Mechanisms of Life History Evolution

Mechanisms of Life History Evolution

The Genetics and Physiology of Life History Traits and Trade-Offs

EDITED BY

Thomas Flatt
*Group Leader at the Institute of Population Genetics
at the Vetmeduni Vienna, Austria*

Andreas Heyland
*Assistant Professor at the Department of Integrative
Biology at the University of Guelph, Canada*

OXFORD
UNIVERSITY PRESS

OXFORD
UNIVERSITY PRESS

Great Clarendon Street, Oxford OX2 6DP

Oxford University Press is a department of the University of Oxford.
It furthers the University's objective of excellence in research, scholarship,
and education by publishing worldwide in

Oxford New York

Auckland Cape Town Dar es Salaam Hong Kong Karachi
Kuala Lumpur Madrid Melbourne Mexico City Nairobi
New Delhi Shanghai Taipei Toronto

With offices in

Argentina Austria Brazil Chile Czech Republic France Greece
Guatemala Hungary Italy Japan Poland Portugal Singapore
South Korea Switzerland Thailand Turkey Ukraine Vietnam

Oxford is a registered trade mark of Oxford University Press
in the UK and in certain other countries

Published in the United States
by Oxford University Press Inc., New York

© Oxford University Press 2011

The moral rights of the authors have been asserted
Database right Oxford University Press (maker)

First published 2011

British Library Cataloguing in Publication Data

Data available

Library of Congress Cataloging in Publication Data

Library of Congress Control Number: 2011920657

Typeset by SPI Publisher Services, Pondicherry, India
Printed in Great Britain
on acid-free paper by
CPI Antony Rowe, Chippenham, Wiltshire

ISBN 978–0–19–956876–5 (Hbk.)
         978–0–19–956877–2 (Pbk.)

1 3 5 7 9 10 8 6 4 2

"…integrating an understanding of mechanisms into life history theory will be one of the most exciting tasks facing evolutionary biologists in the 21st century."

Barnes & Partridge (2003)

Foreword: Harvey's legacy

Graham Bell

"Ex ovo omnia" wrote Harvey (1651), and more than three centuries later his dictum still stands. Many examples of vegetative reproduction in animals have been described since Trembley (1744) astonished the world by describing the asexual budding of *Hydra*, but all lineages, so far as we know, pass through a single-cell stage sooner or later. The only more certain observation is that the individual that develops from the egg will eventually senesce and die. The journey between the two fixed points of egg and corpse has no prescribed route, however, and biologists have struggled to discover and interpret the lives of animals, plants, fungi, and seaweeds since the beginning of scientific biology in Harvey's time.

In very broad terms, two main approaches have been followed. The first is concerned largely with juvenile forms, and specifically with discrete developmental stages such as larvae. This is the older school, and its subject is usually called "life cycles." The second is concerned largely with adults, and specifically with the quantitative schedule of reproduction. This has developed over the last 50 years, and its subject is usually called "life histories." The two have been developed to a large extent independently of one another, and the linkages between them remain few and weak.

The study of life cycles is the older tradition, and dates back to the time when it was first demonstrated by Steenstrup (1845) that very dissimilar individuals could be produced, one from another, as a succession of forms belonging to the same lineage. The only example that is familiar to everyone (now that classical zoology has withered from the curriculum) is the succession of polyp and medusa in some cnidarians, but much more complex life cycles have evolved in groups such as digenean trematodes. It is often difficult to work out how and when one stage gives rise to another, especially when this involves reproduction by eggs. In other cases, however, the developmental sequence is clear even though it passes through larval stages that are very different from the adult, by virtue of the physical continuity of macroscopic individuals. There is no profound difference between the two, however. The juvenile starfish, for example, develops as a miniature version of the adult from a small patch of tissue within the body of a small ciliated-band animal living in the plankton. As the starfish animal enlarges, the ciliated-band animal shrinks, and its remnants are eventually discarded. This process is called "development" because the physical continuity of larva and adult is clear, but it is manifestly the same kind of process as the transition between polyp and medusa, a succession of morphologically distinct phases within a single lineage.

The study of life cycles has been predominantly descriptive, a tradition that has continued, in the form of discovering the regulatory genes responsible for the evolution of body plans, down to the present day. The possibility of a theoretical account of the life cycle was raised by Garstang (1928) nearly a century ago: given that an animal developing from an egg must grow steadily larger, it must be functional first as a small individual and subsequently as a large individual. A ciliary-band animal living at low Reynold's number in the plankton is one possible route to a large hydraulically powered animal living on the sea floor, whereas the reverse route is impracticable. Nevertheless, a systematic theoretical framework capable of interpreting the succession of phases in development has yet to be constructed.

The parallel effort to understand the sexual phases of life cycles is a similar but more complicated story, largely because of the great difficulty of establishing the site and timing of the crucial events of fusion and reduction, even after the advances of microscope design in the later nineteenth century. It was not until the closing decades of the century that the essential distinction between spore and gamete was unequivocally established, and by the end of the first decade of the twentieth century a correct account of the alternation of generations in the life cycle of seaweeds and land plants had been successfully accomplished (see the review by Farley 1982). A theoretical basis for understanding the alternation of generations in terms of the fundamentally different requirements of spores and gametes, and thus the different structure and behavior of spore-producing and gamete-producing individuals, was provided at the same time (Bower 1908). Before it could be extensively developed, however, the field was largely abandoned, swept away by the flood of research into transmission genetics, which had just been provided with the firm theoretical framework of Mendelism. The endlessly varied sexual cycles of eukaryotes have continued to provide rich material for extending our knowledge of natural history, but we cannot yet interpret them within a consistent theoretical framework. In recent years there has been a modest revival of interest in the theory of phenomena such as gamete dimorphism, but this has not yet become firmly coupled to field studies and experimentation.

The field of life histories has developed more recently and in a very different fashion, being highly theoretical from the outset. Simplifying history (as one must in a foreword) it was galvanized by Cole's paradox: a lineage in which females live forever and produce an infinite number of litters of whatever size has the same rate of increase as one in which all females die immediately after producing their first litter, but produce one more offspring (Cole 1954).

Working out the reasons for this counter-intuitive result led to a general interpretation of suicidal versus repeated reproduction that was subsequently elaborated into an account of the schedule of reproduction over the whole of the adult stage. Juvenile stages such as larvae are ignored; alternative adult stages such as sporophytes and gametophytes are not distinguished. Freed from zoology, botany, and genetics a highly abstract and general theory of the life history could be developed.

Much of this theory was based on a principle of optimality: quantities such as the rate of reproduction and the probability of survival are negatively correlated, such that intermediate values of both maximize the overall rate of increase of a lineage. This enabled the schedule of reproduction to be predicted from the costs of reproduction, in terms of reduced growth or survival. This approach has had some brilliant successes, beginning with David Lack's classical interpretation of clutch size in birds (see Lack 1966). It could also be extended to the puzzling phenomenon of senescence, which could now be interpreted as a non-adaptive side effect of selection for early reproductive maturity (Williams 1966a). Moreover, the generality of the predictive framework made it possible to contrive experimental tests in laboratory model systems, which had never been possible for life cycles.

The very generality that gave the theory such power was also a source of weakness, insofar as the sources of the costs of reproduction on which the theory was based did not need to be specified and therefore could not be investigated within the confines of the theory itself. For this reason, attention began to shift towards the nature of the costs themselves. This is not as straightforward as it might appear. It seems obvious that producing larger offspring will mean producing fewer, and almost equally obvious that allocating a greater share of resources to reproduction must deplete the stock available to support maintenance and defense. One function necessarily interferes with the other. It was soon found, however, that in practice the correlations between fitness components are usually positive rather than negative. This was quickly attributed to environmental variance of productivity among sites when comparisons are made in the field, or to genetic variance of overall fitness among strains when comparisons are made in the laboratory. The expected costs are then expressed only at evolutionary equilibrium, when genetic variance for overall fitness has been reduced to a low level by selection. This evolutionary argument, however, implies that costs of reproduction measured as negative genetic correlations need not have any mechanistic basis in terms of functional interference. A simple illustration of this conclusion is to generate n random numbers and label them x_1, x_2 etc; then repeat the process to obtain a second set labeled y_1, y_2 etc. Plotting y_i on x_i produces a cloud of points with zero correlation. Now choose the small percentage of pairs with the highest values of $(x_i + y_i)$; plotting y_i on x_i now produces a graph with a slope of −1 (if the x_i and y_i have equal variance) and $r^2 \approx 0.5$. This striking pattern has been carved out of the original unstructured data by the act of choosing an unrepresentative set of cases, and natural selection will likewise generate negative genetic correlation among components of fitness from random life histories through the propagation of an unrepresentative set of genotypes. No causal connection between the components, for example through pleiotropic gene expression, is necessary for such correlations to arise.

Understanding the mechanistic basis for costs of reproduction is thus an important aspect of understanding life histories as a whole. Moreover, it may contribute directly to human well-being. The evolutionary reasons for senescence can be framed in terms of pleiotropy or delayed gene expression for example, whereas the physiological reasons must be framed in terms of factors such as the irreversible damage caused by reactive oxygen species, the accumulation of somatic mutations, the intrinsically limited metabolic capacity of tissues, and so forth. Much of the recent research into senescence in the *Drosophila* and *Caenorhabditis* model systems has been driven by the physiological agenda, in part because of its potential for identifying palliative therapies. The value of this research is indisputable.

It would be a pity, however, if the evolutionary agenda were to be obscured or lost as a result. Part of the *raison d'être* of this volume is to emphasize that it is important to understand the physiological basis of the costs of reproduction in order to understand how life histories evolve—but conversely that it is equally important to understand how life histories evolve in order to predict how physiological processes are likely to operate. The *clk* genes of *Caenorhabditis*, which slow down vital processes and extend lifespan, are an excellent example of recent research in molecular developmental genetics at the interface between evolution and physiology (Hekimi *et al.* 2001).

A more fundamental task for the future is to build a synthetic evolutionary theory of development that would bring together all the phenomena of life cycles within the same framework. It is not even clear whether this is possible. The classical theory of life histories invokes natural selection, leading to optimal phenotypes. The most successful theories of certain aspects of the life cycle, such as gamete dimorphism and the sex ratio, invoke sexual selection, leading to evolutionary stable states. For some phenomena, such as the alternation of generations, both approaches have been tried without any decisive outcome so far; for others, such as the succession of phases, hardly any formal theory has yet been developed. But the possibility is worth contemplating, and if this volume is not the last word on the subject, it will at any rate be the next word.

Graham Bell
McGill University
23 May 2010

Preface

The major features of a life cycle are shaped by demographic traits—size at birth, growth rate, age and size at maturity, age-specific reproductive investment, number and size of offspring, age-specific survival, and lifespan—connected by constraining trade-offs. Together, these life history traits determine Darwinian fitness by affecting the two most important fitness traits, survival and reproduction. Life history theory seeks to understand the causes and consequences of genetic and environmental variation in life history traits, both within and among species. By combining quantitative genetics, artificial selection, demography, phenotypic manipulations, and optimality modeling, life history theory has had major success in explaining the diversity of life history strategies, as reviewed in three excellent books by Stephen C. Stearns and Derek A. Roff (Stearns 1992, Roff 1992, 2002).

The present book, in contrast to most previous work on life history evolution, emphasizes the mechanistic description, the "molecular natural history," of life history traits and their evolution. Traditionally, life history theory is silent on proximate mechanisms, yet recent advances in mechanistic biology have taught us a great deal about how genetics, development, and physiology affect life history. Although much of this information comes from research outside the realm of evolutionary biology, for example from fields such as the genetics of growth control or the molecular biology of aging, it is often directly relevant for our understanding of the evolution of life histories. To date, however, this mechanistic knowledge has not been adequately integrated into the life history framework. To forge such an integration, and to foster an exchange between scientists who work on organismal versus mechanistic aspects of life histories, this multi-author book brings together leading researchers who share the conviction that many fundamental problems in life history evolution can only be completely understood if we begin to incorporate information on developmental, physiological, and genetic mechanisms into the study of life histories (e.g., Barnes and Partridge 2003, Flatt *et al.* 2005, Heyland *et al.* 2005).

Given the major predictive and explanatory success traditional life history theory has had, why is it important to fill the "black box" of life history evolution with mechanism? A good example is the problem of life history trade-offs. Trade-offs, for example between survival and reproduction, are typically thought to be caused by competitive resource allocation (e.g., Stearns 1992), but whether this physiological explanation is correct is usually unknown. Without detailed knowledge of resource levels, patterns of acquisition and allocation, intermediary metabolism, and endocrine regulation we cannot properly test the assumption that trade-offs are resource based (e.g., Harshman and Zera 2007). Thus, while the existence of trade-offs can often be quite readily established, we do not understand their underlying mechanisms, and this limits our understanding of life history evolution (e.g., Stearns 2000). As several chapters in this book illustrate, the classical assumption of trade-offs being resource based might in fact not always hold.

Another example of how information on mechanisms can illuminate and expand life history theory concerns genes with major effects on life history traits and their integration (e.g., Flatt 2004, Schmidt *et al.* 2008, Paaby *et al.* 2010). Identifying the genes that affect or modulate life history traits will ultimately enable us to answer important evolutionary questions such as: Which genes or alleles affecting life history traits are evolutionarily conserved, and which genes or alleles are lineage-specific? What is the relationship between life history variation segregating within natural populations and genetic differences in life histories among species? Is there standing genetic variation for these genes within populations and are they under selection? What are the genes or alleles that make up genetic correlations and trade-offs? Can the genetic mechanisms that cause trade-offs be uncoupled and how? How do genes interact with the environment to determine life history phenotypes? Several authors in this book review impressive progress in evolutionary quantitative and molecular genetics that has lead to the identification of genes and pathways that are likely to be of major importance in life history evolution.

Although the integration of mechanistic studies into life history evolution is still in its infancy, we believe that—similar to the recent advances made by evolutionary developmental biology (evo-devo), which combines studies of evolution, development, and genetics—future work on life histories will benefit significantly from an explicit consideration of proximate mechanisms. Many examples of such an interdisciplinary approach towards understanding life history evolution can be found throughout this book.

We are targeting this book at advanced undergraduates, graduate students, postdocs, and established researchers in evolution, ecology, evo-devo, development, genetics, physiology, and aging who all aim to understand the mechanisms that shape the expression and evolution of traits that affect Darwinian fitness, including growth, development and maturation, reproduction, and lifespan. In particular, we hope that the chapters in this book will stimulate students and researchers with a strong interest both in organismal biology and molecular biology. For background reading on life histories we refer the reader to the books by Stearns (1992) and Roff (1992, 2002) who cover traditional, non-mechanistic aspects of life history evolution.

The chapters in this book have all been written by leading researchers who use studies of proximate mechanisms to solve fundamental problems in life history biology in a variety of organisms. Their chapters not only represent the current state of the art, but also offer fresh perspectives for future research. In designing the book we have attempted to present a balanced selection of authors (ranging from young to well-established), organismal taxa

(e.g., algae, higher plants, nematodes, insects, echinoderms, fish, amphibians, reptiles, birds, humans), and biological disciplines and approaches relevant to life history biology (e.g., developmental biology, genetics, evo-devo, anthropology, behavior, reproductive biology, aging, phenotypic plasticity, social evolution, immunology, metabolism, and endocrinology). Although we have aimed to cover a lot of ground in this book, many interesting and important subjects had to be omitted due to space limitations. For example, while several chapters discuss the mechanisms that affect lifespan, we did not include a detailed discussion of the evolutionary biology of aging—this has been reviewed extensively elsewhere, for example by Rose (1991) and Flatt and Schmidt (2009).

Chapter authors were asked to write chapters that are equally accessible to evolutionary and mechanistic biologists, to make clear references to fundamental concepts in life history evolution, and to cross-reference other chapters in this volume. Authors obviously differ in their scientific views, the level of their exposition of material, and their writing styles, and we have therefore attempted to make chapters somewhat uniform, for example through author guidelines, editorial and external peer reviews, and several revisions. Nevertheless, differences among the chapters do remain—in fact, we feel that they are desirable since they make the perspectives offered here both more personal and pluralistic. Together with the wide range of topics, scientific approaches, and organisms covered in this volume we hope that this diversity in perspective will be stimulating for the reader.

Although not being a chapter in its own right, the book starts out with a foreword by Graham Bell, who gives a broad historical summary of research on life histories and who emphasizes the need for integrating mechanistic insights into this research area. This sets the stage for the actual book chapters which are grouped into seven parts. Each part of the book, except for the last, is preceded by a brief introduction written by the editors. Part 1 (Integrating mechanisms into life history evolution) consists of two chapters which introduce some of the basic concepts of life history theory and outline the utility of mechanistic approaches for understanding problems in life history evolution. Parts

2–6 consist of 24 chapters that make up the bulk of the book. Since many of these chapters are highly integrative and cover more than one type of organism, we have attempted to group these chapters into conceptual categories. The chapters in Parts 2–4 all address mechanisms that deal with one of the three major phases of an organisms' life cycle (Part 2: growth, development, and maturation, Part 3: reproduction, and Part 4: aging and somatic maintenance). The chapters in Parts 5–6 deal with two major concepts in life history theory, namely phenotypic plasticity and trade-offs (Part 5 life history plasticity and Part 6 life history integration and trade-offs). Despite this conceptual structure, many chapters could have easily been placed into other parts of the book. Thus, the structure of the book is not rigid, and many chapters in one part of the book directly touch on issues discussed in other parts of the book. In fact, we feel that a certain amount of overlap among book parts and chapters is desirable and helps the integration of the diverse subjects we cover. We have also aimed to achieve further integration throughout the book by writing short introductory sections that precede each book part and by asking authors to frequently cross-reference other chapters. Part 7 concludes the volume: in Chapter 27 Stephen C. Stearns summarizes and critically discusses the contributions in the book by asking whether progress on understanding mechanisms forces life history theory to change. Chapter 28 is a postscript that rounds up the book: it consists of an exchange between Stearns and the editors, discussing what mechanistic insights can or cannot contribute to our understanding of life history evolution.

The idea for this volume was conceived at a symposium on molecular mechanisms of life history evolution sponsored by the Society for the Study of Evolution (SSE), which we organized at the Evolution meetings at the University of Minnesota in Minneapolis in 2008. We are grateful to all the speakers and participants of this symposium for many stimulating discussions that have helped to shape some of the ideas in this book. Their enthusiasm convinced us to approach this book project. In particular, we thank Derek Roff for his advice and encouragement.

Each book chapter was reviewed by both editors and in most cases by two external reviewers or chapter contributors. We are extremely grateful to these experts for their help and time: without exception they have provided very thoughtful, critical, and helpful suggestions for improving the chapters. For their timely chapter reviews we are indebted to Gro Amdam, Richard Bribiescas, Goggy Davidowitz, Greg Davis, Tony De Tomaso, David Denlinger, Robert Denver, Michelle Elekonich, Peter Ellison, Caleb Finch, Klaus Fischer, Gary Freeman, Owen Gilbert, Michael Hadfield, Dan Hahn, Larry Harshman, John Hatle, Jason Hodin, Hillard Kaplan, Tad Kawecki, Ellen Ketterson, Teri Markow, Alistair McGregor, Amy Moran, Coleen Murphy, Courtney Murren, Amy Newman, Dan Noble, Mats Olsson, Bruno Pernet, Scott Pletcher, Kim Rewitz, Jens Rolff, Michael Rose, Olav Rueppell, Gerhard Schlosser, Paul Schmidt, David Schneider, Stanley Shostak, Cristian Solari, Gabriele Sorci, Stacia Sower, Michael Stern, Richard Strathmann, Stuart Wigby, Karen Williams, and John Youson, and two reviewers who wished to remain anonymous.

We are also grateful to our editors at Oxford University Press, Helen Eaton and Ian Sherman, for their timely and professional support during all stages of this project; to Laura Rosario Sanchez for major help with manuscript formatting and compiling references; and to Brenda Rascón and April Bojorquez for their work on the cover art. AHs work on this book was supported by a NSERC Discovery Grant.

Last but not least, our largest debt of gratitude is of course to the authors themselves. Despite many other academic and personal commitments, they have all written excellent chapters in a timely fashion and patiently responded to our many requests. We are especially grateful to Steve Stearns for having read and synthesized 26 chapters and for the idea of including an exchange between the three of us at the end of the book. We are also indebted to Graham Bell for contributing the foreword to this volume. All these authors, colleagues, and friends have invested major amounts of their limited time to produce the scientific content of this book: the success is entirely theirs; all failures are ours.

Thomas Flatt (Vienna) and
Andreas Heyland (Guelph)
August 2010

Contents

Contributors

Gro V. Amdam, Arizona State University, School of Life Sciences, PO Box 874501, Tempe, AZ 85287-4501, USA
and
Norwegian University of Life Sciences, Department of Chemistry, Biotechnology and Food Science, PO Box 5003, Aas, N-1432, Norway

Adam Antebi, Baylor College of Medicine, Huffington Center on Aging and Department of Molecular and Cellular Biology, One Baylor Plaza, Houston, TX 77030, USA
and
Max Planck Institute for Biology of Aging, Gleueler Straße 50a, D-50931 Cologne, Germany

Joshua A. Banta, New York University, Department of Biology, 1009 Silver Center, 100 Washington Square East, New York, NY 10003-6688, USA

Johannes H. Bauer, Southern Methodist University, Department of Biological Sciences, 6501 Airline Drive, 238–DLS, Dallas, TX 75275, USA

Alan O. Bergland, Department of Ecology and Evolutionary Biology, Brown University, Box G-W, 80 Waterman Street, Providence, RI 02912, USA
and
Department of Biology, 371 Serra St, Stanford University, Stanford, CA 94305-5020, USA

Christian Braendle, Université Nice Sophia-Antipolis, Institute of Developmental Biology and Cancer (IBDC), Parc Valrose, 06108 Nice cedex 2, France

Paul M. Brakefield, Evolutionary Biology Group Institute of Biology, Leiden University, Postbus 9505, 2300 RA Leiden, The Netherlands
and
University Museum of Zoology, University of Cambridge, Downing Street, Cambridge CB2 3EJ, UK

Anne M. Bronikowski, Iowa State University, Department of Ecology, Evolution & Organismal Biology, 253 Bessey Hall, Ames, IA 50011, USA

Daniel R. Buchholz, University of Cincinnati, Department of Biological Sciences, 832 Rieveschl Hall, 312 Clifton Court, Cincinnati, OH 45221, USA

Tracey Chapman, University of East Anglia, School of Biological Sciences, Norwich, Norfolk, NR4 7TJ, UK

Sandie Degnan, University of Queensland, Integrative Biology School, Goddard Building, Saint Lucia, Queensland 4072, Australia

Dominic A. Edward, University of East Anglia, School of Biological Sciences, Norwich, Norfolk, NR4 7TJ, UK

Deniz F. Erezyilmaz, Princeton University, Department of Ecology & Evolutionary Biology, Guyot Hall Princeton, NJ 08544, USA

John R. Finnerty, Boston University, Biology Department, 5 Cummington Street, Boston, MA 02215, USA

Thomas Flatt, University of Veterinary Medicine Vienna, Department of Biomedical Sciences, Institute for Population Genetics, Veterinärplatz 1, A-1210 Vienna, Austria

Mark V. Flinn, University of Missouri, Department of Anthropology, 107 Swallow Hall, Columbia, MO 65211-1440, USA

Birgit Gerisch, Max-Planck Institute for Biology of Aging, Gleueler Straße 50a, D-50931 Cologne, Germany

Lawrence G. Harshman, University of Nebraska, School of Biological Sciences, 335A Manter Hall, Lincoln, NE 68588-0118, USA

Michaela Hau, Max-Planck Institute for Ornithology, Vogelwarte Radolfzell, Schlossallee 2, D-78315 Radolfzell, Germany

Stephen L. Helfand, Brown University, Divison of Biology and Medicine, Molecular, Cellular Biology Biochemistry, Syndey Frank Hall, 185 Meeting Street, Providence, RI 02912, USA

Andreas Heyland, University of Guelph, Department of Integrative Biology, Rm 1468, Guelph, ON N1G-2W1, Canada

Magdalena Hodkova, Institute of Entomology, Biological Centre, Academy of Sciences, Branisovska 31, 37005 České Budějovice, Czech Republic

Amy Hollar, University of Cincinnati, Department of Biological Sciences, 832 Rieveschl Hall, 312 Clifton Court, Cincinnati, OH 45221, USA

Saurabh Kulkarni, University of Cincinnati, Department of Biological Sciences, 832 Rieveschl Hall, 312 Clifton Court, Cincinnati, OH 45221, USA

Maris Kuningas, Department of Epidemiology, Erasmus Medical Centre, PO Box 2040, 3000 CA Rotterdam, The Netherlands

Lesley T. Lancaster, National Center for Ecological Analysis and Synthesis (NCEAS), University of California, Santa Barbara, 735 State Street, Suite 300, Santa Barbara, CA 93101, USA

Brian P. Lazzaro, Cornell University, Department of Entomology, 3125 Comstock Hall, Ithaca, NY 14853, USA

Richard G. Manzon, University of Regina, Department of Biology, Regina, SK S4S 0A2, Canada

Kurt A. McKean, Department of Biological Sciences, State University of New York (SUNY) at Albany, 1400 Washington Avenue, Albany, NY 12222, USA

Richard E. Michod, University of Arizona, Department of Ecology and Evolutionary Biology, PO Box 210088, Tucson, AZ 85721, USA

Benjamin G. Miner, Western Washington University, Department of Biology, 516 Hight Street, University Mailstop 9160, Bellingham, WA 98225, USA

Christine Moskalik, University of Cincinnati, Department of Biological Sciences, 832 Rieveschl Hall, 312 Clifton Court, Cincinnati, OH 45221, USA

Michael P. Muehlenbein, Indiana University, Department of Anthropology, Student Building 130, 701 E. Kirkwood Avenue, Bloomington, IN 47405-7100, USA

Navdeep S. Mutti, Arizona State University, School of Life Sciences, PO Box 874501, Tempe, AZ 85287-4501, USA

Aurora M. Nedelcu, University of New Brunswick, Department of Biology, PO Box 4400, Fredericton, NB E3B 5A3, Canada

Allison Ng, University of Cincinnati, Department of Biological Sciences, 832 Rieveschl Hall, 312 Clifton Court, Cincinnati, OH 45221, USA

Michael D. Purugganan, New York University, Center for Genomics and Systems Biology, Department of Biology, 1009 Silver Center, 100 Washington Square East, New York, NY 10003-6688, USA

Brenda Rascón, Norwegian University of Life Sciences, Department of Chemistry, Biotechnology and Food Science, P.O. Box 5003, Aas, N-1432, Norway

Adam M. Reitzel, Woods Hole Oceanographic Institution, Redfield 304, MS#32, Woods Hole, MA 02543, USA

Derek A. Roff, University of California at Riverside, Department of Biology, Office 3352, Spieth Hall, Riverside, CA 92521, USA

Paul S. Schmidt, University of Pennsylvania, Department of Biology, Leidy Laboratories, Philadelphia, PA 19104, USA

Tonia S. Schwartz, Iowa State University, Interdeptartmental Genetics Program, Department of Ecology, Evolution & Organismal Biology, 253 Bessey Hall, Ames, Iowa, IA 50011, USA

Alexander W. Shingleton, Michigan State University, Department of Zoology, 203 Natural Science Building, East Lansing, MI 48824-1115, USA

Barry Sinervo, University of California, Department of Ecology & Evolutionary Biology, Earth & Marine Sciences Building, Santa Cruz, CA 95064, USA

Stephen C. Stearns, Yale University, Department of Ecology and Evolutionary Biology, Box 208106, New Haven, CT 06520-8106, USA

Derek Stefanik, Boston University, Biology Department, 5 Cummington Street, Boston, MA 02215, USA

Marc Tatar, Brown University, Division of Biology and Medicine, Department of Ecology and Evolutionary Biology, Box G-W, Providence, RI 02912, USA

Christina Tolfsen, Norwegian University of Life Sciences, Department of Chemistry, Biotechnology and Food Science, PO Box 5003, Aas, N-1432, Norway

Rudi G.J. Westendorp, Leiden University Medical Center (LUMC), Department of Gerontology and Geriatrics C2-R, PO Box 9600, 2300 RC Leiden, The Netherlands

John C. Wingfield, University of California, College of Biological Sciences, Neurobiology, Physiology & Behavior (NPB), 294 Briggs Hall, Davis, CA 95616, USA

Anthony J. Zera, University of Nebraska, School of Biological Sciences, 225 Manter Hall, Lincoln, NE 68588-0118, USA

Bas J. Zwaan, Evolutionary Biology Group Institute of Biology, Leiden University, Postbus 9505, 2300 RA Leiden, The Netherlands
and
Chair in Genetics, Laboratory of Genetics, Plant Sciences Group, Wageningen University and Research Centre, PO Box 309, 6700 AH Wageningen, The Netherlands

PART 1

Integrating mechanisms into life history evolution

Thomas Flatt and Andreas Heyland

Life history evolution seeks to explain the major phenotypic features of organismal life cycles and the diversity of adaptive strategies that different organisms use to optimize their survival and reproductive success in the face of challenges posed by the environment (for an overview, see Roff 1992, Stearns 1992). To understand life histories and their evolution, evolutionary biologists have classically used the tools of demography, quantitative genetics, mathematical modeling, and phylogenetics. While this combination of approaches was—and continues to be—very successful in explaining the broad evolutionary features of life cycles and of variation in Darwinian fitness, the present book aims to draw attention to the functional and mechanistic description and analysis of life history traits and their evolution.

Traditionally, the field of life history evolution, a branch of evolutionary ecology, has focused on analysing the demographic phenotypes that make up the life cycle and which determine fitness: so-called "fitness components." This is even true for quantitative genetic approaches towards understanding life history evolution that are by definition inferring genetic properties by a statistical analysis of phenotypic variation. Thus, within the life history framework, explicit genetic and other mechanistic details are typically treated as a black box. In another area crucial for our understanding of evolutionary processes—population genetics—similar considerations apply. Population genetics typically ignores phenotypes and the complication of explicit genetics and other mechanisms. Although this situation has changed markedly with the

advent of molecular population genetics and modern genotyping techniques such as next generation sequencing, evolutionary geneticists working on molecular variation still rarely attempt to explicitly connect molecular variation—now observable with greatest resolution at the nucleotide level—to the intermediate levels of development and physiology and, ultimately, to the kinds of whole-organism phenotypes that are the real stuff of evolution (e.g., Houle 2010). This is the realm of developmental biology, molecular genetics, physiology, and molecular biology—areas that have typically received little attention from evolutionists. Yet, as pointed out by Lewontin (1974) and many others, our basic understanding of the evolutionary process and its dynamics will remain incomplete until we integrate information on phenotypes with information on genotypes and the intermediate mechanisms that connect them. A notable exception to this pattern is the relatively young field of evolutionary developmental biology, or evo-devo, which investigates the evolution of developmental mechanisms and how they shape the amazing morphological diversity among different organisms (e.g., Carroll 2005, Stern 2010).

Since development and physiology translate genotypes into phenotypes, and since this mapping is complex and not necessarily one-to-one, the mechanistic details of the genotype–phenotype relationship will likely have major effects on evolutionary outcomes. For example, development can bias the kinds of mutations that are available and, as a consequence, the resulting phenotypes that are visible to natural selection. Thus, while evolutionary genetics tells us about the fate of particular muta-tions subject to either selection or drift, develop-mental and physiological genetics tell us how specific genes—and the mechanisms influenced by them—generate particular phenotypes in the first place (e.g., Stern 2010). However, the mechanisms that translate genotypes into phenotypes have largely been ignored in most areas of evolutionary biology, including life history evolution. Historically, this has often been done to keep things manageable and simple, especially in evolutionary theory, but sometimes it has been done simply because the potentially relevant mechanisms were not yet known or amenable to analysis. However, in recent decades, developmental genetics, molecular biol-ogy, and physiology have tremendously advanced our understanding of functional biology, and it is time now to integrate this knowledge into the evo-lutionary framework. Evo-devo has very success-fully started to do exactly this (e.g., Carroll 2008, Stern 2010), and we believe that the time is ripe now to try the same thing for those traits that have the largest impact upon Darwinian fitness: life history traits (e.g., Brakefield 2005).

The authors and editors of the present book share the conviction that a detailed knowledge of the genetic, developmental, and physiological mecha-nisms that affect life history traits is of major impor-tance for understanding, from first principles, how life history traits are expressed, why they vary, and how they evolve. Unravelling such mechanisms has the potential to provide answers to questions such as:

• Which genes and mutations determine specific life history traits and how do they do this?

• Which of these genes are genetically variable and contribute to the evolution of life history traits in natural populations?
• Are there "hot spot" genes for life history evolution?
• Is the tremendous variation in life histories among species explained by variation in the same genes as those that cause variation within a species?
• What are the developmental and physiological mechanisms that integrate life history traits and explain trade-offs between them?
• How do these trade-offs evolve, how do they become fixed, and how do they constrain evolution?
• What are the mechanisms that mediate and mod-ulate life history plasticity?

As the chapters in this volume illustrate, we do not yet know the answers to most of these long-standing questions, but we have got a good start.

The following two chapters set up some of the major themes discussed throughout this volume; together they serve as an introduction to the rest of the book. In Chapter 1, Braendle, Heyland, and Flatt provide some elementary background about the main of concepts life history evolution and briefly sketch some of the key problems that require, and are amenable to, mechanistic analysis. In Chapter 2, Derek Roff exemplifies how genetic and genomic approaches can be used to begin to illuminate some fundamental problems in life history evolution. His discussion focuses on two promising areas of appli-cation: the functional investigation of life history trade-offs, and the question of the relative impor-tance of stochastic versus deterministic factors that affect evolutionary trajectories.

CHAPTER 1

Integrating mechanistic and evolutionary analysis of life history variation

Christian Braendle, Andreas Heyland, and Thomas Flatt

1.1 Introduction

Life histories—describing essential patterns of organismal growth, maturation, reproduction, and survival—show tremendous variation across individuals, populations, species, and environments. Understanding this variation is the goal of life history research. The analytical framework of life history theory focuses on the variation and interaction of different key maturational, reproductive, and other demographic traits, given that natural selection acts to maximize fitness of a life history as a whole (Roff 1992, Stearns 1992). Fitness integrates over the entire reproductive performance of the organism, and life history traits are the major fitness components underlying this integration. However, the investment into alternative life history traits, and thus the possible set of trait combinations, is restricted by genetic, developmental, physiological, and phylogenetic limits. Apart from explaining variation in life history strategies as a result of natural selection, identifying how such trade-offs and constraints shape life histories is the central aim of life history research.

In this chapter we introduce the basic concepts and definitions of life history theory and argue for the importance of integrating a mechanistic perspective into research on life histories. While most traditional life history research is based on mathematical, statistical, and phylogenetic approaches without explicit reference to underlying mechanisms, today's principal research challenge is to fill this gap through experimental characteriza-

tion of the proximate basis of life histories. The analysis of genetic, developmental, and physiological factors that shape life history traits will ultimately allow us to determine how evolutionary changes in such mechanisms generate, facilitate, or constrain the diversification of life histories. Integrating mechanistic and evolutionary analyses of life history variation is part of a global quest in biology that seeks a shared understanding of proximate and ultimate causes of phenotypic variation.

1.2 The life history framework

1.2.1 What is a life history?

A life history encompasses the life of an individual from its birth to its death, describing the age- or stage-specific patterns of maturation, reproduction, survival, and death. The major objective of life history research is to understand how evolution, given selection imposed by ecological challenges, shapes organisms to achieve reproductive success. The second objective of life history research is to understand whether and how, given internal trade-offs and constraints, selection can optimize a set of life history traits to maximize reproductive success. Since organisms dispose of limited resources, which must be competitively allocated to differing functions, such as growth, reproduction, survival, and maintenance, resources invested into one function cannot be invested into another, leading to trade-offs. In addition, life history research explores taxon-specific features of life cycles and life history decisions, including patterns of sex allocation, alternative phenotypes, or larva-to-adult transitions. For in-depth treatments of the evolution of life histories and life history theory see Stearns (1992), Roff (1992, 2002), and Charlesworth (1994).

1.2.2 Life history traits and fitness

Life history traits represent quantitative, demographic properties of organisms that are directly related to the two major components of fitness, i.e., survival and reproduction. Classical life history analysis considers the following to be the principal life history traits (Stearns 1992):

- size at birth
- growth pattern
- age and size at maturity
- number, size, and sex ratio of offspring
- age-and size-specific reproductive investments
- age- and size-specific mortality schedules
- length of life.

These traits essentially represent the demographic parameters required to estimate fitness as defined by the Malthusian parameter (or similar fitness measures). The Malthusian parameter (also called the instantaneous rate of natural increase, r) is the solution to the Euler–Lotka equation, which describes population growth by summing reproductive events and survival probabilities over the entire lifetime of individuals (Stearns 1992). Thus, life history traits are directly linked to fitness, with fitness being defined by population growth models from demography.

In contrast to classical life history traits, morphological, physiological, or behavioral traits are considered to contribute to fitness only indirectly (e.g., Roff 2007b). However, this distinction is somewhat arbitrary. For example, certain morphological traits such as body size or gonad size may correspond to life history traits (or at least are correlates thereof). In the literature, the term "life history trait" is often used interchangeably with fitness components, so that many phenotypic characters with major effects on reproduction and survival have been called life history traits.

Because of their complexity and demographic nature, life history traits are usually treated as quan-

titative, polygenic traits (Falconer and MacKay 1996). The expression of life history traits is also highly contingent on the environment, so that life history research places particular emphasis onto the concept of phenotypic plasticity, i.e., the ability of a single genotype to produce different phenotypes across environments (Stearns 1992). Plasticity is described by "reaction norms", mathematical functions that relate the phenotypic values adopted by a given genotype to changes in the environment. Selection shapes life history plasticity by acting on genetic variation for plasticity, which is present when the reaction norms that represent different genotypes are non-parallel across the same range of environments (so-called genotype by environment interactions, or G × E). Reaction norms (and thus plasticity) are considered to be optimal when they maximize fitness for each of the different environments (Stearns and Koella 1986).

1.2.3 Trade-offs and constraints

A key postulate of life history theory is that the values and combinations of life history traits are limited by factors internal to the organism, namely trade-offs and constraints. These intrinsic factors ultimately limit and direct the evolutionary response to the external force of selection. A life history trade-off occurs when an increased investment in one fitness component causes a reduced investment in another fitness component, i.e., a fitness benefit in one trait exacts a fitness cost in another. Examples of classical life history trade-offs are survival versus reproduction, number versus size of offspring, or current reproduction versus future reproduction (Stearns 1992).

Trade-offs are usually described as phenotypic or genetic covariances or correlations among traits, without reference to their causal relationships. If the relationship can be shown to be genetic, negative genetic covariance among traits is expected to limit the evolution of each of these traits. Such genetic or evolutionary trade-offs are considered at the population level, i.e., as defined by genetic correlations among individuals or correlated phenotypic responses to selection. Genetic trade-offs are traditionally assumed to stem from antagonistic pleiotropy or linkage disequilibrium. These trade-offs also manifest themselves at the physiological or individual level, for example when an individual with increased reproductive effort in one year exhibits a reduction in reproductive output in the next year. Such physiological trade-offs are thought to be due energy limitations, i.e., the allocation of resources among competing functions. Importantly, trade-offs may exist at population level, but not at individual, physiological level (Stearns 1989, Houle 1991, Stearns 1992).

In contrast to trade-offs, the term "constraint" is often used to described *absolute* limits to or biases upon trait expression and combination. Constraints may describe physical factors, developmental properties, or historical contingencies that prevent an organism from expressing a certain phenotype or a population from attaining a certain fitness optimum in response to selection (Maynard Smith *et al.* 1985). The distinction between trade-offs and constraints is not strict, and trade-offs are often regarded as one type of constraint. In the life history context, constraints usually refer to phylogenetic, lineage-specific characteristics that impose *absolute* limits on trait expression in a given organismal group.

1.2.4 Empirical approaches in life history research

Although classic life history analysis has been largely theory-driven, much empirical research has addressed the questions and predictions raised by life history theory, using both non-genetic and genetic approaches (Stearns 1992, Roff 1992, 2002, 2007b; also see Chapter 2). Non-genetic approaches include phenotypic correlations to examine patterns of life history trait covariation among populations and species, experimental phenotypic manipulations, and statistical tools from comparative analysis to control for phylogenetic history. Genetic approaches to the study of life history variation are predominantly based on the framework of quantitative genetics. Most of this work has concentrated on the detection and analysis of genetic trade-offs, either through the study of covariances and correlations among life history traits between relatives (e.g., pedigree analyses) or through correlated responses of life history traits to artificial selection or experimental evolution. This research framework has

generated a substantial body of empirical evidence that has revealed how selection operates on life history traits, contingent on the environment and trade-offs (Stearns 1992, Roff 1992, 2002, 2007a,b). Despite these extensive efforts, very few studies have examined the mechanistic underpinnings of life history traits. For example, inferred interrelationships among life history traits rarely describe more than statistically determined associations. A major limitation common to the classical approaches in life history research is therefore the ignorance of the proximate causes that determine or modulate life histories and their evolution.

1.3 The study of causal mechanisms linking genotype to phenotype

Understanding how a genotype translates into a phenotype is one of the most fundamental problems in biology. In most cases, phenotypes cannot be simply inferred from their underlying genotypes, and vice versa, because the mapping of genotypes onto phenotypes is often a non-linear process, shaped by a multitude of complex genetic and environmental interactions. Moreover, a single genotype may generate multiple phenotypes and, conversely, multiple genotypes may generate a single phenotype. That such properties of the genotype–phenotype map are relevant for our understanding of the evolutionary process has been emphasized for a long time (e.g., Lewontin 1974, Houle 2001), but it is only relatively recently that the causal relationships between genotype and phenotype have received increased attention from evolutionary biologists (e.g., Pigliucci 2010). While research at the interface of development and evolution has begun to tackle the significance of the genotype–phenotype map in morphological evolution, the causal connection between genotypes and phenotypes for fitness components is still extremely rudimentary (e.g., Chapter 2 and Roff 2007b).

Traditionally, attempts to link the genotype with the phenotype have been regarded as the principal task of "reductionist" branches of biology, including molecular, cellular, and developmental biology. Developmental genetics in particular has emerged as the prime discipline in connecting gene function during development with phenotypic outcomes, primarily by relying on mutational analysis and forward genetics. The great power of this approach lies in the typically high degree of causal inference that can be made through carefully controlled manipulation of isolated genetic factors and their phenotypic effects. The general downside of this approach is that such studies are generally limited to the study of single, highly pleiotropic mutations with large phenotypic effects. In addition, developmental genetic analyses are generally limited to the study of a single or or a small number of laboratory populations in highly simplified artificial environments, aiming to reduce variation engendered by genetic background or environmental context as much as possible. This research approach starkly contrasts with that of evolutionary biologists, whose primary concern is the study of quantitative genotypic and phenotypic variation among populations or species. Here, in contrast to developmental genetics, the inferred genotype–phenotype relationships are generally of indirect, associative nature, rarely permitting inferences about the causal connections between genotypic and phenotypic variation.

As advocated in many chapters throughout this book, a better future understanding of many issues in life history evolution will require the integration of evolutionary and organismal biology with molecular and developmental biology (e.g., Dean and Thornton 2007). That unfortunate historical separations between biological disciplines can be overcome is well illustrated by the successful rapprochement of evolutionary and developmental biology (e.g., Raff and Kaufman 1983, Carroll *et al.* 2000, Stern 2010). Although initially mainly concerned with the description of evolutionary diversification or conservation of developmental mechanisms, the central aim of evolutionary developmental biology (evo-devo) has recently shifted to the experimental analysis of how properties of genetic and developmental architecture impact phenotypic evolution. Evo-devo therefore addresses specific issues directly relevant to the understanding of life history evolution, such as the mechanistic basis of developmental biases and constraints or phenotypic plasticity. More generally, as life history traits are high-level phenotypes that depend on the ensemble of morphological and physiological traits, the mechanistic analysis of life

history evolution can consequently be regarded as an extension of the principal objective of evo-devo, namely to understand which developmental and genetic changes underlie phenotypic evolution.

Uncovering the mechanistic basis of life history variation is a non-trivial challenge. Life history traits were defined by evolutionary ecologists with the intent of reducing phenotypic complexity by focusing on a small number of traits that summarize the essential fitness components and by ignoring the underlying genetic, developmental, and physiological mechanisms that govern the expression of these traits. A given life history trait can thus be thought of as a functionally complex phenotype resulting from the integration of a suite of morphological, physiological, or behavioral phenotypes. At the level of the individual, their characteristics have therefore to be understood in terms of both the construction of multiple individual traits as well as their spatial and temporal integration into a higher-level phenotype. As such, life history traits are *a priori* composite, quantitative, polygenic traits whose expression is often highly contingent upon plasticity, pleiotropy, and epistasis. All these properties render the mechanistic analysis of life history traits extremely difficult in practice.

1.4 How can mechanistic insights contribute to understanding life history evolution?

Despite the inherent difficulties in studying the proximate basis of life histories, considerable progress has been made in our mechanistic understanding of life history evolution, with major contributions stemming from molecular genetic studies on experimental model organisms. Here we briefly discuss the importance of integrating such mechanistic information into organismal life history research; many more detailed examples can be found throughout the chapters in this book. For further reading on integrative approaches in life history biology we recommend the reviews by Houle (2001), Leroi (2001), Barnes and Partridge (2003), Harshman and Zera (2007), Chapter 5 in Van Straalen and Roelofs (2006), Roff (2007b), and Flatt and Schmidt (2009).

1.4.1 Why understanding mechanisms is important for answering evolutionary questions

While it is clear that knowledge of the proximate basis of life histories does not provide information about the ecological or evolutionary relevance of such mechanisms, it enables evolutionary biologists to address several fundamental questions about life history evolution, including, for example:

- What is the function of genes that are genetically variable in natural populations and that contribute to ecological adaptation?
- Are major candidate genes, as identified by molecular genetics, variable in natural populations?
- If so, do polymorphisms at these loci actually contribute to the evolution of life history traits in the wild?
- Are the genes that impact life history evolutionarily conserved or lineage-specific?
- What genetic and physiological mechanisms determine or modulate the expression of ecologically and evolutionarily important trade-offs?
- Are such trade-offs, as commonly assumed, resource based, or are they due to mechanisms independent of energy allocation?
- What are the mechanisms that mediate life history plasticity?

1.4.2 The molecular identity and function of genes that affect life history

Studies in molecular and developmental genetics inform us about the molecular identity and function of genes, including those that affect life history traits and other fitness components. The functionally best-understood genes that affect life history traits have been analyzed in model organisms such as *Arabidopsis, Drosophila,* or *C. elegans.* Information about the function of such genes is useful, for example, when evolutionary biologists want to investigate the consequences of allelic variation at such loci in natural populations. Although natural alleles might have much more subtle phenotypes than laboratory induced mutant alleles, detailed knowledge about gene function might help organismal biologists to understand whether and how particular genes contribute to ecologically relevant phenotypes and thus why selection acts on such loci. This does not mean that every gene with a major phenotypic effect on a fitness-related trait, as identified by molecular genetics, is in fact ecologically or evolutionarily relevant in natural populations; many such genes might not harbor standing genetic variation affecting life history phenotypes and might therefore not contribute to evolutionary change in the wild. Yet, it is also clear that loci that do contribute to phenotypic variation in fitness-related traits and thus to ecological adaptation in natural populations are a subset of all genes, including those that have been functionally studied by molecular geneticists (e.g., Stern 2000, Flatt 2004, Flatt and Schmidt 2009).

While developmental and molecular genetic approaches do inform us about the ecological or evolutionary significance of specific genes, they have proved powerful in identifying the molecular mechanisms that affect life history traits, for instance their endocrine regulation (Tatar *et al.* 2003, Fielenbach and Antebi 2008). Perhaps the best examples are genes known to affect adult survival and longevity in the nematode, fruit fly, and mouse; these have received particular attention, not only from biomedical researchers because of their potential implications for human gerontology (see Chapter 16), but also from evolutionary biologists because of their potential relevance for understanding the evolution of aging. During the past 20 years, numerous mutations that extend lifespan have been identified in diverse model organisms (e.g., Kenyon 2010; also see Chapter 14). Many of these mutations were found to affect a key metabolic pathway—the insulin/insulin-like growth factor signaling pathway—indicating that decreased effectiveness of insulin/IGF-like signaling causes lifespan extension, linked to correlated responses in reproduction, growth, and metabolism. These pivotal discoveries, many of which are discussed in this book, not only demonstrate the feasibility of molecular genetic analyses of complex life history traits such as lifespan, but also suggest that certain evolutionarily conserved signaling pathways are potential key regulators of major life history traits (also see Chapters 27 and 28). Many of these findings have also contributed to our understanding of life history trade-offs (see below and Chapters 11 and 13). The molecular genetic analysis of lifespan has thus rapidly become of great interest to many researchers studying life histories, and this interest is now paving the way for an integration of mechanistic and evolutionary approaches towards the understanding of life history variation (e.g., Partridge and Gems 2006, Flatt and Schmidt 2009).

In addition to functional studies of individual mutations, genome-wide gene expression analyses have also been widely used by both molecular and evolutionary biologists to investigate the proximate basis of life history variation (as is discussed in detail in Chapter 2). For example, genome-wide transcriptional profiling has been used to identify candidate genes involved in lifespan regulation (e.g. Murphy *et al.* 2003), or to describe gene expression patterns associated with particular life history stages, for example dauer larva formation in *C. elegans* (Wang and Kim 2003). Many of these studies illustrate the complex and manifold changes in gene expression associated with life history variation and further indicate that life history trade-offs might emerge through "conflicts over gene expression", i.e., antagonistic pleiotropic effects of genes involved in multiple functions (Stearns and Magwene 2003, Bochdanovits and de Jong 2004). However, the functional interpretation of such data remains challenging because the precise causal connections between transcriptional changes and the resulting phenotypes are rarely known. Thus, while it is clear from these few examples that we have learned a great deal about the molecular genetic basis of life history traits, a current key challenge is to integrate such mechanistic insights into the evolutionary framework (also see Chapters 27 and 28). One obvious question for the evolutionary biologist is, for example, whether the candidate genes identified by molecular geneticists actually matter in natural populations.

1.4.3 Are candidate life history genes ecologically and evolutionarily relevant?

Mutational, transgenic, and genomic analyses in model organisms have been successful in identifying at least some of the key mechanisms that affect life history traits. However, while many of these mechanisms show a surprisingly high degree of conservation across widely divergent taxa, their relevance in shaping evolutionary life history variation in natural populations is not yet sufficiently clear. Determining whether and how such mechanisms evolve to generate natural life history variation represents a promising starting point for the integration of functional and evolutionary analysis of life histories. In most cases, however, such studies are limited to model organisms. Such an analysis requires testing of whether the genes involved in these candidate mechanisms show actual variation in natural populations and, as a more challenging step, to functionally demonstrate that this allelic variation impacts the life history trait in question.

Several studies suggest that genes identified through molecular and developmental genetic analyses indeed harbor natural allelic variation that contributes to population variation in life history traits, for example in *Drosophila* (e.g., Schmidt *et al.* 2000, Paaby and Schmidt 2008, Paaby *et al.* 2010; also see Chapter 18), or in *Arabidopsis* (e.g., Todesco *et al.* 2010; also see Chapter 9). Although the screening of natural polymorphisms in candidate life history genes only provides a first glimpse of the molecular basis of life history variation, such initial findings are encouraging since they indicate that developmental and molecular genetic studies indeed generate valuable candidate genes of interest for evolutionary biologists.

In contrast to the analysis of natural allelic variants at major candidate loci identified by molecular and developmental genetics, quantitative trait locus (QTL) mapping provides a less biased, yet technically challenging, approach to the characterization of the genetic basis of polygenic quantitative traits, including life history traits (Falconer and Mackay 1996). While classical QTL mapping approaches have been useful in determining the basic genetic architecture of life history traits (e.g., the number and effect size of the involved loci), they rarely achieve sufficient resolution to pinpoint individual candidate genes (see discussion in Roff 2007b and Mackay *et al.* 2009). However, recent technological advances, such as rapid and cost-effective genotyping methods and refined statistical and mapping methods, have increased the feasibility of high-resolution mapping, now allowing the identification of candidate genes within QTL regions for organisms with well-annotated genomes, in some cases down to the level of single nucleotide polymorphisms (e.g., Mackay *et al.* 2009). High-resolution mapping through recombinant inbred lines and genome-wide association studies have already been successful in characterizing natural polymorphisms underlying genetic variation in complex developmental or life history traits in *C. elegans* (e.g., Kammenga *et al.* 2007, Palopoli *et al.* 2008), *Drosophila* (e.g., De Luca *et al.* 2003, Schmidt *et al.* 2008, also see Flatt and Schmidt 2009 for a recent review), and in *Arabidopsis* (e.g., Atwell *et al.* 2010, also see Chapter 9). Moreover, recent progress in genomic methods now allows the researcher to treat genome-wide expression patterns as complex quantitative traits (e.g., Rockman 2008).

The recent advent of refined QTL and genetical genomics approaches is emblematic for an integrative and novel research program, namely the use of natural genetic variation as a tool to understand the causal connection between genotype and phenotype. By explicitly taking evolutionary variation into account, this approach holds great promise for facilitating the detection of mechanistic features that are involved in phenotype construction. However, the identification of individual genes or nucleotide polymorphisms that contribute to quantitative trait variation remains a major challenge because of subtle phenotypic effects, complex genetic interactions, pleiotropy, and genotype-by-environment interactions (e.g., Weigel and Nordborg 2005, Mackay *et al.* 2009).

1.4.4 How do trade-offs work?

One central and recurring theme in this book is the mechanisms that underlie life history trade-offs (see the chapters in Part 6). Given the central importance of such trade-offs in life history evolution, uncovering their mechanistic basis is one of the most fundamental but unresolved problems in life history research (e.g., Stearns 2000, also see Chapters 27 and 28). Despite numerous and seemingly obvious trade-offs between life history traits in a wide range of taxa, most reported trade-off relationships basically describe no more than a statistically inferred negative correlation. The description of trade-offs

by means of trait correlations or covariances is, however, insufficient for evaluating how genetic architecture influences evolutionary trajectories (e.g., see Chapter 2 and Roff 2007b). Specifically, it remains to be determined to what extent presumptive trade-offs are conclusively due to actual competition for limited resources or caused by alternative mechanisms, such as hormonal signaling independent of resource allocation (see Chapters 11, 13, 27, and 28). The very limited knowledge on the mechanistic underpinnings of trade-offs therefore represents a current key problem in our understanding of life history evolution (e.g., Stearns 2000, Flatt *et al.* 2005, Roff 2007b, Flatt and Schmidt 2009).

Recent progress in this area comes again from the molecular genetic analysis of lifespan. Several studies on the relationship between lifespan and reproduction in worms and flies have challenged the fundamental notion that reproduction exacts an energetic cost in terms of reduced survival (e.g., see Chapter 11, Leroi 2001, Barnes and Partridge 2003). Of particular relevance was the observation of a *C. elegans* insulin receptor mutant with extended lifespan (Kenyon *et al.* 1993). Although this mutant exhibited decreased fecundity—consistent with a resource-allocation trade-off where investment in longevity extension lowers investment in reproduction—detailed experimental analysis of this relationship indicates that decreased reproduction is not the causal agent in extending longevity (e.g., Kenyon *et al.* 1993, Leroi 2001). Therefore, reproductive versus somatic investment may not necessarily be coupled through resource competition but rather via independent underlying signaling processes (see Chapters 11, 13, and 24, and Hsin and Kenyon 1999, Flatt *et al.* 2008b). While these findings do not prove the absence of a cost of reproduction (Barnes and Partridge 2003, Flatt and Schmidt 2009), they underscore the difficulty of inferring resource-allocation trade-offs without a precise understanding of the proximate mechanisms involved. For example, a major technical challenge in demonstrating the resource basis of trade-offs is to experimentally track resource allocation to different organismal functions by detailed measurement of relevant parameters, such as nutrient ingestion and assimilation (see Chapter 24 and O'Brien *et al.* 2008).

Other valuable information on the mechanistic basis of life history trade-offs comes from research exploring the fitness consequences of organismal defensive mechanisms against pathogens, parasites, stresses, or toxins. For example, studies in both vertebrates and invertebrates indicate that elevated immune and other defense functions incur fitness costs in terms of reproduction and survival (see, for example, Chapters 2 and 23, Flatt *et al.* 2005, Harshman and Zera 2007). Similarly, the evolution of pesticide tolerance in insects often results in a fitness cost, which is generally supposed to stem from increased energy allocation to corresponding detoxification mechanisms. Remarkably, however, it turns out that such fitness costs can result from collateral metabolic costs rather than energetic costs due to the detoxification mechanism (Van Straalen and Hoffmann 2000).

Thus, while many observations support the existence and evolutionary relevance of life history trade-offs, their underlying causal mechanisms still remain rather poorly understood. Importantly, one of the central postulates of life history theory, namely that trade-offs are caused by competitive resource allocation, might not necessarily always hold. As discussed in many chapters throughout this book (e.g., Chapters 11, 13, 27, and 28), major efforts are currently under way to dissect the mechanistic basis of life history trade-offs.

1.5 Conclusions

Combining mechanistic and evolutionary analyses of life history variation is a fundamental yet ambitious aim in current biology. On the one hand, there are inherent biological and technical problems with studying complex quantitative phenotypes such as life history traits. On the other hand, there are cultural divides that necessitate a combination of diverse research approaches and concepts from both molecular and organismal biology. Despite these challenges, the chapters in this book illustrate that the successful integration of mechanisms into life history research is fully under way.

CHAPTER 2

Genomic insights into life history evolution

Derek A. Roff

2.1 Introduction

Recent advances in the analysis of genetic variation and genetic control at the level of the genome have contributed significantly to our understanding of the functional basis of life history evolution (Roff 2007a, 2007b). In this paper I address two issues for which genomic analyses are important for our understanding of life history evolution: genomic regulation of trade-offs and the relative importance of stochastic versus deterministic factors.

In life history theory a trade-off is defined as the covariation between two traits where a change in one trait that by itself increases fitness is accompanied by a change in the second trait that results, by itself, in a decrease in fitness. Typically, this trade-off is operationally measured by a negative correlation between the two traits (e.g., survival and reproductive effort). This correlation could arise from linkage disequilibrium, in which case it may only be a transitory phenomenon, or by antagonistically pleiotropic genes (i.e., genes that have opposite fitness effects on the two traits). The existence of trade-offs is a central pillar of evolutionary theory in general and life history theory in particular; it is the interaction of trade-offs that limits and directs evolutionary responses.

Responses to selection can be predicted, at least in the short term, by the multivariate equivalent of the breeder's equation

$$\mathbf{R} = \mathbf{GP}^{-1}\mathbf{S}$$

where $\mathbf{R}$ is the vector of changes in mean trait values, $\mathbf{G}$ is the genetic variance-covariance matrix, $\mathbf{P}$ is the phenotypic variance-covariance matrix, and $\mathbf{S}$ is the vector of selection differentials. Phenotypic trade-offs are represented by the phenotypic covariances in $\mathbf{P}$ but their influence on evolutionary changes depends on the existence of genetic covariances of the appropriate sign. The above equation is general in the sense that the functional source of trade-offs is not considered. Indeed, the strength of this particular approach is that it is based not upon the particular cause of the trade-off, but rather *the emergent properties of the interaction of the functional systems*, namely the means, variances, and covariances. However, understanding the functional underpinnings of a trade-off may facilitate better predictions of evolutionary trajectories by combining functional components into the quantitative genetic framework; for an example in the framework of an experimental evolution study see Roff and Fairbairn (2007). An understanding of the functional basis of trade-offs may also enable us to predict the endpoints of a selective process even though it may not enable the exact trajectory to be plotted. Assuming, as seems reasonable, that similar functional mechanisms will exist, at least among closely related taxa, then an analysis of the functional basis of constraints may help explain the diversity, or lack of diversity, in life history traits among taxa. Thus the approach may prove informative in explaining the present distribution of trait complexes, which in life history theory involve traits such as development time, fecundity, age at maturity, and so forth. These arguments do not merely apply to genomic analyses but to functional analyses in general. Thus an analysis of physiological constraints may elucidate why particular patterns occur. Genomic analyses provide a lower level of analysis and have the potential to account for the apparent constraints at the higher levels (e.g., physiology). Of course, it is possible that as one proceeds down such a hierarchical analysis the biological details may become more and more specific and hence less general. At this point I believe that the jury is still out on the extent to which genomic analyses may provide general answers. It is a question that cannot be answered by theory, but only by observation and experiment.

The multivariate equation described above is a deterministic equation predicting the mean trait values. In the real world, finite population sizes lead to generation-to-generation fluctuations due to random sampling, fluctuations that are superimposed on the mean trajectory and which might even effectively obliterate the deterministic response. The importance of random genetic drift is well recognized in selection experiments, but its relative importance in the evolutionary change in wild populations is poorly studied and its importance in generating variation among taxa is even less well understood. An important question in evolutionary biology is "to what extent are evolutionary trajectories and the endpoints of evolution functions of stochastic or deterministic factors?" The role of stochastic versus deterministic factors in contributing to an evolutionary trajectory might vary. For example, selection for increased fecundity might be facilitated by up-regulation of several factors; which particular factor is up-regulated in the early course of selection might be largely determined by random events. In the later course of selection, however, when the phenotype approaches an extreme, this up-regulation may be possible only by a single combination of factors. Thus stochastic factors may dominate in the early course of evolution but the endpoint will be deterministic. On the other hand, it may be that early divergence may lead to radically different solutions to a selective pressure. This issue is the subject of the second part of this chapter.

For both cases I present illustrative case studies that outline the importance of genomic contributions to our understanding of the evolutionary process. I take a fairly liberal view of life history traits and for the present discussion include morphological variation such as body size as this trait

frequently contributes to reproductive success and developmental duration (Roff 2002).

2.2 Genomic analysis of trade-offs

There are at least four areas in which genomic approaches can advance our understanding of the causal mechanisms of trade-offs:

1) transgenetic analyses
2) quantitative trait locus (QTL) analyses
3) microarray analysis of single gene action
4) microarray analysis of multiple gene action.

For each case I present an illustrative example. In addition to these examples I would draw attention of the reader to three other studies that exemplify the utility of genomic approach to understanding the molecular basis of trade-offs. These studies are:

1) The effect of a point mutation on antagonistically pleiotropic genes in the bacterium *Pseudomonas fluorescens* (Knight *et al.* 2006)
2) Antagonistically pleiotropic genes affected by the FRIGIDA locus in the plant *Arabidopsis thaliana* (Scarcelli *et al.* 2007; see also Chapters 1 and 9)
3) A microarray analysis of genomic variation between dwarf and "normal" whitefish, *Coregonus sp* (St-Cyr *et al.* 2008).

2.2.1 Case Study 1: A transgenic analysis of the cost of resistance in *Arabidopsis thaliana*

Resistance to herbicides, pesticides, and rodenticides is so common that it is generally accepted that any new introduction will be quickly followed by the evolution of resistance. It is also a common observation that once the killing agent is removed resistance equally quickly disappears, or is considerably reduced (for an example in which co-adaptation reduced the fitness cost of resistance see McKenzie (2001)). Life history theory predicts that this cost of resistance will be on some other fitness trait such as development time or fecundity. For example, if resistance, which means increased survival, occurs because of the up-regulation of some agent that metabolizes or otherwise nullifies the effect of the poison then on the assumption that energy is limiting there will be a decline in some other trait, such as a decrease in the rate of growth or production of eggs. Such trade-offs between survival and other life history traits have been observed but the causal genetic basis has not been elucidated in most cases. Demonstrating that the cost is associated with particular resistance genes requires an ability to manipulate the genome, an endeavor that is presently possible in only a few "model" organisms. An excellent example of such a manipulation is given by Tian *et al.* (2003) in their transgenic analysis of resistance in the plant *A. thaliana.*

In plants there is a suite of genes collectively called resistance genes (*R*-genes) that enable the plant system to recognize pathogens and induce a defensive response to these. One particular locus, *RPMl,* in *A. thaliana* codes for a peripheral plasma membrane protein that enables the plant to recognize the pathogen *Pseudomonas syringae* (for a detailed discussion of *R* gene evolution and mode of action see McDowell and Simon 2006). Accessions that are susceptible to this pathogen lack the entire coding region of *RPMl* (Shimizu and Purugganan 2005), thus implicating this locus as a major factor in resistance. Tian *et al* (2003) were able to move only the *RPMl* region into a susceptible ecotype, Bla-2, and created four independent transgenic lines. Next they made use of a Bla-2 line expressing *cre* recombinase to produce matched pairs of the transgenic lines that lacked the *RPMl* region. Thus they were able to compare the response of four genomes, differing only in the presence or absence of the *RPMl* locus, to infection by *P. syringae.*

In all four pairs the resistant genotype showed a statistically significant reduction in total seed production relative to the matched non-resistant genotype. There was no significant difference among the four transgenic lines in the cost of resistance, which amounted to a 9% decrease in seed production. Such a cost would rapidly eliminate the resistant genotypes from a population not faced with continual infestation by the pathogen. This cost can in large part explain the maintenance of *R* gene polymorphisms (Ehrenreich *et al.* 2006). An important aspect of this study is that it was conducted in the field. To assess the actual cost of resistance, experiments should be carried out

under realistic conditions (Zavala *et al.* 2004). Experiments looking for trade-offs under laboratory conditions frequently find no or only small costs, presumably because laboratory conditions are relatively benign (Roff 1992, Roff 2002, Heidel *et al.* 2004).

2.2.2 Case Study 2: A QTL analysis of the cost of resistance to parasite infection in *Tribolium*

QTL analyses have now been extensively carried out on a large array of organisms. While these analyses are valuable in locating a region of the chromosome that is associated with a particular trait they can suffer from a lack of precision if the sample size is insufficiently large or even be misleading (the "Beavis effect": Beavis 1994, Roff 1997, Xu 2003). With respect to trade-offs QTL analysis can be used to locate chromosomal regions that are associated with both the component traits of the trade-off. QTL analysis can be particularly useful in determining if the trade-off is a consequence of linkage disequilibrium, and hence transitory, or pleiotropy. Under the latter hypothesis there should be individual QTLs that are associated with both of the traits. This is well illustrated by the analysis of the cost of resistance in *Tribolium casteneum* to infection by the tapeworm *Hymenolepis diminuta.*

Susceptibility of *Tribolium* to *H. diminuta* varies both between species—*T. casteneum* being more susceptible than *T. confusum*—and among strains (Yan and Norman 1995). The presence of genetic and phenotypic variation for resistance implies that there is likely to be a fitness cost associated with resistance to the parasite. To test this hypothesis Yan and Stevens (1995) compared the relative fitness (i.e., proportion of offspring) of infected and non-infected flour beetles under high and low competitive regimes (Fig. 2-1). Under high competition, male reproductive success was reduced by 42.9% and female reproductive success was reduced by 14.4%, both costs being statistically significant. Only male fitness was estimated under the low competitive regime: the infected males showed a non-significant reduction in fitness (45.6% vs 47.2% for infected and non-infected, respectively). These results indicate that there is a reproductive cost to being infected but that it is a function of the

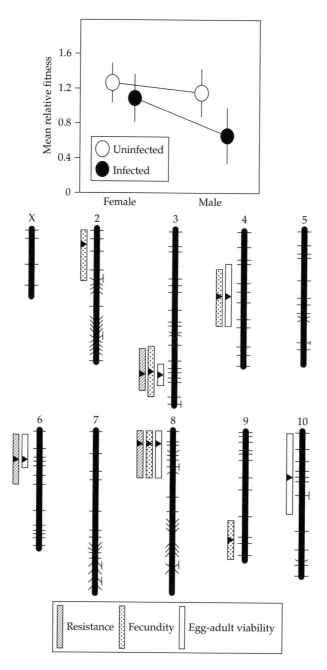

Figure 2-1 The top panel shows the fitness consequences of tapeworm infection in *Tribolium casteneum* under conditions of high competition (Redrawn from Yan and Stevens 1995). The lower panel shows the colocalization of the QTL for resistance to tapeworm infection and two fitness traits in *T. casteneum*. Horizontal lines show positions of the markers. Bar lengths indicate the 95% confidence interval for the QTL and ▶ shows the map position with the highest LOD score (modified from Zhong *et al.* 2005).

environment. Parenthetically it is worth noting that infection increased the likelihood of *T. casteneum* winning in competition with *T. confusum*, though the mechanism underlying this paradoxical result is not known (Yan *et al.* 1998).

Given this background, Zhong *et al.* (2005) used QTL analysis to locate chromosomal regions associated with resistance and the fitness components, fecundity, and egg-to-adult viability. For the QTL analysis a highly resistant strain of *T. casteneum* was crossed with a highly susceptible one and from this cross two independently segregating populations were constructed. Three QTL were located on separate chromosomes, with those from Cross 1 accounting for 15%, 32%, and 11% of the phenotypic variation in resistance and those from Cross 2 accounting for 8%, 21%, and 15%, respectively. All three QTL were associated with QTL for egg-to-adult viability and two were also associated with fecundity (Fig. 2-1). Because the QTL encloses a fairly wide region it is not possible to determine from this analysis if the trade-off is a result of the pleiotropic effects of particular genes or that genes for resistance are linked with genes for the fitness components. Nevertheless the results are certainly suggestive of pleiotropic effects and provide a region that can be dissected by more refined molecular methods.

2.2.3   Case Study 3: A microarray analysis implicating a single gene in the cost of resistance to DDT in *Drosophila melanogaster*

As shown in Case Study 1, a single gene can be a major component of the causal mechanism underlying a trade-off. The present example illustrates the use of microarray analysis to demonstrate that a single gene can be important in reducing male reproductive success while increasing resistance to the pesticide DDT.

In this study Drnevich *et al.* (2004) examined microarray variation among male *Drosophila melanogaster* from a single population that were known to differ in male reproductive success under competitive conditions (MCRS). To assess differences in gene expression they chose three genotypes that showed consistently high MCRS and three genotypes that showed a consistently low MCRS, hereaf-

ter referred to as the H and L genotypes, respectively. As expected from their means of selection, H and L genotypes differed significantly in MCRS values. A "volcano plot" of the probability associated with the difference between the two genotypes and the effect size (the difference in expression between the two genotypes) showed that a single gene, cytochrome P450 (*Cyp6g1*), had the highest level of significance and second-largest effect size (Fig. 2-2). High expression levels of *Cyp6g1* were associated with lowered male reproductive success. Overexpression of this gene is known to confer broad resistance to a variety of insecticides (Le Goff *et al.* 2003) and thus is a prime candidate as an antagonistically pleiotropic gene. This hypothesis was supported by the research of Festucci-Buselli *et al.* (2005), who showed that *Cyp6g1* was relatively over-expressed in fly strains that were resistant to DDT (Fig. 2-2).

While the above data support the hypothesis that a single gene plays a major role in resistance it is also clear from other work that DDT resistance overall is a polymorphic trait (Festucci-Buselli *et al.* 2005). Expression of the gene *Cyp6g1* is itself regulated by a retrotransposon upstream of the transcription site (Chung *et al.* 2007). Maintenance of the gene in the population appears not only to be a function of the antagonistic pleiotropy between male reproductive success and resistance but also a positive pleiotropy in females in which resistant flies are more fecund than susceptibles (McCart and Ffrench-Constant 2005, 2008). Importantly, such a finding may indicate that the resistant gene was already present in the population prior to the use of DDT, a situation found in malathion resistance in the Australian sheep blowfly (Ffrench-Constant 2007).

2.2.4   Case Study 4: A microarray analysis of antagonistic pleiotropy and gene expression in *Drosophila melanogaster*

In the previous case study it was possible to associate a single gene with a particular trade-off. However, because life history traits such as fecundity, development time, and survival show continuous variation, trade-offs between traits will most likely be a result of antagonistic pleiotropy at several genes. One way to approach this problem is to

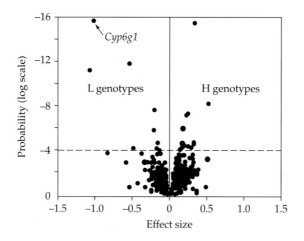

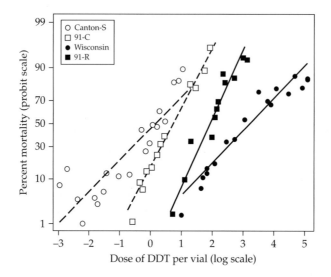

Figure 2-2 "Volcano" plot of microarray results comparing genomes of *Drosophila melanogaster* males with high (H) and low (L) mating success. The *y*-axis shows the log(probability) for each contrast (average expression in H genotype minus average expression in L genotypes). The dashed line indicates the level of significance given the number of contrasts examined. Redrawn from Drnevich *et al.* (2004). The bottom plot shows dose–response curves for DDT-induced mortality in two DDT-susceptible (Canton-S, 91-C) and two DDT-resistant (Wisconsin, 91-R) fly strains that differ in *Cyp6g* levels; both resistant strains were found to have higher expression of *Cyp6g* (not shown). Redrawn from Festucci-Buselli *et al.* (2005).

use microarray analysis, which gives the action of thousands of genes at a particular moment in the life cycle of the organism. While this approach can identify putative antagonistically pleiotropic genes, further experiments are necessary to confirm these as such and to determine the precise mode of interaction.

Bochdanovits and de Jong (2004) used this "shotgun" approach to isolate genes that showed opposite effects in two life history traits in *D. melanogaster*, namely adult male weight and pre-adult survival. Gene expression was determined in third instar larvae of nine isofemale lines (five from Wenatchee, WA, USA and four from San José, Costa Rica). Expression profiles of 1670 genes were assessed and these correlated with the isofemale line means for the two traits. This resulted in two distributions of correlation values from which

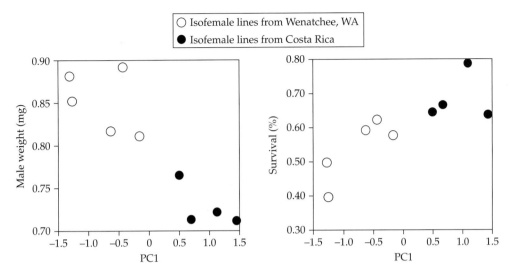

Figure 2-3   Scatter plots between PC1 of the 34 candidate "trade-off" genes and the fitness traits, male weight and pre-adult survival. Redrawn from Bochdanovits and de Jong (2004).

"significant" values were drawn by a cutoff at the lower and upper tails. To reduce the effect of false positives, Bochdanovits and de Jong used a protocol that resulted in the upper and lower 3.5% of the distributions of correlations. Genes that were candidates for being classed as antagonistically pleiotropic were defined as those that appeared in the upper tail of one distribution and the lower tail of the other. Thirty-four antagonistic pleiotropy genes were thus isolated and their joint effects evaluated by considering the first principal component (PC) score (PC1), which accounted for over 70% of the overall variation. There was a highly significant negative correlation between adult male weight and PC1 and a highly significant positive correlation between pre-adult survival and PC1 (Fig. 2-3). A two-way analysis of covariance indicated that PC1 accounted for 86.3% of the observed trade-off, which was more than accounted for by the correlation between the two life history traits. Although the molecular and biological function of some of the genes is known, their individual importance in the trade-off was not (could not) be assessed.

The foregoing four case studies illustrate the potential benefits of a genomic approach to dissecting the functional components of variation in life history traits. In the next section I discuss the question of the extent to which such functional components are determined by drift or selection. The central issue is the question, "To what extent is variation among populations a consequence of random variation among founding individuals constraining the future evolutionary trajectory?"

2.3   To what extent is the phenotype determined by different molecular/developmental mechanisms?

In their review of parallel genotypic adaptation, Wood *et al.* (2005) cite numerous cases in which the same amino acid or nucleotide substitution occurs in different taxa, or have been observed in experimental evolution studies of viruses and bacteria (see their Table 2-1). Additionally, a few studies have demonstrated similar genome composition in multiple lines of eukaryotes under experimental evolution or the same QTLs in different species

(Wood *et al.* 2005). The probability of parallel evolution by mutation at the DNA sequence level has been analysed by Orr (2005), who showed that if there are n beneficial mutations the probability that replicate populations will fix the same mutation is $2/(n + 1)$. Thus the probability of parallel evolution for simple substitutions can be quite high. How often complex developmental pathways are repeated in different lineages is far from clear (for a discussion of how gene networks may arise *de novo*, see Monteiro and Podlaha 2009). Equally unclear is the question of how frequently different developmental pathways produce the same phenotype.

A population that is separated into two populations will over time diverge due to genetic drift, even if subject to the same selective regime. This has been experimentally demonstrated by comparing populations subject to the same artificial selection (see below). Such experiments raise the possibility that the same phenotype may be determined by more than one genotype and that future evolutionary trajectories may be constrained by the initial variation within the founding population. This contrasts with the much-studied phenomenon of phenotypic plasticity, in which a single genotype can give raise to multiple phenotypes (Pigliucci 2001, Whitman and Ananthakrishnan 2009). From a Mendelian geneticist's point of view the production of the same phenotype from several different genotypes is not only unsurprising but is to be expected. Consider, for example, the simple two locus model (each locus labeled, say, A and B), in which each locus comprises two alleles which contribute either 0 or 1 to a phenotypic trait. Now suppose the organism is subject to selection for a phenotypic value of 2: at equilibrium the population will consist of one locus fixed for allele 1 and the other locus fixed for allele 0. If the initial frequencies are 50:50 then in half the cases locus A will be fixed for 1 and in the other half locus B will be fixed for 1. While this simple model demonstrates the expectation of multiple pathways to the same phenotype, it does not demonstrate that such will be the case in more realistic models. There are two perspectives from which this problem can be viewed: that of the population geneticist and that of the molecular geneticist. The former traditionally views genes as somewhat abstract quantities, their

effects being measured in terms of means and variances. The latter is likely to look for causal molecular pathways. The divergence of developmental pathways without phenotypic divergence has been separated from genetic drift and called "developmental system drift" (DSD; True and Haag 2001). I disagree with True and Haag (2001, p. 109) that "the term 'drift' is clearly distinct from genetic drift, but nevertheless is appropriate because, as with genetic drift, chance and not selection determines the details of how developmental systems change under DSD." True and Haag do not make it clear why the phenotypic trait does not change even though the developmental pathway changes. As I argue above, such a process is most likely to result from a combination of drift and selection on the final phenotype. In this manner the separation made by True and Haag between DSD and convergence—in which selection produces the same phenotype in two different lineages, which initially differ in phenotype and developmental pathway, by different developmental pathways—is somewhat artificial as the only difference is the process that generates the initial variation among groups. If this initial separation is due to variation in selection then DSD is not involved, but if the initial separation is a consequence of genetic drift then DSD *is* involved. An interesting question is whether it is possible to have variation in gene frequencies that, from a population geneticist's model, do not lead to a change in phenotype and also do not change the molecular pathway.

That multiple pathways to the same phenotype exist is undoubtedly true but the important evolutionary question is how often do they occur? Equally, is it possible to have different genetic architectures but the same developmental pathway? For example, wing dimorphism, in which there are macropterous (long-winged) and brachypterous (short-winged) morphs in the same population, appears to be determined by a single gene or closely linked set of genes on the X chromosome of the cricket *Gryllus rubens*. However, dimorphism is determined by multiple segregating genes on the autosomes of *Gryllus firmus* (Roff 1990). There is no doubt that the two species differ in their genetic architecture but whether this changes the physiological determination of wing dimorphism is unknown. It is reasonable to assume that a change in genetic architecture will change the developmental pathway, either by altering regulatory genes or by qualitatively altering the developmental pathway. Similarly, if the developmental pathways differ then it is reasonable to assume that the genetic architectures are different. For example, oocyte maturation in some Lepidoptera is controlled by increased production of juvenile hormone (JH), whereas in other species JH plays no role (Hodin 2009). Another perhaps even more

Table 2-1   A brief review of cases in which the same phenotype is or is not produced by different genetic architectures/molecular pathways

Trait	Finding	Species	Reference
Studies among species or sub-species			
Fluctuating asymmetry of morphometric ratios	Hybrids did not differ	*Lepomis macrochirus* (sunfish)	1
Macrochaetae on notum	1) Hybrids differed 2) Hybrids did not substantially differ	1) *D. melanogaster, D. simulans* 2) *D. melanogaster* vs *D. mauritiana*, or *D. sechellia*	2
Morphology of legs	Hybrids had sex combs on the second and third pairs of legs in males	*D. subobscura* and *D. madeirensis*	3
Wing dimorphism	Major gene on X in first sp. but not second	*Gryllus rubens* and *G. firmus*	4
Wings	Interruption of wing development network at different points	Various ant species	5
Sex	Self-fertility achieved by intervention at different points in pathway	*Caenorhabditis elegans* and *C. briggsae*	6
Mating	Different molecular pathways	Baker's yeast and ancestral strain	7
Studies within species			
Resistance to antimalarial drug	Same mutation	*Plasmodium falciparum*	8
Lateral plate number	Same major locus responsible	*Gasterosteus aculeatus* (stickleback)	9
DDT resistance	Different metabolic pathways	*D. melanogaster*	10
Warfarin resistance	Different mutations	Brown rats	11
Wing components	Different genetic architecture	*D. melanogaster*	12
Wing size	Variation in wing components	*D. subobscura*	13
Artificial selection			
Adaptation to laboratory culture	Parallel adaptation	Bacteriophage	14
Adaptation to a novel environment	Various responses (some parallel)	Bacteriophage	15
Adaptation to a novel environment	Different end points to same phenotype	*E. coli*	16
Adaptation to a novel environment	Similar responses	*E. coli*	17
Fluctuating glucose-galactose environment	Same mutation	Yeast	18
Wing dimensions	Different architectures in some races but not others	*D. melanogaster*	19
Learning	Different architectures	*D. melanogaster*	20
Competitive ability	Different mechanisms	*D. melanogaster* and *D. simulans*	21
Weight	Different architectures	Mice	22
Nesting score	Different architectures	Mice	23
Open-field activity	Same QTLs	Mice	24
Fecundity	Mostly same QTLs	*Arabidopsis thaliana*	25

References: 1) Felley (1980); 2) Takano (1998); 3) Papaceit *et al.* (1991); 4) Roff (1990); 5) Abouheif and Wray (2002); 6) Hill *et al* (2006); 7) Tsong *et al.* (2006); 8) Musset *et al.* (2007); 9) Cresko *et al.* (2004), Schluter *et al.* (2004); 10) Pedra *et al.* (2005); 11) Pelz *et al.* (2005); 12) Gilchrist and Partridge (1999); 13) Gilchrist *et al.* (2001, 2004), Calboli *et al.* (2003); 14) Wichman *et al.* (1999); 15) Bollback and Huelsenbeck (2009); 16) Fong *et al.* (2005); Woods *et al.* (2006); Ostrowski *et al.* (2008); 17) Pelosi Tab2 *et al.* (2006); Cooper *et al.* (2008); 18) Segre *et al.* (2006); 19) Weber *et al.* (2008); 20) Kawecki and Mery (2006); 21) Joshi and Thompson (1995); 22) Mohamed *et al.* (2001); 23) Bult and Lynch (1996, 2000); 24) Turri *et al.* (2001); 25) Ungerer *et al.* (2003a,b).

dramatic example is the effect of testosterone on growth rate in the lizards, *Sceloporus undulates, S. virgatus* and *S. jarrovi*: in the first two species elevated testosterone stimulates growth while in the third species growth is inhibited (John-Alder and Cox 2007).

In the absence of a theory to guide us in predicting the prevalence of multiple pathways to a common phenotype we must fall back on empirical evidence: if it is a common occurrence then it should be readily found in experimental studies. In the three next sections I present an overview of the evidence for the widespread existence of multiple pathways to determine if the hypothesis is tenable and worth a more detailed study. Finally, I suggest an artificial experimental approach that would address this issue. The empirical evidence can be divided into three categories:

1) comparisons among species
2) comparisons among natural populations of the same species
3) results from artificial selection experiments (Table 2-1).

2.3.1 Comparisons among species

Seven examples are given in Table 2-1. In three cases variation, or lack of, in the developmental pathway was demonstrated by crossing species. Given that the genetic architecture and physiology among species is likely to be quite substantial and hybrid developmental pathways generally compromised, this type of evidence is not very convincing. Furthermore, it is possible that the genetic architecture of one species determines the outcome of development in the hybrid and thus the lack of a change in the hybrid does not tell us if the developmental pathways within each species differ. Better evidence comes from an analysis of the genetic architecture (e.g., wing dimorphism) or the analysis of the developmental pathway itself (wing development in ants, sex in *Caenorhabditis*, mating in yeast).

The study by Abouheif and Wray (2002) illustrates how the same phenotype can be produced by alteration of the same developmental pathway. The castes of many ant species display variation

in the presence or absence of wings, the origin of which appears to have occurred at least 90 million years ago. The developmental pathway for the production of wings is highly conserved but the disruption of wing development differs among species (Fig. 2-4). The same phenotype, wingless, is thus produced by simple but different changes in the developmental pathways of different species.

Recently, Nahmad *et al* (2008) modeled this developmental system, asking if there are particular points in the network that would result in the same phenotype. Their model did indeed show that there are groups of genes that have similar effects on target gene expression and growth. Furthermore, although genes within these groups occupy different positions in the network their interruption produces vestigial wing discs that are similar in size and shape. Thus this model both supports the concept of DSD and gives a mechanistic explanation for the possible variation in developmental pathways. An important question is what effects, if any, these differences have on other developmental pathways; if the system is entirely divorced from other systems then disruption could presumably occur at any point. However, it is more likely that there are pleiotropic effects associated with disruption at different points and it would be very interesting to investigate how this affects other components of the phenotype, particularly life history traits, which are important components of fitness.

2.3.2 Comparisons among natural populations of the same species

Table 2-1 reports two studies in which the same phenotype was evolved independently in different populations by the same mutation and four studies in which there is evidence that different mutations were responsible for the same phenotype. At first glance these results support the hypothesis that the evolution of different developmental pathways or genetic architectures that produce the same phenotype is common. However, because the exact nature of selection is not known, such a conclusion may be premature. Consider the case of parallel clinal variation in *Drosophila subobscura*.

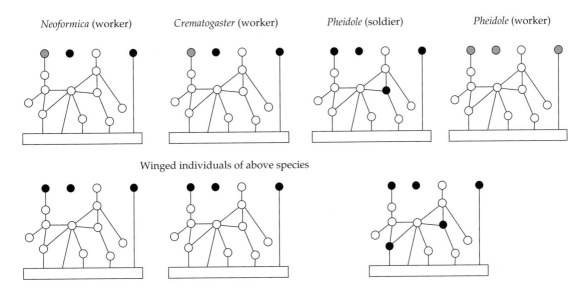

Figure 2-4 An example demonstrating that different molecular mechanisms can produce the same phenotype. Lower section shows the wing patterning network from the late larval stage of three ant species. For simplicity, regulatory interactions have been omitted. ○ Nodes not examine; ● conserved gene expression; ● gene expression interrupted. Modified from Abouheif and Wray (2002).

The fly *D. subobscura* is native to Europe, where it displays clinal variation in wing size (Prevosti 1955, Misra and Reeve 1964). In the late 1970s populations were accidentally introduced into western North America and South America (Huey *et al.* 2000). The species spread southwards in North America and northwards in South America and later collections showed the evolution of clinal variation in wing size in both American regions. In the North American region the cline in female wing size was convergent on that found in Europe (Fig. 2-5, Gilchrist *et al.* 2001), whereas that in the South American region showed the same slope but a different intercept (Gilchrist *et al.* 2004). However, while the overall wing size clines in Europe and North America were congruent, the components of wings were not. In Europe clinal variation is due to the length of the L1 vein and cell number but in North America clinal variation was a consequence of variation in the L2 vein and cell area (Fig. 2-5, Huey *et al.* 2000, Calboli *et al.* 2003). In contrast clinal variation in the South American populations was a consequence of variation in both L1 and L2 and cell number (Calboli *et al.* 2003).

One interpretation of these results is that selection has favored the evolution of wing size but

the manner in which wing size is determined is not critical and hence variation in the wing components among the regions is a result of developmental system drift. On the other hand, one cannot reject the hypothesis that the same cline in wing size was selected for in North America and Europe but that there was also selection on the wing components. This problem faces all comparisons using natural populations: in the absence of information on the factors of selection one cannot conclude that DSD is important, only that it is worth further investigation in any particular case. The solution to this problem is that outlined at the beginning of this section, namely to subject lines taken from a common population to the same selection regime.

2.3.3 Artificial selection experiments

Attention has already been drawn to experimental studies in which the same amino acid or nucleotide substitution has been found (Wood *et al.* 2005). Further studies in bacteriophage, *E. coli*, and yeast, in which similar findings have been made, are given in Table 2-1 in addition to experiments on the same organisms in which different substitutions were

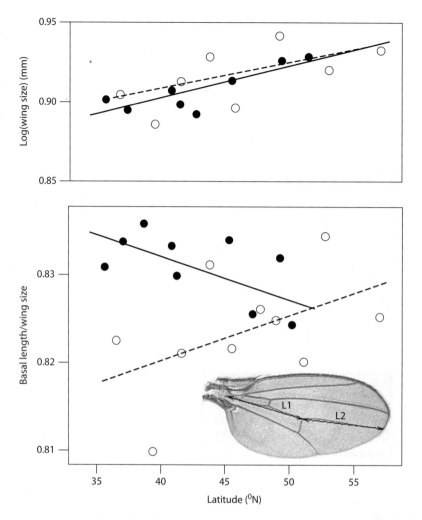

Figure 2-5 A comparison of the latitudinal cline in wing size and relative length of the vein L1 (arcsine square-root transformed) of *Drosophila subobscura* in North America and Europe. Open circle and dashed line show pattern for European populations; closed circle and solid line shows pattern for North American populations. Redrawn from Huey *et al.* (2000).

found. Artificial selection on eukaryotes has demonstrated that selection can produce similar evolutionary trajectories via changes in different genes. These experiments are explained in more detail in Table 2-2. The traits involved are all complex ones that undoubtedly involve the coordinated action of many genes. While selection in all cases was successful in altering the phenotype, the different lines achieved the response via different genetic architectures.

In the microarray analysis of variation associated with selection on wing dimensions in *D. melanogaster*, the results suggested that the same trajectories were followed when the lines were drawn from the same population (Mass) but that populations from different areas had initially different architectures (Weber *et al.* 2008). It is not possible to say if these initial differences were a result of DSD in the past or differences in selection in the two regions. However, the results from selection experi-

Table 2-2 Analyses of the constancy of changes in genetic architecture in artificial selection experiments

Species	Trait	Type of analysis	Finding	Reference
D. melanogaster	Wing dimensions	Microarray analysis of 12 selected lines	Ten lines from within a single population (Mass) showed parallel changes in loci. Populations from Massachusetts showed changes in different loci than those from two lines from California	1
D. melanogaster	Learning	Ten F1 crosses between lines selected for improved learning	Four crosses performed in a manner similar to parentals, six showed reduced or no learning	2
D. melanogaster, D. simulans	Competitive ability	Experimental evolution measured over 11 generations	Considerable variation among environments and among replicates	3
Mouse	8-week weight, 3–5 week weight gain, 5–8 week weight gain	Crosses between replicate lines selected in the same direction	Crosses not intermediate between parents	4
Mouse	Thermoregulatory nest-building behaviour	Crosses between replicate lines selected in the same direction	Crosses not intermediate between parents	5
Mouse	Open-field activity	QTL analysis. Replicated high and low lines crossed	Same four QTL identified in both intercrosses	6
A. thaliana	Three generations of viability and fertility selection	Non-specific molecular markers	Most markers changed in a similar manner between replicates but some did not	7

References: 1) Weber *et al.* (2008); 2) Kawecki and Mery (2006); 3) Joshi and Thompson (1995); 4) Mohamed *et al.* (2001); 5) Bult and Lynch (1996, 2000); 6) Turri *et al.* (2001); 7) Ungerer *et al.* (2003a,b).

ments showing differences for lines originating from a common population (Studies 2–5 in Table 2-2) do indicate the existence of DSD.

2.3.4 A proposed experiment and predictions

In this section I propose an artificial selection experiment to explicitly test for the occurrence of DSD and its impact on evolutionary trajectories. Approaches such as that suggested by Monteiro and Podlaha (2009) identify the occurrence of novel pathways but do not differentiate the origin of such variation, namely the interplay of genetic drift and selection. The experimental protocol, which is specifically designed to address this issue, is as follows:

1) Select a significant number, say ten or more, of gravid females from a single population. These might be selected at random or specifically to encompass the observed range of variation in the trait, such as development time, body size, or fecundity, to be selected. This initial selection generates

the sort of variation that one might find in a set of colonizing episodes. These ten females are the progenitors of the ten selection lines. If the traits under study are likely to be subject to inbreeding depression it may be necessary to begin each line with more than a single female: a reasonable compromise between avoiding inbreeding depression and creating variation would be five females mated to five males.

2) Subsequent to this "colonization", the lines are increased in size to ensure that genetic drift is negligible. This requires that the fecundity of the founding females is large enough to generate populations of at least 100 individuals, which will obviously restrict the type of organism to be used.

3) Following the founding episode the lines are subject to directional selection of sufficient degree that progress is made but that the selected number of parents is sufficient to minimize genetic drift.

4) At particular generations (the first included) the lines are examined for variation both in the selected trait and traits that could be assumed to be

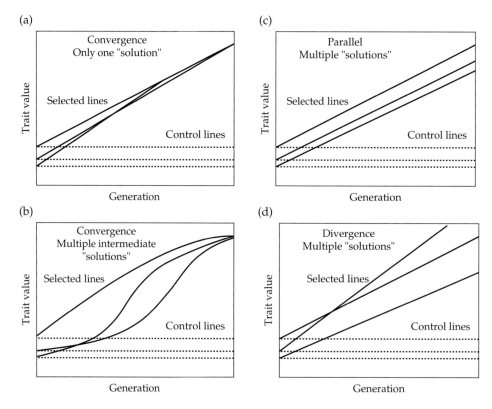

Figure 2-6 Schematic examples of the response to selection as described in the text. Three control and three selected lines are illustrated. The response variable "Trait value" refers to traits that underpin or are correlated with the trait under selection, which, for the purposes of illustration, are assumed to respond along the same trajectory (not shown). Panels a and b illustrate the situations in which there is only a single developmental pathway for the extreme phenotype but possibly several for the intermediate. Panels c and d illustrate the case in which there are multiple pathways (C is unlikely).

correlated with this trait. Examination of variation among lines should be by using a genomic approach such as microarray analysis and/or by a quantitative genetic approach of line cross analysis. Preferably both approaches are used and as many generations as logistically feasible studied.

The initial protocol will generate genetic variation among lines in the genetic and developmental architecture of the trait, which should tend to remain or even increase while the selected trait is not at an "extreme" value. As selection drives the trait to an extreme, either this variation will decrease due to there being only one developmental pathway to produce the extreme phenotype or variation will be maintained because there are several pathways to the extreme phenotype (Fig. 2-6). It is also possible

that in some lines there may not be the required genetic architecture to attain the most extreme phenotype, in which case the response to selection would greatly decrease or cease.

Based on the results given in Table 2-2 the proposed experiment is very likely to show DSD at least in the initial period of selection. What is most interesting is what happens in later generations when the selected and correlated trait values deviate greatly from their initial values.

2.4 Summary

1. The evolution of life history traits clearly depends on the trade-offs inherent in the functional bases of these traits.

2. Equally it is dependent upon the variation in possible mechanisms to achieve optimal combinations of trait values.

3. Genomic analyses coupled with analyses at higher levels (e.g., physiological, ecological) can contribute to our ability to predict variation in specific instances and determine the extent to which information at different levels of resolution is required to make general predictions.

4. Quantitative genetics is based on the emergent properties, namely the means, variances, and covariances of interaction systems and, as such, is not tied to the particular underlying functional mechanism.

5. On the other hand, it may be that broad generalities about these emergent properties might be possible when sufficient cases at the genomic level have been studied.

6. Transgenic, QTL, and microarray analyses have all contributed to our understanding of the functional components of evolutionarily important trade-offs.

7. It is evident from the results of artificial selection and genomic analyses that different genetic architectures and developmental pathways exist that produce the same phenotype.

8. The extent to which such variation is a result of initial variation in founding populations or selection is not, however, known.

9. The analysis of existing variation cannot answer this question because it is not possible to separate drift and selection after the event. Thus we need to take an experimental approach.

10. I have suggested one possible approach in this paper. The results of studies outlined in the first part of this chapter demonstrate that the genomic analysis of the lines in the proposed experiment is both feasible and instructive.

2.5 Acknowledgments

I gratefully thank A. Heyland, T. Flatt, and two anonymous reviewers for their insightful comments on earlier versions of this chapter.

PART 2

Growth, development, and maturation

Thomas Flatt and Andreas Heyland

An organism's fitness depends critically on its patterns of growth and development, in particular on its age and size at maturity. Life history theory has made successful and thoroughly tested predictions about the evolution of these features of the life cycle in terms of the relative costs and benefits of maturing early or late, and of growing large or staying small (Stearns 1992, Roff 1992). In brief, the benefits of maturing early are a shorter generation time and higher survival to maturity because of a shorter duration of the risky developmental and juvenile period. Conversely, the benefits of maturing late include longer growth, leading to a larger body size and thus increased early fecundity, decreased juvenile mortality through improved offspring quality, and higher lifetime fecundity gained through a longer growth period (Stearns 1992, Stearns 2000).

However, the classical life history framework does not usually take into account how developmental and physiological factors determine or modulate the expression of any variation in growth and maturation traits, and future work should therefore make an attempt to integrate such mechanistic information into the life history framework. Moreover, as Graham Bell points out in his foreword to this volume, life history theory traditionally leaves out traits that are involved in major life history transitions, such as changes from simple to complex life cycles, from direct to indirect development, and so forth (see also Stearns 1992, 2000). For example, organisms with indirect development are characterized by a metamorphosis between the larval and juvenile/adult stage—see the special issue of *Integrative and Comparative Biology*, 46(6), 2006—on metamorphosis in animals and on similar life history transitions in non-animals, e.g., Bishop

et al. 2006a). In fact, a majority of phyla (28 out of over 30) have evolved metamorphic life histories, yet life history theory is relatively quiet about the evolution of such indirect life histories and their underlying mechanisms.

Intriguingly, as the chapters in this book section illustrate, a comparison of the mechanisms that affect growth, size, and metamorphosis in organisms with indirect life histories reveals some striking proximate similarities among holometabolous insects (Chapters 4 and 5), amphibians (Chapter 7), fish (Chapter 6), and marine invertebrate phyla (Chapter 8). Perhaps the most obvious similarity is that different organisms use similar hormone systems to orchestrate growth, morphogenesis, and metamorphosis, and to synchronize these processes with the exogenous environment. For example, ecdysteroid and juvenile hormones are critical for the regulation of insect molts and metamorphosis (Chapters 4 and 5). Similarly, as discussed by Buchholz *et al.* (Chapter 7) and Manzon (Chapter 6), thyroid hormones have critical functions in the regulation of amphibian and fish metamorphosis, and there exist remarkable parallels between metamorphic insect hormones and thyroid hormones (Flatt *et al.* 2006). Interestingly, novel evidence now suggests that thyroid hormones can also regulate metamorphic development in invertebrates, for example in basal deuterostomes (echinoderms, cephalochordates, and urochordates), as reviewed by Heyland and colleagues (Chapter 3).

What can we learn from these similarities in metamorphic regulation across vertebrate and invertebrate taxa that is relevant for our understanding of the evolution of life histories? Work summarized by several authors in this part of the book suggests that evolutionary changes in hormonal signal transduction and synthesis may have fitness consequences that contribute to the evolution of alternative life histories. In Chapter 5, Erezyilmaz identifies several major differences in endocrine regulation among ametabolous (no metamorphosis—direct development), hemimetabolous (incomplete metamorphosis), and holometabolous (complete metamorphosis—indirect development) life histories in insects. The transcription factor Broad (Br), a major pupal determinant that has evolved relatively recently within the arthropod lineage, is regulated by two insect hormones and shows temporal shifts in expression between species with direct and indirect life histories. Similarly, the evolution of the larval period in amphibians and echinoderms has likely involved changes in the expression and/or activity of proteins that control the bioavailability of thyroid hormones (Chapters 3 and 7). Moreover, in many animals growth and size are regulated via insulin/insulin-like growth factor signaling and/or growth hormone, hormonal systems that are intimately linked to nutrient uptake, stress, reproduction, growth, and aging, as discussed in several chapters throughout this book (e.g., Chapter 4, and various chapters in Part 4). Physiological changes

in these endocrine pathways are known to have dramatic effects on growth, size, and other life history traits, and it is thus conceivable that variation in these signaling systems contributes to evolutionary changes in metamorphic and reproductive timing.

Although there are many pervasive similarities in the regulation of larval development among holometabolous insects, amphibians, fish, and marine invertebrates, there exist several important differences as well. Indirectly developing insects reproduce almost immediately after metamorphosis, and their growth rates during larval development are therefore highly correlated with fecundity and thus fitness. Similarly, amphibian tadpoles experience lower fecundity and fitness when pre-metamorphic growth rates are low. In contrast, many marine invertebrate larvae reproduce only after considerable periods of juvenile growth. Consequently, post-metamorphic growth patterns appear to have a much stronger impact on fecundity and therefore fitness than larval growth. Unfortunately, little data exist on the mechanisms that regulate post-settlement processes in marine invertebrates. Such differences in life history modes need to be carefully considered in discussions of the mechanisms that underlie the evolution of life histories.

CHAPTER 3

Emerging patterns in the regulation and evolution of marine invertebrate settlement and metamorphosis

Andreas Heyland, Sandie Degnan, and Adam M. Reitzel

3.1 Background

The biophysical properties of seawater and the connectivity of marine habitats impose specific opportunities and constraints on reproduction and dispersal in the ocean. Marine invertebrates have evolved a diversity of reproductive strategies (Fig. 3-1). One distinct mode of reproduction is via dispersing larvae, which must settle out of the plankton and undergo a metamorphic transition into the benthic adult form (Strathmann 1990) Despite the commonality of this life cycle transition in the marine environment, the molecular, cellular, and physiological mechanisms that regulate settlement and metamorphosis as individuals move from the plankton to the benthos are poorly understood. This in turn limits our ability to consider mechanisms and pathways of the evolution of the plankton–benthos transition.

Phenotypes are linked with genotypes via physiological and developmental (proximate) mechanisms (see also Chapter 1). Understanding mechanisms of life history evolution (the ultimate level of inquiry) requires a fundamental understanding of such proximate mechanisms across relatively closely related taxa (Wray 1994). Research of proximate mechanisms of marine invertebrate life histories so far has largely been biased towards early life history stages due to the difficulty of rearing and experimentally manipulating larval and juvenile stages in the laboratory for extensive periods of time. Consequently, life history models and empirical data have focused on the relation-

ship between egg investment and larval type, and how this association affects larval dispersal and survival and subsequent evolutionary changes in life histories (Vance 1973a,b, Christiansen and Fenchel 1979, Caswell 1981, Perron 1986, Roughgarden 1989, McEdward 1997). Specific developmental mechanisms underlying metamorphic competence and settlement have received less attention (but see Hentschel and Emlet 2000, Day and Rowe 2002, Marshall and Keough 2003, Toonen and Tyre 2007).

Understanding mechanisms underlying complex life cycles and their evolution in marine habitats requires an understanding and consideration of both ecology and development (Moran 1994). From an ecological perspective, dispersal potential and the identification of suitable settlement sites are larval life history traits with significant fitness effects (for a review see Chia and Rice 1978). Mechanisms regulating the development of feeding, dispersal and sensory structures that enable larvae to survive in the plankton and to detect suitable settlement sites are likely to be under strong selection, creating an important link between such developmental mechanisms and larval ecology. For example, trade-offs between dispersal potential and survival are mechanistically rooted in processes regulating the development of these larval structures, including mechanisms of plasticity (see Chapter 17). Similarly, the timely recognition of settlement cues, hence metamorphic competence, depends on the differentiation of larval sensory structures. The development of these structures is likely integrated with

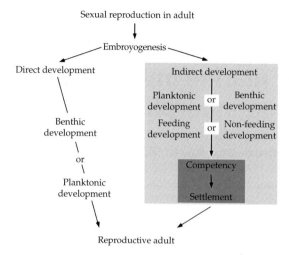

Figure 3-1   A generalized representation of the diversity of life histories in marine invertebrates as well as the emphasis of this chapter: metamorphic development and settlement in marine invertebrate larvae (the shaded area). After fertilization, which can occur via direct mating, spermcast, or broadcast spawning, initial development occurs either in the benthos or in the water column, with or without a larval stage, which may or may not be feeding (planktotrophic—see Box 3-1 for further explanation). Regardless of the pre-settlement developmental strategy, organisms will reach metamorphic *competency* before the transition to the adult life history stage (*settlement*). At this stage, individuals (larvae or pre-adults) begin to sample future benthic habitats for appropriate cues, which, when detected and integrated into the developmental program, elicit settlement and, usually, metamorphosis for indirect developing species (see Fig. 3-3 for further details). For the purpose of this review we focus our discussion on larval development and metamorphosis with specific emphasis on endocrine and neuroendocrine mechanisms regulating this important life history transition.

energetic or biomechanical requirements of the post-settled juvenile and in some cases will be affected by trade-offs between different life cycle stages.

Intraspecific studies of the morphological, physiological, and molecular components regulating life cycle transitions (see above) provide comparative data to enable examination of evolutionary changes *among* species, and for testing of hypotheses on how species may have evolved the diversity of life histories we observe in extant marine invertebrate taxa. Thus, we can use closely related species that vary in particular life history characters and determine how such differences at the level of the life history are regulated at the molecular level.

In this chapter, we discuss proximate mechanisms underlying metamorphic competence and settlement, and how changes in such mechanisms can lead to changes in life histories over evolutionary time. We first describe metamorphic competence and settlement in the context of marine invertebrate life cycles and summarize mechanisms that regulate these processes. We then analyse energy allocation trade-offs between larval and juvenile structures for echinoids (sea urchins and sand dollars). Echinoid larvae are characterized by extreme indirect development (i.e. distinctly separate programs for larval and juvenile morphogenesis and differentiation). Finally we summarize the diversity of chemical cues that are known to modulate the settlement responses among marine invertebrate larvae and discuss hypotheses on the evolution of settlement strategies.

3.2   Introduction to marine invertebrate life histories

Marine invertebrate life histories are diverse and can be categorized based on mode of gamete release (e.g., broadcast or spermcast spawning, copulation, pseudocopulation), location of development (e.g., planktonic, benthic, brooding), presence or absence of larval stages (e.g., indirect or direct), feeding requirements of these larvae (non-feeding or feeding), and associated presence or absence of a settlement and/or metamorphic transition. Marine invertebrate groups have evolved diverse combinations of these broad categories, allowing for some generalizations. In Figure 3-1 we present a simplified diagram of this diversity, while keeping in mind that variation is present in each of these modes. In this chapter, we focus our discussion on indirect developing species that include a distinct feeding or non-feeding planktonic larval stage, and which then undergo a metamorphic transition associated with settlement into the benthic habitat.

Indirect development in the planktonic environment includes a distinct larva, with or without feeding structures. Larvae with the ability to feed are referred to as planktotrophic (see Box 3-1 for definition of feeding development and planktotrophy)

Box 3-1 Explanation of important terms

Feeding development

Developmental mode that is characterized by a larval form that requires exogenous food sources in order to reach metamorphic competence. Feeding larvae always undergo indirect development and have a metamorphic transition in their life cycle. Feeding development is largely synonymous with planktotrophy. Historically, this term has been used in reference to species that produce small eggs.

Non-feeding development

Developmental mode that is characterized by a larval form that does not require exogenous food sources in order to reach metamorphic competence. Non-feeding development is generally considered indirect and includes a metamorphic transition. Non-feeding development is largely synonomous with lecithotrophy. Historically, this term has been used in reference to species that produce large eggs.

Facultative feeding/planktotrophy

Developmental mode that is characterized by a larval form that has the ability to use exogenous nutrition but does not require it in order to reach metamorphic competence. Species with a facultative feeding mode of development have both feeding and digestive structures but develop generally from larger, more energy-rich eggs.

Direct development

This mode of development is characterized by a direct transition of embryogenesis to the juvenile/adult form. By definition, direct development does not involve a larval form.

Indirect development

This mode of development is characterized by a larval form that is distinct from the juvenile/adult body plan. The larva transitions into a juvenile/adult via a more

or less drastic morphological, developmental, and physiological transition called metamorphic transition or metamorphosis.

Metamorphic competence

Developmental capacity to undergo settlement.

Metamorphic transition (metamorphosis)

Life history transition involving changes in morphology, physiology, and development. Can involve fast and drastic morphological changes in a short period of time. Within animals, a metamorphic transition only occurs in those groups with indirect development and a larval form (see above). The metamorphic transition begins with the formation of juvenile structures and ends with the elimination of larval structures. It therefore includes metamorphic competence and settlement (for marine invertebrate species). See also Bishop *et al.* (2006a) for an extensive discussion on defining metamorphosis.

Settlement

Term used primarily for marine invertebrate life cycles. Describes the ecological and behavioral transition from a pelagic larval form to a benthic juvenile/adult form. Usually leads directly on to a metamorphic transition, but not always.

Larva

Post-embryonic life history stage with distinct morphology and physiology from the juvenile and adult organism. Larvae can be feeding or non-feeding and have the capacity to disperse. At the end of the larval period, metamorphosis will transform the larva into a juvenile.

because they require nutrition derived from feeding on plankton in order to reach the juvenile stage. Larvae that can reach metamorphosis without the necessity for exogenous food frequently lack feeding structures; they are called lecithotrophic (see box for definition of non-feeding development and lecithotrophy), as they are sustained instead by maternal nutrients provided in the egg. In some rare cases, larvae have the *ability* to feed but do not *require* food to become juveniles. These larval types are referred to as facultative planktotrophs (Emlet 1986, Hart 1996) and may represent a transitional mode between planktotrophic and lecithotrophic development. Maternal investment in the egg is a relatively strong predictor of developmental mode across invertebrate phyla, where eggs with higher provisioning typically develop into non-feeding larvae (Vance 1973a,b). The relative amount of energy investment varies among lineages, such that an absolute threshold value does not apply broadly (reviewed in Strathmann 1985).

In these biphasic life cycles, selection on the planktonic larval stage may act quite differently from that on the benthic juvenile/adult stages. This situation creates trade-offs between life cycle phases that will ultimately shape the evolution of indirect development. Across all developmental strategies, the larval period may be under strong selection because larvae face various mortality risks including predation and exposure to adverse environmental conditions such as drastic salinity, temperature, and pH changes, all of which can have detrimental effects. Lecithotrophic larvae are, by definition, maternally provisioned with adequate resources to complete the pre-component stage of the larval period. Therefore, the development time of these larvae is typically shorter than for planktotrophic larvae, and environmental factors likely to affect development are not related to food acquisition. Survival and growth of pre-competent planktotrophic larvae are largely dependent on food availability in the environment in addition to abiotic factors. Larval distribution into favorable feeding environments is influenced by adult spawning behavior as well as by larval behavior. In many species, feeding larvae exhibit morphological plasticity to adjust feeding and digestive structures to increase energy acquisition in variable feeding environments (reviewed in Chapter 17).

Metamorphosis is an intricate part of indirect development and so is metamorphic competence, the phase directly preceding settlement (Hadfield *et al.* 2001, Bishop *et al.* 2006a, Heyland and Moroz 2006, Hodin 2006). Competent larvae respond to highly specific environmental cues by rapidly settling into the juvenile habitat (Hadfield *et al.* 2001). In species with indirect development, juvenile structures differentiate during the larval phase and are likely creating energy allocation trade-offs between these two distinct developmental compartments. We will discuss regulatory mechanisms underlying such trade-offs in sea urchin and sand dollar larvae, and the consequences of these trade-offs for life history evolution. It should be noted that although such a developmental pattern is relatively common among marine invertebrates, there are numerous taxa (e.g., many lophotrochozoan groups) without such a clear distinction between juvenile and larval structures. In these groups we generally see a more gradual transition from larva to juvenile (further discussed below).

3.3 Regulation of larval development and the evolution of feeding modes in echinoids: Energy allocation trade-offs during larval development

In echinoids (e.g., sea urchins and sand dollars), bryozoans, nemertean and crustacean larvae, juvenile tissue begins to differentiate inside the larval body as a distinct structure during the pre-competent period. During the entire larval period, larval feeding supports the growth and differentiation of this juvenile tissue in addition to maintaining larval structures; this can create an energy allocation trade-off between larval and juvenile growth and development (Hart and Strathmann 1994, Heyland and Hodin 2004, Strathmann *et al.* 2008). In many of these marine invertebrate groups, growth and development of juvenile structures inside the larval body are extraordinarily plastic. In echinoid and bryozoan larvae, for example, the maturation of the juvenile rudiment can be largely decoupled from larval development, depending on environmental food conditions (Strathmann *et al.* 1992, 2008). Such allocation "decisions" require the integration of ecological with developmental processes that ultimately determine the fitness of the organism. Precisely how endogenous and exogenous energy resources are divided between the development of larval versus juvenile structures will depend on the specific environmental conditions in which a larva finds itself (for some examples see Toonen and Pawlik 2001, Strathmann *et al.* 2008). Moreover, larval life history traits, such as age and size at metamorphosis, as well as juvenile traits such as post-settlement performance and timing of first reproduction, may be linked to maternal investment and availability of nutrition and predation in the larval environment (reviewed in Strathmann 1990, Hanvenhand 1993, Levin and

Bridges 1995, Morgan 1995, Hentschel and Emlet 2000). These relationships are schematically depicted for these examples of indirect development in Figure 3-2. The quality of the settling larva can influence the size of the juvenile, which affects the probability of survival for this stage. Extended larval periods may therefore lead to increased juvenile quality, which creates a trade-off between larval survival and juvenile performance. Note, however, that these trade-offs primarily exist for feeding larvae, because non-feeding larvae are limited by endogenous reserves and therefore more prone to declining juvenile quality by staying longer in the plankton or delaying settlement

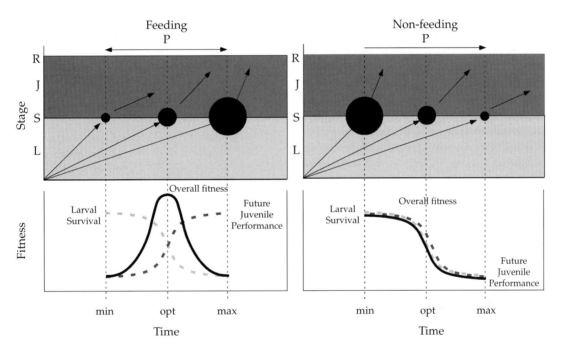

Figure 3-2 Potential fitness consequences of prolonged larval life for species with indirect feeding (left) and non-feeding (right) development. Top: For species with a feeding larva in any given feeding environment (top left), accelerated development to settlement in the larval habitat (L, light grey) results in earlier settlement at a smaller juvenile size (black area), while extended periods of time in the plankton result in later settlement at a larger juvenile size due to additional feeding. Alternatively, for species with non-feeding larval stages (top right), time to competency for settlement is not dependent on exogenous food so extended pelagic periods result in net energy losses, thus smaller juveniles. While reduced juvenile size (J, dark grey) may negatively impact performance and reproductive output, many marine invertebrate species become reproductive after extended periods of time post settlement and such negative effects during the larval period may be compensated during juvenile life (see text for detailed discussion). The slope of arrows reflect developmental rates to settlement (S) and reproduction (R). Bottom: The life history trade-off between larval survival and future juvenile size/performance predicts an optimal timing of settlement in a given environment, which differs between species with feeding and non-feeding larvae. This timing is affected by behavioral patterns of larvae in response to specific environmental and endogenous signals (P for plasticity). For species with feeding larvae (bottom left), the optimal time for settlement balances the decrease in larval survival over time with increases in juvenile performance due to increased size and potentially selection for better benthic habitats. However, for species with non-feeding larvae (bottom right), the shape of the fitness curves for the larval and juvenile stages is similar, with expected decreasing benefits over time. Thus, species with non-feeding larvae would be expected to transition to the juvenile stage with minimal delay.

(Fig. 3-2). Furthermore it should be noted that this situation does not generally apply to marine invertebrate groups. Although extension of larval lifespan can most certainly result in increased larval mortality in the plankton, little evidence exists that changes in developmental rate has long-term consequences on juvenile traits (Miller and Hadfield 1990, Ernande *et al.* 2003) but see (Pechenik *et al.* 1996). Hence, it appears that pre- and post-settlement life history phases can be largely decoupled in marine biphasic life histories.

3.3.1 Hormonal regulation of juvenile development

Many animals with indirect life histories and a drastic metamorphic transition, such as frogs and insects, regulate larval and juvenile development using hormones (reviewed in Chapter 6 for several fish species, in Chapter 7 for amphibians, and in Chapters 5, 13, and 20 for holometabolous insects). The role of hormones in marine invertebrate life histories may be equally critical and widespread, at least for deuterostomes and ecdysozoan larvae (for review see Heyland *et al.* 2005). New data on thyroid hormone (TH) signaling in basal chordates, that is, urochordates and cephalochordates, suggest a regulatory function of THs in the metamorphosis of these groups as well. Paris *et al.* (2008) showed that several types of thyroid hormone (T4, T3, and thyroid hormone triiodothyroacetic acid—TRIAC) can induce metamorphosis by binding to the amphioxus thyroid hormone receptor (Paris *et al.* 2008). Moreover, many TH synthesis and signal transduction genes have been annotated from the recently released *Amphioxus* genome (Holland *et al.* 2008). In the ascidian *Boltenia villosa*, inhibition of TH synthesis leads to a correlated inhibition of adult differentiation after settlement (Davidson *et al.* 2002), and in three other ascidian species thyroxine effects on larval development as well as TH synthesis have been confirmed (D'Agati and Cammarata 2006). Still, detailed information on how THs regulate the energy allocation trade-off between larval and juvenile structures remains to be elucidated in cephalochordates and urochordates.

THs have been identified as regulators of development and metamorphosis in echinoids (sea urchins

and sand dollars) (Chino *et al.* 1994, Saito *et al.* 1998, Hodin *et al.* 2001, Heyland and Hodin 2004, Heyland *et al.* 2005, Heyland and Moroz 2006, Heyland *et al.* 2006a,b, Hodin 2006). Specifically, thyroxine and T3 lead to a significant acceleration of juvenile rudiment development as well as a short-arm larval phenotype, suggesting that thyroxine and other THs may function in the coordination of energy investment between larval and juvenile structures. In one sand dollar species with obligatory feeding larvae, *Leodia sexiesperforata*, TH treatment resulted in larvae metamorphosing in the absence of food (i.e., functionally non-feeding larvae), suggesting that TH may have played an important role in the evolution of non-feeding development (Heyland and Hodin 2004). The molecular mechanisms underlying TH action in echinoids are still poorly understood, but preliminary data suggest that larvae can synthesize TH endogenously (Heyland *et al.* 2006a,b). In addition to these endogenous sources, TH can originate from algal food (Chino *et al.* 1994, Heyland and Moroz 2005). While this has broad implications for larval life history evolution (Heyland *et al.* 2005, Heyland and Moroz 2005, Miller and Heyland 2010) that are further discussed below, the mechanisms of TH synthesis in larvae and the transfer between algae and larva remain to be elucidated.

As in insects, molting cycles in marine arthropods are regulated by hormones (Hartnoll 2001), specifically molt inhibiting hormone (MIH) and various steroid hormones related to 20-hydroxy ecdysone. Crustacean hyperglycaemic hormone (CHH) has a very similar structure to MIH and appears to be involved in the molt regulation via inhibition of ecdysteroid synthesis. Finally, methylfarnesoate can stimulate molting by increasing ecdysteroid production. A review of recent findings suggests that ecdysteroid action in crustaceans is also mediated via nuclear hormone receptors (Hartnoll 2001, Wu *et al.* 2004).

3.3.2 Hormonal signaling and the evolution of alternative life history modes

Life history trade-offs can result from two life history traits with opposite fitness effects that are mechanistically (i.e., physiologically, developmentally, genetically) linked in the same organism.

Mechanisms underlying trade-offs can therefore affect the course of evolutionary change in life histories (see also Chapters 1 and 2 for other examples). Here we will discuss one such evolutionary transition in life history mode in more detail—the evolution of non-feeding development from feeding development—and the role that THs might have played in this process.

Among echinoid groups, non-feeding development has evolved many times independently from feeding development. This evolutionary transition resulted in heterochronic shifts in development as well as the reduction or complete loss of feeding structures (reviewed in Wray 1995). One consequence of feeding loss may have been the loss of exogenous hormone sources, creating a requirement for endogenous hormone synthesis (Heyland and Moroz 2005). Expanding on our data from *Leodia sexiesperforata*, we hypothesize that up-regulation of endogenous TH synthesis may have been sufficient to transform an obligatory feeding larva into a facultative feeding larva (Heyland *et al.* 2004). Therefore the ability of larvae to endogenously synthesize THs may be a pre-adaptation for the evolution of non-feeding development in this group (Heyland and Hodin 2004). Based on that hypothesis, we predict that clades, which frequently underwent the evolutionary transition to non-feeding development (such as the Clypeasteroida) may have a higher capacity to synthesize hormones endogenously in comparison to clades that rarely or never underwent this transition, such as the Diadematoida urchins.

The sand dollar *Dendraster excentricus* produces a feeding larva and responds to the TH synthesis inhibitor thiourea by delaying the metamorphic transition, an effect that can be rescued with exogenous hormone (Heyland and Hodin 2004). Similar responses have been found for facultative planktotrophic larvae of the sea biscuit *Clypeaster rosaceus* (Heyland *et al.* 2006b) and the non-feeding Japanese sand dollar *Peronella japonica* (Saito *et al.* 1998). Together these results indicate that larvae of these three species have the capacity to endogenously synthesize THs, which also regulate larval development and metamorphosis. Preliminary experiments with *Diadema antillarum* larvae show very different results; that is, there is no evidence for endogenous

hormone synthesis (Hodin and Heyland unpublished data). While these results are preliminary and further sampling of taxa is needed, they are consistent with the hypothesis that extant representatives of taxa that frequently evolved non-feeding development are more likely to synthesize THs endogenously.

One important question is whether endogenous and exogenous hormone sources have a) identical structures and b) comparable effects on development. The data so far show that THs in algae are the same as those that larvae endogenously synthesize and TH from endogenous and exogenous sources effect development to metamorphosis in a similar way (Heyland and Hodin 2004). Still, the detailed signaling systems involved in the process have not been elucidated. Studies focusing on this question will have to assess whether larval and juvenile development is regulated by a linked signaling system and whether manipulation of endogenous TH synthesis in larvae can lead to quantitative changes in development comparable to changes induced by nutritionally derived hormone.

A recent study suggests that changes in ecdysteroid signaling may have played a role in the evolution of alternative life histories in crustaceans. As in the majority of marine invertebrate larvae, the larval stage of crustaceans is the dispersal stage whereas the juvenile and adult stage is involved in growth and reproduction. In a parasitic copepod species, the function of these life history phases has been reversed. Metanauplius larvae of *Caribeopsyllus amphiodiae* live parasitically inside burrowing ophiuroids (*Amphiodia urtica*) before they transform into short-lived free-living adults (Hendler and Dojiri 2009). The pedomorphic life cycle of this copepod probably evolved through a delay of metamorphosis regulated by developmental hormones (Hendler and Dojiri 2009).

In summary, hormonal signaling pathways appear to be consistently involved in the differentiation of juvenile structures in marine invertebrate species, primarily in deuterostome and ecdysozoan larvae. The extent to which these hormonal signaling pathways regulate life history trade-offs in other groups such as the lophotrochozoa and non-bilaterian groups remains to be explored.

3.4 Mechanisms underlying larval settlement and the evolution of alternative settlement strategies: Signal detection and modulation during settlement

Regardless of their nutrition source, both feeding and non-feeding larvae eventually reach metamorphic competence, a developmental stage at which they become responsive to species-specific settlement cues that contain information about the suitability of future settlement sites (Fig. 3-1). The regulation of developmental and physiological processes during competence and settlement requires the coordination of a complex set of exogenous and endogenous factors. For example, echinoid larvae are sensing and processing information about their current environment, while trading off the allocation of energy into larval versus anticipatory juvenile structures (see above). Marine larvae perceive information about future habitats via chemical signals detected

either in the water or via direct contact with the substratum. Responses to these exogenous cues need to be coordinated with endogenous signals that regulate the development of the organisms, including hormones, peptides and neurotransmitters. Understanding the source, nature, and transduction of these signals, and their integration with morphogenesis and differentiation, can provide critical insights into the regulation and evolution of marine life histories (Zimmer and Butman 2000).

Figure 3-3 provides a model of how settlement signals may be processed by the larva and how changes in transduction mechanisms may contribute to the evolution of alternative settlement strategies. For the purpose of this discussion we distinguish between two main components: the larval sensory system and the competence system (Fig. 3-3A). Receptors expressed by cells in the sensory system are capable of detecting specific settlement cues and therefore establish the link between the larval nervous system and the environment. The

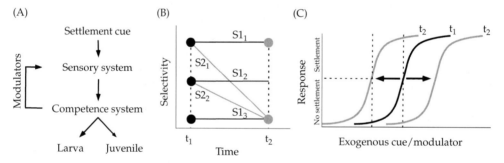

Figure 3-3 A hypothesis for proximate mechanisms regulating metamorphic competence in marine invertebrate larvae. A) Metamorphically competent larvae enter a state of developmental arrest during which little to no morphological differentiation and growth occurs. The metamorphically competent state is regulated by a hypothetical competence signaling system (*competence system*) that regulates both larval and juvenile structures. Differentiation will progress once larvae are exposed to highly specific signals from the environment that are detected by the larval *sensory system* and convey reliable information about the future post-settlement habitat (see Fig. 3-1). Modulators of the settlement response feedback to the sensory system are likely a component of the larval nervous and/or endocrine system. B) Settlement strategies are diverse among marine invertebrate groups and they are characterized by how the selectivity of a larva to cues from the post-metamorphic habitat changes over time. Such larval strategies can be extremely fixed ($S1_1$–$S1_3$, i.e., the selectivity does not change with extended periods spent in the competent stage—t_1 to t_2) or relatively flexible (i.e., the selectivity decreases with extended periods spent in the competent stage—$S2_1$, $S2_2$). Note that some larvae can have different selectivity at the beginning of the competent period and this figure primarily emphasizes how selectivity changes over time. C) The mechanistic basis underlying these strategies is unknown but could be based on changes in response curves to endogenous and exogenous factors (i.e., settlement cues or endogenous modulators). For example, a larva that changes its specificity to a settlement cue as a function of larval age may be able to do so by shifting the affinity of receptor systems to these cues (S1). Alternatively, these larvae may be able to adjust endogenous modulators of the settlement response. Note that it is unclear whether such responses are genetically determined or phenotypic plasticity and need to be investigated on a case by case basis. Still, the evolution of settlement strategies likely involves modifications in the sensory system, the competence system and the interaction between them, via synthesis and release mechanisms of specific metamorphic modulators. $S1_{1–3}$, settlement strategy independent of larval age (also known as death before dishonor); $S2_{1–2}$, settlement strategy dependent on larval age; t_1: early time point during metamorphic competence; t_2, late time point during metamorphic competence.

competence system is responsible for maintaining metamorphic competence and hence integrates information from the external environment (sensory system) with the internal environment (i.e., factors maintaining metamorphic competence). In the following sections, we discuss the regulation and induction of the settlement process in the bentho-planktonic life cycle, as well as specific settlement strategies that have evolved in marine invertebrate taxa.

3.4.1 The sensory system: Cues, receptors, and signal transduction mechanisms

Exogenous cues capable of inducing settlement are detected by sensory components of the larval nervous system. These cues are diverse and convey information about food availability, physical or chemical substrate composition, or other biotic and abiotic factors in the post-settlement habitat (reviewed in Chia and Rice 1978, Burke 1983b, Bishop *et al.* 2006b). Cues can be released by complex biofilms, predators, potential prey species, conspecifics, algae, or sediment. Detection of these compounds by competent larvae requires a high degree of specificity, because the consequences of settling into sub-optimal habitat can be detrimental (reviewed in Levin and Bridges 1995). Still, the exact chemical identity of the settlement-inducing cue is rarely known, further complicating studies on signaling transduction systems of the settlement response in marine invertebrate larvae (but see Hadfield and Paul 2001).

The apical sensory organ (ASO) is the central component of the larval nervous system and its involvement in the settlement response has been discussed for a diversity of marine invertebrate species, including mollusks, cnidarians, and echinoderms (Chia and Rice 1978, Hadfield *et al.* 2001, Page 2002, Kempf and Page 2005, Bishop *et al.* 2006a, Bishop and Hall 2009). For example, in gastropod larvae the ASO has motor and sensory functions and coordinates the activity of swimming and feeding structures in the larval stage (Page 2002). It also has a critical function in perceiving and modulating sensory input during the competent stage leading to the metamorphic transition (Burke 1980, Couper and Leise 1996, Hadfield *et al.* 2000, 2001). Using

excision experiments in sand dollar larvae, Burke (1983b) was able to show that the apical neuropile and oral ganglion mediate the perception of the natural cue and control the initiation of metamorphosis. Ablation experiments on the tropical nudibranch *Phestilla sibogae* conclusively showed that the ASO is an important structure for the ability of larvae to metamorphose (Hadfield *et al.* 2000). Recently, the adoral lobe has been discussed as an important structure in echinoid larval olfaction, specifically in the detection of settlement cues, and preliminary data suggest that chemosensory receptors are involved in the detection of these cues (Bishop and Hall 2009).

The repertoire of signaling molecules used by larvae to transduce settlement cues includes both activating and repressing neurotransmitter actions. For example, serotonin (5HT) has been detected in neurons of larval nervous systems across diverse marine invertebrate phyla (Burke 1983b, Heyland and Moroz 2006) and abundant evidence suggests a modulatory function. Inhibitory effects of 5HT on settlement have been discovered in cnidarians (Zega *et al.* 2007), ascidians (Zega *et al.* 2005), and bryozoans (Shimizu *et al.* 2000, Yu *et al.* 2007b). In contrast to these findings, 5HT has settlement-inducing effects in several mollusk species (Couper and Leise 1996, Satuito *et al.* 1999, Urrutia *et al.* 2004) and one cnidarian species (McCauley 1997). Intriguingly, settlement in barnacle larvae was inhibited by 5HT-like barettin-related compounds isolated from the marine sponge *Geodia barretti*, suggesting that 5HT-related compounds of naturally exogenous origin may also be involved in interspecies interactions (Hedner *et al.* 2006). Still, neurotransmitter action in settlement is not restricted to 5HT. A diversity of other amino acids and transmitters appear to modulate this process, including dopamine, noradrenalin, GABA, histamine, glutamine, and many others (reviewed in Hadfield and Paul 2001, Paul and Ritson-Williams 2008). However, it appears that many such compounds act as "artificial" modulators of the settlement process; that is, they are likely acting downstream of the receptor activation and do not represent a specific exogenous signal.

Biofilms are an important feature of marine ecosystems, consisting of communities of bacterial, fungal, or algal species. Bacteria use a specific group of signaling molecules for inter- and intra-specific communication, called quorum sensing. Although quorum sensing compounds are restricted to prokaryotic species, increasing evidence suggests that eukaryotes can respond to these compounds, as exemplified by various chemical interactions of bacteria with their eukaryotic hosts (Keller and Surette 2006) as well as the settlement response of algal dispersal stages in response to quorum sensing compounds (Joint 2006). A signaling function of such compounds, especially acylhomoserine lactones (AHSLs), in the settlement response of marine invertebrate species has been previously proposed (Dobretsov *et al.* 2009). However, their low stability in seawater make experiments with such compounds difficult and convincing evidence for any biological function related to marine invertebrate settlement is lacking (M. Hadfield, pers. comm.). Finally, several recent studies have implicated genes of the innate immune pathway in signal transduction mechanisms of settlement (Davidson and Swalla 2002, Heyland and Moroz 2006, Roberts *et al.* 2007, Meyer *et al.* 2009, Williams *et al.* 2009a). The analysis of such signal transduction pathways in the context of the settlement response may provide interesting new insights.

G-protein coupled receptors (GPCRs) are important transducers of exogenous signals in animals and plants (Schoneberg *et al.* 2007). They have been abundantly discussed in the context of vertebrate and insect olfaction, and increasing evidence suggests that these seven trans-membrane domain-spanning cell-surface receptors fulfill important functions during the induction of metamorphosis in at least some marine invertebrates. GPCR involvement in metamorphic signal transduction has been demonstrated for hydrozoans, barnacles, and mollusks (Baxter and Morse 1987, Schneider and Leitz 1994, Clare 1996b). Similarly, signal transduction during metamorphic induction frequently involves protein kinase activity, as shown in a wide variety of marine invertebrate species, including cnidarians, crustaceans, and polychaetes (Freeman and Ridgway 1990, Yamamoto *et al.* 1996, Biggers and Laufer 1999, Siefker *et al.* 2000). These receptor systems likely act together and may involve calcium

signaling (e.g., sea urchin *Strongylocentrotus purpuratus* (Amador-Cano *et al.* 2006)). Calcium signaling also appears to be critical for barnacle metamorphosis (Clare 1996a) and cAMP dependent protein kinase activity has been shown to mediate settlement in the polychaete *Hydroides elegans*. In contrast, data on bryozoans, gastropods, and polychaetes (*H. elegans*) suggest no involvement of GPCR activity in the settlement process (Holm *et al.* 1998, Bertrand and Woollacott 2003), further emphasizing the variability of transduction events in marine invertebrate species.

3.4.2 The competence system

Two important factors affect the transition between plankton and benthos: the availability of suitable settlement sites (an ecological consideration) and the morphogenetic progression of juvenile structures inside the larval body (a developmental consideration). Larvae will not be responsive to specific settlement cues before they reach metamorphic competence (by definition, the developmental maturity to respond to these settlement cues). During competence, larvae have to balance the benefit of finding the "perfect" habitat with the risk of mortality. Based on our simplified model (Fig. 3-3), we suggest that larvae have a competence system that regulates these "decisions" and integrates them with signals from the external environment. For the purpose of this discussion, we propose that this competence system has the ability to modulate the sensitivity of the sensory system and regulate larval and juvenile morphogenesis differentially. We will first discuss some known components of this system, and then speculate on how it may be different in species with different settlement strategies.

Several lines of evidence implicate endocrine and neuroendocrine signaling in the regulation of metamorphic competence. For example, EGF-like signaling regulates programmed cell death (PCD) associated with metamorphic competence in the metamorphic transition of the ascidian *Herdmania curvata* (Eri *et al.* 1999). A new study on the sea hare *Aplysia californica* shows that transcripts related to hormone synthesis and signaling, specifically several nuclear hormone receptor genes, are differentially expressed pre- and post-settlement (Heyland *et al.* 2010). Moreover, over 50 secretory products and peptides show increased gene expression levels during the metamorphic transition, and preliminary data suggest that some of these gene expression changes may be linked to behavioral changes (Heyland *et al.* 2010).

Both nitric oxide (NO) and thyroid hormones (THs) have been shown to regulate metamorphic competence in echinoids (Bishop *et al.* 2006a). Specifically, NO, an inhibitory signal on the progression of metamorphosis) is antagonized by TH (a metamorphosis-promoting signal). Evidence for such a regulatory mechanism originates from combined treatments of TH and NO as well as histochemical analysis of the larval nervous system during metamorphic competence (Bishop *et al.* 2006a).

One transmitter that has been recently identified as a natural inducer of settlement in sea urchin larvae is histamine (Swanson *et al.* 2006, 2007). Swanson and colleagues (2006) first identified histamine from the host plant of newly recruited sea urchin larvae (*Holopneustes purpurascens*) using gas chromatography-mass spectrometry (GC-MS) and nuclear magnetic resonance (NMR). They also noted that different algae species contain different amounts of histamine. Next, they tested histamine experimentally on metamorphically competent larvae and showed that it can induce settlement. Finally, they were able to provide evidence that the sensitivity of larvae to histamine varies with the length of the competent period (Swanson *et al.* 2007). Specifically, older larvae showed higher sensitivity to histamine than competent larvae, suggesting that histamine concentration is used as a proxy for larvae to assess their future habitat and that larvae of this sea urchin species expand the range of potential settlement sites by changing their sensitivity to histamine.

Several recent studies have employed genomic, transcriptomic, and even proteomic approaches to elucidate mechanisms underlying metamorphic competence and settlement. Although these studies have produced a large number of candidate mechanisms, they frequently lack data confirming a functional involvement. In a recent review, Heyland and Moroz (2006) found evidence for signaling pathways related to the stress response, immunity, and apoptosis, which appear to have been co-opted for signaling events during settlement and metamorphosis. Some very recent studies provide evidence from a genomics level that an independent developmental program is activated during the metamorphic transition. Williams *et al.* (2009a) analysed gene expression changes in the abalone *Haliotis asinina* using a microarray, and found evidence for marked temporal changes in transcription associated with the attainment of competency, and again with the metamorphic transition of this mollusk species. Specifically, their data suggest that gene activity associated with endogenous attainment of competence is somewhat independent from gene activity associated with exogenous induction of settlement. Jacobs *et al.* (2006) analysed gene expression changes as a consequence of delayed metamorphosis in the ascidian species *Herdmania momus*. They discovered that independent of larval age, an autonomous developmental program is activated in response to settlement cues. The fact that both mollusks and ascidians appear to recruit independent developmental regulatory mechanisms during metamorphosis further emphasizes the modularity of this process and has important evolutionary implications, which are further discussed below.

Although settlement involves a drastic change in habitat among marine invertebrate species, morphological changes associated with this transition are not always so radical. Some groups, such as barnacles and echinoids, undergo drastic changes when transferring to the benthos, losing the larval body almost entirely. In these groups, juvenile structures generally become functional only after settlement. In other groups, such as mollusks and polychaetes, the transition from the larva to a juvenile is more gradual, in that functional larval structures are carried over to the juvenile. A recent comparison between the proteome of a barnacle and a polychaete species (Mok *et al.* 2009), representing two extremes of this spectrum, revealed that these differences are also reflected in protein expression patterns. Barnacle larvae show a much more drastic change in protein expression during the metamorphic transition.

3.5 Settlement strategies: Evolution of sensory structures and signaling networks

Competent larvae generally respond to species-specific settlement cues, yet important differences exist among species in how they adjust their responsiveness to these cues over the duration of larval competence (reviewed in Bishop *et al.* 2006a). Some larvae may lower their sensitivity to a specific settlement cue as a function of time spent in the competent period, whereas other types will not change their specificity over time and explore putative juvenile habitats until their requirements are met. These two examples are extremes of a spectrum of settlement strategies, as illustrated graphically in Figure 3-3B.

The evolution of settlement strategies is likely influenced by ecological factors such as predation, food availability, and competition in a given pre- and post-settlement habitat. We predict that differences between such strategies are correlated with changes in the larval sensory and competence system outlined above. Possible causes of changes relevant to evolutionary changes are illustrated in Figure 3-3C. For example, a larva with an opportunistic settlement strategy (Fig. 3-3B, $S2_{1-2}$) may have the ability to shift the affinity of receptor systems to specific settlement cues as a function of the time that it spent in the competent state. Alternatively, opportunistic larvae may be able to adjust endogenous modulators of the settlement response. The evolution of larval selectivity likely depends on the relative frequency at which putative habitats of variable qualities are encountered, as well as the length of the larval search period (Toonen and Tyre 2007). Bishop *et al.* (2006a) also hypothesized that changes in regulatory mechanisms underlying the decision process are likely linked to the metabolism of the larva.

While direct interactions between the competence and sensory systems have rarely been studied in detail, two recent examples help to illustrate the potential importance of this interaction. *Holopneustes purpurascens* larvae increase their sensitivity in response to histamine as a function of time spent in the competent stage (Swanson *et al.* 2006, Swanson *et al.* 2007). Based on our model in Figure 3-3, this shift in sensitivity corresponds to a shift in the response curve to the left, so that lower concentrations of an exogenous modulator (settlement cue) can illicit the settlement response. Note that in this example, the shift in sensitivity could either result from changes in histamine receptor affinity to its ligand (change in settlement system only) or from interactions between the competence system and the settlement system. Another example for a potential interaction between the competence and settlement system was illustrated by Bishop *et al.* (2006a). Thyroid hormones, which are synthesized endogenously by competent larvae (Chino *et al.* 1994, Heyland and Moroz 2005, Heyland *et al.* 2004, 2006a,b), can inhibit the formation of neurons containing nitric oxide synthase (NOS) in the apical neuropile of sea urchin larvae, thus counteracting the inhibitory role of NO in competent larvae.

It is unlikely that competent larvae have fixed responses to specific environmental stimuli. Instead, one would predict that settlement behavior will vary depending on the internal environment of the organism as well as the external environment under which it encounters the signal. Developmental plasticity that allows larvae to tolerate and colonize variable environments may play a crucial role in determining both the distribution and evolution of marine invertebrate species. In particular, transcriptional plasticity is a potentially powerful mechanism for fast phenotypic responses without trade-offs that compromise fitness under different ecological conditions, because the phenotypic variability is genome-encoded, yet can be switched to suit on a generational basis (Levine and Davidson 2005). For larvae of the tropical abalone *Haliotis asinina*, for example, variation in the settlement cue induces differential transcriptional responses in a whole suite of genes and tissues that are involved in both attainment of competency and metamorphosis into the post-larval form (Williams and Degnan 2009). The widespread nature of the differences in response suggests a hormonal role via the neuroendocrine system, because hormones can act simultaneously on numerous genes and tissues of an organism (hormonal pleiotropy); the nervous system can respond directly to signals from the environment by producing chemical signals that in turn lead to the production of hormones that can later affect gene expression

(Kucharski *et al.* 2008). We suggest this as a valuable focus for further study of the mechanisms and evolution of marine invertebrate settlement and metamorphosis, both within and between species. Comparative studies among species have already demonstrated variation in life history strategies (Krug 2009), but the mechanisms underlying specific changes remain elusive. Several studies have looked at the nervous systems of planktotrophic and lecithotrophic gastropod larvae and found interesting clade-specific differences between both sensory and non-sensory components (Page 2002). Still, a close comparison of asteroid larvae with different developmental modes (feeding and non-feeding) provides evidence for a high degree of conservation in the neurogenesis and the organization of the nervous system (Elia *et al.* 2009).

In summary, available data suggest that settlement strategies are variable, and presumably evolutionarily labile, as a consequence of specific ecological factors in the marine environment. Although mechanisms underlying such strategies are variable, some lineages appear to have evolved hormonal signaling mechanisms independently. Future studies of such mechanisms will therefore be valuable in contributing to our understanding of specific constraints and opportunities for the evolution of complex developmental programs.

3.6 Future directions

The relative paucity of data describing the mechanisms underlying settlement and metamorphosis for most marine invertebrates provides the most critical hurdle towards generating a broad and comprehensive picture for our understanding of the development, ecology, and evolution of these critical events in life histories. Some preliminary data suggest that hormonal signaling systems may play a role in the energy allocation between larval and juvenile structures and that these hormones may originate from nutritional sources in some cases. We have also described settlement strategies pursued by various marine invertebrate species. While chemo-reception on the level of the larval nervous system appears to play a pivotal role in the transduction of environmental signals, the actual cues used in this process appear to be almost as diverse as the larvae responding to them. Some modulators such as NO and 5HT appear to be the same in different groups and it will be interesting to further investigate these cases of parallel or convergent evolution. For the most part, however, more detailed information on these mechanisms is needed.

Larval development, ecology, and evolution have been most thoroughly studied in echinoderm larvae. For example, comparative analysis of two Australian sea urchin species with dramatically different developmental modes revealed that major developmental changes occurred before the trophic transition to non-feeding development (Sly *et al.* 2003). Some of these evolutionary differences in development correlate with larval ecology, suggesting that larval ecology represents an important factor shaping the evolution of development in echinoderms (Wray 1994). Unfortunately, these two species differ in a multitude of ways (e.g., developmental and morphological), precluding a clear approach for discerning how or why one strategy diverged from the other. Another productive use of the comparative approach could involve study of closely related species with more subtle differences in particular aspects of the life history (e.g., egg size range within obligate planktotrophic species). Within-species studies dissecting the signaling network and developmental architecture underlying plastic responses will also likely provide a productive means of identifying mechanisms for generating phenotypic diversity.

3.7 Summary

1. Marine invertebrate life histories are diverse and frequently involve microscopic dispersal stages (larvae) that are morphologically, physiologically, developmentally and ecologically different from the adult stage. Such larval stages can be either feeding or non-feeding, and undergo a drastic metamorphosis at the end of the planktonic period that transforms them into the juvenile stage.
2. Mechanisms underlying larval development and metamorphosis are poorly understood for marine invertebrate species. Available data suggests that both neuronal and hormonal signaling systems are

involved in the modulation of environmental input. Both systems overlap during larval development. Neuronal signaling involves a wide variety of neurotransmitters and is functional on the level of the metamorphically competent larva responding to specific settlement cues. Hormonal signaling appears to be functional in coordinating the larval and juvenile developmental program. Novel genomics and proteomics approaches reveal a diversity of signaling pathways awaiting functional characterization.

3. Thyroid hormone signaling in echinoid larvae is involved in the regulation of pre-metamorphic development by coordinating energy allocation between larval and juvenile structures. Changes in this signaling system, including switching from an exogenous to an endogenous signal, may have facilitated the evolution of non-feeding development.

4. Understanding the mechanisms underlying life history evolution in marine invertebrates will require detailed mechanistic studies on regulatory mechanisms of larval development and metamorphosis. Larval ecology is one, if not the most, important factor shaping the evolution of these mechanisms.

3.8 Acknowledgments

We would like to thank Drs Michael Hadfield and Bruno Pernet for insightful comments on previous versions of the manuscript. We would also like to thank NSERC Discovery Grant to AH for financial support.

CHAPTER 4

Evolution and the regulation of growth and body size

Alexander W. Shingleton

4.1 Introduction

The relationship between body size and life history parameters has been long recognized. Correlates between body size and clutch size, lifespan, gestation time, and age at maturation have been well studied in myriad taxa. Similarly, changes in relative organ size are associated with major life history transitions: the decision to grow horns in beetles, become a worker or queen ant, or develop a defensive helmet in *Daphnia* all involve changes in organ growth and have important life history implications. Developmentally, final body and organ size is regulated by mechanisms that control the rate and duration of growth, two primary life history parameters. At many levels, therefore, the genetic, developmental, and physiological mechanisms that regulate body and organ size are intimately involved with those that regulate life history. Changes in body and organ size will consequently impact other life history traits, and vice versa, and bias their collective evolution (Stearns 1992, Roff 2002). Work over the last 30 years has seen a dramatic deepening of our understanding of the mechanisms that regulate growth and body size, and how these mechanisms influence other aspects of life history. The challenge for evolutionary biologists is to integrate this new mechanistic information with an evolutionary perspective to understand how growth and body size evolves. In other words, we need to link the proximate with the ultimate causes of body size evolution.

This chapter is divided into two parts. In the first part I will concentrate on the regulation of body size. I will describe the molecular and physiological mechanisms that regulate the duration and rate of growth, explore how these mechanisms work together to create variation in body size, and review how they influence other aspects of life history. This first part will concentrate on body size regulation from an environmental rather than genetic perspective. This is because most of our mechanistic understanding of body size regulation comes primarily from studies of how body size changes in response to environmental factors. In the second part I will focus on body size evolution. I will consider observed patterns of body size evolution in response to natural and artificial selection, and ask whether these patterns can be explained in terms of the genetic, developmental, and physiological mechanisms that regulate body size. I will then explore the relationship between environmental and evolutionary changes in body size, before asking whether we can predict which size-regulatory mechanisms are the target for selection on body size.

I will largely, but not exclusively, concentrate on growth regulation in holometabolous insects. This is not solely a personal bias. First, we perhaps know more about the physiological and molecular regulation of growth in insects than in any other animal. Second, as will become clear, these mechanisms are highly conserved evolutionarily and aspects of them are likely shared among all animals.

4.2 The regulation of body size in insects

Holometabolous insects, as their name suggests, undergo complete metamorphosis, molting through a series of grub-like larval stages before pupating and metamorphosing into their final adult form. Adult insects, like all arthropods, have a stiff growth-restricting exoskeleton, so adult size is entirely regulated by growth during the premetamorphic larval stages. This is true both for the body as a whole and for the external organs (wings, legs, eyes, etc.), which grow as imaginal discs within the developing larva. Maximal body and organ size is therefore fixed at the point of metamorphosis.

The physiological processes that control the initiation of metamorphosis, and hence define the duration of growth, have been best described in large lepidopteran such as the tobacco hornworm *Manduca sexta* and the silkworm *Bombyx mori* (Nijhout and Williams 1974a,b, Truman and Riddiford 1974, Ishizaki and Suzuki 1994). These physiological processes take the form of a hormone cascade, illustrated in Figure 4-1. At some point in the final larval instar of a lepidopteran, attainment of a particular body size, referred to as *crtical size* or *critical weight*, is associated with a drop in the level of circulating juvenile hormone (JH). This is a consequence of a reduction in the production of JH by the corpora allata (Nijhout and Williams 1974a) and an increase in the production of JH esterase (Sparks *et al.* 1983, Roe *et al.* 1993, Browder *et al.* 2001). Once JH levels fall below a certain threshold, this de-inhibits the release of prothoracicotropic hormone (PTTH) (Nijhout and Williams 1974a), which in turn

stimulates the synthesis of ecdysone by the prothoracic gland (PG). The subsequent peaks in ecdysteroid levels (the metabolic derivatives of ecdysone) ultimately cause the cessation of feeding, then, after a period of larval wandering, initiate pupation itself. Growth of the body stops at the cessation of larval feeding. However, the imaginal discs continue to grow until the beginning of the pupal period, when even higher levels of circulating ecdysteroids cause them to stop growing and differentiate into their respective adult structures (Chihara *et al.* 1972, Champlin and Truman 1998a,b). While the hormonal regulation of growth cessation has been less well elucidated in other holometabolous insects, the processes are likely similar to those of the lepidopterans, albeit with some key differences, as I will describe below.

This cascade of hormonal events means that there is a delay between the attainment of critical size and the termination of body and organ growth, called the terminal growth period (TGP) (Shingleton *et al.* 2007). Final body size in holometabolous insects is thus regulated by three basic mechanisms (Davidowitz and Nijhout 2004, Nijhout *et al.* 2006, Shingleton *et al.* 2007, Shingleton *et al.* 2008):

1) critical size
2) the duration of the TGP
3) growth rate during the TGP.

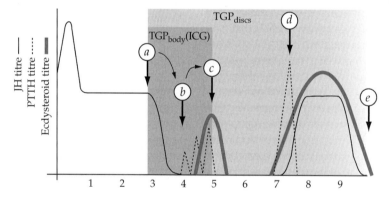

Figure 4-1 The physiological regulation of metamorphosis in *M. sexta*. Attainment of critical size (*a*) is associated with decline in the JH titre. The elimination of JH from the haemolymph de-inhibits PTTH secretion (*b*), which stimulates ecydsteroidgenesis by the prothoracic gland. The subsequent rise in the ecdysteroid titre causes the end of feeding (*c*), delimiting the TGP of the body (also called the interval to cessation of growth, ICG). A second peak in PTTH (*d*) causes another rise in the ecdysteroid titre. This, combined with a rise in JH, initiates pupation (*e*). Subsequent peaks in ecdysone during pupation (not shown) initiate imaginal disc differentiation, ending growth and delimiting the TGP of the imaginal discs. Redrawn after Nijhout 1994.

The same is true for the organs, although because they continue to grow after the cessation of body growth, their TGPs are longer than the TGP of the body. The regulation of body size in holometabolous insects can therefore be approached by considering these three developmental processes separately.

4.2.1 The regulation of critical size

While the hormonal link between attainment of critical size and the initiation of pupariation was elucidated in *M. sexta* (Nijhout and Williams 1974a,b), critical size was first described in *Drosophila* (Beadle *et al.* 1938). It was recognized as the minimal size at which transient starvation no longer delays metamorphosis: withdrawing food and then re-feeding larvae smaller than critical size delayed metamorphosis but did not affect final adult size, while the same manipulation in larvae larger than critical size did not delay metamorphosis but reduced final adult size. Thus critical size represents a size checkpoint, attainment of which is dependent on nutritionally regulated growth. How does an insect know when it has achieved critical size? Three mechanisms that regulate critical size have been identified (Fig. 4-2): insulin-signaling in the PG, the production of PTTH by neurosecretory cells in the brain, and growth of the imaginal discs.

Work in *Drosophila* indicate that nutritional regulation of critical size is via insulin/insulin-like growth factor signaling (IIS) (Shingleton *et al.* 2005). The IIS signaling pathway is the major molecular mechanism regulating growth with respect to nutrition, in all animals (see below). It is activated by the nutritional release of insulin-like peptides (ILPs) from insulin-producing cells in the brain (IPCs) and from endocrine tissue around the body (Fig. 4-3). As with transient starvation, suppression of the IIS pathway early in development delays attainment of critical size. Conversely, activation of the IIS pathway by increasing the expression of ILPs causes precocious metamorphosis (Walkiewicz and Stern 2009). These effects appear to be controlled by IIS signaling in the PG. An increase in IIS signaling in the PG alone is sufficient to accelerate the attainment of critical size, resulting in a premature release of ecdysone and precocious metamorphosis in larvae of a smaller size (Caldwell *et al.* 2005, Mirth *et al.*

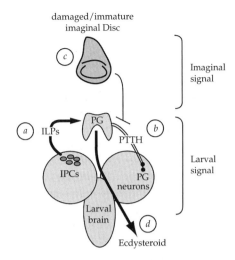

Figure 4-2 A model of critical size regulation in *Drosophila*. Critical size is regulated by larval and imaginal signals. The larvae signals comprise (a) a nutritional/size signal from the IPCs, and (b) a temporal signal from the PTTH-producing neurons. The imaginal signal (c) is inhibitory and affects the synthesis of PTTH, in part in a retinoid-dependent manner. It is the balance of these signals that ultimately regulates the release of ecdysteroids (d).

2005). Suppressing the IIS in the PG has the opposite effect.

More recently, ablation of PTTH-producing neurosecretory cells in the *Drosophila* brain has been shown to severely delay the release of ecdysteroids, retarding metamorphosis and increasing adult size (McBrayer *et al.* 2007). PTTH promotes ecdysteroidogenesis in the PG by acting as a ligand for the receptor *torso* and activating the Ras-Raf-MAPK pathway. Changes in *torso*, *Ras* and *Raf* expression in the PG consequently alter critical size and final body size (Caldwell *et al.* 2005, Rewitz *et al.* 2009). In lepidopterans, PTTH also appears to activate the synthesis of ecdysone via the Ras-Raf-MAPK pathway (Rybczynski *et al.* 2001, Smith *et al.* 2003). Interestingly, PTTH transcription in wild-type flies shows a cyclic profile during the third and final larva instar (McBrayer *et al.* 2007). PTTH synthesis may therefore impose circadian rhythm on the attainment of critical size, the release of ecdysone, and the timing of the resulting developmental events (Walkiewicz and Stern 2009).

Finally, there is increasing evidence that the imaginal discs also regulate critical size. Slowing the growth of the imaginal discs alone, by inducing DNA damage using X-irradiation or by repressing the expression of ribosomal proteins using RNAi, delays attainment of critical size and retards metamorphosis in *Drosophila* (Poodry and Woods 1990, Stieper *et al.* 2008). This seems to be a repair mechanism that ensures damaged or slow-growing imaginal discs are given additional developmental time to recover to their normal size before differentiating into their adult structures (Simpson *et al.* 1980). Damage to the imaginal disc delays critical size in part by inhibiting the transcription of the gene encoding PTTH, via retinoid signaling—inhibiting retinoid signaling prevents larvae with damaged discs from delaying metamorphosis (Halme *et al.* 2010). The same may be true for larvae with slow-growing discs. Intriguingly, however, larvae with slow-growing or damaged discs pupariate at their normal size, unlike larvae in which PTTH production is directly inhibited. Thus, either the PTTH-inhibiting signaling or some other signal produced by slow-growing/damaged imaginal discs must also slow growth, or prevent overgrowth, of the rest of the body.

4.2.2 The regulation of TGP

In lepidopterans, the duration of the TGP is regulated by the timing of the hormonal cascade that is initiated at attainment of critical size (Fig. 4-1; Browder *et al.* 2001). Two factors affect the timing of this cascade:

1) the speed at which JH is cleared from the haemolymph
2) the time between the clearing of JH to the synthesis of PTTH (Davidowitz *et al.* 2002).

The rate at which JH declines is in part regulated by the activity of JH esterase (Roe *et al.* 1993). Once JH has been cleared from the haemolymph, the precise timing of PTTH release is regulated by photoperiod: PTTH is only released during the first night after JH has been cleared from the haemolymph (Truman 1972).

The hormonal dynamics that follow attainment of critical size in *Drosophila* are less well elucidated and appear to be somewhat different than in *M. sexta.* The ecdysone titre shows multiple peaks throughout the third larval instar in *Drosophila*, each

of which initiates a particular developmental process. The first peak just precedes attainment of critical size, the second peak occurs just before the synthesis of glue protein by the salivary glands, the third precedes larval wandering and the cessation of body growth, while the fourth initiates pupariation itself (Berreur *et al.* 1979, Warren *et al.* 2006). All of these peaks are preceded by a spike in PTTH synthesis (McBrayer *et al.* 2007), suggesting that the control of the body and organs' TGPs is also regulated by PTTH action in *Drosophila.* Furthermore, there is some evidence that the timing of development is influenced by photoperiod in *Drosophila* (Mirth *et al.* 2005), suggesting that the cycles of PTTH transcription may also be photoperiodically regulated. However, loss of PTTH (McBrayer *et al.* 2007) or application of compounds that mimic JH (Riddiford and Ashburner 1991) delays but does not prevent pupariation in *Drosophila,* so there must be additional mechanisms that regulate the TGP apart from JH and PTTH.

It is not only the temporal dynamics of JH, PTTH, and ecdysone that define the body's and organs' TGPs, however, but also the sensitivity of growing tissues to these hormones. For example, in *M. sexta* body growth stops with the cessation of feeding, when ecdysteroids rise above a minimum threshold in the absence of JH and act on the nervous system to initiate larval wandering (Dominick and Truman 1986a,b). In contrast, the eye-antennal imaginal disc continues to grow until ecdysteroid levels rise above a higher threshold during pupal development (Champlin and Truman 1998a,b). The cessation of growth and the initiation of differentiation also seem to be dependent on levels of ecdysteroids in developing *Drosophila* (Chihara *et al.* 1972, Currie *et al.* 1988, Cullen and Milner 1991, Peel and Milner 1992, Mirth 2005). Consequently, different organs may have different sensitivities to the circulating hormone that regulates developmental timing, and this may account for observed differences in their periods of growth and the timing of growth cessation (Emlen and Allen 2003).

4.3 The regulation of growth rate

The second determinant of body size in animals is growth rate. A key regulator of growth rate is metabolic rate, and much has been written about broad interspecific relationships between growth rate and metabolic rate, and metabolic rate and body size (Calder 1984, Schmidt-Nielsen 1984, West *et al.* 1997, 2001, 2003). Nevertheless, among individuals within a population there may be considerable variation in growth rate, much (if not most) of which is due to environmental factors such as nutrition, temperature, and oxygen level. Research over the last decade has made substantial progress in elucidating how these factors influence growth rate at a molecular level. In doing so they provide a deeper understanding of how growth rate is controlled and how this control may change through evolutionary time.

In almost all animals, reduced developmental nutrition lowers growth rate and reduces final adult size. This relationship is intuitive: growth takes place through the conversion of nutrition to tissue, and the more limited the nutrition, the more limited the growth. Remarkably, in all animals this relationship appears to be regulated by the same signaling pathways: the insulin/insulin-like growth factor signaling (IIS) pathway, the target of rapamycin (TOR) signaling pathway, and the AMP-dependent kinase (AMPK) pathway, collectively called the IIS/TOR system (Fig. 4-3; Conlon and Raff 1999, Edgar 1999, 2006, Oldham *et al.* 2000, Shingleton 2005). The IIS/TOR system ultimately regulates growth rate by regulating transcription, translation, endocytosis, autophagy, and mitosis, and is essentially identical in insects and vertebrates.

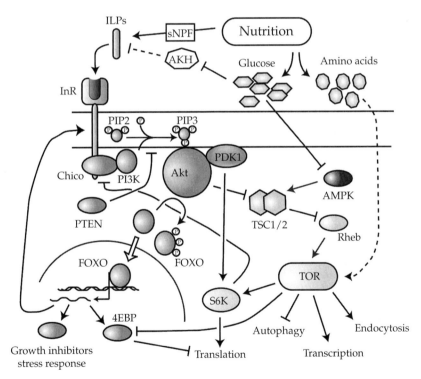

Figure 4-3 The insulin/IGF-signaling system in *Drosophila*. Variation in nutrition influences the release of ILPs, possibly by the action of sNPFs and AKH. ILPs bind to InR which initiates a signal transduction cascade involving the phosphorylation of multiple intermediate proteins. Downstream growth regulators include FOXO, which is deactivated by IIS via phosphorylation by AKT, and S6K, which is activated by IIS via PDK1. S6K is also a target of TOR, which additionally restricts the effects of FOXO by inhibiting one of FOXO's transcriptional targets, 4EBP. TOR is regulated indirectly by IIS via the action of AKT on TSC1/2. TOR also responds to amino acids, by an unknown mechanism, and glucose, via the AMPK pathway. Both FOXO and TOR regulate the activity of multiple growth inhibitors and promoters, respectively. Data from (Rintelen *et al.* 2001, Radimerski *et al.* 2002, Miron *et al.* 2003, Oldham and Hafen 2003, Harrington *et al.* 2004, Edgar 2006). Dotted lines are un-elucidate/putative relationships. Reproduced from Shingleton (2010).

The IIS/TOR system responds to nutrition in two ways (Fig. 4-2). The first is indirect, through the nutritionally dependent release of insulin-like peptides (ILPs) and their subsequent binding to the insulin receptor (InR). The second is more direct, through the cell-autonomous response of TOR to cellular levels of amino acids and glucose, the latter via the AMPK pathway. Suppression of the IIS system, for example by mutation of *InR* or *TOR*, genocopies starvation and results in a reduction in growth rate and final adult size, in insects (Chen *et al.* 1996), nematodes (Kimura *et al.* 1997), and vertebrates (Baker *et al.* 1993).

For ectothermic animals, growth rate is also regulated by temperature: an increase in temperature results in an increase in growth rate. Canonically, this is thought to be a consequence of the effect of temperature on biochemical kinetics (Gillooly *et al.* 2002). However, as I discuss below, while biokinetics undoubtedly has a substantial influence on growth rate and hence body size, there is compelling evidence that animals can regulate the effect of temperature on the rate of growth, even at the level of individual tissues. The molecular mechanisms by which this is achieved remain unknown, leaving a conspicuous hole in our understanding of the molecular regulation of body size.

Finally, a reduction in oxygen reduces growth rate and decreases final adult size in many animals. For example, *Drosophila* reared under hypoxic conditions are smaller than when reared under normoxic conditions (Peck and Maddrell 2005). The trend for amphipods to be larger at increasing latitude, once seen to be a result of decreasing temperature, is now thought to be dependent on oxygen availability in the ambient water (Chapelle and Peck 1999). Similar trends are seen in deep-sea turrid gastropods (McClain and Rex 2001) and oceanic nematodes (Soetaert *et al.* 2002). Much research has been directed to the molecular and physiological response to hypoxia in mammals, although this has only recently been extended to *Drosophila*. In mammals, the transcriptional response to hypoxia is mediated by the hypoxia-inducible factors, Hif-1 and Hif-2 (Déry *et al.* 2005). In *Drosophila*, overexpression of Sima protein, the homolog of Hif-1, genocopies hypoxia and causes an autonomous reduction in cell size (Centanin *et al.* 2005), in part

by suppressing TOR signaling (Reiling and Hafen 2004). In mammals, Hif-1 activates FOXO3a (Bakker *et al.* 2007), and the same may be true for FOXO in *Drosophila*. Thus the mechanisms that regulate body and organ size in response to oxygen concentration appear to converge on the mechanisms that regulate size in response to nutrition.

4.4 Environmental variation in body size: The functional interaction between critical size, TGP, and growth rate in insect size regulation

Understanding the individual mechanisms that control growth rate and growth duration is necessary but not sufficient to explain the regulation of body and organ size. For example, changes in growth rate need not result in a change in final body size if there is a compensatory increase in developmental time. Therefore, we must consider the functional interaction between the mechanisms that regulate growth rate and duration if we are to understand the regulation and evolution of body and organ size. This interaction can be best appreciated by examining how size variation is achieved in response to environmental regulators of size; that is, the phenotypic plasticity of body size.

As discussed above, a reduction in developmental nutrition reduces adult body size in almost all animals. In *Drosophila* this is primarily achieved through the negative effects of reduced IIS/TOR signaling on growth rate. Malnourished flies do not adjust either their critical size or lengthen their TGP to compensate for their reduced growth rate (Shingleton *et al.* 2005, although see Layalle *et al.* 2008). Consequently, such flies grow more slowly during their TGP, resulting in a reduction in final body and organ size (Fig. 4-4; Shingleton *et al.* 2005, 2008). In *M. sexta*, the size response to developmental nutrition is mechanistically similar, but not identical, to *Drosophila*. As for *Drosophila*, a reduction in size through nutritional deprivation is largely through a reduction in growth rate during a nutritionally-insensitive TGP (Davidowitz *et al.* 2004). However, dietary restriction also reduces critical size in *M. sexta*, providing a second mechanism by which nutrition affects final body size (Davidowitz *et al.* 2004).

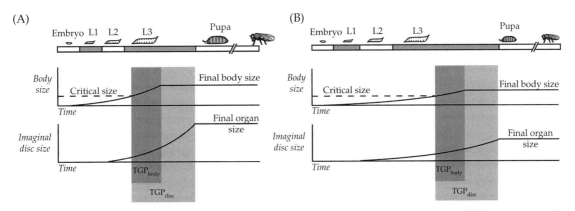

Figure 4-4   The physiological regulation of body and organ size in *Drosophila*. (A) Larvae grow until they reach a critical size at the beginning of the third larval instar. Attainment of critical size begins a terminal growth period for the body (TGP_{body}) and the imaginal discs (TGP_{disc}). Final body and organ size is controlled by the size of the body and discs at critical size, plus the amount of growth achieved during the subsequent TGPs. (B) Reduced nutrition slows growth rate but does not lower critical size nor extend the TGPs. Dietary restriction *before* critical size delays attainment of critical size and extends total developmental time, but does not reduce final body size. Dietary restriction *after* critical size slows growth during the TGP_{body} and TGP_{disc} and reduces final body and organ size, but does not increase total developmental time. Redrawn after Shingleton (2008).

The thermal response of the mechanisms that regulate growth rate and duration also explain why *M. sexta* larvae reared at higher temperatures produce smaller adults. Critical size is unaffected by temperature in *M. sexta*, but both the duration of the TGP and growth rate during the TGP are temperature-sensitive (Davidowitz *et al.* 2004). However, because an increase in temperature shortens the duration of the TGP more than it increases growth rate, the result is an overall decrease in body size. It is unclear whether the same is true for *Drosophila*.

In the majority of ectotherms hitherto studied, an increase in rearing temperature leads to a decrease in body size, an effect dubbed the temperature-size rule (Atkinson 1994). One explanation for this rule is that growth duration/developmental rate is more sensitive to changes in temperature than growth rate, as observed in *M. sexta* (van der Have and de Jong 1996). Furthermore, these same principles can explain why, in many animals, individuals reared at higher temperatures also have smaller cells: temperature may have a greater effect on the rate of cell division than on the rate of cell growth (Angilletta *et al.* 2004). The result is a decrease in cell size as temperature increases to maintain energy budget.

It is tempting to view the thermal response of growth and differentiation, and hence the effect of temperature on size, as simply being a result of the underlying biochemical kinetics, and consequently non-adaptive. However, there is extensive evidence for population and individual variation in thermal sensitivity for body size (Angilletta *et al.* 2004). Thus animals appear to be able to modulate how temperature effects their growth and development rate. Intriguingly, this control of thermal sensitivity can be seen at the level of different organs within an individual. For example, in *Drosophila* the size of the wing is much more thermally sensitive than the size of other organs and the size of the body as a whole, and means that individuals reared at lower temperatures have proportionally larger wings relative to body size (Shingleton *et al.* 2009). A recent study has shown that flies with lower wing loadings (wing area divided by body area) perform better at lower temperatures, so the high thermal sensitivity of the wings appears to be an adaptation to flight at different temperatures (Frazier *et al.* 2008). Because the TGP of the wing is regulated by the same hormonal cascade that regulates the TGP of the body and all imaginal discs, this difference in thermal sensitivity likely reflects a difference in the way temperature affects growth rate rather than growth duration.

In general, which of the three major regulators of body size in holometabolous insects (critical size, TGP, or growth rate) underlies size plasticity varies between environmental factors and between species. Consequently, we might expect the evolutionary response of insects to selection on body size to be similarly variable.

4.5   Trade-offs between body size and other traits

The mechanisms used to regulate body size in insects also regulate other life history traits. These interactions have the potential to create trade-offs, such that an increase in one life history trait negatively impacts another life history trait. In some cases, these trade-offs are an inevitable consequence of the way that body size is physiologically regulated. In other cases, trade-offs exist because the same genes and signaling pathways are co-opted to regulate multiple developmental and physiological processes. Consequently, these trade-off may be apparent when body size changes in response to one factor, for example developmental temperature, but not another, for example direct selection on body size.

An example of a life history trait that is inevitably affected by the physiological mechanisms that regulate body size in holometabolous insects is developmental time. All other things being equal, natural selection favors an increase in body size (larger females tend to be more fecund) and a decrease in developmental time (a shorter generate time increases reproductive rate) (Roff 2002). Increasing body size by increasing critical size or the duration of the TGP alone will coincidentally increase developmental time, potentially ameliorating any fitness benefits gained from a larger adult size (Davidowitz et al. 2005). In contrast, increasing body size by increasing growth rate will shorten developmental time by accelerating attainment of critical size. Which size-regulatory mechanism is targeted by natural selection to increase body size presumably accommodates these potential trade-offs in developmental time and reflects the ecological context in which selection is acting.

An example of a signaling pathway that is utilized in the regulation of multiple life history traits is the IIS/TOR system. IIS signaling not only affects growth rate (and hence developmental time), but also aging and egg production. A reduction in IIS causes an increase in longevity in Drosophila (Tatar et al. 2001b), nematodes Caenorhabdites elegans (Dillin et al. 2002,

Lee et al. 2003), and mouse Mus musculus (Holzenberger et al. 2003). In Drosophila this is mediated through the action of FOXO in the adult fat body, the insect equivalent of the mammalian liver and the major nutritional storage organ. Constitutive activation of FOXO in the fat body leads to an increase in median lifespan by as much as 35% (Hwangbo et al. 2004). Furthermore, in the adult ovary, suppression of IIS in follicle cells reduces their proliferation and inhibits the ability of ovarian cells to enter vitellogenesis (Drummond-Barbosa and Spradling 2001, LaFever and Drummond-Barbosa 2005). This, combined with a decrease in proliferation of both germline and somatic stem cells in the ovary, results in a 60-fold reduction in the rate of egg production in protein-starved adult female flies (Drummond-Barbosa and Spradling 2001). These pleiotropic effects of IIS mean that any changes in systemic IIS activity will not only influence body size and developmental time, but may also coincidentally affect fecundity and longevity.

Changes in the expression and activity of the hormones that regulate critical size and the timing of metamorphosis may also impact adult phenotype. Both JH and ecdysteroids are implicated in the regulation of longevity and in egg production. In monarch butterflies and several grasshopper species (Herman and Tatar 2001, Pener 1972), removal of adult JH synthetic tissue arrests reproduction and increases longevity. In Drosophila these same phenotypic effects are seen in adults undergoing reproductive diapause, which is marked by a decline in JH (Tatar et al. 2001a). Flies that are heterozygous for a mutation of the ecdysone receptor (EcR) are also long-lived but do not show a reduction in fecundity (Simon et al. 2003), although flies with a heteroallelic combination of two EcR mutations do show reduced oogenesis (Carney and Bender 2000). There is also evidence that ecdysone and IIS interact antagonistically during development (Colombani et al. 2005). Up-regulating ecdysteroidgenesis during the final larval instar not only accelerates metamorphosis but also reduces growth rate, by reducing IIS in the fat body. Systemic changes in the synthesis, release, and response to both JH and ecdysone may therefore not only impact the timing of larval developmental events, but also influence growth rate, longevity, and fecundity.

Further details of the molecular and physiological regulation of reproduction (Chapter 11), metamorphosis (Chapters 5, 6, and 7), and aging (Chapters in Part 4 of this volume) are described elsewhere in this book. These chapters, along with the chapters on diapause (Chapter 18), social interactions (Chapter 25), and immunity (Chapter 23), serve to emphasize that the mechanisms that regulate growth and body size also regulate multiple life history traits. This creates potential trade-offs among life history traits that, as I discuss below, have important implications for the evolution of individual traits and for the integrated phenotype as a whole. An open question is why developmental and physiological systems have evolved so that these trade-off exist in the first place. In some cases, the trade-offs may be an inevitable consequence of processes that depend upon each other (West-Eberhard 2003), for example the interaction between developmental time and growth in holometabolous insects. However, why the same signaling systems, for example the ecdysone-signaling pathway, should be used to regulate processes as disparate as the timing of critical size and egg production is less clear. Such pleiotropy may be a consequence of natural selection linking developmental and physiological processes that maximize fitness in a particular environmental context. Alternatively, it may reflect the general pattern of evolution co-opting pre-existing signaling pathways for novel developmental functions (True and Carroll 2002).

4.6   The evolution of body size

From the discussion above it is evident that we have a good (and improving) understanding of the developmental and physiological mechanisms that regulate body size in insects, particularly in response to environmental variation. It is also clear that there are multiple mechanisms by which body size can change, each of which may trade-off with other aspects of life history. What, if anything, does this knowledge add to our understanding of how body size evolves? To answer this question I will briefly review some of the evolutionary trends in body size variation, and the selective forces that are thought to underlie them, as well as studies that have used artificial selection and experimental evolution to

alter body size in a population. I will then examine some of the few examples where we know the developmental mechanisms that underlie evolved changes in body size, and explore whether these are the same as the mechanisms that control environmental variation in body size. Finally I will pose the question "Can we predict which size-regulatory mechanisms are the target for selection on body size?". The goal of this section is to begin to clarify how the proximate (mechanistic) causes of body size evolution interact with the ultimate (selective) causes of body size evolution.

4.6.1   Evolutionary trends

There is a tendency for body size within a lineage to increase over evolutionary time. This macroevolutionary trend, referred to as Cope's rule, has been documented in a wide variety of animal and plant taxa and appears to be driven by individual selection on size within populations (Kingsolver and Pfennig 2004). There are a number of hypotheses as to the selective pressures that drive this trend. Larger individuals may utilize a larger share of resources within local ecosystems, can more easily avoid predation whilst being more effective predators, are more successful in mating and intraspecific competition, and are more resistant to environmental perturbation (Hone and Benton 2005). Although Cope's rule does not apply to insects throughout their evolutionary history (gigantism was widespread amongst the insects during the Paleozoic, see below), it does appear to apply to insects after the Permian mass extinction, at least in the Coleoptera, Hymenoptera, Diptera, and Lepidoptera (Chown and Gaston 2010).

A second pattern is for species to become larger with latitude and altitude, known as Bergmann's rule. Although the rule was first applied to interspecific patterns in endotherms, it also applies to intraspecific patterns (where it is also referred to as the James' rule) and to ectotherms (Chown and Gaston 2010). The trend has been particularly well studied in *Drosophila*. For example, a genetic increase in body size with latitude is observed in *D. melanogaster* from western Europe and Africa (Capy *et al.* 1993), North America (Coyne and Beecham 1987), South America (Van't Land *et al.* 1995) and Australia

(James *et al.* 1997). Similar clines are seen in other *Drosophila* species (James *et al.* 1997). These clines have apparently evolved quickly: *D. subobscura* populations introduced into North and South America in the 1970s have evolved genetic wing size clines that are similar to those observed in endemic European populations (Calboli *et al.* 2003). The nature of the selective pressures that underlie Bergman's rule remains controversial, but they include changes in temperature, oxygen level, starvation resistance, and duration of growing season with latitude (Chown and Gaston 2010). It should be noted, however, that that Bergman's rule is not immutable. Many species show a negative correlation between body size and latitude (Mousseau 1997).

Finally, for insects at least, there is also a pattern of increased body size in species living in the late Paleozoic. Gigantism was common among insect taxa during this period, with dragonflies growing wing spans as wide as 710 mm (Wootton and Kukalova-Peck 2000). A common explanation for this phenomenon is increased oxygen levels and atmospheric pressure during the Paleozoic (Dudley 1998). Support for this hypothesis is the loss of gigantism with decreased hyperoxia during the late Permian and a second spike in gigantism coinciding with an oxygen peak during the Cretaceous. However, increased body size in the Paleozoic may also have been in response to an increase in predation by vertebrates (Chown and Gaston 2010). The phenomenon of gigantism may therefore be another example of Cope's rule.

4.6.2 Artificial selection

Artificial selection has been most commonly used to alter body size in domesticated animals and plants. Examples include toy and giant breeds of dogs, and increased size in cattle (Yerex *et al.* 1988) and swine (Partridge *et al.* 1999). There have also been numerous studies in which insect body size has been changed through artificial selection on body size itself (Partridge *et al.* 1999) or as a correlated response to selection on another trait, for example developmental time (Pijpe *et al.* 2006) or stress resistance (Bubliy and Loeschcke 2005). Finally, body size has evolved through experimental evolution, where laboratory populations

are reared in environmental conditions thought to select for changes in body size. Examples include rearing populations of *Drosophila* in low-temperature or low-nutrition environments (Partridge *et al.* 1995, Bochdanovits and de Jong 2003) and bacteria in a low-glucose environment, which resulted in an increase in cell size (Mongold and Lenski 1996).

4.6.3 The developmental mechanisms underlying the evolution of body size

Despite the amount of experimental and observational evidence that body size can and does evolve, the developmental mechanisms underlying evolved changes in body size have generally been poorly elucidated. Nevertheless, there are a few examples where we have some idea of the proximate mechanisms that underlie evolved changes in body size.

4.6.3.1 Evolution of body size in Manduca sexta

Manduca sexta have been used as a model for insect physiology for over 30 years. Consequently, colonies have been reared under laboratory conditions for extended periods of time. One such colony, maintained for over 30 years, has shown an increase in average body size of 50%. Subsequent investigation revealed that this increase in body size was due to an increase in critical size, an increase in the delay between attainment of critical size and the release of PTTH (that is the TGP), and an increase in growth rate (D'Amico *et al.* 2001); that is, all three regulators of body size in holometabolous insects.

The precise genetic basis for these physiological changes is unknown. However, the experimental manipulations of the mechanisms that regulate critical size, TGP, and growth rate described above hint at what some of these changes might be. For example, in *Manduca sexta* (unlike *Drosophila*) critical size is increased in well-fed animals. The evolved changes in critical size might be a consequence of modifications in the production of ILPs and/or the autonomous response of the PG to these peptides. Alternatively, changes in critical size may be a result of changes in the synthesis of, or response to, PTTH. Changes in TGP could be due to alterations in the expression and/or efficacy of JH esterase, in the suppressive effect of JH on PTTH synthesis, in the effect of PTTH on ecdysteroidgenesis, or the sensitivity of growth tissue to changes in the ecdysteroid titre. Finally, whole-body growth rates may evolve through changes in the production of, and/or response to, ILPs.

4.6.3.2 *Evolution of body size in Drosophila*

There have been several studies where *Drosophila* body size has been subjected to artificial selection. In one such study (Partridge *et al.* 1999), direct selection for increased body size (thorax length) was associated with an increase in larval development time, and an increase in the mean viable weight for pupariation (MVW), but no change in growth rate. In contrast, selection for a decrease in body size was associated with a reduced growth rate and a decrease in MVW but no change in developmental time. The MVW is the point in development at which starvation no longer prevents a larva from pupariation, and occurs at approximately the same time as critical size in *Drosophila*. Consequently, it is seems likely that the observed changes in body size were due to changes in critical size. The developmental response of *Drosophila* to selection on body size therefore uses all of the mechanisms that account for changes in body size in *M. sexta*, but in a different pattern depending on the direction of selection. Furthermore, because there are many processes that regulate growth rate, critical size, and the duration of growth in insects, the molecular-genetic response to selection on any one aspect of size regulation may be very different in the two species.

4.6.3.3 *Evolution of body size in domestic dogs*

Domestic animals are commonly subjected to artificial selection on size. In the case of dogs, this selection appears to have targeted the IIS/TOR system (Sutter *et al.* 2007). Within a breed (e.g., Portuguese Water Dogs) variation in body size is associated with variation at the Insulin Growth Factor 1 (IGF-1) allele, the mammalian homolog of insulin-like peptides in insects (Sutter *et al.* 2007). Like dILPs, IGF-1 promotes growth and its serum levels are regulated by nutritional status (Donahue and Phillips 1989). Variation in IGF-1 serum levels correlates with variation in body size both within and between breeds (Eigenmann *et al.* 1984a,b). Among breeds, there is evidence of a selective sweep at the IGF-1 allele, with single IGF-1 SNP haplotype common to nearly all small breeds but generally absent in large breeds (Sutter *et al.* 2007).

4.6.4 The relationship between evolutionary and environmental variation in body size

The examples above suggest that the mechanisms that account for evolved genetic variation in body size converge on the mechanisms that control environmental variation in body size. To a certain extent this may be because body size is primarily regulated by critical size, TGP, and growth rate in insects, and so both natural selection and phenotypic plasticity are constrained to target these mechanisms. Consequently, there will inevitably be overlap between the developmental and physiological mechanisms responsible for evolutionary and environmental variation in body size. However, critical size, TGP, and growth rate are regulated through multiple processes, so evolution need not affect these in the same way as environmental factors. Nevertheless, because the processes involved in the environmental regulation of size by definition affect body size, they may provide additional targets for selection on body size and facilitate its evolution.

Alternatively, phenotypic plasticity may play a more central role in evolutionary change, through the process of genetic assimilation (West-Eberhard 2003). Genetic assimilation occurs where a phenotype formerly produced only in response to a particular environment becomes stably expressed independent of the environmental effect (Flatt 2005). During genetic assimilation, the phenotype therefore loses its plasticity and becomes canalized. For this to occur there must be genetic variation in phenotypic plasticity, that is, there is a genotype by environment (G × E) interaction for a trait. There is abundant evidence for G × E interactions for traits in general (Schlichting and Pigliucci 1998) and body size in particular (Bergland *et al.* 2008). Consequently, genetic assimilation is an appealing mechanism for body size evolution.

Genetic assimilation should be evident from examining the evolved developmental response of flies to a particular environmental pressure. If body size has evolved through genetic assimilation then, where evolution is a response to a particular environmental pressure, for example low nutrition, temperature, or oxygen levels, the mechanisms targeted by natural selection should be the same as those utilized in the plastic response to that environmental pressure. More specifically, evolution should target those components of the plasticity mechanism that regulate the extent of the phenotypic plasticity.

Experimental evolution of *Drosophila* reared at different temperatures provides only partial support for body-size evolution by genetic assimilation. Both the evolved and plastic response to higher temperatures is a reduction in body size mediated primarily through changes in cell size, at least in the wing (Partridge *et al.* 1994). This is unlike the plastic response to nutrition, which involves changes in both cell size and cell number. However, the effect of thermal selection on other regulators of body size shows a different pattern. The plastic response to increased temperature results from an increase in growth rate but a decrease in developmental time, while the evolved response is exactly the opposite (when comparing high- and low-temperature lines reared at the same temperature) (Partridge *et al.* 1994). Thus thermal selection appears to target some but not all aspects of the mechanisms that underlie the thermal plasticity of body size.

The developmental response of body size to temperature has been hypothesized to underlie Bergman's rule in *Drosophila*, whereby individuals get larger with increasing latitude, and presumably decreasing temperature (James *et al.* 1997, Zwaan *et al.* 2000). Again, this trend may arise through genetic assimilation of an initially plastic response to temperature. However, in *D. melanogaster* the plastic response of the wing size to temperature is mediated primarily through changes in cell size while the latitudinal cline in wing size is a result of changes in both cell size and cell number (James *et al.* 1997). Furthermore, the relative importance of cell size versus cell number in explaining the cline differs between different continents (Zwaan *et al.* 2000). The same is true for *D. obscura* (Gilchrist *et al.* 2004). Thus the same selective pressure can result in the same gross morphological response but for different developmental reasons. The incongruence between the developmental mechanisms that underlie the latitudinal cline and the thermal plasticity of body size suggest that either the

latitudinal cline is not driven solely by thermal selection, or that thermal selection does not affect body size solely through genetic assimilation (or both).

Despite the lack of support for the role of genetic assimilation in the evolution of body size in *Drosophila*, there is evidence that it has occurred in snakes. Head size is smaller but more plastic in tiger snake populations isolated on islands for the last 30 years than it is in populations isolated for 6000–9000 years (Aubert and Shine 2009). Nevertheless, the details of the size-regulatory mechanisms that account for this pattern are unknown. In general, additional studies elucidating the molecular-genetic basis for body size plasticity and evolution are necessary before firm conclusions on the relationship between the two can be drawn. Experimental evolution and artificial selection provides a particularly powerful method for testing theories of genetic assimilation (Frankino *et al.* 2009) and for understanding body size evolution in general.

4.6.5 Can we predict which size-regulatory mechanisms are the target for selection on body size?

The extensive pleiotropic effects that changes in different developmental regulators of body size have on other life history traits will determine how body size evolves through space and time. They also make it very difficult to predict *a priori* which mechanisms will be targeted by which selective pressures. In general, we might expect that changes in those size-regulatory mechanisms that have the fewest pleiotropic effects will occur first, since pleiotropy is thought to constrain evolutionary change (Hansen and Houle 2004). For example, changes in cell size should have fewer pleiotropic effects than changes in cell number because the former can, in principle, occur at the very end of development, while the latter require alterations in the rate and duration of cell proliferation during development. However, there are also likely limits on the extent to which body size can be affected solely through changes in cell size, due to functional constraints on cell size itself. Consequently, short-term thermal selection may initially affect body size through changes in cell size, while longer-term thermal selection along a latitudinal cline may subsequently break life history trade-offs and affect body size further through changes in cell number also.

Trade-offs between different size-regulatory mechanisms and other life history traits need not always constrain evolution. Moderate levels of pleiotropy may enhance the evolvability of a trait through an increase in mutational target, as long as directional selection on the target trait, in this case body size, is sufficiently large relative to stabilizing selection on the pleiotropic traits, for example developmental time (Hansen 2003). Furthermore, variation among pleiotropic traits need not always be antagonistic. For example, certain environmental conditions may select for both a reduction in body size and a reduction in developmental time, in which case selection may target growth period rather than growth rate as the mechanism for reduced body size.

The picture that emerges of the developmental mechanisms that underlie body size evolution is a complex one. In general, it would be surprising if the evolution of body size did not reflect the many different physiological and molecular mechanisms by which body size is regulated. Just as the mechanisms insects use to adjust body size in response to the environment varies with environmental factors, so too might we expect the mechanistic targets for natural selection on body size to vary with selective pressure. Life history trade-offs, general pleiotropy, and possibly genetic assimilation will play important roles in determining precisely what these targets are. Consequently, we should expect considerable variation among insects specifically, and animals in general, in the mechanisms used by natural selection to alter body size. Only with a more profound understanding of the mechanisms that regulate body size will we be able to better explain why, and predict which, mechanisms are targets for selection on body size.

4.7   Summary

1. Body size is regulated by mechanisms that control the duration and rate of growth. In holometabolous insects these mechanisms regulate body size through their influence on critical size, TGP, and growth rate.

2. We have a good and improving understanding of the molecular and physiological regulation of critical size, TGP, and growth rate, and in particular how this regulation generates variation in body size in response to environmental factors, that is, phenotypic plasticity. Different environmental factors influence body size through different mechanisms in different species.

3. The molecular and physiological regulators of body size are also intimately involved in regulating other life history traits, for example lifespan and fecundity. This creates life history trade-offs that will influence how an animal responds to an environmental factor that favors change in body size, on both developmental and evolutionary time-scales.

4. Body size is evolutionarily labile and responds rapidly to natural and artificial selection. Which size-regulatory mechanisms are the proximate target for selection varies between species and with the nature of the selective agent. Nevertheless, selection appears to frequently target mechanisms involved in the environmental regulation of size. This may be because these plasticity mechanisms provide additional targets for selection on size. Alternatively, body size may evolve through the process of genetic assimilation.

5. A deeper understanding of the mechanisms that regulate body size and how these mechanisms regulate other life history traits is essential if we are to understand intra- and interspecific patterns of body size variation through space and time.

4.8   Acknowledgments

I thank the editors for the invitation to contribute to this volume, two anonymous readers for their helpful comments on early versions of this manuscript, and Ian Dworkin for inspiring conversations. The National Science Foundation (USA) supported the writing of the chapter through grants IOS-0919855 and IOS-0945847.

CHAPTER 5

The genetic and endocrine basis for the evolution of metamorphosis in insects

Deniz F. Erezyilmaz

5.1 Introduction

Within 100 million years of their emergence, insects progressed from an exclusive reliance on direct development to a dramatic metamorphosis (Grimaldi and Engels 2005). Today, the extant representatives of ametabolous (direct-developing), hemimetabolous (incomplete metamorphosing), and holometabolous (with complete metamorphosis) insects permit comparison between these diverse developmental strategies. Moreover, the extensive knowledge of the endocrine mechanisms that regulate insect molting and metamorphosis make this class an ideal comparative model for the study of life history evolution.

The earliest insects were ametabolous (Fig. 5-1; see also Box 5-1 for an explanation of important terms). These primitively wingless insects emerge from embryonic development as miniature adults, increase in size during the juvenile molts, and continue to molt as sexually mature adults (Sehnal *et al.* 1996). The primitively wingless insects are represented today by just two insect orders: Archaeognatha, or bristletails, and Zygentomea, which includes the firebrat (Grimaldi and Engels 2005).

With the advent of wings, postembryonic development became more complex. In hemimetabolous insects, embryonic development also produces a miniature version of the adult. However, hemimetabolous insects posses external wing pads during the immature stages, which encase the wing primordia (Fig. 5-1). This innovation occurred early in the history of insects, and was followed by a increase in the number of insect orders shortly thereafter (Kristensen 1999). Although the majority of the hemimetabolous body plan grows isometrically, the wing pads undergo a period of anisometric growth at some point during the nymphal stages. At the final molt to the adult, the animal acquires genitalia, and the wings are allowed to spread from within the wing pad cuticle (Sehnal *et al.* 1996). Cockroaches, crickets, mantids, and the true bugs are all common examples of hemimetabolous insects.

Complete metamorphosis in insects has evolved just once, approximately 300 million years ago and from a hemimetabolous ancestor (Kristensen 1999, Wheeler *et al.* 2001). The fossil record gives little insight into either the earliest holometabolous insects or their immediate predecessors (Grimaldi and Engels 2005). The ecological advantage of holometaboly is clear: where hemimetabolous nymphs competed with the adult stage for resources, larvae and adults of holometabolous insects are able to occupy vastly different niches. The monophyletic Holometabola are distinguished by a larval stage that is morphologically distinct from the adult form, and a typically quiescent pupal stage that bridges the broad developmental gap between larva and adult in an intense bout of morphogenesis (Fig. 5-1). In the most dramatic examples of complete metamorphosis, the nervous system is completely rewired, the musculature is digested and rebuilt, the gut is replaced with a new epithelium, and the epidermis of the adult displaces the larval epidermal cells (Truman and Riddiford 2002, Erezyilmaz

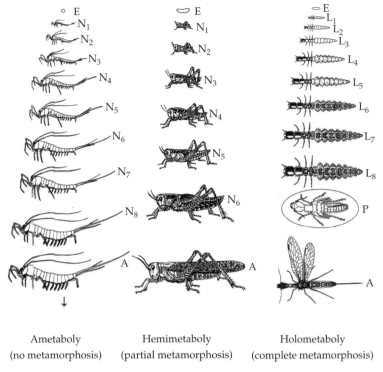

Ametaboly   Hemimetaboly   Holometaboly
(no metamorphosis)   (partial metamorphosis)   (complete metamorphosis)

Figure 5-1 The three general modes of insect life-history strategy. N, nymph; L, larva; P, pupa; A, adult. Ametabolous insects hatch as miniature versions of the adult form, obtaining genitalia at the adult stage. The first-stage nymph of hemimetabolous insects also resembles the adult, except that the wing primordia grow during the nymphal stages and are freed from the wing pads at the molt to the adult stage. The form of the immature stages of holometabolous insects is distinct from the adult form. Adapted from Sehnal and colleagues (1996), with permission from Elsevier.

2006). In many cases, the primordia for the adult epidermis develop during the larval stages. In the highly derived fruit fly, *Drosophila melanogaster*, imaginal cells (that will give rise to adult structures) are specified during embryogenesis and remain diploid while the larval epidermis becomes polyploid. The imaginal primordia grow as either:

1) imaginal discs of the head, thorax and genitalia, which pattern and proliferate throughout larval development
2) abdominal histoblasts that divide and pattern during pupal and adult development (Madhavan and Schneiderman 1977).

In other holometabolous insects, however, imaginal development does not occur as the larva grows, but only begins at the onset of metamorphosis in the final larval stage. This mode of imaginal development appears to be the ancestral condition, and

an early "telescoping" of imaginal development into the early larval instars has arisen independently at least six times in the Holometabola (Truman and Riddiford 1999). Therefore, the first holometabolous insects probably initiated and completed imaginal development during the final two larval instars.

The aim of this review is to compare the molecular mechanisms that regulate direct development with those that regulate metamorphic development in insects. To accomplish this, I will briefly describe what is known of the major endocrine factors derived from insect models. I will then outline differences between the major endocrine events in the three types of insect life history strategies: ametaboly, hemimetaboly, and holometaboly. I will next discuss recent findings from loss-of-function studies of endocrine effectors. Finally, I will discuss the *broad* gene, a master regulator of insect

---

Box 5-1 Explanation of some important terms

Pronymph

The stage that follows the second cuticle produced in embryonic development of insects, used by Truman and Riddiford (1999). Other authors use the term prolarva to describe the equivalent stage of holometabolous embryos.

Imaginal

Pertaining to the primordia that will give rise to adult structures.

Direct development

A general term to describe the life-history strategy that lacks a metamorphosis, or when the adult resembles the product of embryogenesis. In insects, direct development best describes the ametabolous insects. However, it can also be used to describe hemimetabolous insects in contrast to insects of the Holometabola, which have a radical metamorphosis.

Ecdysis

The emergence of an insect, or any member of the Ecdysozoa, from its previous cuticle to accommodate growth or a change in form.

Instar

The duration between molts. The name of each instar is given by the cuticle that was most recently deposited. For instance, "fifth instar" describes the stage after deposition of the fifth nymphal or larval cuticle.

Larva

A general term to describe the juvenile form of a metamorphosing animal. In insects, the term *nymph* is often used to distinguish the larvae of ametabolous and hemimetabolous insects from the juvenile stages of the holometabolous insects.

---

polymorphisms, its mode of action, and its prevalence in published genomes.

5.2 Endocrine regulation of metamorphosis

Postembryonic development of insects is constrained by the exoskeleton. This cuticle is shed periodically to accommodate growth or to reveal a new form. Therefore, development of the epidermis, which secretes cuticle, is episodic; each period between molts is called an *instar*. Hormones regulate these bouts of development. V.B. Wigglesworth first identified the roles of the two most significant hormones in a series of experiments on the hemimetabolous blood-sucking bug, *Rhodnius prolixus* (Wigglesworth 1934, 1936). Using decapitation and parabiosis experiments at different stages of postembryonic development, he showed that molting was initially induced by a factor released from the brain. After a period, the onset of a molt became independent of the signal

from the brain, but was conveyed by a factor circulating in the insect's blood. In addition to the factors that induced the *onset* of molts, Wigglesworth detected the presence of a factor that controlled the *identity* of the next stage (Wigglesworth 1936). Specifically, he showed that the blood of a young nymph could inhibit the progression of a last stage nymph to adult. This indicated that an inhibitory factor was circulating in the blood of an immature stage and that this prevented progression to the adult stage. Subsequent experiments have revealed that the hormone from the brain is prothoracicotropic hormone (PTTH), which stimulates release of the "molting hormone," or ecdysone, a steroid hormone produced in the prothoracic glands during the immature stages, and which is converted to 20-hydroxyecdysone in the cells of target tissues. The inhibitory hormone is known as juvenile hormone, or JH, a lipid-soluble sesquiterpenoid that is produced by the corpora allata, glands that are housed in the neck region of most insects (Nijhout 1994).

In these early experiments, Wigglesworth articulated the two most important facts of insect life history regulation: *peaks of ecdysone determine the timing of the molt, and JH determines the nature of the molt.* The two hormones act in concert during "critical periods," which typically occur during the earliest stages of a molt, when the ecdysteroid titers begin to rise. During these periods, the presence of JH selects for one of two optional fates. In the holometabolous moth, for instance, the presence of JH during a larval molt promotes the larval fate, and when JH is present at the pupal stage, JH promotes pupal development over adult (Riddiford 1994). While these properties are true of most insects, many aspects of JH regulation have been lost in the highly derived fruit fly *Drosophila*, with only vestiges of JH action remaining

(Zhou and Riddiford 2002). Therefore, mechanisms derived from *Drosophila* development are not always considered generalizable to other holometabolous insects throughout this review.

The presence of JH does not always promote the status quo, and JH secretion is not tethered to a single developmental outcome. During embryonic development of a number of hemimetabolous insects, for instance, JH promotes differentiation during embryonic molts (Sbrenna-Micciarelli 1977, Truman and Riddiford 1999, Erezyilmaz et al. 2004). In these cases, the presence of JH at an early molt will redirect the ensuing stage so that the features of a terminal stage appear precociously. Therefore, each insect molt, which is triggered by ecdysone, can be seen as a binary switching mechanism where

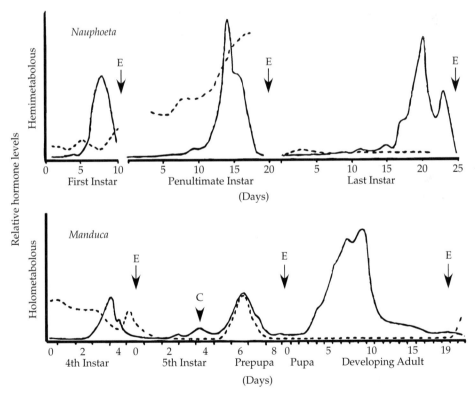

Figure 5-2 20-hydroxy-ecdysone and JH titers during the metamorphic molts of the hemimetabolous cockroach and the holometabolous hawk moth. 20-hydroxy-ecdysone titers are shown as a solid line, JH titers as a dashed line. Top, hemimetabolous cockroach, *Nauphoeta cinerea*; bottom, holometabolous hawk moth, *Manduca sexta*. E, ecdysis; H, hatching. The commitment peak of ecdysone is indicated by (C). After Truman and Riddiford (2002), reproduced with permission of Annual Reviews, Inc.

one of two alternate fates is determined by the presence or absence of JH.

5.3 Comparative endocrinology across insect life history strategies

Because JH and ecdysone have been shown to regulate molting and metamorphosis in both hemi- and holometabolous insects, the differences in their timing or action are likely to be a proximate cause of the differences in life history strategies. Currently, the majority of data come from studies of holometabolous insects, particularly of the endocrine models, *Manduca sexta*, and the silkmoth, *Bombyx mori* (Bergot *et al.* 1981, Riddiford 1994; Fig. 5-2, bottom). Among the hemimetabolous insects, complete profiles of endocrine levels are available for the cockroach *Nauphaeota* (Bruning *et al.* 1985, Lanzrein *et al.* 1985; Fig. 5-2, top) and embryonic titers of JH and ecdysone are available for the locust, *Schistocerca gregaria* (Lagueux *et al.* 1979, Temin *et al.* 1986). Unfortunately, hormone profiles during development do not exist for any ametabolous insect. As a result, studies of the endocrine basis for the evolution of complex life history in insects have been largely restricted to the acquisition of holometaboly.

A comparison of JH and ecdysone titers during the nymphal and larval molts of hemi- and holometabolous insects reveals no significant differences (Fig. 5-2). In both cases, the peaks of ecdysteroids trigger molts and the presence of JH at these molts maintains the juvenile form. In contrast, significant differences in these two hormones exist:

1) for embryonic development, when either the larval or nymphal form is generated
2) during the last larval instar, where the holometabolous insects have acquired new endocrine features (Fig. 5-2).

5.4 Endocrine titers and cuticle progression during embryonic development

Insects pass through molts while developing within the eggshell, and appearance of these cuticles coincides with elevation of ecdysteroid levels. Although data is not available for the embryonic titers of JH

or ecdysteroids for any ametabolous insect, embryonic cuticle formation has been described for the firebrat, *Thermobia domestica*, using light and electron microscopy. These studies describe two embryonic cuticles: the E1, which is deposited at about 22% of embryonic development, and the second embryonic cuticle (also called the pronymphal cuticle), which is deposited during mid-embryogenesis following dorsal closure (Konopova and Zrzavy 2005). Truman and Riddiford (1999, 2002) use morphometric analysis to show that in the ametabolous insect, *Ctenolepisma longicaudata*, transformation from the pronymphal to the nymphal stage occurs after hatching during the first postembryonic instar. After the first instar, the growth that occurs between stages is isometric.

The acquisition of wings in insects correlates with the appearance of an additional cuticle during embryogenesis. Analysis of a number of embryos of hemimetabolous insects shows that three cuticles are produced prior to hatching. The first of these is also called the E1 cuticle, and it coincides with an elevation of ecdysteroids (Edwards and Chen 1979, Lagueux *et al.* 1979). In the milkweed bug, *Oncopeltus*, a hemimetabolous insect, ecdysteroid conjugates are maternally provided in yolk, and released at the time of E1 cuticle formation (Dorn 1983). A peak of ecdysone appears at about 50% of embryonic development, with the formation of the pronymphal cuticle. A final embryonic peak occurs with the appearance of JH, to induce the first nymphal cuticle (Bruning *et al.* 1985, Temin *et al.* 1986). This cuticle possesses differentiated structures such as sensory bristles, sclerotized mandibles, and other specializations for life outside of the eggshell (Edwards and Chen 1979).

Until recently, holometabolous embryos were thought to hatch after producing only two embryonic cuticles, the E1 and the pronymphal (or prolarval) (Truman and Riddiford 1999). Data of the embryonic cuticles in this group were restricted to the more derived groups, and had been examined with light microscopy only. However, a survey of embryos from four holometabolous insect orders showed that three embryonic cuticles represent the ancestral condition of this group as well (Konopova and Zrzavy 2005). As it turns out, the moth *Manduca sexta* has a highly reduced prolarval, or pronymphal cuticle, and the higher flies may have lost this cuticle entirely, since only two cuticles have been detected: the E1 and the first larval (Hillman and Lesnik 1970, Callaini and Dallai 1987, Konopova and Zrzavy 2005). Correspondingly, *Drosophila* produces only one peak of ecdysone, which coincides with the single true cuticle that is deposited during embryonic development (Maroy *et al.* 1988). JH levels in *Manduca sexta* peak by 50% of embryonic development (Bergot *et al.* 1981), and JH is present as the pronymphal (or prolarval) cuticle is formed. JH treatment of holometabolous embryos has little effect (Truman and Riddiford 1999), presumably because endogenous levels are already high at the E1 and pronymphal molts.

In summary, the current data on embryonic cuticle formation suggests that an additional embryonic cuticle emerged as insects evolved wings and hemimetaboly. However, this comparison is based upon the study of a single ametabolous insect, the firebrat. In this insect, differential growth extends into postembryonic life. Erezyilmaz *et al.* (2004) suggest that the postembryonic pronymphal cuticle of ametabolous insects was embryonized, allowing an extended period of JH-free differential growth. Hemimetabolous and holometabolous insects (except for the higher Diptera) are now known to hatch after producing three embryonic cuticles, although the timing of JH production differs between the two groups. In the holometabolous insects, JH is present for the formation of the pronymphal/prolarval cuticle. However, JH is absent from the molt to the pronymphal stage of hemimetabolous insects, but present at the final molt, which produces the first nymphal cuticle.

Since the differences between the hemimetabolous nymph and the holometabolous larva are produced as the insect proceeds from the phylotypic germ band stage to the later stages of embryogenesis, differences in the morphogenetic hormones that organize these stages could be the cause of differences between the two forms. This is the basis for the pronymph hypothesis (Truman and Riddiford 1999), which accounts for the origins of the larval stage as the result of an advancement of JH production into earlier stages of embryonic development. The holometabolous larval form is developmentally simpler than the adult form or hemimetabolous

nymphal form. The larval nervous system, for instance, is little more than a truncated version of the adult or the nymphal counterpart (Truman and Riddiford 1999, 2002). When hemimetabolous insects are provided with JH at the pronymphal stage to simulate the holometabolous JH profile, anisometric growth is blocked, and the pronymphal cuticle produces terminal structures of the first nymphal cuticle precociously (Sbrenna-Micciarelli 1977, Truman and Riddiford 1999, Erezyilmaz *et al.* 2004). These data suggest that the advancement of JH into earlier stages of embryonic development, which has occurred in the evolution of holometaboly, would similarly freeze the present developmental form and endow the pronymphal cuticle with terminally differentiated features. In this scenario, the pronymph would hatch precociously, thereby accounting for the loss of an embryonic cuticle.

The realization that holometabolous and hemimetabolous insects have an equivalent number of embryonic instars shows that "de-embryonization" did not occur to produce the Holometabola (Konopova and Zrzavy 2005). However, this does not rule out the possibility that the earlier appearance of JH triggered formation of the holometabolous larva. This advancement must have had a profound effect upon the morphogenesis of the mid-stage embryo. JH treatment of early embryos of a number of hemimetabolous insects shows that JH given prior to the pronymphal molt, to simulate the holometabolous endocrine situation, blocks differential growth (Fig. 5-3). For instance, legs from cricket embryos treated with JH prior to the pronymphal molt grew, but this growth became isometric. Leg segments that grew at the greatest rate were stunted to the greatest extent (Erezyilmaz *et al.* 2004).

The Czech physiologist V. Novak first described the inhibitory effect of JH upon morphogenesis. Novak discriminated two types of growth: gradient growth and isometric growth. During nymphal development of *Oncopeltus*, for instance, the wing pads undergo gradient growth while the remainder of the body experiences isometric growth (Novak 1966). JH, he suggested, blocks gradient growth but allows isometric growth to occur. This would account for the isometric increase that occurs between nymphal or larval instars while JH is

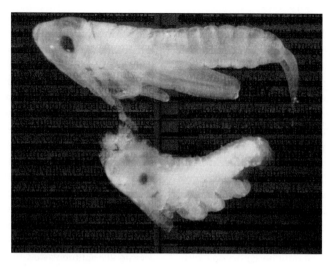

Figure 5-3 The morphostatic action of JH. An embryonic first-stage cricket nymph, just prior to hatching (above). The nymph below was treated with juvenile hormone prior to the molt to the pronymphal stage. Growth continued, but became more isometric, while nymphal features appeared precociously (Erezyilmaz et al. 2004).

present, and the induced isometric growth observed in embryos. Some tissues, such as the nymphal wing pads, Novak speculated, are endowed with the ability to grow anisometrically during the larval stages. These tissues, he suggested, possess a "gradient factor" (Novak 1966).

Truman et al. (2006, Truman and Riddiford 2007) similarly discriminate between extrinsic growth factors, such as JH, and intrinsic growth factors, like the proposed "gradient factor." Intrinsic factors include morphogens such as decapentaplegic (Dpp), wingless (Wg), or epidermal growth factor signaling systems, while JH or insulin are extrinsic factors that interact with these signals during normal development. Tanaka and Truman (2007) studied the expression of Bric-a-brac (Bab), which is regulated by the Wg and Dpp gradients in the leg proximal-distal axis of the holometabolous moth, Manduca. In the developing limbs of hemimetabolous embryos, Bab resolves from a thick band in the medial tarsus, to subdivide into tarsal rings that correspond to the nascent tarsal segments (Erezyilmaz et al. 2004). The leg of the Manduca larva is a truncated version of the adult limb, and has a reduced number of tarsal segments. Tanaka and Truman (2007) found that the initial stages of Bab expression in Manduca embryos resembles the

pattern seen during development of the hemimetabolous limb. As JH levels rise, however, the resolution of Bab into tarsal rings does not occur. These data suggest that JH, an extrinsic factor, acts upon, the intrinsic factor, Bab, to inhibit morphogenesis during metamorphosis. For a more complete discussion of insect growth regulation, see Chapter 4.

5.5 Comparison of hemi- and holometabolous endocrine events during postembryonic development

The endocrine profile of the penultimate nymphal instar in hemimetabolous insects resembles the endocrine profile of any nymphal stage (Fig. 5-2, top). The penultimate instar of holometabolous insects, by contrast, has acquired additional features to accommodate the transition to the pupal stage (Fig. 5-2, bottom). These features probably evolved through a combination of transposition of existing features of hemimetabolous development and the emergence of novel mechanisms. The first evolved feature is considered a safety mechanism, whereby the presence of JH inhibits the release of ecdysteroids from the prothoracic glands (Rountree and Bollenbacher 1986). This failsafe mechanism prevents the onset of a metamorphic molt before the appropriate size threshold and developmental stage is achieved (Nijhout 1994). The decline in JH titers that occurs during the last larval instar is triggered by the acquisition of "critical weight," the minimum viable size at which starvation can no longer delay metamorphosis (Mirth and Riddiford 2007). This occurs as the corpora allata, the glands that produce JH, are shut off through an unknown mechanism. This inhibition of ecdysone release by JH emerges only in the last larval instar, since JH is present during ecdysone pulses of earlier larval stages, and exogenous JH has no effect upon the timing of these molts. In the final larval stage, however, exogenous JH delays the onset of metamorphosis (Nijhout and Williams 1974a).

The acquisition of critical weight culminates in ecdysone-induced epidermal cell commitment. This phenomenon was first described for holometabolous insects in the moth, *Manduca sexta*. Once critical weight is achieved, and the corpora allata are inactivated, the prothoracic glands produce a small peak of ecdysone during the last larval instar (Watson *et al.* 1987, Nijhout 1994). This peak (marked "C" in Fig. 5-2, bottom) occurs in the absence of JH, and does not trigger cuticle formation. Instead, it triggers two events:

1) cessation of feeding, and the onset of wandering behavior, when the larva searches for an appropriate site for metamorphosis
2) a switch in epidermal cell fate.

Treatment with JH prior to the production of this peak prevents formation of pupal cuticle and causes re-induction of the larval cuticle. However, exogenous JH given after the small surge of ecdysteroid does not alter the fate of the pupally committed cells. Therefore, this small ecdysone peak causes a switch in commitment, from larval cells to pupal cells (Riddiford 1976).

A similar size-based safety mechanism may also occur in the transition to incomplete metamorphosis in the hemimetabolous insects. Instead of occurring at the penultimate instar, however, this small rise in ecdysteroids appears during the final instar, after JH levels decline, but before the molt to the adult stage. For instance, a size-based threshold mechanism has been demonstrated for progression to incomplete metamorphosis in a hemimetabolous

insect, since a mere injection with saline is able to induce the molt to the adult stage in final instar nymphs (Nijhout 1979). In addition, a small peak of ecdysteroid appears in the middle of the final larval instar of the hemimetabolous desert locust, *Locusta migratoria* (Hirn *et al.* 1979), and the dragonfly, *Aeshna cyanea* (Andries 1979, Schaller and Charlet 1980). In both cases, this peak coincides with the loss of JH sensitivity, as exogenous JH given before this small peak prevents adult differentiation, while treatment afterwards does not. Therefore, hemimetabolous insects probably use low levels of ecdysteroid to promote a switchover in cell fate as well.

The commitment peak of holometabolous insects resembles the peak observed in hemimetabolous insects in that they both:

- regulate a turnover in cell fate
- appear after the initial decline of JH.

These similarities suggest that the mechanisms used to regulate the transition from nymph to adult in hemimetabolous insects may have been transposed from the last instar to the penultimate instar of holometabolous insects to facilitate the transition from larva to pupa. Therefore, the appearance of a critical weight mechanism, which inhibits JH release to allow commitment, in the penultimate instar may have been a key event in the evolution of holometaboly.

A second distinguishing endocrine feature of the final larval instar of holometabolous insects is the reappearance of JH at the molt to the pupal stage (Fig. 5-2, bottom). The corpora allata are shut off during the final larval stage, once critical weight is obtained. The "*status-quo*" action of JH reappears after pupal commitment at the molt to the pupal stage, when JH is required to prevent precocious adult development. This was shown in *Manduca* final instar larvae by allatectomy—surgical removal of the corpora allata. Allatectomy during the final larval instar results in pupal–adult intermediates at the ensuing molt. In these animals, the imaginal tissues produce adult-like cuticle, and normal pupal development is rescued by exogenous JH treatment (Kiguchi and Riddiford 1978). Moreover, loss of the JH-effectors *broad* (*br*) and *Kruppel-homolog*-1 (*Kr-h*1) in the flour beetle *Tribolium castaneum* also produces precocious adult features (Konopova and Jindra 2007,

Parthasarathy *et al.* 2008, Minakuchi *et al.* 2009). Therefore, the *status-quo* action of the JH pathway at the pupal molt is probably a common feature of holometabolous metamorphosis.

The mechanism that accounts for the reappearance of JH at the pupal molt has not been identified. However, analysis of JH biosynthesis in *Manduca* shows that corpora allata activity is altered at this stage, such that the glands produce JH acid (JHA), an inactive prohormone (Bhaskaran *et al.* 1986). Kinjoh *et al.* (2007) followed the expression of enzymes involved in JH biosynthesis during silk moth metamorphosis. The initial decline in JH production by the corpora allata occurs as expression of each of eleven transcripts of JH-biosynthesis enzymes is reduced. Ten of these are re-expressed at the prepupal stage, accounting for the production of JH acid. The final enzyme, JH acid O-methyl transferase, which catalyses the formation of JH from JHA, is expressed in the imaginal tissues but it is not detected in the corpora allata, where it is normally expressed during the earlier larval instars (Kinjoh *et al.* 2007). The situation in *Bombyx* therefore suggests that JH acid is produced in the corpora allata, by the normal complement of enzymes, but the rate-limiting step occurs in the tissues. It is possible that acquisition of new stage- and tissue-specific regulatory enhancers in the genes of these biosynthetic enzymes may contribute to the evolution of the pupal stage.

5.6 The "status-quo" action of juvenile hormone and its signal transduction

Historically, the identity of the JH receptor and its signal transduction pathway has been refractory to biochemical and genetic research. A relatively large amount has been learned in the past few years, however, and comparative studies in non-model insects have played a large part in the discovery (Erezyilmaz *et al.* 2006, Konopova and Jindra 2007, Parthasarathy *et al.* 2008, Suzuki *et al.* 2008, Minakuchi *et al.* 2009). Although many aspects of JH signaling are conserved between hemi- and holometabolous insects, differences in the pathways point to the causal changes in the evolution of complete metamorphosis.

The effect of JH during *Drosophila* metamorphosis is clear, but less dramatic than in other holometabolous insects (Riddiford and Ashburner 1991), and this has complicated the use of this genetic model to identify genes involved in JH signal transduction. In a clever genetic screen, Wilson and Fabian (1986) made use of pesticide resistance by selecting flies that were resistant to methoprene, an analog of JH that is used as an insecticide. This process implicated the gene *Methoprene-tolerant* (*Met*), a bHLH-PAS protein (Ashok *et al.* 1998). *Met* mutants are 100 times less sensitive to methoprene treatments (Wilson and Fabian 1986), and JH binding to a cytosolic JH protein had 10-fold lower binding affinity in *Met* flies than in wild-type flies (Shemshedini and Wilson 1990). Moreover, JH binds Met protein directly at physiological levels and activation of transcription of a reporter gene in *Drosophila* S2 cells by Met required JH (Miura *et al.* 2005).

Despite the clear case for a role for *Met* in JH sensitivity in *Drosophila*, loss of *Met* did not disrupt larval development or metamorphosis in the fruit fly (Wilson and Ashok 1998), and a function for *Met* in these aspects of JH action remained debatable. However, with the emergence of double-stranded RNA-mediated interference as a viable tool for testing gene function in non-model insects, the role of *Met* in JH signaling at metamorphosis was firmly established. In contrast to *Drosophila*, the red flour beetle *Tribolium castaneum* is highly sensitive to JH during metamorphosis. Konopova and Jindra (2007) isolated *Tc'Met* from this insect to test its role at metamorphosis. Depletion of *Tc'Met* prevented a JH-induced second pupa, and *Tc'Met* knockdown during the larval stages resulted in precocious metamorphosis.

Met mediates JH signaling through a second factor, *Kruppel-homolog*-1 (*Kr-h1*), a C_2H_2 zinc finger gene that is broadly expressed during the late larval and prepupal stages of the holometabolous insects, *Drosophila* and *Tribolium* (Minakuchi *et al.* 2008, 2009, Konopova and Jindra 2009). The role of this protein in regulating metamorphosis was first discovered in *Drosophila*, when flies with P-element-induced mutations in *Kr-h1* failed to complete prepupal development (Pecasse *et al.* 2000). *Kr-h1* was subsequently identified in a microarray experiment, where it was significantly induced in *Drosophila* prepupae 12 h after treatment with a JH mimic (Minakuchi *et al.* 2008). Misexpression of the gene in the abdominal histoblasts during adult development causes irregular or missing bristles along the dorsal midline, a very similar phenotype to that found after treatment of adults with JH. In *Tribolium*, which has a canonical *status-quo* response to JH, loss of *Kr-h1* causes metamorphosis to occur at the sixth larval instar, instead of the eighth (Konopova and Jindra 2009, Minakuchi *et al.* 2009). In this insect, *Kr-h1* is not expressed during the pupal stage, when JH is absent and the pharate adult is developing. However, topical application of JH during the pupal stage re-induces prepupal development, which causes re-induction of *Kr-h1*. Moreover, *Tc'Met* expression is required for re-induction of *Tc'Kr-h1* (Konopova and Jindra 2009, Minakuchi *et al.* 2009). In this situation therefore, *Tc'Met* acts upstream of *Tc'Kr-h1* expression.

A role for *Met* has also been established in regulating the incomplete metamorphoses of a hemimetabolous insect. Konopova and Jindra (2009) have shown that depletion of *Met* phenocopies the effect of chemical allatectomy in the hemimetabolous firebug, *Pyrrhocoris apterus* (*Pa'Met*). Specifically, when *Pa'Met*[RNAi] was injected into firebug nymphs, these insects passed through one normal molt, but then molted to precocious adults with miniature wings, adult pigmentation patterns, and genitalia (Konopova and Jindra 2009).

5.7 The *broad* gene and specification of the pupal stage

If the *status quo* roles of the JH signal transduction pathway are conserved between hemi- and holometabolous insects, then how could alterations to this pathway account for emergence of a morphologically divergent larval stage and the novel pupal stage? The answer lies, in large part, with the temporal deployment of the *broad* gene during ontogeny.

The *br* gene was first identified in *Drosophila*, in a screen of mutants that fail to pupariate. Gynandromorph flies, which are a mosaic of wild-type and *br*-mutant tissue, form mosaics of pupal and larval cuticle at the end of larval life (Kiss 1976). Molecular studies show that *br* is a complex gene that produces four isoforms created when a Broad-Tramtrack-Bric-a brac (BTB) domain-containing

core region is fused to one of four different zinc fingers, Z1–Z4, to produce four different transcription factors (DiBello *et al.* 1991, Bayer *et al.* 1996). Four different complementation groups have been identified that remove Z1, Z2, Z3 or complete *br* functions (Belyaeva *et al.* 1980, Kiss *et al.* 1988). The expression pattern of *br* in flies (Karim *et al.* 1993) reveals that *br* expression is prominent during the larval–pupal transition but is absent as the pupa initiates adult differentiation (Zhou and Riddiford 2002). The significance of this pattern was seen when heat-shock driven expression of Br isoforms during the adult molt induced a second pupal cuticle, while heat-shock driven expression of Br during a larval molt resulted in the precocious expression of pupal genes (Zhou and Riddiford 2002). Therefore, *br* seems to specify the pupal stage in *Drosophila*.

The tight correlation between *br* expression and pupal development of holometabolous insects has been confirmed wherever it has been examined. *br* expression is also strongly up-regulated at the larval-pupal transition in two lepidopterans (Zhou and Riddiford 2001, Uhlirova *et al.* 2003), *Tribolium* (Konopova and Jindra 2007, Parthasarathy *et al.* 2008, Suzuki *et al.* 2008), and the neuropteran, *Chrysopa perla* (Konopova and Jindra 2007). In *Manduca*, where the endocrine titers are well defined, *br* is first expressed at the commitment peak, when a small pulse of ecdysone occurs in the absence of juvenile hormone (Fig. 5-4, bottom; Zhou *et al.* 1998, 2001). It is then maintained at the molt to the pupal stage, when the ecdysone peak is accompanied by JH (Zhou *et al.* 1998, 2002). Treatment of *Manduca* pupae with JH redirects the pupal-adult molt to produce a second pupal stage. In this situation, *br* is re-induced when it would normally be absent during adult development. This re-induction after JH treatment is also observed during the adult molt of *Tribolium* (Suzuki *et al.* 2008), and this re-induction requires *Met* (Konopova and Jindra 2007). Therefore, the onset of *br* expression is inhibited by JH during the larval molts, but once it is induced at commitment, the presence of JH does not inhibit *br* expression. In a sense, JH is maintaining the *status quo* of *br* expression.

Loss of function studies from other insects confirm the role of *br* in regulating metamorphosis

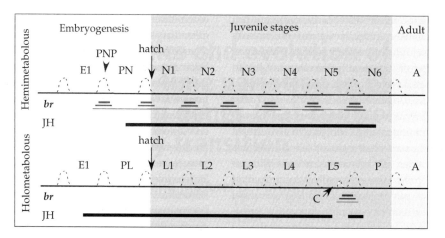

Figure 5-4 The regulation of *br* expression by ecdysone and JH. Regulation of *br* expression by ecdysone and JH follows two general rules in both hemi- (top) and holometabolous insects (bottom): 1) *br* expression is activated by ecdysone in the absence of JH, and 2) once activated, *br* expression is maintained by the presence of JH at ecdysone peaks. Ecdysone pulses are shown as dashed lines, the presence of JH is represented by the solid black bars, and *broad* expression is denoted by grey underlines. In hemimetabolous embryos *br* is activated during the pronymphal peak (PNP) of ecdysone in the absence of JH, and its expression is maintained at the ensuing nymphal molts. In holometabolous insects, *br* first appears at the commitment peak (C), as JH titers decline. It is then maintained at the molt to the pupal stage, which occurs in the presence of JH. L, larval stage; N, nymphal stage; E1, stage that follows E1 cuticle deposition; A, adult stage. Data summarized from Zhou and Riddiford 1999, Zhou and Riddiford 2001, Uhlirova *et al.* 2003, Erezyilmaz *et al.* 2006, Konopova and Jindra 2007, Parthasarathy *et al.* 2008, Suzuki *et al.* 2008, Erezyilmaz *et al.* 2010).

throughout the Holometabola. In the lepidopteran, *Bombyx mori*, loss of *br* in animals transformed with Sindbus virus driving *br*^RNAi blocks the larval–pupal transition (Uhlirova *et al.* 2003). This role in specifying the pupal stage is probably ancestral within the Holometabola, since RNAi directed against the common region of *br* from *Chrysopa*, a more basal holometabolous insect, also results in an arrest at the pupal stage (Konopova and Jindra 2007). Loss of *br* in *Tribolium* also has no effect prior to the pupal stage, but produces mosaic pupae comprising larval abdomens in combination with adult structures, such as the appendages. Analysis of the function of each individual isoform reveals that the timely progression through metamorphosis occurs by collaboration of each of the Br proteins (Suzuki *et al.* 2008). For instance, pupae formed after injection with the dsRNA, which knocks down the Z2 or Z3 isoforms, have shorter wings, and malformed claws, but are otherwise normal looking. Loss of the Z4 isoform, however, causes a ballooning of the wings, adult-like legs during the pupal stage, and patches of pupal cuticle during the adult stage. Therefore, entry to and exit from the pupal stage requires combined input

from multiple Br isoforms. Interestingly, in the most basal group of holometabolous insects, the snake flies, pupae have larval-like abdomens and adult-like appendages. Given this, Suzuki *et al.* (2008) speculate that the first holometabolous insects formed pupae in the absence of *br* function.

An analysis of the temporal expression and functions of *br* reveals a great deal of variation across the three types of developmental strategies. In the ametabolous insect, *Thermobia domestica*, an analysis of the common region of Br shows constitutive expression on each day of embryonic development (Erezyilmaz *et al.* 2010). Unfortunately, the function or postembryonic expression of *br* has not been described for this or any other ametabolous insect.

In a coarse analysis of embryos of the hemimetabolous cricket, *Acheta domesticus*, *Ach'br* is found in the earliest stages of segmentation, and it is strongly expressed in the last quarter of embryonic development (D.F. Erezyilmaz, unpublished data). In the milkweed bug *Oncopeltus*, which is also hemimetabolous, *Of'br* is first detected in early embryonic development, with transient expression appearing on day one (Erezyilmaz *et al.* 2010). This transient expression roughly coincides with segmentation and germ band invagination, a morphogenetic movement that positions the embryo inside the yolk, and knockdown of *Of'br* disrupts morphogenesis at this stage. No further *Of'br* expression is then seen until just before the appearance of the pronymphal cuticle, after which *Of'br* expression remains high throughout the remainder of embryonic development (Fig. 5-4).

In holometabolous insects, *br* expression has been detected in whole extracts from holometabolous embryos of *Tribolium* and *Chrysopa* and in the first larval stages of *Chrysopa*. However, despite injection with *Cp'br* dsRNA in the first instar, knockdown of this gene had no effect until pupal development. In addition, loss of *Tc'br* in the fifth larval stage had no effect upon the subsequent three larval stages, although it did affect the ensuing metamorphic molt (Konopova and Jindra 2007). A more detailed analysis of Br protein production during the immature stages of holometabolous insects with an antibody directed against the Br core region in whole *Drosophila* and *Manduca* embryos showed that Br is restricted to a handful of neurons

in the central nervous system, and subsequent analysis with isoform-selective antisera in *Drosophila* revealed that only BR-Z3 was produced in these neurons (Zhou *et al.* 2009). Therefore, a role for *br* production in the epidermis of holometabolous insects has only been shown for the metamorphic stages.

The role of *br* during holometabolous development reveals a fundamental divergence from its role during hemimetabolous development. Whereas *br* does not appear to have a function in the epidermis during the immature stages of the Holometabola, loss of *Of'br* was shown to have a profound effect upon the nymphal stages of *Oncopeltus*, a hemimetabolous insect that molts to an adult after five nymphal stages (Erezyilmaz *et al.* 2006, 2010). Specifically, *Of'br* is induced at the end of each instar as the animal prepares to molt to the next nymphal stage. *Of'br* is absent as the last stage nymph molts to the adult stage (Erezyilmaz *et al.* 2006, Fig. 5-4). These data suggest that *Of'br* is not inhibited by the presence of JH during the immature stages as it is in holometabolous insects.

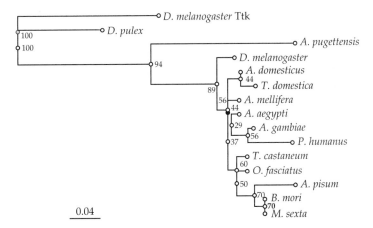

Figure 5-5 Gene tree generated from amino acid sequences from 12 different insect Br BTB domains aligned against 49 amino acids of a putative Br sequence found in the amphipod crustacean, *Apopayhyale pugettensis* (N. Goehner. J.W. Truman and L.M. Riddiford, unpublished data). The *Daphnia pulex* sequence was taken from the published *Daphnia* genome, and had the greatest similarity to *D. melanogaster* Br. The *A. pugettensis* sequence clusters more closely with the insect Br BTB domains than with the *D. pulex* sequence, which shares more identity with other *D. melanogaster* BTB domains. The BTB domain of *D. melanogaster* Tramtrack was used as an outgroup. Holometabolous Br sequences include the mosquitoes *Aedes aegypti* and *Anopheles gambiae*, the house cricket, *Acheta domesticus*, the moths, *Manduca sexta*, and *Bombyx mori*, the bee, *Apis mellifera*, and the fruit fly, *Drosophila melanogaster*. Br BTB domains for hemimetabolous insects include the hemipterans *Oncopeltus fasciatus* and *Acythosiphum pisum*, the cricket, *Acheta domesticus* and the louse, *Pediculus humanus*. The sequence of the single ametabolous insect, *Thermobia domestica*, is missing one residue. Amino acid sequences were aligned using CLUSTALW, and Geneious Tree Builder software (Drummond *et al.* 2010) was used to create the neighbor-joining tree. The values at nodes are per cent consensus support and the scale bar gives the number of amino acid substitutions per site.

To test this, Erezyilmaz *et al.* (2006) induced a supernumerary larval stage by applying JH to a final-stage nymph. In this situation, *Of'br* is re-induced at the end of the fifth instar. They next asked if *Of'br* expression was lost after loss of a nymphal molt. Nymphs that were treated with precocene, a drug that destroys the corpora allata, which produces JH, lost expression of *Of'br* in the fourth instar as they molted to a precocious adult (Erezyilmaz *et al.* 2006). Therefore, *Of'br* expression requires JH, and expression of *Of'br* correlates tightly with nymphal development.

To test the function of *br* expression during the nymphal molts, *Of'br* dsRNA was injected into third-stage nymphs. The five nymphal stages of *Oncopeltus* can be distinguished by the pattern of melanin on the thorax and by the size of the wing pads. Loss of *Of'br* did not prevent the growth of the nymph or the molt itself. Instead, *Of'br*[RNAi] prevented changes in proportion and the sequence of nymphal identity. For instance, after injection with *Of'Br* dsRNA during the third instar, nymphs molted to a larger version of the third nymphal stage (a 3′ nymph), instead of to a fourth-stage nymph; additional injections induced a 3′-stage result during the fifth instar—an additional repetition of the third nymphal pattern (Erezyilmaz *et al.* 2006). Therefore, *Of'br* is required for changes in identity between instars. Without *Of'br*, it is as if the genome remains locked, and only simple isometric growth and molting from one stage to the next are permitted.

5.7.1   Molecular aspects of Broad action

A comparison of the role of *br* in regulating changes between nymphal instars in a hemimetabolous insect with its role in regulating change at the pupal stage of the Holometabola suggests that the salient feature of *br* action is to convey competence for change. The fact that knockdown of *br* function prevents transitions in identity between any nymphal instar suggests that *br* does not specify the genes of any particular stage, but instead permits changes in gene expression required for changes in form between instars. The role of Br isoforms in regulating the expression of *fat body protein* 1 (*fbp*-1) in the fat body at the onset of metamorphosis in *Drosophila*

(Mugat *et al.* 2000) may provide insight into how *br* regulates changes between instars without specifying pupal- or nymphal-specific genes. In this tissue, *Br* isoforms alone are not able to induce expression from an *fbp*-1 minimal promoter and they are unable to regulate its expression outside the fat body. Instead, a switch in isoform expression in this tissue appears to regulate the timing of *fbp*-1 expression. *fbp*-1 is expressed in the latter half of the third larval instar. The Z2 isoform is expressed until the mid-third instar, when Z3 expression accumulates, followed by Z1 and Z4. Loss of the Z2 isoform allows precocious expression from the minimal promoter, while loss of any of the other isoforms reduces expression (Mugat *et al.* 2000). These results suggest that expression of this gene during metamorphosis is gated by the switch from Z2 to Z1/Z3/Z4, and suggest an explanation for the phenotypes that are observed after knockdown of *Of'br* during *Oncopeltus* development. By extrapolation, the changes in nymphal identity are not specified by *Of'br*, but are instead gated by the turnover in isoforms. Loss of *Of'br* then prevents these gated changes, the genome remains locked, and the status quo is maintained.

5.7.2   A *broad*-based view of the pronymph hypothesis

The patterns of *br* expression and *br*'s roles in hemi- and holometabolous insects indicate that a heterochronic shift in the expression of this factor, which conveys mutability to insect molts, has been restricted to the penultimate postembryonic molt in the metamorphosing insects (Fig. 5-4). The endocrine events that are unique to holometabolous development, including the inhibition of PTTH by JH at the last larval stage and the re-induction of JH at the pupal molt may have evolved in tandem. The observed heterochronic shift follows a change in the appearance of *br* and JH during the immature stages, since *br* is suppressed by JH during the larval stages of the Holometabola, but activated by JH during the nymphal stages of hemimetabolous insects. This could have evolved through one of two mechanisms:

1) *Acquisition of inhibitory regulation by JH*. This may have occurred through the emergence of an

inhibitory JH response element in the *br*-regulatory region in the line of insects that led to the Holometabola. This would have repressed *br* expression as soon as JH appeared, late in embryonic development. The coincidence of JH and *br* expression during pupal development would have evolved secondarily.

2) *An earlier onset of JH production inhibits the initial activation of br.* This idea was suggested by Lynn Riddiford (Erezyilmaz *et al.* 2010). This scenario does not require any evolved changes in the regulatory circuitry between *br* and JH. From the extensive work on *br* during holometabolous metamorphosis, endocrine assays have established that *br* transcription is induced by the commitment peak of ecdysone, which occurs in the absence of JH (Fig. 5-4, bottom). Once it is activated, the presence of JH during the ecdysone peak is no longer inhibitory. During embryonic development of hemimetabolous insects, *br* expression first appears coincident with the peak of ecdysone that triggers pronymphal cuticle formation. Like the commitment peak, the pronymphal peak of hemimetabolous insects occurs in the absence of JH, while the following nymphal molts occur in the presence of JH, like the molt to the pupal stage. Therefore, the ancestral commitment peak of hemimetabolous insects may have been the pronymphal peak. The causal event in the evolution of metamorphosis would then be the encroachment of JH into pronymphal development, thereby suppressing the onset of *br* transcription until JH titers subsided at the penultimate larval instar.

Regardless of the mechanism that caused the temporal shift in *br* expression in the Holometabola, the expression profiles of *br* support the suggested homology between the hemimetabolous nymph and holometabolous pupa that was suggested by the pronymph hypothesis (Truman and Riddiford 1999). The appearance of *br* during the pronymphal stage, and its presence at the molt to the nymphal stage, strongly resembles the appearance of *br* during the final larval stage of holometabolous insects and its presence at the molt to the pupal stage (Fig. 5-4). The function of *br* during the pronymphal stage of hemimetabolous embryos is not known.

During the later nymphal stages, however, it is required for changes in form, or mutability (Erezyilmaz *et al.* 2006). If *br* plays a similar role at the pronymphal to nymphal transition, then a suppression of *br* expression caused by an earlier appearance of JH would suppress the transition in form between the pronymphal and nymphal stages. This would suspend progression to the nymphal form, and fix the pronymphal proportions during the postembryonic instars until JH titers disappeared at the end of postembryonic development. The de-repression of pronymphal proportions would allow the suspended nymphal body plan to be deployed, in a process called complete metamorphosis.

5.7.3 The appearance of *broad* in the arthropods

A database search of Br orthologs does not reveal any sequences outside the arthropods, including the nematode *Caenorhabditis elegans*. One sequence from the crustacean water flea, *Daphnia pulex*, has 69% amino acid identity to *Drosophila* Br, but is more similar to other *Drosophila* BTB-domain proteins (Fig. 5-5, and data not shown). On the other hand, a sequence of a fragment of Br obtained from the direct-developing amphipod crustacean, *Apohyale pugettensis* (N. Goehner, J.W. Truman and L.M. Riddiford, unpublished data) segregates more strongly with the insect Br BTB sequences (about 75% amino acid identity, Fig. 5-5 and data not shown). The amphipod and water flea are from two sister taxa, the Malacostraca and Branchiopoda, respectively, and this clade is often considered a sister clade to the Insecta (Cook *et al.* 2005). Therefore, Br seems to have arisen within the arthropods, but before the split between insects and crustaceans.

In insects, *br* has emerged as a creative force, which regulates the nymphal heteromorphosis of hemimetabolous insects as well as the more dramatic metamorphoses of Holometabola. Because these two disparate processes occur as JH-regulated switches at molts, the *br* gene is likely to also function at the JH-regulated polymorphisms and polyphenisms. Caste determination in social insects and the environmentally induced reproductive morphs of aphids are two examples (see Chapter 20 on honey-bee life-history plasticity). Another example is the extreme life-history adaptations observed in parasitic insects. In the parasitic order Strepsiptera, for example, females from the genus *Xenos* remain within the host in a reproductive larval state. The male, however, metamorphoses and emerges from within its host. Multiple *br* isoforms have recently been isolated from the metamorphosing *Xenos* male, and studies are underway to determine whether females have reduced or altered *br* expression (A. Hayward, D. Erezyilmaz and J. Kathiritamby, unpublished data). If so, this would underscore the pivotal use of this protein in generating the fantastic diversity of life-history transformations found within the insects.

5.8 Summary

1. Endocrine events during embryonic development differ between the three major insect life-history strategies. Ametabolous insects hatch after formation of just two embryonic cuticles, but nothing is known of their JH or ecdysone profiles. Both hemi- and holometabolous embryos hatch after producing three embryonic cuticles: the E1, pronymphal/prolarval, and first nymphal/larval. However, the two groups differ in the onset of embryonic JH production. JH appears with formation of the prolarval cuticle in holometabolous insects, but not until the first nymphal cuticle in the hemimetabolous insects. Experiments that simulate the early appearance of JH by exogenous JH treatment of the pronymphal stage of hemimetabolous insects show that JH has a profound effect upon embryonic growth and differentiation: JH treatment of pronymphs alters growth so that it becomes more isometric and terminal features appear precociously.

2. Two major endocrine events of the penultimate instar distinguish holometabolous insects from their hemimetabolous ancestors. In the first, the acquisition of critical weight, the subsequent decline in JH titers, and the following small peak of ecdysteroid probably occur in hemimetabolous insects as well as holometabolous insects. What differs is that these events have been transposed to the penultimate instar

of Holometabola, although they occur in the last instar as they do in hemimetabolous insects. The second difference involves the restarting of production of JH at the pupal molt, which suspends the precocious development of adult structures.

3. After over 70 years of research on JH signaling, two proteins have recently been established in transduction of the status quo response. Loss of the bHLH-PAS protein Met, or the C_2H_2 zinc finger *Kr-h1* results in precocious metamorphosis in *Tribolium*, and *Met* is required for formation of a JH-induced second pupa. *Met* is also required for the status quo action of JH in hemimetabolous insects, since loss of *Met* in the firebug, *P. apterus* resulted in formation of precocious adults. These data suggest that the initial steps of JH signal transduction may be conserved between hemi- and holometabolous insects.

4. Numerous studies of the role of *br* in holometabolous insects have established this gene as a pupal determinant. They have also shown that epidermal expression of this gene is largely restricted to the metamorphic stages. Analysis of *br* expression of hemimetabolous insects shows that *br* appears during embryonic development and persists during the nymphal stages. The postembryonic expression of *br* is not known for any ametabolous insect, although *br* is expressed throughout embryonic development of the direct-developing *Thermobia domestica*. Therefore, *br* expression has been transposed from embryonic and juvenile stages in the holometabolous lineage.

5. A comparison of *br* expression and function has suggested that the temporal shift in *br* expression could account for the evolution of insect metamorphosis. What, then, caused the shift in *br* expression? The most likely explanation involves the shift in JH production to earlier stages of embryonic development that has occurred in holometabolous insects. In both hemi- and holometabolous insects, *br* expression is activated by ecdysone in the absence of JH, but once activated, its expression is maintained by JH and ecdysone. Given these two properties, an earlier appearance of JH would suppress the onset of *br* expression during embryogenesis of hemimetabolous insects, and subsequently delay the onset of *br* expression until

JH titers declined at the penultimate instar, creating a metamorphosis.

6. Br orthologs have been found in both direct-developing and metamorphosing insects, but not outside of the arthropods. Within the arthropods, a short sequence with high amino-acid identity to the insect Br sequences has been obtained from the amphipod crustacean, *A. pugettensis*, although a Br ortholog does not exist in the genome of the crustacean, *D. pulex*. These data suggest that *br* arose relatively recently within the arthropods.

5.9 Acknowledgments

I thank Professor Lynn Riddiford for her helpful comments on this manuscript and for the use of unpublished data. I additionally thank Professor Marek Jindra and Barbora Konopova for their helpful comments and use of unpublished data as well as Drs Jeya Kathiritamby and Alex Hayward for allowing me to cite their unpublished data. This work was supported by an NIH Kirschstein Fellowship to me, D.F.E.

CHAPTER 6

Thyroidal regulation of life history transitions in fish

Richard G. Manzon

6.1 Introduction

Fish are the largest and most diverse group of vertebrates with an estimated 27,977 described species (Nelson 2006). Although at present there is still some controversy over precise fish taxonomy and formal classification, there is a consensus as to the major descriptive groupings used in fish biology. This essay will conform most closely to the taxonomy presented by Nelson, who recognizes five extant fish classes—the agnathans Myxini (hagfish) and Petromyzontida (lampreys) and the gnathostomes Chondrichthyes (cartilaginous fish), Actinopterygii (ray-finned fish), and Sarcopterygii (lobe-finned fish and tetrapods)—and 515 fish families (Nelson 2006). Diversity is also evident in the areas of fish ontogeny, and the nature, number, and frequency of fish life history transitions. Interestingly, amid this variation, thyroid hormones (THs) appear to function in some capacity in the ontogeny and life history transitions of most fish, including basal (i.e., lampreys) and derived (i.e., flatfish) species. This suggests that THs are central ancestral signaling molecules in fish ontogeny and may provide a mechanism through which natural selection might act. Minor endogenous or exogenous perturbations of THs at critical times in development could result in more substantial changes in the timing of developmental events (heterochrony) and the subsequent evolution of diverse life history strategies.

The suggestion that THs are ancestral signaling molecules is not novel. Iodinated compounds (i.e., iodohistidine and iodotyrosine) are present in representatives from most metazoan groups; THs and other iodothyronine derivatives are synthesized by all chordates (reviewed in Eales 1997); all vertebrate taxa contain the requisite TH signaling components (reviewed in Paris and Laudet 2008). Furthermore, THs (or their derivatives) are important during the development of most vertebrates studied to date (reviewed in Hulbert 2000), including fish (Youson 2007, Dufour and Rousseau 2007), amphibians (see Chapter 7), birds (McNabb 2007), and mammals (Yen 2001, Galton 2005), and also in some non-vertebrates such as the cephalochordate amphioxus (Paris *et al.* 2008) and the echinoderms (see Chapter 3). The involvement of THs and TH-derivatives in chordate development and the conservation of the TH-signaling system (i.e., nuclear receptors) prompted Paris and Laudet (2008) to propose that all chordates, and perhaps all deuterostomes, undergo a metamorphosis that is regulated by THs or their derivatives. According to their new definition for chordate metamorphosis, which is based on the conservation of TH signaling and a function in development, they further hypothesize that metamorphosis evolved once in chordates and that the chordate ancestors followed an indirect pathway of development (Paris and Laudet 2008).

The alternative (traditional) view is that the ancestral chordates were direct developers and that metamorphosis (see also Box 6-1 for important definitions and explanations of terms) evolved several times independently in some chordates (see below, and Nielsen 1998, Hadfield *et al.* 2001). If we consider both of these hypotheses collectively, it is plausible that THs and their derivatives are ancient signaling molecules that act via conserved nuclear

Box 6-1 Definitions and major characteristics of periods in fish ontogeny and fish life history transitions

Embryonic period

The embryonic period begins with egg activation (i.e., fertilization) and ends with independent exogenous feeding and digestion regardless of whether or not a yolk sac is still present (Balon 1999).

Larval period

The larval period begins with the onset of exogenous feeding and ends following the completion of a first or "true" metamorphosis (Just *et al.* 1981, Youson 1988) and thus should be reserved for fish that follow indirect development. Most fish species are direct developers and have no "true" larval period, thus it is the juvenile period that begins with the onset of exogenous feeding (Balon 1999). However, to facilitate the broadest discussion of early life history transitions in fish and avoid the introduction of more terminology, I will adopt a more liberal use of the term "larva". This will allow the discussion of the less dramatic transitions (i.e., larval to juvenile) that occur early in the post-embryonic period.

Juvenile period

The juvenile period commences when the fish has lost most of its larval (or embryonic) characteristics and acquired most of its adult characteristics and ends at sexual maturation (Balon 1999).

Adult period

The adult period begins with sexual maturation (i.e., spermatogenesis or vitellogenesis) and ends with senescence or death (Penaz 2001).

Embryo to larval transitions

These transitions mark the end of the embryonic period and the start of the larval period in the case of indirect developers or juvenile period in the case of direct developers. During this time fish acquire the ability to obtain nutrients from exogenous sources and process these endogenously. There is also an increase in motility and a loss of most embryonic characteristics.

Larval to juvenile transitions

Larval to juvenile transitions will refer to all transitions in fish that occur shortly after the onset of exogenous feeding and do not meet the criteria of a "true" metamorphosis. These transitions usually occur early in the post-embryonic period and involve progressive changes that began in the embryonic period; in the literature these larval to juvenile transitions are often inappropriately referred to as "metamorphoses." Examples of fish that undergo a larval to juvenile transition include grouper, milkfish, sea bass, sea bream, tilapia, and zebrafish.

Metamorphosis ("true" or first metamorphosis)

Among fish, a "true" metamorphosis is restricted to the Petromyzontiformes, members of the subdivision Elopomorpha (i.e., eel, tarpon, bichir) and the order Pleuronectiformes (i.e., flounder, sole, halibut) of the subdivision Euteleostei (Youson 1988). A "true" metamorphosis must include several of the following: 1) a post-embryonic change in non-reproductive structures which are unrelated to embryogenesis, 2) an ecological niche that differs from that of the embryo and juvenile/adult, 3) a marked change in body form such that larvae and juveniles/adults do not look alike, 4) no growth, and 5) regulation by exogenous and/or endogenous (hormonal) signals (Just *et al.* 1981, Youson 1988).

Juvenile transitions

Juvenile transitions are those that occur during the juvenile period and can involve changes in body morphology and coloration, physiology (i.e., change in osmoregulatory ability), behavior (i.e., benthic versus pelagic), and migration. The best studied of these transitions is the parr to smolt transformation (smoltification) that occurs in salmonids (Hoar 1988, McCormick *et al.* 1998, Stefansson *et al.* 2008). Although often referred to as a metamorphosis, in my opinion these are not "true" fish metamorphoses (see above and text). Smoltification, other similar transitions, and those associated with sexual maturation have been referred to as a second metamorphosis or second type of metamorphosis (reviewed in Youson 2008).

receptors to regulate development and thus represent a mechanism for the evolution of metamorphosis. Accordingly, metamorphosis may have evolved independently in a variety of taxa following a perturbation (i.e., change in environment, nutrient supply, competitors, or predators) that altered TH signaling, resulting in a change in developmental timing (heterochrony) and the appearance of a larval period (i.e., a delay in sexual maturation). This chapter will provide an overview of the different life history transitions in fish, review data on the fish thyroid system, and discuss the conserved role of the thyroid system in fish ontogeny as a potential mechanism for the evolution of life history transitions.

6.2 Fish ontogeny and life history transitions

The diversity in fish ontogeny has generated both controversy and communication difficulties regarding such concepts as:

- gradual versus saltatory ontogeny
- direct versus indirect development
- the onset of larval and juvenile periods
- the existence of a larval period in all fish
- the appropriate use of the term "larva"
- what constitutes a metamorphosis in fish.

Discussion of the aforementioned debates is beyond the scope of this chapter; I refer those interested to other essays on the topic (Youson 1988, Balon 1999, Penaz 2001, Bishop *et al.* 2006a). Given the disparity in opinions regarding fish ontogeny, an overview of the terminology and definitions used in this chapter are detailed in Box 6-1. A broad approach will be taken when discussing the topic of fish life history transitions, such that both the subtle and dramatic transitions beginning with embryonic development and ending with the sexually immature juvenile will be considered (Box 6-1). Although sexual maturation is an important life history transition, it is beyond the scope of the current chapter and readers are referred to other reviews on the topic and the overlap between the reproductive and thyroid axes (Youson and Sower 2001, Blanton and Specker 2007, Dufour and Rousseau 2007, Sower *et al.* 2009).

6.3 Overview of the hypothalamic–pituitary–thyroid axis

The importance of the THs 3,5,3′,5′-tetraiodothyronine (thyroxine; T_4) and 3,5,3′-triiodothyronine (T_3) is evident in chordates, but they also function in numerous invertebrates and even some plants and prokaryotes (Eales 1997). In vertebrates, THs influence thermoregulation, many aspects of metabolism, growth, immune and antiviral defenses, cardiovascular function, reproduction, development, and life history transitions (Hulbert 2000, Yen 2001). Gudernatsch was the first to note the influence of thyroid tissue on vertebrate development; this seminal work on anurans ultimately led to the identification of THs as the first developmental morphogens (Gudernatsch 1912). Beyond their role in anuran metamorphosis (see Chapter 7), THs influence embryonic and/or post-embryonic development in all vertebrate taxa studied to date. TH activity can be modulated both centrally by the HPT axis, which regulates their synthesis and secretion by thyroid tissue, and peripherally by a number of other mechanisms, including serum TH distributor proteins (THDPs), transmembrane cellular uptake transporters, cytosolic TH binding (transport) proteins (CTBPs), TH deiodinases, TH nuclear receptors (TRs) and their heterodimeric partners, and transmembrane receptors.

6.4 The hypothalamic–pituitary axis

In mammals and other amniotes, hypothalamic thyrotropin-releasing hormone (TRH) is a key stimulator of thyrotropin (thyroid stimulating hormone; TSH) synthesis and secretion from the thyrotrophs of the adenohypophysis, which in turn stimulates the synthesis of THs from follicular thyroid tissue (Yen 2001). In the late 1980s, a second hypothalamic neuropeptide with thyrotrophic activity was discovered when corticotrophin releasing hormone (CRH) was shown to be a more potent stimulator of northern leopard frog (*Rana pipiens*) pituitary TSH secretion than either TRH or gonadotrophin-releasing hormone (Denver 1988). Subsequently, CRH has been found to stimulate pituitary TSH synthesis and secretion in many non-mammalian vertebrates, including birds (Meeuwis *et al.* 1989, De Groef *et al.*

2003), reptiles (Denver and Licht 1989b), amphibians (Denver 1988, Denver and Licht 1989a), and fish (Larsen *et al.* 1998). Some reports suggest that TRH may have thyrotrophic activity in fish, but it does not appear to be a primary regulator of TSH synthesis and secretion (Larsen *et al.* 1998, Mackenzie *et al.* 2009), and it appears that agnathans may use pituitary trophic hormones other than TSH to regulate TH synthesis (Sower *et al.* 2009).

As in other vertebrates, the hypothalamic–pituitary axis of fish consists of a hypothalamus, neurohypophysis, and adenohypophysis. However, the organization, relative size of each division, innervation, and vascularization can vary (reviewed in Lagios 1982, Sower 1998, Norris 2007). For instance, agnathans and teleosts lack a median eminence and hypothalamic–pituitary blood portal system for transport of regulatory molecules from the hypothalamus in the brain to the pituitary (Sower 1998). Teleosts have hypothalamic neurosecretory neurons that directly innervate adenohypophysial cells (Peter *et al.* 1990), but agnathans lack both neuronal and vascular communication. Regulation of the agnathan adenohypophysis is achieved by the diffusion of brain peptides ("the diffusional median eminence") from the adjacent neurohypophysis, which forms the floor of the third ventricle of the brain (Sower 1998). Chondrichthyes (i.e., sharks) and ancient non-teleost ray-finned fish (i.e., bichirs, bowfins) have both a median eminence and short-portal blood vessels (Lagios 1982).

6.5 Thyroid tissue and hormone synthesis

With the exception of the larval lamprey, thyroid follicles or thyroid tubules are the site of TH synthesis, storage, and secretion in fish. In the larval sea lamprey (*Petromyzon marinus*), THs are produced by the subpharyngeal endostyle, which transforms into typical vertebrate follicular tissue during metamorphosis (Wright and Youson 1980). The thyroid tissue of most teleosts is arranged as scattered follicles (or tubules) associated with vascular tissue in the subpharyngeal or basibranchial region, and is not encapsulated into a discrete gland as in the more derived vertebrates (tuna, swordfish, and parrotfish have an encapsulated gland) (Leatherland

1994). Moreover, ectopic thyroid follicles have been localized most commonly to the ventral aorta (e.g., medaka (*Oryzias latipes*); Raine *et al.* 2001) and kidneys (e.g., carp, *Cyprinus carpio*; Geven *et al.* 2007), but also the brain, intestine, and other regions (Leatherland 1994, Youson 2007). The thyroid tissue of both Chondrichthyes and the Sarcopterygii fish is encapsulated, as is case for tetrapods (Leatherland 1994, Youson 2007), and it has been suggested that the non-teleost, ray-finned fish Polypteriformes (i.e., Senegal bichir, *Polypterus senegalus*) and Acipenseriformes (i.e., lake sturgeon, *Acipenser fulvescens*) may have an encapsulated thyroid as well (Youson 2007).

Although it has not been studied in great depth, the cellular biosynthesis of THs by fish is thought to be similar to that observed in other vertebrates (reviewed in Eales and Brown 1993, Leatherland 1994, Yen 2001). As is the case with most vertebrates, the teleost thyroid secretes predominately T_4. However, unlike mammals, where circulating T_4 is often two orders of magnitude higher than T_3, fish serum T_3 is routinely comparable to and can even exceed T_4 concentrations (Eales and Brown 1993).

6.5.1 Serum thyroid hormone distributor proteins, cellular uptake, and cytosolic transport

Free THs are hydrophobic and although their solubility in aqueous solutions generally exceeds serum concentrations, almost all THs in the circulation are non-covalently and reversibly bound to serum TH distributor proteins (THDPs). Serum THDPs are required to ensure an adequate distribution of THs to all cells and tissues; in their absence THs tend to partition into the first cells that they encounter (Mendel *et al.* 1987). Human serum contains three dominant THDPs: thyroxine-binding globulin (TBG), transthyretin (TTR), and albumin (ALB). Each of these proteins is present at a different concentration and has a different affinity for T_4 and T_3. Although this redundancy is present in many other vertebrates, not all vertebrates have all three THDPs, nor are all present during all phases of the life cycle (Richardson *et al.* 2005). For instance, TTR is absent in adult anurans but appears transiently during metamorphosis when TH concentrations peak (Yamauchi *et al.* 1993).

TH kinetics in fish has been studied quite extensively, but surprisingly little work has focused on the identification and function of THDPs. Despite conflicting reports on the presence of ALB in fish, there is sufficient evidence to suggest that at least some ray-finned (i.e., salmonids) and lobe-finned (i.e., lungfish) fish possess ALB or ALB-like serum proteins capable of binding fatty acids (Metcalf *et al.* 2007). However, a comprehensive literature search failed to reveal evidence that ALBs from these fish bind either THs or steroids. In contrast, the sea lamprey ALBs (SDS-1 and AS) bind both T_4 and T_3 (Gross and Manzon 2011). A recent survey indicates that the TTR gene is present in at least six different species of Perciformes, three species of Cypriniformes, two species of Salmoniformes, and two species of Petromyzontiformes. Although recombinant sea bream (*Sparus aurata*) TTR binds THs, and TTR is present in the serum of three species of Salmoniformes, its precise function in these fish is unclear (Santos and Power 1999, Richardson *et al.* 2002). Lastly, lipoproteins appear to be of particular importance with respect to TH distribution and homeostasis in some teleosts. Lipoproteins in rainbow trout (*Oncorhynchus mykiss*) serum bind 67% and 89% of T_4 and T_3, respectively (Babin 1992).

Contrary to the once-held belief that, due to their lipophilic nature, THs diffuse freely into cells, it is now understood that saturable transport mechanisms mediate cellular TH import and export (Abe *et al.* 2002). Five groups of transmembrane TH transport molecules have been identified: organic anion transporter proteins, Na^{2+}/taurocholate co-transporting polypeptides, fatty acid translocases, system I-leucine-preferring amino acid transporters, and system T–tryptophan preferring amino acid transporters (reviewed in Visser *et al.* 2008). Once inside the cell, THs do not diffuse into and out of the nucleus. The nucleus concentrates THs and CTBPs and TRs are important mediators of nuclear TH transport and retention (Yamauchi and Tata 1997, Mori *et al.* 2002). Transmembrane and cytosolic TH transporters and transmembrane TH receptors have been studied mainly in model mammalian and anuran systems. Investigation of these receptors and transporters in fish should significantly enhance our understanding of the regulation of TH action in fish.

6.5.2 Thyroid hormone deiodinases

Once imported into the cell, THs can be modified in several ways, including deamination, decarboxylation, sulphate or gluconuride conjugation, and ether-link cleavage. In most vertebrates these reactions produce inactive TH metabolites destined for excretion. Deiodination of THs by a class of enzymes known as TH deiodinases represents another important means of regulating the action of THs (reviewed in Bianco *et al.* 2002). Three different deiodinases have been identified in most vertebrate taxa and are designated deiodinase types 1, 2, and 3 (D1, D2, and D3, respectively). Deiodinases contain a selenocysteine in their catalytic site, and each type is encoded by a separate gene whose expression is regulated by THs. Deiodination reactions can be broadly classed as either inner-ring or outer-ring deiodinations (IRD and ORD, respectively; Fig. 6-1). Generally, ORD is an activation reaction that converts T_4 to the more biologically active T_3 and IRD is an inactivation reaction that converts T_4 and T_3 to reverse T_3 (rT_3) and 3,3'-diiodothyronine (3,3'-T_2), respectively (Fig. 6-1). Both rT_3 and T_2 have negligible activity in terms of modulating gene expression via TRs. In tetrapods, D2 and D3 are ORD and IRD deiodinases, respectively, while D1 is capable of both IRD and ORD.

The ability to modulate the bioavailability of active THs at the cellular level has significant implications for developmental processes as it facilitates the differential regulation of organ systems despite exposure to the same circulating TH concentrations. For example, during the early stages of anuran metamorphosis, when serum TH levels are low, elevated D2 activity in the limb buds promotes their development, and elevated D3 activity in the tail prevents precocious tail regression until all other systems for terrestrial life have developed (Becker *et al.* 1997, Brown 2005).

Fish deiodinases have been studied intensively over the past three decades. Historically, much of this work has focused on characterizing the nature of deiodination reactions *in vitro* using tissue extracts (Eales and Brown 1993, Orozco and Valverde 2005). The metabolism of various iodothyronine substrates by IRD and ORD has been examined in a wide variety of fish, including agnatha

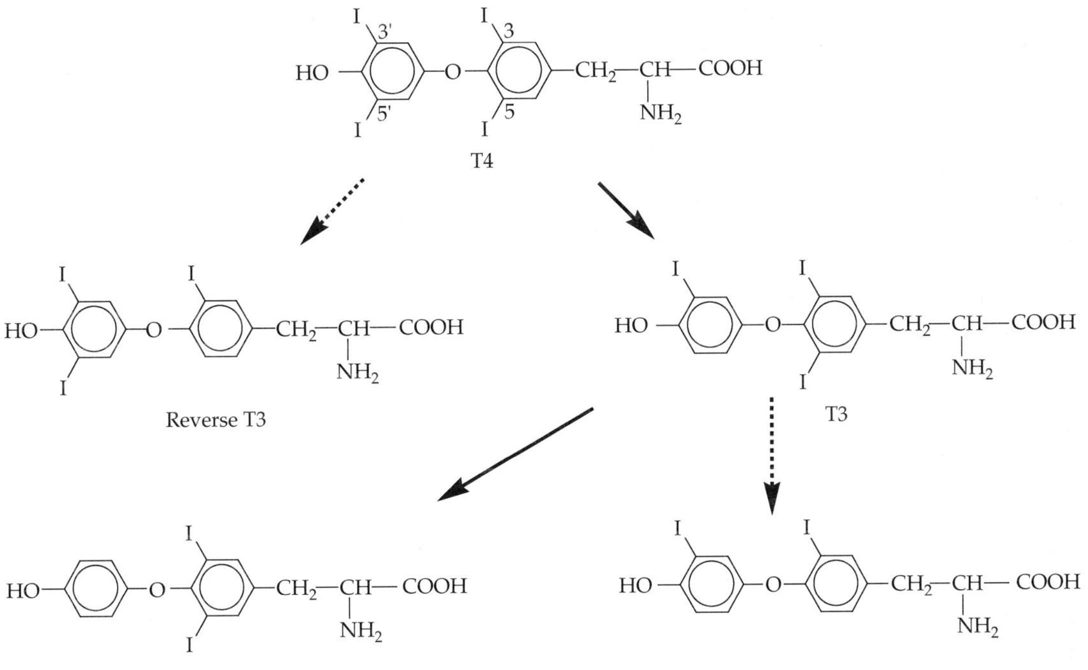

Figure 6-1 Common pathways of monodeiodination of various iodothyronines. Solid arrows represent outer-ring monodeiodination reactions, broken arrows represent inner-ring monodeiodination reactions. Outer-ring deiodination of T4 (3,5,3'5'-tetraiodothyronine) produces the more biologically active T3 (3,5,3'-triiodothyronine). Inner-ring deiodination of T4 or T3 produces reverse T3 (3,3',5'-triiodothyronine) and T2 (3,3'-diiodothyronine), respectively, both of which have negligible activity. T3 can also undergo inactivation via a second outer-ring deiodination (e.g., 3') producing the inactive 3,5-T2 (3,5-diiodothyronine).

(i.e., hagfish and lampreys), Chondrichthyes (i.e., dogfish shark, *Squalus acanthias*), Actinopterygii (i.e., lake sturgeon, and both basal (eels) and derived (salmonids and flatfish) teleosts), and Sarcopterygii (i.e., Australian lungfish, *Neoceratodus forsteri*) (reviewed in Orozco and Valverde 2005, Youson 2007). Furthermore, D1, D2, and D3 cDNAs have been isolated and/or characterized from a variety of the aforementioned fish (Orozco and Valverde 2005), and the cDNAs for D2 and D3 have recently been isolated from the gut of the sea lamprey (R.G. Manzon, unpublished data). These fish deiodinases have a conserved selenocysteine in the putative catalytic site and have been classified as types 1, 2, or 3 based on sequence similarity to the other vertebrate deiodinases. However, the *in vitro* biochemical characteristics of fish deiodinases do not necessarily correlate with their mammalian counterparts, perhaps due to different cofactor require-ments or the presence of deiodinase inhibitors in *in vitro* assays.

6.5.3 Thyroid hormone nuclear receptors

Downstream of the aforementioned regulatory points, TH action on cellular function is largely mediated by TRs, which act as ligand-regulated transcription factors (reviewed in Lazar 1993, Yen 2001, Lazar 2003), although other non-genomic, receptor-mediated mechanisms of TH action have been identified (reviewed in Davis *et al.* 2008). The retinoid-X-receptor (RXR) is the most common het-erodimeric partner of TRs—together this pair, in the presence or absence of THs, will bind to TH response elements (TREs) located in the promoters of TH responsive genes. When bound to a positive TRE (pTRE), the RXR–TR heterodimer associates with a co-repressor complex and histone deacetylases

(HDACs) and represses gene transcription in the absence of ligand (Fig. 6-2). Once bound to ligand there is a conformational change in the RXR–TR heterodimer, resulting in the release of co-repressors, the recruitment of co-activators and histone acetyltransferases, and the upregulation of gene transcription (Fig. 6-2). Conversely, gene expression is repressed in the presence and activated in the absence of THs in those genes, such as TRH and TSHβ, that contain negative TREs (nTREs). The mechanisms responsible for gene regulation by nTREs remain to be fully elucidated, but several different models have been suggested (reviewed in Lazar 2003, Oetting and Yen 2007).

All gnathostome genomes that have been mapped contain two TR genes, each of which encodes for a unique TR subtype. With the exception of sea lampreys, these vertebrate subtypes are denoted TRα and TRβ and encode for proteins that are homologous to the mammalian TRα and TRβ (Lazar 1993).

Although sea lampreys also contain two distinct TR subtypes, these have been denoted TR1 and TR2 (Manzon 2006), since molecular phylogenetic analyses indicate that the sea lamprey TRs diverged from the gnathostome lineage prior to the lamprey–gnathostome split (Escriva *et al.* 2002, Manzon 2006). Not only are there two TR subtypes, but several gnathostome TR and TRβ isoforms have been identified that are generated by differential splicing, differential promoter usage, or species-specific gene duplications (Lazar 1993). Full-length TR cDNAs have been isolated from a variety of fish species, including sea lamprey, eels, zebrafish (*Danio rerio*), goldfish (*Carassius auratus*), salmonids, sea bream, and flatfish. Numerous studies on the various TRs have shown that differences in both function, and temporal and spatial expression play important roles in modulating TH action during developmental processes in a variety of vertebrates, from fish to humans (Tata 1996, Brown and Cai 2007, Nunez

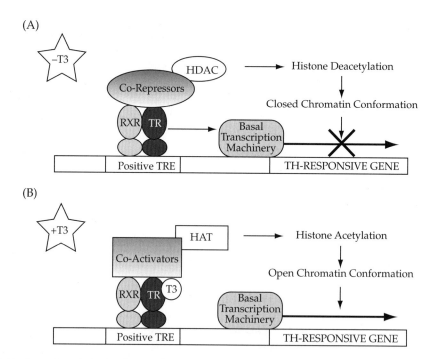

Figure 6-2 Models of repression and activation of genes with positive thyroid hormone response elements (pTRE) by thyroid hormone (TRs) and retinoid-X (RXRs) receptors in the presence and absence of T3 (3,5,3′-triiodothyronine). A, repression of basal gene expression by the RXR-TR heterodimer—in the absence of T3, RXR-TR interact with a series of co-repressor proteins and histone deacetlyase (HDAC), resulting in the tight packaging of chromatin and the repression of basal gene expression. B, T3 binding causes the release of co-repressors and the recruitment of co-activators and histone acetyltransferase (HAT) that relax the chromatin structure and activate gene transcription. (Figure modified from Manzon 2006).

et al. 2008, Nelson and Habibi 2009, Chapter 7). The fact that TRs can act as either homodimers or het-erodimers (with RXRs and other nuclear receptors) adds to the diversity of cellular responses.

The cellular regulatory components of the thy-roid-signaling system are multi-tiered and include the cell- and tissue-specific regulation of TH bioa-vailability (TH uptake, cytosolic transporters, and deiodinases) and a variety of functionally distinct TRs and RXRs (as well as dimer variations) that dif-fer in their abilities to regulate gene expression. Collectively, this cellular machinery offers an ele-gant and complex system with vast potential for the differential coordination of developmental proc-esses by a single molecule. Clearly it represents a powerful and flexible system, which allows for response to environmental perturbations that could alter developmental timing (heterochrony). Thus, the differential regulation of the thyroid system during development represents a mechanism through which developmental timing could be altered and selected for, ultimately leading to a diversification of life history strategies.

6.6   Thyroidal regulation of fish ontogeny and life history transitions

6.6.1   Embryogenesis and embryo to larval transitions

The influence of THs on various aspects of the embryonic development of mammals, birds, and amphibians is well established and supported by several lines of evidence:

1) THs are available to the developing embryo, either via the placental blood supply or by maternal deposition in egg yolk.
2) There are spatial and temporal differences in deiodinase gene expression and activity in embryos, thus indicating the ability to regulate the bioavaila-bility of active hormone at the embryonic cellular level.
3) The presence and differential expression of TRs, both spatially and temporally, indicates that embryos have the capacity to respond to THs.
4) Cause-and-effect relationships between THs and embryonic development of the brain, small intes-

tine, cochlea, and bones have been established (Brent 2000, Hulbert 2000, Nunez *et al.* 2008, Mcnabb 2007, Brown and Cai 2007, Morvan-Dubois *et al.* 2008).

A role for THs in fish embryogenesis was first suggested by the discovery that teleost eggs contain THs (Kobuke *et al.* 1987, Tagawa and Hirano 1987). In fact, THs are present in significant quantities in the eggs of all teleosts studied to date, with concen-trations ranging from 0.04–15 ng/g and 0.07–9.9 ng/g for T_4 and T_3, respectively (reviewed in Tagawa *et al.* 1990b, Leatherland 1994). In rainbow trout, THs move passively on their concentration gradi-ents between the ovarian fluid and the oocyte. But movement between these two compartments is not directly proportional, as the high lipid content of yolk tends to retain THs within the oocyte (Raine and Leatherland 2003). Additionally, egg TH con-centrations vary with maternal thyroid status and can be elevated or lowered by treatment with exog-enous THs or antithyroid agents (i.e., goitrogens such as perchlorate, thiourea, or propylthiouracil), respectively (Brown *et al.* 1988, Tagawa and Hirano 1991, Ayson and Lam 1993).

Although there is maternal deposition of THs in yolk, the function of these maternally-derived THs in embryogenesis is uncertain. Studies of the tilapia (*Oreochromis mossambicus* Ruppell), striped bass (*Morone saxatilis*), and rabbitfish (*Siganus guttatus*) show that the treatment of females with exogenous THs has a positive effect on yolk absorption, growth, time to hatching, and survival (Lam 1980, Brown *et al.* 1988, Ayson and Lam 1993), but delays the secretion of hatching enzymes in tilapia (Reddy and Lam 1991). Conversely, treatment with goitrogens does not affect embryogenesis in medaka or rabbit-fish, despite lowering egg TH concentrations to only 10% of controls (Tagawa and Hirano 1991, Leatherland 1994). The variable results of these studies are related in part to different treatment approaches and doses. The application of physio-logical concentrations of THs to rainbow trout eggs has no influence on embryonic development, growth, or survival, suggesting that any positive effects of THs on embryogenesis are likely due to high pharmacological doses (Raine *et al.* 2004). These latter data suggest that either there is no absolute requirement for THs during fish embryogenesis and their presence in yolk is simply a metabolic accident or coincidence (Leatherland 1994), or that oocytes contain an excess of yolk THs of which only a small fraction is required for normal embryogenesis.

More recent research has examined the expression and regulation of TH deiodinases and TRs during fish embryogenesis, and these studies provide both correlative and causal evidence that strongly support a role for THs in this developmental process. Deiodinase expression and activity are present and regulated in zebrafish embryos (Thisse et al. 2003, Walpita et al. 2009), and both TRα and TRβ mRNAs are present during all stages of embryogenesis, commencing with the early blastula stage, in a variety of species, including zebrafish, sea bream, rainbow trout, and orange spotted grouper (Epinephelus coioides) (Essner et al. 1997, Liu et al. 2000, Power et al. 2001, Tang et al. 2008). TRα and TRβ expression in rainbow trout embryos is reduced in response to T_3 treatments of 30 ng/egg, but not 110 ng/egg (Raine et al. 2004). Rearing zebrafish embryos in 5 nm T_3 increases TRα, but not TRβ expression, downregulates D2 expression, reduces the time to hatching, accelerates pigmentation rate, and stimulates the embryo–larval transition (Walpita et al. 2007). Furthermore, transgenic loss and gain of function studies show that an overexpression of TRα is detrimental to zebrafish development, resulting in loss of the midbrain–hindbrain boundary, swelling in the rostral hindbrain, loss of rhombomere borders, severe disorganization of the hindbrain, reduced eye size, and delayed pigmentation (Essner et al. 1999). The use of antisense oligonucleotides that target D2 mRNA and knockdown D2 proteins significantly delay zebrafish embryogenesis and pigmentation; these effects are reversed by treatment with exogenous T_3 (Walpita et al. 2009). The results of this knockdown study suggest that the observed developmental delays were in response to a lack or depletion of intracellular T_3 due to the decrease in D2 availability. Collectively, these overexpression and transgenic studies indicate that embryos are capable of modulating TH action and provide definitive evidence that THs are essential for zebrafish embryogenesis.

6.6.2 Larval to juvenile transitions

THs are involved in some, but not all, aspects of early post-embryonic development and larval to juvenile transitions in fish. A rise in serum or whole-body TH concentrations correlates with the larval to juvenile transition in tilapia, sea bass (Lates calcarifer), milkfish (Chanos chanos), and grouper (Reddy et al. 1992, de Jesus 1994, Nugegoda et al. 1994, de Jesus et al. 1998). Similarly, the larval to juvenile transition in zebrafish coincides with an increase in thyroid function, as determined by an increase in ^{125}I uptake by thyroid tissue (Brown 1997). Exogenous TH and goitrogen treatments (see below) support the aforementioned results. However, these studies are difficult to interpret because the doses are often above the normal physiological range and are administered to animals that continue to synthesize endogenous THs. Furthermore, serum/tissue TH concentrations are not always reported. Nevertheless, the resulting data provide sufficient evidence to stimulate further investigation aimed at identifying a causal relationship between THs and larval to juvenile transitions of fish.

Treatment of larvae with exogenous THs hastens fin and gut development and the acquisition of the juvenile pigmentation patterns in several teleosts. There is precocious development of the adult fan-shaped pectoral fins in zebrafish and a resorption of the dorsal fin rays in grouper following TH treatment (Brown 1997, de Jesus et al. 1998). Treatment of striped bass with 1 ppm (1.53 μm) T_3 stimulates thickening of the muscle layers of the digestive tract, elongation of the fundus, and development of the gastric blind sac (Huang et al. 1998a). If either treatment is prolonged or higher concentrations of T_3 are used, normal stomach development is disrupted (Huang et al. 1998a). Exogenous THs also elicit pigmentation changes such as a proliferation and redistribution of melanophores, xanthophores, and/or iridophores, resulting in the adult striped pattern in zebrafish, sea bream, striped bass, and grouper (Hirata et al. 1989, Brown 1997, de Jesus et al. 1998, Huang et al. 1998a, Kawakami et al. 2008), and silvering in milkfish (de Jesus 1994).

An ablation-and-replacement study by Brown (1997) provides conclusive evidence that THs are involved in the larval to juvenile transition of zebrafish. Treatment of zebrafish larvae with various goitrogens reduces thyroid gland activity (as measured by ^{125}I uptake) and inhibits the acquisition of certain adult features such as pigmentation, and paired pectoral and pelvic fins. Importantly, the simultaneous treatment of these larvae with exogenous T_3 prevents these goitrogenic effects. However, goitrogens have no effect on several aspects of the larval to juvenile transition in zebrafish, including development of the unpaired fins, attainment of the adult body shape or ossification of the skull and spine. Collectively it appears that THs influence some, but not all aspects of the larval to juvenile transition in fish, which is similar to their function in smoltification (discussed below). To date, the function of THs in larval to juvenile transitions cannot be considered homologous or even directly analogous to the scenario in anurans, where TH is the one obligate morphogen that initiates virtually all the signaling cascades associated with metamorphosis, but it is apparent that THs have a conserved role in this life history transition, as with chordate development.

6.6.3 First or "true" metamorphoses in ray-finned fish

Among the bony fish, a "true" metamorphosis is restricted to members of the subdivision Elopomorpha (tarpons, ten pounders, true eels) and the order Pleuronectiformes (flatfish) of the subdivision Euteleostei. The members of these two groups of ray-finned fish are widely spaced (i.e., basal versus derived) on the taxonomic scale of fish. This apparent absence of a taxonomical relationship among teleosts that metamorphose suggests that the use of metamorphosis as a developmental strategy must have evolved more than once in this diverse group of gnathostomes (Youson 2004). The hormonal control of metamorphosis in bony fish has been studied in the Japanese (*Paralichthys olivaceus*; Inui and Miwa 1985) and summer (*P. dentatus*; Keefe and Able 1993) flounders, Senegalese sole (*Solea senegalesis*; Isorna *et al.* 2009), and conger (*Conger myriaster*; Yamano *et al.* 1991b), European (*Anguilla anguilla*; Edeline *et al.* 2005) and Japanese eels (*Anguilla bicolor pacifica* and *Anguilla obscura*;

Ozaki *et al.* 2000). The metamorphosis of Atlantic halibut (*Hippoglossus hippoglossus*) has also been described and characterized in detail and additional studies on hormonal control are likely forthcoming (Power *et al.* 2008).

Substantial species variability exists, but in general flatfish metamorphosis transforms a pelagic, bilaterally symmetrical larva to a benthic, asymmetrical juvenile. This process involves, among other internal and external morphological and physiological changes, the translocation of one eye across the dorsal midline to the opposite side of the body (Keefe and Able 1993). THs play an important role in flatfish metamorphosis, similar to that described for anurans. Exogenous THs accelerate fin resorption, eye translocation, and the rate of settling associated with the metamorphoses of both the Japanese and summer flounders (Miwa and Inui 1987, Schreiber and Specker 1998). Interestingly, the influence of THs appears to be stage-specific in the summer flounder, with the greatest effects occurring between prometamorphosis and mid-climax (Schreiber and Specker 1998). One explanation for these stage-specific effects is that treatment with higher exogenous TH concentrations may be required to accelerate the final stages of metamorphosis. A hastening of other tissue-specific changes is also seen with exogenous TH treatment, such as the shift from larval to adult myosin light chain troponin T isoforms (Yamano *et al.* 1991a), the development of the gastric glands in the stomach (Miwa *et al.* 1992, Huang *et al.* 1998b), the shift to juvenile erythrocytes (Miwa and Inui 1991), the appearance of the two types of juvenile mitochondria-rich cells in the gills, and salinity tolerance (Schreiber and Specker 1999b,c). In many of the aforementioned studies, chemical ablation of the thyroid with thiourea results in metamorphic stasis and oversized, pelagic larvae. Significantly, thiourea-induced metamorphic stasis is alleviated in a dose-dependent manner by treatment with either T_4 or T_3, with T_3 being several fold more potent than T_4 (Miwa and Inui 1987, Yamano *et al.* 1994, Schreiber and Specker 1999a).

The influence of THs on flatfish metamorphosis is further demonstrated by the correlation of whole-body T_4 concentrations and TR expression levels with tissue morphogenesis. Whole-body T_4 levels are generally low during pre- and pro-metamorphosis, rise rapidly at the onset of metamorphic climax, peak at climax, and decline to 50% of peak concentrations during post climax (Miwa *et al.* 1988, Tagawa *et al.* 1990a, Schreiber and Specker 1998). The differential expression of TRα and TRβ during Japanese flounder metamorphosis is consistent with that observed in anuran metamorphosis. The expression of TRα (rather than TRβ as in anurans) correlates with TH concentrations and metamorphic change in the Japanese flounder (Yamano and Miwa 1998), but TRβ levels are low throughout metamorphosis, peak post climax, and remain elevated in juveniles (Yamano and Miwa 1998). These changes in whole-body TR expression are further supported by *in situ* hybridization studies showing that TR transcript levels are elevated in skeletal muscle and stomach mucosae during metamorphosis and decrease to low levels in post-climax fish (Yamano and Miwa 1998). In the Senegalese sole, both TRα2 and TRβ2 expression levels correlate with metamorphosis and TH levels, D2 activity and expression are upregulated during metamorphosis, and D3 activity and expression decrease at late metamorphosis (Isorna *et al.* 2009). Although data are still lacking on the specific gene-expression cascades regulated by THs, there is little doubt that THs are essential for flatfish metamorphosis.

All members of the subdivision Elopomorpha undergo a "true" metamorphosis (Youson 2004). Data on the endocrine regulation of metamorphosis in this group of basal teleosts is most extensive for eels, where THs promote metamorphosis in a manner similar to anurans and flatfish. In conger eels, whole-body T_4 levels begin to increase early in metamorphosis and decline late in metamorphosis, at which time whole-body T_3 levels increase rapidly (Yamano *et al.* 1991b) and exogenous THs stimulate precocious metamorphosis (Kitajima *et al.* 1967). Moreover, TR expression correlates with the aforementioned data: TRβ levels increase during metamorphic prophase and remain elevated in elvers, and TR levels decrease following metamorphic climax (Kawakami *et al.* 2003). The role of THs in the metamorphosis of eels has not been tested directly using TH ablation and replacement experiments. However, exogenous THs and the goitrogen, thiourea, stimulate and inhibit, respectively, locomotor

activity associated with migration in European glass eels (Edeline *et al.* 2005).

Recent data also suggest that THs are required for the metamorphosis of tarpons (*Megalops cyprinoids*) (Shiao and Hwang 2006). Exposure to exogenous T_4 and T_3 results in a slight acceleration of metamorphic development and thiourea inhibits the metamorphosis of tarpon leptocephali (larvae) (Shiao and Hwang 2006). Together these studies indicate that THs are required for the metamorphoses of both derived and basal teleosts, for which there is likely no taxonomic relationship. They therefore support the notion that the role of THs in the metamorphoses of bony fish is ancestral and conserved.

6.6.4 First or "true" metamorphosis in lampreys (Agnatha)

All lampreys undergo a first or "true" metamorphosis (Youson 1988) from a blind, sedentary, filter-feeding larva to a free-swimming, sexually immature juvenile that has acquired the adult body form. Half of the approximately 40 species of lampreys have a postmetamorphic parasitic (or predatory) feeding phase as part of their life history, while the others are non-parasitic. Non-parasitic lampreys begin their sexual maturation shortly after the completion of metamorphosis, spawn the following spring and, like all lampreys, die shortly thereafter. In contrast, parasitic species feed for a period of 1–21 months following metamorphosis, but prior to sexual maturation. Some juvenile lampreys feed in their natal freshwater streams, while others (like the sea lamprey) migrate downstream to lakes or oceans for their feeding period, then migrate back upstream to spawn.

The induction of precocious metamorphosis in the non-parasitic lamprey (*Lampetra planeri*) following treatment with the goitrogen, potassium perchlorate (KClO$_4$), provided the first evidence that THs might be involved in lamprey metamorphosis (Hoheisel and Sterba 1963). However, this result suggests that the function of THs in lamprey metamorphosis may differ from that in other vertebrates; an idea that is supported by serum TH levels. In sea lamprey, serum TH levels gradually increase throughout the protracted larval period and peak prior to the onset of metamorphosis (Youson *et al.*

1994). Coincident with the first external signs of metamorphosis, serum TH levels decrease rapidly and remain depressed for the remainder of the life cycle in sea lamprey, the American brook lamprey (*Lampetra appendix*), and the pouched lamprey (*Geotria australis*) (Wright and Youson 1977, Lintlop and Youson 1983, Leatherland *et al.* 1990, Holmes *et al.* 1999).

The importance of a peak in THs prior to metamorphosis and a decline in THs at the onset of metamorphosis is demonstrated by ablation and replacement experiments in sea lampreys. Several studies show that the treatment of larval sea lampreys with $KClO_4$ suppresses serum TH concentrations and induces precocious metamorphosis (Youson *et al.* 1995, Manzon and Youson 1997, Manzon *et al.* 1998). The fact that treatment with exogenous T_4 or T_3 completely blocks $KClO_4$-induced metamorphosis provides convincing evidence that the decline in serum TH concentrations is required for sea lamprey metamorphosis (Manzon *et al.* 1998). Furthermore, the ability to induce metamorphosis is not a response to $KClO_4$ specifically, as several different goitrogens can elicit the same effect and the induction of precocious metamorphosis is correlated with the magnitude of the decline in serum TH concentrations (Manzon *et al.* 2001). Finally, treatment of immediately premetamorphic sea lampreys with exogenous THs (primarily T_3) disrupts the normal metamorphic process (Youson *et al.* 1997). Taken together these data clearly show that the metamorphosis of sea lampreys and at least four other lamprey species require a decline in serum TH levels. However, the precise function of other regulatory components of the thyroid axis in lamprey metamorphosis remains to be fully elucidated.

Most of the data regarding the influences of THDPs, TH deiodinases, TRs, and RXRs on lamprey metamorphosis are consistent with the requirement for a decline in THs during metamorphosis, but these data also come largely from a single species, the sea lamprey. The primary site of deiodinase activity in sea lampreys, unlike most other vertebrates, is the intestine (Eales *et al.* 1997, 2000). T_4 ORD activity, which is responsible for converting T_4 to the more biologically active T_3, is low in young larvae but is elevated in immediately

premetamorphic larvae and during stages 1 and 2 of metamorphosis (staging according to Youson and Potter 1979). This transient elevation is followed in stage 3 by a rapid decrease to very low levels, which are maintained until the upstream-migrant phase (Eales *et al.* 2000). Coincident with this decrease in T_4 ORD is a significant increase in T_4 IRD activity (Eales *et al.* 2000). The decline in T_4 ORD by stage 3 and elevation in T_4 IRD between stages 3 and 7 of metamorphosis indicates the system is functioning to minimize hormone activation and maximize hormone inactivation during the metamorphic process, further confirming the need for a decline in TH action during lamprey metamorphosis. Consistent with these findings are the observations that the overall total TH-binding capacity of sea lamprey serum decreases only slightly, if at all, during metamorphosis, as sea lamprey transition from AS to SDS-1 and CBIII as the main serum THDPs (Gross and Manzon 2011). Thus, it is unlikely that a dramatic increase in hormone availability is compensating for the observed decrease in total serum TH concentrations at the onset of metamorphosis (Gross and Manzon 2011).

Even though a decline in THs appears to be necessary for the onset of lamprey metamorphosis, this does not mean that THs are not important in lamprey development and the metamorphic process. Lampreys, like all other vertebrates, require THs at some point in their life cycle for metamorphosis to occur, and this is consistent with an evolutionarily conserved role for THs in chordate development (Youson 2004). There is speculation that the first lampreys, much like the presumed common ancestors of vertebrates and protochordates, were pelagic and free-swimming, dwelled in marine estuaries, and were larval-like in character and underwent direct development (i.e., they were paedomorphic; Youson 2004). The ability of the lamprey endostyle to bind iodide, and perhaps synthesize and store large quantities of iodothyronines (i.e., THs), could represent a mechanism to initiate either a major developmental event or a change in developmental timing (i.e., heterochrony) in response to environmental stressors such as a change in temperature or food availability (Youson 2004). Ultimately, through natural selection, a "true" metamorphosis was incorporated into the life cycle and encoded for in the lamprey genome (Youson 2004).

If we accept a conserved role for THs in chordate development and thus lamprey development, one can further speculate that despite the requirement for a decline in TH levels at the onset of metamorphosis in lampreys, THs could still function in a stimulatory fashion for the remainder of the metamorphic process; these two ideas need not be mutually exclusive. Although serum TH levels are low during lamprey metamorphosis when compared to larval concentrations, these lower levels are likely similar to those observed during anuran metamorphic climax (i.e., when TH levels are maximal and significant morphological changes are occurring) and therefore may be sufficient to drive metamorphic change in lampreys (reviewed in Manzon 2006). It is noteworthy that serum TH concentrations in larval lampreys are among the highest measured in any vertebrate (7–10-fold higher), thus even low TH levels during metamorphosis are still substantial. Although these data must be interpreted with caution, as comparisons between different assays, laboratories, and organisms are not necessarily direct and linear, they do suggest that metamorphic lampreys have sufficient hormone to drive gene expression. Lintlop and Youson (1983) demonstrated that the binding of T_3 by nuclear hepatocytes is elevated during metamorphosis, suggesting that the pool of serum THs during metamorphosis is sufficient to allow for an increase in cellular uptake and nuclear binding, and a subsequent increase in gene expression. Furthermore, preliminary TR developmental expression analyses indicate that sea lamprey TR1 is elevated during tissue morphogenesis and TR2 is expressed constitutively (Manzon 2006). These sea lamprey TR expression data are similar to what is observed in anuran and Japanese flounder metamorphoses, where the expression of one TR subtype varies during metamorphosis while another is expressed at relatively constant levels (Wong and Shi 1995, Yamano and Miwa 1998). Although data are still lacking, one possible hypothesis to explain the sea lamprey TH-metamorphosis paradox is that THs have a dual role in sea lamprey development. High TH levels are required to promote larval feeding, growth, and lipid accumulation, while inhibiting

metamorphosis. Upon initiation of metamorphosis by some endogenous or exogenous (environmental) signal, lower TH levels are important (required) for tissue morphogenesis, in a fashion similar to other metamorphosing vertebrates.

6.6.5 Smoltification: A juvenile transition in salmonids

The parr-smolt transformation (smoltification) occurs in anticipation of downstream migration to (rather than acclimation to) a marine environment and is a significant life history transition that involves changes in morphology, physiology, biochemistry, and behavior that are influenced by both endogenous and exogenous signals (reviewed in Hoar and Randall 1988, McCormick *et al.* 1998, Stefansson *et al.* 2008). Some of these changes associated with smoltification include the development of a more streamlined body in response to alterations in metabolism (i.e., increased lipid mobilization and protein accretion), loss of parr marks and increased silvering in response to the deposition of purines in scales and skin, the development of hypoosmoregulatory abilities, and a shift from being territorial and benthic to schooling and pelagic.

In my opinion smoltification is not a "true" or first metamorphosis (Box 6-1; Just *et al.* 1981, Youson 1988) for several reasons:

1) Unlike a first metamorphosis, smoltification is the coordination of a series of independent developmental events each regulated by different signaling systems and these events can be uncoupled, for instance body silvering and hypoosmoregulatory adaptation.
2) Aspects of smoltification are reversible whereas a first metamorphosis, once initiated, is irreversible.
3) Smoltification is not required for sexual maturation and spawning, as resident parr can spawn either with each other or with upstream migrants (also see Stefansson *et al.* 2008, Dufour and Rousseau 2007).

The influence of THs on smoltification was first reported more than 45 years ago (Gorbman and Bern 1962) and since then many studies have suggested that THs regulate various aspects of this process. For instance, treatment with exogenous

THs stimulates purine deposition and silvering (Gorbman and Bern 1962), but the extent of silvering following TH treatment is not as pronounced as that observed during natural smoltification and silvering can be induced at any stage of development, even early embryogenesis. The precise role of THs in silvering during smoltification is therefore unclear (Stefansson *et al.* 2008). The observations that serum TH concentrations increase at the time of smoltification and that a surge in THs correlates with downstream migration and the lunar phase further suggest that THs function to regulate smoltification (reviewed in Hoar and Randall 1988). THs have also been implicated in olfactory imprinting and behavioral changes (Morin *et al.* 1997, Specker *et al.* 2000). Although there are sufficient data to suggest that THs influence salmonid smoltification, much of these data are correlative and indirect.

THs may exert their effects on smoltification via synergistic actions with other key endocrine regulators such as growth hormone (GH), cortisol, and prolactin. In particular, GH and the insulin-like growth factors (IGFs) are potent regulators of numerous aspects of smoltification. GH and IGFs strongly influence the development of a more streamlined body by stimulating food intake, protein accretion, and lipid mobilization, which collectively result in more rapid increases in length than weight (Johnsson and Bjornsson 1994, Martin-Smith *et al.* 2004). The development of hypoosmoregulatory ability is also influenced by GH and IGFs, as they enhance the proliferation and differentiation of gill chloride cells, Na^+/K^+-ATPase activity and Na^+/K^+ cotransporter expression. These effects of GH and IGFs are independent of their effects on growth (Bolton *et al.* 1987, McCormick *et al.* 1991, Sakamoto and Hirano 1993, Sakamoto *et al.* 1993). Likewise, cortisol plays a key role in the regulation of metabolism, protein turnover, and lipid mobilization, and synergizes with GH, IGF, and prolactin in the development of salinity tolerance (McCormick 1996, 2001). As with other life history transitions, it is clear that THs influence smoltification in salmonids, but given the lack of direct experimental evidence it appears likely that their role is to synergize with, and modulate the action of, other key regulators, perhaps by increasing the number of receptors for these other regulatory molecules (see McCormick 2001).

In conclusion, THs and their derivatives are ancient signaling molecules with a well-conserved molecular mechanism of action, and which are involved in a diverse range of biological processes in all chordates studied to date. Particularly noteworthy is their ubiquitous involvement in chordate development. Fish are no exception to this trend, as THs influence various aspects of both embryonic and post-embryonic life history transitions. Given the diversity in fish species and variety of life history transitions in fish, there is great opportunity for expanding our understanding of both the evolution of life history transitions and the function and molecular mechanisms of TH action. As our toolboxes continue to grow with the sequencing of fish genomes, our ability to examine the molecular mechanisms of TH action during fish ontogeny is increasing rapidly. Data from such studies should aid our ability to develop synthetic hypotheses on evolution of fish life history transitions and the molecular mechanisms that regulate these developmental events.

6.7 Summary

1. Fish ontogeny includes several different life history transitions. Included among these are embryo to larval transitions, larval to juvenile transitions, "true" or first metamorphoses, juvenile transitions (i.e., smoltification), and transitions associated with sexual maturation (Box 6-1).

2. Despite minor variations, both the central and peripheral components and molecular signaling cascades of the thyroid system in fish are similar to those observed in other more derived vertebrates. When coupled with data from non-vertebrate chordates and other deuterostomes they strongly support the notion that the thyroid system is ancestral and highly conserved. Given the diversity of fish species, future efforts should be focused on further characterizing these systems in fish.

3. Thyroid hormones (THs) are critical for some aspects of all fish life history transitions and these findings are consistent with data from most chordates. Given there are several aspects of fish development associated with the embryo to larval, larval to juvenile, and juvenile transitions that are not absolutely dependent on THs, it does not appear

that there are sufficient data to support the hypothesis that their role in these transitions is homologous to their role in anuran metamorphosis. However, it must be emphasized that the aforementioned statement does not argue against the idea that THs are conserved and ancestral developmental morphogens.

4. The role of THs in the "true" or first metamorphosis of teleosts is likely homologous to anuran metamorphosis. The taxonomic distance between the two groups of teleosts that metamorphose (i.e., basal Elopomorpha and derived Pleuronectiformes) suggests that metamorphosis evolved separately in each of these two groups of divergent fish. The central role of THs in metamorphosis likely stems from their ancestral function in development as a whole.

5. Lamprey metamorphosis, unlike the metamorphoses of other vertebrates, appears to require a decline in TH levels. However, it is clear that THs are involved in lamprey ontogeny and this is consistent with their role in chordate development. The hypothesis is put forth that THs have a dual role in lamprey metamorphosis, where high TH levels promote larval growth and inhibit metamorphosis and lower TH levels are stimulatory during metamorphosis.

6. THs are ancient signaling molecules with a ubiquitous role in vertebrate, and perhaps deuterostome, development. The complex and multi-tiered regulatory checkpoints in the vertebrate thyroid system provide the potential to respond to environmental perturbations and result in an alteration of developmental timing at both the organismal and evolutionary levels.

6.8 Acknowledgments

I am most grateful for the careful reading of earlier versions of this manuscript and the valuable suggestions provided by Lori A. Manzon. I also thank John H. Youson and another anonymous reviewer for their helpful comments. Financial support was provided by the National Science and Engineering Council of Canada and the Canada Foundation for Innovation.

CHAPTER 7

Hormone regulation and the evolution of frog metamorphic diversity

Daniel R. Buchholz, Christine L. Moskalik, Saurabh S. Kulkarni, Amy R. Hollar, and Allison Ng

7.1 Introduction

A vast body of work exists on the endocrine regulation of frog metamorphosis (Dodd and Dodd 1976, Shi 1999, Denver *et al.* 2002, Buchholz *et al.* 2006, Furlow and Neff 2006, Brown and Cai 2007), and a separate set of literature describes the dramatic larval period diversity found among amphibians (Duellman and Trueb 1994, McDiarmid and Altig 1999). This chapter begins to unite these topics by addressing endocrine and molecular mechanisms underlying life history evolution in larval amphibians. Amphibian metamorphic life history diversity includes species differences in time to, and size at, metamorphosis, evolution of direct developing forms from typical free-living larvae, and neoteny. Ecological aspects of this diversity are not covered, nor is the diversity of reproductive modes. Metamorphic life history parameters are under neuroendocrine control, including both central and peripheral control mechanisms (Fig. 7-1). Despite limited evidence, theory suggests mechanistic changes at the central level of control, whereas evolution of control at the peripheral level has been more adequately demonstrated. We present mechanisms of tissue responsiveness to thyroid hormone from model frog species in order to shed light on potential sites of evolutionary change in endocrine physiology that may underlie life history differences found within amphibians.

The larval period of the amphibian life cycle occurs from the beginning of post-embryogenesis (i.e., completion of organogenesis) through metamorphosis to the production of a juvenile frog. Within tadpoles, life history variation includes larval period duration (ranging from 8 days in *Scaphiopus couchii* to 2–3 years in *Heleophryne* sp., *Ascaphus*, and some *Rana*) and size at metamorphosis (ranging from <1 cm in many species to adult size in *Pseudis* sp.). Even though virtually every environmental factor affects larval period duration and size at metamorphosis (Wilbur and Collins 1973, Denver 2009), species diversity in larval period and metamorph size is not completely accounted for by phenotypic plasticity, as determined by rearing different species under identical laboratory conditions (Lieps and Travis 1994, Buchholz and Hayes 2002). Direct development and neoteny are dramatic evolutionary departures from the free-living transitory larva. Direct developers lack a free-living larval period and hatch from the egg as a juvenile. Evolutionary loss of larval features in direct developers varies widely across species, from *Eleutherodactylus*, with its vestigial and highly modified larval structures, to other species which hatch from the egg with the appearance of a tadpole but have enough nutrition from yolk to complete metamorphosis (Callery *et al.* 2001, Thibaudeau and Altig 1999). Conversely, neoteny is reproductive maturation in the larval form, found in salamanders but not frogs or caecilians (Dent *et al.* 1968). Depending on the species, neotenic salamanders do or do not undergo metamorphosis in nature, and can or cannot be induced to undergo metamorphosis by hormone injections.

7.2 Ecological context of metamorphic life history evolution

Both time to and size at metamorphosis are important life history traits (Smith 1987). Short larval period durations are often favored to increase larval survival from aquatic challenges such as predators and pond drying. Larger metamorph sizes are often favored to increase overwinter survival rates and adult reproductive fitness (larger males are often preferred by females and larger females have larger clutch sizes). However, evolutionary changes in larval period may impact metamorph size and vice versa due to an apparent trade-off between time to, and size at, metamorphosis. For example, if pond drying is a relatively large threat then evolution of a short larval period may be accompanied by a decrease in metamorph size because:

1. energy allocation is shifted away from growth to enable more rapid development
2. less time for growth reduces metamorph size (assuming ancestral rates of growth).

The impact of this trade-off could be reduced via evolutionary changes in rates of growth and development (such that size at metamorphosis could be maintained at a shorter larval period). However, the relationship between time to and size at metamorphosis varies widely across amphibians, in that not all rapidly developing species are small, and some slowly developing species are quite small. Thus, knowledge of mechanisms underlying growth and development and the mechanistic relationship between them will help explain the diversity of larval periods and metamorph sizes observed.

7.2.1 Escape from the growth versus development trade-off

Some lineages have radically departed from this paradigm by evolving direct development. Direct developers have "chosen sides" of the trade-off in which relatively small size was an acceptable consequence of eliminating the uncertainties of the aquatic habitat. In contrast to direct development, where development is favored over growth, salamander neoteny is a major life history paradigm, where metamorphic development is abrogated and larval forms become reproductively mature. Neoteny often occurs in cases where the terrestrial habitat is inhospitable or non-existent (as in caves).

7.3 Key concepts in the endocrinology of metamorphosis

7.3.1 Overview of the endocrinology of metamorphosis

Hormones control the morphological and physiological transitions during metamorphosis, including limb elongation, gill and tail resorption, and remodeling of most other organs from the larval to adult versions (Dodd and Dodd 1976). Of predominant importance are thyroid hormones (THs), which induce a gene regulation cascade leading to metamorphic changes (Shi 1999). THs are small molecules derived from two tyrosine residues. The predominant form of TH released into the blood by the thyroid is thyroxine (T4), which has two iodines on each tyrosine moiety, and needs to be altered by the removal of an iodine to become the active form of TH, triiodothyronine (T3). T4 can be considered a hormone precursor compared to T3 because T4 has a 10-fold lower affinity for the TH receptor (TR) compared to T3. Thus, T3 is the form that controls initiation and rate of transformation for each tissue, as well as the overall duration of the larval period. Both T4 and T3 travel in the blood mostly bound to plasma proteins such as transthyretin and albumin. THs enter the cell and bind TR in the nucleus to interact with tissue-specific transcription factors and alter gene transcription associated with metamorphosis. Other hormones, such as corticosterone and prolactin, modulate but cannot replace TH in the control of progress through metamorphosis (Kaltenbach 1996).

All frogs, and indeed all vertebrates, have a peak in TH at some point during their life cycle (White and Nicoll 1981, Buchholz *et al.* 2006). In frogs, this peak in TH occurs at climax of metamorphosis and its regulation is under central control via production of corticotropin releasing factor (CRF), secreted from axon terminals in the median eminence into the pituitary portal circulation and acting on the pituitary (Denver 1996). CRF induces release of two hormones from the pituitary:

1) adrenocorticotropic hormone (ACTH) from corticotropes, which stimulates interrenal glands to produce corticosterone

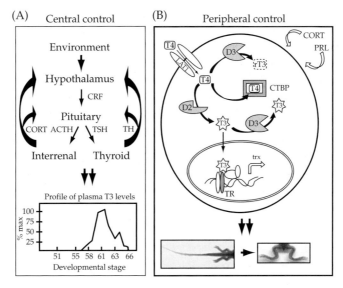

Figure 7-1 Central and peripheral control in thyroid hormone physiology. (A) Brain processing of environmental signals regulates hypothalamic neurosecretion of corticotropic releasing factor (CRF). CRF enters the pituitary portal vein to induce release of adrenocorticotropic hormone (ACTH) and thyroid-stimulating hormone (TSH). ACTH induces the interrenal glands (homologous to mammalian adrenal glands) to synthesize and release corticosterone (CORT), and TSH acts on the thyroid gland to stimulate release of thyroid hormone (TH). Both CORT and TH exert maturational and feedback effects on the brain, hypothalamus, and pituitary. These regulatory interrelationships constitute central control and give rise to the developmental profile of blood TH titre. The profile for CORT (not shown) resembles that for TH. (B) Circulating TH (mostly in the form of T4) enters the cell via any number of TH transporters (the L-type amino acid transporter LAT1 is shown). Once inside the cell, T4 may be (i) degraded by deiodinase D3 to the inactive TH, reverse T3 (rT3), (ii) sequestered by cytoplasmic thyroid hormone binding proteins (CTBP), or (iii) converted by deiodinase D2 to the active TH, T3. T3 may then enter the nucleus to bind to TR on promoters or enhancers of TH-response genes. T3 binding to TR causes displacement of corepressors for coactivators and subseqeunt transcriptional activation (trx). CORT and prolactin (PRL) signaling mechanisms (not shown) may affect one or more of the TH transport, metabolism, and transcription processes, thereby modulating gene induction by TH. These mechanisms constitute peripheral control of TH signaling, ultimately resulting in metamorphic changes, exemplified by tail resorption.

2) thyroid-stimulating hormone (TSH) from thyrotropes, which stimulates the thyroid gland to produce TH.

Corticosterone and TH affect this process by maturational actions on the brain and pituitary and by feedback control on the hypothalamus and pituitary. Thus secretion of hormones by the thyroid and interrenal glands is controlled centrally by the hypothalamus and pituitary. Peripheral control refers to cellular and molecular mechanisms regulating the ability of TH/TR complexes to alter gene expression. TRs are ligand-activated transcription factors constitutively bound to enhancers and promoters of TH-response genes. Gene expression and subsequent developmental changes are induced by TH only if free active forms of TH (i.e., T3) can enter the nucleus and bind TR. For cells to respond, the hormone must first enter the cell from the blood via

membrane transporters. Then, deiodinases metabolize THs to active or inactive forms. Cytoplasmic TH-binding proteins can bind T3 in the cytoplasm, perhaps preventing it from entering the nucleus. Peripheral control determines tissue sensitivity and responsivity (see Section 7.3.2) to circulating TH and corticosterone and thereby contributes to the timing of tissue transformation. Because TH physiology regulates developmental timing during metamorphosis, evolutionary changes in central and/or peripheral control may contribute to metamorphic diversity.

7.3.2 Tissue sensitivity and tissue-specific responses to thyroid hormones

Tissue sensitivity refers to whether or not a threshold concentration of TH has been achieved to elicit a response by the tissue. Tissues differ in their sensitivity to TH at specific developmental stages (e.g., the hind limbs require lower TH levels to initiate transformation compared to the tail). Individual tissues may also vary in sensitivity to TH across development (e.g., at the climax of metamorphosis, the hind limbs are no longer affected by TH). Tissue sensitivity differences may also be found in the same tissue across species. Tissue-specific responses to TH refer to the tissue-specific gene expression and morphological responses to TH, and the rate at which these changes occur. The two concepts of sensitivity and tissue-specific responses are interrelated because some genes that determine TH sensitivity are themselves induced by TH in a tissue-specific manner.

Whether or not a tissue is sensitive to TH resulting in altered gene expression is determined by:

1) the amount of TH in the cell nucleus required to achieve a threshold number of hormone-bound TH receptors (TRs)
2) the expression TRs, coregulatory factors, other transcription factors, and the activity of other hormonal systems (such as activated corticosteroid receptors).

The threshold effective cellular TH concentration that determines tissue sensitivity involves TH transporters, TH metabolizing enzymes, cytoplasmic TH binding proteins, and TRs and associated transcriptional cofactors (Section 7.5). Above the threshold TH level, higher TH concentrations increase the rate of gene induction and consequent morphological change, up to a maximum level.

7.3.3 Tissue developmental asynchrony

During metamorphosis, each tissue is exposed to the same blood levels of TH yet undergoes remodeling at different times. This tissue asynchrony, where limbs develop first, then intestine remodels, then tail resorbs, is due to tissue-specific initiation and/or rate of transformation in response to TH. Differences in tissue sensitivity and responsivity among tissues, the gradual increase in TH up to metamorphosis, and the tissue-specific alterations of tissue responses to TH by other hormones such

as corticosterone and prolactin account for tissue asynchrony. Little coordination of timing between tissues occurs during metamorphosis (i.e., tissues do not communicate to coordinate timing). Rather, tissue autonomy in response to TH is the rule. Indeed, organs, including tail, hind limb, and intestine, dissected from the tadpole and cultured *in vitro*, transform in response to TH similarly to their *in situ* counterparts (Derby 1975, Ishizuya-Oka and Shimozawa 1991, Tata *et al.* 1991). In addition, transgenic studies show striking tissue developmental autonomy of tissues, with nerves, muscle, cartilage, and skin of the hind limb transforming independently of one another (Das *et al.* 2002, Marsh-Armstrong *et al.* 2004, Brown *et al.* 2005).

Tissue asynchrony can break down upon treatment with exogenous TH. Lower concentrations of exogenous TH allow for a normal sequence of developmental events, but at an accelerated pace. On the other hand, high levels of exogenous TH lead to initiation of all tissue developmental programs at the same time, even though they would not normally develop synchronously. With high TH, tissue asynchrony is also severely disrupted because some metamorphic events take more time than others. For example, limb elongation, which takes a long time, normally ends before opercular window formation, which occurs quickly. In the presence of high TH, however, opercular window formation is complete before the limbs have completely elongated. The next sections detail endocrine and molecular mechanisms underlying the control of initiation of tissue sensitivity/responsivity associated with derived life histories.

7.4 Endocrine basis of amphibian life history evolution

7.4.1 Larval period duration

The time to metamorphosis (from beginning of feeding to tail resorption) is determined by the time of initiation and rate of metamorphosis. Because TH is necessary and sufficient to initiate tissue remodeling during metamorphosis, much of amphibian metamorphic diversity in larval period duration can be explained by evolutionary changes in central and/or peripheral control mechanisms of TH physiology. The initiation of metamorphosis depends on developmental maturation of the neuroendocrine system, tissue TR expression, and environmental factors that affect central control (such as pond duration, food availability, and/or predators). Because TR expression occurs very early in the larval period prior to the presence of circulating TH (Baker and Tata 1990), maturation of central control determines the timing of metamorphic initiation. However, the timing of metamorphic initiation is a plastic trait within species, which is subject to brain processing of environmental signals and body condition cues leading to activation of the metamorphic process (Denver 2009). The initiation of metamorphosis varies across species due in part to the rate of embryonic development to achieve maturation of the neuroendocrine system, as well as changes in central control of growth and development rates. Once initiated, the rate of metamorphosis depends both on circulating levels of TH, which are determined by central control, and tissue sensitivity/responsivity to TH, which is determined by peripheral control (see below).

Evolutionarily, it would seem easiest to achieve species differences in metamorphic rate and larval period duration by a single change at the central level (such as a change in CRF production or signaling) to affect the developmental profile of TH and corticosterone. Such a change would make the hormone peaks occur earlier in time and would perhaps achieve higher levels, which would then act globally on all tissues to increase development rate and reduce the larval period. Indeed, hormone levels within tissues have been measured in tail and liver across species in spadefoot toads, showing that species with faster rates of metamorphosis have a higher TH tissue content during metamorphosis (Buchholz and Hayes 2005). However, it is also possible that hormonal differences seen within tissues may not reflect differences in blood concentrations. For example, tissue-specific expression levels of TH-degrading enzymes could result in different amounts of cellular hormone retention despite similar blood TH levels (see below). Thus, the role of central control underlying species differences in larval period duration is not clear. On the other hand, evidence

exists for altered peripheral control as a contributor of species differences in developmental rate. Tail tips of three spadefoot toad species in culture shrank at different rates after exposure to T3, consistent with their differences in larval period (Buchholz and Hayes 2005). Altered larval periods due to changes in peripheral control would seem to require changes in each tissue to avoid disruption of normal tissue developmental asynchrony. Thus, changes in peripheral control are expected to comprise many small evolutionary changes over time.

7.4.2 Size at metamorphosis

Growth hormone (GH) and insulin-like growth factors control size in all vertebrates and transgenic overexpression of GH results in larger tadpoles and adults in the African clawed frog *Xenopus laevis* (Huang and Brown 2000a). Size at metamorphosis may also be controlled indirectly by larval period duration, where environmental signals may alter initiation and/or rate of development, thereby changing the overall time for growth and resulting in different metamorph sizes. Tadpole size and developmental stage are highly correlated, but the extent to which mechanistic interactions between growth and development exist is virtually unknown. However, larval period duration was not affected by transgenic GH overexpression. Thus, any trade-offs that may exist between growth and development are in need of further study.

7.4.3 Direct development

The biphasic amphibian life history of larvae metamorphosing to the juvenile, found in many frog families and plethodontid salamanders, is ancestral to direct developers, which lack free-living feeding larvae. Endocrine studies on the frog *Eleutherodactylus coqui* have shown:

- similarities in TH titre profile
- CRF release of TSH and ACTH
- lack of TSH release by TRH.

These observations are all consistent with conservation of central control of metamorphosis (Jennings and Hanken 1998, Callery and Elinson

2000, Kulkarni *et al.* 2010). Conserved larval neuroendocrine control mechanims may suggest that direct development evolved from ancestral indirect development via changes in the preipheral regulation of metamorphosis. Importantly, loss of requirement for feeding in aquatic larval form (except in viviparous caecilians and the viviparous toad *Nimbaphrynoides occidentalis*, where fetal feeding occurs) likely led to the lack of functional larval organs such as larval mouth parts, gills, and intestine. Despite this deletion of a functional larval phase, the ontogeny of direct developers still contains vestigial and non-functional larval organs. Surprisingly, treatment with methimazole, which inhibits TH production, blocked development to the juvenile form of limbs, skin, intestine, muscle, and tail in *E. coqui* (Callery and Elinson 2000). Thus, tissues are still sensitive to TH and depend on TH for development to the adult versions. Therefore, neither central nor peripheral control seems to have changed in a way that would explain the evolution of direct development. Embryonic and tissue-specific development prior to a response to TH seem to be changed, instead of TH physiology itself. Although extensive research in embryonic modifications to deal with a large yolky egg has been done (Elinson 2001), evolutionary changes in the mechanisms of post-embryonic development (i.e., after all major organ systems have been developed), but prior to TH involvement, are unknown.

7.4.4 Neoteny

Reproductive maturation in the larval form is not found in frogs or caecilians but is widespread among salamanders (Duellman and Trueb 1994). Some neotenic salamanders are capable of undergoing metamorphosis in nature under certain environmental conditions. Thus, because peripheral responses to TH leading to metamorphosis are intact, mechanisms of neoteny are to be found in central control. The Mexican axolotl undergoes metamorphosis only when induced by T3 in the laboratory. Surprisingly, the axolotl has a peak in T4 early in life, when toes differentiate, but metamorphosis fails to occur (Rosenkilde *et al.* 1982). Lack of a peripheral deiodinase conversion of T4 to T3, in combination with

low TR levels, may explain neoteny in axolotls (Galton 1992, Rosenkilde and Ussing 1996).

The obligate neotenic salamander *Necturus maculosus* cannot undergo metamorphosis even when treated with TH although TRs are present, functional *in vitro*, and expressed *in vivo* (Vlaeminck-Guillem *et al.* 2006). In *Necturus*, loss of TH-induction of key genes involved in metamorphosis may explain its neoteny. Thus, the molecular mechanisms underlying neoteny vary with the type of neoteny and may be found at central and/or peripheral levels.

7.5 Molecular mechanisms of peripheral control: Potential evolutionary targets underlying diversity in larval period diversity

Even though changes in central control may have contributed to amphibian metamorphic diversity, we focus here on peripheral control (Fig. 7-1B), where potential evolutionary changes are better known. The timing of transformation for each tissue is determined by the combinatorial effect of TH signaling proteins expressed at their tissue-specific level. Evolutionary changes in the expression and activity levels of these proteins can change tissue sensitivity/responsivity to TH and can thereby alter the timing of initiation and/or rate of tissue transformation, leading to species differences in larval period duration. Below, proteins involved in TH transport, metabolism, and gene regulation in frogs are described, including their potential role underlying amphibian life history diversity.

7.5.1 Thyroid hormone transporters

Many transport proteins that allow TH entrance into the cells have been identified. These have varying efficiency and specificity for TH and have been sorted into different families (Heuer and Visser 2009). These proteins transport THs as well as other small molecules, often amino acids. MCT8 and OATP1c1 (monocarboxylate anion transporter 8 and organic anion transport protein 1c1) described in mammals have the highest specificity for TH among all TH transporters and preferentially transport THs across cell membranes. Importantly, TH transport has the potential to regulate TH bioavailability in the nucleus because TH transport is specific, subject to regulation, and rate-limiting for intracellular TH metabolism (Hennemann *et al.* 2001). The strongest evidence for an *in vivo* role of TH transport comes from the association of mutations in MCT8 with severe X-linked mental retardation and elevated circulating T3 levels (Jansen *et al.* 2008). Most TH transporters are expressed in varying patterns in different types of tissues (Heuer and Visser 2009). *In-vitro* studies have shown that the activity of these transporters affects tissue sensitivity to TH and how the cell responds to competing substrates like amino acids (Shi *et al.* 2002).

The only TH transporter studied in frogs is the L-type amino acid transporter 1 (LAT1). Expression of LAT1 varies across tissues and correlates with tissue transformation during metamorphosis (Shi *et al.* 2002). An *in-vivo* role for LAT1 is suggested by its strong up-regulation during TH-dependent development (Liang *et al.* 1997). In addition, LAT1 overexpression in *Xenopus laevis* oocytes affected TH-response gene transcription (Ritchie *et al.* 2003). Because the free serum TH concentration is 1000 times lower than the receptor binding constant (Km) for LAT1, TH bioavailability via LAT1 is not regulated by substrate saturation, but rather by the number of transporters in the cell membrane. Thus, changes in LAT1 expression across species may affect tissue sensitivity, thereby potentially altering metamorphic initiation. However, at least two features of LAT1 function may limit the potential role of LAT1 in metamorphic diversity. First, LAT1 cannot function without its heterodimerization partner, 4F2 (Friesema *et al.* 2001), and thus, simultaneous and/or sequential evolutionary changes in both LAT1 and 4F2 may be required for this TH transporter to affect TH bioavailability across species. Second, ligands for LAT1 include not only TH, but also leucine and large aromatic amino acids (tryptophan, phenylalanine, tyrosine), such that changes in the flux of these amino acids could occur and may disrupt normal cellular function.

7.5.2   Thyroid hormone metabolizing enzymes

Like mammals, three deiodinases, D1–3, have been identified in amphibians (St Germain *et al.* 1994, Davey *et al.* 1995, Kuiper *et al.* 2006). Developmental studies in amphibians focused on D2 and D3 because frogs were not known to have D1 until recently (Kuiper *et al.* 2006). D1 activity in frogs had not been detected biochemically because, unlike the mammalian version, it is not inhibited by propylthiouracil. D1 converts T4 to T3, T4 to reverse T3 (rT3, an inactive form), and T3 and rT3 to T2 (another inactive form). D2 activates T4 to make T3, and D3 inactivates T4 and T3 to rT3 and T2, respectively. These enzymatic activities are significant because only T3 has biologically significant affinity to TR.

D2 and D3 act to regulate tissue developmental asynchrony during metamorphosis (Brown 2005). Studies measuring deiodinase activity in *Rana catesbeiana* showed that high rates of T4 to T3 conversion correlated with the timing of transformation of limbs, intestine, skin, and eye (Becker *et al.* 1997). *In-situ* hybridization in *X. laevis* confirmed the presence of D2 in transforming tissues (Cai and Brown 2004). Experimental evidence for the importance of D2 in regulating the timing of differentiation was shown by use of the TH synthesis blocker, methimazole, and the pan-deiodinase inhibitor, iopanoic acid (Becker *et al.* 1997). When tadpoles of *R. catesbeiana* were treated with methimazole and iopanoic acid, hind-limb growth and change in body morphology were inhibited. Hind-limb growth was restored when tadpoles were treated with T3 but not with T4. This confirms that the D2 activity to convert T4 to T3 can control the timing of tissue transformation. In addition, overexpression of D3 using transgenesis has confirmed the ability of D3 to inhibit metamorphic changes in *X. laevis* (Huang *et al.* 1999, Marsh-Armstrong *et al.* 1999).

Species differences in expression and activity of deiodinases could account for differences in larval period duration. For example, tails might shrink faster by expressing more D2 and less D3 in one species compared to another. Also, unlike several other factors that affect TH signaling, deiodinases are specific to TH physiology and would not be burdened by pleiotropic effects on other physiological systems.

7.5.3   Cytosolic thyroid hormone binding proteins

After first being distinguished from serum TH binding proteins (Tata 1958), the existence of cytosolic thyroid-hormone binding proteins (CTBPs) is now well established and many have been reported in mammals and amphibians (Kato *et al.* 1989, Ishigaki *et al.* 1989, Ashizawa *et al.* 1991, Shi *et al.* 1994, Yamauchi and Tata 1994). Multiple types of CTBPs exist in a given species and are expressed in a variety of tissues. Surprisingly, CTBPs do not represent a specific protein family; they are enzymes of diverse function that seem to have acquired TH binding capacity independently. In frogs, three CTBPs have been identified: aldhehyde dehydrogenase 1, pyruvate kinase subtype M2, and protein disulfide isomerase (Shi *et al.* 1994, Yamauchi and Tata 1994, 1997). Another CTBP in frogs was detected by photoaffinity labeling and SDS-PAGE, but was not identified (Yamauchi and Tata 1997). Yet another CTBP has been reported in *Rana catesbiana* and was characterized as metal-ion-dependent but its identity was not determined (Yoshizato *et al.* 1975).

CTBPs have been shown to affect TH-regulated gene transcription, most likely by modulating intracellular bioavailability of TH (Ashizawa and Cheng 1992, Mori *et al.* 2002) and/or transporting TH to the nucleus (Ishigaki *et al.* 1989). In turn, CTBPs may contribute to developmental tissue asynchrony via decreased tissue-specific expression levels of CTBPs in tissues actively undergoing TH-dependent remodeling (Shi *et al.* 1994). Also, high ALDH1 mRNA expression was observed in the liver and smaller amounts were detected in head, intestine, and tail during metamorphosis (Yamauchi and Tata 1997).

Despite these potential roles in TH physiology, CTBPs are multifunctional proteins, and the primary role of TH binding may be to modulate the activity of the enzyme rather than modulate TH bioavailability. Importantly, the enzymatic function and TH binding of CTBPs are mutually exclusive, i.e., when TH is bound the enzyme is inactive. Furthermore, the equilibrium dissociation constants (Kd) among known CTBPs for TH is often above the concentration of TH found in the blood, indicating that their identification as CTBPs may be of only pharmacological importance and not relevant for physiological TH levels. Thus, the evolution of CTBP expression and function may have been constrained by their enzymatic functions, suggesting they may not be likely candidates for the evolution of differential tissue sensitivity to TH. While *in-vitro*

studies have confirmed TH binding in cytosolic extracts of various tissues (Yamauchi and Tata 1997), and correlations exist between TH binding activity and expression levels of TH response genes (i.e., TRβ) (Ashizawa and Cheng 1992, Mori *et al.* 2002), future studies are required to clarify the role of CTBPs in TH-dependent development.

7.5.4 Thyroid hormone receptors

Frogs, like all vertebrates, have two types of TR, TRα and TRβ, coded for by two separate genes (Yaoita *et al.* 1990). TRs belong to the nuclear receptor superfamily of transcription factors and influence transcription by binding to TH-response elements (TREs) in DNA. TRs can bind to TREs weakly as monomers or homodimers, but with a much higher affinity as a heterodimer with retinoic X receptors (RXRs) (Wong and Shi 1995). Gene regulation by TRs is hormone dependent, such that in the absence of TH, TRs function to repress expression of TH-response genes, while in the presence of TH they activate those same genes (Buchholz *et al.* 2006). This dual function is achieved through the interaction of TRs with cofactors. Unliganded TRs are bound to corepressors, such as N-CoR (nuclear receptor corepressors) and SMRT (silencing mediator for retinoid and thyroid hormone receptors). These corepressors form complexes with other proteins that function to prevent transcription at promoters of TH-response genes via histone deacetylation. The presence of TH causes a conformational change in TR favoring binding of coactivators, such as steroid receptor coactivator (SRC) 1–3 and p300, leading to the recruitment of additional cofactors, causing changes in histone acetylation and gene induction.

The importance of TRs and co-factors in tadpole metamorphosis is supported by studies using transgenic overexpression of mutant receptors and co-factors (Shi 2009). Metamorphic transformation can be initiated in the absence of TH in tadpoles that overexpress a constitutively active TR (Buchholz *et al.* 2004), and TH-induced morphological changes are accelerated by overexpression of wild-type co-activator protein arginine methyltransferase 1 (PRMT1), which methylates histones and other nuclear proteins (Matsuda *et al.*

2009). On the other hand, none of these developmental events occur in tadpoles overexpressing dominant negative TRs or cofactors, that is, mutant versions that block function of endogenous TRs (Buchholz *et al.* 2003, Paul *et al.* 2005a, 2007). Transgenic tadpoles that overexpressed the dominant negative corepressor NCoR experienced derepression of TH-response genes as well as accelerated premetamorphic development (Sato *et al.* 2007). These studies provide *in-vivo* evidence for the profound role that TRs and co-factors play in gene regulation *in vivo* and in regulating timing of tissue transformation.

The dual molecular function of TRs coupled with their peak in expression during metamorphosis has the potential to impact developmental timing in frogs. TRα, with significant premetamorphic expression levels, is important for repressing TH-response genes to ensure the tadpole does not develop precociously (Sato *et al.* 2007). At the same time, TRα expression levels allow tissues to establish tissue competence to respond to the hormone signal. Because TRβ is a direct TH-response gene, it requires TH for significant expression and consequently may be less important than TRα for repression during premetamorphosis and controlling tissue sensitivity. However, premetamorphic tissues contain limiting amounts of TR, and thus autoregulation of TRβ expression is necessary for tissues to respond fully to circulating TH (Buchholz *et al.* 2005). Therefore, both TRs participate in the activation of TH-response genes and thus responsivity to TH. Consequently, evolutionary changes in TRα and TRβ expression levels across development and tissues are expected to underlie species differences in metamorphic timing.

7.5.5   Modulation of thyroid hormone responsiveness by corticosterone and prolactin

While TH may be necessary and sufficient to initiate tissue transformation, other hormones interact with TH to affect the timing of developmental events. The blood titre of corticosterone (CORT) and prolactin (PRL) peak at metamorphic climax, like TH (Clemons and Nicoll 1977, Krug *et al.* 1983, Jolivet-Jaudet and Leloup-Hatey 1984), and these hormones can influence the initiation and rate of tissue

transformation (Kaltenbach 1996). However, neither hormone by itself can initiate metamorphic progress. Rather, CORT and PRL have a complex and tissue-specific role, interacting with TH, and underlying tissue developmental asynchrony.

7.5.5.1   Corticosterone

Although TH is sufficient to initiate metamorphosis, TH alone cannot complete metamorphosis in the absence of CORT. Because CORT does not have metamorphic actions independent of TH, the role of CORT is likely to permit a maximal level of TH signaling required to complete metamorphosis. Hypophysectomized tadpoles treated with T4 do not completely resorb the tail nor leave the water, whereas injection of ACTH to induce CORT production allows complete metamorphosis of T4-treated hypophysecotomized tadpoles (Remy and Bounhiol 1971). Similarly, bullfrog tadpoles treated with amphenone B, a corticoid synthesis blocker, show inhibited metamorphosis (Kikuyama *et al.* 1982), but a 50% reduction of CORT whole-body levels by the 11β-hydroxylase inhibitor, metyrapone, does not affect developmental timing (Glennemeier and Denver 2002).

On the other hand, CORT will either inhibit metamorphosis (if present in early stages) or accelerate metamorphosis (if present during later stages). CORT levels increase naturally in the presence of stress and studies have examined the affect of exogenous CORT on developmental timing of amphibian metamorphosis (Leloup-Hatey *et al.* 1990). However, this acceleration by CORT occurs in some tissues, such as tail, but not others, such as hind limbs (Hayes 1995). Generally, CORT has inhibitory actions on growth in all vertebrates, even in prometamorphic tadpoles. CORT's early inhibition of development is unlikely to be via negative feedback on CRH because the receptors for CRF on thyrotropes are not expressed until prometamorphosis (Manzon and Denver 2004). However, by prometamorphosis CORT's role as a facilitator of TH action becomes significant because TH levels are rising. In particular, CORT increases sensitivity and responsivity to TH in tail tips (Kikuyama *et al.* 1993). CORT affects TH physiology by at least two mechanisms: the regulation of deiodinases and the presence of a glucocorticoid response element (GRE) in the promoter of TRβ and perhaps other TH response genes. Because nearly all cell types express the CORT receptor (GR), the tissue-specific actions of CORT are likely explained by tissue-specific factors affecting gene regulation by GR. CORT enhances T4-to-T3 conversion by increasing D2 activity and decreases the degradation of T3 by reducing D3 activity (Galton 1990), thereby increasing the intracellular concentration of T3. The mechanisms by which CORT affects D2 and D3 are not known, but could be direct via gene regulation or indirect via regulation of genes that affect deiodinase enzyme stability or activity. A GRE may be present in the TRß promoter because CORT upregulates TRß in the intestine (Krain and Denver 2004), so TRß upregulation in the presence of TH and CORT could be due to a combination of CORT's action on TH signaling via D2 and D3 and direct transcriptional regulation at the TRβ promoter.

7.5.5.2 Prolactin

At one time PRL was believed to be a juvenile hormone analogous to juvenile hormone in insects because of the anti-metamorphic effects of PRL. However, the developmental expression profile of PRL mRNA in pituitary (Takahashi *et al.* 1990, Buckbinder and Brown 1993), and blood levels of PRL that peak at climax, like TH, in *Bufo japonica* (Niinuma *et al.* 1991) and bullfrog (Yamamoto and Kikuyama 1982) put this interpretation of the role of PRL into question. Functional analysis of PRL via overexpression in transgenic animals shows that PRL indeed antagonizes TH action in transgenic animals, resulting in a tailed juvenile frog, where muscle still resorbs but notochord, fibroblasts, and epithelium remain (Huang and Brown 2000b). However, the transgenic animals otherwise metamorphose at the same time, with no known delays in other organs. PRL receptors are detected in most tadpole tissues tested, with high expression in tail and low expression in liver and intestine (Hasunuma *et al.* 2004). As for CORT, these data argue for tissue-specific action of PRL because, even though PRL receptors are present in many tissues, delays in tissue transformation in PRL overexpressing transgenic tadpoles are observed only in the tail. One mechanism by which PRL antagonizes TH action is upregulation of D3 mRNA in the tail (Shintani *et al.* 2002). The effect of PRL on T3-induced tail shrink-

age is reversed by iopanoic acid, which blocks deiodinase activity. PRL blocks TRβ upregulation by T3 (Baker and Tata 1992), where TRβ is a direct response gene in no need of protein synthesis, which suggests that PRL signaling can act upstream of TR-mediated gene regulation, in addition to D3 gene regulation.

7.6 Conclusions

Amphibian larval period diversity includes time to and size at metamorphosis, direct development, and neoteny. TR and associated cofactors function as a metamorphic switch and depend upon upstream TH transport and metabolism to regulate the timing and rate of transformation in a tissue-specific manner, contributing to the total larval period duration. Few studies have attempted to identify evolutionary changes in TH signaling that may serve as a mechanistic basis for life history differences among species. Research to understand the control of metamorphosis in model species reveals a large complexity of potential mechanisms from which evolution could "choose" to generate species life history differences. Changes at the central and/or peripheral levels could underlie life history differences, but the available evidence so far has provided more support for peripheral mechanisms. In theory, a change in a single protein that influences TH bioavailability could underlie differences between species, but species differences are more likely controlled by a number of different mechanisms, such as expression and activity changes in TH transporters, CTBPs, TRs, and/or deiodinases. For instance, higher TH sensitivity could be due to increased LAT1, TR, and/or D2 expression or decreased D3 and/or CTBP levels. The dual functionality of CTBPs and multiple substrates for LAT1 may limit evolutionary flexibility in their expression levels and thus may limit their role in contributing to metamorphic diversity. On the other hand, TRs and deiodinases are specific to TH physiology, such that changes in their expression levels might not affect other aspects of development. Future comparative studies to identify specific differences in these proteins between species will require analysis of protein activities and/or cloning relevant genes from each of the derived species for analysis of gene expression patterns. However, such comparative analysis will at best provide correlations between life history traits and expression-level differences of TH signaling genes. Strong inferences about the mechanistic basis of life history divergence will come from experimental tests of these correlations by studying the effects of altered gene expression levels on metamorphosis in model species.

7.7 Summary

1. Amphibian larval period diversity includes time to and size at metamorphosis, direct development, and neoteny.

2. Thyroid hormone physiology regulates the timing and rate of metamorphic transformation on a tissue-by-tissue basis via thyroid hormone receptors and associated cofactors, which together function as a developmental switch dependent upon bioavailability of thyroid hormone in the nucleus.

3. Thyroid hormone bioavailability in the cell nucleus is controlled by cellular mechanisms of thyroid hormone transport and metabolism involving thyroid hormone transporters, cytosolic thyroid hormone binding proteins, thyroid hormone receptors, and iodothyronine deiodinases.

4. Larval period evolution likely involved changes in expression level or activity of proteins that control thyroid hormone bioavailability.

5. Thyroid hormone receptors and deiodinases are specific to thyroid physiology, such that changes in their expression or activity levels might not affect other aspects of development.

6. On the other hand, the dual functionality of cytosolic thyroid hormone binding proteins and multiple substrates for thyroid hormone transporters may limit evolutionary flexibility in their expression levels and thus may limit their role in contributing to metamorphic diversity.

7. Comparative studies to identify specific differences in proteins controlling thyroid hormone bioavailability between species will require analysis of protein activities and/or cloning relevant genes from each of the derived species for analysis of gene expression patterns.

8. Comparative analyses should be complemented with functional studies on the role of these genes in development using model species.

PART 3

Reproduction

Thomas Flatt and Andreas Heyland

The major objective of life history theory is to explain how natural selection designs organisms to achieve reproductive success (Stearns 1992, Roff 1992; see also Chapter 1). Thus, traits that more or less directly determine or affect reproductive output, such as fecundity and fertility, are clearly the most important life history traits, and insights into the mechanisms underlying the timing, rate, and duration of reproductive processes are key to our understanding of life history evolution. The chapters in this part of the book give a glimpse of the various mechanisms that determine or affect reproductive traits in a broad range of organisms, including plants, cnidarians, insects, and humans.

In Chapter 8, Reitzel and colleagues discuss the diversity of asexual reproduction modes in cnidarians and the developmental mechanisms that underlie this diversity. Because asexual reproduction involves cell proliferation, death, and differentiation, the interplay of cellular signaling events is critical for successful deployment of these developmental processes, and the authors discuss recent comparative genomic surveys for candidate genes and studies of transcriptional regulation in cnidarians and other animals that utilize asexual reproduction and which elucidate these cellular signaling events. As Reitzel *et al.* illustrate, cnidarians provide a powerful comparative system for studying the mechanisms underlying asexual reproduction and how these gene networks evolve to produce the diversity of asexual reproductive mechanisms.

In Chapter 9, Banta and Purugganan use findings from the model organism *Arabidopsis thaliana* to illustrate how insights from evolutionary ecology, quantitative genetics, and developmental genetics can be combined to advance our understanding of the evolution of a major life history trait in plants: flowering time. In the main part of their chapter, the authors describe the molecular mechanisms that control this life history transition and the methods for detecting genetic polymorphisms accounting for flowering time variation, including association mapping and QTL analyses. They close by discussing the genes that have been identified through these approaches, how they affect flowering time, and how selection acts at these loci.

Chapter 10 by Bergland reviews the physiological mechanisms that underlie the effects of nutrition on reproduction in Dipteran insects. The author synthesizes a large body of work describing how nutritional resources acquired during larval and adult stages affect various reproductive processes and how these effects are mediated by nutrient-dependent hormonal signals provided by insulin, ecdysone, juvenile hormone, and biogenic amine signaling. He notes that many studies have found genetic variation for reproduction and concludes that the identification of natural alleles will help our understanding of how these pathways evolve to shape reproductive success.

Edward and Chapman in Chapter 11 look at another major aspect of reproduction, the so-called survival cost of reproduction, a classic example of a life history trade-off (see also Chapters 13, 27, and 28). The authors review the available evidence for the notion that survival–reproduction trade-offs are physiologically caused by competitive resource allocation, so that increased investment of resources into reproductive processes leads to fewer resources available for longevity assurance, thus causing decreased lifespan. While this resource allocation model has been broadly confirmed by phenotypic and genetic manipulative experiments, Edward and Chapman discuss several examples that suggest that this intuitively appealing idea might in fact not always be correct; instead, trade-offs might often occur because of the integration of molecular signals from different sources.

In the final chapter in this section, Chapter 12, Muehlenbein and Flinn examine human life histories in the context of reproductive maturation and senescence. As the authors discuss, humans, like many other organisms described throughout this volume, are required to allocate physiological resources between reproduction and a number of competing functions, particularly growth and survivorship. Muehlenbein and Flinn discuss several major aspects of human reproduction, including reproductive maturation, male and female reproductive ecologies, mating/parental behaviors, and reproductive senescence, with a particular emphasis on the key endocrinological mediators of life history trade-offs, especially those involving reproduction.

There are several major recurring themes among the chapters outlined above. For example, as several authors discuss, the coordination of reproductive processes often relies on similar regulatory principles, in particular involving the action of endocrine and neuroendocrine signaling pathways, and future work in this area is likely to yield important insights into life history regulation. Similarly, as suggested by Edward and Chapman in Chapter 11 and Muehlenbein and Flinn in Chapter 12, the study of social interactions holds great promise for a better understanding of reproductive conflicts and associated life history trade-offs. This theme is also taken up again in Part 6, by Rascon et al. (Chapter 20) and Lancaster and Sinervo (Chapter 25).

The perhaps most important area for future research is the study of natural genetic variation for the mechanisms that affect reproductive processes. As discussed by Banta and Purugganan in Chapter 9 and Bergland in Chapter 10, QTL mapping approaches and genome-wide association studies have already begun to further our understanding of how natural variation affects reproductive traits (see also Chapters 1, 2, and 18). While such methods are most readily applicable to model systems such as *Arabidopsis* and *Drosophila*, the integration of mechanisms into the life history theory framework will hopefully guide the expansion of such inquiries to comparisons among closely related non-model species with different reproductive modes.

CHAPTER 8

Asexual reproduction in Cnidaria: Comparative developmental processes and candidate mechanisms

Adam M. Reitzel, Derek Stefanik, and John R. Finnerty

8.1 Introduction

Animals display a remarkable diversity of life histories, each of which is composed of successive, often highly distinctive, stages. Studies of molecular mechanisms linking the stages following sexual reproduction elucidate the evolution of developmental transitions between stages of embryogenesis, larval development, and metamorphosis. Previous examples of these studies include evolution of direct development in ascidians and echinoderms (Jeffery and Swalla 1992, Raff and Byrne 2006, see also Chapter 3), insect metamorphosis (Truman and Riddiford 1999, see also Chapter 5), and neoteny in salamanders (Bonett and Chippindale 2004, see also Chapter 7). It is clear that sexual life histories can evolve rapidly, so that even closely related taxa exhibit distinctive life cycles. A general finding of these studies is that conserved molecular mechanisms (e.g., transcription factors, signaling pathways, and hormones) can play critical roles in diversification of animal life histories (see Chapter 3, 5, 6, and 7).

In many animals, sexual reproduction is accompanied by one or more forms of asexual reproduction. Asexual reproduction occurs in most every animal phylum (reviewed by Blackstone and Jasker 2003) and at every stage of invertebrate life histories (Jaeckle 1994, Craig *et al.* 1997). Despite the near ubiquity of asexual reproduction, the mechanisms underlying the generation of diverse asexual modes are largely unknown (Ferretti and Geraudie 2001) and the potential selection factors leading to different strategies are often not considered in traditional life history theory (Roff 1992, Stearns 1992).

In this chapter we explore the diversity of asexual reproduction strategies within and among species to discuss the question: "*How* and *why* are there so many methods of asexual reproduction?" While the molecular mechanisms underlying sexual life histories are known in exquisite detail for a handful of model systems, the mechanisms underlying asexual reproduction, from developmental patterning, to morphogenesis, to environmental sensing and fitness consequences are not well understood in any animal model. With few exceptions, our knowledge amounts to descriptive studies at the organismal or tissue level. Indeed, molecular mechanisms of asexual reproduction are considerably less understood than for regeneration, for which broad, comparative data exist (Bimbaum and Sánchez Alvarado 2008). The relative paucity of molecular data from animals undergoing diverse forms of asexual reproduction and the potential fitness consequences of one form over another represent significant hurdles to understanding how these life history strategies evolve in particular lineages. We propose in this chapter that the diversity of asexual life histories can be largely explained by the reordering of developmental modules and their plastic expression in response to different environmental cues. This chapter focuses on one phylum of animals, the Cnidaria, where the life history frequently includes sexual reproduction and one or more form of asexual reproduction, but our conclusions are applicable to other organisms where life histories include facultative asexual life histories.

8.2 Diversification of clonal reproduction in cnidarians

The phylum Cnidaria is an ancient lineage of metazoans composed of more than 10 000 species, including corals, sea anemones, hydrozoans (e.g., *Hydra*), and true jellyfish. Cnidarians exhibit two principal body plans as adults, the polyp and the medusa (Fig. 8-1). The two major cnidarian lineages—the Anthozoa and the Medusozoa—differ with respect to the possession of the medusa (Anthozoans lack a medusa) and, correspondingly, they differ with respect to their life histories.

Despite being limited to two basic body plans, cnidarians exhibit a tremendous degree of morphological and ecological diversity achieved, in large part, through modes of asexual reproduction (Shostak 1993). Indeed, the ecological impact of cnidarians is partly attributable to their rapid population increases through clonal growth, and this, in turn, is partly attributable to their ability to reproduce asexually. For example, sea anemones are densely distributed on the available substrate on temperate rocky shores (Francis 1988), jellyfish blooms can profoundly impact pelagic food webs (Moller 1984), and the growth of colonial corals supports the ocean's most species-rich ecosystem, the coral reef (Moberg and Folke 1999). In addition, asexual reproduction of colony-forming cnidarians contributes to the fine-tuning of ecomorphological responses, including distinct colony shapes matched to particular environments (Francis 1979, Jackson and Coates 1986, Sebens 1982a, Hughes 1989).

For this discussion, we focus on agametic clonal reproduction of cnidarian polyps. We have chosen the polyp and not other forms of asexual reproduction (Shostak 1993) because this stage is likely the ancestral body plan and occurs throughout the phylum (Bridge *et al.* 1992). Clonal reproduction of polyps occurs in both colonial and solitary species. In colonial cnidarians (e.g., hard corals, colonial hydroids), clonal reproduction adds new polyps to an interconnected colony. In solitary polypoid cnidarians (e.g., sea anemones, hydras), clonal reproduction generates two or more solitary polyps from a single parent individual. The key distinction in clonal

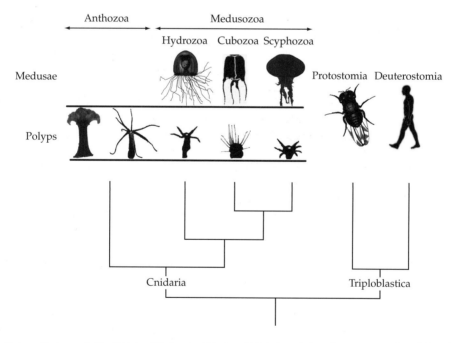

Figure 8-1 Phylogenetic placement of the Cnidaria within the animal kingdom. Cnidarians are the immediate outgroup to the triploblasts (deuterostomes and protostomes). Within the Cnidaria, there are two clades: the Anthozoa possess only a polyp stage in the life history and the Medusazoa typically have a biphasic life history comprising both polyp and medusa stages.

reproduction between colonial and solitary cnidarians is whether the new polyp is anatomically cleaved from the parent polyp, but the developmental processes at the organismal level are comparable.

8.2.1 Diversity of clonal reproduction modes in cnidarian polyps

Clonal reproduction of polyps can vary according to:

• the axial orientation of the daughter polyp relative to the parental polyp
• the temporal and spatial sequence of developmental events by which the new polyp forms.

In solitary cnidarians, clonal growth may be broadly categorized into three modes: fission, fragmentation, or budding (Shick 1991, Arai 1997). Fission occurs when the body column of the parent polyp is cleaved in two (Fig. 8-2A–D). The plane of division may be perpendicular to the primary body axis, the oral-aboral (OA) axis (transverse fission; Fig. 8-2A–C), or it may pass along the OA axis (longitudinal fission; Fig. 8-2D). If the plane of division is perpendicular to the OA axis, it may be located at different points along the axis (compare Fig. 8-2A–C). Furthermore, the daughter polyp may be anatomically complete at the time of fission (Fig. 8-2A,C), or it may develop missing body regions and structures after division has occurred (Fig. 8-2B). Finally, the initial polarity of the daughter polyp may mirror that of the parent, or it may be reversed 180° relative to the parent. Fragmentation occurs when a small piece of differentiated tissue detaches from the parental polyp via a tangential fission plane that does not pass through the entire body column (Fig. 8-2E). The newly separated fragment grows and differentiates to yield a complete and independent daughter polyp. Budding occurs when lateral outgrowth from the parental polyp gives rise to a daughter polyp. The outgrowth may occur in a relatively undifferentiated region, such as a budding zone (Fig. 8-2F, Bosch 2003), or it may occur in a differentiated region of the body such as a tentacle (Fig. 8-2G, Pearse 2002). Among colonial cnidarians, the addition of polyps to the colony occurs by a process of lateral outgrowth (Fig. 8-2H), which in colonial hydrozoans involves connecting structures called stolons. We point interested readers to more comprehensive discussions of particular modes and axonomic distribution of asexual reproductive strategies (Shick 1991, Arai 1997, Fautin 2002).

As the incomplete survey presented here reflects, clonal reproduction appears to be of an extremely labile character over the evolutionary history of the Cnidaria (Stephenson 1935, Geller and Walton 2001, Fautin 2002). Simply categorizing these diverse reproductive modes does not provide much insight into possible homology between modes, their evolutionary origins, or the underlying developmental mechanisms. We could pinpoint the origin(s) of a particular mode of asexual reproduction on a phylogeny of the Cnidaria and determine whether it predates or postdates the origins of other modes of asexual reproduction in a given lineage. However, we can gain more insight if, rather than treating mechanisms of asexual reproduction as indivisible characters (in the phylogenetic sense), we consider that each mode is composed of constituent developmental modules. Indeed, the diversity of agametic reproductive strategies both within and between cnidarian species reveals an underlying modularity of their developmental programs, which in turn provides an approach for explaining the diversification of asexual reproduction to altered developmental mechanisms.

8.2.2 The role of developmental modularity in life history diversification

The modular organization of organisms has become an overarching theme in evolutionary developmental biology (Schlosser and Wagner 2004, Wagner et al. 2007). Modules are generally described as networks of interacting elements that behave in a semi-autonomous manner during the life history of an individual, and among related taxa in a particular lineage (von Dassow et al. 2000). Under this description, modularity applies equally well to many levels of biological organization including gene networks, tissue interactions, and morphological structures (Bolker 2000, Smith and Krupina 2001, Winther 2001). The origin of biological novelty can be understood, in part, by investigating how modules are constituted, how their constitution and interactions with other modules may be altered, and how their expression may change over the

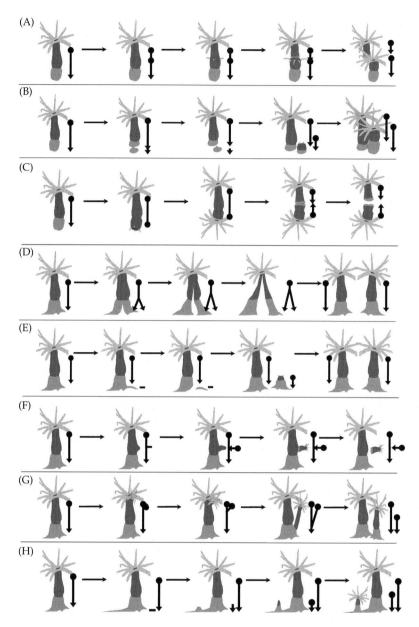

Figure 8-2 Diagrammatic representation of the diverse modes of asexual reproduction in cnidarian polyps. The three major body regions are delineated by shades of gray. A, intracolumnar transverse fission with conserved axial polarity (e.g., *Anthopleura stellula* in Schmidt 1970); B, physal pinching (*e.g.,* *Nematostella vectensis* in Reitzel 2007); C, transverse fission with polarity reversal (e.g, *Nematostella vectensis* in Reitzel 2007); D, longitudinal fission (e.g., *Anthopleura elegantissima* in Geller 2001); E, pedal laceration (e.g., *Aiptasia pallida* in Clayton 1985); F, lateral budding (e.g., *Hydra* species, various hard corals, in Soong 1992, Bode 2003); G, intratentacular budding (e.g., *Favia* species, in Gateño 2003); H, stolon outgrowth (e.g., *Hydractinia*, in Cartwright 2003). The stick figure to the right of each diagram indicates the axial polarity (circle, oral region; vertical line, body column; arrowhead, foot).

course of evolution (Dover 2000). Because modules are semiautonomous, they should exhibit cohesion over evolutionary time, both within and across lineages. Therefore, the identification of modules has frequently relied on interspecific comparisons (Harris *et al.* 2002, Mabee *et al.* 2002, Friedman and Williams 2003, Poe 2004).

However, while comparisons between distantly related organisms are useful for identifying highly conserved developmental modules, closer taxonomic comparisons are more powerful for revealing the mechanisms of diversification (Jeffery and Swalla 1992, Averof and Patel 1997, Mabee *et al.* 2000, Schram and Koenemann 2001). A complementary approach to comparisons between distantly related species is provided by species where alternate developmental trajectories converge on the same adult phenotype. Species with polyphenic larval or adult morphologies have been used to study how divergent gene regulation can generate unique phenotypes from identical zygotes (Pfennig 1990, Nijhout 1999, 2003, Abouheif and Wray 2002). Understanding developmental plasticity within species is particularly important because it may provide insight into the mechanisms underlying the generation of diversity at higher taxonomic levels while avoiding the confounding effects inherent in interspecies comparisons (West-Eberhard 2003).

8.2.3 Evidence of modularity in cnidarian developmental programs

Developmental modules exhibit the following properties:

- autonomous or semi-autonomous genetic regulatory networks
- hierarchical composition
- discrete physical location (Bolker 2000).

Furthermore, modules interact in networks with one another, and they may undergo evolutionary and developmental transformations (Dover 2000). Comparisons among cnidarians reveal asexual reproduction to be a highly modular developmental process, consisting of discrete, recognizable components.

Based on published observations of several forms of clonal reproduction in cnidarian polyps, we define 13 distinct developmental modules (Table 8-1). These provisional definitions are necessarily broad as they are based on developmental processes at the gross morphological level, and also on a simplified model of polyp anatomy. We divide the polyp into three main body regions (oral region, column, and aboral region), and we recognize three distinct structures (the "mouth," the tentacles, and the "foot," which may be either a flat "pedal disc" or a bulbous extensible "physa"). A detailed examination of any particular taxon would allow much finer anatomical resolution, but this rudimentary model applies equally well to polyps from all four classes of cnidarians. Furthermore, despite the simplicity of the anatomical model, and the fact that our survey of reproductive modes is undoubtedly incomplete given the paucity of cnidarian diversity that has been characterized (interested readers are directed to reviews by Fautin 2002, Shostak 1993), we observe a number of qualitative shifts in the deployment of individual modules.

We hypothesize that the diversity in modes of clonal reproduction (Fig. 8-2) can be explained by changes in the temporal order and/or the spatial deployment of these modules (Fig. 8-3). The shifting of modules can be compared between morphogenesis following sexual reproduction and various modes of clonal reproduction. In larval development, the primary (oral–aboral) and secondary (dorsal–ventral) body axes must be established (Fig. 8-3A). By contrast, at the outset of physal pinching—a type of transverse fission—there is no need to establish the body axes *de novo* (Fig. 8-3B), and with polarity reversal—another type of transverse fission (Fig. 8-3C)—a new primary axis must be established oriented at 180° to the original axis. All three modes require the patterning and subsequent morphogenesis of the oral region, but physal pinching alone does not require the patterning and subsequent morphogenesis of the aboral end. Fission occurs at the outset of physal pinching, but at the close of polarity reversal and not at all during larval development. Longitudinal fission (as seen in anemones such as *Anthopleura elegantissima*) is markedly different still from all three of these developmental modes (Fig. 8-3D). Unlike larval development, physal pinching, or polarity reversal, longitudinal fission does not require axial patterning or morphogenesis of any of the three primary axial regions

Table 8-1   Hypothesized developmental modules deployed in asexual reproduction by cnidarians

Module	Symbol	Description
Establishment or re-establishment of primary (oral–aboral) body axis	1°	Establishment of polarity along the oral–aboral axis
Establishment or re-establishment of secondary (transverse) body axis	2°	Establishment of polarity along the transverse or "directive" axis
Oral patterning	OR	Delineation of oral region
Column patterning	CO	Delineation of column region
Foot patterning	FO	Delineation of foot region
Oral morphogenesis	OR	Formation of oral morphology (e.g., mouth and tentacles)
Column morphogenesis	CO	Formation of column morphology
Foot morphogenesis	FO	Formation of foot morphology (e.g., pedal disc or physa)
Growth along primary body axis	1°	Elongation of animal along the oral-aboral axis
Growth along primary body axis	2°	Widening of animal along the transverse axis
Lateral outgrowth	OUT	A protuberance extends outward from the body column
Fission	CUT	Cleavage of body column
Epidermal healing	HEAL	Healing of cut edges of body column

These categorizations largely divide into two types of processes observed in the diverse modes of asexual reproduction: patterning of portions of the individual and differentiation of structures. Although these two processes are necessarily linked, we suggest that they represent different and discrete stages of clonal reproduction.

(oral region, body column, or aboral region). These four examples illustrate that equivalent developmental processes can undergo differential spatial and temporal deployments and result in identical end products (i.e., two complete individuals).

Further evidence for discrete developmental modules can be observed by examining morphogenesis of the oral region across different modes of asexual reproduction. For example, morphogenesis of the oral region can occur at the outset of asexual reproduction (Fig. 8-3C), at the close of asexual reproduction (Fig. 8-3B), or not at all (Fig. 8-3D). Such temporally labile developments of the same morphological structure suggests that morphogenesis of the oral region is coordinated by a semi-autonomous genetic architecture.

8.2.4   Elucidating the genetic architecture of cnidarian modules

Because morphology is a phenotypic manifestation of molecular processes, the anatomical structures (e.g., oral opening, tentacles) that form during development are ultimately driven by changes in gene expression and post-transcriptional processes (e.g., alternative splicing, protein–protein interactions). By definition, homologous anatomical modules should involve the same, or at least broadly similar, gene networks regardless of the context in which they are deployed. Therefore, a potentially insightful method for discovering the genetic architecture underlying development of putative anatomical modules is to identify genes that are

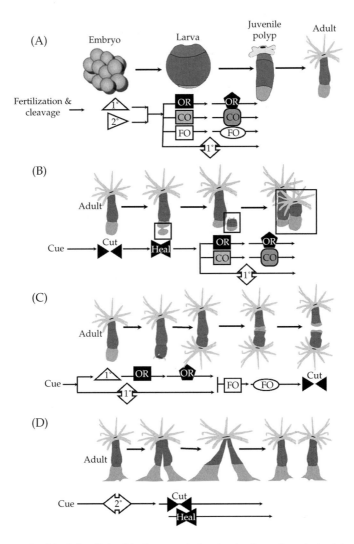

Figure 8-3 Temporal sequence of modules deployed in larval development and selected modes of asexual reproduction. A, larval development; B, transverse fission with physal pinching; C, transverse fission with polarity reversal; D, longitudinal fission. Edwardsiid anemones such as *Nematostella vectensis* and *Edwardsiella lineata* can undergo A, B, and C (Reitzel *et al.* 2007, 2009), and thus may be useful experimental systems for studying the mechanisms and environmental conditions for initiation among modes of asexual reproduction. *Anthopleura elegantissima* is an example of a species that can undergo both A and D. Symbols for modules, along with their definitions, are given in Table 8–1.

specifically associated with the development of a particular structure in multiple developmental modes. We discuss morphogenesis of the oral region of cnidarian polyps as an example.

Studies of developmental patterning of adult polyps from two primary cnidarian models (*Hydra*, *Nematostella*) suggest that the primary body axis is determined through an oral organizer (Matus *et al.* 2006, Fritzenwanker *et al.* 2007, Bode 2009). Similar to

developmental organizers from other taxa (e.g., Spemann's organizer in amphibians), the cnidarian oral organizer is a region at one pole of the developing animal that expresses a suite of genes that convey axial identity through gene activation and repression throughout the body column. The organizer also influences expression of downstream genes specific to particular regions or morphological structures, for example tentacles. In *Hydra*, a plethora of research has shown that the organizer works by a suite of activators and inhibitors that form a gradient along the body column (reviewed in Bode 2003, 2009). This morphogenetic gradient confers the developmental signal to the body-column tissue and maintains the oral–aboral axis by continuous active patterning. The molecular basis of these two gradients is not known, but it is speculated that they are established through the diffusion and transport of small molecules (Meinhardt 2009). On the other hand, it is uncertain if all cnidarians, particularly members of the class Anthozoa (corals, sea anemones), utilize a morphogenetic gradient akin to that described in *Hydra*. The signaling in *Hydra*, and potentially other members of the Medusazoa, depends upon cell communication via gap junctions (Fraser *et al.* 1987), present throughout the Medusozoa but potentially not the Anthozoa (Magie and Martindale 2008). However, a recent study reported the identification of a gene for the innexin gap junction in the anthozoan, *Nematostella vectensis*, suggesting that a new search for gap junctions with ultrastructural analyses may be warranted (Chapman *et al.* 2010). In addition to axial pattering, the interaction between the activator/inhibitor gradient in *Hydra* determines the placement of asexual buds along the body column. At present, it is not known what molecular mechanisms other species use to specify the location of asexual reproduction or to initiate it, although some studies have indicated relative spacing of polyps (in colonial species) (Gateño and Rinkevich 2003) and expression of developmental regulatory genes involved in embryogenesis (Burton and Finnerty 2009).

Although data on the molecular mechanisms driving asexual reproduction lags far behind our understanding of embryogenesis, recent work has begun to elucidate the genes and molecular interactions involved in axial patterning in embryogenesis, and some forms of asexual reproduction, in a number of different cnidarians. In recent years, the diversity of genes known to be involved in axial specification has increased considerably, including the Hox and Wnt families, genes instrumental in axis patterning in other animals.

Most work on comparative developmental gene expression in the Cnidaria has investigated the diversity and expression of Hox genes (Gauchat *et al.* 2000, Finnerty *et al.* 2003, 2004, Kamm *et al.*

2006, Ryan *et al.* 2007, Chiori *et al.* 2009). In many but certainly not all cases, Hox genes are spatially restricted along the oral–aboral axis (Shenk *et al.* 1993, Masuda-Nakagawa *et al.* 2000, Ryan *et al.* 2007), suggesting that they have a role in specification of the primary axis and the diagnostic structures. RNAi gene suppression studies in one species, the hydrozoan *Eleutheria dichotoma*, have shown that knockdowns of Hox genes in the Antp class substantially alter the primary axis by producing additional oral regions, deformation of the axis, and tentacle duplication (Jakob and Schierwater 2007). However, it is presently unclear how generally conserved the role of Hox genes in axial specification is in the Cnidaria. Comparisons among species have shown that the spatial expression for Hox genes differs among species (Kamm *et al.* 2006, Chiori *et al.* 2009). These divergences suggest that a Hox-based gene network may not be a broadly conserved mechanism for axial pattering in the Cnidaria.

More recently, a rapidly expanding literature has developed that describes the conservation of genes in the Wnt signaling pathway in the Cnidaria, and the expression of Wnt genes and other members of this gene network (reviewed in Lee *et al.* 2006b, Ryan and Baxevanis 2007). Together, these data suggest that Wnt signaling appears to be highly conserved among the studied cnidarians and that Wnt/ß-catenin signaling is involved in specification of the oral region. In *Hydra*, canonical Wnt/ß-catenin signaling is involved in specification of oral identity in regenerating tissue and in asexual reproduction via lateral budding (Lee *et al.* 2006b). Investigations into Wnt/ß-catenin signaling during embryogenesis in *Nematostella* and *Clytia* have identified a critical role for this pathway in specifying axial polarity during embryo development (Kusserow *et al.* 2005, Momose *et al.* 2008). Similarly, canonical Wnt/ß-catenin signaling appears to be instrumental in oral determination and axis patterning in the hydroid *Hydractinia* (Purcell 2007, Liu *et al.* 2009). From this comparison, Wnt/ß-catenin signaling specifies the axial identity of tissue that will develop into the oral crown of the adult polyp in embryogenesis, asexual reproduction, and regeneration. In addition, chemical disruption of the Wnt/ß-catenin pathway by lithium and paullones results in severe disruption of the primary axis in multiple cnidarian species (Matus *et al.* 2006, Plickert *et al.* 2006, Lengfeld *et al.* 2009). Thus, Wnt/ß-catenin signaling is likely to be a well-conserved component of the developmental module that coordinates oral identity during cnidarian development.

Studying Wnt/ß-catenin signaling could offer additional insights into the molecular mechanisms underlying modularity in axial patterning during asexual reproduction. Perhaps most interesting would be the study of Wnt signaling in diverse asexual reproductive strategies where differentiated tissue is transformed into tissue of a different axial fate, as in the case of intracolumnar fission, where a new oral region forms in the column (Fig. 8-2A), or in transverse fission via polarity reversal, where existing aboral tissue at the physa is converted into oral tissue (Fig. 8-2C). Under the modular model of axial patterning, the change of axial identity in polarity reversal, for example, likely occurs through activation of the oral developmental module at the aboral end of the animal. One potential mechanism by which the oral module could be activated in these modes of asexual reproduction is through ectopic induction of Wnt/ß-catenin signaling at the site of future tissue transformation.

Recent work by Philipp and colleagues has shed some light on additional players that may be involved in defining tissue fate at finer morphological resolution within the oral developmental module of cnidarians (Philipp *et al.* 2009). Canonical Wnt/ß-catenin signaling induces expression of Wnt genes that interact with the non-canonical Wnt/PCP pathway to signal the location and evagination of tentacles in *Hydra* polyps. Given other similarities between gene expression of the Wnt pathway among cnidarians, it is possible that this mechanism of tentacle development may be conserved more broadly in cnidarians. Thus, the interaction between canonical Wnt/ß-catenin and non-canonical Wnt/PCP signaling may represent another candidate molecular mechanism by which development is coordinated in a modular fashion at successively finer morphological scales.

8.3 Trade-offs and environmental signaling in asexual reproduction

The wide taxonomic distribution and apparent evolutionary stability of asexual reproduction may be attributable to a number of factors. Traditional models exploring the evolution and maintenance of mixed reproductive strategies have historically taken two general approaches that compare sexual and asexual reproduction:

- a classic genetic approach balancing mutation rate and fitness within a population
- an energy-intake optimality approach.

When considering potential genetic consequences, infrequent sexual reproduction is likely sufficient to offset the absence of recombination from asexual propagation (Green and Noakes 1995). Because the genetic costs of sexual reproduction increase with a higher proportion of offspring produced sexually, there is an optimal proportion of offspring produced sexually or asexually (Kondrashov 1993). The deleterious genetic effects of asexual reproduction are likely offset by occasional sexual recombination for the vast diversity of organisms, which utilize both within their life history, thus allowing a "best of both worlds" strategy. Optimal energy-intake approaches emphasize the selective benefits of asexual reproduction from the perspective of growth of genetically identical units (Pearse *et al.* 1989). In this way, asexual propagation is not solely a mode of reproduction, but also increases an individual genotype's ability to produce gametes through growth. For example, Sebens (1980, 1982b) developed and empirically tested a model to quantify when organisms (in his case sea anemones) with indeterminate growth should asexually reproduce to maximize fitness. By asexually propagating, organisms can maintain an optimal body size for higher energy intake rates, which will maximize growth, and thus likely fitness. Sebens' model successfully predicted when adults should asexually reproduce in order to maximize energy intake rates and thereby growth rates.

In addition to genetic and growth hypothesis outlined above, other hypotheses have considered the roles of developmental processes for the evolution and retention of clonal reproduction. In a recent review of regeneration in animals, Bely and Nyberg (2010) suggest that the ability to regenerate may be:

- maintained through a selective advantage (e.g., more beneficial to invest in regrowth than to live without lost structure)

- conserved due to pleiotropy of shared developmental processes with embryogenesis
- neutrally retained from phylogenetic inertia.

We suggest that similar hypotheses apply to asexual reproduction because each involves the re-development of morphological features at a post-embryonic stage. Asexual reproduction, like regeneration, also shows a complex evolutionary history as a primitive feature in early animals as well as independently evolving (Delmotte *et al.* 2001) and being lost (e.g., most vertebrates) in lineages. Dynamic evolutionary patterns of asexual reproduction extend to the Cnidaria, where a molecular phylogenetic study indicated that clonal reproduction has been lost and regained multiple times within the sea anemone genus *Anthopleura* (Geller and Walton 2001).

8.4 Trade-offs between methods of asexual reproduction

Like sexual reproduction, initiation of asexual reproduction by an individual is likely dependent on a combination of exogenous signals from the environment and endogenous condition of the individual (Minasian and Mariscal 1979, Clayton 1985, Rodolfo-Metalpa *et al.* 2008). Because many cnidarians have multiple modes of asexual reproduction, the deployment of a particular type is plastic and may be dependent on more specific conditions, which may differ among modes. In certain sea anemones, different modes of asexual reproduction can be employed by single individuals, that is, a single polyp is capable of two forms of transverse fission (Reitzel *et al.* 2007, 2009). Similarly, the jellyfishes *Aurelia aurita* and *Sanderia malayensis* display a variety of clonal reproduction strategies in particular stages of the life cycle (Vagelli 2007, Adler and Jarms 2009).

Fitness advantages for one mode of asexual reproduction over another could be predicted from trade-offs characteristic for each mode of reproduction, although we know of no experimental study that has actually measured relative fitness benefits or costs. One predicted trade-off is a balance between producing many small and few large asexual offspring. A number of cnidarian species with multiple methods for asexual reproduction have modes that produce either small, relatively undifferentiated products (e.g., pedal laceration in *Metridium*, physal pinching in *Nematostella*) or nearly complete individuals (e.g., longitudinal fission in *Metridium*, polarity reversal in *Nematostella*) at the time of separation. The former strategy would be favored in more stable environments with less predation, such that these propagules would have high chance of developing successfully to the adult stage independent of the parent. Conversely, the latter selection would be favored in more variable environments where the separation of a fully developed daughter polyp would increase the probability for a successful clonal offspring. Utilizing one form over another would then optimize growth of a particular individual genotype given the present, yet potentially changing, environmental conditions (Sebens 1982b, 2002).

Utilizing different modes of asexual reproduction may also represent a trade-off for the relative dispersal potential of clonally produced individuals. Just as larval dispersal is influenced by relative maternal investment and pelagic duration (see Chapter 3), products of asexual reproduction may differ substantially in their likelihood for retention or migration from the parent habitat. An extreme case is exemplified by scyphozoan jellyfish, where asexually produced individuals from a parent can be either the pelagic medusa stage, which will have large dispersal potential, or a sessile, budded polyp stage, which will have effectively zero dispersal. More subtle differences in dispersal potential may occur in other species with mixed asexual strategies, like the "many small" or "few large" products discussed above. In these species, the relatively undifferentiated asexual products may disperse in a manner more like embryos due to the lack of attachment structures and overall small size. These individuals may be subjected to passive dispersal from currents, which could result in moderate dispersal distances, like those observed in other species (Wulff 1991, Lirman 2000). For asexual products that are largely differentiated individuals, we would expect no or low realized dispersal from the parent location, although some studies have found that even minimal dispersal can be ecologically significant (Ayre 1983, Wahl 1985, Sherman and Ayre 2008).

8.5 Environmental signals and reception in cnidarian asexual reproduction

A variety of environmental variables influence the timing and frequency of asexual reproduction in cnidarians. Some cues result in divergent responses between species, indicating that common environmental signals have different effects on asexual reproduction, which suggests a divergence of reception and transduction mechanisms. Light–dark cycles (photoperiod, season) have been shown to increase the incidence or pace of asexual reproduction in some species (Purcell 2007, Liu *et al.* 2009) but not others (Rodolfo-Metalpa *et al.* 2008). Increased temperatures have a positive correlation with asexual reproduction for many species (Willcox *et al.* 2007, Rodolfo-Metalpa *et al.* 2008). Asexual reproduction has been reported to increase with high (Minasian and Mariscal 1979, Clayton 1985, Rodolfo-Metalpa *et al.* 2008) and low (Spangenberg 1965, 1967, 1972, Smith and Lenhoff 1976, Sebens 1980) food environments. In addition, physical conditions, including higher-energy wave environments and intertidal habitats, also result in increased asexual reproduction (Francis 1979, Shick *et al.* 1979, Geller *et al.* 2005). Finally, substrate availability and contact with conspecifics can be an apparent inducer for asexual reproduction in species occupying densely-occupied habitats (Ayre 1983, Francis 1988). These signals may not be mutually exclusive (e.g., season, temperature), so that individuals may integrate a combination of signals from the local environment.

Despite the number of observational and experimental studies showing how particular environmental signals influence the frequency of asexual reproduction, there have been few studies elucidating how these signals are "received," how these external signals are transduced to internal signaling gene networks, or the particular molecular mechanisms involved. The lack of data represents a tremendous opportunity for ascertaining how organisms make decisions about initiating asexual reproduction and identifying particular mechanisms. The only specific molecular work we are aware of aimed at identifying particular genes is a study by Geller *et al.* (2005) which reported the results of transcriptome survey of tissues at the fission site for the anemone

Anthopleura elegantissima. Even from this relatively small data set, the authors identified a number of candidate genes (e.g., cell death and proliferation genes, cell signaling receptors) as targets for future study. Two of the best-studied examples for physiological mechanisms to integrate external signals and initiate asexual reproductive decisions are the potential role of iodide in scyphozoan strobilation and the reactive oxygen species (ROS)-induced signaling in colonial hydrozoans.

Particular compounds or chemicals have been shown to either initiate or inhibit asexual reproduction in some cnidarian species. The best characterized of these chemical inducers is the role of iodide as an inducer for a particular mode of transverse fission (strobilation) in scyphozoan polyps. Originally described by Spangenberg (Spangenberg 1965, 1967, 1972) in a series of experiments with *Aurelia*, the addition of iodide to seawater induced initiation of strobilation. The mechanism through which iodide induces strobilation appears to be related to the production of ROS, which oxidize iodide to iodine, which is then incorporated into an unidentified compound for inducing strobilation (Berking *et al.* 2005). The importance of ROS in the initiation of strobilation has also been supported by additional observations by culturing polyps on polystyrene dishes (Stampar *et al.* 2007).

The importance of ROS in mediating the transition to an asexually producing polyp is one example of the potentially larger role that ROS may play in cnidarian developmental and clonal growth strategies (Blackstone 1999, 2001, Blackstone and Bridge 2005). Through chemical manipulation of components of mitochondrial redox signaling, ROS can directly influence the growth strategies of colonial hydroids (Blackstone 2003, 2006, Blackstone *et al.* 2004a,b. Increasing the relative oxidation of the electron transport chain creates "runner-like" colony growth, in which the polyps are spaced widely apart along stolons with few branches. However, reduction in relative oxidation results in another colony morphology, termed "sheet-like" growth, in which polyps are closely spaced along stolons with many branches (Blackstone 2003). These physiological changes may be linked to the expression of specific developmental patterning genes. For example, the cnidarian homeobox gene, Cnox-2, is localized to the tips of elongating stolons, newly forming stolon buds, and developing polyps (Cartwright *et al.* 2006), suggesting a role in patterning stolon growth and colony morphology in *Hydractinia*. Because ROS concentrations are related to food ingestion rates, this mechanism provides a potential link between observations of the role of food environment and asexual reproduction decisions, and potentially to the expression of a transcription factor. It is yet unclear how widespread is the role of ROS as a signaler for asexual reproduction throughout the Cnidaria. The role of ROS in eliciting asexual reproduction in some scyphozoans and hydrozoans and the potential relationship with feeding suggest that future studies in species with diverse modes of asexual reproduction may be informative. In addition, redox state can be easily manipulated with chemicals that target different steps of ROS signaling, so that experimental studies will be comparable across species and asexual reproductive modes.

From the described abiotic and biotic variables that influence the frequency of asexual reproduction, we can hypothesize additional genes or pathways involved in sensing or transducing signals. Molecular mechanisms for detecting and responding to photoperiod and seasonal changes likely involve the circadian clock, a gene network well conserved in bilaterian animal models (Dunlap 1999). Cnidarians appear to have orthologs to many of the molecular components of the circadian clock (Vize 2009), and one component, the cryptochromes (photoreceptive proteins), respond to diurnal photoperiods in corals (Levy *et al.* 2007). For temperature signals, cnidarians have a diverse set of heat shock proteins (Black and Bloom 1984, Bosch *et al.* 1988, Kingsley *et al.* 2003), some of which may be involved in mediating fluctuating environmental signals for initiating reproductive decisions. As mentioned above, metabolic rate is related to the concentration of ROS, which may be a common signal for initiating asexual reproduction in cnidarians (Blackstone and Bridge 2005, Blackstone 2008). A recent survey of *Nematostella's* genome identified a diverse set of genes important in ROS signaling in other animals (Reitzel *et al.* 2008), but their role in cnidarians is unknown. The identification of these genes in combination with experimental evidence for ROS signaling in cnidarian asexual reproduction

may provide productive links for future investigation. Potential genes and molecular mechanisms for shear or physical stress, the fourth environmental signal, have been discussed recently by Blackstone and Bridge (2005). Among the discussed genes are integrins, a group of membrane proteins involved in cell–cell adhesion and signal transduction (Takada *et al.* 2007). Integrins have been identified in two distantly related cnidarians, the hydrozoan *Podocoryne* (Reber-Muller *et al.* 2001) and the anthozoan *Nematostella* (Reitzel *et al.* 2008), suggesting that these genes are present widely in the phylum. Finally, environmental signals for substrate availability and allorecognition of conspecifics may be facilitated by a diverse set of mechanoreceptors (Watson and Hessinger 1989, Watson and Mire 2004) and compatibility loci (Lakkis *et al.* 2008).

8.6 Looking ahead: Combining signaling with developmental mechanisms

Developmental processes are the observable output from networks of developmental regulatory genes. From a mechanistic standpoint, the gene networks themselves and the specific environmental signals, and not the observable phenotypic processes, are the relevant units. The module hypothesis implies that conserved suites of genes will underlie homologous developmental processes regardless of cellular context. The validity and genetic make-up of provisional modules can be tested empirically by studying the genes that underlie the relevant developmental processes in different spatial and temporal contexts within species, and in homologous pathways and structures between species. If conserved suites of coordinately regulated genes ("synexpression groups"; Niehrs 2004) are found to underlie the same process, then the mechanistic basis of a developmental module has been identified. Efforts can then be directed towards understanding the changing deployment of modules during evolution.

The availability of genomic and larger-scale transcriptional profiling techniques is rapidly expanding to traditionally non-model organisms (Wilson *et al.* 2005, Dupont *et al.* 2007, Travers *et al.* 2007). Within the Cnidaria there are currently two species with completely sequenced genomes (*Nematostella,*

Hydra) and a handful of other species with transcriptome data sets. These genomic datasets have provided the necessary data for a variety of studies searching for cnidarian orthologs to genes important in developmental and ecological responses of triploblast species, like vertebrates and insects (Kortschak *et al.* 2003, Reitzel *et al.* 2008, Schwarz *et al.* 2008, Rosenstiel *et al.* 2009). In addition, a number of studies using cnidarians have successfully used small- and moderate-scale microarrays to assay changes in transcriptional profiles in response to ecological variables and during development (Morgan and Snell 2002, Desalvo *et al.* 2008, Grasso *et al.* 2008). The extension of these combined approaches of gene identification and transcriptional profiling will result in dramatic advances in identifying genes involved in the mechanisms of asexual reproduction.

In addition, and importantly, elucidating the intersection of environmental signals with activation of these gene networks will provide a synthetic understanding of how organisms make particular developmental "decisions" in nature and how these are initiated at the molecular level. The link between environmental signaling and initiation of asexual reproduction is almost completely unknown and needs to be approached to understand the mechanisms for how organisms interpret and respond to their environment. The emergence of the interdisciplinary field of ecological and evolutionary developmental biology (Sultan 2007, Gilbert and Epel 2009) provides a firm conceptual basis for future studies linking known inducers of asexual reproduction with the underlying developmental and genetic mechanisms. The confluence of genetic tools with the diverse asexual reproductive modes in the Cnidaria will provide the tools and experimental systems to address the conservation and divergence of mechanisms for asexual reproduction.

8.7 Summary

1. Cnidarian life histories typically include one or more distinct modes of asexual reproduction. In many cases, asexual reproduction represents the developmental mechanism for producing the diversity of morphologies stemming from a relatively simple body plan, particularly in colonial species.

2. Organismal-level comparisons of the developmental pathways among modes of asexual reproduction in cnidarian polyps suggest conserved morphological components forming discrete modules. The diversity of modes of clonal reproduction can be explained in large part by a reordering of the spatial and temporal expression of these modules.

3. Molecular mechanisms underlying asexual reproduction in the Cnidaria, and most other taxa, are poorly understood or completely unknown. A recent surge in gene expression data for transcription factors and other developmental regulatory genes during embryogenesis from a handful of species provides a set of candidate genes for exploring in the context of asexual reproduction. The Wnt-signaling pathway appears to be the most conserved axial patterning mechanism across sampled cnidarians, and thus a promising avenue for future research.

4. Ecological factors for initiating asexual reproduction among species, including abiotic and biotic cues, appear to be diverse. Genomic approaches comparing cnidarians and bilaterians suggest a suite of conserved genes and signaling pathways that may be involved in environmental sensing. The expression of these genes and potential role in asexual reproduction await experimental characterization. Additionally, we lack any experimental data comparing potential fitness consequences of the diverse modes of asexual reproduction in laboratory or natural settings.

5. New advances in transcriptional profiling will result in the identification of genes expressed during different stages of asexual reproduction, thereby providing candidate markers for comparing modes of asexual reproduction within and between species.

8.8 Acknowledgments

We would like to acknowledge Dr Jonathan Geller for discussions of the developmental modularity of asexual reproduction in cnidarians. During the writing of this chapter, AMR was supported by a postdoctoral scholar program at the Woods Hole Oceanographic Institution, with funding provided by The Beacon Institute for Rivers and Estuaries, and the J. Seward Johnson Fund.

CHAPTER 9

The genetics and evolution of flowering time variation in plants: Identifying genes that control a key life history transition

Joshua A. Banta and Michael D. Purugganan

9.1 Introduction

The study of life history evolution has traditionally focused on animals (Stearns 1992, Roff 2002). This is due to historical factors, such as the fact that the pioneers in the field happened to be zoologists, and probably also due to logistics, such as the fact that vegetative reproduction in plants makes "individuals" harder to identify and that modular growth complicates the measurement of life history traits (Vourisalo and Muitkainen 1999). Whatever the reason, the study of life history evolution and ecology in animals is relatively advanced compared to that in plants, although this gap is narrowing (Vourisalo and Muitkainen 1999).

Plants are particularly interesting organisms for the study of life history evolution because they are sessile and must match their phenotype and phenology to their local environment. The developmental transition to flowering in angiosperms illustrates this challenge. Flowering is analogous to crucial life history transitions in animals, like metamorphosis and diapause (see Chapters 2, 3, 5, 6, 7, 18, and 22), in that flowering is a discrete and costly morphological transition to another stage of the life cycle. Plants must carefully integrate genetic and environmental signals before committing themselves to flower. Plants that flower too soon may lack sufficient material resources or favorable environmental conditions to complete reproduction (Elzinga et al. 2007). Plants that flower too late may be out-com-

peted by their earlier-flowering neighbors (Schmitt and Wulff 1993) or may miss out on the window of favorable environmental conditions for their offspring (Donohue et al. 2005). Plants that flower at the "wrong" time may be out of sync with their pollinators (Knight et al. 2005), have more inbred offspring (Kitamoto et al. 2006), or be more susceptible to herbivores and pathogens (Elzinga et al. 2007).

Flowers are a key innovation in evolution, and the differences in morphology, growth, and life history among angiosperms are so vast that they astonished Darwin (Darwin 1859). The timing of flowering, both within and among species, varies widely in nature (Rathcke and Lacey 2003) and is determined by genetic and environmental factors, and by their interaction (Rathcke and Lacey 2003, Putterill et al. 2004, Engelmann and Purugganan 2006). Angiosperm life histories include both iteroparity ("polycarps," also known as "perennials") and semelparity ("monocarps"). Some monocarps may complete their entire reproductive cycle within one year ("annuals") while others may accrue resources for one, or several, years before flowering ("biennials"). Different populations of the same species may have different life histories (e.g., Koch et al. 1999). Within populations, flowering time in nature is often found to be under strong local selection (Elzinga et al. 2007).

The ecological and evolutionary significance of flowering time, and the ample variation commonly observed within and among species, has led to growing interest over the last two decades in elucidating the genetic network responsible for natural variation in flowering phenology (Mitchell-Olds and Schmitt 2006). As Stinchcombe and Hoekstra (2008) point out, identifying the genes accounting for ecologically important trait variations, such as life history variation, is the first step in answering a host of longstanding ecological and evolutionary questions. Identifying the genes controlling natural trait variation can also lead to wholly unanticipated advances in evolutionary theory (e.g., Hittinger *et al.* 2010).

Much of the interest in studying the mechanistic basis of flowering time variation has focused on the model species *Arabidopsis thaliana*. In this chapter, we describe some of the genes identified for this key life history trait. We then illustrate how this molecular information can be used to better understand the evolution of flowering time in nature. We then look at how the genetics of flowering time variation in *A. thaliana* translates to other species, both those closely related and those more anciently diverged. The picture that emerges is that many of the genes in the flowering time genetic network are conserved over much of angiosperm evolution, and some genes accounting for flowering time variation in *A. thaliana* are implicated in other species as well.

9.2 The natural and laboratory history of *Arabidopsis*

Arabidopsis thaliana (L.) Heynh. (Brassicaceae) is uniquely well suited for the molecular characterization of flowering time variation. With a long history as a research organism tracing back over 60 years (summarized in Leonelli 2007), it is already one of the best-studied organisms on a molecular genetic level. Moreover, large numbers of "immortal" inbred *A. thaliana* accessions are maintained and freely available, and can be, or have been, genotyped once and then phenotyped repeatedly. *A. thaliana* has a wide natural range throughout Eurasia (Hoffmann 2002) and displays extensive phenotypic variation and a wide niche breadth (Nordborg and Bergelson 1999, Mitchell-Olds and Schmitt 2006). Connecting the molecular level to the phenotypes of *A. thaliana* has thus been a major goal of both geneticists and evolutionary biologists (Mitchell-Olds and Schmitt 2006, Wilczek *et al.* 2009).

A. thaliana is a relatively short-lived, highly selfing (Abbott and Gomes 1989), and annual plant found in ruderal habitats (Napp-Zinn 1985). It is characterized by a small size and rapid growth—it is able to complete its life cycle in less than six weeks depending on the genetic background and environmental conditions. Moreover, *A. thaliana* has a low outcrossing rate (approximately 1%; Hoffmann *et al.* 2003), which allows one to easily collect genetically uniform seeds and grow multiple genetically uniform replicates. *A. thaliana* is estimated to have diverged from other *Arabidopsis* species 5–6 million years ago, and ranges across Eurasia and North Africa as well as being widely introduced in North America and Japan (Hoffmann 2002).

A. thaliana has an apically dominant type of above-ground architecture, resulting in a rosette shoot. The shoot apical meristem initially produces rosette leaves until, elicited by endogenous genetic factors and exogenous environmental factors, it is converted to a reproductive meristem that produces the inflorescence and the elongation of the main shoot (bolting). Bolting occurs a few days prior to flowering, which is quickly followed by seed development and fruit elongation. While the time from germination to flowering can be quite rapid, there is considerable genetic diversity, which interacts with environmental variation in natural populations to produce a wide range of times until initiation of the reproductive phase (Nordborg and Bergelson 1999, Wilczek *et al.* 2009).

This wide range results in a multimodal distribution of flowering times in this species, which contributes to life history differences among populations (Fig. 9-1; Nordborg and Bergelson 1999, Weinig and Schmitt 2004, Elzinga *et al.* 2007, Wilczek *et al.* 2009). Winter annuals, for example, germinate in the fall, overwinter as rosettes, where they experience cold temperatures and short day lengths, and flower in early spring (Nordborg and Bergelson 1999). Rapid cyclers germinate in the early autumn, spring, or summer and quickly grow to maturity, flowering and setting seed prior to the onset of winter, thus flowering without rosette vernalization (Engelmann and Purugganan 2006, Wilczek *et al.* 2009). The life history adopted by *A. thaliana* varies with latitude, which is presumably driven by variation in climate. At higher latitudes and altitudes, plants have either

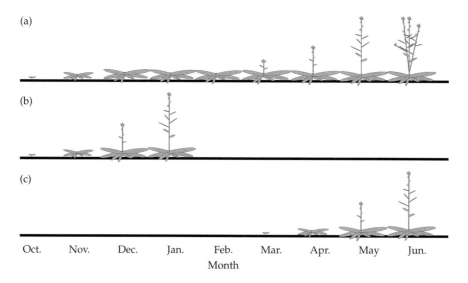

Figure 9-1 Life-history strategies of *Arabidopsis thaliana* in the wild: (a) winter annual; (b) rapid cycling—fall germination; (c) rapid-cycling—spring germination. Adapted from Weinig and Schmitt 2004.

a rapid-cycling or winter annual life history (Fig. 9-1b and c), whereas at low latitudes plants tend to have a winter annual life history (Fig. 9-1a, Nordborg and Bergelson 1999, Weinig and Schmitt 2004, Wilczek *et al.* 2009). Winter annual plants are often observed to flower at a later developmental stage, with a larger number of rosette leaves, and to have higher fitness as compared to rapid-cycling plants (Weinig and Schmitt 2004, Engelmann and Purugganan 2006), suggesting that a winter annual life history strategy involves fewer generations and more progeny per generation, whereas a rapid-cycling strategy involves more generations and fewer progeny per generation. See Stearns (1992) for more information on the general phenomenon of generation time versus brood size.

9.3 The molecular genetics of flowering time

There are extensive data on the molecular genetics of flowering time in *A. thaliana*, mostly obtained through forward genetic screens (i.e., identifying genes through the phenotypic and downstream molecular effects of laboratory-generated knockout mutations). Flowering time in this species is determined in part by expression of genes in at least four main genetic pathways (Komeda 2004, Fig. 9-2).

The autonomous pathway channels signals to *FLOWERING LOCUS C* (*FLC*), which represses flowering (Michaels and Amasino 1999). These signals come from genes such as *FCA* and *LUMINIDEPENDENS* (*LD*), which suppress *FLC* and thereby promote flowering (Koornneef *et al.* 1991). Conversely, *FRIGIDA* (*FRI*) and the *FRIGIDA LIKE* genes (*FRL1*, *FRL2*), which are not part of the autonomous pathway, are positive regulators of *FLC*, which delay flowering (Johanson *et al.* 2000, Michaels *et al.* 2004, Schläppi *et al.* 2006).

The vernalization pathway responds to temperature cues to repress *FLC* through genes such as *VERNALIZATION 1* (*VRN1*) and *VERNALIZATION 2* (*VRN2*), thereby promoting flowering in a temperature regime-dependent manner (Chandler *et al.* 1996). The light-dependent pathway channels signals to *CONSTANS* (*CO*), which promotes flowering (Suarez-Lopez *et al.* 2001). These signals come from phytochromes (e.g., *PHYTOCHROME A; PHYA*) and cryptochromes (*CRYPTOCHROME 1* and *CRYPTOCHROME 2; CRY1*, and *CRY2*), which respond to enhanced light quality by repressing *CO* activity and thereby delaying flowering (Levy and Dean 1998), and also from circadian clock genes such as *FLAVIN-BINDING KELCH REPEAT F-BOX 1* (*FKF1*), which respond to enhanced light quantity by degrading the *CYCLING DOF FACTOR*

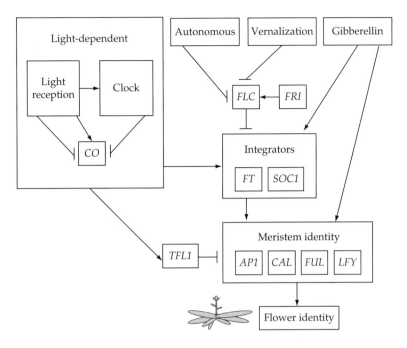

Figure 9-2   Summarized version of the known flowering time genetic network in *Arabidopsis thaliana*.

1 (CDF1) protein, a transcriptional repressor of *CO*, thereby promoting flowering. Finally, there is a flowering pathway that encompasses genes in the biosynthesis and signaling of the plant hormone gibberellic acid, which promotes flowering (Dill and Sun 2001).

The four flowering pathways channel into a set of floral integrator genes, which receive the upstream signals and integrate them into the decision to flower by activating downstream genes that engage the floral morphogenic process (Komeda 2004). Two of these genes in particular—*SUPPRESSOR OF OVEREXPRESSION OF CONSTANS (SOC1)* and *FLOWERING LOCUS T (FT)*—serve as the crucial floral switches initiating the flowering developmental program (Caicedo *et al.* 2009).

9.3.1   Getting at the mechanistic basis: Genes controlling flowering time variation and what they do

While there is a wealth of information available about how flowering time in *A. thaliana* is determined on a molecular genetic level (see Komeda

(2004) for a review), we know much less about which molecular polymorphisms account for *natural variation* in flowering time. Nevertheless, substantial progress has been made in identifying the allelic variants at flowering time genes that lead to standing genetic variation (Alonso-Blanco *et al.* 2009, Ehrenreich *et al.* 2009). Despite the identification of many promising candidate genes, however, the list of genes that have been confirmed to control natural variation in flowering time is quite small (Table 9-1). This is probably because follow-up studies on promising genes identified by mapping methods take a significant amount of work. Smaller yet is the list of genes for which ecologically significant polymorphisms have been described, as opposed to polymorphisms that have only been described under artificial laboratory conditions. In this section, we will present a few of the genes controlling variation in flowering time that have been isolated thus far, together with what is known about their effects on flowering time in natural populations under natural conditions. Furthermore, we will illustrate how these genes can have ecologically significant pleiotropic effects on other traits, even

Table 9-1 Summary of the genes that have been confirmed to account for natural variation in flowering time in *Arabidopsis thaliana*, and the molecular characteristics of the natural alleles implicated in the definitive studies.

Gene	Molecular function	Functional polymorphism	Functional alteration	References
CRY2	Photoreceptor	SNP	AA substitution	El-Assal *et al.* 2001
FLC	MADS-box transcription factor	INDELs	Expression level	Gazzani *et al.* 2003 Michaels *et al.* 2003
FLM	MADS-box transcription factor	INDEL	Deleted gene	Werner *et al.* 2005
FRI	Post-transcriptional regulator (?)	INDELs	Truncated protein	Kowalski *et al.* 1994 Gazzani *et al.* 2003 Johanson *et al.* 2000
FRL1	Post-transcriptional regulator (?)	SNP	AA substitution	Schläppi *et al.* 2006 Michaels *et al.* 2004
FRL2	Post-transcriptional regulator (?)	SNP	AA substitution	Schläppi *et al.* 2006 Michaels *et al.* 2004
FT	Phosphatidylethanolamine binding/protein binding	*cis*-regulatory element	Expression level	Swartz *et al.* 2009
HUA2	RNA processing	INDEL	Truncated protein	Doyle *et al.* 2005
		SNP	AA substitution	Wang *et al.* 2007
PHYC	Photoreceptor	SNP	AA substitution	Balasubramanian *et al.* 2006
PHYD	Photoreceptor	INDEL	Truncated protein	Aukerman *et al.* 1997

Adapted from Koornneef *et al.* 2004 and updated.

with seemingly little, if any, physiological or biochemical connection to flowering.

9.3.2 *CRY2*

One of the first genes controlling natural variation in flowering to be cloned was *CRY2*, in the light-dependent flowering pathway. It had been shown previously by quantitative trait locus (QTL) analysis between individuals from Germany (*Ler*) and the Cape Verde Islands (*Cvi*) that one locus at the top of chromosome 1 was controlling the timing of flowering under short-day but not long-day photoperiods (named *EDI*;Alonso-Blanco *et al.* 1998). A map-based positional cloning strategy determined that *CRY2* was a likely candidate gene within this region, and a transgenic approach revealed that this gene was indeed responsible for the *EDI* QTL (El-Assal *et al.* 2001). It turns out, however, that a rare allele of *CRY2* (*CRY2EDI*), found exclusively in the *Cvi* ecotype of *A. thaliana*, is responsible for this observed phenotype in this accession.

Following up on this work, Olsen *et al.* (2004) studied a large set of naturally occurring accessions and identified two other haplotypes in the *CRY2* genomic region that are associated with flowering time variation under short-day photoperiods (although, interestingly, the way in which the haplotypes effect flowering depends on whether the

plants were grown in a growth chamber or with overwintering conditions in the field; Fig. 9-3). These haplotypes, unlike the one containing the *CRY2EDI* allele, were found to have a widespread distribution and therefore may play an important role in controlling natural variation in flowering time in this species. This is supported by the fact that one allele they identified, *HAP AS*, is significantly more prevalent in environments with colder mean January temperatures. Furthermore, *HAP AS* is particularly interesting because one of its amino acid replacements is a radical change that is otherwise conserved across vascular plants.

The haplotypes identified by Olsen *et al.* (2004) appear to underlie QTL of much smaller effect than those of *CRY2EDI*, even though they may be more important in accounting for patterns in the field. This shows that effect size in any laboratory study should not be taken by itself as evidence of the importance a particular locus in nature and that, conversely, a locus with a small effect size should not be deemed unimportant for controlling natural variation. A locus with a small effect size could play an important role in controlling natural variation in a trait if the alleles are each at high enough frequency throughout the natural landscape. It is also important to note that the effects of the *CRY2* polymorphisms in all of the studies were found only under short day lengths. This illustrates the

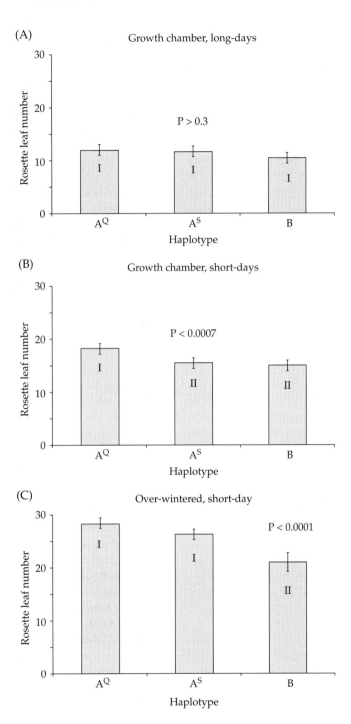

Figure 9-3 Associations between *CRY2* haplogroups and rosette leaf number at bolting, a conservative estimate of flowering time. A, long-day photoperiod, growth chamber; B, short-day photoperiod, growth chamber; C, short-day (overwintering) photoperiod, field. *P* values indicate significance in one-way analyses of variance. Bars that share roman numerals are not significantly different from each other at *P* = 0.05. From Olsen *et al.* 2004, with permission from the Genetics Society of America.

importance of studying life history variation, such as flowering time, under a variety of conditions; the effect of *CRY2* would have been invisible under more standard laboratory conditions with long (indeed, sometimes 24-h) photoperiods.

9.3.3 *PHYC*

Evidence has been accumulating that genes from the light-dependent pathway are major factors in natural variation of flowering time in *A. thaliana* (Aukerman *et al.* 1997, El-Assal *et al.* 2001, Olsen *et al.* 2004). Adding to this evidence, Balasubramanian *et al.* (2006) found that naturally occurring variation at *PHYTOCHROME C* (*PHYC*) affects flowering time in F_2 crosses under short-day photoperiods. *PHYC* has opposing effects on flowering under long and short days, with functional *PHYC* alleles inhibiting flowering under short days and promoting flowering under long days (Monte *et al.* 2003).

Balasubramanian *et al.* (2006) identified two common haplotypes of *PHYC* that occur in natural populations and that differ in gene expression levels and phytochrome activity levels; a QTL region containing *PHYC* was controlling flowering time variation in F_2 crosses. By analysing the patterns of gene expression within the QTL region, it was shown that *PHYC*, rather than flanking regions/genes, is controlling flowering time. The phenotypic effect appears to be due to a nonsense polymorphism in the first exon, which converts the Lys299 codon to a stop codon. Ecologically, the authors found significant latitudinal differentiation among the *PHYC* haplotypes, above what could be explained by neutral processes alone, such that the more *PHYC*-active haplotype was more common at low latitudes, and the less-active haplotype was more common at high latitudes, suggesting the action of clinal selection on *PHYC*.

Based on the fact that genes in the light-dependent pathway sense photoperiod (among other aspects of the light environment), and transmit this information through the genetic pathways leading to flowering (Fig. 9-2), Samis *et al.* (2008) tested the hypothesis that the natural *PHYC* polymorphism identified by Balasubramanian *et al.* (2006) also affects the *plasticity* of flowering time to photoperiod. They used a large set of naturally occurring accessions from Eurasia, with a structured associa-

tion-mapping approach targeted at *PHYC*. They found that, indeed, the *PHYC* polymorphism affects flowering time plasticity, although the situation is complicated by other unknown loci that change the effect of *PHYC* on flowering time plasticity, depending on the longitude of collection of the plants. They also found a similar contingent effect of *PHYC* on flowering under short days, suggesting that there are also unknown loci modulating the effect of *PHYC* on flowering time itself, and which have a clinal distribution. Samis *et al.* (2008) make the case that these apparent, but as of yet uncharacterized, epistatic interactions make sense ecologically because the information conveyed by the photoperiod changes depending on where the plant is geographically, and thus the response elicited by *PHYC* should be different in different areas.

9.3.4 *FRI*

The floral repressor gene *FRIGIDA* (*FRI*) was the first flowering time gene to be cloned. It has long been known that the vernalization-mediated induction of flowering was controlled by *FRI* (Burn *et al.* 1993), and that individuals with a dominant *FRI* allele would only flower after a prolonged cold period at the rosette stage, whereas individuals with the recessive *fri* allele would flower early regardless of cold treatment. Johanson *et al.* (2000) used map-based positional cloning to characterize the gene. They also analyzed a large set of early flowering ecotypes and found that most of them contained one of two different deletions within *FRI*, which disrupt the open reading frame, and which evolved independently. While the molecular details of *FRI* function are not well understood, recent work suggests that *FRI* binds to the nuclear cap binding complex, and thus increases the transcript abundance (and possibly splicing) of another important flowering time gene, *FLOWERING LOCUS C* (see below) (Geraldo *et al.* 2009).

The functional importance of *FRI* was supported by Le Corre *et al.* (2002), who sequenced 25 ecotypes of *A. thaliana* from Western Europe and found evidence that the *FRI* deletions arose due to recent positive selection for early flowering (perhaps due to the spread of agricultural and other man-made disturbances), and also by Stinchcombe *et al.* (2004), who found evidence of a latitudinal cline in flowering time, but only in accessions with a functional *FRI* allele. This was followed up by Toomajian *et al.* (2006), who studied more than 1000 short DNA fragments throughout the *A. thaliana* genome, in 96 different accessions. They determined that the linkage disequilibrium surrounding early-flowering (nonfunctional) *FRI* alleles was significantly larger than those observed in the rest of the genome, and it is likely these *FRI* alleles have increased in frequency due to recent positive selection. They estimate the age of the proliferation of these alleles at less than 13,000 years, suggesting a possible link of selection in this weedy species to the spread of human colonization and agriculture in Eurasia. Adding even further weight to the importance of *FRI* is the fact that it is found to explain a large proportion of the natural variation in flowering time in *A. thaliana* (between 12.6% and 70%, depending on the study; Shindo *et al.* 2005, Scarcelli *et al.* 2007). Interestingly, since the *FRI* deletions arose multiple times independently, and since it is so important for accounting for flowering time variation, it is likely that parallel evolution of flowering time occurred in *A. thaliana*; that is, the same locus evolved in response to selection in multiple independent instances (see also Chapters 2, 3, 4, and 8 for similar examples of independent evolution). Thus *FRI* appears to be an important locus of evolutionary change in this species.

9.3.5 *FLC*

It turns out the *FRI* story is intertwined with that of another gene, *FLOWERING LOCUS C* (*FLC*) (see the next subsection for details about this interaction). As with *FRI*, *FLC* was first characterized using map-based positional cloning (Michaels and Amasino 1999). It encodes a MADS-box protein (a family of anciently conserved transcription factors regulating flowering) that represses the flowering time integrator genes (Michaels and Amasino 1999). Michaels and Amasino (1999) found that the loss of *FLC* functionality resulted in early flowering, whereas ectopic expression of *FLC* delayed, or arrested, the developmental transition to flowering. Furthermore, the amount of *FLC* transcript was subsequently found to be correlated with time to flowering in mutants from several different genetic backgrounds (Sheldon

et al. 2000) and in natural populations (Michaels *et al.* 2003). The ecological importance of *FLC*-mediated early flowering is also supported by the independent evolution of different early-flowering *FLC* alleles (Gazzani *et al.* 2003, Michaels *et al.* 2003). Thus the picture painted by the studies initially suggested an ecologically important and additive effect of *FLC* transcript on flowering in nature.

In addition to the importance of *FLC* itself, recent work suggests that closely related homologues of this gene also account for natural variation in flowering time in *A. thaliana*. Caicedo *et al.* (2009) studied a set of four genes from within the same monophyletic clade of MADS-box genes as *FLC* and *FLM*, known as the *MAF* MADS-box family of genes. Because of their sequence similarity to *FLC* and *FLM*, Caicedo *et al.* (2009) hypothesized that other *MAF* genes would also be involved in controlling flowering time variation in nature. They found evidence that this is indeed the case: using a panel of 169 natural accessions with a structured association mapping approach, they found significant association between *MAF2* and *MAF3* polymorphisms and flowering time. These genes had high levels of nonsynonymous single nucleotide polymorphisms, insertion-deletions, and rearrangements, as well as novel gene fusions that persist as moderate-frequency polymorphisms. The fused genes involve *MAF2* and portions of *MAF3*, and result in chimeric, alternately spliced transcripts. The story involving the *MAF2* and *MAF3* genes, while requiring further confirmation of a causal link to natural variation, illustrates that clusters of genes related to known genes, already established as accounting for flowering time variation, are a good place to focus efforts aimed at finding more such loci.

9.4 Epistatic effects among *FRI* and *FLC*

When natural variation at the *FLC* locus was examined in conjunction with variation at *FRI*, it became apparent that, rather than having straightforward additive effects, *FRI* and *FLC* interact non-additively (epistatically, in the quantitative genetic sense) with each other. Caicedo *et al.* (2004) studied several hundred naturally occurring accessions of *A. thaliana* and documented two major *FLC* haplogroups that are associated with flowering time variation. This polymorphism, however, only affected flowering time in the presence of a functional *FRI* allele. The overall finding of *FRI–FLC* epistasis was later supported by Korves *et al.* (2007) and Scarcelli *et al.* (2007), although the latter two studies both found that the *FLC*[A] allele confers later flowering, whereas Caicedo *et al.* (2004) found that it confers earlier flowering, relative to the alternative *FLC*[B] allele. Furthermore, while Caicedo *et al.* (2004) found non-additive *FRI–FLC* effects only under environmental conditions experienced by winter annuals, and not under conditions experienced by spring annuals, Scarcelli *et al.* (2007) found the opposite.

Regardless of which *FLC* allele confers later flowering, and which specific environments elicit that effect, the repeated finding that *FLC* has an effect on natural flowering time variation, and that its effects are contingent on a functional *FRI* background and on the environmental circumstances, is corroborated by other sources of evidence. First of all, Caicedo *et al.* (2004) found that the *FLC* haplogroups differed in their latitudinal distributions, indicative of selection for different *FLC* haplogroups at different latitudes, but only in the presence of a functional *FRI* allele. Secondly, they found population genetic evidence that selection acts on particular *FRI–FLC* genotype combinations: the loci were in strong linkage disequilibrium, despite the fact that they occur on different chromosomes.

The ecological and microevolutionary context of the *FRI–FLC* interaction was put into sharper focus by Korves *et al.* (2007). Using a collection of 136 European natural accessions of *A. thaliana*, Korves *et al.* (2007) documented that *FRI* is associated with fitness variation under field conditions, but that which type of *FRI* allele (functional or non-functional) is favored depends upon the seasonal environment and the genotype at the *FLC* locus. Specifically, they found that accessions with functional *FRI* alleles had higher winter survival in one *FLC* background in a fall-germinating cohort, but that accessions with non-functional *FRI* alleles had greater seed production in the other *FLC* background in a spring-germinating cohort. Overall, their study suggests that seasonally varying selection and epistasis between *FRI* and *FLC* could explain the maintenance of variation at *FRI* and, more generally, may be important in the evolution

of genes underlying complex traits such as life history (see Chapter 27 for more examples).

9.5 Pleiotropic effects of genes controlling flowering time variation

Trade-offs are an important aspect of life history theory (Stearns 1989). They are represented as the fitness cost of increasing a beneficial trait at the expense of decreasing another beneficial trait (Stearns 1989, 1992, Roff 2002). Life history transitions involve energetically costly decisions regarding reproduction, that is, strategically postponing, hastening, or changing the form of it. Life history theory postulates that reproduction is limited by trade-offs, since traits correlated with fitness seem to be below the limits they could achieve given organisms' designs and the laws of physics (Stearns 1989). But the relationship between molecular machinery and trade-offs is not straightforward (Stearns 1992). To determine the mechanistic scenario governing a trade-off, one must understand which particular genes, and which particular polymorphisms in those genes, are causing the observed genetic correlations. As Roff points out in Chapter 2, trade-offs at a quantitative genetic level can be caused by many different phenomena that can only be distinguished from each other by the use of molecular genetic methods. If a trade-off between flowering time and another trait is caused by antagonistic pleiotropy, the trade-off may be harder to escape than one caused by more transient phenomena such as linkage disequilibrium. Thus the molecular mechanism accounting for the trade-off may affect how constrained the evolution of the trait is.

There is evidence that polymorphisms affecting natural flowering time variation also have pleiotropic effects. *FRI* and *FLC* polymorphisms are implicated in controlling nitrogen content, inflorescence architecture, and water use efficiency (Loudet *et al.* 2003, McKay *et al.* 2003, 2008, Scarcelli *et al.* 2007), *FLC* polymorphisms were found to be involved in circadian leaf movements and seed germination (Swarup *et al.* 1999; Chiang *et al.* 2009), and *CRY2* polymorphisms have been shown to affect fruit length, the number of ovules per fruit, and the percentage of unfertilized ovules (El-Assal *et al.* 2004).

Only one study, to our knowledge, has explicitly studied the role of *antagonistic* pleiotropic effects—opposite effects of different traits on fitness (Roff 2002)—involving flowering time and other traits. Scarcelli *et al.* (2007) used an outbred population of *A. thaliana*, derived from the intermating of 19 natural accessions from throughout the species' natural range, and confirmed the findings, mentioned previously, that *FRI* accounts for variation in flowering time and that it interacts epistatically with *FLC*. Yet, interestingly, they found no association between *FRI* and fitness, under either spring or fall simulated field conditions. A path analytical approach proved useful in understanding the source of this problem: the authors found that earlier-flowering plants had higher fitness, and that non-functional alleles at *FRI* were associated with early flowering, but that variation at *FRI* was not associated with fitness, regardless of the simulated field conditions. Instead, the nonfunctional *FRI* alleles had negative pleiotropic effects on fitness, by reducing the numbers and nodes of branches on the inflorescence (Scarcelli *et al.* 2007; Fig. 9-4). Thus, Scarcelli *et al.* (2007) propose that these antagonistic pleiotropic effects reduce the adaptive value of *FRI* and help to explain the maintenance of alternative life history strategies of *A. thaliana* populations in nature.

Not all pleiotropic effects relevant to life history evolution are necessarily antagonistic, however. Chiang *et al.* (2009) documented what appears to be adaptive pleiotropy between flowering time and another important life history trait. They used several lines that were near-isogenic for natural alleles at *FLC* in different genetic backgrounds to test whether natural variation at the *FLC* locus is associated with natural variation in temperature-dependent germination as well. Their hypothesis was based on the fact that germination and flowering are both fundamental life history stages that require precise environmental sensing and responses to multiple seasonal cues to accurately match developmental timing to appropriate seasonal conditions, and that they respond to similar seasonal cues. Therefore, they reasoned, it seems plausible that the genetic pathways partially overlap. Furthermore, they pointed out that the timing of germination appears to be a stronger factor in determining the flowering time of *A. thaliana* in the field than the effects of genes acting directly on flowering time itself (Wilczek *et al.* 2009).

Chiang *et al.* (2009) found that the natural *FLC* polymorphisms in the near-isogenic lines did, in fact, account for germination variation, under both laboratory and field conditions. Further corroboration of this finding came from the fact that transgenic lines overexpressing *FLC* had much higher germination than the background control line, and that *FLC* expression was significantly associated with increased germination in 52 natural accessions of *A. thaliana*, but only at low temperatures. Based

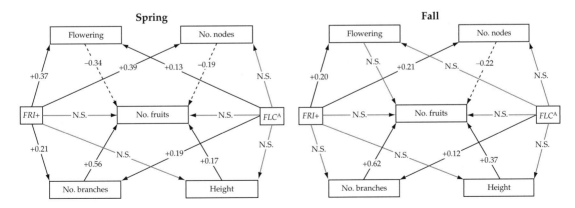

Figure 9-4 Path coefficients (standardized partial regression coefficients) from *FRI* and *FLC* to flowering time, inflorescence traits, and fruit production. Solid lines indicate a path was significantly positive and dashed lines indicate it was significantly negative. Gray lines indicate a path was not significant. (Adapted from Scarcelli *et al.* 2007)

on the role of *FLC* in germination that they documented, Chiang *et al.* (2009) believe that a reevaluation of the adaptive significance of natural variation at *FLC* is required. They argue that selection on *FLC* and other flowering time genes may reflect a history of selection on germination as much as, if not more than, selection on flowering. Furthermore, they make the case that *FLC*-mediated selection on germination and flowering time are not necessarily at odds with one another; the natural accessions with *FLC* alleles conferring high *FLC* expression tend to originate from higher latitudes, where later flowering, conferred by higher *FLC* expression, is favored and where higher *FLC* expression will increase seed germinability.

Another example of putatively adaptive pleiotropy in *A. thaliana* involving flowering time and another trait comes from two studies by McKay *et al.* (2003, 2008). These studies show how pleiotropic effects of natural alleles can be quite surprising, tying together traits that might superficially seem to be unrelated to each other on a developmental or physiological level. They tested the hypothesis that an observed correlation involving flowering time and water-use efficiency was due, at least in part, to molecular genetic pleiotropy. In the first of their two studies, McKay *et al.* (2003) found a positive genetic correlation between flowering time and ^{13}C, a proxy for water-use efficiency, in 39 natural accessions originating from locations spanning a wide range of climactic conditions. Water-use efficiency is achieved in plants primarily by closing the stomata, which slows the rate of carbon assimilation as a cost and consequence of sparing water evapotranspiration. Using lines that were near-isogenic for natural polymorphisms at *FRI* and *FLC*, they found that these two flowering time genes appear to account for the genetic correlation Specifically, the *FRI* and *FLC* alleles that increased the interval until flowering also increased water-use efficiency. More evidence for these pleiotropic effects came from McKay *et al.* (2008), where the authors created a recombinant inbred line population created from two naturally collected accessions previously found to represent different ends of the range of water-use efficiencies observed in *A. thaliana*. The two QTL they detected for water-use efficiency mapped near the locations of *FRI* and *FLC*. Mckay *et al.* (2003) claim that this

pleiotropic effect is adaptive because plants that flower early "escape drought" by flowering before desiccation and withering becomes a major problem, whereas plants that flower later must endure dry summer conditions before flowering; the pleiotropic effects of *FRI* and *FLC* allow plants that are flowering early to avoid paying the cost of being drought-tolerant (decreased growth rate), while allowing plants that flower later to endure the dry conditions they will experience.

Rather than representing an interesting, but isolated and idiosyncratic, example of the intersection of two genetic pathways, McKay *et al.* (2008) make the case that flowering time genes are rife with pleiotropic effects on other traits. This could also explain the pleiotropy observed in the study by Chiang *et al.* (2009), above. Their reasoning is as follows: the work on flowering time genetics was originally based on a model where loci involved in flowering were envisioned as flowering time switches, whose functions were to turn flowering on or off. This metaphor is problematic, however, because, as they point out, the reality is that flowering time is a very "downstream" trait, and therefore any polymorphisms that affect the ability to sense the environment, or to acquire and shunt resources, may affect flowering time. In other words, far from being highly specified "floral switches," flowering time genes may generally be involved in very basic metabolic processes that affect flowering time as well as a host of other traits.

9.6 Comparative functional genomics: The genetics of flowering time in other species

While this chapter has focused on the elucidation and characterization of genes controlling flowering time in *Arabidopsis thaliana*, parallel work, similar to that described here, has been going on with a number of different species, although the knowledge of the genetic architecture is much less extensive. Interestingly, in other members of Brassicaceae, the genes accounting for natural variation in flowering time identified so far have orthologues that account for natural flowering time variation in *A. thaliana* (Table 9-1). In *A. thaliana*'s closest relative, *A. lyrata*, a *FRI* polymorphism involving a functional, but truncated, version of the protein accounts for natural variation in flowering time (Kuittinen *et al.* 2008). In field mustard (*Brassica rapa*) and in shepherd's purse (*Capsella bursa-pastoris*), polymorphisms at *FLC* orthologues have been implicated (Kole *et al.* 2001, Slotte *et al.* 2009). In a more distantly related species also from the eudicot branch of angiosperms, the pea (*Pisum sativum*), a polymorphism at *TFL1* accounts for flowering time variation (Foucher *et al.* 2003). Even in this instance, the *TFL1* locus has population genetic characteristics that make it a candidate gene for natural flowering time variation in *A. thaliana* as well (Olsen *et al.* 2002), although this role has not yet been documented.

In species more distantly related to *A. thaliana*, most of the polymorphisms accounting for natural flowering time variation occur in genes that have orthologues in *A. thaliana* that are part of the known flowering time genetic network (summarized in Alonso-Blanco *et al.* 2009). This is not surprising because, generally speaking, the molecular genetic flowering time network in other species involves the same or similar players as those in *A. thaliana* (Tan and Swain 2006). But there are also differences. A particularly interesting comparison in this regard is between *A. thaliana* and rice (*Oryza sativa*). Rice, which diverged from a common ancestor with *A. thaliana* about 200 million years ago (Wolfe *et al.* 1989), is a monocot, which represents a different branch of an ancient split in the angiosperm lineage from *A. thaliana*. In addition to being far-removed evolutionarily, *Arabidopsis* and rice are also very different ecologically. For example, they have opposite responses to photoperiod. *A. thaliana* is a facultatively long-day plant (Napp-Zinn 1985), meaning that it flowers more rapidly in response to long day lengths. Rice, on the other hand, is a short-day plant, meaning that flowering is triggered by short day lengths, so that it can synchronize sexual reproduction with the rainy season (Putterill *et al.* 2004).

Despite the differences between *A. thaliana* and rice, it appears that many aspects of the genetic network controlling flowering are the same in these two lineages. This suggests that many of the genes involved in flowering are conserved through evolutionary history (see Izawa *et al.* (2003) for a full list). For instance, rice has phytochromes and cryptochromes that comprise a light-dependent

pathway of flowering signals, as they do in *A. thaliana* (Izawa *et al.* 2003). Furthermore, the *A. thaliana* floral integrators, *FT* and *SOC1*, have orthologues in rice that serve the same functions downstream of the various flowering genetic pathways (Kojima *et al.* 2002, Andersen *et al.* 2004).

On the other hand, there are also some key differences between the genetic networks in these species. For instance, although orthologues of some phytochromes exist in rice (*PHYA*, *PHYB*, *PHYC*), there do not appear to be any *PHYD* or *PHYE* orthologues, suggesting that those functions are maintained by other loci or are absent in rice. Most significantly, the gene that acts downstream of the light-dependent pathway in *A. thaliana* to promote flowering, *CO*, appears to have the opposite effect in rice, suppressing flowering under long days (Yano *et al.* 2000). While it makes sense, from an ecological perspective, that signals would be channeled to the floral integrators to suppress flowering under long days in rice, it is interesting that the *A. thaliana* orthologue is used for this function, at the same point in the flowering time signaling pathway but with opposite effect. It is also interesting that *FRI*, a key regulator of flowering in *A. thaliana*, is not found in rice, or in any species outside the Brassicaeae for that matter, and that the clade containing *FLC* has not had the sort of diversification that has been documented in *A. thaliana*.

9.7 Synthesis and prospectus

There are only a few genes that have been confirmed to control natural variation in flowering time, and much of the variation has yet to be explained (Ehrenreich *et al.* 2009). We know even less about the effects of gene–gene interactions on this variation (with the notable exception of *FRIGIDA* and *FLOWERING LOCUS C*). Furthermore, we know little about the pleiotropic effects of these functional polymorphisms on other traits. Clearly there is much work left to be done to understand the genetic basis of flowering time variation.

Arabidopsis thaliana remains the main focus of research on flowering and its variation in nature (Alonso-Blanco *et al.* 2009), but the situation is slowly changing, especially for agricultural plants, where there is interest in manipulating flowering time to increase yield (Tan and Swain 2006). While *A. thaliana* is an invaluable resource, there are limits to what we can learn about the genetics of flowering time from studying *A. thaliana* alone. It is a very small annual plant that has a natural range confined to temperate areas of Europe and Asia, and it has a genome that is small and has other unusual characteristics (Schranz *et al.* 2007). Even within this one species, our chapter demonstrates that the genetic control of flowering time changes depending on the environmental conditions, so it is reasonable to expect different species with different niches to have differences in the genetic mechanisms underlying flowering variation. It is also worth noting that the bulk of our knowledge about the genetic control of variation in flowering time comes from plants with annual life histories, as opposed to biennials or perennials, although this is also slowly changing (Tan and Swain 2006). We believe the trend towards studying flowering in a larger diversity of species, with different life histories and occupying different ranges of the natural environment, is a promising one, which will create a fuller picture of the genetic mechanisms underlying natural variation in flowering time in angiosperms.

While there is much work left to be done, we believe this work is important to begin to answer longstanding questions about life history evolution. We need to first know the genes governing variation in life history before we can fully address questions such as:

- How many genes influence intraspecific variation in life histories, and what is the distribution of their effect sizes (Orr and Coyne 1992)?
- Which evolutionary processes influence life history variation (Mitchell-Olds *et al.* 2007)?
- Do important life history shifts (such as changes in flowering phenology) arise due to structural changes or to changes in regulatory elements (Hoekstra and Coyne 2007)?
- Does selection for certain life histories in ecologically similar environments operate on the same genes, or do the same life histories arise independently due to different genetic mechanisms?

These questions are also fundamental to evolutionary theory in general, and as such illustrate how

the molecular genetic study of life history evolution in plants can shed light on the nature of evolution as a whole.

9.8 Summary

1. Flowers are a key innovation in evolution, and flowering is a key life history transition in angiosperms. There is wide variation in the timing of flowering within and among species, and flowering is determined mechanistically by complex interplay of genetic and environmental signals.

2. Much of the interest in the mechanistic basis of flowering has focused on the model species *Arabidopsis thaliana*. A few genes, such as *CRYPTOCHROME 2*, *FLOWERING LOCUS C*, *FRIGIDA*, and *PHYTOCHROME C*, have been cloned and shown to have complex, environmentally contingent effects on flowering time, as well as pleiotropic effects on other traits. This has ramifications for the study of life history evolution.

3. The comparative functional genomics of *A. thaliana* and other species illustrate that many of the genes involved in flowering are conserved throughout a large part of angiosperm evolution, although the comparison of *A. thaliana* and rice shows that the same genes can affect flowering by very different mechanisms in different species.

4. In future work, it is important that the genetic basis of flowering time variation be understood in a variety of angiosperms with different niches and life histories. Studying *A. thaliana* advances this goal by suggesting particular genes and genetic pathways for further study in other species.

9.9 Acknowledgments

We would like to thank Yoshie Hanzawa for many helpful discussions about this chapter, and Andreas Heyland and Thomas Flatt for their specific editorial suggestions. Rocky Graziose helped design Figure 9.1. This work was funded in part by grants from the National Science Foundation *Arabidopsis* 2010, Population and Evolutionary Processes and Plant Genome Research Programs to M.D.P.

CHAPTER 10

Mechanisms of nutrient-dependent reproduction in dipteran insects

Alan O. Bergland

10.1 Introduction

Reproductive output, along with development time and lifespan, are the core parameters of an organism's life history. Together, these three parameters allow us to predict an individual's fitness and, by extrapolation, the growth rate of a population. Ultimately, natural selection should act to maximize fitness (Fisher 1930) and as a consequence these life history traits will respond in a correlated fashion (Robertson 1968).

The study of life history traits has repeatedly demonstrated that they show reduced genetic variation relative to putatively neutral characters such as morphological traits (Roff 2002). Presumably, reduced genetic variation in fitness components is due to the constant action of natural selection to maximize fitness. Although genetic variation in life history traits is generally low, it is still present in many populations. The presence of genetic variation for life history traits is possibly due to mutation–selection balance (discussed in Charlesworth and Hughes 2000), life history trade-offs (see several chapters in Part 6 of this volume), or some form of balancing selection such as genetic overdominance or, more likely, environmentally dependent marginal overdominance (e.g., Chapter 18). In this last scenario, environmental variation affects life history traits but not all genotypes are affected the same way. For instance, one genotype may do very well in one environment but very poorly in another; an alternative genotype may have the opposite pattern. Such a scenario can lead to the stable persistence of these two, hypothetical, genotypes. If these environments affect genotypic performance by altering life history traits then genetic variation will likely be observed.

The persistence of genetic variation in life history traits due to marginal overdominance is plausible given the sensitivity of many life history traits to the environment (Roff 2002, Hodin 2009). In general, the three major environmental variables affecting life history traits are photoperiod, temperature, and nutrition. Photoperiod, at least for many organisms living in seasonal environments (e.g., Chapter 9), plays a major role in determining the timing of development and reproduction (cf., Chapter 18). For these species, alterations in the timing of life history transitions affect the length of the growing season and consequently the number of reproductive cycles per year. In many organisms, notably ectothermic animals, exposure to variable temperatures affects development time, reproductive output, and lifespan. This effect is mediated by changes in the rate of metabolic and catabolic processes that are direct functions of temperature. Finally, nutrition affects life history traits by altering the rate and duration of larval growth and by directly limiting resources available for reproduction and somatic maintenance (Chapter 11). The biology of nutrient-dependent reproduction is reasonably well understood and is mediated by a complex set of interactions between molecular processes, morphology, and intraspecific interactions.

The goal of this chapter is to integrate what is known about the molecular, morphological, and behavioral basis of one aspect of an organism's life history: reproduction. In particular, I will focus on recent developments that have been made in understanding the mechanistic basis of nutrient-de-pendent reproduction in dipteran insects. Although I will be focusing on dipteran insects, it is reasonable to hypothesize that many of these mechanisms are shared among more divergent animals.

This chapter will be divided into three main sections. The first section will examine the relationship between larval nutrition and adult reproduction via growth and will focus on two related processes. First, I will discuss how nutrition affects the physical size of the adult and how body size and other allometric correlates directly relate to reproductive capacity. Second, I will discuss how larval nutrition provisions the adult with nutritive resources necessary for reproduction.

In the second section of this chapter, I will focus on the relationship between adult nutrition and reproduction. In this section, I will cover the role of the sensory, digestive, and endocrine systems and their role in reproduction. Finally, I will conclude the chapter by highlighting recent findings that integrate advances in the mechanistic basis of nutrient-dependent reproduction with the predictions that evolutionary theory makes about the dynamics of life history evolution.

10.2 Larval nutrition and reproduction

For Dipterans, larval nutrition affects female reproduction by mediating various aspects of adult body size. Environmentally induced variation in body size and other allometric correlates affect fecundity through at least four mechanisms. First, larval nutrition affects ovary size, which is a direct determinant of reproductive capacity. This effect appears to be universal amongst Dipterans. Second, larvally acquired nutrients are often necessary for adult reproduction and thus the extent of larval nutrition directly affects the number of eggs that can be provisioned. Third, adult body size is directly related to adult meal size in many Dipterans. Meal size affects the ability of adults to provision eggs, and thus affects fecundity. Finally, because larval nutrition affects body size it consequently affects male–female interactions, which are size-dependent for many Diptera. Size-dependent mating can negatively affect fecundity because larger males can cause greater physical harm to females during cop-

ulation and potentially invest fewer resources into each female. Below, I discuss these four mechanisms in detail.

10.2.1 Ovary size

The insect ovary is composed of repeated structures called ovarioles (reviewed in Hodin 2009) and each ovariole is capable of simultaneously producing an egg. Therefore, given sufficient adult nutrition, ovariole number sets an upper limit on reproductive rate and capacity (David 1970).

The positive relationship between ovariole number and fecundity is present within and amongst species of Diptera (reviewed in Honek 1993). For instance, variation in ovariole number within populations of *D. melanogaster* is positively correlated with fecundity (David 1970, Bergland and Tatar, unpublished data). Likewise, variation in ovariole number amongst populations of *D. melanogaster* is correlated with fecundity (e.g., Boulétreau-Merle *et al.* 1982). Finally, variation in ovariole number amongst closely related Drosopholids (reviewed in Hodin 2009) is positively correlated with fecundity.

While there is abundant genetic variation in ovariole number, it is also highly sensitive to the larval environment (e.g., reviewed in Hodin 2009) and in particular to larval nutrition (e.g., Bergland *et al.* 2008 and references therein). Hodin and Riddiford (2000) showed that for *Drosophila melanogaster* larval nutrition affects ovariole number by modifying the rate of differentiation of a specialized set of cells at the anterior tip of the ovariole—the terminal filament cells. Interestingly, this variable rate of differentiation occurs during the wandering stage, a period when larvae are no longer feeding. This observation suggests that the ovariole number is set by endocrine or paracrine signals from another tissue that is growing in direct response to larval nutrition. It is plausible that this tissue may be either the larval fat body or the prothoracic gland, two organs that have recently been identified as regulators of body size in insects (reviewed in Mirth and Riddiford 2005).

This hypothesis is substantiated by the results of Bergland *et al.* (2008), who performed a quantitative trait locus (QTL) mapping study of nutritional plasticity of ovariole number in *D. melanogaster*. They identified at least nine QTL affecting ovariole number and ovariole number plasticity in response to larval nutrition. One pair of epistatically interacting QTL contains the imaginal disc growth factors, *Idgf1*, *Idgf2*, *Idgf3*, and *Ras85D*. The IDGFs are secreted by the larval fat body and affect imaginal disc growth (Kawamura *et al.* 1999). *Ras85D* is a member of the RAS signaling pathway, which, in conjunction with *PI3K* activity in the prothoracic gland, affects nutrient-dependent insulin signaling in the larval fat body (Mirth and Riddiford 2005). Thus, it is plausible that nutrient-dependent insulin signaling alters IDGF secretion, which in turn affects the rate of terminal filament differentiation during the wandering stage. Further work is needed to test this hypothesis.

Orgogozo *et al.* (2006) have also suggested that genetic variation in insulin signaling underlies differences in ovariole number between two closely related species, *D. simulans* and *D. sechellia*. The latter species is endemic to the Seychelles Islands and has evolved resistance to the toxic *Morinda citrifolia* fruit, which it uses as its exclusive nutritive resource (discussed in Orgogozo *et al.* 2006). For unknown reasons, *D. sechellia* has evolved a nearly two-fold reduction in ovariole number (Orgogozo *et al.* 2006) and concomitant reduction in fecundity (R'Kha *et al.* 1997). Line crosses between these two species initially identified autosomal factors underlying differences in ovariole number (discussed in Orgogozo *et al.* 2006). Fine-scale QTL mapping has localized these loci to a region on chromosome 2 and to two epistatically interacting regions on chromosome 3. Orgogozo *et al.* (2006) point out that the insulin receptor (*InR*) lies within one of these epistatic QTL on chromosome 3 and hypothesize that genetic variation at this locus has contributed to differences in ovariole number. This hypothesis is consistent with the observation that mutants in the insulin signaling pathway have decreased ovariole number (e.g., Tu and Tatar 2003).

10.2.2   Meal size

Larval nutrition affects adult body size, which subsequently determines how much food adults can ingest. Adult meal size, of course, will determine the amount of resources available for reproduction.

The relationship between body size, meal size, and fecundity is best understood in blood feeding mosquitoes such as *Anopheles* spp. where blood meal size is directly mediated by abdomen size (e.g., Roitberg and Gordon 2005).

In mosquitoes, the relationship between meal size and body size is determined by sensory mechanisms in the abdomen (Gwadz 1969). Mosquitoes, being relatively opportunistic feeders, will gorge themselves on blood when they find a suitable host. Individuals gauge how much they have eaten by the expansion of their abdomen, the extent of which is sensed by a set of nerves that innervate the abdomen. When these nerves are severed, adults will feed until their abdomen explodes (Gwadz 1969). This rather dramatic behavior demonstrates that the control of meal size in mosquitoes is purely physical. However, physiological mechanisms will determine how many meals are necessary before sufficient resources are attained for reproduction in anautogenous species.

10.2.3   The effects of mate size

Nutrient-induced plasticity in body size affects males as well as females (Chapter 4). For some dipteran species, male body size affects access to females through competition (e.g., McLachlan and Allen 1987) and may also directly affect female fecundity. The best-known examples of the latter are in *D. melanogaster*, where a negative correlation between male body size and female reproductive output exists, particularly for reproductive output during early and mid life (Pitnick 1991, Pitnick and García-González 2002). In these studies, male body size was experimentally altered by manipulating larval density, thus larval nutrition may have indirect effects on female fecundity through male body size.

Pitnick (1991) suggested a behavioral mechanism for this phenomenon. For many insect species, males compete with each other for access to mates and typically larger males win. Because large males might have access to multiple females in a short period of time, and for many male insects sperm is limited (discussed in Bretman *et al.* 2009, Wigby *et al.* 2009), it could be beneficial for large males to distribute relatively smaller ejaculate to each female, thereby increasing the male's total reproductive fitness but potentially decreasing an individual female's reproductive potential. Small males, in contrast, might only have access to females on rare occasions. Therefore, it is plausible that small males take advantage of their infrequent mating and release most of their stored ejaculate at a single time. Small males copulate (Pitnick 1991, Kelic *et al.* 2007) for a longer period of time than large males and thus it is possible that a longer copulation duration results in a larger ejaculate, thereby increasing reproductive success per female. This hypothesis is plausible given recent evidence (Bretman *et al.* 2009, Wigby *et al.* 2009) that copulation duration is positively correlated with the quantity of at least one oviposition stimulating protein, sex peptide, which is transferred from males to females upon mating (see Section 10.3.1 for a longer discussion of the mode of action of sex peptide, SP).

The relationship between male size and female reproductive output in *D. melanogaster* may be mediated by phenotypic plasticity in female size (Lefranc and Bundgaard 2000, but see Pitnick and González 2002). Lefranc and Bundgaard (2000) found that male body size was negatively correlated with female fecundity only in large and medium-sized females. However, the size-dependent interactions between males and females with respect to fecundity has not been confirmed by independent replication (Pitnick and González 2002). Given that size-dependent assortative mating occurs in some Diptera (e.g., Sisodia and Singh 2004 and references therein), further research on the relationship between male and female size with respect to fecundity is warranted.

10.2.4 Larval nutrition and teneral reserves

It would be safe to say that larvally acquired resources are used for reproduction in all Diptera, yet the extent to which they are necessary for reproduction varies amongst different lineages. The quantity and quality of adult nutrition certainly determines how important larvally acquired resources are. For instance, many species of mosquitoes do not feed on blood during adulthood or prior to the first ovarian cycle. For these mosquitoes, larval nutrition totally determines the extent of reproduction.

This is the case for some populations of the pitcher-plant mosquito, *Wyeomyia smithii*, which are completely autogenous (non-blood-feeding). These populations are found at northern latitudes, where larval resources are not limited due to low intraspecific competition (Lounibos *et al.* 1982) and abundant larval resources (Bergland *et al.* 2005). For larvae that experience these relatively benign conditions, adults emerge with enough resources to reproduce and hematophagy is not observed. However, in southern populations larval densities can be much higher and adults that experience intense intraspecific competition as larvae are unable to provision eggs without a blood meal. Thus, for this species autogeny is both genetically fixed (north versus south) and plastic (nutrient-dependent in southern populations).

Autogenous–anautogenous polymorphisms are present in many other species of mosquito (Attardo *et al.* 2005). In some species, this polymorphism is environmentally induced, but in others it is under strict genetic control. For example, in the Asian tiger mosquito, *Aedes albopictus*, researchers were able to artificially select for autogeny from a stock laboratory population (Mori *et al.* 2008). From that selection experiment, Mori *et al.* (2008) were able to produce a completely autogenous strain, generate a mapping population by intercrossing with an anautogenous strain, and map three to four QTL affecting this autogenous behavior. These loci each contribute less than 10% of the total genetic variation within this mapping population, suggesting that other loci of small effect or epistatically interacting loci also contribute to autogeny. Although the position and effect of loci controlling autogeny/anautogeny polymorphisms have not been determined for other mosquito species, most evidence suggests that in each species many loci segregate alleles that can confer autogeny in primarily anautogenous species (Attardo *et al.* 2005).

Dipterans that must feed as adults in order to reproduce nonetheless utilize larvally acquired resources for adult reproduction. In blood-feeding mosquitoes, larvally acquired resources are an important component of reproductive output. This is especially true during the first gonotropic cycle, which is prior to blood feeding in some anautogenous species. Zhou *et al.* (2004 and references therein) found that for *A. aegypti*, larvally derived carbohydrates, lipids and amino acids make up roughly 70–99% of egg protein and lipids found in the first clutch of eggs. The remainder came from sugars acquired during early adult feeding. While subsequent gonotropic cycles will utilize blood-meal-derived resources, it has not been determined how long teneral reserves will persist in adults. Furthermore, it has not been determined whether components of larvally acquired resources are non-renewable in anautogenous mosquitoes.

There do not appear to be any non-renewable, larvally acquired resources in *D. melanogaster*. Furthermore, larvally acquired resources make a very small contribution to total reproductive output. A set of studies by Min *et al.* (2006) and O'Brien *et al.* (2008) found that larvally acquired carbon and nitrogen are used for early life reproduction (i.e., prior to day 10) and these larvally acquired resources only make up less than 15% of the total carbon and nitrogen in the eggs during this short timespan. After day 10, there were virtually no larvally derived resources found in eggs. Thus, in contrast to other insects such as Lepidoptera (e.g., O'Brien *et al.* 2002) and Dipterans mentioned above, teneral reserves in this species do not play an important role in provisioning eggs nor do there seem to be any irreplaceable, larvally acquired resources.

10.3 Adult-acquired resources

For anautogenous species of Diptera, variation in adult resources directly affects the extent of egg production (reviewed in Hodin 2009). Variation in the acquisition of adult resources is determined by present nutritional status (i.e., hunger), her ability to find food, and the conversion of those resources into the proteins used to provision eggs. Hunger is mediated by neuroendocrine signals, which stimulate the drive to locate resources. The ability to locate food as an adult is controlled by sets of olfactory and gustatory receptors located in the sensory neurons spread throughout the adult body. Once food has been identified, enzymes throughout the gut digest these complex proteins and sugars, which are then transported to the haemolymph. Changes in sugar, fat, and protein in the haemolymph ultimately affect egg production by altering hormonal

balances that alter the production of yolk proteins. In this section, I will discuss recent work identifying the specific molecular bases for each of these behavioral and physiological processes and their relation to fecundity.

10.3.1 Hunger

Presumably, the feeling of hunger is required in order to initiate the search for food. In adult Diptera, hunger is mediated by two factors. First, and quite obviously, hunger is mediated by nutritional status. To date, the specific physiological mechanisms mediating the relationship between hunger and nutritional status have been elucidated in larval *D. melanogaster*. It is likely, however, that the same mechanisms mediate the relationship between nutritional status and hunger in adults. In larvae, response to hunger (as measured by feeding rate after bouts of starvation) is suppressed by the release of *Drosophila* insulin-like peptides (*dIlps*) from the brain (Wu *et al.* 2005). Wu *et al.* (2005) further demonstrate that high levels of *dIlps* maintain food selectivity, whereas low levels of *dIlps* (i.e., conditions of starvation) promote feeding on suboptimal media. The effect of *dIlps* on food preference is caused by interactions with set neurons in the brain, which are sensitive to neuropeptide F (NPF, Wu *et al.* 2005). Thus, through *dIlps* and NPF signaling, food acquisition rates and food preference are altered in response to hunger.

Second, sexual status (i.e., mated versus virgin) alters food acquisition rates. During copulation, males transfer SP in the seminal fluid and this peptide stimulates adult feeding (Carvalho *et al.* 2006), along with other post-mating behaviors such as oviposition and mating refractoriness. Presently, it is unclear how sex peptide induces these coordinated behaviors. Specifically, it is unknown if sex peptide affects all of these processes directly or if sex peptide initiates one of these behaviors (e.g., feeding), which in turn stimulates another (e.g., oviposition). To date, data suggest both of these general processes cause the coordinated action of reproduction and feeding in response to sex peptide.

Sex peptide is thought to cause post-mating behaviors through two general processes. First, neurons that terminate in the reproductive tract are responsive to sex peptide through the sex peptide receptor (*SPR*; discussed in Clyne and Miesenböck 2009). Genetic ablation of *SPR* in these neurons inhibits oviposition after mating compared to wild-type controls. The neuronal action of sex peptide on reproduction could be subsequently mediated by neuronal connections in the abdominal ganglion. Some of the *SPR* neurons terminate in the abdominal ganglion, which is known to affect ovulation (the release of mature eggs) through the production of octopamine (see below; Monastirioti 2003). However, flies lacking *SPR* in their reproductive tract neurons lay more eggs than wild-type females mated to males lacking sex peptide. These results suggest other modes of action for sex peptide with respect to reproduction.

Sex peptide could also stimulate post-mating behaviors through action on other tissues or neurons elsewhere in the adult. Sex peptide is present in the haemolymph following mating and is known to stimulate the production of juvenile hormone (JH) by the corpora allata *in vitro* (reviewed in Kubli 2003). JH is well known to positively regulate egg production (Flatt *et al.* 2005, see below and Chapter 13). Thus, differences in reproduction between females lacking *SPR* in the reproductive tract and those mated to males without sex peptide could be due to the absence of JH induction by sex peptide.

Current evidence suggests that increased feeding following exposure to sex peptide (Carvalho *et al.* 2006) is most likely a functional consequence of reproductive activity (or, at least, of oogenesis). These data come from experiments that have examined the role of sex peptide on feeding in genetically sterile *ovo*[D1] flies. *ovo*[D1] is a dominant allele that causes sterility by inhibiting germline stem cell (GSC) maintenance; after the first division of the GSC, no new oocytes are produced and thus, in *ovo*[D1] flies, the ovary does not act as a nutrient sink. Barnes *et al.* (2008) demonstrated that *ovo*[D1] flies exposed to sex peptide do not feed after mating. This result suggests that production of vitellogenic oocytes are required for post-mating induced feeding. This interpretation implies that reproduction (or ovulation) depletes nutritional reserves, inducing hunger and subsequently feeding.

However, adult feeding might also be directly affected by JH. Recently, a putative JH binding protein encoded by the gene *takeout* (*to*) has been associated with adult feeding rate and other phenotypes (Meunier *et al.* 2007 and references therein). Reductions in *to* levels increase feeding rate and locomotion in males and females (Meunier *et al.* 2007). Meunier *et al.* (2007) also demonstrated that, in males, exogenous application of the JH analog, methoprene, rescues wild-type locomotor behaviors. While this effect was not observed in females, it does not preclude an interaction between *to* and JH in regulating feeding in females. *to* is also expressed in the gustatory neurons located in the mouthparts of *D. melanogaster* and has been implicated in taste perception after bouts of starvation. In contrast to wild-type flies, *to* mutants do not display increased activity of sugar-sensitive neurons following starvation (Meunier *et al.* 2007). Thus, *to* might be an important integrator of mating-status-induced feeding (via JH via sex peptide), taste perception, and reproductive output.

10.3.2 Finding nutrition

Olfactory and gustatory receptors are necessary for reproduction because they allow females to identify resources used for egg production and to identify oviposition substrates. For many dipteran species, identification of adult resources and oviposition substrates is clearly differentiated in time and space. For instance, in anautogenous mosquitoes, females must find a suitable mammalian host, obtain a blood meal, digest it, and then find a location to oviposit. Thus, the chemical cues that signify food are distinct from those that identify oviposition substrate. For other dipteran species, such as those in Drosophilidae, this distinction is not as clear. Adults feed at the same locale as they oviposit and this raises the possibility that the chemical cues that allow adults to identify food sources are the same as those that identify oviposition substrate. However, recent work indicates that these cues are perceived by female *D. melanogaster* through distinct chemosensory mechanisms. How, then, do Dipterans find food and oviposition substrate?

Dipterans sense chemical cues through taste and smell receptors located in sensory neurons throughout the head, legs, and abdomen. Two large gene families, the gustatory receptors (GRs) and the odorant receptors (ORs), enable individuals to identify taste and smell. In dipteran species studied to date, ORs contain roughly 60–70 members and the GRs contain roughly 70–80 members (reviewed in Touhara and Vosshall 2009). Each receptor has a high affinity for a small set of ligands (Touhara and Vosshall 2009), and through combinations of receptors, individuals are able to perceive complex chemical environments. The ability to perceive such complex chemical cues enables species to display very subtle host and oviposition preferences.

For instance, in mosquitoes, CO_2 in combination with other perspired chemicals allows female mosquitoes to find hosts. Host specificity is determined by sensitivity to particular combinations of emanations released by a host. For instance, anthropophilic mosquitoes are attracted to lactic acid, which is a major component of human sweat (reviewed in Takken and Knols 1999). Attraction to lactic acid is potentiated by CO_2 (Dekker *et al.* 2005), which is sensed by two GRs, *GPRGR*22 and *GPRGR*24. To date, the OR or GR that senses lactic acid has not been identified.

Female mosquitoes use very different chemical cues to identify oviposition substrate. First, females utilize volatile chemicals released by bacterial decomposition of organic matter to identify suitable oviposition sites (Millar *et al.* 1992). However, the specific receptors sensing these chemicals have not been identified. Female mosquitoes are also attracted to chemical cues deposited by conspecific females upon oviposition, but the mechanisms regulating this type of aggregation have not been identified. Although the mechanisms for host-seeking and oviposition behavior have not been fully worked out, it is clear that the sensory mechanisms mediating these behaviors are biologically distinct.

For species of Diptera that utilize the same resources for adult nutrition and oviposition, such as members of the Drosophilidae, it is not clear *a priori* that females use distinct chemical cues in order to identify adult resources and oviposition sites. However, recent work suggests that different sensory mechanisms are used in these two processes. Work in *D. melanogaster* and related species demonstrates that females are attracted to by-products of yeast fermentation, such as acetic acid, ethyl acetate, and ethanol (e.g., Ruebenbauer *et al.* 2008). However, yeast odor *per se* does not influence oviposition (Libert *et al.* 2007). Rather, it appears that the choice to oviposit on a particular substrate is governed by chemosensory neurons located in the ovipositor (Yang *et al.* 2008). This choice is mediated, in part, by the expression of an insulin like peptide, *dIlp7*. The role of *dIlp7* in mediating the choice to oviposit suggests that oviposition is a secondary process following nutrient identification, acquisition, and digestion.

10.3.3   Oogenesis and ovulation

Given that a female Diptera has sufficient nutrition, and has found a mate and oviposition substrate, how does she make and release eggs? The process of egg production, or oogenesis, is primarily controlled by the action of three classes of hormones,: JH, 20-hydroxyecdysone (20E), and the *insulin-like* peptides (*dIlps*). These hormones have well-known, coordinated effects on metamorphosis (reviewed by Mirth and Riddiford 2005) and there is a growing body of literature demonstrating that the proper balance of at least JH and 20E is necessary for oogenesis. The coordinated actions of JH and ecdysone are thought to also be mediated by at least two biogenic amines, dopamine and octopamine. Octopamine is also known to play a crucial role in ovulation. Below, I shall briefly discuss some recent work elucidating the relationship between these four hormones and oogenesis and ovulation; further discussion of the role of JH on insect life histories can be found in Chapters 4, 5, and 13.

Oogenesis is a complex physiological process that is composed of several distinct processes, including germline stem-cell division and differentiation, yolk production (vitellogenesis), and the construction of the eggshell itself (chorionogenesis). Of these three processes, nutrition is known to play a major role in germline stem-cell division, differentiation, and vitellogenesis. Adult-acquired nutrition alters the secretion of neuronally derived *dIlps*, which directly stimulate the asymmetric division of germline stem cells (i.e., division into one self-renewing germline stem cell and one cell, called a germline cyst, destined to become an egg) in a niche-independent fashion (reviewed in Drummond-Barbosa 2008). The germline cyst must undergo four rounds of cell division to become a 16-cell cyst, at which point it acquires yolk and develops into a mature egg. Acquisition of yolk by the 16-cell cyst is mediated by cyst autonomous insulin signaling, that is, cysts lacking functional *InR* fail to acquire yolk (reviewed in Drummond-Barbosa 2008).

While the uptake of yolk by germline cysts is mediated by autonomous insulin signaling, nutrient-dependent production of yolk itself appears to be regulated in a more complex fashion. Yolk proteins (either vitellogenin or yolk peptides, YP) are upregulated upon feeding (discussed in Attardo *et al.* 2005) and the upregulation of these genes is mediated by the joint actions of 20E and JH. For instance, nutritionally deprived *D. melanogaster*, which normally have low levels of YP synthesis, increase YP production upon exogenous application of 20E and methoprene, a JH analog (reviewed in Postlethwait and Shirk 1981). 20E increases YP transcription through the action of ecdysone receptor/ultra spiricle transcription factor complex, which directly binds to the ecdysone response element 5′ of the YPs (Bownes *et al.* 1996). The action of JH on YP production is more elusive, partly due to the general difficulty identifying the molecular modes of JH action (Jones and Jones 2007). However, it is generally thought that JH, synthesized in the corpra allata upon nutrient acquisition, directly stimulates early fat body and ovarian YP production, while concomitantly stimulating 20E production by the ovary. This ultimately leads to sustained YP production in the fat body (reviewed in Gruntenko and Rauschenbach 2008).

The levels of 20E and JH must be appropriately balanced, however, to successfully stimulate oogenesis (Soller *et al.* 1999). For instance, in *D. melanogaster* under starvation conditions, 20E titers rise (Terashima *et al.* 2005) and JH titers fall (Tu and Tatar 2003, cf. Rauschenbach *et al.* 2004). These changes in hormonal titers are correlated with apoptosis of early oocytes (Terashima *et al.* 2005) and with the repression of YP production. Exogenous application of 20E without simultaneous application of JH under normal nutritive conditions also reduces egg production (Soller *et al.* 1999, Terashima *et al.* 2005). Additionally, exogenous application of JH under starvation conditions elevates 20E titres

and increases the number of mid-stage oocytes compared to non-JH-treated flies (Terashima *et al.* 2005).

The interaction between 20E and JH on vitellogenesis may be modulated by the action of two biogenic amines, dopamine and octopamine (reviewed in Gruntenko and Rauschenbach 2008), which are thought to regulate the metabolism of 20E and JH in an environment and age-dependent fashion. Evidence to date suggests that dopamine inhibits JH degradation in early life and under normal environmental conditions. As an individual ages or when she experiences unfavorable environments, Gruntenko and Rauschenbach (2008) propose that dopamine ceases its inhibitory role in JH degradation and starts stimulating JH degradation. The net effect of this switch in dopamine action would be to reduce egg production. Gruntenko and Rauschenbach (2008) further suggest that this switch is mediated by a transition from inhibitory to stimulatory dopamine receptors in the fat body. However, to date there is no evidence that these receptors are expressed in the fat body (Draper *et al.* 2007). Exogenous application of octopamine has also been noted to affect JH degradation (reviewed in Gruntenko and Rauschenbach 2008), but its modes of action are not known.

Octopamine has another very important role in reproduction in that it acts to stimulate ovulation, or the movement of the egg from the ovary through the oviduct. Initially, the role of octopamine on ovulation was observed by noting that *D. melanogaster* with octopamine deficiencies produce eggs but do not release them (Monastirioti 2003). When these flies are fed octopamine or when octopamine synthesis is rescued in neurons that innervate the ovary, ovulation proceeds normally (Monastirioti 2003). Ovulation, stimulated by octopamine, is thought to occur via contractions of the muscles surrounding each ovariole and by relaxation of the muscles surrounding the oviduct (Middleton *et al.* 2006).

Qazi *et al.* (2003) have also suggested that the process of ovulation *per se* induces oogenesis. This process would also feedback on the endocrine system because it would deplete nutritional reserves and would thus alter behavior, inducing foraging. These coordinated processes are also induced by accessory gland proteins (ACPs, and notably SP) transferred from males to females during mating (see above). Further work is certainly needed to examine the interrelations of these ACPs, ovulation, oogenesis, and behavior in the Diptera.

10.4 The evolutionary genetics of reproduction: Future prospects

Many ecological, behavioral, morphological, and physiological factors affect reproduction in Diptera. This chapter has highlighted several of these key components, but there are still large gaps in our general understanding of dipteran reproduction. One particularly promising area of study will be identifying how genetic variation in the processes underlying reproduction vary amongst species (reviewed in Markow and O'Grady 2005, 2008) and amongst populations within species. Such information will be particularly interesting because it can be tied to environmental differences between populations and species, and potentially to the selective pressures causing phenotypic differentiation.

Work in this area is underway. For example, naturally occurring polymorphisms in the *D. melanogaster*'s *InR* have been associated with female fecundity as well as other life history traits (Paaby *et al.* 2010). These polymorphisms vary in a clinal manner, suggesting that this gene, and the associated phenotypes, are under contemporary natural selection. A similar story has been documented for reproductive diapause in *D. melanogaster* (see Chapter 18). Reproductive diapause in *D. melanogaster* is characterized by a strong resorption of vitellogenic oocytes upon exposure to cold temperatures. The ability of an individual to enter into reproductive diapause is controlled by naturally occurring polymorphisms in the *couch potato* (*cpo*) gene. These polymorphisms show latitudinal clines consistent with the action of natural selection (Chapters 13 and 18).

Identification of naturally occurring genetic polymorphisms that affect reproduction and other life history traits is a crucial step in understanding their short- and long-term evolutionary dynamics. For instance, application of population genetic tests to the natural polymorphisms in *InR* (Paaby *et al.* 2010)

suggests that this gene is evolving under directional selection. However, this is just one example and it would be wrong to assume that other naturally occurring polymorphisms affecting life history traits are evolving by similar mechanisms. Thus, in order to understand the evolutionary history of genes affecting fitness-related traits much broader genomic-level perspectives are needed.

Fortunately, such opportunities are becoming available. The advent of high-throughput sequencing technologies will allow for the rapid and cheap genotyping of many individuals. This will facilitate more efficient association mapping of complex traits and will hopefully lead to the identification of point mutations affecting life history traits. With such association maps in hand, we will be able to further interrogate the population genetic patterns of these loci that are putatively under natural selection. This type of investigation may finally provide insight into a longstanding problem in life history evolution, namely why genetic variation in life history traits exists even though we expect natural selection to efficiently remove it.

10.5 Summary

1. Reproduction in Dipterans is an environmentally labile trait, which is strongly influenced by many environmental factors, including nutrition.

2. Nutrition affects reproduction in Dipterans by altering larval growth, and this affects body size and other allometric correlates. In females, larval growth conditions limit reproduction by altering ovariole number, a direct determinant of fecundity. Also in females, body size determines meal size and thus the availability of resources to provision eggs. In males, body size affects competitive ability and thus access to females. Competitive dynamics between males may alter a male's reproductive investment in females and subsequently alter his mate's reproductive output.

3. Adult-acquired resources affect reproductive output by limiting resources available for reproduction. Hunger, the ability to locate resources, and conversion of those resources into yolk mediate the relationship between adult-acquired resources and reproduction. Insulin, ecdysone, juvenile hormone,

and biogenic amine signaling underlie these behavioral and physiological processes.

4. The genetic and physiological determinants of nutrient-dependent reproduction are likely subject to natural selection. Most likely, this will erode genetic variation in these pathways. However, many studies have shown that genetic variation, while limited, does exist for life history traits, including reproduction. The identification of natural alleles conferring genetic variation in reproduction will help characterize the molecular evolution of these pathways and will help in our understanding of the processes that create and maintain genetic variation.

10.6 Acknowledgments

I would like to thank the editors, Thomas Flatt and Andreas Heyland, for inviting me to write this chapter and for their thoughtful comments on it. I would also like to thank Stuart Wigby and an anonymous reviewer for helpful suggestions. Finally, I would like to thank Marc Tatar for guidance and insightful discussion. Because of space limitations, I could not always cite primary sources and instead cited other review articles. I apologize to any researchers who have published relevant research that I was not able to discuss.

CHAPTER 11

Mechanisms underlying reproductive trade-offs: Costs of reproduction

Dominic A. Edward and Tracey Chapman

11.1   Introduction

The extraordinary variation in reproductive life histories is often overlooked. We tend to be more familiar with traits studied under the umbrella of sexual selection, such as astonishing plumage patterns and weaponry, than we are with the equally striking variability in how often, when, and over what period organisms reproduce. Understanding how this diversity of reproductive life histories evolves is a key challenge in evolutionary biology.

Fisher (1930) was the first to formulate mathematically the important tenets of life histories. He suggested that individuals of a certain age would have a "reproductive value," defined as the mean amount of expected future reproductive success for individuals of that age and sex in a population. Natural selection will act to maximize the reproductive value of an organism at each age by balancing growth, maintenance, and reproduction (e.g., Fisher 1930, Charlesworth 1980), and theory shows that the fitness of a particular life history is strongly linked to r, the intrinsic rate of increase for a population or the "Malthusian parameter" (Charlesworth 1980, Lande 1982). However, an "ideal" life history (e.g., one that maximizes both reproductive rate and survival) is constrained by reproductive trade-offs or "costs of reproduction" (a negative relationship between reproductive activity and future reproduction or lifespan; Williams 1966b). There can be ecological trade-offs in which elevated reproductive activity renders organisms more vulnerable to predation or parasitism, but also intrinsic trade-offs, which limit the reproductive output that can be achieved   at each age because of competition between life history traits for a share of a finite resource pool (Van Noordwijk and De Jong 1986). Elucidating the nature of these costs of reproduction is central to understanding the diversity of life histories (Charlesworth 1980, Stearns 1992).

In this chapter, we first set the scene by defining some of the key life history traits that are observed to trade-off with one another and to result in costs of reproduction. Included is a broad but brief summary of reproductive trade-offs and how to measure them, to illustrate the full breadth over which reproductive costs can occur (for in-depth reviews on this topic, see Stearns 1989, 1992, Roff 1992, 2007b). The main focus of this review then describes recent advances in elucidating the mechanisms that underlie the physiological and evolutionary costs of reproduction. We then highlight the importance of deriving an understanding of reproductive costs in a fitness-based framework. Finally, some gaps that remain in our mechanistic understanding of reproductive costs are identified, and we discuss new, and potentially fruitful, avenues for investigation.

11.2   Key life history traits and costs of reproduction

Key life history traits (e.g., Stearns 1992) include:

- size at birth
- growth rate
- age and size at sexual maturity
- number, size, and sex ratio of offspring produced
- reproductive schedule and age-specific reproductive investment
- age-specific mortality
- lifespan.

Over 40 different trade-offs between these life history traits have been identified (Stearns 1989), including those between:

- current reproductive rate and survival
- current and future reproductive rate
- the number and size of offspring

The "cost of reproduction" describes the trade-offs in the first two of these major categories. The majority of mechanistic research has focused more narrowly on the trade-offs between reproduction and lifespan, or "survival costs of reproduction." We define lifespan as a potentially important life history trait, as it is significantly influenced by survival probability. However, it is important to use a fitness-based framework in order to correctly assess the information emerging from mechanistic studies of lifespan in long-lived model organisms (see Section 11.5). For example, extended post-reproductive lifespan may confer limited fitness benefits. However, the processes involved in extending lifespan, identified via the study of long-lived mutants, may also be significant modulators of life history traits throughout life.

We focus on defining and describing the *intrinsic* costs of reproduction that arise because of trade-offs between reproductive rate and future survival or lifespan, and between current and future reproduction. However, there is also a growing literature on the *ecological* costs of reproduction, where elevated reproductive activity alters the susceptibility of organisms to extrinsic threats such as increased predation, disease, or parasitism (Sheldon and Verhulst 1996).

11.3 Intrinsic costs of reproduction: Trade-offs between reproductive activity and survival or future reproductive rate

Reproductive trade-offs fall into two categories: physiological and evolutionary (Stearns 1989, 1992; Flatt and Schmidt 2009). To date, the overwhelming research effort has focused on investigating the mechanisms underlying physiological trade-offs. For a summary of methods used to measure physiological and evolutionary costs of reproduction, see Boxes 11–1 and 11–2.

Box 11–1 Methods for measuring costs of reproduction

1. Measuring physiological costs of reproduction

(i) Phenotypic manipulations

Phenotypic manipulations can demonstrate how the physiological costs of reproduction are manifested in real organisms and give some indication of the magnitude of the cost. For example, manipulating the reproductive rates of organisms assigned randomly to groups in similar environments can successfully reveal the costs of reproduction in decreased future survival and fertility (Reznick 1985, Bell and Koufopanou 1986, Partridge and Harvey 1988). This technique has been increasingly employed in measurements of the responses of organisms to different diets (e.g., Chapman and Partridge 1996, Skorupa *et al.* 2008, Grandison *et al.* 2009; see review by Partridge *et al.* 2005a). Such studies can reveal how elevated reproductive rates lead to shortened lifespan, and how extended longevity is often associated with lowered age-specific fertility (for further discussion see below). The

extent of reproductive costs will often depend upon the environment (Reznick 1985, Fricke *et al.* 2009), which highlights a great need for studies employing a much broader range of environmental conditions (Cornwallis and Uller 2010). For example, in many cases trade-offs are only seen under stressful conditions (e.g. Stearns 1989, Marden *et al.* 2003). Field studies on birds have also revealed the costs of reproduction using phenotypic manipulations of brood size (e.g. Gustafsson and Sutherland 1988).

(ii) Genetic manipulations

In genetic manipulations reproductive trade-offs are identified through direct manipulation of the genetic pathways that are predicted to be involved. The idea is to manipulate components of trade-offs by using loss of function or over-expression mutants and study their phenotypic effects on reproductive rate or survival. This has been an enormous growth area for research over the last 10–15 years, as genes that are important in determining trade-offs between reproduction and longevity have been identified (see main text). Within this category fall the many kinds of experiments that manipulate components of reproductive pathways, for example germ line removal (Maynard Smith 1958, Barnes *et al.* 2006, Flatt *et al.* 2008b), nutrient sensing pathways (Libert *et al.* 2007, Partridge *et al.* 2005a), and genetic manipulations of heat shock chaperone genes (Tatar 1999, Silbermann and Tatar 2000). A particularly powerful approach is to combine phenotypic and genetic manipulations to test the effects of, for example, the dietary components involved in trade-offs along with the genes that respond to those components.

(iii) Phenotypic correlations

Physiological costs of reproduction can also be measured by testing for phenotypic correlations, where fertility and survival are measure in organisms allowed to reproduce at their normal rate. The limitations are that it can be difficult to determine the causal relationships involved because any observed correlation may be caused by a common correlation with another uncontrolled factor (Reznick *et al.* 2000). However, this approach has been used successfully to gather evidence for costs of reproduction in human populations (e.g., Westendorp and Kirkwood 1998, Helle *et al.* 2002; see also Chapter 10). The value of correlative techniques can be substantially increased by combining it with approaches (i) and (ii) above to demonstrate causal links between the life history traits predicted to show trade-offs.

2. Measuring evolutionary costs of reproduction

(i) Genetic correlations and correlated responses to selection

The measurement of genetic correlations between life history variables can give strong evidence for evolutionary costs of reproduction (Rose and Charlesworth 1981, Lande 1982, Reznick 1985) and can be derived directly from artificial selection, experimental evolution, or from breeding experiments (Falconer 1981). Genetic correlations between life history variables can indicate the presence and extent of antagonistic pleiotropy. A negative genetic correlation between early fecundity and longevity would imply that, on average, mutations that increase fecundity also decrease longevity (Reznick 1985). Correlated responses to artificial selection are measured in artificial selection or experimental evolution. For example, Rose and Charlesworth (1981) reported a decrease in

the early fecundity of lines of *D. melanogaster* that were selected for late-age reproduction by using older adults as parents in successive generations. Reduced fecundity is also commonly observed in response to selection for increased lifespan (e.g., Zwaan 1999).

Selection experiments (and inbred lines) can also be combined with QTL analysis for a useful way of determining genes with major effects on life history traits (e.g., Leips and Mackay 2000). Such techniques have been used to determine genes that affect longevity (e.g., Lai *et al.* 2007), and there is, in principle, no reason why such methods could not be used to detect genes with major effects upon life history trade-offs. Evolutionary trade-offs can also come from the characterization of isogenic lines for evolved differences in the shape of trade-offs. These approaches have so far been underused in the study of reproductive trade-offs and represent a potentially useful avenue for future work.

There are various caveats about the measurement of genetic correlations from artificial selection and breeding experiments: (i) even in the laboratory, the measurement of quantitative traits can be imprecise (Falconer 1981), (ii) genetic variation in the rate of reproduction is generally much lower than can be produced by phenotypic manipulations, hence the power to detect genetic correlations may be lower than for phenotypic correlations, (iii) genetic correlations may not remain constant over time or during artificial selection. Indeed, in some selection experiments the sign of the genetic correlation between two fitness-related traits changed from positive to negative (e.g., Archer *et al.* 2003). It is also important to consider whether trade-offs that evolve under benign and constant environments in the laboratory will reflect those seen under natural conditions. An advantage is to combine evolutionary manipulations and measurements of genetic correlations with the physiological techniques described in 1 above. Concordance between the results of studies using these varied techniques allows powerful inferences to be made.

(ii) Population and species comparisons

Comparative data can indicate broad-scale evidence for costs of reproduction. For example, within several different groups of animals, there are negative correlations between high reproductive output and repeated breeding, giving evidence for a trade-off between current and future reproductive output (e.g., in several species of triclads, mites, many species of lizards, and birds; Roff 1992, Stearns 1992). Caveats include the influence of gene–environment interactions, which may confound when comparing the reproductive rates of different populations or species in

continues

Box 11-1 (*continued*)

environments to which they are unequally adapted. Another potential drawback is the impact of ecological variables and population dynamics (e.g., Gustafsson and Sutherland 1988). Correlated life history traits may also be independent adaptations to different environments and their association

need not, therefore, imply a constraint or trade-off. Combining the results of comparative data with experimental studies is an advantage, if it is possible, because, as mentioned earlier, it allows the causal relationships to be elucidated with greater confidence.

Box 11-2 New directions in measuring physiological and evolutionary costs: Genomic approaches

Genomic approaches can measure both physiological and evolutionary trade-offs and there has been a rapid increase in their deployment (e.g., Bochdanvovits and De Jong 2004). Included in this category are the determination of expression profiles by microarray or deep sequencing technology, sequencing of candidate genes involved in determining lifespan across different populations (Schmidt *et al.* 2000), detecting signatures of selection in candidate genes, and sequencing entire genomes. A promising approach is the full genomic characterization of lines selected for reproductive trade-offs, to test for evolved

differences in gene sequences and the shape of such relationships. These methods cut across phenotypic and evolutionary trade-offs because they can measure the downstream responses to phenotypic and genetic manipulations to selection for different reproductive strategies (e.g., McElwee *et al.* 2007). They can also measure the expression or sequence of genes or genomes across populations or species. Genomic, and also the currently underutilized proteomic, analyses are set to provide an increasingly important global and tissue-specific signature of the impact of reproductive costs.

11.3.1   Physiological costs of reproduction

Physiological trade-offs occur within individuals and can represent plastic responses to resource levels (Stearns 1992). For example, an organism experiencing an overabundance of resources might elevate current reproductive rate at the expense of future survival or reproduction. Conversely, it might be beneficial to conserve energy reserves if resources are scarce, and increase current and future survival probabilities until resource levels increase. Organisms are faced with these allocation "decisions" because current and future reproductive rates, along with survival, cannot all be maximized. Plastic responses allow organisms to adjust reproductive rates to prevailing conditions. The result is that current reproductive rate may trade-off with future reproductive rate and survival. Even though this kind of trade-off is contrasted with evolutionary trade-offs (see below), the plasticity is itself, of course, an evolved strategy.

There is evidence that the physiological costs of reproduction can be influenced by hormones and/or by nutrient-sensing pathways, and this evidence is discussed in section 11.4. However, the nature of physiological costs will depend upon resource acquisition modes. So-called "income breeders" are those organisms that maintain no energy reserves. Their current reproductive rate depends entirely on current food intake and such organisms have "direct costing" of reproduction (Sibly and Calow 1986). For income breeders, trade-offs between different life history traits will depend largely on current foraging strategies rather than physiological trade-offs *per se*. On the other hand, "capital breeders" are those able to stockpile energy reserves, either during development or through times of glut. Capital breeders exhibit "absorption costing" (Sibly and Calow 1986), i.e., where costs can be buffered. Reproductive rate can therefore be maintained at a level that is not directly linked to the prevailing conditions. In this case, decisions concerning which resource allocation strategy to adopt are more complex. Many species will fall somewhere between income and capital breeders. To date, the majority of theoretical and empirical investigation has been into systems that exhibit capital breeding. Interestingly, even in organisms such as *D. melanogaster* that can maintain some energy reserves, reproductive activity is tightly linked to external resource levels. This can be seen in the remarkably tight temporal correlation between mortality rate and reproductive schedule in *D. melanogaster* switched between good- and poor-quality diets (Mair *et al.* 2003).

11.3.2   Evolutionary costs of reproduction

Evolutionary trade-offs are revealed by the existence of fixed life history strategies that differ between individuals. For example, artificial selection for high early reproductive rate often results in a correlated response in shortened lifespan, and *vice versa* (e.g., Rose and Charlesworth 1981, Partridge and Fowler 1992, Sgrò and Partridge 1999). These observations reveal survival costs of reproduction because individuals cannot normally show high reproductive rate *and* long lifespan. Such data provide important evidence for antagonistic pleiotropy (Williams 1957). Mechanisms underlying evolutionary costs of reproduction have been revealed by testing the effects of abolishing egg production in lines artificially selected for early and late age reproduction (Sgrò and Partridge 1999), and by assessing hormone titers in wing polymorphic crickets (Harshman and Zera 2007; Chapter 24 and Sections 11.3.5 and 11.4 below). Nevertheless, there is relatively little mechanistic work in this area so far. This is an important oversight because it is not yet clear whether physiological and evolutionary trade-offs occur via the same underlying mechanisms. It would be interesting to know, for example, whether individuals selected for early- or late-age reproduction retain equal capacity to express physiological trade-offs; that is, whether the effects underlying these different kinds of trade-offs are additive. It would also be useful to know whether nutrient signaling evolves during artificial selection for early- and late-age reproduction.

A further type of evolutionary trade-off is found in the broad-scale differences between reproductive output and fecundity that occur between species, as identified in comparative analyses (Partridge and Gems 2006). Such patterns must be products of selection for different reproductive strategies, but very little mechanistic work has yet been conducted. It would greatly illuminate the study of reproductive costs to test for differences in the expression or sequence of genes influencing trade-offs in different populations or species. The study of evolutionary gerontology or "Evo-Gero" (Partridge and Gems 2006) is therefore a promising new field. One of the few examples of this kind of mechanistic work to date concerns the lifespan extending *Methuselah* gene in *D. melanogaster* (Lin *et al.* 1998), which shows an intraspecific geographic cline in sequence variation (Schmidt *et al.* 2000).

11.3.3   Mechanisms underlying reproductive costs

Current reproduction could trade off with future reproduction or with lifespan, either because it diverts resources away from somatic maintenance or because it causes damage. Both explanations predict a causal and negative relationship between reproductive rate and future reproduction/survival. Long-lived animals often show increased resistance to heat and other stresses (see Section 11.7 and review by Partridge *et al.* 2005a). However, increasing evidence from studies of model organisms suggests that reproductive trade-offs arise because of links between resource acquisition (diet and nutrients), metabolism, lifespan, and reproduction (Flatt 2009). The main lines of evidence for this conclusion are as follows (adapted from Flatt and Schmidt 2009):

• The existence and extent of trade-offs between reproduction and lifespan depend upon nutrient levels (e.g., Chapman and Partridge 1996, Marden *et al.* 2003).
• The increased lifespan seen in *D. melanogaster* lacking a germ line is accompanied by reduced levels of glucose and trehalose (Flatt *et al.* 2008b), suggesting that the germ line is intimately involved in nutrient signaling.

• The increased lifespan seen in *C. elegans* mutants without a germ line depends upon the presence of a downstream component (Daf-16/FOXO) of the insulin/IGF-like signaling (IIS) pathway (Arantes-Oliveira *et al.* 2002), showing that the extension of lifespan is dependent on intact nutrient sensing pathways.

• The removal of germ cells in *C. elegans* leads to fat deposition throughout the body and modulation of the extent of fat deposition can itself alter lifespan (Wang *et al.* 2008). This again suggests a link between the germ line and nutrient metabolism.

• *C. elegans* that lack germ cells (but not the gonad itself) show increased lifespan that cannot be further extended by dietary restriction (DR). Hence the removal of the germ line and the effects of DR appear to be non-additive and may represent part of the same mechanism (Crawford *et al.* 2007).

Although the evidence for a role of nutrients in mediating reproductive costs is increasing, the details of the links between all these processes are not yet well understood (Flatt 2009). The majority of the mechanistic data come from intraspecific phenotypic or genetic manipulations resulting in lifespan extension. These data therefore help us to understand the trade-off between reproduction and lifespan. There are, however, few mechanistic data on the mechanisms underlying trade-offs between current and future reproductive rate, or on mechanisms underlying evolutionary trade-offs in general.

In the following sections we outline some of the known mechanisms that influence the trade-off between current reproductive rate and lifespan: nutrients and nutrient sensing, hormones, immunity, and damage repair mechanisms.

11.3.4 Nutrients, nutrient sensing, and costs of reproduction between reproductive rate and lifespan

Nutrient signaling can influence the extent and shape of reproductive trade-offs (Fig. 11–1) and may allow individuals to exhibit plastic trade-offs by shifting resources from reproduction to somatic maintenance. Whether these trade-offs are mediated by organisms physically shifting resources from one process to another or by molecular signaling is

not yet known (see Section 11.5). It is now well established that DR leads to extended longevity in yeast, flies, worms, rodents, and perhaps primates, as well as resulting in beneficial effects on health in humans (reviewed in Partridge *et al.* 2005b). These phenotypic manipulations are sometimes associated with decreased age-specific reproductive output (e.g., Chapman and Partridge 1996, Toivonen and Partridge 2009), which implies that how nutrients are sensed and used may underlie the trade-off between reproductive rate and survival. Consistent with this, there are many studies in which genetic manipulations have been used to alter expression in components of nutrient sensing pathways such as in the IIS and target of rapamycin (TOR) pathways. These manipulations usually result in increased lifespan and are sometimes (although not always, Grandison *et al.* 2009), associated with decreased age-specific reproduction (Clancy *et al.* 2001, reviewed in Toivonen and Partridge 2009).

Great strides have recently been made in identifying the specific nutrients that influence reproductive trade-offs (Skorupa *et al.* 2008). An emerging theme is that such trade-offs are not simply founded upon variation in resource levels (e.g., calories) but by the balance of nutrients available (Mair *et al.* 2005, Grandison *et al.* 2009). Studies in *D. melanogaster* (Lee *et al.* 2008b) and in field crickets *Teleogryllus commodus* (Maklakov *et al.* 2008b) reveal that lifespan and reproductive rate can be maximized by different diets. Components of the diet such as casein, in *D. melanogaster* (Min and Tatar 2006), and essential amino acids such as methionine, in *D. melanogaster* and rodents (Miller *et al.* 2005, Zid *et al.* 2009), alter lifespan. In a series of ingenious dietary component add-back experiments, methionine has been identified as the amino acid that promotes longer life in *D. melanogaster* (Grandison *et al.* 2009). What is intriguing is that the addition of methionine to the diet was able to restore egg production in long-lived flies on an otherwise restricted diet. This finding apparently indicates that the trade-off between reproductive rate and lifespan can be abolished if dietary nutrients are finely tuned (Grandison *et al.* 2009, see Section 11.5). The known roles of the two main nutrient signaling pathways in influencing trade-offs between reproductive rate and lifespan are outlined below. Much remains to be discovered about the

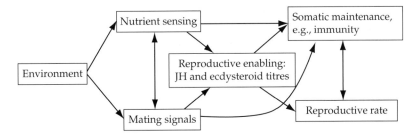

Figure 11-1 Potential links between the major players that determine reproductive costs. Nutrient sensing pathways detect nutrient levels/condition determined by the local environment and may determine the extent of reproductive activation (enabling) that can occur, or may have only direct effects on somatic maintenance. The quality or quantity of mating signals (and responses to them) may depend on the environment directly or indirectly, via nutrient sensing. Mating signals can determine the extent of reproduction, but may also have direct effects on somatic maintenance that bypass reproductive enabling. Information on the level of reproduction is integrated and inputs into the level of reproduction and somatic maintenance. Reproductive hormones may themselves directly suppress somatic processes such as the immune system. The challenge is to determine the existence and relative importance of these links.

exact role of these pathways in mediating the costs of reproduction more generally.

11.3.4.1   Insulin signaling

Elegant experiments employing single gene mutations in the nutrient sensing IIS pathway have shown that reduced insulin signaling can lead to extended longevity (e.g., by manipulation of *chico, InR, Lnk* and *dilp* 2,3, and 5; Clancy *et al.* 2001, Tatar *et al.* 2001b, Ikeya *et al.* 2002, 2009, Broughton *et al.* 2005, Grönke *et al.* 2010, Slack *et al.* 2010). This effect is sometimes also associated with reduced fertility or female sterility (e.g., Clancy *et al.* 2001, Tatar *et al.* 2001b) and the effects of insulin signaling appear to be evolutionarily conserved (Partridge *et al.* 2005a). These findings support the general idea of a trade-off between longevity and reproductive rate that is mediated at least in part by insulin signaling. However, not all of the genes in the insulin pathway whose manipulation leads to extended longevity also lead to reduced fecundity. For example, *chico* heterozygotes and certain long-lived flies with altered dFOXO signaling do not have impaired fertility (reviewed in Toivonen and Partridge 2009). Nevertheless, reproductive trade-offs may sometimes be apparent only under stressful conditions such as food limitation, so may be overlooked. For example, the fertility of *Indy* long-lived mutants appears normal under food abundance, but is reduced under food limitation (Marden *et al.* 2003). This is not a *Drosophila*-specific phenomenon. There

are similar inconsistencies in *C. elegans* (e.g., for the *Daf*-2 gene) and in mice, where long life is associated with decreased fertility in some but not all strains that are long-lived as a result of manipulations to insulin signaling (Toivonen and Partridge 2009). It seems that there is no obligate relationship between increased lifespan and decreased fertility mediated by insulin signaling, and reasons for these discrepancies are discussed below (see Chapter 13).

Insulin signaling may influence reproductive costs because this pathway conveys information about the total resource budget. Alternatively, it may be the case that insulin signaling genes themselves alter reproductive hormone levels. For example, mutations in the *Drosophila* insulin receptor (*DInR*) are associated with both reduced levels of ecdysteroid release from ovaries *in vitro* and with reduced juvenile hormone (JH) biosynthesis (Tatar *et al.* 2001b). The latter effect was seen *in vitro* using the *chico* mutation in some experiments but not others (Richard *et al.* 2005).

11.3.4.2   TOR signaling

The target of rapamycin (TOR) nutrient sensing pathway is responsive to amino acid levels and also interacts with insulin signaling. As for IIS, mutations in members of the TOR pathway also extend lifespan in *Drosophila* (Kapahi *et al.* 2004) and in yeast. It is not yet known whether lifespan extension via TOR signaling also reduces fertility, so it is not yet clear the extent to which TOR influences fertility and survival trade-offs. The impact of the cross-talk between the IIS and TOR pathways on reproductive rate is also not yet known (Flatt 2009).

11.3.5 The presence of a germ line and costs of reproduction between reproductive rate and lifespan

Evidence for the importance of the germ line in mediating trade-offs between reproduction and lifespan comes from classic studies in which extended lifespan was seen in animals that lay no, or reduced numbers of, eggs (Maynard Smith 1958). The underlying hypothesis is that in intact animals, nutrients are signaled to the germ line to match reproductive rate to nutrient levels. In the absence of the germ line, resources are not allocated to reproduction and stay in longevity assurance, leading to longer lifespan.

Early studies showed increased lifespan in *grand-childless* females of *Drosophila subobscura* that lacked a germ line (e.g., Maynard Smith 1958). Sterilization of females by low doses of X-irradiation can also produce extended lifespan (e.g., in *Ceratitis capitata*; Chapman *et al.* 1998). These results are consistent with the idea that some aspect of reproductive activity leading to egg production is costly. Similar results have been obtained from studies of germ line ablation in *C. elegans* (e.g., Leroi 2001). Elevated rates of egg-production and exposure to males cause a drop in *D. melanogaster* female lifespan (Partridge *et al.* 1987); these effects could be due to the costs of egg production or costs of mating. However, elevated egg production by itself can lead to decreased lifespan in both phenotypic manipulations and in selection experiments (e.g., Partridge *et al.* 1987, Sgrò and Partridge 1999). The latter study provides one of the few pieces of mechanistic evidence for an evolutionary trade-off between current reproductive rate (egg production) and lifespan. The ovo^{D1} mutation conferring sterility on females was crossed into lines of flies selected for early- and late-age reproduction. The subsequent patterns of mortality in egg laying and non-laying females suggested that reproduction (egg laying) caused a delayed wave of increased mortality. This suggests that similar pathways can operate in both physiological and evolutionary trade-offs.

Recent studies have highlighted that it is the developmental stage at which the germ line is inactivated that is important. For example, early ablation of the germ line during development leads to no effect on lifespan (Barnes *et al.* 2006), possibly because of increased signaling of the somatic gonad that can proliferate in the absence of the germ line. However, later-acting germ line silencing had the predicted effect of increasing lifespan (Flatt *et al.* 2008b). Germline removal extended lifespan in both sexes. However, the nature of any trade-off in males is not yet known. Germline-lacking males will court and mate at levels similar to those of control males (e.g., Chapman *et al.* 1993), but the investment of such males in courtship and mating, and their reproductive hormone titers, have not yet been tested.

11.4 Reproductive hormones as mediators of trade-offs between reproductive rate and lifespan

Hormones have disparate effects that underlie many different life history traits and therefore they have long been thought to play a central role in mediating life history trade-offs (Harshman and Zera 2007). In insects, activation of reproduction is influenced by the balance of JH and ecdysteroid signaling (Nijhout 1994). For example, in *D. melanogaster* females, reproduction is controlled by the actions of 20-hydroxyecdysone (20E) and JH. The balance between 20E and JH determines whether oocytes undergo vitellogenesis (the uptake of yolk proteins, stimulated by JH) or apoptosis (stimulated by 20E) (Soller *et al.* 1999). In males, JH is essential for the formation and function of the accessory glands, which synthesize much of the non-sperm part of the ejaculate. 20E is also important; it is synthesized in the male prothoracic gland and regulates spermatogenesis and accessory gland development.

Evidence for the involvement of hormones in life history trade-offs between reproduction and dispersal comes from selection experiments in wing-polymorphic crickets. Elevated early reproduction is linked with high JH and ecdysteroid titres in short-wing morphs. This elevated early reproductive rate is associated with a low dispersal ability (i.e., the presence of short wings). Consistent with this, the application of a JH mimic to long-winged morphs of *Gryllus firmus* produces changes in lipid metabolism, ovarian growth, and flight muscle to levels more characteristic of short-winged morphs (Harshman and Zera 2007). The role of JH on lifespan and in mediating trade-offs has been investigated in *Drosophila* (Flatt and Kawecki 2007), and there appears to be no obligate trade-off with fecundity. Similarly, reductions in signaling by the ecdysteroid pathway (achieved via mutations to ecdysone receptor, *EcR*) cause an increase in lifespan with no apparent decrease in fecundity (Simon *et al.* 2003). Further support for a role of JH in influencing lifespan comes from studies of diapausing insects. Elevated JH is associated with reduced lifespan in the Monarch butterfly, and levels of JH are also low in the diapause stage (in which aging is reduced) of several invertebrate species (reviewed by Flatt *et al.* 2005). However, it is not yet clear whether these associations covary with reproductive costs. JH is essential for vitellogenesis, which can itself be costly (e.g., Partridge *et al.* 1987, Sgrò and Partridge 1999), and mating also causes a significant increase in JH levels in females of many insects.

11.5 Male seminal fluid proteins as mediators of trade-offs between reproduction and lifespan in females

There are considerable costs for females in mating itself (Fowler and Partridge 1989). Using transgenic males in which a population of seminal fluid producing cells is genetically ablated (Chapman *et al.* 1995) shows that the cost of mating in female *D. melanogaster* is explained by the transfer of seminal fluid proteins. Furthermore, there is a dose–response effect of seminal fluid proteins on lifespan and lifetime reproductive success. Whether there is one seminal fluid protein or many that are responsible for this cost of mating remains an open question. There are four seminal fluid proteins that are toxic when ectopically over-expressed (Mueller *et al.* 2007). In addition, there are associations between sequence variation at two seminal fluid protein loci and the differences in lifespan between virgin and singly mated *D. melanogaster* females (Fiumera *et al.* 2006).

One of the candidate costly seminal fluid proteins is the sex peptide (SP; Wigby and Chapman 2005, Fricke *et al.* 2010). The potential involvement of SP in mating costs is intriguing because it also activates systems known to be costly. For example, likely candidates for SP-mediated mating costs in female *D. melanogaster* are JH, the immune system, and/or nutrient intake. SP causes the release of JH-BIII from the corpora allata, which stimulates vitellogenesis and hence oocyte progression in the ovary (Soller *et al.* 1999). High levels of JH could decrease lifespan directly and are already found to be negatively associated with length of life in other insects (Flatt *et al.* 2005). JH might also be costly because it increases vitellogenesis, i.e., egg production, which has itself been shown to be costly (e.g., Sgrò and Partridge 1999). JH could also incur costs because it suppresses the immune system, as shown in *Tenebrio molitor* (Rolff and Siva-Jothy 2002). Alternatively, mating costs in females may result from effects of increased SP on immunity, independent of JH levels. Although SP could be costly via its effects on female feeding rate, behavioral observations have correlated feeding with the ability to lay eggs and not with female survival *per se* (Barnes *et al.* 2008). What is now needed are measurements of the potential reproductive costs following manipulations of immune and hormonal pathways, independent of the receipt of SP and the presence or absence of a germ line.

11.6 The immune system as a mediator of costs between current reproductive rate and survival

The idea that maintaining the immune system, mounting an effective immune response, or evolving an immune response may be costly and may be traded off against other life history components, such as reproduction, has gained increasing credence (reviewed in Lawniczak *et al.* 2007). Much work has been done in birds and has focused on males and their responses to testosterone. In *D. melanogaster*, constitutive expression of antimicrobial peptides via the Toll (Tl) pathway results in female sterility because Tl, as well as being a switch to upregulate immunity, is also involved in dorso–ventral pattern formation in eggs. This suggests that there are likely to be costs of antibacterial immune defense over and above those of antibacterial peptide production itself (reviewed in Lawniczak *et al.* 2007). This resembles a "design" (rather than "allocation") trade-off, i.e., where genes are selected in one context because of evolutionary history but come to fulfill other new functions, sometimes with lowered efficiency.

Reproductive activity in *D. melanogaster* males is also traded off against the ability to clear bacteria following an experimental injection (McKean and Nunney 2001). The potential trade-offs between immunity and sexually selected traits in males became of interest because of the idea that hormone-dependent traits are "honest" signals of male quality under the immunocompetence handicap hypothesis (Sheldon and Verhulst 1996, Lawniczak *et al.* 2007). Mating-induced declines in immune function have been shown in damselflies and *D. melanogaster* males and immunity-related genes show significant changes in expression following mating in *D. melanogaster* females (reviewed in Lawniczak *et al.* 2007). Taken together, the data suggest that mating costs could be incurred because mating suppresses immunity, which then leads to a decline in fitness. This could be because of allocation trade-offs, limited resource pools, or a type of "design" trade-off as mentioned above (e.g., perhaps males suppress female immunity because high immunity otherwise impairs fertilization).

Despite the hypothesis that trade-offs between reproduction and the immune system are likely to be mediated by hormones (Flatt *et al.* 2008a), there are relatively few experimental data so far on the mechanistic links between JH and the immune system (Flatt *et al.* 2005, see Chapter 13). In the mealworm beetle, *Tenebrio molitor*, mating decreases the activity of phenoloxidase (PO), a major humoral effector system, in both sexes (Rolff and Siva-Jothy 2002). Furthermore, the downregulation of PO is mediated by JH. Experimental injections of JH into male *T. molitor* also increase the attractiveness of pheromones produced by males, whilst simultaneously suppressing immune function (affecting both PO activity and ability to encapsulate non-self). Hence, the costs of mating could arise because mating produces JH, which suppresses immunity, leading to a decline in fitness. A caveat is that JH titer was not measured, and accurate measurement of JH titers *in vivo* remains an empirical hurdle in the study of insect endocrinology. However, new techniques based on mass spectrometry are emerging. In summary, it is not yet clear whether the links between immunity and reproductive rate occur because of resource trade-offs or because elevated reproductive rate leads to increased damage, perhaps leading to disease susceptibility.

11.7 Damage as a mediator of trade-offs between current reproductive rate and survival

The possibility that trade-offs between reproduction and survival are influenced by resource allocation is described above, but it is also possible that such trade-offs occur because of increased levels of damage. Such damage could occur because reproductive processes cause direct damage to the soma or alternatively suppress repair or protection mechanisms (Salmon *et al.* 2001, Wang *et al.* 2001). Consistent with the idea of direct damage to the soma are the positive associations between the level of reproductive activity and the level of reactive oxygen species (ROS; e.g., Dowling and Simmons 2009). Many long-lived strains of *D. melanogaster* and *C. elegans* are also resistant to heat stress and to oxidative stress (reviewed in Partridge *et al.* 2005a). Female *D. melanogaster* that overexpress a heat shock chaperone (Hsp70, which protects proteins from the effects of misfolding at high temperatures) extends lifespan but reduces egg hatchability (Silbermann and Tatar 2000). This is consistent with the idea that protection mechanisms such as Hsp70 expression are costly and can trade-off with life history traits. Further evidence for this view comes from the links between elevated reproductive rate and decreased immunity (see Chapter 23). Data on whether trade-offs between reproduction and immune function result from resource allocation decisions may help to resolve whether damage is manifested directly because of the effects of elevated reproduction, or indirectly via suppression of protection mechanisms. It will be interesting to discover how often elevated damage (e.g., ROS) features in pathways that mediate reproductive costs.

11.8   Resource allocation: Allocation versus adaptive signaling

A common theme of much of the experimental work described above is the lack of consistency with which extended lifespan leads to decreased fertility, especially in studies that test the effects of single gene mutations. This observation is important because it challenges the view that "Y" resource allocation models are an appropriate framework in which to study trade-offs. Such models refer to the situation where a set pool of resources are allocated to reproduction or somatic maintenance but not both (Van Noordwijk and De Jong 1986). In this final section we explore these ideas and possible explanations for these discrepancies (Flatt and Promislow 2007).

There are three main lines of evidence to suggest that the literal application of "Y" models may not fully explain the proximate mechanisms underlying the relationships between life history traits. Firstly, the elimination of reproduction does not necessarily extend life span. In *C. elegans* ablation of the whole gonad, genetic sterilization or chemical inhibition of egg production all fail to extend lifespan (Hsin and Kenyon 1999, Arantes-Oliveira *et al.* 2002, reviewed in Barnes and Partridge 2003). Also, *daf*-2 gene mutants of *C. elegans* have a significantly increased longevity over the wild type, but ablation of the wild-type germ line does not extend lifespan more than in *daf*-2 mutants. This suggests that *daf*-2 does not extend lifespan simply through re-allocating resources from reproductive processes (Leroi 2001). Secondly, although mutations that extend lifespan generally cause reduced fecundity, some apparently also increase fecundity (e.g., *daf*-2 in worms, *EcR* in flies), or have no effect on fecundity (see Barnes and Partridge 2003). Recent findings using *D. melanogaster* suggest that, given a carefully calibrated balance of dietary components, trade-offs between reproduction and lifespan can be avoided (Grandison *et al.* 2009). Thirdly, there are marked sex differences in the response of males and females to interventions that increase longevity. A mutation in the IGF-1 receptor in the mouse is found to extend female but not male lifespan. Similarly, lifespan extension in both *chico* and *EcR* mutant females is much greater than seen in males (Clancy *et al.* 2001, Simon *et al.* 2003). Such sex differences require an explanation.

A proximate explanation for this challenge to traditional life history theory is that trade-offs are mediated by molecular signals (Leroi 2001). This hypothesis is derived from observations in *C. elegans*, where ablation of the gonad fails to extend lifespan, but ablation of the germ line alone (prior to proliferation) causes lifespan to double (Hsin and Kenyon 1999). This may indicate that reproductive ability *per se* is not the key to determining longevity. Instead, what is important is the presence or absence of specific reproductive tissues. A model compatible with these findings is that the proliferating germ line produces a signal that down-regulates life span, which is counter-balanced by an equal and opposite signal from the somatic gonad (Fig. 11-2, adapted from Barnes and Partridge 2003). Ablation of the whole gonad leaves longevity unaffected (because both signals are removed) but ablation of the germ line eliminates the negative signal, extending lifespan. The putative molecular signals represent arbitrary connections between life history traits and may be independent of resource availability. Importantly, the hormonal effects of such gonad ablations in the worm are now being elucidated (Gerisch *et al.* 2007, see Chapter 22). An alternative view is that resource allocation is important in the evolution of life histories, but that life history traits do not trade-off via a literal apportioning of resources (Fig. 11-2). For example, as described above, the use of nutrients in reproduction itself might generate damage, rather than simply diverting resources from somatic protection (i.e., longevity; Barnes and Partridge 2003). An interesting empirical avenue is to track the movement of resources between different tissues. Studies that track the passage of amino and fatty acids between tissues in the wing dimorphic cricket *Gryllus firmus* provide evidence that the literal diversion of resources can occur. Long-winged morphs divert more resources to the production of triglycerides used for flight and less to ovarian protein than the short-winged morph (reviewed in Harshman and Zera 2007).

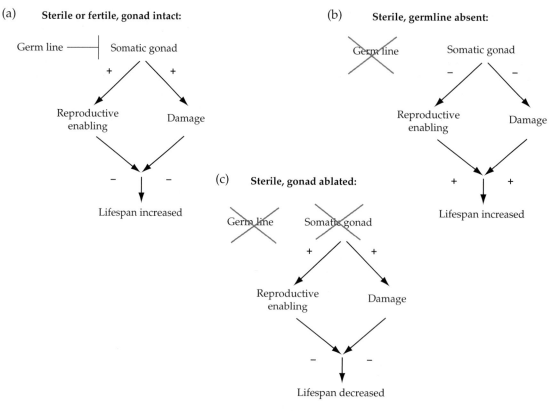

Figure 11-2 A model for adaptive reproductive signaling. Here, signals from the germ line and somatic gonad co-ordinate reproduction and life span adaptively. An intact germ line (a) causes allocation to reproduction even if the organism is otherwise sterile. In (b), the absence of a germ line allows signaling from the somatic gonad to cause investment in somatic processes. Whole gonad ablations (c) knock out the sources of all signals, mimicking a default state where investment in reproduction is high at the cost of lifespan. Reproduction and longevity can be uncoupled by manipulations to germ line or somatic gonad signaling, but the connections between the traits are adaptive because in an unmanipulated state, reproductive state is matched to local environmental conditions. Longevity and reproduction can be unlinked because costs of reproduction may still be maintained in non-reproductive individuals by the allocation of resources to reproductive function where this allocation is not blocked by sterilization. Adapted from Barnes and Partridge (2003).

The apparent absence of trade-offs between reproduction and survival could also arise because of the methodology employed. For example, gonad ablation is an extreme manipulation, so perhaps it is not surprising that resources are not shifted from reproduction to maintenance under these conditions. In addition, mutational analysis of the cost of reproduction may not be an appropriate method for confirming the absence of trade-offs, because mutations of large effect (such as those in the insulin signaling pathway) could have pleiotropic effects that obscure the normal relationships between different life history traits (Harshman and Zera 2007). The existence and shape of many trade-offs observed in the laboratory may also be significantly altered under stressful conditions and in situations where there are interactions with other ecological costs.

11.9   Costs of reproduction in a fitness-based framework

Central to studies of costs of reproduction should be an accurate quantitative assessment of key components of life history in terms of fitness—the sum of the products of individual age-specific survival probabilities times age-specific reproductive output (Charlesworth 1980). As described above, many phenotypic and genetic manipulations can extend longevity, but do so by rescheduling reproductive rate. Therefore it is necessary to be cautious and avoid viewing all manipulations that increase longevity as beneficial in fitness terms. This issue is acute, particularly in the experimental study of life history, because it is essential to assess the extent to which we can learn about fitness and fitness trade-offs from long-lived mutants. To assess fitness accurately, consideration is needed of the effects of manipulations on

both survival and the timing of reproduction (i.e., to measure the effect of manipulations on current versus future or residual reproductive output; see Fig. 11-3 for a useful way to present such data). However, fitness can be defined in several ways (Endler 1986, Clutton-Brock 1988) and the strategy that maximizes fitness may be different for decreasing, stable, or expanding populations.

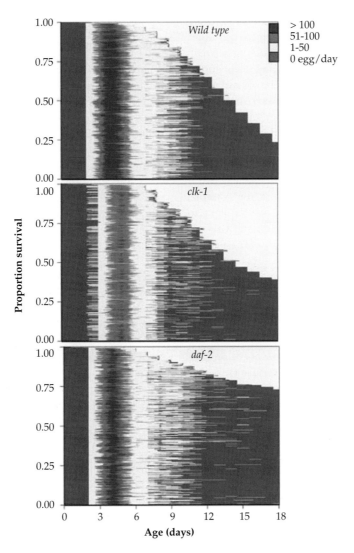

Figure 11-3   Event history diagrams. Individual event histories can help determine when in the lifetime of an organism the most important reproductive events occur, with respect to fitness. Here, the event history diagrams depict individual life histories of *C. elegans* wild type, *clk-1* and *daf-2* strains. Each row represents a single individual in increasing order of lifespan (for *n* = 1000, 800, and 800 individuals, respectively). The degree of shading depicts the amount of daily egg production. Despite increased lifespan for both *clk-1* and *daf-2* strains, a high level of early reproduction in the wild type confers a greater intrinsic rate of increase. Reproduced, with permission, from Chen *et al.* (2007).

A frequently adopted approach is to correlate changes in life history traits with lifetime reproductive success (LRS). However, a drawback to the use of LRS is that it fails to account for differences in the individual value of offspring (Brommer *et al.* 2002). In addition to genotypic and phenotypic variation in the quality of offspring, there is strong selection on the timing of offspring production from a purely demographic standpoint. Two individuals with identical LRS and lifespan can have different fitness due to the way that they partition current to future reproduction. Fitness may be maximized at intermediate levels of LRS due to a trade-off between offspring production and the timing of offspring production (Brommer *et al.* 2002). The costs of reproduction may also be age-dependent so that the trade-off might only be detected at some ages (or across some age classes), but not at others. The use of rate-sensitive measures of fitness may well be an advantage in this context (for a review see Metcalf and Pavard 2007). The important message is that it is not enough to consider effects of, for example, single gene mutations on lifespan or LRS alone to make robust conclusions about ultimate costs. To do this, findings should be placed in a fitness-based framework to investigate trade-offs between current and future reproduction (residual reproductive value).

11.10 New directions

In this final section we discuss a few of the promising, and perhaps underutilized, avenues for future research, which together will generate a richer and deeper mechanistic understanding of reproductive costs.

11.10.1 Mechanistic data are incomplete

The majority of mechanistic studies to date have used phenotypic or single gene manipulations to examine phenotypic responses, and most of these have examined the relationship between reproductive rate and lifespan. There are few mechanistic data on the reproductive trade-offs between current and future reproduction. This leaves open the opportunity for a much broader investigation of the mechanisms underlying the responses in selected lines, and of the extent to which they overlap with the responses seen upon phenotypic manipulations, as well as of the differences between populations and species. A stronger integration of evolutionary and molecular genetics of life history trade-offs is also necessary and we look forward to the comparative study of mechanisms and trade-offs that is now possible with the advent of new genomic technologies (Box 11-2).

11.10.2 The evolution and conditional economics of reproductive costs

As has been noted by many, current trade-offs are not necessarily indicative of the existence or level of costs that existed earlier in evolutionary time (Stearns 1992, Rowe and Day 2006). However, using experimental evolution and model systems with short generation times it should be possible to track the evolutionary trajectory of life history costs. For example, female costs of reproduction can indeed evolve even over relatively modest evolutionary timescales (Linklater *et al.* 2007). The evolution and mating economy of costs of reproduction is of crucial importance in understanding how mating systems are shaped by selection (Fricke *et al.* 2009). The role of the environment in determining the magnitude of costs of reproduction has long been realized, but tests of the effects of systematic manipulations of the local environment on the magnitude of reproductive costs are scant (Cornwallis and Uller 2010, Fricke *et al.* 2010). Manipulations of the local environment can include resource levels, but also disease levels and access to the opposite sex (Kokko and Rankin 2006). Another potentially promising area is the analysis of comparative demography, specifically the idea that there may be predictable differences between species in the shape of the cost of reproduction with age. Related to the evolution of reproductive trade-offs is also whether their effects are reversible (e.g., Mair *et al.* 2003) and whether they alter the rate of aging itself as well as lifespan (e.g., Priest *et al.* 2008). In addition, trans-generational reproductive trade-offs may have significant effects on inclusive fitness, but as yet there are no mechanistic data to probe the potentially important mechanisms involved.

11.10.3 Integration of life history data from social species

There is a gap in integrating information from the study of life histories from social and non-social species (see Chapter 20). There is a rich source of novel mechanistic data emerging from the study of social species, and there are several ways in which the study of life history and reproductive costs under sociality can provide new insights. Three examples are briefly discussed here.

11.10.3.1 *Novel insights into soma (growth) and reproductive trade-offs*

Social colonies often have an initial growth phase in which only workers are produced, followed by a reproductive phase in which sexuals (new queens and males) emerge. There may also be an orphan stage in which the colony continues after the death of the queen, with workers sometimes producing males. There is therefore a clear and easily measurable resource allocation to soma (growth of worker numbers) and reproduction (sexuals). This offers new and experimentally tractable ways in which to investigate trade-offs above the level of the individual, using principles that pertain at the individual level in non-social species (Bourke and Franks 1995).

11.10.3.2 *Reproductive conflicts*

A fundamental difference between eusocial and non-social species is the division between germ line and soma (Bourke and Franks 1995). In eusocial species, the germ line resides in the reproducing individuals (queens and males), with most of the soma in the sterile workers. Because of asymmetries in relatedness between sexuals and workers in a social colony, the interests of the germ line and soma are not perfectly aligned as they are within non-social reproducing individuals. This opens up novel sources of conflict between germ line and soma, and new opportunities for the expression of reproductive costs.

11.10.3.3 *Maximizing reproductive rate and longevity*

If ever there was a "Darwinian Demon" that could reproduce at high rate over a long lifespan, then it is likely to be found within social insect queens, some species of which can live and reproduce for decades (Keller and Genoud 1997). The extreme longevity of queens over that of workers is achieved in individuals that bear identical genotypes. This indicates, therefore, that differential gene expression can, at a stroke, produce startlingly different life histories through phenotypic plasticity. The gene expression differences that underlie the queen–worker divide are now being characterized (e.g., Keller and Jemielity 2006). There is much to gain from the study of the mechanisms by which genes can be turned on and off to produce such extremes in lifespan and reproductive output (Gräff *et al.* 2007), and in searching for parallels in non-social systems.

11.11   Summary

1. The costs of reproduction are a superlative example of a life history phenomenon amenable to mechanistic analysis at all levels and great progress has been made in understanding the underlying physiological and genetic basis of such costs.

2. It is important to consider physiological and evolutionary trade-offs in the study of reproductive costs and to determine whether there are common underlying mechanisms.

3. Our brief review of the current state of knowledge of the mechanisms underlying reproductive costs highlights that several pathways are emerging as important. These include nutrients and nutrient sensing mechanisms, reproductive hormones, immunity, and reproductive signals from males. However, we still lack knowledge of how, and in which order, these pathways are connected.

4. The application of new genomic technologies will yield new insights into the genomic signature of reproductive costs both within and between individuals and populations, and between species.

5. It is important to measure reproductive costs in a fitness-based framework. A wider application of such approaches may resolve some of the inconsistencies in the effects of manipulations of pathways implicated in causing reproductive costs.

6. Several existing lines of research are underutilized. Benefit will be derived from a full understanding of the conditional economics of reproductive costs, as this tells us about the magnitude of reproductive costs, the evolution of the mechanisms by which they occur, and also the evolutionary dynamics which shape them. In addition, the effective integration of

data from the study of life histories in social and non-social systems will reap great rewards.

11.12 Acknowledgments

We thank the editors for inviting us to make this contribution, and apologize to authors whose work we did not have room to cite. We thank Thomas Flatt, Andreas Heyland, Tad Kawecki, and an anonymous reviewer for their constructive comments on a previous draft. We thank the Natural Environment Research Council and the University of East Anglia for supporting our research.

CHAPTER 12

Patterns and processes of human life history evolution

Michael P. Muehlenbein and Mark V. Flinn

Phenotypic plasticity, or the ability to alter one's morphological, physiological, and behavior phenotype in response to environmental change (Pigliucci 2001), is central to organismal life histories. Plasticity in reproductive physiologies and behaviors in response to stochastic environmental signals (i.e., diet, activity, stress, disease, availability of mates) is hypothesized to represent a suite of complex adaptations, reaction norms produced by natural and sexual selection and constrained by trade-offs under conditions of resource restriction (Schlichting and Pigliucci 1998, Sinervo and Svensson 1998). This is to be expected given the central role of reproduction in life history evolution. Like other organisms described throughout this volume, humans are required to allocate physiological resources between reproduction and a number of competing functions, particularly growth and survivorship (Stearns 1992). Humans are capital breeding, iteroparous organisms that budget time and stored energy over a number of reproductive events within a lifetime.

Humans have unusual life history characteristics: we are born helpless, take a long time to mature, and can live for 70 years or more, including a female post-reproductive period following menopause (Mace 2000). These life history traits may be linked with our remarkable cognitive and social abilities compared to other species: humans have very large brains that have complex and lengthy patterns of psychological development (Geary 2005). The interactive development of human life history traits may be understood by considering the context in which they could have evolved. Here we review both theory and mecha-

nisms of human life history evolution. We emphasize the key endocrinological mediators of life history trade-offs, particularly those involving reproduction. Specific topics include reproductive maturation, male and female reproductive ecologies, mating/parental behaviors, and reproductive senescence.

12.1 The evolution of human life histories

Humans are characterized by a unique combination of traits, including:

- large brains
- habitual bipedal locomotion
- use of the upper limbs for tool-use including projectile weapons
- reduced anterior dentition with molar dominance (Lovejoy 1981) relative to other hominoids
- concealed or "cryptic" ovulation
- high fertility rates relative to other hominoids
- physically altricial but mentally precocial infancy
- long periods of juvenile dependence
- an adolescent growth spurt (Bogin 1994)
- an unusual rate of dental development (Smith 1992)
- extensive biparental and alloparental care, including large transfers of information
- menopause
- long lifespans
- multi-generational bilateral kin networks
- lethal competition among kin-based coalitions
- culture, including language.

A few other species exhibit several of these traits, but only humans are characterized by the entire combination (Alexander 2005).

The selective pressures responsible for this unique suite of characteristics and life history patterns appear central to understanding human evolution (Alexander 1990, Kaplan *et al.* 2000, Bjorklund and Pellegrini 2002, Rosenberg 2004). Here we examine how, when, and why the patterns and processes of human life history emerged. A likely scenario involves interactive changes in our brains and gonads, including changes in complex cognitive processes and reproduction/developmental physiologies.

12.1.1 Ecological dominance: Lowered mortality, better food and tools, and increased sociality

Low extrinsic mortality and subsequent slower senescence are traits of the Primate order in general (Austad and Fischer 1992, Charnov and Berrigan 1992) and humans in particular. Possibly the most important initial variable in determining mammalian life history parameters (Charnov 1993), extrinsic mortality rates could have decreased in hominin ancestors through a combination of events, including the development of flake stone tools thought to be concomitant with the emergence of our genus *Homo* (Klein 1984), or perhaps even earlier (*Australopithecus garhi*: de Heinzelin *et al.* 1999). Mortality rates may have also decreased as a consequence of increased feeding and foraging efficiency, particularly with specialization on high-quality, nutrient-dense, and difficult-to-acquire food resources (Kaplan *et al.* 2000). Such practices could have preceded or co-evolved with other traits that further promoted encephalization and complex cognitive processes. Food sharing and provisioning of juveniles could have selected for extended periods of juvenile learning, with increased capacity for learning and expansion of social networks (Kaplan *et al.* 2000, Flinn *et al.* 2005). These advantages could have outweighed the costs associated with delayed reproduction (Williams 1966b) in our hominin ancestors.

Social networking and feeding efficiency are slowly acquired skills in primates (Janson and van Schaik 1993). Slower physical development and extended juvenile dependency would allow for increased brain growth and learning of complex sexual and social behaviors (Bogin 1994, Joffe 1997). Human children are especially tuned to their social worlds and the information that it provides. The social world is a rich source of useful information for cognitive development. The human brain appears designed by natural selection to take advantage of this bonanza of data (Tooby and Cosmides 1992, Bjorklund and Pellegrini 2002, Belsky 2005). "Culture" may be viewed as a highly dynamic information pool that co-evolved with the extensive information processing abilities associated with our flexible communicative and socio-cognitive competencies (Alexander 1979). With the increasing importance and power of information in hominin social interaction, culture and tradition may have become an arena of social cooperation and competition (Baumeister 2005, Flinn 2006) with important consequences for the evolution of human life history.

12.1.2 Human cognitive evolution

The human brain has high metabolic costs: about 50% of an infant's, and 20% of an adult's, energetic resources are used to support neurological activity (Aiello and Wheeler 1995). Although the increase in energetic resources allocated to the brain was accompanied by a corresponding decrease in digestive tissue, this does not explain what the selective pressures for enhanced information processing were, nor why the resources were not reallocated to direct reproductive function. The obstetric difficulties associated with birthing a large-headed infant generate additional problems (Rosenberg and Trevathan 2002). The selective advantages of increased intelligence must have been high to overcome these costs.

The human brain, in short, is a big evolutionary puzzle. It is developmentally and metabolically expensive, evolved rapidly, and enables unusual human cognitive abilities such as language, empathy, consciousness, mental time-travel, creativity, and theory of mind. Advantages of a larger brain may include enhanced information processing capacities to contend with ecological pressures that involve sexually dimorphic activities, such as hunting and complex foraging (Kaplan and Robson 2002). There is little evidence, however, of domain-specific enlargement of those parts of the brain associated with selective pressures from the physical environment (Geary and Huffman 2002, Adolphs 2003). Indeed, human cognition has little to distinguish itself in the way of specialized ecological talents. A large brain may have been sexually selected because it was an attractive trait for mate choice (Miller 2000). However, there is little sexual dimorphism in encephalization quotient or intelligence psychometrics (Jensen 1998), nor is there a clear reason why brains would have been a target for sexual selection driven by mate choice uniquely among hominins.

The human brain did not evolve as an isolated trait; concomitant changes in other traits may provide clues to what selective pressures were important during hominin evolution. Changes in life history patterns accompanied the evident increases in information processing and communication during the Pleistocene (Dean *et al.* 2001). Gestation (pregnancy) was lengthened, but the resultant infant was even more altricial (Rosenberg 2004). Human infants must be carried, fed, and protected for a long period in comparison with those of other primates. Human childhood and adolescence are also exceptionally lengthy (Smith 1992, Bogin 1999, Leigh 2004). An extension of the juvenile period appears costly in evolutionary terms. The delay of reproduction until at least 15 years of age involves prolonged exposure to extrinsic causes of mortality and longer generation intervals. Parental and other kin investment continues for an unusually long time, often well into adulthood and perhaps even after the death of the parents. Like the big brain, human life history is an evolutionary puzzle.

Of course the child must accumulate the energetic resources necessary for physical growth. Whether the lengthening of the human juvenile period was an unavoidable response to an increasing shortage of calories, however, is uncertain. Other hominoids (chimpanzees, gorillas, orangutans) grow at similar overall rates, but mature earlier (Leigh 2004). Paradoxically, some data suggest that increased body fat is associated with earlier

puberty for girls but not for boys (Lee *et al.* 2010), and that low birth weight is actually associated with earlier puberty (Karaolis-Danckert *et al.* 2009). Better nutrition actually leads to shorter periods of juvenility, suggesting that human childhood, although possibly evolved to increase learning experiences, is still constrained by energetic requirements.

The peculiarities of the human growth curve are, however, difficult to explain from a simple model of food scarcity—the general timing of growth spurts does not appear linked to a pattern of caloric surpluses. Hence, although it is clear that human female growth and development are sensitive to nutritional constraints (Donovan and van der Werff ten Bosch 1965), the lengthening of the juvenile period during human evolution seems likely to have involved more than simple energetic constraints on growth.

One of the most difficult challenges in understanding human cognitive evolution and its handmaiden culture is the informational arms race that underlies human behavior. The reaction norms posited by evolutionary psychology to guide evoked culture within specific domains may be necessary but insufficient (Chiappe and MacDonald 2005). The mind does not appear limited to a pre-determined Pleistocene set of options—such as choosing mate A if in environment X, but choosing mate B if in environment Y—analogous to examples of simple canalized phenotypic plasticity (MacDonald and Hershberger 2005).

Keeping up in the hominin social chess game required imitation. Getting ahead favored creativity to produce new solutions to beat the current winning strategies. Random changes, however, are risky and ineffective. Hence the importance of cognitive abilities to hone choices among imagined innovations in ever more complex social scenarios. The theater of the mind that allows humans to "understand other persons as intentional agents" (Tomasello 1999, p. 526) provides the basis for the evaluation and refinement of creative solutions to the never-ending novelty of the social arms race. This process of filtering the riot of novel information generated by the creative mind favored the cognitive mechanisms for recursive pattern recognition in the "open" domains of both language

(Pinker 1994, Nowak *et al.* 2001) and social dynamics (Geary 2005, Flinn 2006). Cultural "traditions" passed down through the generations also help constrain the creative mind (Flinn and Coe 2007). The evolutionary basis for these psychological mechanisms underlying the importance of social learning and culture appears rooted in a process of "runaway social selection" (Alexander 2005, Flinn and Alexander 2007).

Selection that occurs as a consequence of interactions between species can be intense and unending, for example with parasite–host red queen evolution (Hamilton *et al.* 1990) and other biotic arms races. Intra-specific social competition may generate selective pressures that cause even more rapid and dramatic evolutionary changes. Relative to natural selection, social selection has the following characteristics (West-Eberhard 1983):

• The intensity of social selection (and consequent genetic changes) can be very high because competition among conspecifics can have especially strong effects on differential reproduction.
• Because the salient selective pressures involve competition among members of the same species, the normal ecological constraints are often relaxed for social selection. Hence traits can evolve in seemingly extreme and bizarre directions before counterbalancing natural selection slows the process.
• Because social competition involves relative superiority among conspecifics, the bar can be constantly raised in a consistent direction, generation after generation in an unending arms race.
• Because social competition can involve multiple iterations of linked strategy and counter-strategy among interacting individuals, the process of social selection can become autocatalytic, its pace and directions partly determined from within, generating what might be termed "secondary red queens." For example, reoccurrence of social competition over lifetimes and generations can favor flexible phenotypic responses such as social learning that enable constantly changing strategies. Phenotypic flexibility of learned behavior to contend with a dynamic target may benefit from enhanced information-processing capacities, especially in regard to foresight and scenario building.

Human evolution appears characterized by these circumstances, generating a process of runaway social selection (Alexander 2005, Flinn and Alexander 2007). Humans, more so than any other species, appear to have become their own most potent selective pressure, via social competition involving coalitions (Alexander 1989, Geary and Flinn 2001) and dominance of their ecologies involving niche construction (Laland *et al.* 2000). The primary functions of the most extraordinary and distinctive human mental abilities—language, imagination, self-awareness, theory of mind, mental time travel, including foresight, and consciousness—involve the negotiation of social relationships (Siegal and Varley 2002, Tulving 2002, Flinn *et al.* 2005). The multiple-party reciprocity and shifting nested sub-coalitions characteristic of human sociality generate especially difficult information processing demands for these cognitive facilities that underlie social competency. Hominin social competition involved increasing amounts of novel information and creative strategies. Culture emerged as a new selective pressure on the evolving brain.

Some of the standout features of the human brain that distinguish us from our primate relatives are asymmetrically localized in the prefrontal cortex, including especially the dorsolateral prefrontal cortex and frontal pole (Ghazanfar and Santos 2004; for review see Geary 2005). These areas appear to be involved with "social scenario building" or the ability to "see ourselves as others see us so that we may cause competitive others to see us as we wish them to" (Alexander 1990, p. 7), and are linked to specific social abilities such as understanding sarcasm (Shamay-Tsoory *et al.* 2005) and morality (Moll *et al.* 2005). An extended childhood seems to enable the development of these unusual social skills (Joffe 1997).

12.1.3 Prolonged development

The human brain exhibits high amounts of pre- and post-natal growth (Harvey and Clutton-Brock 1985). Corresponding cognitive competencies throughout development in modern humans largely direct attention toward the social environment. Plastic neural systems adapt to the nuances of the local community, such as its language (Alexander 1990,

Geary and Bjorklund 2000, Bjorklund and Pellegrini 2002, Fisher 2005). In contrast to the slow development of ecological skills of movement, fighting, and feeding, the human infant rapidly acquires skill with the complex communication system of human language (Pinker 1994, Sakai 2005). The extraordinary information-transfer abilities enabled by linguistic competency provide a conduit to the knowledge available in other human minds. This emergent capability for intensive and extensive communication potentiates the social dynamics characteristic of human groups (Deacon 1997, Dunbar 1998) and provides a new mechanism for social learning and culture.

An extended childhood appears useful for acquiring the knowledge and practice to hone social skills and to build coalitional relationships necessary for successful negotiation of the increasingly intense social competition of adolescence and adulthood. Modern rates of development were obtained by at least 150 000 years ago (Guatelli-Steinberg *et al.* 2005, see also Dean *et al.* 2001), although some differences from chimpanzee life history were already apparent among *Australopithecus afarensis* by 3.3 million years ago (Alemseged *et al.* 2006). Significant increases in encephalization occured by 1.6 million years ago (Walker and Leakey 1993) and were associated with a change in pelvic structure permitting the birth of larger-brained infants (Ruff 2002). Change in body size and pelvic structure, however, may have permitted initial brain-size increases without altering selection on timing of birth or rate of brain growth. Subsequent substantial brain-size increases would have necessitated more altricial infants, with a greater percentage of brain growth occurring postnatally (Portmann 1941). This, and the slowing rates of development during this period, may reflect intensified selection for a long period of dependency and learning throughout the evolution of *Homo erectus*, appearing in roughly modern human form by the appearance of anatomically modern *Homo sapiens*. Selection for larger females over the past 2.5 million years may have been due to increasing fecundity and the ability to produce larger, higher quality infants, perhaps facilitated by increasing ability to extract higher quality resources from the environment more consistently (Ungar *et al.* 2006).

Social and ecological competencies are developmentally expensive in time, instruction, and parental care. But when fitness is determined largely through the accumulation of resources (including body size) and skills (including social skills), delaying reproduction is appropriate (Stearns 1992, Roff 2002). For example, larger men and women of the Aché Amerindian group of Paraguay exhibit higher fertility rates compared to smaller Aché adults (Hill and Hurtado 1996). Delayed maturity may also lead to higher quality of offspring with reduced mortality rates (Stearns 1992, Roff 2002). These benefits could outweigh any costs of a longer generation time, mortality exposure during development, and a shortened reproductive span. The unusual scheduling of human reproductive maturity, including an "adrenarche" (patterned increases in adrenal activities preceding puberty), bimaturism, and a delay in direct mate competition among males (Bogin 1994, see below) appears to extend social ontogeny and increase fertility.

12.1.4 High fertility, biparental and alloparental care

Despite slow growth, a long juvenile period and late age at first reproduction, women achieve very higher lifetime fertility compared to even the closely related apes (Hill 1993) (Fig. 12-1). Humans accomplish this by shortening the length of the interbirth interval via a decrease in necessary workload and food acquisition during lactation. This could be accomplished through the exploitation of new food sources, including meat and weaning foods, the development of new food-processing methods like cooking, and other technological innovations, as well as non-maternal (paternal, kin, grandparent) childcare and provisioning of lactating and pregnant women. The altricial human infant is indicative of a protective environment provided by intense parenting and alloparental care in the context of kin groups (Hrdy 2005). But the advantages of intensive parenting, including paternal protection and other care, require a most unusual pattern of mating relationships: moderately exclusive pair bonding in multiple-male groups (Chapais 2008). No other primate (or mammal) that lives in large, cooperative multiple-reproductive-male groups has extensive

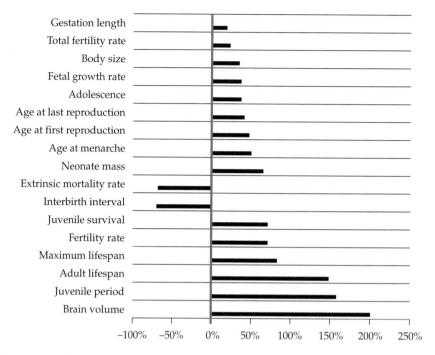

Figure 12-1 Life history traits of humans compared with chimpanzees. Horizontal bars represent the percentage differences of life history traits of modern-day natural-fertility human populations compared with wild chimpanzees. For example, the extrinsic mortality rate of humans is approximately 75% lower than that of chimpanzees, and the brain size of humans is approximately 200% greater than that of chimpanzees. Data courtesy of Robert O. Walker, Department of Anthropology, University of Missouri.

male parental care, although some protection by males is evident in baboons (Buchan *et al.* 2003).

Humans exhibit a unique "nested-family" social structure, involving complex reciprocity among males and females to restrict direct competition for mates among group members. Ensuring biparental care would be particularly important during the period of lactation that coincides with attachment, a key component in the development of social competence. Among the Hadza foragers of Tanzania, Marlowe (2003) found that husbands appear to compensate for their wives' diminished foraging return when they have young children. Similarly, among the Ache and Hiwi foragers women's time spent foraging and in childcare were inversely related; nursing women spent less time foraging than did non-nursing women, and women's foraging time was inversely related to their husbands' foraging (Hurtado *et al.* 1992). Based on these findings Marlowe (2003) suggests that pair-bonds in human evolution may function to provision a mate

and offspring during a "critical period" coinciding with lactation.

Human lactation is very energetically expensive (Prentice and Prentice 1988) and can be time consuming. Therefore there is an expected trade-off between amount of time and energy invested in breastfeeding versus women's work (Arlotti *et al.* 1998). But the benefits of breastfeeding demand investment in the task. For example, prolonged nursing likely has significant positive influences on long-term psychomotor and neural development in well-nourished populations (Horwood *et al.* 2001, Clark *et al.* 2006). Breastfeeding duration has also been associated with long-term reduction in children's stress hormone levels (Quinlan *et al.* 2003) and increased "developmental stability" (Leone *et al.* 2004). Nursing can be important to the mother–child bond, associated with positive emotions and attachment linked to maternal hormones, including prolactin and oxytocin (Ellison 2001). Maternal responsiveness, related to nursing, appears to influence the development of children's attachment styles and later sexual relations (Belsky 1997, Britton *et al.* 2006). Fathers (as well as grandparents and other kin) likely directly contributed to the evolution of high fertility in humans through various tasks, particularly the provisioning of mates and other activities that reduce maternal workload. One might say that ecological exigencies for social competence drove the evolution of human pairbonds and presently regulate pair-bond stability in contemporary populations.

It is difficult to imagine how this system could be maintained in the absence of another unusual human trait: concealed or "cryptic" ovulation (Alexander and Noonan 1979). Human groups tend to be male philopatric (males tending to remain in their natal groups), resulting in extensive male kin alliances, useful for competing against other groups of male kin (Wrangham and Peterson 1996, LeBlanc 2003). Females also have complex alliances, but usually are not involved directly in the overt physical aggression characteristic of intergroup relations (Campbell 2002, Geary and Flinn 2002). Parents and other kin may be especially important for the child's mental development of social and cultural maps because they can be relied upon as landmarks who provide relatively honest information. From this perspective, the evolutionary significance of the human family in regard to child development is viewed more as a nest from which social skills may be acquired than just an economic unit centered on the sexual division of labor (Flinn *et al.* 2005).

An increase in infant altriciality necessitates greater social support for females, and much of this can come from parents and grandparents. The fact that human women cease to reproduce when they still have a high probability of surviving for many years is a rare pattern among mammals and one that probably is absent among all other wild primates (Hawkes *et al.* 1998). However, post-menopausal women could assist in the provisioning, care, and education of dependent children and grandchildren. Stopping reproduction early before the risks of pregnancy complications are too great could potentially function to increase inclusive fitness (Williams 1957, Hamilton 1966). The evolution of large kin networks with intergenerational flow of resources would place further selection pressure on

long juvenile dependency with resulting social competence, encephalization, late age at sexual maturation followed by high fertility rates, and the other complex life history patterns of humans.

Human childhood appears to be a life history stage that is useful for acquiring the information and practice to build and refine the mental algorithms critical for negotiating the social coalitions that are key to success in our species (Bogin 1999). Slow, steady growth throughout childhood would also result in larger adult body size (ibid), which again would reflect investment in physical embodied capital capable of better protection and provisioning of offspring. All of this would have been accompanied by concomitant changes in our reproductive biology and behavior. These functional attributes have been and are being shaped through various life history mechanisms, most notably through the functions of hormones.

12.2 Proximate mechanisms of human life history patterns

It is inherently difficult to measure life history mechanisms and quantify trade-offs in humans as we are unable to intentionally manipulate the system to produce genetically evolved response patterns so as to witness unequivocally the production of phenotypic variation keyed to specific environmental variations. But, as in most other organisms examined to date, the neuroendocrine system is undoubtedly central to mediating phenotypic variation. That is, human reproduction, including its correlated physiological and behavioral functions, has evolved into a complex system of components whose proper performance is regulated in part by the neuroendocrine system. Both from a macro- and microevolutionary perspective, hormones are central mechanisms that contribute to the onset and timing of key life history events, the optimal allocation of time and energy between competing functions, and the general modulation of phenotypic and genotypic expression in response to environmental signals (Ketterson and Nolan 1992, Zera and Harshman 2001, Muehlenbein and Bribiescas 2005, Bribiescas and Ellison 2008). Their functions in reproductive development, gonadal function, mating/parenting behaviors, and reproductive senescence are both similar and different in males and females.

12.2.1 Reproductive development

Puberty marks a fundamental life history shift: from investment in growth to investment in reproductive effort. Organisms become subjected to a different set of life history trade-offs now involving reproduction rather than growth. Organisms are selected to begin this process at a size and age that will maximize survivorship and lifetime reproductive success (Charnov 1991). Late ages at puberty and first reproduction in humans allow for increased learning as discussed above, as well as investment in embodied capital through a sustained growth period. Body size is directly correlated with fertility in some human populations (Hill and Hurtado 1996), probably due to greater net energy-production capabilities. Increased cognitive and general somatic tissue investment would further function to decrease extrinsic mortality, allowing for still longer lifespans and periods of juvenile dependency.

12.2.1.1 Patterns and processes of development
Humans have a rather unusual growth curve compared to all other animals: we exhibit a combination of very rapid pre- and post-natal brain growth (resulting in a very large adult brain relative to body size) up to about one year postpartum, very slow somatic postnatal growth rates, a long juvenile period with delayed puberty, and an adolescent growth spurt (Bogin 1999). That is, our secondary altriciality is followed by a nadir in growth, allowing for a learning phase, and then followed by a growth spurt in both sexes in which rates of growth of both bone and sexually dimorphic tissue rise dramatically, allowing body size to essentially "catch up" with brain size. Throughout human evolution this pattern was likely made possible only via biparental and alloparental childcare, since otherwise child foraging returns are quite low (Hewlett and Lamb 2005).

The actual mechanisms that allow for the production of such an unusual pattern and process are incompletely understood. A number of hormones are responsible for growth and differentia-

tion (Grumbach and Styne 1998). Changes in the frequency and amplitude of gonadotropin production and release occur throughout pubarche (Ojeda 2004). A myriad of other hormones play central roles in bone and tissue proliferation, including growth hormone and insulin-like growth factors (Reiter and Rosenfeld 1998), androgens (Jepson et al. 1973), estrogens (Juul 2001), glucocorticoids (Ballard 1979), thyroid hormones (Steinacker et al. 2005), and even leptin and ghrelin (Klok et al. 2007).

Adrenal androgens (androstenedione, dehydroepiandrosterone, and dehydroepiandrosterone sulfate) may also be central determinants in the timing of pubarche. Adrenarche, or the onset of adrenal androgen secretion, precedes the pubertal growth spurt by several years (Odell and Parker 1985) and may be a life history trait of humans and only a few other species (Muehlenbein et al. 2001, Arlt et al. 2002). It has been suggested that adrenarche alters hypothalamic sensitivity to steroids and thus plays a determinant role in the timing of puberty (Havelock et al. 2004). Adrenarche may also play an important role in human brain maturation and thus cognitive development (Campbell 2006).

12.2.1.2 Timing of reproductive development
An organism should undergo sexual maturation at a time when the environment is conducive to offspring survival (low mortality), when mates are available, and when one can afford the enormous energetic costs associated with the pubertal process and subsequent reproductive events. The timing of reproductive maturation may be dependent on resource availability in the environment, as reflected in some by the accumulation of sufficient energy stores. Frisch and others suggested that a minimum amount of body fat is necessary for adolescent onset and the maintenance of regular menstruation, although the relationship between weight and age at menarche was actually non-significant in their studies (Frisch and Revelle 1971, Frisch and McArthur 1974). However, undernourishment is associated with delayed menarche in females and even later puberty in males (Campbell et al. 2004, 2005). Conversely, obese girls often reach menarche earlier than the average (Donovan and van der Werff ten Bosch 1965). The secular trend of earlier

age at maturation in industrialized populations throughout the 20th century is also indicative of the effects of better nutrition on development (Marshall and Tanner 1986). If reproductive physiology is heavily dependent upon energy availability, then it should be sensitive to ecological variables.

Adipose tissue may "communicate" relative energetic status to the brain, operating to trigger a delay in gonadarche. Adrenal androgens may be aromatized in adipose tissue and the resultant steroids could alter hypothalamic release of gonadotropins (Katz *et al.* 1985). Leptin produced by adipocytes could perform a similar function, although leptin levels accounted for only 3% of variation in age at menarche in a sample of young women (Matkovic *et al.* 1997). Furthermore, females with a genetic mutation that prohibits them from depositing white adipose tissue still undergo menarche (Andreelli *et al.* 2000).

Whatever the endocrine mechanisms may be, it appears that skeletal maturation actually exerts a much stronger influence, accounting for three times as much menarcheal age variance as weight or relative weight (Ellison 1982). In fact, mean weight at menarche has remained fairly stable over the past 100 years in industrialized populations while mean height at menarche has increased, indicating there is an appropriate physical size (specifically a biiliac diameter of approximate 25 cm) for reproduction (ibid). From a life history perspective, reproduction should not begin (or at least function properly) until growth has essentially stopped, as indicated through skeletal maturation in this case. And even after reaching this minimal skeletal development, young women still display diminished capacity for secretion of progesterone from the corpus luteum, exhibiting high anovulatory and luteally insufficient cycles in the first few years after menarche (Ellison 2001).

In general, these young females demonstrate delayed follicular development (Apter *et al.* 1987) and increased risk of low birth weight, prematurity, and higher infant mortality rates, even when controlling for sociodemographic factors such as partner support, prenatal care, education, and income (Fraser *et al.* 1995). These gynecologically young women are still growing and thus competing with their developing fetus. Biological immaturity is not

conducive to maximizing investments in somatic growth and reproductive effort at the same time.

In addition to nutrition, social factors may also influence variation in human developmental timing, including bimaturation between the sexes. In some rodents, exposure of female pups to unrelated adult males may result in accelerated maturation (Vandenbergh 1967, Vandenbergh 1973). Males from a variety of species can exhibit alternative adult phenotypes, thought to be adaptations to the social environment and reproductive opportunities. In humans, early maturation has been reported in females raised in households characterized by high levels of stress (e.g., history of maternal mood disorders, lack of resources, stepfather presence, etc.) compared to less stressful households (Ellis and Garber 2000). The mechanisms may or may not be similar for human and nonhuman animals: develop quickly to either mate with the unrelated males or disperse quickly from potentially dangerous environments.

Earlier development in women relative to men may also be an adaptation to human social context. In this case, girls may be afforded the opportunity to learn important parenting and social skills while they are technically infertile (due to adolescent subfecundity discussed above) but are perceived as mature adults (Bogin 1999). In contrast, young boys who are still reproductively potent during early adolescence (Bogin 1994) could learn their social roles when they are not yet perceived as adults, postponing direct competition with other males for access to mates and allowing for the attainment of larger body size (Bogin 1999). The social context may set the stage for bimaturation in humans, and bimaturation may further reinforce the social context of typified sex differences in behaviors.

12.2.2   Ovarian and testicular functions

Iteroparous organisms must carefully balance investments between current and future reproductive events. Current reproductive events may be curtailed if the natural and social environments are not conducive to the successful production and survival of offspring (Wasser and Barash 1983). Additionally, investment in reproduction can compromise survivorship, so organisms (including humans) are equipped with endocrinological mechanisms capable of augmenting or suppressing reproduction physiologies and behaviors in response to environmental cues, including energy balance, disease, and availability of mates. Variation in hormone levels, as well as receptor location, number, and sensitivity, allows for the alteration of target tissues and ultimately a modulation of reproductive effort relative to other investments such as growth or maintenance.

A trade-off between reproductive effort and survivorship is evident in contemporary human populations and was likely a salient feature well before the evolution of modern humans. Women in poor energetic condition often produce children of poor health (Pike 2000). Multiparous women also exhibit shorter lifespans than women with fewer children (Jasienska et al. 2006, Pike 1999). Women with many children are also at higher risk of developing cardiovascular disease, diabetes, and stroke (Ness et al. 1993, Kington et al. 1997, Qureshi et al. 1997, Simmons et al. 2006), suggesting that there is a "maternal depletion" affect with negative consequences for survivorship. Reproductive suppression in response to certain environments may function to protect maternal condition.

12.2.2.1 Female reproductive ecology

Changes in human female reproductive endocrinology in response to diet, activity, disease, stress, and other factors are now viewed by evolutionary anthropologists as adaptations, rather than pathological conditions of the female, which function to optimize lifetime reproductive success (Ellison 1990, 2003). Depressed estradiol and progesterone levels often result from:

- high energy expenditure due to elevated workload (Panter-Brick et al. 1993, Jasienska 2001, 2003) or exercise (Bullen et al. 1985, Ellison and Lager 1986, Prior et al. 1992)
- low nutritional intake (Pirke et al. 1985, Lager and Ellison 1990)
- psychological stress (Nepomnaschy et al. 2004).

Estradiol produced by the thecal cells of the ovaries prepares the endometrium for implantation. Progesterone produced by the corpus luteum supports zygote implantation and maintenance of the endometrium and myometrium, mucosal development, and glandular development in the breast (Carr 1998, Casey and MacDonald 1998), and inhibits smooth muscle contraction of the myometrium by inhibiting prostaglandin synthesis (Sfakianaki and Norwitz 2006). Estradiol and progesterone are necessary for a successful pregnancy (Lipson and Ellison 1996), but these hormones vary significantly between women and within the same woman over time. Reproductive suppression as the result of environmental stress usually exhibits a dose–response relationship, ranging from small fluctuations in cycle length due to minor perturbations, to complete amenorrhea due to major insults (Ellison 1990).

It is no wonder female reproductive physiology is sensitive to perceived environmental conditions: pregnancies are extremely costly and risky. Even menstruation is energetically costly, although it would probably be more energetically taxing to continuously maintain the uterine lining rather than cycle regularly (Strassman 1996). Nursing is also extremely nutritionally demanding (Prentice and Prentice 1988, Oftedal and Iverson 1995) and, as such, is one of the largest contributors to interbirth interval variation within and between populations (Howie and McNeilly 1982, Wood 1994). Lactation suppresses ovarian function postpartum, primarily through the effects of prolactin, resulting in "lactational amenorrhea" or "postpartum anovulation": prolactin directly inhibits follicular development and steroidogenesis in the ovary (McNeilly et al. 1982). The length of lactational amenorrhea is largely dependent upon the frequency and intensity of nursing, in addition to general maternal condition (Ellison and Valeggia 2003). A long period of breastfeeding may be taxing, but necessary to produce an offspring with such a large brain.

Another mechanism for decreasing reproductive effort at any given time is fetal loss. The probability of fetal loss per conception is as great as 80% (Holman and Wood 2001), with the majority happening within three weeks of conception (Wilcox et al. 1999). Pregnancies characterized by elevated maternal cortisol levels (and therefore possibly low maternal mood) are at increased risk of early spontaneous abortion (Nepomnaschy et al. 2006). In this case, a potentially expensive reproductive event could be terminated early to restrict pregnancy to better conditions.

12.2.2.2 *Male reproductive ecology*

Like women, men are subject to life history trade-offs, primarily that between reproductive effort and survivorship. Men and women differ in their reproductive physiologies and behaviors, and thus the physiological mechanisms regulating their life history trade-offs are different. For men, gametogenesis is energetically inexpensive (Elia 1992). As a result of this, sperm quality and quantity is relatively unaffected by energetic output except under the most taxing circumstances (Bagatell and Bremner 1990). In response to seasonal workload or exercise, testosterone levels vary little (Ellison and Panter-Brick 1996, Bribiescas 2001) or only modestly (Bentley *et al.* 1993). High alcohol consumption (Muller *et al.* 2003), low zinc intake (Abbasi *et al.* 1980), and a low carbohydrate diet (Anderson *et al.* 1987) have all been associated with lower testosterone levels. Depressed testosterone levels can also result from acute or chronic exposure to psychological stressors, including military training (Gomez-Merino *et al.* 2005) and skydiving (Chatterton *et al.* 1997). But there is little evidence that these lower testosterone levels would compromise male fecundity (Guzick *et al.* 2001).

Lowered testosterone levels could be advantageous under certain conditions. Androgens can facilitate mammalian male reproductive effort through the modifications of behaviors (confidence, dominance, etc.) and physical attributes, including secondary sexual characteristics (i.e., signals of fitness) and musculoskeletal functions (e.g., skeletal muscle mass, red blood cells, cortical bone density, etc.). These would augment inter- and intrasexual competition (i.e., male–male conflict and female sexual coercion), mate attraction, and protection of mates and offspring. However, androgens can also compromise survivorship by:

- increasing the risk of negative energy balance (particularly under conditions of resource restriction; Bribiescas 1996, 2001)
- increasing the risk of prostate cancer (Carter *et al.* 1995, Soronen *et al.* 2004)

- facilitating the production of oxygen radicals (Zirkin and Chen 2000)
- reducing resistance against oxidative damage (Alonso-Alvarez *et al.* 2007)
- increasing risk of injury due to hormonally-augmented behaviors such as aggression, violence, and risk-taking (Wilson and Daly 1985, Dabbs and Dabbs 2000)
- directly causing immunosuppression (Muehlenbein and Bribiescas 2005, Muehlenbein 2008)
- reducing the amount of energy and nutrients available for tissue repair and the maintenance and activation of immune responses (Wedekind and Folstad 1994, Sheldon and Verhulst 1996, Muehlenbein and Bribiescas 2005, Muehlenbein 2008).

Developing, maintaining, and activating immune responses generates a substantial energetic burden (see Demas 2004 for review). Despite the dearth of information on human ecological immunology relative to other species, immunocompetence should be an integral part of human life history trade-offs, as it is in other species (Lochmiller and Deerenberg 2000; also see Chapter 23). Resting metabolic rates are elevated by more than 8% in young adult men during acute respiratory tract infections, even in the absence of fever (Muehlenbein *et al.* 2010). Testosterone levels are also depressed during infection, as is usually the case. Honduran men infected with *Plasmodium vivax* exhibit significantly lower testosterone levels during infection compared to post-recovery samples as well as age-matched healthy controls (Muehlenbein *et al.* 2005). Such a response could function to avoid some of the costs associated with higher androgen levels.

Changes in reproductive endocrine function throughout the normal range of variation in men functions is a basic aspect of phenotypic plasticity. High testosterone levels reflect investment in work capacity and sexual selection whereas low testosterone levels reflect investment in survivorship and parental effort (Bribiescas 1996, 2001). In both men and women, hormones are important information transducers that facilitate integrated reproductive and survivorship responses. See Chapters

24 and 26 for more details on this topic in avian and insect species.

12.2.3 Reproductive behaviors

As with other species, hormones facilitate and inhibit our behaviors, and the actions of these hormones vary according to the social and ecological contexts we find ourselves in. They may affect our perception and interpretation of events and our motivation to perform certain actions or elicit emotional responses. Some of these actions relevant to the present discussion include mating, attachment, and parental behaviors, although the relationships between hormones and these behaviors are rather equivocal.

12.2.3.1 Libido

Acute variation in testosterone levels does not correlate well with sexual motivation in healthy, eugonadal men. Testosterone and other androgens may increase the likelihood that a stimulus will elicit a sexual response, but these hormones are not absolutely necessary to produce such feelings or activities, nor are they necessary to produce or maintain an erection in most cases (Buena *et al.* 1993, Caretta *et al.* 2006). In contrast, testosterone supplementation in hypogonadal men does increase the frequencies of erotic thoughts, erections, and sexual activity (Davidson *et al.* 1978). However, in the absence of clinical deficiencies or prolonged abstinence, testosterone is not necessary given the right stimuli.

Testosterone levels do respond to perceived coital opportunities, which could facilitate sexual motivation, confidence, and competitive abilities to attract and maintain mates. Heterosexual men exhibit increased testosterone levels during short conversations with women (whom they identify as potentially sexually receptive) but not other men (Roney *et al.* 2003). In response to potential mates, salivary testosterone levels increase even more in men who self-identify as having an aggressively dominant personality (van der Meij *et al.* 2008). Mobilization of reproductive functioning during the presence of a potential mate would be advantageous, and suppression of these androgenic hormones in the absence of a mate could function to curb some of the costs associated with elevated androgen levels

(Muehlenbein 2008). Testosterone levels also increase in response to coitus (Dabbs and Mohammed 1992) and erotic stimuli (Carani *et al.* 1990). Testosterone levels increase in preparation for competitions (Booth *et al.* 1989, Mazur 1992), which would be useful for attracting and retaining a mate.

There is a significant percentage of men who are willing to consent to sexual intercourse with a potential partner they have only recently met (Clark and Hatfield 1989). A majority of young men surveyed prefer short-term over long-term mating relationships with a great number of desired partners compared to women (Buss and Schmitt 1993). They also express more interest in extra-marital mating (Wiederman 1997), pornography (Malamuth 1996), and prostitution (McGuire and Gruter 2003). These reproductive preferences in men are not necessarily the result of high testosterone levels, but rather reflect a different reproductive strategy for men, which originates from paternal uncertainty (the result of internal gestation) combined with the low energetic costs of spermatogenesis.

For healthy young women, testosterone levels are not associated with the frequency of sexual behaviors, although testosterone supplementation elevates sexual functioning and mood in women with natural- or surgically-induced menopause (Shifren *et al.* 2000). Estradiol is also not necessary for receptivity in human females, and estradiol levels are not usually associated with sexual appetite or frequency of masturbation (Cutler *et al.* 1986). However, libido likely increases around the time of ovulation when hormone levels are changing significantly (Matteo and Rissman 1984).

12.2.3.2 Attachment

Some of the most precious of all our human feelings are stimulated by close social relationships: a mother holding her newborn infant for the first time, brothers reunited after a long absence, or lovers entangled in each other's arms. Natural selection has produced neuroendocrinological mechanisms that generate potent sensations during our interactions with these most evolutionarily significant individuals. We share with our primate relatives the same basic hormones and neurotransmitters that underlie these mental gifts, but our unique evolutionary history has modified us to respond to different circumstances and situations. For example, the endocrine, neurological, and associated emotional responses of a human father to the birth of his child (see below) are likely to be quite different from the responses of a chimpanzee male.

Attachments are central in the lives of the social mammals. Basic to survival and reproduction, these interdependent relationships are the fabric of the social networks that permit individuals to maintain cooperative relationships over time. Interestingly, biochemical mechanisms appear to regulate bonds between mates, parents and offspring, the family group, and even larger social networks (Fisher 2002, Bridges 2008). Oxytocin (OXT) and arginine-vasopressin (AVP) are closely related chains of nine amino acids that are produced within the hypothalamus and appear to be primary regulators of attachment (Carter 2002, Young and Insel 2002, Curtis and Wang 2003). OXT and AVP act on a wide range of central and peripheral neurological systems, and their influence varies among mammalian species and stage of development. Within humans, the neurological effects of OXT and AVP are likely key mechanisms (e.g., Bartels and Zeki 2004) involved in the evolution of family behaviors. The effects of OXT and AVP in humans are also likely to be particularly context-dependent because of the variable and complex nature of family relationships.

12.2.3.3 *Maternal behaviors*
Women have evolved to incur the significant costs of menstruation, internal fertilization, placentation, gestation, parturition, and lactation, so it is not surprising that their contribution to child development is extensive. Among mammals in general, OXT and AVP, prolactin, estradiol, and progesterone appear involved in parental care, although the involvement of these hormones varies across species and between males and females (Insel and Young 2001). Within females, estrogens and progesterone likely prepare the brain during pregnancy for parental behavior. For example, estrogen activates the expression of genes that increase the receptor density for OXT and prolactin, thus increasing their influence (Young and Insel 2002). OXT increases just prior to birth may also prime maternal care (Fleming *et al.* 1999,

Carter 2002). OXT also increases during breastfeeding, and this increase may function to inhibit limbic hypothalamic-anterior pituitary-adrenal cortex system activity and shift the autonomic nervous system from a sympathetic tone to a parasympathetic tone (Uvnas-Moberg 1998). This results in calmness conducive to remaining in contact with the infant. It also results in a shift from external-directed energy toward the internal activity of nutrient storage and growth (Uvnas-Moberg 1998). In humans, adrenal hormones and maternal behavior are related during the postpartum period. Cortisol appears to have an arousal effect, focusing attention on infant bonding. Mothers with higher cortisol levels are more affectionate, more attracted to their infant's odor, and better at recognizing their infant's cry during the postpartum period (Fleming *et al.* 1997).

Functional magnetic resonance imaging (fMRI) studies of brain activity involved in maternal attachment in humans indicate that the activated regions are part of the reward system and contain a high density of receptors for OXT and AVP (Bartels and Zeki 2004, Fisher 2005). These studies also demonstrate that the neural regions involved in attachment activated in humans are similar to those activated in nonhuman animals. Among humans, however, neural regions associated with social judgment and assessment of the intentions and emotions of others exhibit some deactivation during attachment activities, suggesting possible links between psychological mechanisms for attachment and management of social relationships. Falling in love with a mate and offspring may involve temporary deactivation of psychological mechanisms for maintaining an individual's "social guard" in the complex reciprocity of human social networks. Dopamine levels are likely to be important for both types of relationship but may involve some distinct neural sites. It will be interesting to see what fMRI studies of attachment in human males indicate, since father–offspring, male–female mating, and male–male coalitionary relationships are where the most substantial differences from other mammals would be expected. Similarly, fMRI studies of attachment to mothers, fathers, and alloparental caretakers in human children may provide important insights into the other side of parent–offspring bonding.

12.2.3.4 *Paternal behaviors*

Paternal care is uncommon among mammals and the extent and types of paternal care vary among species. In general, the mammalian male physiological investment in offspring production is relatively small. Paternal care is more common in monogamous than polygamous mammals and is often related to hormonal and behavioral stimuli from the female. In the monogamous California mouse, disruption of the pair bond does not affect maternal care but does diminish paternal care (Gubernick 1990). In other species with biparental care, however, paternal care is not as dependent on the presence of the female (Young and Insel 2002).

Experience plays a role in influencing hormonal activation and subsequently paternal behavior in some species. Among tamarins, experienced fathers have higher levels of prolactin than first-time fathers (Ziegler and Snowdon 1997). The receptor density for OXT and AVP in specific brain regions might provide the basis for mechanisms underlying other social behaviors. Other neurotransmitters, hormones, and social cues also are likely to be involved, but slight changes in gene expression for receptor density, such as those between the ventral palladiums of meadow and prairie voles (located near the nucleus accumbens, an important component of the brain's reward system), might demonstrate how such mechanisms could be modified by selection (Lim *et al.* 2004). The dopamine D2 receptors in the nucleus accumbens appear to link the affiliative OXT and AVP pair-bonding mechanisms with positive rewarding mental states (Aragona *et al.* 2003). The combination results in the powerful addiction that many parents have for their offspring.

In humans cross-culturally, men invest less in parental effort than do women (Munroe and Munroe 1997). However, men can contribute significantly to offspring via protecting, provisioning, teaching social skills, and nurturing emotionally. That is, although one could argue that men have been selected to invest more heavily in mating than parental effort, we contend that pair-bonding and intensive paternal investment may have been important determinants of human life history evolution. In men, hormones could facilitate a shift from mating effort to parental effort, as evidenced by decreases in testosterone and increases in prolactin following pair-bonding, fatherhood, and holding a baby (Storey *et al.*, 2000, Delahunty *et al.* 2007, Gray and Campbell 2009). Cross-culturally, both marriage and fatherhood are associated with lower testosterone levels compared to unmarried men (Gray *et al.* 2002, 2006, 2007, Gray 2003). Along with elevated prolactin and oxytocin and decreased testosterone levels, increased vasopressin production may also prepare the male to be receptive to, and care for, infants (Bales *et al.* 2004). The same general neurohormonal systems active in pair bonding in other species appear to exist in humans (Wynne-Edwards 2003). The challenge is to understand how these general systems have been modified and linked with other special human cognitive systems (e.g., Allman *et al.* 1998, Blakemore *et al.* 2004) to produce the unique suite of human family behaviors and associated life history transitions.

12.2.4 Reproductive senescence

Among mammals, humans are one of the longest-lived species with a relatively late age at first reproduction (Austad 1997). Humans typically live more than four times as long as predicted by body size compared to other mammals (Austad and Fischer 1992). Adult human mortality rate doubles approximately every eight years (Austad 1999) and this is accompanied by progressive loss of somatic functional capacity and efficiency. See Chapter 16 for more details on general human senescence and longevity.

12.2.4.1 *Menopause*

Unlike gradual somatic senescence, human female reproductive senescence happens abruptly around age 45–50, well short of the usual life expectancy for women. We are afforded the luxury of a long post-menopausal life because of a combination of low extrinsic mortality (which extends the lifespan) and the loss of fertility in later life. All viable oocytes are actually produced before a female's birth and stored in arrested metaphase until use each cycle, with a resultant depleted supply of oocytes with age. Ovarian steroid production decreases, with an eventual loss in hypothalamic–pituitary–ovarian feedback control. These changes in oocyte count and endocrine function are actually similar to those in other large-bodied mammalian species (Leidy 1994, Packer *et al.* 1998, Videan *et al.* 2006), although few other female mammals demonstrate post-menopausal life as they do not usually survive that long (Crews 2003).

Human pregnancies become increasingly hazardous with age (Williams 1957), and maternal death is associated with significant increases in mortality of their children (Hill and Hurtado 1996). Women who produce children very late in life might also not live long enough to assure the survival of such altricial offspring (Alexander 1974). However, women could still increase their reproductive effort in supporting kin during advanced age, when mortality rates increase and extractive efficiency decreases. So women of a certain age could theoretically benefit more through investment in existing children and grandchildren rather than through the production of additional offspring themselves (Hawkes *et al.* 1989, 1998).

Our unique life history pattern allows for the potential of significant generational overlaps. Like few other species, humans live in groups with multiple overlapping generations of kin and are exceptional in maintaining significant social relationships among individuals two or more generations apart. Grandparenting is cross-culturally ubiquitous and pervasive (Murdock 1967, Sear *et al.* 2000). Grandmothers could increase their foraging time after their grandchildren are born to allow their daughters to decrease their own workload and thus invest more time and energy in pregnancy and lactation (Hawkes *et al.* 1998). However, it has yet to be determined unequivocally whether or not assistance from post-reproductive adults significantly increases the likelihood of offspring survival (specifically the survival of their own grandchildren; Hill and Hurtado 1991, 1996; cf. Hawkes *et al.* 1998, Hawkes 2003).

We contend that older females may have important effects on the success of their developing children, perhaps in part because of the importance of their accumulated knowledge for negotiating the social environment (Hrdy 2009). Socially skilled and well-connected older mothers and grandmothers may have been especially valuable teachers of social and political wisdom, with associated reproductive benefits (Alexander 1990, cf. O'Connell *et al.*

1999). In short, the doubling of the maximum lifespan of humans, involving an increased period of pre-reproductive development on the one hand and an increased period of post-reproductive parental and kin investment on the other, suggests the importance of parent–offspring relationships for acquiring and mastering sociocompetitive information (Bjorklund and Pellegrini 2002, Flinn and Ward 2005, Geary 2005).

12.2.4.2 *Andropause*

Reproductive senescence in men is very different from menopause in women. Male reproductive senescence may involve gradual declines in fertility (de La Rochebrochard *et al.* 2006) with the accumulation of genetic abnormalities in sperm with age (Plas *et al.* 2000). Hormone levels appear to change in Western populations (those with low energetic output and high caloric availability) with age, most notably declines in testosterone (Ellison *et al.* 2002, Feldman *et al.* 2002) and increases in sex hormone binding globulin (Gray *et al.* 1991, Harman *et al.* 2001). Elevated gonadotropin levels are also indicative of declines in Sertoli and Leydig cell sensitivities (Tennekoon and Karunanayake 1993). The circadian rhythm in testosterone production and secretion becomes impaired with age (Bremner *et al.* 1983). All of these changes may be reflective of a compromised ability to adjust reproductive function in response to energy availability (Bribiescas 2006). Somatic changes such as decreases in strength and muscle mass (van den Beld *et al.* 2000, Walker and Hill 2003) may also result in a "social andropause," with decreased ability to attract mates and compete with conspecifics (Bribiescas 2006). In this manner, men may have been selected to adjust reproductive behavior from mating to parenting with increased age (Draper and Harpending 1988).

12.3 Summary

1. Humans have a unique suite of traits that appear key to understanding our unusual life history. Foremost is our large brain, with its extraordinary socio-cognitive processes, including consciousness, empathy, and creative language. The human infant is physically altricial, but mentally precocial.

Human childhood and adolescence are unusually long, enabling extensive learning.

2. Humans are highly parental and reside in complex, multigenerational social groups with networks of families linked by kinship. Humans are also highly alloparental, with especially high effort by grandmothers facilitated by menopause.

3. Human life histories are flexible and sensitive to the social environment. Human reproduction, including its correlated physiological and behavioral functions, has evolved into a complex system of components whose proper performance is regulated in part by the neuroendocrine system.

4. Hormones are central mechanisms that contribute to the onset and timing of key life history events, the optimal allocation of time and energy between competing functions, and the general modulation of phenotypic and genotypic expression in response to environmental signals. Their functions in reproductive development, gonadal function, mating/parenting behaviors, and reproductive senescence are both similar and different in males and females.

5. Reproductive suppression in response to certain environments may function to protect maternal condition and maximize lifetime reproductive success. Depressed estradiol and progesterone levels often result from i) high energy expenditure due to elevated workload or exercise, ii) low nutritional intake, and iii) psychological stress.

6. Androgens can facilitate male reproductive effort through the modifications of behaviors and physical attributes, including secondary sexual characteristics and musculoskeletal functions. However, androgens can also compromise survivorship by i) increasing the risk of negative energy balance, ii) increasing the risk of prostate cancer, iii) facilitating the production of oxygen radicals, iv) reducing resistance against oxidative damage, v) increasing risk of injury due to hormonally-augmented behaviors such as aggression, violence, and risk taking, vi) directly causing immunosuppression, and vii) reducing the amount of energy and nutrients available for tissue repair and the maintenance and activation of immune responses.

7. Relationships between hormones and libido in humans remain equivocal, but surely differ between men and women as well as healthy eugonadal individuals and clinically deficient or menopausal subjects.

8. Several hormones, most notably oxytocin, arginine-vasopressin, prolactin, cortisol, estradiol, and progesterone, may play important roles in mediating attachment and parental behaviors.

9. Menopause is characterized by a loss of oocytes and an eventual loss in hypothalamic–pituitary–ovarian feedback control followed by a lengthy post-reproductive life during which these older females may impart important accumulated knowledge for negotiating the social environment. Male reproductive senescence may involve changes in hormones, fertility, and somatic composition, and an important shift from mating to parental effort.

PART 4

Lifespan, aging, and somatic maintenance

Thomas Flatt and Andreas Heyland

This part of the book deals with the last part of an organism's life cycle, its lifespan, the process of aging, and the maintenance of the soma in the face of aging. The length of the reproductive period of an organism, its reproductive lifespan, is determined by selection for increased reproductive success. On the one hand, since a longer reproductive period typically translates into increased reproductive success, selection will tend to lengthen lifespan; on the other hand, this increase is balanced by reproductive trade-offs, so-called "costs of reproduction," that increase intrinsic mortality with age, and this limits the length of life. Related to this concept of lifespan is that of aging (or senescence), the age-dependent decline in physiological function, which is demographically manifest as decreased survival and fecundity with increasing age (Rose 1991, Stearns 1992, 2000).

In the 1940s and 1950s, based on early insights by Fisher and Haldane, evolutionary geneticists set out to explain why aging, a disadvantageous, maladaptive trait, evolves in the first place (for reviews see Rose 1991, Flatt and Schmidt 2009). Their key insight was that aging evolves as a byproduct of selection for increased fitness early in life, with the decreasing force of natural selection being inefficient at maintaining function late in life (Medawar 1952, Williams 1957). Medawar and Williams developed two conceptually similar theories that posit that aging either evolves through the accumulation of effectively neutral mutations with deleterious effects only late in life (Medawar 1952), or via the accumulation of mutations with strong beneficial early-life effects that outweigh any deleterious late-life effects (Williams 1957), with the negative effects late in life being unchecked by the declining strength of selection late in life. By the 1980s and 1990s the cornerstones of this evolutionary theory of aging had been well confirmed empirically by evolutionary biologists (Rose 1991, Stearns 1992), yet only very little was understood about the mechanisms whereby organisms age. This situation changed dramatically with the application of developmental and molecular genetics approaches to aging, starting in the 1980s, leading to the discovery of numerous long-lived mutants in model organisms (e.g., Tatar *et al.* 2003, Flatt and Schmidt 2009, Kenyon 2010; see also Chapter 1).

Recent mechanistic work in aging has not only advanced our understanding of the fundamental molecular mechanisms that affect aging and lifespan, but has also lead to many insights that touch upon evolutionary aspects of aging, for example the genetic basis of the evolution of lifespan and the mechanisms underlying "cost of reproduction" trade-offs. Despite this progress, there has been little cross-talk between researchers studying the mechanisms of aging and those interested in evolutionary aspects of aging. As has been argued, for example by Barnes and Partridge (2003), Partridge and Gems (2006), and Flatt and Schmidt (2009), this molecular body of knowledge on aging has to be integrated into the evolutionary framework (see also Chapters 1 and 28). The chapters in this part of the volume document some of the recent developments in our understanding of the proximate mechanisms that affect aging; for an overview of evolutionary aspects of aging we refer the reader to Rose (1991) and Flatt and Schmidt (2009).

Chapter 13 by Hodkova and Tatar discusses their recent work on the endocrine mechanisms underlying the commonly observed trade-off between lifespan and reproduction in two insect models, the bug *Pyrrhocoris apterus* and the fruit fly *Drosophila melanogaster*. By focusing on the induction of reproductive diapause, their work and review of the literature suggests that juvenile hormone is a key mediator of this trade-off in insects (see also Chapters 11 and 18).

In Chapter 14, Bauer and Helfand review what has recently been learned about the effects of dietary or caloric restriction on lifespan. Reducing food intake without malnutrition extends animal lifespan nearly universally across taxa, and it is thought that this evolutionarily conserved plastic response helps organisms to survive poor diet conditions until the environment becomes more favorable again for reproduction. As Bauer and Helfand discuss, major progress has been made in uncovering the genetic and physiological mechanisms underlying this diet-induced plasticity. Based on these insights, it will become possible in future work to address the proximate basis of the life history strategies that organisms have evolved to cope with environmental stresses such as poor nutrition.

Chapter 15 by Schwartz and Bronikowski further elaborates on this theme of environmental stress and how organisms cope with it. The authors synthesize recent evidence from reptiles on how environmental stresses (e.g., thermal, oxidative, caloric, predation, social) induce molecular stress pathways, and how such pathways can respond to natural selection to shape life history strategies in reptiles.

Chapter 16 by Kuningas and Westendorp closes the section by reviewing the genetics of aging in humans. The success of molecular geneticists in identifying genes that affect lifespan in model organisms has prompted gerontologists and geneticists to look at the homologues of these genes in humans, especially in centenarians. One of the most remarkable findings in this research area is that particular mutations in genes involved in insulin/IGF-1 signaling (IIS) not only affect lifespan in model organisms such as worms, flies, and mice, but are also associated with exceptional longevity in humans. Together with work reviewed in other chapters in this book, these findings indicate that IIS might be an evolutionarily conserved master regulatory pathway of life history traits, and it will thus be interesting to learn in the future whether genetic variation in this pathway is important in shaping life history traits in natural populations (e.g., Paaby *et al.* 2010; see also Chapters 27 and 28).

While it is true that "…aging, in the evolutionary view, is a byproduct of selection for reproductive performance, no matter what the molecular mechanisms" (Stearns 2000), the chapters in this part of the book clearly illustrate that mechanistic work on aging is now paving the way for an improved understanding of the evolutionary genetics of aging in natural populations and the nature of life history trade-offs (see also Chapter 28).

CHAPTER 13

Parallels in understanding the endocrine control of lifespan with the firebug *Pyrrhocoris apterus* and the fruit fly *Drosophila melanogaster*

Magdalena Hodkova and Marc Tatar

13.1 Introduction

In many organisms reproduction is negatively correlated with lifespan (Williams 1966b, Roff 1992), but little is known about the proximate mechanisms of this trade-off (Barnes and Partridge 2003, Harshman and Zera 2007, Flatt and Kawecki 2007, Flatt and Schmidt 2009). According to the traditional view, internal energy reserves are limited, giving rise to the trade-off of their use (de Jong and van Noordwijk 1992, Stearns 1992, Rose and Bradley 1998). More recently, this idea was challenged by the observation that survival and reproduction can be experimentally decoupled (Hsin and Kenyon 1999, Tu and Tatar 2003, Partridge *et al.* 2005a, Flatt and Kawecki 2007). Pleiotropic effects of endocrine signals are currently considered to play a key role in the modulation of life history trade-offs, but the way these signals connect reproduction (or other fitness traits) and longevity still remains elusive (Finch and Rose 1995, Barnes and Partridge 2003, Tatar *et al.* 2003, Flatt *et al.* 2005, Harshman and Zera 2007, Russell and Kahn 2007, Kleeman and Murphy 2009, Toivonen and Partridge 2009). Beneficial effects of endocrine signals (reproductive hormones) on the reproductive output, which is the subject of natural selection, may be constrained by the costs caused by these signals. Determining the nature of these costs is therefore central to the study of the endocrine effects on life history evolution (Flatt and Schmidt 2009, Chapter 11).

Insects, like vertebrates, have discrete endocrine tissues that control a broad diversity of physiological functions and life history traits (Nijhout 1994). The principal endocrine tissues include the conventional endocrine glands, the prothoracic glands producing larval ecdysteroids and the corpora allata (CA) producing the lipid-like juvenile hormone (JH). In adult females, ecdysteroids are produced in the ovarian follicle cells (somatic gonad). As in the case of vertebrate hypothalamic–pituitary axis, the secretory activity of the conventional glands is controlled via secretion of tropic or inhibitory neuropeptides produced by the neurosecretory cells of the pars intercerebralis of the brain (PI), which is also a major source of the insulin-like peptides.

An extensive body of work has shown that mutations in many genes in the insulin-like signaling pathway prolong the lifespan of *Drosophila melanogaster* (Tatar *et al.* 2003, Partridge *et al.* 2005a, Toivonen and Partridge 2009). JH is a major developmental and reproductive hormone affecting a remarkable number of physiological processes (Nijhout 1994), and is therefore considered a key mediator of life history trade-offs (Flatt *et al.* 2005). Unfortunately, surgical removal of the CA in *D. melanogaster* is difficult because of its small size, and mutations directly affecting JH production are limited, which makes this powerful genetic model species a challenge for testing life history effects of JH (Flatt and Kawecki 2007).

In relatively large species, surgical removal of insect adult CA (allatectomy) prevents the synthesis of JH and thus reveals life history and physiological traits regulated by this terpenoid hormone. Notably, lifespan has been extended in a number of insects by allatectomy (Pener 1972, Herman and Tatar 2001), suggesting that JH plays a role in the control of insect longevity. However, the way it does so remains largely unknown. Complementary approaches in two models, the firebug *Pyrrhocoris apterus* (Heteroptera) and the genetic workhorse *D. melanogaster*, help to uncover these mechanisms. These systems represent opposite ends of how we typically do experiments with insects: surgical analysis with the relatively large but genetically intractable *P. apterus* and engineered genetic manipulations in the surgically inaccessible fruit fly. Technical advances are now permitting us to approximate both approaches in each model such that we are now making progress in understanding how JH modulates insect life history traits.

A common thread in *P. apterus* and *D. melanogaster* is adult reproductive diapause associated with a lack of JH and improved survival. In this chapter we will primarily focus on the role of gonad in the regulation of lifespan by JH, but will also discuss other interactions potentially involved in the trade-off between reproduction and lifespan, namely those among JH, insulin-like peptides, germ-line stem cells, and fat body.

13.2 Reproductive diapause

Appropriate environmental cues induce adults of some species, including those of *Drosophila* (Tatar and Yin 2001) and *P. apterus* (Hodek 1983), to enter a state of reproductive arrest associated with somatic endurance. This state permits survival through winter, extreme summer conditions, lack of food, or migration until the diapause is terminated by environmental cues or by spontaneous physiological processes (diapause development), and adults initiate reproduction (Hodek 1983, Saunders *et al.* 1990, Pener 1992, Tatar *et al.* 2001a, Brakefield *et al.* 2007). Diapause-inducing cues vary among species and include factors such as day length, humidity, and temperature. In each case, reproductive diapause is associated with low levels of JH (Denlinger 2002,

Saunders 2002). A causal role for reduced JH is demonstrated in a number of studies, where allatectomy induces reproductive arrest and prolongs lifespan under diapause preventing conditions, while treating allatectomized adults with JH induces reproduction and shortens lifespan (Pener 1972, Herman and Tatar 2001, Hodkova 2008). Clearly, JH is a proximal endocrine mediator of insect reproductive diapause. Our goal at this point is to understand how JH modulates the suite of diapause-associated life history traits making up a characteristic diapause "syndrome"(Saunders 2002). Animals in reproductive diapause are stress resistant and use specific metabolic pathways and gene expression patterns (Denlinger 2002), as seen for instance in the distinctive profiles of hemolymph proteins (Sula *et al.* 1995), phospholipids molecular species (Hodkova *et al.* 2002) and mRNA transcripts of genes implemented in energy metabolism and cryoprotectant biosynthesis (Kostal *et al.* 2008) in diapause *P. apterus*, and in the energy stores and heat shock proteins of diapause endemic *Drosophila* (Ohtsu *et al.* 1992; Goto *et al.* 1998). Adults can survive much longer than normal in such physiological states, even at a high temperature.

An initial problem is to understand the character of improved survival during diapause. Diapause is a dynamic state and its intensity gradually decreases due to yet unidentified processes of diapause development. In *P. apterus*, similar to many other insect species, diapause development is associated with a decrease in responsiveness to diapause-terminating environmental signals, but metabolic rates remain suppressed (e.g., Hodek 1983, Kalushkov *et al.* 2001). At the molecular level, changes in the expression patterns of many genes during diapause development have been described (Denlinger 2002). Given that somatic persistence is a hallmark of diapause, how can these various traits of diapause development relate to aging? In particular, does reproductive diapause increase survival because it retards the progressive degeneration (senescence) of aging or because it reduces some source of age-independent mortality risk? A design to address this question with *D. melanogaster* must also control for the fact that its diapause is induced by low temperature, which could confound interpretations about lifespan. To circumvent this problem, the mortality rates of cohorts of *Drosophila* that had previously been in diapause were subsequently compared in a common environment to the mortality rates of newly eclosed, young flies (Tatar *et al.* 2001a). The rates and patterns of mortality were indistinguishable between young and post-diapause adults, suggesting that diapause suspended the process of aging (Fig. 13-1A). This conclusion was further supported by studies of temperate endemic *Drosophila* where photoperiod alone is sufficient to induce diapause (Tatar 2004).

A similar conclusion emerges from robust analysis with *P. apterus*, where diapause is induced by short day-length even at relatively high temperature. Diapausing adults live more than twice as long as their reproductive counterparts (Hodkova 2008). Importantly, if diapause *P. apterus* are transferred to non-diapause conditions after various durations, the post-diapause lifespan is similar to that of newly eclosed non-diapause females (Fig. 13-1B). Diapausing females appear to not senesce during at least the first 2.5 months of adult life, in spite of

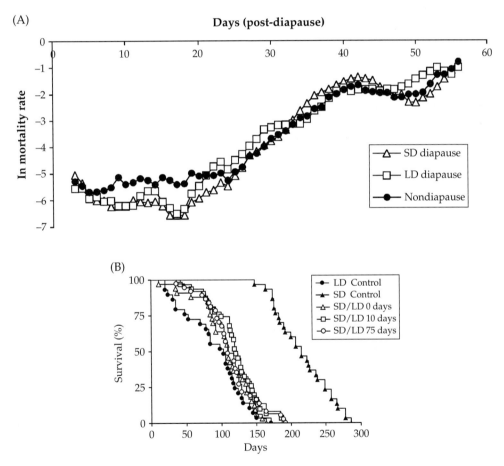

Figure 13-1 Survival and mortality under diapause and non-diapause conditions. (A) Mortality rate of female *D. melanogaster* at 25 °C to compare demographic aging of post-diapause adults that had previously experienced 3 weeks of reproductive diapause (11°C with either long or short day) relative to young, newly eclosed adults (non-diapause). Results are representative of data from Tatar *et al.* (2001a). (B) Survival of diapause, non-diapause and post-diapause females of *P. apterus*. Diapause was induced in short-day photoperiod (SD, 12L:12D), 26°C. Adult females were either kept under these conditions until death (SD control), or transferred to long-day photoperiod (LD, 18L:6D), 26°C on the day of adult ecdysis (SD/LD 0 days), 10 days after adult ecdysis (SD/LD 10 days), or 75 days after adult edcysis (SD/LD 75 days). Survival of SD/LD-females after the transfer was not significantly different from the total adult lifespan of non-diapause females always kept in LD, 26°C (LD control) (Hodkova 2008 and unpublished data).

high temperature. Unlike *D. melanogaster*, where cool temperature might contribute to the absence of aging during diapause, *P. apterus* shows that the retarding effect of diapause upon aging is temperature-independent.

The relationship between adult reproductive diapause and aging has also been studied with *Drosophila melanogaster* in natural populations of a latitudinal cline along the eastern USA (Schmidt and Paaby 2008). The frequency of genotypes with propensity to express diapause increases from southern, to mid-Atlantic, to northern populations, likely due to selection on this behavior to ensure adult survival through the winter. A number of traits correlate with this diapause trend when measured in a common, benign laboratory environment: with increasing latitude of origin there was an increase in mean lifespan, a decrease in reproductive output, an increase in cold shock tolerance, and a decrease in heat stress resistance. Diapause genotype explains a large portion of the variance for each of these correlated phenotypes except that of heat stress resistance, suggesting that genetic variance for diapause expression may directly or indirectly affect the selection and expression of longevity and fecundity. A potential molecular keystone for these interrelated life history traits was uncovered through mapping of genetic markers segregating among the diapause genotypes. The gene *couch potato* (*cpo*) strongly influences the propensity for *D. melanogaster* diapause, and also the expression of correlated life history traits such as longevity (Schmidt *et al.* 2008; also Chapter 18). Remarkably, a difference at a single nucleotide resulting in a Lys/Ile substitution in exon 5 of one *cpo* transcript produces the diapause regulatory allelic variation. To date, the function of *cpo* is largely unknown. It was first discovered as a mutant affecting sensory-organ development and has been subsequently characterized as a RNA binding protein (Bellen *et al.* 1992). It is expressed in numerous tissues, notably in the ring gland, which is the larval site of JH and ecdysone synthesis (Harvie *et al.* 1998). Analysis of the molecular targets and affected cellular and physiological function of *cpo* may provide a major breakthrough in how we understand the way diapause affects aging via the endocrine system.

13.3 Reproduction and its trade-offs

Given the well-documented trade-off between reproductive activity and survival (Bell and Koufopanou 1986), reproductive diapause might slow aging as an indirect effect of repressed reproduction. *P. apterus* is particularly suited to addressing this problem because one can study the contribution of multiple specific tissues by a combination of surgical interventions (Hodkova 2008). Allatectomized females are non-reproductive and long-lived (Fig. 13-2). If loss of egg production were responsible for extended lifespan we might expect ablation of the ovary to increase survival, but it does not and yet allatectomy extends the lifespan of ovariectomized females. These results show that egg production itself is not costly to female longevity, and suggests that the lifespan-reducing effect of JH is not mediated through the ovary.

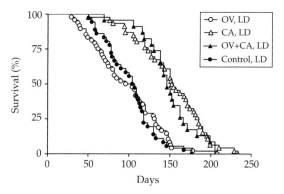

Figure 13-2 Survivorship of *P. apterus* females following ablation of corpora allata (CA), ovary, both CA and ovary, relative to shame operated controls (Data from Hodkova 2008).

The first of these conclusions may come as a surprise, since trade-offs between reproductive activity and survival appear to be ubiquitous. However, there is an emerging precedent for uncoupling this relationship at a physiological level. For instance, in the nematode *Caenorhabditis elegans*, ablation of the entire gonad does not increase lifespan (Hsin and Kenyon 1999). Aging is slowed, however, by ablation of the germ-line progenitor cells alone, suggesting that the germ line produces a signal that accelerates aging while the somatic gonad produces a balancing signal that favors longevity assurance; these signals appear to control longevity independently in *C. elegans* (Kleemann and Murphy 2009) (Fig. 13-3A). A variant of this model requires a single endocrine longevity assurance signal from the somatic gonad and a paracrine signal from the germ line that represses the somatic longevity assurance signal (Fig. 13-3B). Loss or repression of the germ line releases the longevity assurance signal and thus extends lifespan. Loss of the whole gonad eliminates both signals, and the survival of the adult reverts to the default duration. While the postulated longevity assurance signal may affect lifespan independently of the negative effects of JH upon longevity, the longevity assurance signal might function as a negative regulator of JH (Fig. 13-3C).

If loss of the whole gonad in the model 3C eliminates the JH-repressive signal and thus favors JH production, then:

- ovariectomy should reduce lifespan
- the effect of allatectomy on lifespan should be more pronounced in females lacking the ovary (with enhanced JH levels) than in normal females (Table 13-1, alternative (a)).

In *P. apterus*, however:

- ovariectomy has no effect on lifespan
- lifespan extension by allatectomy is similar in ovariectomized and normal females (Fig. 13-2).

Therefore, the model 3C would be consistent with data on *P. apterus* if loss of the whole gonad has no effect on the rate of JH synthesis and thus on the intensity of lifespan reduction by JH (Table 13-1, alternative (b)). However, this may not be the

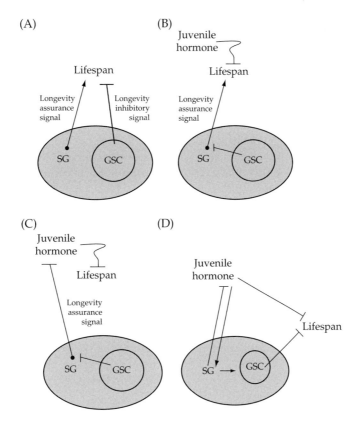

Figure 13-3   Hypothetical models for the potential interactions of the gonad and juvenile hormone on invertebrate aging. SG, somatic gonad; GSC, germ-line stem cells. Juvenile hormone is produced in the adult corpora allata. For other explanations see the text and Table 13-1.

Table 13-1 Hypothetical responses of longevity and JH levels to the loss of whole gonad or germ line cells predicted in models

Intervention	Longevity				JH levels		
	Model B	Model C	Model D	Model B	Model C	Model D
Loss of the	No change	(a) decrease	No	No	(a) increase	Increase
whole gonad		(b) no change	change	change	(b) no change	
Loss of the germ line	Increase	Increase	Increase	No change	Decrease	No change

As presented in Fig. 13-3.

case. *P. apterus* shows cyclic changes in CA volume (Hodkova 1999) and JH synthetic activity (Hodkova and Okuda, unpublished data) that correlate with the cycle of egg maturation. Likewise, ovariectomy in cockroaches results in high rates of JH synthesis due to the absence of negative feedback between mature ovary and CA (Gadot *et al.* 1991, Chiang and Schal 1994), and an injection of ecdysterone (product of somatic gonad follicular epithelium) represses this elevated JH synthesis (e.g., Stay *et al*, 1980, Chiang *et al.* 1991). Interestingly, *adipose female sterile* mutants of *D. melanogaster* with a reduced number of oocytes show hypertrophy of the CA (Doane 1960). Figure 13-3D combines models 13-3A and 13-3C to incorporate potential differences in JH levels that are seen between ovariectomized and intact females. JH in this case reduces lifespan via a "non-ovarian" pathway, but also stimulates the germ line to produce an additional longevity inhibitory signal (either directly or via somatic gonad). JH also acts on the follicle cells (somatic gonad) surrounding each oocyte. This results in yolk uptake and subsequent inhibition of JH synthesis via a feedback mechanism such as ecdysteroid production from the follicle. The loss of a longevity-limiting germ-line signal after ablation of the whole gonad may be counterbalanced by a higher rate of JH production because the CA is no longer inhibited by somatic gonad. In the net effect, lifespan is not changed. On the other hand, when the CA is ablated this loss eliminates both JH and the longevity-limiting signal from active germ-line stem cells, and lifespan is prolonged. Responses of longevity and JH synthesis to the loss of whole gonad or germ-line cells that are predicted in the proposed models (Fig. 13-3 B, C, D) are summarized in Table 13-1.

Genetic approaches with *D. melanogaster* also elucidate the relationship between reproduction and aging. Fly lifespan is extended by many physical and behavioral manipulations that repress egg production in normal females. Sterile mutants provide new insights on this trade-off. Mutants of *ovo*[D1] have a somatic ovary and intact germ-line stem cells but egg chambers fail at an early stage, resulting in adult sterility (Oliver *et al.* 1987). In some conditions, *ovo*[D1] extends female lifespan (Sgrò and Partridge 1999, Mair *et al.* 2004). To resolve the question of whether such longevity benefits are modulated from activity of germ-line stem cells or from reduced egg production, genetic manipulation was used to eliminate the germ-line stem cells during early adulthood (Flatt *et al.* 2008b). Adult lifespan was substantially increased. Importantly, lifespan was not increased in flies made sterile by eliminating germ-line progenitor cells of the embryo (see also Barnes *et al.* 2006) or by a mutation that inhibits late egg development without affecting the germ-line stem cells (Flatt *et al.* 2008b). Consistent with data from *C. elegans* and with whole gonad ablation in *P. apterus*, loss of egg production itself does not confer longevity; this benefit appears to require loss or inhibition of the adult germ-line stem cells.

13.4 Endocrine regulation

Because the endocrine regulation of life history traits such as longevity, reproduction, and diapause is complex, the combinatorial surgical approach with *P. apterus* is remarkably informative. The impact of photoperiod on reproduction (Hodkova 1976) and lifespan (Hodkova 2008) is mediated through brain neurosecretory cells of the PI and the CA. Based on a strong immunoreactivity of the PI and aorta wall (a neurohemal tissue) to bombyxin antibody, the PI of *P. apterus* is a likely source of insulin-like peptides (Hodkova 2008) much as the brain PI of adult *D. melanogaster* produces four forms of insulin (Brogiolo *et al.* 2001, Ikeya *et al.* 2002). In non-diapause conditions of long day, surgical ablation of both the PI and CA increases *P. apterus* lifespan by 96% relative to intact controls, which is almost exactly the additive benefit on median lifespan produced by only ablating the PI (32%) or the CA (60%) (Fig. 13-4). Thus, signaling from the PI and CA may repress lifespan through independent pathways, or within a common network where maximal effects on longevity are complementary (Gems *et al.* 2002). For instance, independent of how insulin-like peptides may directly affect somatic aging, some proportion of the influence on lifespan from the PI may act through its documented effect on CA growth and synthetic activity (Hodkova

et al. 2001). A similar condition could be at play in *D. melanogaster*. Mutants of the insulin receptor (*InR*) and its receptor substrate (*chico*) are long-lived, sterile, and have reduced JH synthesis, while treatment with a JH analog partially restores longevity and induces egg production (Tatar *et al.* 2001b, Tu *et al.* 2005). Genetically directed ablation of the *D. melanogaster* PI also extends lifespan and impairs reproduction (Wessells *et al.* 2004, Broughton *et al.* 2005). It remains to be studied whether these effects of PI ablation occur through downregulation of JH.

Observations made with short-day treatments of *P. apterus* help resolve the dependent and independent effects of PI-signaling (presumably, insulin-like peptides) and JH (Fig. 13-4). The CA does not produce JH under diapause conditions of short photoperiod. In this case it is not surprising that CA ablation has no effect on lifespan. But in contrast to the situation with long days, ablation of the PI

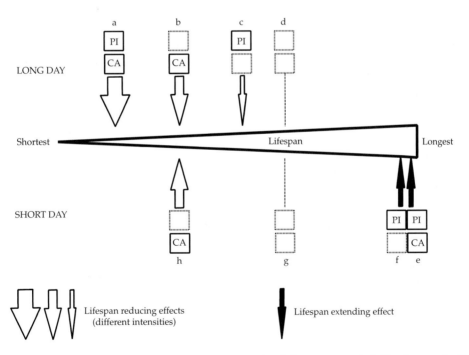

Figure 13-4 Model of photoperiodic regulation of lifespan in *P. apterus* by signals from the brain pars intercerebralis (PI) and corpora allata (CA). LD, long day; SD, short day (a) LD, intact high fecundity; (b) LD, PI removed: low fecundity; (c) LD, CA removed: no reproduction; (d) LD, both PI and CA removed: no reproduction; (e) SD, intact: no reproduction; (f) SD, CA removed: no reproduction; (g) SD, both PI and CA removed: no reproduction; (h) SD, PI removed: low fecundity. In LD, negative signals from PI and CA to lifespan are additive, but the CA plays a dominant role. In SD, diapause extension of lifespan results not only from the absence of negative signals, but also from the presence of positive signals from the PI. Modified from Hodkova 2008.

reduces median lifespan by 41% in females maintained on short photoperiod. The lifespan of females lacking their PI is prolonged by CA-ablation, but it is still shorter than lifespan of intact females maintained on short days. These results imply there is a factor from the PI involved in extending lifespan of short-day females, and this factor acts by downregulating CA-signaling (JH) and in a CA-independent manner. Positive signals from the PI to lifespan have not yet been reported for other insect species, although *C. elegans* insulin-like peptide-1, one of 37 insulin-like genes in this nematode, is thought to be an insulin-receptor antagonist (Pierce *et al.* 2001). In theory, the production of an antagonistic insulin-like peptide by the insect PI could positively regulate lifespan. A future challenge is to sort out the interactive and independent roles of the several insulin-like peptides produced by insect PI.

The combinatory surgical approach also reveals that the lifespan extension by PI-ablation in non-diapause *P. apterus* need not be a consequence of reduced fecundity. First, ovariectomized females of *P. apterus*, despite their sterility, do not live as long as PI-ablated females showing at least a low rate of reproduction (Fig. 13-4). Second, the lifespan of allatectomized females can be prolonged if allatectomy is combined with ablation of the PI (Fig. 13-4). This effect of PI ablation is not attributable to reduced fecundity because allatectomized females have already blocked both the synthesis of yolk proteins by the fat body and their uptake by the ovary (Socha *et al.* 1991). This resembles the situation in *D. melanogaster*, where lifespan of sterile *ovo*[D1] females of *D. melanogaster* is prolonged if *ovo*[D1] is combined with *chico* (Clancy *et al.* 2001). In addition, there is no obligatory association between increased lifespan and reduced fecundity in mutants in the insulin-like signaling pathway, suggesting independent control of the two life history traits (Toivonen and Partridge 2009).

A tissue that may potentially link together the endocrine signaling and the physiology of insect life history is the fat body, which has both adipose and liver-like functions. The lifespan of *D. melanogaster* is extended by transgenic over-expression of the insulin-signaling responsive transcription factor encoded by *dfoxo* in the fat body (Hwangbo *et al.* 2004, Giannakou *et al.* 2004). This manipulation reveals a feedback system to the PI because expression of *dfoxo* in the fat body suppresses the transcription of *ilp* in the brain, which presumably reduces levels of circulating insulin. This reduction of insulin is thought to account for the observed increase in lifespan, reduced reproduction, and activated FOXO protein in fat body. Although not investigated, *dfoxo* expressed in the fat body may indirectly reduce JH synthesis through its reduction of insulin-like peptides and thus further benefit longevity. JH could in turn regulate genes expressed in the fat body that are important for reproduction and metabolism, including those encoding yolk-protein vitellogenin and hexameric storage proteins (Sula *et al.* 1995). Notably, DAF-16 of *C. elegans* (the homolog of dFOXO) represses vitellogenin mRNA, and knockout of the vitellogenin genes prolongs lifespan (Murphy *et al.* 2003). In many insects, including *D. melanogaster* and *P. apterus*, JH promotes vitellogenin synthesis in the fat body and uptake of the protein by the egg chamber (Socha *et al.* 1991, Wyatt 1997, Soller *et al.* 1999). However, the absence of vitellogenins cannot explain either lifespan extension by the CA-ablation in male *P. apterus* (Hodkova and Provaznik, unpublished data), because the male does not produce this protein (Socha *et al.* 1991), or lifespan extension by PI-ablation in females of *P. apterus* that lack the CA (Fig. 13-4). The road from germ-line stem cells to aging may also go through the fat body. The brains of flies without germ line stem cells produce excess *ilp* mRNA and yet these adults are long-lived, potentially because they also transcribe an abundance of message for the insulin-like binding protein ImpL2 (Flatt *et al.* 2008b). ImpL2 has the capacity to inactivate *Drosophila* insulin-like peptides, and the fat body is a primary source for ImpL2 in the adult fly (Arquier *et al.* 2008, Honegger *et al.* 2008). How germ-line stem cells might communicate with the fat body and with the PI are open and important questions.

13.5 Conclusion

Understanding how hormones control life history provides insights into how evolution resolves trade-offs and shapes demographic fitness traits, including lifespan. The insects *P. apterus* and *D. melanogaster* provide complementary data on the roles and interactions of JH, insulin-like peptides, reproduction, and fat body.

13.6 Summary

1. Both insects show adult reproductive diapause associated with a lack of JH and retarded aging.

2. Based on different modes of study, both insects suggest that JH accelerates aging while it promotes reproduction.

3. Egg production itself, however, is not the likely mechanism by which JH accelerates aging, again based on different types of evidence from each species. Processes outside the gonad, most probably in the fat body, seem to mediate the effect of JH on longevity.

4. Yet reproduction still seems to be central to the hormonal control of aging, potentially through hypothesized endocrine signals produced by the somatic gonad that are regulated by germ line stem cells.

5. A postulated longevity assurance signal from the gonad may directly or indirectly modulate insulin-like peptides and JH.

6. Both insulin-like peptides and JH are pro-reproductive and both have multiple documented actions on somatic tissues. Insulin-producing neurosecretory cells of the PI and JH influence lifespan via at least partially independent pathways.

7. Continued surgical investigations with *P. apterus* may further resolve these mechanisms by adding measurements of components of insulin signaling pathway (InR, FOXO) and JH replacement to its technical repertoire. Moreover, it may be possible to induce transient gene knockout in *P. apterus* by injecting RNAi, as used in honey bees (Amdam *et al.* 2006), *Tribolium* (Belles 2010) and *Culex pipiens* (Sim and Denlinger 2008).

8. Progress with *D. melanogaster* could be advanced by a technique to reduce JH in the adult, potentially by directed transgene ablation of the CA. Analysis of adult *D. melanogaster* with reduced JH would facilitate understanding of lifespan through analyses in sterile backgrounds, by genetic epistasis analysis, and by genomic scans to identify JH-regulated pathways.

13.7 Acknowledgments

The study on *Pyrrhocoris apterus* was supported by the Grant Agency of the Czech Republic (Grants 206/05/2222 and P502/10/1612).

CHAPTER 14

The genetics of dietary modulation of lifespan

Johannes H. Bauer and Stephen L. Helfand

14.1 Introduction

In 1935, McCay *et al.* published the account of their experiments on the growth-retarding effects of lowering the number of calories consumed by laboratory rats. What they stumbled upon, however, was much more than that: the rats that consumed lower calories were also living exceptionally long lives (McCay *et al.* 1935). McCay *et al.* thus described for the first time what is now generally known as calorie restriction (CR)—the reduction of calorie intake without the dilution of essential nutritional content, in essence undernutrition without malnutrition. Even in this relatively crude experiment, rats on the lower calorie diet had a median lifespan that was extended by ~76% over the control group. This extended longevity was explained as mostly due to the growth-retarding effects of the lower calorie diet, and thus the field of CR research languished for the most part of the next quarter century. Only when evidence accumulated that CR extends lifespans even when started in adulthood (Weindruch and Walford 1982), and at the same time provides substantial health benefits to the animals, did research into the mechanisms of CR take off. This chapter attempts to give a brief overview of the current state of knowledge of the genes and molecular mechanisms associated with the lifespan-extending effects of CR.

14.2 Calorie restriction as a modulator of life history traits

14.2.1 Is lifespan extension due to calorie restriction universal?

After the initial findings that reduction of caloric intake extends longevity in rats, experiments in other species followed. The list of species that respond to CR with lifespan extension is impressive, and includes unicellular organisms such as protozoa (Rudzinska 1951) and yeast (Jiang *et al.* 2000), invertebrates such as spiders (Austad 1989), nematodes (Klass 1977), and fruit flies (Chapman and Partridge 1996), and even complex vertebrates such as fish (McCay *et al.* 1929), mice (Goodrick 1978), and dogs (Kealy *et al.* 2002). These data seem to imply that CR is a universal phenomenon that extends lifespan in any species, even humans. The universality of CR, however, is not universally accepted (Mockett *et al.* 2006). Houseflies exposed to CR, for example, do not show extended lifespans (Cooper *et al.* 2004). Even within a species, CR can have inconsistent results. Certain *Drosophila* strains respond exceedingly well to CR, while other strains have only small increases in lifespan (Libert *et al.* 2007). Mice caught in the wild do not respond to CR (Harper *et al.* 2006), and the effects in laboratory mice differ from strain to strain (Goodrick 1978). Studies in rhesus monkeys are thus far inconclusive. Preliminary data indicate that monkeys on CR have a lower mortality rate (Colman *et al.* 2009, Roth *et al.* 1999), and several studies have reported an increase in health parameters (Colman *et al.* 2009; for a review, see Lane *et al.* 2002). Nonetheless, these studies await completion (the lifespan of a rhesus monkey is about 40 years) and are thus too premature to make statements on lifespan extension.

It has been argued that primates, including humans, will not benefit from CR (Shanley and Kirkwood 2006). According to this view, animals have evolved a variety of different coping mechanisms to survive adverse and changing environmental conditions. In some animals this may take the form of a general reduction in metabolic activity, as observed in hibernating animals or the *dauer* stage of *C. elegans*. Other animals may cope with environmental stresses simply by migrating away from the affected areas or by increasing the energy stored in their bodies in times of food abundance for later use in times of crisis. Therefore, animals that are geographically restricted or have evolved a state of metabolic stasis are predicted to benefit from CR, while migratory animals are not. As for humans, they can rely longer on energy stores than smaller animals, are omnivores, toolmakers (i.e., able to use even difficult to access food sources), and migratory. Therefore, nutrient reduction may still elicit physiological changes in humans, but may have only minor effects on longevity (Shanley and Kirkwood 2000).

The example of the residents of Okinawa might serve to illustrate this point. It is estimated that the calorie intake of Okinawans is about 40% of the intake of Americans, yet their life expectancy is only about 5% greater (Everitt and Le Couteur 2007). Of course, direct comparison of these numbers is difficult, as lifestyle and general nutrition patterns are different as well. However, a 5% greater life expectancy is nowhere near the 25–40% longevity increase observed in animal models of CR. These data might thus serve as a caveat when considering the longevity effects of CR on humans.

14.2.2 The difficulty of defining what constitutes calorie restriction

Some of the problems in assessing whether CR is a universal intervention to extend animal lifespan stems from the difficulty in precisely defining what CR actually means and how to carry out CR experiments. While "undernutrition without malnutrition" appears at first glance a useful definition, it is still unclear in its definition of undernutrition and of "normal" food conditions.

One important complication facing CR researchers is the question of whether it is a reduction in calories, or rather a reduction in specific nutrients that drives longevity extension. Mice and rats live longer when the amount of protein in their diet is reduced (Goodrick 1978). Flies respond with increased lifespan when the amount of yeast extract, a protein source, but also containing lipids and other micronutrients, is restricted. However, reduction of sucrose content has only minor effects on lifespan (Mair *et al.* 2005, Min and Tatar 2006). Other researchers have found that yeast extract and sucrose in the fly diet may have antagonistic effects on lifespan, and that therefore dietary balance may be a better indicator of longevity than energetic content (Lee *et al.* 2008b, Skorupa *et al.* 2008). Furthermore, when the diet of flies raised on restricted yeast extract was supplemented with extra carbohydrates, lipids, vitamins, or amino acids, only supplementation with essential amino acids blocked the lifespan-extending effects (Bjedov *et al.* 2010). Similarly, restriction of methionine in the food has been shown to increase longevity in rats (Orentreich *et al.* 1993), mice (Miller *et al.* 2005), and flies (Troen *et al.* 2007). These results suggest that specific dietary components may act as signals to turn on a "low-calorie state," leading to changes in physiology and metabolism and/or extended lifespan. CR, the reduction of total calorie content, may therefore at least in part be explained by dietary restriction (DR), the reduction of specific nutrients. Not surprisingly, these signals may be amino acids.

Throughout this review article we will use the mechanistic term DR, rather than the historical term CR, even when discussing studies that were conducted using CR conditions. In this way, we hope to avoid confusion through frequent switching between CR and DR, and to indicate that the mechanistic explanation of the CR phenotype may be found in the DR effect, possibly even in amino acid restriction (see above).

Another consideration for the practical application of DR is that laboratory animal culture usually aims for optimization of reproductive output rather than long lifespans, and therefore tends to use very nutrient-rich diets. In fruit flies, for example, there is a linear relationship between food content and fertility. However, if lifespan is plotted against nutrient concentration, a parabolic curve is observed. Under nutrient-poor conditions, animals are starving and have shortened lifespan, while in nutrient-rich conditions animals are overfed and also have shortened lifespan. The lifespan maximum would thus delineate the point of optimal DR (Fig. 14-1).

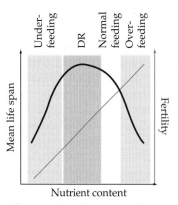

Figure 14-1 The relationship between nutrient content and mean lifespan. Mean lifespan (black line) and fertility (gray line) are dependent on the nutrient content of the food source. Increasing the dietary content leads to an increase in fertility. However, high-nutrition food leads to shortened lifespan, possibly due to overfeeding, while low nutrition food also leads to shortened lifespan due to underfeeding and/or starvation (light gray areas). Lifespan optimization can be achieved under conditions of reduced food intake and reduced fertility (dark gray area), compared to what is considered "normal" (white area).

Interestingly, the exact shape of this curve is different for different fly strains and even different between males and females of the same strain (Libert *et al.* 2007), indicating different nutritional requirements for each strain, as well as for males and females, respectively. Therefore, the failure to observe lifespan extension using a particular DR regimen does not necessarily mean DR *per se* does not extend lifespan, the conditions might merely not be optimized. On the other hand, lifespan extension observed using a DR regimen might simply reflect the switch from overnutrition to a healthier number of calories. Therefore, DR studies need to clearly show the lifespan optimum on the food-response curve; any study using only two food conditions (which includes the majority of worm and fly studies until the publication of Mair *et al.* (2005), and most of the mammalian data) must be carefully interpreted in this light.

14.3 The evolution of dietary restriction and its lifespan-extending effect

DR, as it is most commonly practiced in laboratory experiments, involves the reduction of caloric intake by about 30–40% of *ad libitum* while leaving the essential nutritional content (vitamins etc.) unchanged. This is a harsh regimen that leads to reduced body mass, mostly due to decreases in fat content (Berg and Simms 1960, Skorupa *et al.* 2008) and to growth deficiencies when started early in life, before animals are fully grown (McCay *et al.* 1935). Even when started later in life, animals on DR have reduced body mass (Weindruch and Walford 1982). Importantly, DR is accompanied by marked declines in fertility (Holehan and Merry 1985, Chapman and Partridge 1996). On the other hand, DR treatment leads to an increase in overall health parameters, and delays deterioration of general health. For example, DR has been shown to delay tumor onset and occurrence (Weindruch 1989), and improve cardiovascular health (Herlihy *et al.* 1992) and immune function (Good and Lorenz 1992). Memory function is positively influenced by DR in several studies (Ingram *et al.* 1987, Hashimoto and Watanabe 2005, Witte *et al.* 2009), even though negative effects have been reported as well (Yanai *et al.* 2004). These data suggest that DR may involve trade-offs between physiological systems, especially between reproductive and maintenance systems.

Little is known about how this state of improved health may have evolved (see Flatt and Schmidt 2009 for a recent discussion). It has been hypothesized that DR may be a case of phenotypic plasticity—that is, the ability to change a phenotype (in this case: lifespan) in response to variations in environmental conditions (food shortage). In flies and mice, DR can be instituted later in life, with a corresponding shift in the rate of mortality. Importantly, the mortality rate of flies shifted from a DR regimen to normal food conditions reverts to that of flies raised continuously on normal food (Mair *et al.* 2003, Bross *et al.* 2005). These data suggest that DR is a plastic, adaptive response that induces a reversible alteration of physiology (see Flatt and Schmidt 2009; for a discussion on life history modulation, please refer to Chapter 22). Therefore, animals encountering food shortage in the wild might be able to adjust their physiology, metabolism, growth rate, and fertility to account for these altered environmental conditions. The "disposable soma" theory explains this adjustment as a reallocation of resources from costly reproduction to improved maintenance of somatic cells (Kirkwood and Holliday 1979), which may explain the observed trade-off between fertility and reproduction under DR conditions (for a discussion of fertility trade-off, see Chapter 22). This shift would ensure that animals survive successfully until environmental conditions improve, at which point resource re-allocation increases fertility and decreases somatic maintenance. This mechanism would furthermore guarantee that reproduction only occurs under favorable food conditions, which in turn will also be beneficial for offspring survival (Holliday 1989).

Interestingly, this theory predicts that the two most important life history traits, survival and reproduction, are so intricately linked that they may not be separated. While animals on a DR regimen exhibit lower fertility (Chapman and Partridge 1996), the discovery of single-gene mutations linked to DR without fertility defects has challenged this view. Furthermore, flies in which vitellogenesis has been blocked still display the shift in mortality rate when switched from normal to DR food (Mair *et al.* 2004), and the fecundity, but not the lifespan, of flies on a DR regimen can be reverted to normal by adding methionine to the food (Grandison *et al.* 2009, see Flatt 2009 for a discussion). These data indicate that the resource allocation model may not adequately explain the longevity-increasing effects of DR.

Nonetheless, even though these caveats need to be considered when evaluating DR experiments, they also serve to develop a better understanding of DR. DR may be considered a signaling mechanism that gets activated when certain signals, such as amino acids, are missing from the food. Therefore, it can be dissected on a molecular level. The outcome of this signaling may vary between species and may not always be increased longevity: It may be a state of environmental stasis, like the *dauer* state in *C. elegans*, diapause in certain insects or hibernation in some mammals. Or it may lead to behavioral adaptations, such as increased foraging behavior. However, the most important feature of DR, improved health parameters, appears to be conserved between species, even in rhesus monkeys.

The evolutionary origins of this signaling are unclear, but recent research suggests nutrient-sensitive signaling pathways as crucial mediators of DR-induced longevity. It is no surprise that these pathways are evolutionary conserved, and often affect systemic signaling, for example through neuroendocrine mechanisms. Interestingly, these pathways may serve a dual role by functioning in, or crosstalking with, stress-related signaling pathways. Some of the signaling pathways and molecules will be discussed in the following section.

14.4 Dietary restriction in lower organisms

14.4.1 *C. elegans*

A big hurdle facing the use of *C. elegans* as a model for DR research is the definition of DR in the context of the nematode system and correspondingly the nature of the protocols used for DR. A variety of different regimens are used, which can yield conflicting answers. The classic method is to dilute the amount of bacteria fed to the worms (Klass 1977). However, since bacteria are toxic to *C. elegans* (Garigan *et al.* 2002), increased longevity may merely be a reflection of the lowered toxicity as a consequence of bacterial dilution. Attempts have been made to develop entirely synthetic food sources for *C. elegans* (axenic and synthetic food), or to raise worms on plates entirely devoid of bacteria (Kaeberlein *et al.* 2006). The latter approach suggests more of a starvation response than DR in the conventional sense. However, the experiments using axenic (Houthoofd *et al.* 2002) or synthetic media (Szewczyk *et al.* 2006) have yielded promising results. Another method that has been employed to calorically restrict *C. elegans* utilizes the *eat*-2 mutation, which reduces the rate of pharyngeal pumping (Lakowski and Hekimi 1998). It is assumed that this pumping reduction in turn leads to decreased food intake. Therefore, this approach may be problematic, especially since decreased food intake occurs from the day of hatching, with possibly detrimental side effects on animal growth and development.

Recently, an elegant paper has compared some of these DR regimens with special emphasis on the genes involved in the response to each of those different methods. Interestingly, different ways of inducing DR in worms had additive effects on longevity extension. At the same time, the authors found that different DR methods employed different molecular signaling pathways (Greer and Brunet 2009). Thus, these results may explain some of the conflicting reports in the literature, and clearly indicate that, at least in nematodes, different DR signaling pathways exist.

One of the most important signaling pathways to modulate aging in the worm is the insulin/insulin-like growth factor signaling pathway (IIS), decreasing activity of which leads to exceptional lifespan extension. This signaling pathway is conserved between species, and controls cell growth and proliferation. IIS activity reduction has been observed to lead to extended lifespans in a variety of model systems (Tatar *et al.* 2003). In worms, deactivating mutations in the insulin receptor *daf-2* (Kimura *et al.* 1997), the PI3-Kinase *age-1* (Friedman and Johnson 1988), or activation of their downstream target, the fox-o type forkhead transcription factor *daf-16* (Lin *et al.* 1997), are known to extend lifespan. Since IIS is a nutrient-dependent pathway that is downregulated in response to nutrient depletion, it is reasonable to hypothesize that the effects of DR may be mediated via downregulation of IIS. However, experiments using either bacterial dilution (Houthoofd *et al.* 2003, Panowski *et al.* 2007), axenic media (Houthoofd *et al.* 2003), or the *eat-2* mutation (Lakowski and Hekimi 1998) in a *daf-16* mutant background have still yielded animals with extended longevity. Furthermore, lifespan extension by either of these regimens was additive to the lifespan extension observed by a *daf-2* mutant. In addition to confirming these results, one study found that a special form of bacterial dilution did require *daf-16* for extended worm lifespan (Greer and Brunet 2009). Thus, at least in worms, IIS and most DR-signaling pathways constitute two separate signaling networks.

Another candidate nutrient sensing signaling pathway to extend *C. elegans* lifespan is the TOR pathway (Vellai *et al.* 2003). This amino acid sensing signaling pathway is named after the target of rapamycin (TOR) kinase and controls protein synthesis. Downregulation of TOR through the use of the *let-363* mutant has been shown to extend lifespan in the worm (Henderson *et al.* 2006). However, lifespan extension by the *eat-2* mutation is unaffected by TOR RNAi (Hansen *et al.* 2007). More

extensive analysis of TOR mutants using different levels and regimes of DR has not been performed yet, leaving open the question of whether the TOR pathway mediates the effects of DR in the worm.

The sirtuin family of NAD-dependent deacetylases has been implicated as a major regulator of lifespan. Sirtuins are named after the *sir* (silent information regulator) proteins and were first linked to lifespan regulation in yeast (Kennedy *et al.* 1995). In yeast, *sir2* overexpression extends lifespan (Kaeberlein *et al.* 1999), which has been linked to the ability of *sir2* to limit the appearance of extrachromosomal rDNA circles (Guarente 2000). In *C. elegans*, overexpression of *sir2.1*, one of four *sir2* homologs in the nematode, leads to lifespan extension (Tissenbaum and Guarente 2001). The lifespan-extending effects of *sir2.1* overexpression are dependent on *daf-16*, as worms lacking *daf-16* do not show extended longevity when *sir2.1* is expressed, thus linking *sir2.1* to IIS. The relationship between DR and *sir2.1*, however, is unclear, as conflicting results have been presented in the literature. *Sir2.1* mutants have been shown to block the lifespan extension of an *eat-2* mutant (Wang and Tissenbaum 2006), while other DR regimens were not inhibited by *sir2.1* mutations (Lee *et al.* 2006a, Hansen *et al.* 2007, Greer and Brunet 2009). This is reminiscent of the situation in yeast, where *sir2*-dependent and -independent DR mechanisms have been identified (Kaeberlein and Powers 2007). These results may be the consequence of multiple sirtuins with overlapping and redundant functionalities.

While the results with IIS, TOR signaling, or sirtuins in worms are either negative or unclear and possibly in contradiction to what is observed in other model systems, two candidate transcription factors have emerged that might mediate the DR effects in worms.

Pha-4 is a member of the fox-a family of forkhead transcription factors, has overlapping DNA binding sites with *daf-16*, and is upregulated in response to DR. Importantly, the lifespan extending effects of the *eat-2* mutation and of bacterial dilution, but not of IIS mutants, are lost in *pha-4* mutants (Panowski *et al.* 2007). Interestingly, *pha-4* mutants also suppress the increased lifespan caused by downregulation of the TOR pathway (Sheaffer *et al.* 2008). These data suggest that *pha-4* is a crucial downstream component of the nematodes' DR response, and, furthermore, that the TOR signaling pathway is a crucial mediator of the worms' DR response.

Skn-1, a member of the NF-E2 family of stress-related transcription factors, is increased in response to DR by bacterial dilution, and loss-of-function *skn-1* mutants block DR-dependent lifespan extension. Interestingly, this possible function of *skn-1* resides in the two ASI neurons of the worm, neuroendocrine cells that have been implicated in food sensing and lifespan regulation. *skn-1* reduction specifically in these two neurons is sufficient to block the effects of DR (Bishop and Guarente 2007b). Recent evidence suggests that *skn-1* is a downstream target of and inhibited by IIS (Tullet *et al.* 2008).

Thus, in *C. elegans*, depending on the method used to induce DR, different molecular pathways are activated that lead to extended longevity (Greer and Brunet 2009), some of which are *daf-16*- and IIS-dependent, while others are not. It remains to be seen whether different pathways for DR exist in other animals as well.

14.4.2 *D. melanogaster*

Just as with the nematode model system, knowing if a fly is indeed calorie restricted is difficult. Flies are usually fed a mixture of sucrose and yeast extract, which itself is a complex mixture of different nutrients, mostly amino acids, but also contains lipids, vitamins, and other small molecules. Most standard protocols for *Drosophila* DR employ dilution of either of these components. Since flies live in their food source, it is unclear if merely diluting the nutritional content of the food source indeed leads to DR. Measurements of food consumption have revealed some compensatory feeding in response to DR, but this is probably not enough to make up for the overall loss of nutrients (Carvalho *et al.* 2005, Ja *et al.* 2007, Min *et al.* 2007). Even though direct measurements of food intake and utilization are currently not easily performed on flies, *Drosophila* can conveniently be grown in large numbers, so that experiments comparing a multitude of experimental conditions can be carried out simultaneously. In this way it is possible to optimize nutrient conditions for each experimental fly strain used, and thus avoid the pitfalls associated with choosing experimental DR conditions that do not optimize lifespan (cf. Fig. 14-1). Different *Drosophila* strains, as well as males and females of the same strain, have lifespan optima at different food concentrations and thus do not display identical nutritional requirements (Libert *et al.* 2007). It is therefore highly unlikely that DR in the *Drosophila* model merely reflects a shift from unhealthy overfeeding to a more healthy food regimen. It will be interesting to see what the genetic differences between strains are that allow one strain to respond to DR more than the other.

In flies, IIS activity is known to regulate lifespan, as well as growth and fertility. Mutations in the single *Drosophila* insulin receptor (*InR*) (Tatar *et al.* 2001b) or the insulin-receptor substrate (IRS) homologue *chico* (Clancy *et al.* 2001) have been shown to extend lifespan. Overexpression of the downstream forkhead transcription factor dFoxO (Giannakou *et al.* 2004, Hwangbo *et al.* 2004), which leads to its nuclear accumulation and thus represents overactive dFoxO, also results in extended lifespan. Unfortunately, the experiments to determine whether dFoxO is required for longevity extension by *InR* or *chico* mutations have not yet been carried out in the fly. The signaling mechanism for *InR* or *chico* lifespan extension is therefore largely unclear, as is the role of dFoxO in this pathway. The role of dFoxO itself is better understood. Overexpression of dFoxO specifically in the fat body, a tissue that combines functionality of liver and adipose tissue, extends lifespan (Giannakou *et al.* 2004, Hwangbo *et al.* 2004). This lifespan extension is not observed when dFoxO is expressed in other tissues. Importantly, when flies lacking a functional version of dFoxO are exposed to DR conditions, they still have extended lifespans (Giannakou *et al.* 2008, Min *et al.* 2008). Therefore, just as in some situations in *C. elegans*, DR lifespan extension in flies is independent of FoxO. On the other hand, lifespan extension by *chico* mutations (Clancy *et al.* 2002), or even dFoxO overexpression (Giannakou *et al.* 2008), are not additive to the lifespan-extending effects of DR, suggesting a shared pathway. While similar experiments have not been performed with lifespan-extending *InR* mutations, experiments using a dominant-negative (DN) InR construct to inhibit IIS demonstrate that DN-InR can increase lifespan beyond what is seen with DR alone. However, this lifespan increase on DR is much smaller than what is seen with fully fed flies and was only performed using two food conditions (Bjedov *et al.* 2010). These data present a somewhat confusing picture of the involvement of IIS in the DR lifespan-extension pathway. The synthesis of these conflicting results could be that any DR signal might still employ upstream IIS components (such as InR), but then branches off before reaching dFoxO. dFoxO might therefore not be required for the complete longevity response, but may act to modulate it or to mediate other DR phenotypes.

The TOR signaling pathway might be a promising candidate for such a branch point. Protein kinase B/Akt not only phosphorylates FoxO, but also the Tsc1/Tsc2 kinase complex, which controls activity of the TOR pathway. Inhibition of dTOR pathway signaling activity, specifically in the fat body through overexpression of dTsc1, dTsc2, or a DN version of S6-kinase, has been shown to extend *Drosophila* lifespan (Kapahi *et al.* 2004). Importantly, lifespan extension by dTsc overexpression was not additive to the lifespan-extending effects of decreasing yeast extract content. The TOR signaling pathway controls protein synthesis through activation of 4E-BP, a protein that binds to and inhibits the translational regulator eIF4E. Interestingly, flies lacking 4E-BP do not respond to DR with extended lifespan, and overexpression of 4E-BP in the fat body extends longevity (Zid *et al.* 2009, Bauer *et al.* 2010). Thus, the TOR signaling pathway may be responsible for mediating the lifespan-extending effects of DR.

Histone deacetylases also play a role in *Drosophila* longevity regulation. Under DR conditions, transcript levels of the histone deacetylase Rpd3 are lowered. Accordingly, genetic reduction of *rpd*3 levels extends *Drosophila*'s lifespan (Rogina *et al.* 2002). Interestingly, both DR and lowered Rpd3 levels result in higher levels of dSir2, the fly homologue of *C. elegans sir2.1*. Accordingly, overexpression of dSir2 does lead to lifespan extension. Interestingly, after exposing flies without a functional copy of dSir2 to DR using two food conditions that encompass the point of maximal DR-induced longevity for this fly strain, no increased longevity is observed (Rogina and Helfand 2004). Thus, at least in flies, some of the lifespan-extending effects of DR are

mediated by dSir2. Although in mammals Sir2 affects FoxO activity (Brunet *et al.* 2004, Daitoku *et al.* 2004), the role of dFoxO in longevity extension caused by dSir2 overexpression has not been assessed yet, and it is still unclear how DR leads to Rpd3 reduction or an increase in dSir2 activity.

The sirtuin family of histone deacetylase was originally described in yeast, where they were shown to be upregulated in response to glucose deprivation (the yeast version of DR) and to extend lifespan when overexpressed (Kim *et al.* 1999). It is thought that in yeast sirtuins act to extend yeast mother-cell lifespan by reducing the numbers of extrachromosomal ribosomal DNA circles (ERC), which contribute to yeast mother-cell aging. However, no role has been established so far for ERC in *Drosophila* aging. Histones are only one of the targets of the deacetylation activity of Sir2. In flies, dSir2 interacts with and deacetylates Dmp53, the fly ortholog of the mammalian tumor suppressor p53. This deacetylation inhibits Dmp53 activity (Bauer *et al.* 2009). Accordingly, inhibition of Dmp53 activity through the use of DN-Dmp53 overexpression extends *Drosophila* lifespan (Bauer *et al.* 2005). Lifespan extension by DN-Dmp53 expression is not additive to the lifespan-extending effects of dSir2 expression or DR, suggesting that Dmp53 is a downstream target of dSir2 in the DR lifespan-extending pathway (Bauer *et al.* 2009). These results were confirmed by similar experiments in *C. elegans* (Arum and Johnson 2007). Importantly, the lifespan-extending effects of dSir2 and DN-Dmp53 are observed when they are expressed in the fly nervous system, suggesting an important role of the nervous system in responding to a state of reduced nutrients.

How does a reduction in p53 activity lead to longevity? If p53 activity is reduced in non-post-mitotic cells, aberrant cell growth and subsequent tumor development is almost unavoidable (Donehower *et al.* 1992). However, fully differentiated neurons are post-mitotic and therefore do not give rise to tumors. The strategy to reduce p53 function in neurons may therefore be applicable to mammalian models as well. Interestingly, it is sufficient to reduce Dmp53 activity in a specialized subset of fly neurons to achieve the same measure of increased longevity as with pan-neuronal activity reduction.

These cells are the fourteen insulin-producing cells (IPC) (Bauer *et al.* 2007), ablation of which has been demonstrated to increase lifespan (Broughton *et al.* 2005). IPC share functional similarities with the ASI neurons in worms, which have previously been implicated in lifespan regulation (Bishop and Guarente 2007b), and with the b-cells of the mammalian pancreas (Rulifson *et al.* 2002). Therefore, the main activity of p53 in these cells is most likely not related to its "normal" tumor-suppressor activity (induction of apoptosis or replicative senescence). This is demonstrated by the fact that merely decreasing apoptosis in the fly nervous system does not lead to increased longevity (Bauer *et al.* 2005, Zheng *et al.* 2005). Instead, a clue as to what p53's function in these cells might be comes from the observation that mammalian p53 is capable of regulating the rate of glycolysis by controlling the transcription of TP53-inducible glycolysis and apoptosis regulator, as well as several other metabolic enzymes (Bensaad *et al.* 2006). However, the exact mode of action of p53 in these neuroendocrine cells still remains to be determined.

What is clear is that reduction of Dmp53 activity in the IPC leads to a reduction in *Drosophila* insulin-like peptide 2 (dILP2) mRNA levels and a concomitant reduction in IIS activity in the fat body (Bauer *et al.* 2007). Furthermore, expression of DN-Dmp53 in flies also overexpressing dILP2 does not lead to increased longevity, suggesting that IIS is involved in the DR response in *Drosophila* (Bauer *et al.* 2010). Whether downregulation of dILP2 causes longer lifespans is currently unclear. Recent results suggest that individual dILPs have very specific functions. dILP2, for example, is required for lowering the levels of trehalose, the major circulating sugar in the fly (Broughton *et al.* 2008). However, dILP2 RNAi does not influence lifespan, perhaps due to compensatory upregulation of dILP3 (Broughton *et al.* 2008). Flies in which dILPs1–5 have been knocked out are small, with severe growth phenotypes. Interestingly, these phenotypes can be rescued through overexpression of dILP2 (Zhang *et al.* 2009). Furthermore, DR has been shown to lower the levels of dILP5, while dILP2 and dILP3 remain unchanged (Min *et al.* 2008). Flies with a reduction of dILP2, 3 and 5 levels through RNAi have extended lifespans, even under DR conditions (Min *et al.*

2008). Thus, specific dILP-mediated signaling pathways might be required to differentially regulate lifespan, metabolism, and the different branches of IIS.

Downstream of dILPs, DN-Dmp53-induced longevity requires 4E-BP, since flies lacking 4E-BP do not live longer upon DN-Dmp53 expression (Bauer *et al.* 2010). Taken together, these data suggest a DR signaling pathway that involves downregulation of Dmp53 activity in IPC, followed by dILP reduction. Reduced dILPs may then inhibit IIS; the longevity signal would, however, be transmitted through the TOR branch, while other phenotypes may utilize the FoxO branch of IIS.

The source of the IPC input is unclear. One possibility is that information about nutrient availability may originate in the digestive system and is transferred to the IPC. Generation of this signal may require INDY (I'm not dead yet), reduction of which has been shown to extend longevity (Rogina *et al.* 2000). INDY encodes for a transporter of Krebs-cycle intermediates, and, interestingly, is expressed in the plasma membrane of cells in the midgut region, as well as the fat body and oenocytes (Knauf *et al.* 2002). Reduced expression of INDY extends lifespan and this lifespan extension is not additive to the effects of DR (Wang *et al.* 2009b) or a low-calorie diet (Toivonen *et al.* 2007). Furthermore, INDY long-lived flies display a reduction in IIS (Wang *et al.* 2009b), and a downregulation of components of the electron-transport chain (Neretti *et al.* 2009). These data indicate that INDY may perform a critical nutrient sensing function and thus may mediate the first stages of the DR response.

Olfaction may also play a role in the initial, nutrient sensing stages of the DR response. The addition of yeast-derived odorant shortens the lifespan of flies on a DR regimen, while reduction of the *Drosophila* odorant receptor Or83b has been shown to increase longevity (Libert *et al.* 2007). However, Or83b mutants live longer under DR conditions, which suggests that odorants and their receptors may employ DR-dependent and -independent mechanisms.

These initial nutrient-sensing stages need to relay their signals to effector cells, such as IPCs. This signaling may be mediated through the *Drosophila* homolog of neuropeptide Y, the short neuropeptide

F (sNPF). sNPF is expressed in neurons adjacent to IPC, while sNPF receptors are present in IPC. Interestingly, sNPF inhibition by RNAi increases fly longevity (Lee *et al.* 2008c). Since sNPF is involved in the regulation of food intake (Lee *et al.* 2004), the signaling from sNPF-ergic neurons to IPC may be a critical step in translating information about nutrient conditions into physiological adjustments.

14.5 Dietary restriction in higher organisms

In contrast to the situation in worms and flies, where exact measurements of food intake and nutrient utilization are difficult to obtain, mammalian models allow for the exact quantification of food intake and utilization. The downside, however, is that lifespan is almost prohibitively long (in the case of rhesus monkeys on the order of a human adult lifespan), genetic manipulation somewhat more difficult, and maintenance costs are much higher. Nonetheless, the closer genetic kinship of rodent and monkey models to humans, as well as the ease with which physiologic measurements can be obtained, makes them valuable tools to complement the research in the more genetically malleable lower organisms.

14.5.1 Rodents

Ever since the pioneering work in rats by McCay *et al.* (1935), countless studies have confirmed the beneficial effects of reducing calorie intake on the health and lifespan of laboratory rodents. DR affects longevity even when applied later in life (Weindruch and Walford 1982). This observation is consistent with data observed in fruit flies, where it was shown that shifting flies even as late as 50% maximum lifespan to a DR regimen leads to beneficial effects on mortality rate and survival (Mair *et al.* 2003). Furthermore, DR in rodents delays the onset of cancers (Weindruch 1989), neurodegenerative (Chen *et al.* 2008, Wu *et al.* 2008), and cardiovascular diseases (Good and Lorenz 1992), increases immune function and preserves it longer (Ingram *et al.* 1987, Hashimoto and Watanabe 2005, Witte *et al.* 2009), lowers insulin and IGF (insulin-like growth

factor) levels, and preserves insulin-sensitivity (Bartke 2008c).

Just as in flies, some mouse strains respond well to DR, while others show very little effect (Goodrick 1978). This might be due to the different genetic makeup of these strains leading to different nutritional requirements. For example, wild-caught mice subjected to DR do not show an increase in median lifespan, even though they show a slightly increased maximum lifespan (Harper *et al.* 2006). This experiment seems to suggest that DR does not necessarily work in all mouse strains. However, it has been suggested that the DR regimen for these particular mice was a bit too extreme, as evidenced by an increase in early mortality, thus obscuring the beneficial effects of DR. In fact, the wild-caught mice do show an increase in maximum lifespan and a slope change in their mortality rate, indicating that DR does have a substantial effect even in wild animals (Harper *et al.* 2006).

Interestingly, reducing the levels of the amino acid methionine leads to similar effects on longevity as a reduction in calories alone (Miller *et al.* 2005). These data underscore the importance of specific micronutrients for the beneficial effects of DR.

Signaling pathways that were found to have an important role in DR in lower organisms also play an important role in mediating the effects of DR in higher organisms. Ames dwarf mice that carry a spontaneously arisen mutation in a gene controlling the early stage differentiation of pituitary cells are deficient in growth hormone (GH), prolactin, and thyrotropin, but are longer lived (Bartke *et al.* 2008). Consistent with the hypothesis that neuroendocrine signaling via the growth hormone–insulin-like growth factor (GH–IGF) signaling pathways can regulate lifespan (Holzenberger *et al.* 2004), GH receptor knockout mice (GHRKO) (Coschigano *et al.* 2000), fat-specific InR knockout mice (FIRKO) (Bluher *et al.* 2003), and IRS-1 (Selman *et al.* 2008) and IRS-2 (Taguchi *et al.* 2007) knockout mice are long-lived. Both Ames and GHRKO mice have reduced levels of insulin and increased insulin sensitivity (Dominici *et al.* 2002). However, Ames mice subjected to DR show further lifespan extension beyond DR alone (Bartke *et al.* 2001), while GHRKO subjected to the same treatment show no further effect on longevity (Bonkowski *et al.* 2006). Unfortunately, the other mouse models with lowered insulin levels have not yet been tested on DR regimes. Thus, the role of lowered IIS in DR-dependent lifespan extension in rodents remains unresolved.

The role of the TOR signaling pathway is likewise poorly understood. The activity of both Akt and TOR is lowered in response to DR in epithelial tissues (Moore *et al.* 2008), while Akt activity is unchanged in the liver and elevated in the muscle (Al-Regaiey *et al.* 2007, Hayashi *et al.* 2008). Mice fed the TOR inhibitor rapamycin late in life have extended longevity (Harrison *et al.* 2009), suggesting that, as in flies, lifespan regulation in mammals involves the TOR pathway.

The role of sirtuins in mammalian aging is currently a topic of much research. The levels of SirT1, the mammalian homolog of *C. elegans sir2.1* and *D. melanogaster* dSir2, are elevated in kidney, brain, fat, and liver in response to DR (Cohen *et al.* 2004), similar to the situation in flies (Rogina *et al.* 2002). Furthermore, upregulation of SirT1 leads to the mobilization of fat storage (Picard *et al.* 2004) and phenocopies some aspects of DR (Bordone *et al.* 2007). SirT1 has been shown to deacetylate the transcription factor FoxO (Brunet *et al.* 2004, Daitoku *et al.* 2004), the mitochondrial biogenesis controlling factor PGC1α (Nemoto *et al.* 2005, Rodgers *et al.* 2005), and PPARγ (Picard *et al.*, 2004), as well as p53 (Luo *et al.* 2001, Vaziri *et al.* 2001, Langley *et al.* 2002, Bauer *et al.* 2009). Finally, SirT1 has been shown to localize to sites of damaged DNA to promote repair (Oberdoerffer *et al.* 2008). Moderate SirT1 overexpression in the mouse heart protects the heart from oxidative stress and delayed heart aging. In contrast, high levels of SirT1 overexpression induce cardiomyopathy (Alcendor *et al.* 2007). When overexpressed in the pancreatic β-cells, SirT1 promotes insulin secretion in response to glucose stimulation (Moynihan *et al.* 2005). It is unclear how an increase in SirT1 levels leads to increased glucose-stimulated insulin secretion, but one possibility is via the observed repression of UCP2 (Bordone *et al.* 2006). UCP2 is a protein of the inner mitochondrial membrane that uncouples respiration from ATP generation and leads to lower ATP levels when present and higher ATP levels when repressed. Interestingly, overexpression of human UCP2 in the *Drosophila* IPC leads to longer lifespans and lowered dILP3 levels (Sanchez-Blanco *et al.* 2005). It has been proposed that the different *Drosophila* dILP have different functionalities (Broughton *et al.* 2008), and may thus functionally be more related to IGF-1 than insulin itself. This may explain the apparent contradiction between the mouse and the fly model. However, even though SirT1 expression in the pancreatic β-cells remains at high levels as the mice age, they lose their increased ability to stimulate insulin secretion, apparently due to a decrease in NAD biosynthetic pathway activity (Ramsey *et al.* 2008). These models of increased SirT1 expression have not yet been evaluated for effects on longevity. However, SirT1 null mice did not respond with an increase in activity when exposed to DR (Chen *et al.* 2005, Boily *et al.* 2008). Preliminary evaluation of one of the SirT1 null mouse models showed that SirT1 mice are shorter-lived than control animals, and may not have extended longevity when exposed to a two-food condition DR regimen (Boily *et al.* 2008). These data suggest that sirtuins play an important role in behavioral and physiologic adaptation to DR, and possibly the longevity response.

14.5.2 Primates

As mentioned above, it is unclear whether DR will extend lifespan in humans. One reason may be the different evolutionary path the human lineage took. On the other hand, as animals get more complex, the lifespan-extending effects of single gene mutations become less pronounced (Kuningas *et al.* 2008). Therefore, it may be difficult to extrapolate the results from worms, flies, and rodents directly to humans. A possible way out of this conundrum is to employ non-human primates as aging model systems. Non-human primates resemble humans much more closely than rodents in terms of genome, physiology, and even age-related diseases. Unfortunately, non-human primate studies are extremely expensive, in terms of both acquisition and maintenance, which makes them cost-prohibitive for individual researchers. In addition, non-human primates have long lifespans, on the order of 40 years for rhesus monkeys (Lane 2000).

Studies on the effects of DR on rhesus monkeys have been ongoing since 1987 (Mattison *et al.* 2003). In these studies, the DR regimen involves a reduction in food intake by 30%, initiated pre-puberty (1–2 years of age), during puberty (3–5 years of age), or as adults (8–15 years). Preliminary data from these studies reveales that DR elicits the same physiological changes in monkeys as it does in the rodent model. For example, monkeys on DR generally weigh less than controls and have lower body temperature and less body fat. In addition, disease risk factors are changed in the same direction in monkeys as well, as both insulin and IGF-1 levels are reduced and insulin sensitivity is increased (Mattison *et al.* 2003). Furthermore, blood pressure, triglycerides, and cholesterol are lower in DR monkeys. The rate and age of onset of cancer formation, cardiovascular disease, or metabolic defects is greatly reduced in animals exposed to DR, as well as the occurrence of age-related diseases (Colman *et al.* 2009). Importantly, animals on DR show reduced mortality, especially when non-age-related deaths are excluded, and are projected to have increased longevity (Colman *et al.* 2009). Despite these promising results, it is still too early to be confident whether monkeys on DR will actually have an increase in overall longevity or if the mortality rates will remain different during the entire lifespan. Nonetheless, even if DR does not lead to an increase in longevity, it certainly shows beneficial effects on monkey healthspan.

Studies in squirrel monkeys demonstrated beneficial effects of a 30% DR regimen on the development of Alzheimer's disease, confirming earlier studies in mice. Squirrel monkeys on DR had reduced levels of Aβ peptides, and, interestingly, increased levels of SirT1 protein in the same brain region (Qin *et al.* 2006). While this suggests that, just as in the lower organisms, Sir2 may play a role in mediating the beneficial effects of DR, further work is needed to clarify these issues. Furthermore, preliminary data suggest that squirrel monkeys on DR have a slightly lowered mortality compared to control animals (Lane *et al.* 2002).

Studies in humans have found similar patterns (for a detailed discussion, refer to Chapter 16). During the Biosphere 2 experiment, residents experienced DR conditions due to the constraints of the Biosphere habitat. Inhabitants had reduced body weight and body fat, as well as lowered insulin levels (Walford *et al.* 1992). In addition, their biomarkers resembled those of rodents and monkeys on DR (Walford *et al.* 2002).

Short-term DR (3 months) has furthermore been shown to be effective in increasing memory function in the elderly (average age ~60 years) (Witte *et al.* 2009). Serum obtained from donors on long-term DR led to the increased expression of SirT1 and PGC1α when applied to a cell line (Allard *et al.* 2008). In contrast, in another study, short-term (1 year) and even long-term (6 years) DR, did not lower IGF levels. However, insulin levels were still markedly decreased (Fontana *et al.* 2008). Importantly, this same study found that IGF levels could be lowered if protein levels were reduced. These data are reminiscent of the situation in mice, where reduction of protein or methionine content in the food is sufficient to achieve the same lifespan-extending effects as observed with total CR. Taken together, these data suggest that DR has positive effects on healthspan in primates, and even humans.

14.6   Concluding remarks

After over 70 years of research on DR in model systems, as well as humans, some patterns emerge. It appears that DR actively employs cellular signaling pathways to inform the organism at large to change stress response and metabolic parameters to a different set point. It is still largely unclear what these new parameters are. However, the signaling pathways involved are starting to become clearer. While the specifics of each DR signaling mechanism might differ somewhat from organism to organism, it is reasonably clear that certain commonalities are conserved (Fig. 14-2). Nutrient sensing is performed by specialized neuroendocrine cells, like the ASI neurons in *C. elegans*, the IPC in *D. melanogaster*, and possibly the hypothalamus and pancreatic β-cells in mammals, with their input likely stemming from nutrient sensors in the brain or the digestive system. The exact signaling mechanisms remain to be worked out, but presumably involve the action of Sir2 or related sirtuins, p53, janus kinase (JNK), or

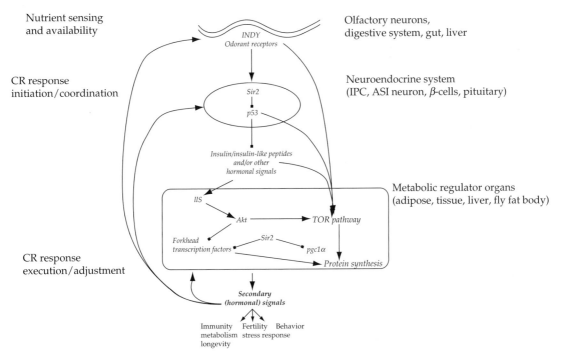

Figure 14-2   The DR response involves complex inter- and intracellular signaling mechanisms. The organismal response to DR may begin with an assessment of nutrient conditions. The digestive tract and the gut are promising candidates for this sensing stage as they are directly involved in nutrient uptake. This information is then relayed to tissues that can coordinate an organism-wide response to changes in nutrient status. These tissues likely include secretory organs, such as the neuroendocrine system, but may also include secretory tissue such as adipose tissue. Hormones and/or other signaling molecules secreted from these "secretory" organs may provide feedback to the upstream elements of this response system. In addition, they provide the crucial signals for the downstream tissues to initiate the tissue-specific DR response. In the final stage, the organism adjusts to the change in nutrient status with tissue-specific response patterns. Molecules and signaling pathways mentioned in the text are in italics, tissues likely involved in the DR response are in grey. Signaling connections that are inferred, or the nature of which is still unclear, are in light grey.

forkhead transcription factors. The activation of these signaling molecules in turn may lead to a change in the secretion status of neuroendocrine hormones, possibly insulin, IGF, or ILPs. The targets of these hormones are the metabolic tissues of the organism, such as the gut, adipose tissue, and/or liver. In these tissues, inhibition of nutrient-responsive pathways, like IIS or TOR signaling, may then result in a systemic reprogramming of the physiology, metabolic state, stress response, detoxification response etc., that ultimately leads to an increase in health parameters and may even extend longevity. This response may be mediated by secretion of hormones and other messengers from those metabolic tissues and may involve feedback loops to the neuroendocrine system.

The details of these signaling mechanisms, how they differ, and how they are conserved between different organisms, remain to be determined, as do the ultimate effector genes that mediate the multitude of physiologic changes observed during DR. However, this signaling framework provides a good starting point for further dissection and hypothesis testing of the DR signaling pathway.

In addition, the framework also provides a starting point for drug discovery of DR mimetics. DR mimetics could be used instead of the demanding DR regimen itself (Ingram *et al.* 2004) to increase healthspan or treat metabolic diseases, such as diabetes and obesity. Depending at which point of the pathway they act, they could potentially also have substantially fewer side effects than DR itself. One such mimetic might be resveratrol. Naturally found in red grapes, this polyphenol has many beneficial properties, one of which has been described as the activation of Sir2 (Howitz *et al.* 2003). Subsequently, resveratrol has been reported to extend the lifespan of yeast, worms, and flies (Howitz *et al.* 2003, Wood *et al.* 2004). However, not all researchers have found beneficial effects of resveratrol (Bass *et al.* 2007), but this may reflect the difficulties inherent in working with this drug or a food regimen that may be insufficiently suited to observing any effects of resveratrol on lifespan. Importantly, resveratrol did extend *Drosophila* longevity in a dSir2-dependent fashion (Rogina and Helfand 2004). Interestingly, in *C. elegans*, resveratrol's effect on lifespan was independent of daf-16 (Greer and Brunet 2009). In addition, resveratrol was found to improve health parameters and lifespan of mice fed a high fat/high calorie diet (Baur *et al.* 2006), but failed to increase longevity of mice on a normal diet (Pearson *et al.* 2008). Nonetheless, it was reported that the transcriptional changes elicited by treatment of mice using resveratrol closely resembled the transcriptional profile of mice on a DR regimen (Barger *et al.* 2008, Pearson *et al.* 2008). Furthermore, resveratrol was found to elicit similar changes in signaling pathways as DR (Barger *et al.* 2008, Smith *et al.* 2009). In light of data obtained in the yeast and the nematode systems (Kaeberlein and Powers 2007), it remains to be seen whether these beneficial effects of resveratrol are indeed mediated through the activation of Sir2, or are the consequence of other effects, such as its function as an antioxidant, an anti-inflammatory drug, or a combination of the two.

Another potential candidate may be rapamycin, an immunosuppressant that targets the TOR kinase. Treatment of mice late in life with rapamycin leads to impressive effects on longevity (Harrison *et al.* 2009), suggesting that inhibition of the TOR pathway may be a promising target for pharmacological aging interventions. To avoid detrimental side effects of rapamycin administration, more specific TOR pathway inhibitors need to be discovered—these must only target the longevity effects of the TOR pathway or only inhibit TOR signaling in tissues linked to longevity regulation.

14.7 Summary

The field of DR research has come a long way from its early days. A variety of model systems have been developed that each excels at addressing particular aspects of the DR question. Novel, short-lived vertebrate model systems, such as killifish (Genade *et al.* 2005), will further contribute to solving the question of DR. The first drugs are about to emerge from this research, which have the potential to not only treat metabolic diseases, but may also be able to prolong human healthspan.

Nonetheless, important questions and challenges remain:

1. Is DR a universal intervention that can extend healthspan (and lifespan) even in primates and humans? If so, is there one conserved core signaling pathway, or are there several pathways that mediate the DR response?
2. If different DR signaling pathways exist, as the data from worms and yeast would imply, what exactly are the different molecular cues activating these different pathways?
3. What are the evolutionary origins of these different pathways, what is their original purpose? Is there mechanistic cross-talk between them?
4. What are the ultimate effector molecules leading to better health and extended longevity upon DR treatment?
5. And, finally, the most important question: Can knowledge of the effector molecules and the signaling pathways leading to them be utilized to develop treatment strategies for age-related diseases and to increase human healthspan?

14.8 Acknowledgments

The authors would like to apologize to all colleagues whose work we could not mention due to space constraints. This work was supported by NIA grants AG029723 to JHB and AG16667, AG24353, and AG25277 to SLH. SLH is an Ellison Medical Foundation Senior investigator and recipient of a Glenn Award for Research in Biological Mechanisms of Aging.

CHAPTER 15

Molecular stress pathways and the evolution of life histories in reptiles

Tonia S. Schwartz and Anne M. Bronikowski

15.1 Reptiles possess remarkable variation and plasticity in life history

Reptiles are extraordinarily diverse in their life history traits (reviewed in: Shine 2005). Post-embryonic life ranges from a few months (Labord's chameleon, *Furcifer labordi*) to centuries (tortoises, sea turtles), with correlated completion of growth, maturation, and reproduction within these highly variable time frames. These life history traits are environmentally plastic but have an underlying pleiotropic genetic basis. Thus, selection on any of these life history traits should have accompanying evolution across the life history. Within the amniote vertebrates, the reptilian lineage is characterized by a high degree of flexibility in metabolic and physiological processes on one hand (excepting *Aves*), and remarkable adaptations on the other. This combination of variability and novelty has led to in-depth studies of the life histories of avian and non-avian reptiles. A relatively smaller molecular tool kit for application to wild-living species, however, has resulted in far fewer studies of molecular pathways in reptiles. Here, we focus on the recent literature of molecular pathways in ectothermic lineages of reptiles (crocodilians, turtles, tuatara, lizards, and snakes: Fig. 15-1). These pathways are highly conserved across a diversity of organisms; for many model organisms, the links between molecular pathway and life history have been, or are being, identified. By analogy, we hypothesize that these molecular pathways will be found to determine reptilian life history as well (see Chapter 26 for examples for birds), with the exceptions being informative to reptilian adaptations.

Evolutionary theory posits the life history as a set of co-evolved traits that have responded to natural selection operating through mortality and reproductive success (Stearns 1992, Roff 2002). From life history theory, it follows that the continuum of fast- to slow- "pace-of-life" (i.e., short-lived with fast growth, early maturation, rapid and high effort reproduction, versus long-lived with slow growth, late maturation, and extended reproduction, potentially with a post-reproductive stage) results from variation in time and space in the sources of mortality that shape the life history (Williams 1957). Thus, evolution is expected to occur due to changes in mortality environments (Hamilton 1966, Charlesworth 1994). More recently, it has been proposed that the source of the extrinsic mortality (e.g., predation versus pathogens) and condition-dependent mortality may also be an important factor in how evolution can shape the life history (Bronikowski and Promislow 2005, Williams *et al.* 2006b). Reptiles possess many adaptations related to mortality selection that recommend them for studies that link morphological and physiological evolution with underlying molecular events and pathways. These include:

- an external ribcage (turtles)
- venom (snakes)
- limblessness (snakes and some lizards)
- extended metabolic shut-down (all groups)
- starvation resistance, including remodeling of the digestive tract (snakes)
- supercooling, freeze tolerance, and heat tolerance (all groups)
- extended hypoxia resistance (turtles, crocodilians, lizards).

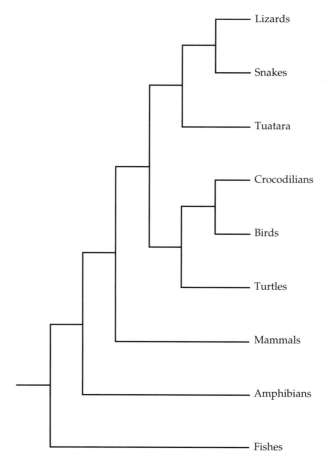

Figure 15-1 Simplified vertebrate phylogeny illustrating the relationships of reptiles relative to other major vertebrate groups. Branch lengths have no meaning.

Although there are definable lifespans and reproductive schedules within each species, many may have remarkably little physiological deterioration with advancing age. Thus, some reptile species may indeed exhibit the phenomenon known as negligible senescence—the lack of age-related deterioration (Finch 1990). Furthermore, many reptile species have indeterminate growth and fecundity, such that the costs of reproduction (Reznick 1985)—at least with respect to the lifespan—are not apparent at the level of the external phenotype (*sensu* Harshman and Zera 2007). In these species, the oldest individuals in the natural populations are often the most fecund and robust (Sparkman *et al.* 2007). Therefore, in many reptile species, strong selection against del-

eterious mutations may be characteristic of older individuals, leading to increased longevity and longer reproductive lifespans.

Reptiles are proposed as a model for the trade-off between lifespan and reproduction (Bronikowski 2008 and references therein) because they have evolved plastic responses to external stresses and, putatively, plastic modulation of cell signaling pathways. Ectothermic reptiles have different physiological and cellular responses to environmental and metabolic stress, relative to endotherms. This is mostly driven by the reptilian ability to regulate metabolic function by behaviorally modulating their body temperature, which results in lower energy requirements than birds and mammals that must use their metabolism to maintain higher body temperatures. This underlying metabolic difference drives the evolution of many reptilian life history traits and allows for plasticity and the ability to acclimate to environmental conditions (Shine 2005). Understanding the molecular pathways that have been conserved versus those that have evolved in reptile lineages will lead to an understanding of how these pathways and interactions evolve under different selective pressures. Many of these reptilian adaptations to environmental stress are known to activate molecular pathways linked to mechanistic theories of aging (e.g., the free-radical theory of aging (and its derivations)), which provides *a priori* predictions of outcomes for stress-response modulation (Finch 1990).

In this chapter, we propose that not only the source of extrinsic mortality matters, but also the class of organism in which it is occurring. Reptiles have evolved extreme metabolic and physiological plasticity in response to environmental factors such as temperature and food availability, and may be one of a few classes of organisms to display negligible senescence across their phylogeny. Thus, we propose that, for reptiles in particular, mortality due to incorrect or imprecise responses to these types of environmental stresses could drive the evolution of stress-response networks that, in turn, have pleiotropic effects on the life history (Fig. 15-2). Thus, we propose a "physiology evolves first" scenario of life history evolution for reptiles. It is becoming increasingly apparent that the molecular mechanisms underlying the complex traits of life history, stress response, and metabolism are controlled by evolutionarily conserved, and equally complex, molecular networks (Flatt and Schmidt 2009). Here we focus on the molecular level to highlight the best-characterized metabolic and stress response pathways in reptiles. In doing so, we explore support for our guiding hypothesis, and suggest causal hypotheses of how these molecular networks modulate the trade-off between reproduction (including growth and maturation rates) and longevity, when these hypotheses are supported in other organisms (e.g., Chapters 11 and 13). Therefore, the central theme of this chapter is that in reptiles, past and current natural selection on environmental stress response (e.g., thermal, oxygen,

and dietary stresses) has resulted in exceptional molecular stress responses that impact many life history traits. We summarize what is known about these molecular networks in reptiles, describe how three types of environmental stresses affect these networks, and point out potential consequences for life history evolution. Additionally, we provide a case study on garter snake life history evolution (Box 15-1) to exemplify how selection on the physiology from different environmental stresses may drive the evolution of life histories in opposite directions.

15.2 The molecular stress networks: What is known in reptiles?

Many molecular networks can affect the reproduction/longevity life history trade-off (reviewed in Harshman and Zera 2007, Flatt and Schmidt 2009; see also Chapter 12). Experiments on laboratory animals have clearly illustrated the importance for longevity of single allelic differences in specific pathways. Gene-network analyses indicate that these longevity genes are both highly conserved and highly connected nodes (Bell *et al.* 2009). Thus the networks described below may be shaped in wild populations through selection on genetic variation in stress-response, and likely have pleiotropic effects on the trade-offs among growth, reproduction, and longevity.

15.2.1 Metabolic pathways

Metabolic pathways produce energy, which is essential for growth, reproduction, response to stress, and cellular maintenance/repair; concordantly they also produce deleterious byproducts that can contribute to cellular dysfunction and aging. Under normal conditions, most of the energy production within an organism is through oxidative phosphorylation by the electron transport chain (ETC) in the mitochondria, resulting in ATP production. The core of the ETC consists of five protein complexes (I–V) in the inner-membrane of the mitochondria (Fig. 15-2). Electrons, from dietary carbohydrates and fats oxidized by the tricarboxylic acid (TCA) cycle and β-oxidation, are donated to the ETC, which uses them to produce a

Box 15-1   Case study on garter snake life history ecotypes

While natural selection acts on the individual, evolution occurs at the level of the population. Therefore, studies on diverging populations of the same species are particularly informative to ask how genomes can evolve under differing selection pressures and what the pleiotropic consequences are on life history traits. This opportunity has been provided by a unique natural study system, 25 populations of the western terrestrial garter snake, *Thamnophis elegans*, arrayed over a 100 km[2] study area at 1555–2055 m at the northern end of the Sierra Nevada mountains in northeastern California. These populations consist of two distinct garter snake ecotypes (slow- versus fast-living) that inhabit contrasting environments (mountain meadow versus lakeshore). The garter snake ecotypes have evolved different morphologies and life history strategies along the growth/reproduction/longevity continuum (Table 15-1; Fig. 15-2), which have been revealed by systematic mark/recapture studies since 1976, and laboratory breeding and common-garden experiments (Bronikowski and Arnold 1999, Bronikowski 2000, Sparkman *et al.* 2007). Studies of mitochondrial DNA and nuclear markers (microsatellites) show that the ecotypic differentiation is driven by selection that counteracts moderate gene flow between ecotypes (Bronikowski and Arnold 2001, Manier *et al.* 2007).

The contrasting habitat selection pressures and the evolved differences between the ecotypes match in ways predicted by evolutionary theory. The fast-living ecotype, found at widely-spaced intervals along the shore of Eagle Lake, has higher extrinsic mortality (due to predation) and lower dietary restriction (due to constant prey availability). The slow-living ecotype, meanwhile, found in the surrounding mountain meadows, has lower predation and probability of mortality, high dietary stress due to annual variation in prey availability, and cooler temperatures at the higher elevations.

One goal of ongoing research is to identify the underlying cellular and molecular bases for the evolved ecotype differences. In doing so, we have identified key differences in how the ecotypes differ in stress response—in their behavior, hormone levels, free radical production, and DNA damage and repair (Robert *et al.* 2007, 2009, 2010, Bronikowski 2008, Sparkman *et al.* 2009). In summary, the fast-living ecotype has higher ROS production during stress, less efficient DNA repair mechanisms, and a stronger innate immune response (Sparkman *et al.* 2009), which may include the use of ROS to kill pathogens (Nappi and Ottaviani 2000). Ongoing experiments will continue to characterize differences between stress response in these ecotypes, including whole transcriptome gene expression and additional mechanisms to regulate ROS and oxidative stress. Thus far, evidence suggests that the slow-living ecotype has a more efficient stress response and less potential for oxidative damage.

Table 15-1   Summary of differences between slow-living and fast-living garter snake ecotypes in the vicinity of Eagle Lake, Lassen County, California, USA.

Trait	Slow-living ecotype	Fast-living ecotype
Habitat[1]		
Substrate	Grassy meadow	Rocky lakeshore
Elevation	1630–2055 m	1555 m
Summer daytime temperature	15–30°C	20–34°C
Avian predators	Medium bodied raptors, robin	Eagle, osprey, robin
Food/water availability	Variable across years	Continuous
Major prey types	Anurans, leech	Fish, leech (anurans in flood years)
Most common parasites	Tail trematodes	Mites

Morphology/life history[2,3]

Color and stripe patterns	Black with bright yellow stripe	Checkered, muted grays, browns
Adult body size (mean)	538 mm (range: 370–598)	660 mm (range: 425–876 mm)
Female maturation size/age	400 mm/5–7 years	450 mm/3 years
Reproductive rate	Infrequent/resource dependent	Annual
Litter size (mean)	4.3 liveborn (range: 1–6)	8.8 liveborn (range: 1–21)
Baby mass	2.85 g (range: 1.8–3.5 g)	3.27 g (range: 2.5–4.2 g)
Annual adult survival probability	0.77/year	0.48/year
Median life span (years)	8	4

Cellular physiology[4,5,6,7]

Mean metabolic rate at 28°C at 1 mo of age, both sexes	0.52 ± 0.02 ml O_2/h	Statistically equivalent
H_2O_2 production under stress	56 pmol/min × mg mitochondria	240 pmol/min × mg mitochondria
DNA repair efficiency to UV damage	73%	35%
Field baseline corticosterone levels	50 ± 8 ng/ml plasma	7.7 ± 12 ng/ml plasma
IGF-1 levels	Lower/resource dependent	Consistently higher
Innate immune response	Low: natural antibodies (sheep RBC), complement-mediated lysis, and bactericidal competence	Higher for all three immune measures

Sources: [1] Bronikowski and Arnold 1999; [2] Bronikowski 2000; [3] Sparkman *et al.* 2007; [4] Sparkman and Palacios 2009; [5] Robert and Bronikowski 2010; [6] Sparkman *et al.* 2009; [7] Bronikowski, AM and D. Vleck 2010. Metabolism, body size and life span: A case study in evolutionarily divergent populations of the garter snake (*Thamnophis elegans*) Integrative and Comparative Biology 50: 880-887.

proton gradient by pumping hydrogen atoms from the matrix to the inter-membrane space of the mitochondria. The subsequent movement of these protons down this gradient into the matrix, by means of ATP-synthase (complex V), drives production of ATP. This process of oxidative phosphorylation is a major source of reactive oxygen species (ROS), as electrons "waiting" during the early stages of the ETC (particularly complex I and III) can interact with oxygen, resulting in the production of ROS—specifically superoxide (O_2^-), and subsequently hydroxyl radicals (OH^-), nitric oxide (NO), hydrogen peroxide (H_2O_2), and others. In this way, ROS are produced continuously throughout life and may be related to species-specific longevity (Barja 2004, Lambert *et al.* 2010).

ROS have unbalanced electrons and can pull electrons from other molecules including DNA, proteins, and lipids, thereby damaging these molecules (reviewed in Pamplona and Barja 2007). Although the deleterious effects of ROS are typically studied with respect to life history traits, they are also essential for many cellular and physiological processes, including cell signaling, immune function, and nor-mal reproduction. Therefore, cells need to manage ROS levels to maintain their essential functions while minimizing their potential deleterious effects. ROS levels can be regulated through the balance between:

• the production of ROS
• the levels of antioxidants that can neutralize ROS.

Oxidative stress results if this balance is disrupted, causing an overabundance of ROS and ultimately oxidative damage. Oxidative stress and oxidative damage have established roles in reproductive pathologies, particularly with age, and ROS have been implicated in causing age-related decline in oocyte quality and hormone biosynthesis, and the degradation of reproductive tissues (reviewed in Martin and Grotewiel 2006). Additionally, the accumulation of DNA mutations and damaged proteins and lipid membranes, due to ROS can lead to cellular senescence, and may influence organismal aging and lifespan—the premise of the free-radical theory of aging (Finkel and Holbrook 2000 and references therein).

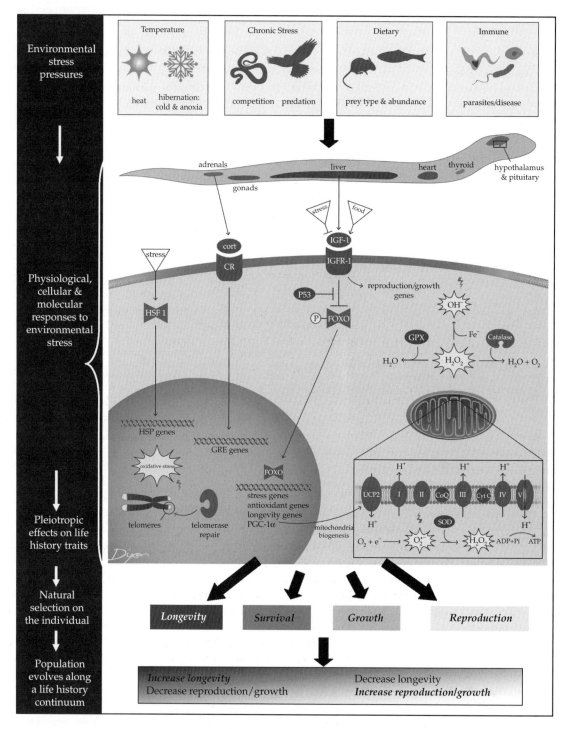

Figure 15-2 Conceptual diagram illustrating environmental stresses as selective pressures affecting the organism through physiology and molecular stress response pathways that have pleiotropic effects on multiple life history traits. Fitness consequences of being able to modulate these pathways appropriately allow for natural selection on the individual, and cause the population to evolve along the trade-off continuum between longevity and growth/reproduction. The simplified molecular pathways depicted in the cell represent stress responses (or signaling cascades) causing cytoplasmic transcription factors (HSF-1, CR, FOXO) to enter the nucleus upon perceived stress and induce transcription of genes involved in stress response and longevity. The mitochondria is represented as the major source of energy and ROS production, which antioxidants neutralize. Abbreviations:

The cell can defend itself from oxidative-stress–induced damage by increasing the tolerance of cellular components to oxidation by ROS and/or by activating mechanisms to repair the oxidative damage they cause (reviewed in Monaghan *et al.* 2009). Thus, there are four components to managing oxidative stress and how it affects cells:

- ROS production
- ROS conversion or neutralization through antioxidants
- cellular protection from and tolerance of ROS
- repair of damage due to ROS.

The networks underlying these four components of oxidative stress regulation can contribute to the trade-off of whether to invest in regulating oxidative stress for cell maintenance and longevity or to invest in growth and reproduction (Monaghan *et al.* 2009).

15.2.2 Molecular mechanisms to regulate the production of reactive oxygen species

Although the relationship is complex, the production of ROS is often associated with an increased rate of aging and decreased lifespan. For example, in comparing snake species of varying lifespans, long-lived species have lower H_2O_2 production than short-lived species (Robert *et al.* 2007). As most ROS production occurs in the mitochondria, this is where most of the oxidative damage ensues, causing the mitochondrial genome to accumulate mutations with age (Barja 2004). As damage accumulates, energy production becomes increasingly inefficient, which in turn causes more ROS production. In this way, mitochondrial function plays an important role in the aging process (Balaban *et al.* 2005). Understanding the evolution of mitochondrial

function is essential for understanding the evolutionary relationship of energy production, reproduction, and aging.

The rate of metabolism is not a cause of aging in mammals, but this has not been well explored in reptiles (but see Robert *et al.* 2007), which can have metabolic rates that are extremely plastic with environmental conditions. Typically, reptile metabolic rate (as measured by mitochondrial O_2 consumption) is 5–10 times lower than similar sized endotherms of the same body temperature (Brand *et al.* 1991). Ultimately, this is due to the low energy requirements of reptiles. Proximately, this may be due to differences in lipid membrane properties, remodeling of the ETC (see below), and/or mitochondrial function. The types of phospholipids in the plasma and mitochondrial membranes and their degree of saturation (i.e., the number of double bonds in the fatty acid chains) vary in their susceptibility to peroxidation. Furthermore, the process of lipid peroxidation creates additional ROS, which can further damage the cell (Hulbert 2008). Moreover, different types of phospholipids can alter membrane fluidity and metabolic rate. Reptile species typically have membranes with lower polyunsaturated fatty acid content relative to endotherms (birds and mammals) at the same body temperature—this is consistent with less leaky membranes and increased resistance to peroxidation (Brand *et al.* 1991; Brookes *et al.* 1998). Since body temperature also affects membrane fluidity, understanding how reptile membrane fatty acid composition changes with metabolic rate and body temperature would be fruitful to understanding the role of the lipid membrane in reptile metabolic plasticity.

Recent molecular studies on reptiles have proposed a remodeling of the reptile ETC, which may have drastic effects on metabolic rate, energy production, and ROS production. The genes that code

Extracellular: CORT (corticosterone) CR (corticosterone receptor); IGF-1 (insulin-like growth factor-1); IGFR-1 (insulin-like growth factor-1 receptor).

Cytoplasmic: HSF-1 (heat shock factor-1); P-FOXO (Fork-head transcription factor—homologous to DAF-16 in C. elegans), that is phosphorylated; P53 (tumor protein 53); H_2O_2 (hydrogen peroxide); GPX (glutathione peroxidase-1); Fe- (iron); OH- (hydroxyl radical).

Nucleus: HSP genes (genes encoding heat shock proteins); GRE genes (genes encoding glucocorticoid response elements); PGC-1α (peroxisome activated receptor γ coactivator 1- α).

Mitochondria: UCP2 (uncoupling protein 2); I (ETC complex I, NADH dehydrogenase); CoQ (co-enzyme Q_{10}); II (ETC complex II, succinate dehydrogenase); III (ETC complex III, bc_1 complex); CytC (cytochrome c); IV (ETC complex IV, cytochrome c oxidase); V (ETC complex V, ATP synthase); e- (electrons); O_2^- (superoxide); SOD (superoxide dismutase); H_2O_2 (hydrogen peroxide).

for the protein subunits of the ETC are located in either the nuclear or mitochondrial genomes. Allelic variants (mutations) at many of these genes have been shown to affect not only mitochondrial function but also lifespan in mice and *C. elegans* under laboratory conditions (Greer and Brunet 2008b). Research on reptile ETC has focused on genes encoded by the mitochondrial genome. Comparative studies reveal varying rates of mitochondrial genome evolution among the different reptile lineages. While turtles and some lizards appear to have relatively slow rates of mitochondrial evolution, crocodilian (Janke *et al.* 2001), tuatara (Hay *et al.* 2008), and snake lineages (and some lizards) demonstrate accelerated evolution relative to other vertebrates (Jiang *et al.* 2007, Castoe *et al.* 2009).

Further evidence of reptile mitochondrial evolution is seen in the rapid evolution of gene sequences encoding the ETC protein subunits (e.g., COXI, COX2, ATP6, ATP8, ND5, ND6, CytB) in the lineage leading to snakes, and unprecedented convergent evolution among agamid lizards and snakes (e.g., COXI, NDI) (Castoe *et al.* 2009 and references within). The rapid adaptive evolution of the snake mitochondrial proteome is predicted to affect mitochondrial function and metabolism, including the production of ROS and ultimately influencing how energy production, aging, and stress-response will evolve (Castoe *et al.* 2009). Interestingly, birds (having evolved from reptiles) have lower rates of ROS production than do mammals of the same size and metabolic rate (reviewed in Pamplona and Barja 2007). Changes in mitochondrial genomic architecture have also occurred in reptiles, the most striking of which is the duplicated control region (D-loop) found in tuatara, most snakes, and some lizards, as well as rRNA gene rearrangements. It is hypothesized that these changes may be adaptive for energy conservation and increased plasticity in mitochondrial function and proliferation (Jiang *et al.* 2007).

Proteins in the mitochondrial membrane, such as uncoupling proteins (UCP), can influence the production of ROS. Members of this protein family reside in the inner membrane of the mitochondria and enable protons to leak across the inner membrane into the matrix without passing through ATP-synthase. This uncoupling of the ETC from ATP production can result in the production of heat (UCP1—only in mammals), and a decrease in ROS production (UCP2 and/or UCP3) (Andrews and Horvath 2009). Additionally, these proteins can be activated by ROS to increase the proton leak. Because of this ability to regulate and be regulated by ROS, UCPs may provide a mechanistic link between metabolic function, lifespan, and evolutionary fitness (Brand 2000, Criscuolo *et al.* 2005). All reptiles have lost UCP1, but the ectothermic reptiles still have and express UCP2 and UCP3; birds have lost UCP2 (Mezentseva *et al.* 2008, Schwartz *et al.* 2008). Interestingly, *peroxisome activated receptor coactivator 1-α* (PGC-1α) has been found to regulate production of UCPs and mitochondria biogenesis in mammals (reviewed in Scarpulla 2002), both of which can lead to increased energy production with decreased ROS production. Comparative studies of ectothermic reptiles and endotherms regarding the role of UCPs in regulating ROS in response to metabolic stress would be highly informative to our fundamental understanding of metabolism and the functional constraints on metabolic evolution.

15.2.3 Molecular mechanisms to neutralize reactive oxygen species

Antioxidants are molecules that neutralize ROS. These include endogenously produced enzymes, such as superoxide dismutase (SOD), catalase, and glutathione peroxidase (GPX), which act on H_2O_2, as well as antioxidants obtained from the diet, such as vitamin E (e.g., tocopherols and tocotrienols) and carotenoids (Catoni *et al.* 2008). Metabolically active tissues generally produce the most antioxidants. For example, liver, kidney, and muscle have the highest levels of SOD, catalase, and GPX in green sea turtles (Valdivia *et al.* 2007). Additionally, the production of endogenous antioxidants can be induced through stress response pathways, including the insulin/insulin-like signaling pathway (IIS) discussed below.

Dietary antioxidants such as carotenoids contribute to bright coloration in plants and some animals. It has been hypothesized that these antioxidants may represent an honest signal for health (via levels of oxidative stress and immune function) and are thereby used in mate choice and sexual selection. Although this hypothesis has received support in some birds and fish it has yet to be supported in the few studies using reptiles (reviewed in Catoni *et al.* 2008, Olsson *et al.* 2008a).

Antioxidant levels are affected by reproduction and aging. In birds, reproduction causes a decrease in total antioxidant defenses in blood (Alonso-Alvarez *et al.* 2004). Interestingly, vitellogenin, a precursor to yolk, a biomarker of reproduction, and part of the IIS pathway (see below), is known to shorten lifespan in *C. elegans*. In contrast, it is an antioxidant and correlated with stress resistance and survival in honeybees (Seehuus *et al.* 2006b; see also Rascón *et al.* Chapter 20 of this volume). A recent study suggests vitellogenin may function as an antioxidant in lizards as well. Olsson *et al.* (2009) determined that, in painted dragon (*Ctenopherus pictus*), juveniles from large clutches, in which mothers produced more vitellogenin, had lower levels of O_2^- in their circulating blood relative to juveniles from small clutches. Studies on another antioxidant, catalase, in the garden lizard (*Calotes versicolor*), demonstrated that catalase activity decreases (brain) or stays the same (kidney and liver) during maturation and then increases with age in all these tissues (Jena *et al.* 1998 and references therein). Lipid peroxidation also increases with age in these same tissues of the garden lizard, suggesting that antioxidant activities increase in response to oxidative stress (Majhi *et al.* 2000 and references therein).

15.2.4 Tolerance and resistance to reactive oxygen species

Not all types of proteins and fatty acids are equally susceptible to oxidation, thus the composition of macromolecules can determine the susceptibility of cells to damage by ROS (Pamplona and Barja 2007). Proteins with methionine residues and membranes with a higher proportion of polyunsaturated fatty acids are most susceptible to ROS, and both of these types of macromolecules are found in lower concentrations in long-lived mammals and birds. Lipid

membranes with more saturated fatty acids are more resistant to peroxidation. Hulbert *et al.* (2006) have proposed membrane composition in the naked mole rat as the reason this species can tolerate high levels of ROS and have an extended lifespan relative to other rodents of the same size. Based on the few reptile species examined, compared to mammals, reptiles have relatively low levels of polyunsaturated fatty acids in their membranes, and are thus predicted to have high tolerance to ROS. Lipid peroxidation potential increases with age in many tissues in mammals and short-lived lizards (Majhi *et al.* 2000). More research is necessary to understand how these fatty acids change under environmental stress and with age across reptiles with different rates of aging.

Heat-shock proteins (HSPs) are molecular chaperones found in the cytoplasm and mitochondria. They are responsible for maintaining the conformational structure of cellular proteins, particularly under stress conditions. Different HSPs (from different genes and/or splice variants) are utilized under different stress conditions. The transcription of all HSPs is controlled by transcription factor HSF-1, which is normally present in the cytoplasm, and which, when activated by stress, moves into the nucleus to bind the promoters of HSP genes. HSPs have been identified in lizards and turtles and play important roles in temperature and hypoxia-stress-response (discussed below).

DNA damage leads to cellular apoptosis unless it can be repaired, therefore long-lived organisms must invest in either preventative measures to protect the DNA or repair mechanisms. Telomeres, consisting of repetitive DNA and protective proteins, are found at the ends of linear chromosomes and protect their ends from degradation during replication and mitotic division. They are considered a biomarker of biological aging because they typically decrease in length with every cell division. Telomeres can be further eroded with oxidative stress, chronic stress, reproduction, infection, and increased growth rate. Thus telomeres provide a clear link between stress and aging, which may reflect differences in life histories (Monaghan and Haussmann 2006 and references therein, Ilmonen *et al.* 2008).

Studies on mammals and birds have documented decreased telomere length with age, although this may not be the case for at least one species of long-lived bird (petrels), which also had enhanced telomerase expression (Haussmann *et al.* 2007). In reptiles, alligators have long telomeres (~31 Kb) relative to mammals, birds, and snakes (14–25 Kb), but not as long as turtles (>60 Kb) (Paitz *et al.* 2004). Furthermore, in both alligators and snakes, telomeres decrease with age (or size as a proxy for age), although hatchling pythons seem to be able to increase their telomeres in their first year of life (Scott *et al.* 2006, Bronikowski 2008, Ujvari and Madsen 2009). A study on loggerhead sea turtles found no significant association between age and telomere length, although this study had low sample sizes (Hatase *et al.* 2008). Overall, telomere dynamics in ectothermic reptiles has been little studied.

15.2.5 Molecular pathways for repair

ROS can damage proteins, lipids, and DNA. Although there are methods to remove and repair these damaged proteins and lipids (reviewed in Pamplona and Barja 2007) these mechanisms have not been explored in reptiles and therefore will not be discussed further here. Mechanisms for DNA repair in reptiles have been explored indirectly. Robert and Bronikowski (2010) demonstrate that neonate snakes can repair DNA damage from UV exposure, and that there is variation in repair efficiency across populations. Another aspect of DNA repair is through the enzyme telomerase, which can actively increase the length of the telomeres and thereby repair damage from cell division and oxidation. Conversely, hyperactive telomerase can increase the risks of cancer in some species. Telomerase activity is tissue and species specific—in most tissues, adult cells have low levels of telomerase activity, whereas the gonads and germ cells display high levels.

Telomerase has a highly conserved structure that consists of protein and an RNA that serves as a template to elongate the telomeres (reviewed in Monaghan and Haussmann 2006). Although there are no peer-reviewed publications that directly measure telomerase activity in reptiles, a recent study by Ujvari and Madsen (2009) documents the increase in telomere length within individual hatchling pythons during their first year of life, thus suggesting that telomerase is active during that time. Studies on other ectotherms (e.g., fish, lobster; Klapper *et al.* 1998), as well as a long-lived species of bird (Haussmann *et al.* 2007), have found telomerase to be continuously active, which is in contrast to most mammalian tissues. Considering the extreme longevity of some reptile clades and their dynamic temperature range, understanding how such an important enzyme functions in reptiles is an obvious area that needs further exploration.

15.2.6 Insulin/insulin-like growth factor signaling pathway

The insulin/insulin-like growth factor signaling pathway is well characterized and evolutionarily conserved and interacts with multiple signaling networks to integrate nutritional and external stress signals to regulate reproduction, growth, stress response, and aging (Tatar *et al.* 2003, Greer and Brunet 2008b, Piper *et al.* 2008). Indeed, allelic variants (or knock-out mutations) in this molecular network that cause decreased signaling can increase longevity in diverse organisms (nematodes—Kenyon *et al.* 1993, mice—Liu *et al.* 2005, Suh *et al.* 2008, humans—Willcox *et al.* 2008). Some, but not all of these mutations also decrease reproductive output (nematodes—Friedman and Johnson 1988, Ghazi *et al.* 2009). In invertebrates, such as *Drosophila* and *C. elegans,* there is one IIS pathway. In vertebrates, components of this pathway have duplicated and resulted in two pathways: the insulin pathway and the insulin-like growth factor (IGF) pathway. Key regulators of these pathways are FOXO transcription factors. Information on the oxidative stress and nutritional state of the cell can be integrated through post-translational modifications (e.g., phosphorylation, acetylation) of FOXO by other signaling proteins that regulate the behavior of FOXO, that is, its location in the cell and transcription-regulatory activity (Hedrick 2009). In vertebrates, the FOXO gene family has expanded via gene duplications (Wang *et al.* 2009a), and at least three of the FOXO proteins are known to be regulated in response to oxidative stress and metabolism

(Hedrick 2009). For vertebrates in general, under non-stress, high-nutrient conditions, growth hormone from the hypothalamus–pituitary axis stimulates production of IGF-1 from the liver. The binding of IGF-1 to its cell membrane receptors in multiple tissues activates a signaling cascade that modifies FOXO. This modification prevents FOXO from entering the nucleus, which ultimately activates genes for growth and reproduction (including vitellogenin) and prevents the activation of genes for stress response/longevity (Fig. 15-2). Decreased IGF-1 signaling allows FOXO entrance to the nucleus, which results in the transcription of genes for antioxidants (e.g., SOD), DNA repair, and other stress-response genes (Greer and Brunet 2008a). Concordant with model organisms, studies measuring circulating levels of IGF-1 blood samples from reptiles (sea turtles, garter snakes, and alligators) indicate that:

- IGF-1 levels change seasonally with food availability
- IGF-l increases during reproduction
- IGF-1 is positively correlated with reproductive output (Crain *et al.* 1995, Guillette *et al.* 1996, Sparkman *et al.* 2009).

15.3 Environmental stress and evolving molecular pathways: Evidence in reptiles

To integrate evolution with molecular mechanisms, we discuss three major environmental selective pressures: temperature, anoxia, and caloric restriction. We summarize the empirical evidence in reptiles for how these environmental stressors affect the underlying molecular mechanisms and pathways discussed above, and their implied pleiotropic effects on aging and reproduction based on evidence in other lineages. This approach illustrates the integrated molecular basis for responses to multiple environmental stressors and the consequential pleiotropic effect these responses have on many life history traits and their evolution.

To discuss the evolution of life history traits through selective forces on stress-response pathways, we need clear definitions of the time frame and the hierarchical level (cell, tissue, individual, population, species) upon which the different proc-

esses are acting. Selective force on metabolic and molecular stress pathways would focus on the reaction norms of individuals in two time frames:

- an individual's *ability to respond* to daily changes in stressors, including extreme/immediate stress (e.g., heat shock)
- an individual's *ability to acclimate* their metabolic networks over longer periods (e.g., weeks to months) and potentially predictable (seasonal) fluctuations.

The fitness consequences for individuals that adjust their physiologies in response to these environmental stress pressures will drive the evolution of these molecular networks at the population level, leading to local adaptation (Fig. 15-2). How particular reptile species (or populations) have evolved the ability to respond to stress may define their current location in life history trait space (e.g., Fig. 15-2, a simplified, two-dimensional continuum trade-off in aging and reproduction).

15.3.1 Temperature (heat) stress

Thermodynamics determine the biochemical reaction rates that underlie an organism's ability to function. Temperature is thus a driving force in reptile life histories. In contrast to most mammals and birds, reptiles do not use metabolic processes to maintain a constant body temperature; rather, the ambient temperature and organismal thermoregulatory behavior together determine body temperature, which can be incredibly precise. Despite this, reptiles are found in extremely diverse thermal habitats ranging across altitude and latitude. While each species has different life histories, they likely have adapted to their local temperature profiles using the same underlying molecular pathways, which could be selected for in a multitude of directions, provided that genetic variation exists in those pathways. This would result in the variation in resting metabolic rate and in the plastic responses (reaction norms) of metabolic rate to temperature that are found within and among reptile populations.

Both metabolic rate and growth rate increase with body temperature in reptiles. Faster growth rate leads to higher mortality in lizards (Olsson and Shine 2002). Although this may be due to increased predation associated with higher foraging rates, there is also evidence that fast growth rate is metabolically costly (Peterson *et al.* 1999) and is linked to increased age-related mortality (Ricklefs 2006). In comparing wild-caught caimans of different size (age) classes, most oxidative damage seemed to occur within the first years of life, possibly due to fast growth (Furtado *et al.* 2007). Furthermore, studies on snakes and lizards indicate that local adaptations to cooler temperatures include slow growth, low reproduction, and longer lifespan (Bronikowski 2000, Ibarguengoytia and Casalins 2007), all of which are predicted with a decrease in IIS.

Daily and seasonal temperature fluctuations can dramatically affect energy metabolism, metabolic rate, and the production of free radicals, unless offset by thermoregulatory behavior and/or thermal compensation in physiological and molecular pathways (Seebacher 2005). This requires the integration of temperature sensors, cellular signaling pathways, and ultimately behavioral responses. Thus far, two transient receptor potential ion channels (TRPs), which are temperature sensitive cell-membrane channels associated with nerve endings, have been identified as internal temperature sensors in crocodiles and lizards: the heat sensing TRPV1 and the cold sensing TRPM8 (Seebacher and Murray 2007). The signals from these TRPs are likely passed through the hypothalamus and activate neuronal β-adrenergic receptors on cell membranes. These latter receptors are related to cardiovascular response and PGC-1α expression, which in birds and mammals causes an up-regulation of metabolic enzymes (CCO and ATPase), UCPs, and mitochondrial biogenesis, suggesting the possibility of increasing energy production with less ROS production.

Heat stress causes metabolic rate to increase, ROS to be overproduced, and denaturation of cellular proteins. During heat stress, heat shock proteins (e.g., HSP-70) can protect other cellular proteins. Studies on the master transcription factor HSF-1 and on HSP-70 in lizards report that both temperate and desert lizards express HSP-70 at high levels under heat stress. Interestingly, under normal conditions desert lizards constitutively express HSP-70, whereas temperate species maintain higher concentrations of the transcription factor HSF-1. This sug-

gests that within these two species, selection on different components in the same pathway has allowed each species to become adapted to the thermal fluctuations in their respective habitats. Diurnal desert lizards may maintain a high level of thermal tolerance that can be induced further under extreme stress, whereas the temperate species may respond faster to moderate heat stress through HSP-70 synthesis, allowing for more temperature sensitivity (as reviewed in Evgen'ev *et al.* 2007). Heat-stress treatment can also increase activity of the antioxidant, catalase, in garden lizards (Jena and Patnaik 1992).

Physiological and phenotypic flexibility (i.e., reversible phenotypic plasticity, acclimation) of reptiles to seasonal temperature changes has recently been reviewed (Seebacher 2005). Generally, winter-active reptiles that experience seasonal temperature fluctuations use metabolic compensation during cooler temperatures to maintain performance. They accomplish this by increasing thermosensitivity, increasing metabolic enzyme activities (via transcriptional up-regulation), and changing the membrane fatty acid composition to increase their metabolic potential (Seebacher 2005 and references therein). They may also decrease the negative effects of metabolic up-regulation (ROS production) by increasing UCP activity to prevent ROS production (Schwartz *et al.* 2008) and/or increasing antioxidant levels to neutralize ROS.

Activating these pathways to assist in heat shock or acclimation may also promote overall cellular maintenance and longevity, and once these pathways are activated they may have a hormetic effect on longevity, possibly through epigenetic modifications (reviewed in Arking and Giroux 2001). The selective advantage of precise acclimation and balanced metabolic up-regulation is likely profound. Furthermore, evolving the ability to acclimate thermally may select for allelic variants in regulating pathways that also promote longevity.

15.3.2 Hibernation: Supercooling, freeze tolerance, and anoxia tolerance

Hibernation is a strong selective force on life history traits because the ability to hibernate allows reptile populations to persist, and individuals to reproduce for multiple years, in colder climates. Hibernation is a drastic response to seasonal temperature fluctuation, in which rather than attempting to compensate for the thermal (or dry) conditions the organism further intensifies the down-regulation of metabolic rates to a level of dormancy. The overall decrease in metabolism for an extended period of time has prompted the suggestion that hibernation or torpor could increase lifespan through "suspended animation." In addition to this, the evolution of the ability to hibernate (and recover) may select for molecular mechanisms that also increase lifespan as a by-product.

Three types of stresses associated with hibernation would be selective forces at the molecular level:

- cold stress
- anoxia and the build-up of toxic anaerobic by-products
- oxidative stress due to reperfusion during rewarming.

Reptile species that experience freezing or nearly freezing temperatures have evolved mechanisms for freeze avoidance through supercooling, preventing freezing of tissues, and/or through freeze tolerance—the ability to recover from being frozen. Decreased blood circulation is concomitant with these conditions, resulting in tissue anoxia and a switch to anaerobic metabolism. This is most extreme in turtles, as they typically hibernate under the water. Consequently, turtles have evolved exceptional ability to deal with anoxia as well as cold stress (Storey 2006, Bickler and Buck 2007). Post-hibernation rewarming or thawing has immense potential for injury due to oxidative stress, as the tissues are reperfused with oxygenated blood and aerobic metabolism is reestablished.

At the molecular level, it is difficult to separate the response to these stresses because they have coevolved as an integrated hibernation response. In this way, some environmental stressors may provide cues for the stresses the organism must prepare for at the cellular level. Furthermore, a response to a particular stress treatment may also activate molecular responses for other naturally occurring stresses. For example, some freeze-tolerance genes are known to be regulated by the same signal transduction pathways as anoxia (e.g., the transcription factor hypoxia-inducible factor: HIF-1; Cowan 2003). Therefore, an experimental anoxia treatment would also activate pathways for cold tolerance, as natural selection acts on the organismal level to react to these stresses concordantly.

Appropriate anticipatory responses to these environmental cues are important to prepare the organism for hibernation at a cellular level. Under extended anoxia, turtles go into a state of suspended animation such that essentially all protein degradation pathways and transcription/translation are silenced (except for a subset of genes essential for hibernation survival), thus no new proteins are made and no proteins are broken down (Storey 2007). Therefore, to survive the impending oxidative stress during the rewarming process after hibernation, they need an anticipatory response to prior cues before they enter anoxia and/or hibernation.

Metabolic regulation is obviously important for hibernation. Freeze tolerance in reptiles is intimately related to water content in the cell as well as oxidative stress and metabolic regulation. For example, in a study on turtle hatchlings placed at –3°C for 72 h, survival was predicted by plasma levels of lactate dehydrogenase, and cryoinjury was predicted by body size and water content (Costanzo et al. 2006). Anoxia and freezing studies on turtles and frogs consistently detect the up-regulation of mitochondrial genes (COX1, NAD5, NAD4, CytB), although different genes have different patterns across tissues (Storey 2006, Storey 2007). Experiments on freezing and thawing (live) lizards demonstrated an up-regulation of the mitochondrial protein UCP3 (along with PGC-1α and PPAR—regulators of UCP expression in mammals) and a consequential decrease in mitochondrial superoxide production during thawing (Rey et al. 2008).

Protection of cellular proteins during anoxia is provided by the up-regulation of particular heat-shock proteins; most notable of which is HSP70–9b (also called mortalin-2). Mortalin-2 is normally found in mitochondria and is associated with cell survival, possibly through interaction with tumo protein, p53 (reviewed in Storey 2007). Increasing antioxidant defense is necessary to neutralize ROS and prevent oxidative damage during rewarming. Gene expression for iron-binding proteins (e.g., hemoglobin, ferritin, and transferrin) is up-regulated during hibernation. This is hypothesized to protect against oxidative damage, since free iron in the cell can interact with H_2O_2 and lipid peroxides, producing highly reactive free radicals. It is thus important to keep free iron bound until needed for cellular processes. Many of these genes are under the control of the transcription factor HIF-1, which is also up-regulated in response to freezing and hypoxia in hatchling, but not adult, turtles (reviewed in Storey 2006). Studies across species vary widely as to which antioxidants (SOD, GPX, catalase, etc.) are up-regulated, in which tissues, and at which life history stages, but the overall consensus is that antioxidants are up-regulated during cold stress and/or anoxia, most likely in preparation for free-radical generation when recovering from hibernation (Storey 2006). Additionally, the up-regulation is more extreme in species that have experienced less frequent cold stress/anoxia, such as garter snakes and hatchling turtles. In contrast, adult turtles, which have repeated cold stress/anoxia exposure due to diving and underwater hibernation, have constitutive levels of antioxidants that are high relative to other ectothermic vertebrates, but similar to the levels in endotherms (reviewed in Storey 2006, 2007).

Lutz *et al.* (2003) proposed that hibernation adaptations may have a pleiotropic effect on longevity, as the turtle brain has evolved mechanisms to increase protection from oxidative stress and repair mechanisms to combat damage during anoxia. For example, anoxic turtles have increased GABA, which protects neurons, and the loss of which is associated with aging diseases. In addition, binding of transcription factor NF-kappaB typically declines with age in mammals, but increases in turtle brains under hypoxia (reviewed in Lutz *et al.* 2003). Thus, the first hibernation event may be a strong selective force for maintaining and activating these molecular pathways. It may also provide a hormetic experience, allowing these pathways to have an increased basal activity and/or be activated faster under other types of cellular stress.

15.3.3 Dietary stress: Availability and type of food

Many animals have adaptive physiological and biochemical responses to fluctuations in diet (prey availability and quality). In many animals systems (*C. elegans*, *Drosophila*, fish, rodents, primates), dietary restriction (DR)—particularly protein restriction—decreases growth rates, maximum body size, and reproductive output, and conversely increases stress resistance, immune response, and lifespan (Pamplona and Barja 2007, Chapter 14). Additionally, DR generally results in an overall decrease in IIS and oxidative damage (Caro *et al.* 2008). Genetic experiments on lab models have implicated multiple pathways likely involved in this life history trade-off response to DR, including metabolic, mTOR, and particularly the IIS pathway. In rodents, the effect of DR is not due to a decrease in mass-specific metabolic rate. Rather, it is likely due to a decrease in ROS production and/or increase in resistance to ROS due to changes in lipid membrane composition (Faulks *et al.* 2006). Additionally, under DR there is also mitochondrial biogenesis and up-regulation of UCPs, which are associated with increased mitochondrial efficiency and decreased ROS production (reviewed in Pamplona and Barja 2007).

Our focus here is on the adaptations in reptiles to caloric stress that utilize pleiotropic molecular pathways for the evolution of growth, reproduction, and aging. Like laboratory model systems, reptiles show a correlation between reproductive rate and prey abundance (Shine and Madsen 1997) and quality (Warner *et al.* 2007). Beyond that, comparing caloric studies between endotherms and reptiles is difficult because the latter have much lower resting metabolic rates, can lower their metabolic rates even further during fasting, and have extraordinary abilities to go without any food for long periods of time—up to two years in some snakes—whereas endotherms need daily dietary intake (Wang *et al.* 2006a). Consequently, even the terms "fasting" and "starvation" do not have the same meaning across birds/mammals and reptiles. While we have found no studies on caloric restriction at the molecular level in reptiles, there are many studies on food deprivation in reptiles, particularly snakes. During fasting and starvation, reptiles can down-regulate their resting metabolic rate (RMR; up to 70% in some snakes), and drastically increase aerobic metabolic rate upon food intake (reviewed in Secor 2009). Different species have evolved different levels of plasticity of metabolic rate with feeding—very infrequent feeders such as pythons have the largest fluctuations. Interestingly, in many snake species, either gravid or pregnant females stop eating voluntarily to allow for the more precise thermoregulation that is necessary for embryogenesis. In contrast to DR and starvation in mammals, these gravid females typically have increased levels of IGF-1 (Sparkman et al. 2009). With starvation in snakes and lizards, there is an increase in the proportion of polyunsaturated fatty acids in the total body lipid content (McCue 2008 and references therein). Additionally, DR in male, but not female, lizards causes a decrease in selected body temperature (Brown and Griffin 2005). It is interesting that both starvation/DR and decreased body temperature characterize hibernation in both mammals and reptiles, suggesting these correlated responses may have integrated molecular networks.

15.3.4 Type of food

The composition of the diet can impact reproduction and aging. For example, higher protein diet increases reproductive output in lizards (Warner et al. 2007), ingesting phytochemicals produced by stressed plants can affect an animal's metabolism and cellular defense (reviewed in Baur and Sinclair 2008), decreasing protein intake can decrease ROS production and increase lifespan in rats (Pamplona and Barja 2007), and dietary antioxidants may negate oxidative stress (Olsson et al. 2008a). Most studies on reptiles have focused on the fatty acid composition (polyunsaturated versus saturated fats) of the diet. Collectively, these studies indicate that changes in fatty acid composition of the diet can change the composition in tissues, blood plasma, and the plasma membranes, but different patterns across species and across tissues are reported (Cartland-Shaw et al. 1998, Simandle et al. 2001). Furthermore, diets rich in polyunsaturated fats cause a decrease in selected body temperature, an ability to decrease RMR with temperature, and a

decrease in lipid membrane rigidity (Geiser et al. 1992, Geiser and Learmonth 1994), which is consistent with polyunsaturated fats being important energy stores for hibernation.

It is apparent that local diet restrictions (seasonal fluctuations in prey type and availability) can impact individual physiology, reproduction, and longevity. A heritable difference in prey preference has been documented in snakes (Arnold 1981). Thus, natural selection on prey preference has the potential to drive populations to evolve along the reproduction/longevity continuum (Fig. 15-2).

15.4 Perspective

In this chapter, we have written from the position that reptiles are organisms that have been challenged by myriad environmental stresses, whose physiologies (perhaps more than other classes of organisms) have often adapted to such stresses through plastic regulation, and whose life histories have followed suit in correlated evolutionary fashion. This is a departure in assumption from the more frequently envisioned model of selection acting directly on the life history (through extrinsic mortality from predation, for example) with concomitant evolution of physiology or other "mechanism" following. In a "Which came first?" argument, we do not have conclusive data, although it is certainly the case that both scenarios are possible and may vary across populations. Experiments using artificial selection or experimental evolution approaches can speed up or slow down life history (reviewed in Rose 1991, Stearns et al. 2000, Rose et al. 2002) through largely uncharacterized mechanisms (e.g., Harshman 1999). Likewise, in model laboratory systems, the opposite pattern has also been reported. For example, in classic studies of the nematode (C. elegans), mutations in age-1 or daf-2 loci (Friedman and Johnson 1988, Kenyon et al. 1993) caused altered life history phenotypes through changes in signaling pathways. Unless life history variation is a manifestation of existing plasticity, some mechanistic basis to life history evolution at the cellular level is likely present, which leads to our hypothesis of "physiology evolves first." Tests of this assertion could include novel experimental evolution studies on, for example, short-lived reptiles—likely lizards—where both response to selection and physiological evolution could be monitored at each generation.

We have highlighted a number of compelling areas of research into the physiological mechanisms underlying life history variation and their molecular bases. We note that in almost each section, we have surmised by analogy to other amniote vertebrates what may be similar molecular induction in reptiles. Data are sorely lacking in these ectothermic vertebrates, which makes it an exciting time to study the molecular bases of life history evolution in reptiles. Although difficult to study in a lifespan and aging context due to their often long lives, several areas where immediate progress can be made include telomere biology, ectothermic mitochondrial function, and the roles of uncoupling proteins and lipid membrane dynamics in stress response. Experiments on oxidative stress need to look at all aspects of oxidative stress (free radical production, antioxidant defense, protection, repair, and overall damage accumulation) to make predictions of how these molecular events can influence the evolution of life history traits. More detailed experiments in reptiles (both in natural populations and under controlled laboratory conditions) are necessary to tease apart these components of oxidative-stress regulation in response to different environmental stresses and Monaghan *et al.* (2009) provide a nice review of procedures for measuring these components.

Recent advances in technologies such as RNAi are being used on chickens (*in ovo* and post-hatching) to knockdown gene expression of targeted genes (Das *et al.* 2006, Chen *et al.* 2009). Undoubtedly, with optimization, these technologies will be applicable to future experimental studies on reptiles, to investigate specific nodes in complicated gene networks underlying life history evolution. Additionally, the sequencing of two reptile genomes (*Anolis* lizard and painted turtle) are currently underway, and transcriptomes from various reptile species will be available by the time this book is published. These advances in the molecular tool kit of reptile genetics introduce the entirely realistic possibility that molecular network regulation, including epigenetic modification, gene expression, and proteome dynamics, will be revealed both in the laboratory setting and in natural populations of diverse reptiles.

While the molecular tool kit is rapidly expanding, there will always be difficulty in doing laboratory-based molecular studies on life history evolution on most reptiles due to their long generation time. Although some reptiles—such as *anolis* lizards, which reach sexual maturity in under a year and reproduce readily in captivity—will be good laboratory models, they represent only one end of the life history continuum. Therefore, more studies will need to focus on wild populations, capitalizing on the natural genetic variation available, to achieve comprehensive understanding of the evolution of life histories on both ends of the continuum. Future studies on natural populations should include measuring heritable genetic variation, which is necessary for selection to act. Very few studies have measured genetic variation in stress response or ROS regulation in wild populations, or heritability of these traits. One exception in reptiles has demonstrated that genetic variation in ROS production (measured O_2^-) is heritable (Olsson *et al.* 2008b). Thus, at least in this lizard species, ROS production has the ability to respond to selection. Predictably, advances in genomic and computational technologies will allow the reconstruction of population/pedigree history, with genomic data and detection of heritability of traits in wild populations along with detailed association studies (Stinchcombe and Hoekstra 2008). In addition, studies on whole-system physiology, including transcriptomics and metabolomics, would greatly assist in our understanding of the molecular networks underlying stress response and life history trade-offs. These types of studies on wild reptile populations may allow for further discovery of genes and pathways important for these life history trade-offs that would not otherwise be indentified in inbred laboratory animal models selected for laboratory conditions.

15.5 Summary

1. The life history trade-off between reproduction and aging has been well established. Because of

their unique metabolic and physiological adaptations, which suggest remodeling of the molecular network underlying both stress-response and life history trade-offs, reptiles should be excellent model systems to study the trade-off between reproduction and aging and the role of stress response pathways in the evolution of this trade-off.

2. Many types of stress response use integrated networks, which may constrain or accelerate their ability to evolve. Genes known to be involved in longevity are also incorporated in these stress-response networks—often as a major node (highly connected).

3. Natural selection on the molecular networks underlying stress response will have a pleiotropic effect on longevity and the trade-off between longevity and reproduction. External environmental stresses can be a source of mortality and they are thus agents of selection on the pathways that underlie the ability to respond appropriately to environmental cues.

4. Rate of aging has a genetic basis and longevity can evolve under direct selection. Longevity may also be able to evolve as a pleiotropic response to selection on the physiological basis of stress response.

5. Future research needs to utilise populations from both the laboratory and nature in order to advance our understanding of underlying molecular networks, and to elucidate how evolution has shaped these conserved networks into unique evolutionary adaptations in reptiles.

15.6   Acknowledgments

Our thanks to the Bronikowski Lab members, D. Warner, M. Olsson, the editors, and an anonymous reviewer for their constructive comments and efforts to improve this chapter. Also our thanks to Amy Dixon and Katelyn McDonald for the artwork. TSS acknowledges support from a US NSF-IGERT fellowship 0504304 in Computational Molecular Biology. The garter snake case study was supported by US NSF DEB-0323379 and IOS-0922528.

CHAPTER 16

Mechanisms of aging in human populations

Maris Kuningas and Rudi G.J. Westendorp

16.1 Introduction

Lifespan (the length of life) is an integral part of life history, and is influenced by a multitude of factors, including genetic and environmental. In humans, it has been observed that mean life expectancy in Western societies has increased dramatically over the last century (Oeppen and Vaupel 2002). In Japan, for instance, the mean life expectancy has increased from 50 years to 80 years in no more than six decades. It is unlikely that changes in population genome over this time-period can explain the observed increase in lifespan, which is more likely because of improvement of environmental conditions and medical care. The increase in mean life expectancy of the total population, however, has left the marked inter-individual variance in lifespan unaltered. Studies of twins and long-lived families have estimated that 20–30% of the variation in human lifespan is determined by genetic factors, which become more important for survival at older ages (Herskind *et al.* 1996, Mitchell *et al.* 2001, Hjelmborg *et al.* 2006). Siblings of centenarians have a significantly higher chance of becoming a centenarian themselves when compared to other members of their birth cohort (Perls *et al.* 2002). In addition, it has been shown that offspring of long-lived sibling pairs already have a lower mortality risk at middle age, whereas their spouses, with whom they have shared a common environment, do not show this survival benefit (Schoenmaker *et al.* 2006).

Over the years several evolutionary theories of aging have been proposed, which help to explain why there are genetic factors that influence aging

and lifespan (Austad and Finch 2008). Aging or senescence is the progressive loss of function accompanied by decreasing fertility and increasing mortality with advancing age. It has been proposed that aging has evolved as a by-product of natural selection for the optimization of reproductive schedules. Genes with beneficial effects early in life, but detrimental effects on fitness late in life, would be selected because of the diminished power of natural selection with age. This antagonistic pleiotropy theory of aging (Williams 1957) was further developed in the disposable soma theory, which emphasizes the trade-off between reproduction and somatic repair and maintenance (Kirkwood 1977). The existence of such trade-offs has been demonstrated in various studies and the most common trade-off that has been observed is between lifespan and fertility (see also other contributions in Part 4 of this volume). The latter trade-off has also been observed in humans (Westendorp and Kirkwood 1998).

Studies with model organisms have considerably contributed to the identification of genes and physiological processes that influence lifespan. The most prominent example includes the *Caenorhabditis elegans daf*-2 and *clk* double mutants, which live nearly five times longer than wild-type worms (Lakowski and Hekimi 1996). Most of the genes indentified to influence lifespan in model organisms are evolutionarily conserved and present in a variety of organisms, including humans. In the following sections, we briefly review these genes and their role in human health and longevity. In addition, we discuss the integration of genetic pathways and the environment.

16.2   Mechanisms of aging

16.2.1   Insulin/IGF-1 signaling

The first evidence that genes can influence lifespan came from studies with *C. elegans*. It was discovered that mutations in dauer formation (Daf) genes, such as *age-1* and *daf-2* lead to lifespan extension (Larsen 2001). Molecular characterization of these genes revealed that they belong to the evolutionarily conserved insulin and insulin-like growth factor (IGF) signaling (IIS) pathway (Bartke 2008a). In the following years it was shown that, similar to *C. elegans*, reduced insulin signaling extends lifespan in *Drosophila melanogaster* (Giannakou and Partridge 2007). However, in the case of *C. elegans* it was also noted that whereas moderately decreased signaling through the IIS pathway prolongs lifespan, severe reduction in IIS during development can result in the formation of dauer larvae and failure to complete normal development.

In *C. elegans* and *D. melanogaster* the IIS pathway functions via a single receptor and a host of ligands. In vertebrates, this has changed during the course of evolution. The IIS pathway has split into two pathways, with separate receptors for insulin (IR) and IGF (IGF-1R; IGF-2R). The insulin and IGF-1 pathways have overlapping, but also distinct, functions, where insulin mainly controls metabolism, while IGF-1 controls growth and development. Data from mouse models indicate that both these pathways influence lifespan (Bartke 2008a). Reduced signaling through the IGF-1 pathway, which acts through growth-hormone-releasing hormone (GHRH), growth hormone (GH), and IGF-1, leads to reduced fertility but also to longer life (Rincon *et al.* 2005). In contrast, reduced signaling through IR, as demonstrated by complete- or tissue-specific disruption of the *IR* gene, leads to insulin resistance, impaired glucose regulation, diabetes, and shortened lifespan (Rincon *et al.* 2005). The only exception is the fat-specific *IR* knockout mouse models (FIRKO), which have reduced fat mass, are protected against age-related obesity, and live longer than their littermates (Bluher *et al.* 2003). In the case of insulin signaling substrates (IRS), it has been observed that reduced IRS-2 signaling in the brain promotes healthy metabolism, attenuates meal-induced oxidative stress, and extends the lifespan of overweight and insulin-resistant mice (Taguchi *et al.* 2007). This suggests that by directly attenuating brain IRS-2 signaling, an aging brain can be shielded from the negative effects of hyper-insulinemia that ordinarily develops with overweight and advancing age (Taguchi *et al.* 2007).

In humans, it has been observed that an optimal regulation of IIS is necessary for survival. This regulation prevents morbidities associated with either deficient or excessive IGF-1 and/or GH levels. For instance, reduced levels of IGF-1 and/or GH have been associated with protection against cancer (Yang *et al.* 2005) and with longevity, as centenarians have decreased plasma IGF-1 levels and preserved insulin action (Barbieri *et al.* 2003), but also with cardiovascular disease, diabetes, and reduced life-expectancy (Yang *et al.* 2005). Several studies have tried to identify genetic factors that contribute to this regulation, and to assess whether genetic variation in the evolutionarily conserved IIS pathway influences lifespan in humans, as observed in model organisms. To date, no genetic variants have been identified in the coding region of *IGF-1* gene, suggesting that the *IGF-1* locus is under strong selection. On the other hand, for the *IGF-1R* locus, it has been observed that centenarians are enriched for genetic variants that lead to reduced IGF-1 signaling (Bonafe *et al.* 2003, Suh *et al.* 2008). Similar findings have been observed in a prospective follow-up study of elderly Dutch subjects, where a polymorphism in the *GH*1 gene, which controls IGF-1 activity, was associated with longevity in females (van Heemst *et al.* 2005a). In the same study, a combined effect of variations at the *GH*1, *IGF-1* and *IRS*1 loci was observed, suggesting an additive effect of multiple variants that reduce IIS on human longevity. In contrast to the common variants, mutations in the *GH* receptor have been show to lead to high levels of GH, with low IGF-1 levels, and to cause GH-insensitive or Laron syndrome, which has a phenotype of short stature, obesity, mental retardation, and glucose intolerance. Likewise, mutations in the *IR* have been associated different degrees of insulin resistance and diabetes, and with pathogenesis of polycystic ovary syndrome (Lee *et al.* 2008a). These detrimental effects are observed as a result of severe function-disrupting mutations.

Therefore, it might be argued that reduction in IIS signaling probably has to be rather mild in order to have beneficial effects on lifespan.

Taken together, the human IIS pathway contains several genes that influence growth, reproduction, and lifespan, although differential biological effects of IGF-1 and insulin are observed, which are likely related to differences in the engagement of other signaling pathways. This might also explain why variation in the IIS does not reveal a modulation of human lifespan of a magnitude that would resemble what is seen by analogous defects in some of the model organisms. It could also be that in vertebrate models the effects are too strong or more pleiotropic, causing deleterious side effects that are not present in invertebrate models. On the other hand, it might also be argued that, even in invertebrate models, the reduction in IIS signaling has to be rather mild in order to have beneficial effects on lifespan.

16.2.2 Lipid metabolism

In *C. elegans*, DAF-12 is a member of an evolutionarily conserved nuclear hormone receptor (NHR) superfamily, and has been implicated in dauer diapause, developmental timing, metabolism, fertility, and longevity (Rottiers and Antebi 2006). In humans, the NHRs most similar to DAF-12 are the liver X receptors alpha and beta (LXRs), which have cholesterol breakdown products (oxysterols) as ligands. Upon activation, LXRs regulate processes that result in cholesterol excretion from the body. Recently, a common *LXRA* haplotype was associated with increased survival, predominantly because of lower mortality from cardiovascular disease and infection (Mooijaart *et al.* 2007). A possible mechanism through which LXRA could lead to the observed beneficial effects includes involvement of its target gene apolipoprotein E (*APOE*). APOE transports cholesterol, fat-soluble vitamins, and lipoproteins to the blood. It has three major isoforms, *apoE2*, *apoE3* and *apoE4*, of which *apoE2* is the rarest and *apoE3* the most common allele. *ApoE2* allele has been associated with low cholesterol levels, whereas *apoE3* and *apoE4* are associated with intermediate and the highest cholesterol levels, respectively. High plasma APOE levels, and also the *apoE4* allele, have

consistently been associated with cognitive decline and cardiovascular disease mortality (Panza *et al.* 2007). It has also been observed that the *apoE4* allele is depleted whereas *apoE2* allele is enriched among centenarians in a variety of populations. Furthermore, the *apoE2* allele has been associated with reduced fertility, suggesting an antagonistic pleiotropic effect for the *APOE* gene (Corbo *et al.* 2008). Similar associations, although not as consistent as with *APOE* gene, have been observed for other genes in the apolipoprotein family and for genes involved in lipid metabolism (e.g., *APOA1*, *APOA4*, *APOB*, *APOC3*, *ACE*, *CETP*, *Lp(a)*, *PPARgamma*). These findings highlight the influence of lipoprotein metabolism on lifespan, which is further reinforced by studies on lipoprotein particle size in relation to age. It has been shown that families of Ashkenazi Jewish centenarians have larger particles of both high-density lipoproteins (HDL) and low-density lipoproteins (LDL), which are associated with a decreased incidence of metabolic syndrome, cardiovascular disease, and hypertension (Panza *et al.* 2007). Also, in a Dutch population, offspring of long-lived sibling pairs have larger LDL particles than their age-matched partners, again suggesting that larger LDL particles confer a survival benefit (Panza *et al.* 2007). Taken together, in humans, there is enough evidence to suggest that lipid profiles and genes involved in lipid metabolism are associated with longevity.

16.2.3 Antioxidant enzymes

Antioxidant enzymes, such as catalase (CAT) and superoxide dismutase (SOD), prevent damage to cellular lipids, proteins, and DNA from reactive oxygen species (ROS). The evidence from model organisms on the beneficial effect of these enzymes on lifespan has been controversial. In *D. melanogaster*, there are two forms of SOD: Cu/ZnSOD (SOD1), which is located in the cytoplasm and the outer mitochondrial space, and MnSOD (SOD2), which resides in the inner mitochondrial space. It has been observed that the extended lifespan of several long-lived *D. melanogaster* populations is correlated with increased expression of *SOD1* and *CAT*, and associated with increased oxidative stress resistance. In line with these observations, loss of

CAT and/or *SOD1* activity by mutations has been shown to decrease the resistance of *D. melanogaster* to oxidative stress and dramatically reduce lifespan (Phillips *et al.* 1989). On the other hand, over-expression of *SOD1*, *SOD2* and *CAT* has been shown to increase lifespan (Sohal *et al.* 2002, Tower 2004). Although these findings are not undisputed, they provide support for the idea that the oxidative damage theory of aging holds, at least in *D. melanogaster*.

In mammals, in addition to SOD1 and SOD2, there is an extra SOD (SOD3) in the extracellular space. Studies with mouse models have shown that mice without a *SOD1* gene are viable and appear normal at birth, but there are also studies reporting a number of moderate-to-severe pathologies and reduced lifespan for *SOD1* null mice. On the other hand, mice over-expressing *SOD1* are more resistant to oxidative stress and have lower levels of lipid peroxidation, with lifespans similar to wild-type mice (Perez *et al.* 2009). Disruption of the *SOD2* gene has been reported to be lethal because of neurodegeneration and damage to the heart (Jang and Remmen 2009), whereas over-expression of *SOD2* leads to increased lifespan, as does over-expression of *CAT* targeted to mitochondria (Jang and Remmen 2009). Mice heterozygous for the mitochondrial form of *SOD2* have high levels of DNA oxidation in multiple organs. In spite of their abnormally oxidized DNA, these animals show no decline in lifespan and no acceleration in hallmarks of aging, such as cataracts, immune dysfunction, and protein modifications (Jang and Remmen 2009). These data suggest that mice can live reasonably long and healthy lives despite unusually high levels of oxidative damage.

The evidence for the role of antioxidative enzymes in the preservation of human health is not well established. Mutations in the *SOD1* gene lead to familial amyotrophic lateral sclerosis, but there are no studies reporting on the influence of genetic variation in the *SOD1* gene on lifespan. For the *SOD2* gene, genetic variants have been associated with a number of phenotypes, including increased risk for prostate and breast cancer, Alzheimer's disease, and mortality. Findings also indicate that epigenetic silencing of *SOD2* constitutes a mechanism that leads to decreased expression of *SOD2*, as observed

in many breast cancers (Hitchler *et al.* 2006). In addition, variants in both *SOD2* and *SOD3* have been associated with reduced lung function and susceptibility to chronic obstructive pulmonary disease (COPD; Siedlinski *et al.* 2009). One *SOD3* gene variant has also been implicated in pre-eclampsia complicated by severe fetal growth restriction (Rosta *et al.* 2009). Similar to the *SOD1* gene, there are no studies that have found an effect of genetic variation in the *SOD3* and *CAT* genes on lifespan in humans.

16.2.4   Macromolecule repair mechanisms

Defects in the mechanisms that repair damage to DNA, proteins, and membranes have been shown to reduce lifespan in various model organisms, but beneficial effects of increased repair capacity on lifespan have been demonstrated in only a few studies. In *D. melanogaster* it has been shown that the absence of mei-41 excision repair reduces lifespan, but the presence of one or two extra copies significantly increases lifespan. Likewise, over-expression of protein carboxyl methyltransferase (PCMT), which is a protein repair enzyme, has been correlated with increased lifespan. Both of these genes, *mei-41* and *pcmt*, have homologues in mammals, which are ataxia telangiectasia and Rad3 related (ATR), and *PCMT*, respectively. In mice, the disruption of the *ATR* gene leads to chromosomal fragmentation and embryonic lethality, and the disruption of *PCMT1* to fatal seizure disorder and retarded growth. In humans, there are data only for the *ATR* gene, in which mutations lead to a rare Seckel syndrome.

In a similar fashion, increased expression of heat shock proteins, which are key mediators of the organism's resistance to stress by assisting in the establishment of proper protein conformation and prevention of unwanted protein aggregation, has been implicated in lifespan extension in *C. elegans* and *D. melanogaster* (Zhao *et al.* 2005, Terry *et al.* 2006). In mouse models, over-expression or exogenous delivery of heat shock proteins have been shown to prolong survival of a transgenic mice with Huntington's disease (Vacher *et al.* 2005) and with amyotrophic lateral sclerosis (Gifondorwa *et al.* 2007). In humans it has been observed that a metionine/threonine polymorphism located in HSP70-related gene (HSP70-hom),which is in tight linkage disequilibrium with *HSP70-1*, is associated with longevity (Bonafe and Olivieri 2009), but there are also studies finding opposite or no associations between longevity and the various heat shock proteins (Terry *et al.* 2006, Bonafe and Olivieri 2009). Therefore, in humans it still needs to be elucidated whether heat shock proteins and their ability to maintain proteostasis protect against age-associated pathology caused by protein malfolding, and influence lifespan.

To study the influence of differences in DNA repair mechanisms on lifespan, DNA-repair-deficient mouse models have been generated, which display a common phenotype of progeria, or cancer predisposition, or both, and have a reduced lifespan (Hasty *et al.* 2003). Similarly, in humans, all mutations identified in DNA repair genes severely compromise health. For instance, mutations in transcription-coupled repair components have been associated with the progeroid syndromes of Cokayne and trichothiodystrophy (Hasty *et al.* 2003). Mutations in the RecQ-like DNA helicase genes, *WRN*, *BLM* and *RecQ4*, lead to progeroid syndromes of Werner, Bloom, and Rothmund–Thomson, respectively. In contrast to the strong phenotypes associated with loss-of-function mutations in the RecQ helicases, common variants with subtle effects on the functionality of these genes do not seem to influence aging trajectories and survival in the general population. This could be because of the non-overlapping functions of the different RecQ helicases, meaning that failure of one given RecQ gene cannot be complemented by another. These observations underpin the importance of DNA repair in all organisms.

The key questions that have yet to be answered are whether subtle variants in macromolecule repair genes contribute to variation in lifespan, and whether above-average repair contributes to lifespan extension in humans. In addition, it is still under question whether subtle variation in genes that assure genomic integrity influence disease susceptibility and lifespan. Their study is important, as defective or reduced DNA damage recognition and repair capacity, together with decreased ROS scavenging, can greatly compromise human health.

16.2.5 Cellular responses to damage

In response to unrepaired damage, cells trigger either apoptosis, which is a programmed cell death, or cell cycle arrest. The most well-known protein implicated in the maintenance of genomic stability is p53. Recently, p53 homologues were identified in *C. elegans* and *D. melanogaster*. In contrast to mammalian p53, which elicits apoptosis or cell cycle arrest, the p53 of *C. elegans* and *D. melanogaster* affects only apoptosis. Despite these differences, reduced p53 activity has been associated with increased lifespan across a variety of organisms (Bauer and Helfand 2006). In humans, this extension comes at the cost of increased risk for cancer. It has been shown that carriers of the *Arg72Pro* genetic variant in the *p53* gene, which leads to lower apoptotic potential, have increased survival at old age, despite an increased risk for mortality from cancer (van Heemst *et al.* 2005b). This observation supports the hypothesis that reduced p53-mediated induction of apoptosis can have beneficial effects on lifespan if tumor formation can be avoided.

Similar to p53, p16 is active in cell senescence, aging, and tumor suppression, and its expression is triggered by a variety of major biological stresses, including DNA damage and oxidative stress. In humans it has been found that a deletion of the p16^{INK4a}/ARF/p15^{INK4b} region occurs in many cancers, and that a genetic variant in this region associates with reduced physical impairment in old age (Melzer *et al.* 2007).

16.3 Convergence of longevity signals

Many longevity signals converge on members of the evolutionarily conserved forkhead transcription factor (FOXO) and silent information regulator 2 (Sir2) NAD-dependent protein deacetylase (Sirtuin) protein families, which interact with each other. FOXOs are human homologues to *C. elegans'* DAF-16, which is the main downstream target of the IIS pathway, which negatively regulates the activity of DAF-16. In *C. elegans*, it has been shown that reduced signaling through the IIS pathway relieves the repression of DAF-16, thereby promoting an anti-aging pattern of gene expression. Similar effects are observed when *daf-16* is over-expressed. The majority of these findings have also been corroborated in *D. melanogaster* (Russell and Kahn 2007, Toivonen and Partridge 2009). In addition, over-expression of dFOXO in the *Drosophila* adult fat body increases the lifespan and reduces the fecundity of female flies, whereas no effect is observed for males (Giannakou *et al.* 2004).

In mammals, four FOXO proteins have been identified: FOXO1a, FOXO3a, FOXO4, and FOXO6. Similar to DAF-16/dFOXO in invertebrates, the mammalian FOXOs relay the effects of insulin on lifespan, influence fertility, and play an additional role in tumor development and in diabetes (Carter and Brunet 2007). However, besides insulin and IGF-1, FOXOs activity is regulated by neuro-trophins, growth factors, nutrients, cytokines, and oxidative stress (Salih and Brunet 2008). These stimuli control FOXOs expression, subcellular localization, or DNA binding and transcriptional activity. In turn FOXOs can regulate a broad array of cellular processes, such as apoptosis, glucose metabolism, cell cycle progression, and differentiation. In humans, studies analysing the role of genetic variation in *FOXOs* in relation to age-related diseases, fertility, and lifespan have found that FOXO1a and FOXO3a influence these phenotypes. Genetic variants in *FOXO1a* have been associated with increased glucose levels, increased risk of diabetes, and decreased lifespan, but not with fertility (Kuningas *et al.* 2007a). In the same study, *FOXO3a* was associated with increased risk of stroke and mortality. In contrast, different genetic variants in the *FOXO3a* gene have been shown to contribute to longevity, as these variants were enriched in centenarians compared to middle-aged participants (Willcox *et al.* 2008; Flachsbart *et al.* 2009). In addition, carriers of these variants had lower plasma insulin levels, lower cancer and cardiovascular disease incidence, and a good self-reported health. Taken together, even though, to date, no studies have assessed the role of other FOXO family remember on human lifespan, the results with FOXO1 and FOXO3a are concordant with findings in model organisms, suggesting a high level of conservation of FOXO effects on aging across species.

Sirtuins represent another family of proteins that have been identified as interacting with and influencing the activity of various transcription factors and co-regulators. Through these interactions, sirtuins affect many metabolic and stress-resistance pathways, including those involved in DNA repair, apoptosis, fertility, glucose, and fat metabolism (Haigis and Guarente 2006). It has been shown that increased expression of *Sir2*, either because of an extra copy of the gene or caloric restriction, prolongs lifespan in various model organisms (Haigis and Guarente 2006). Similar effects, although not undisputed, have been observed after administration of resveratrol, which is normally synthesized by plants in response to stress and which increases the activity of Sir2.

Mammals have seven sirtuins: an intranuclear SIRT1, a cytoplasmic SIRT2, three mitochondrial sirtuins (SIRT3, SIRT4, and SIRT5), a heterochromatin-associated nuclear SIRT6, and a nucleolar SIRT7. From these genes, *SIRT1* and *SIRT3* have been studied the most in relation to phenotypes in humans. It has been found that genetic variants in the *SIRT1* gene are associated with lower risk for cardiovascular mortality and with better cognitive functioning (Kuningas *et al.* 2007b). In addition, it has been shown that *SIRT1* variants are associated with decreased basal energy expenditure and a lower lipid-oxidation rate (Weyrich *et al.* 2008). Therefore it has been proposed that genetic variation in *SIRT1* may determine the response rates of individuals undergoing caloric restriction and increased physical activity. However, there are also studies showing that *SIRT1* variants are associated with increased risk for obesity (Peeters *et al.* 2008), and studies finding no associations with *SIRT1* variants and lifespan. For *SIRT3*, associations with genetic variants and increased lifespan have been reported, but again there are also studies finding no evidence for such an association (Bellizzi *et al.* 2005, Lescai *et al.* 2009). For other human sirtuins, no data implicating a role in lifespan are yet available.

16.3.1 Dietary restriction

Dietary restriction (DR) without malnutrition has been shown to increase lifespan in a variety of organisms (see Chapter 14), although the precise molecular mechanisms for this action remain controversial. Currently there are several ongoing studies on primates to evaluate the effects of DR. The first results have revealed that an adult-onset DR delays the onset of age-associated pathologies and promotes survival in rhesus monkeys (Colman *et al.* 2009). These monkeys had improved metabolic function, lower incidence of diabetes, cancer, and cardiovascular disease. In two other studies, conducted with rhesus monkeys and rhesus macaques, the benefits of DR for health and longevity were less obvious, which could be because of differences in study design (Bodkin *et al.* 2003, Mattison *et al.* 2003, Spindler 2010). Even though studies with primates show positive indications it is still not certain if DR has a positive effect on longevity in humans. There are a number of studies that report on similar effects of DR in humans as in primates. In humans, DR has been shown to contribute to lower cholesterol, fasting glucose, and blood pressure and to reduce systemic inflammation and other risk factors for cardiovascular disease (Spindler 2010). Similar effects have been observed in weight-loss studies, suggesting that DR contributes to better survival in humans. However, this conclusion is not supported by prospective cohort studies of the relationship between body mass index (BMI) and longevity in humans (Spindler 2010). Low body-weight in middle aged and elderly humans has been associated with increased mortality. This suggests that in humans, a caloric intake to maintain an optimum BMI is necessary.

16.4   Integration of genetic pathways and the environment

It is well known that environment has a vast influence on the expression of genetic information. This is also the case for lifespan. Studies with model organisms have demonstrated that lifespan is very plastic, that is, different environmental factors (temperature, nutrition, population density) can have a considerable effect on lifespan. The phenotypic plasticity, which is defined as the ability of a genotype to change phenotypically when exposed to environmental change, allows organisms to adjust their life history in response to environmental cues. For instance, in optimal environments, organisms might invest in reproductive success at the expense of future survival, whereas in stressful environments organisms might switch to increased

investment in survival, until conditions for reproduction have improved (Flatt and Schmidt 2009). An example for such action is the dauer diapause in larval nematodes in response to low temperatures and short day-length.

Testing model organisms in laboratory conditions has provided valuable information on genetic and environmental factors that influence lifespan. However, in most cases laboratory conditions poorly mimic the evolutionary niche in which these genes come to expression. It is largely unknown to what extent mutations that influence lifespan in laboratory conditions affect fitness in natural environments. Furthermore, it is unknown whether in natural populations these candidate loci contain genetic variation, which would contribute to phenotypic variance for lifespan. These questions are of importance because not all candidate loci with major effects on longevity in laboratory conditions may exhibit variation in natural populations. Already, there are a few studies that have addressed these questions. For instance, it has been shown that the long-lived mutant fruit fly methuselah (mth) underperforms in most cases under conditions that resemble natural situations (Baldal *et al.* 2006). On the other hand, there is also evidence that the *mth* locus is genetically variable in natural populations and that these variants do contribute to variation in lifespan (Paaby and Schmidt 2008). These studies are of importance, since such results can be more easily interpreted for studies with humans, where naturally occurring genetic variants in candidate genes identified in model organisms are studied in relation to lifespan.

Even for humans, the environment in which the genome effectively evolved has changed. Many aspects of our modern environment and lifestyle, including diet, exercise, exposure to chemicals, and hygienic practices are mismatched to our bodies' evolutionary state. Genes that were originally selected for survival in adverse environments are now expressed under completely new, affluent environmental conditions. Such mismatch is believed to underlie many currently prevalent diseases such as diabetes, obesity, and cardiovascular disease (Stearns *et al.* 2008). For instance, genes that increased the efficiency to store energy in times of abundance and use these stores in times of famine contributed to a survival advantage. In modern Western societies, where food is constantly abundant, these genes are thought to underlie the increased prevalence of storage diseases, such as obesity and diabetes. This reinforces the idea that our genomes have been optimized to increase fitness under adverse environmental conditions and not under modern affluent conditions, resulting in new interactions with outcomes that are both unknown and unpredictable. For lifespan, we have observed an increase in Western societies, but such increase is likely to come at a cost. Despite the accumulating evidence from model organisms and humans, it still remains unclear how exactly lifespan evolves when populations adapt to novel diets or other environments and whether there are long-term benefits and costs associated with responding to such environmental change.

16.5 Summary

1. Lifespan is influenced by a multitude of factors, including genetic and environmental factors.
2. Studies with model organisms have identified evolutionarily conserved mechanisms that influence lifespan across species, including humans.
3. Mutations in genes that lengthen lifespan in model organisms have many detrimental side effects in humans. It is likely that genetic effects have to be mild in order to contribute to beneficial effects on lifespan in humans, but also in model organisms.
4. In model organisms, trade-offs between lifespan and other life history traits are commonly observed. The most common trade-off is between lifespan and fertility.
5. Environmental factors can have a considerable effect on lifespan and influence the expression of genetic information. Studies with model organisms have provided valuable information on understanding how some genotypes can lead to different phenotypes with changes in the environment (phenotypic plasticity).
6. In humans, mismatch to modernity is believed to underlie many currently prevalent diseases such as diabetes, obesity, and cardiovascular disease.

16.6 Acknowledgments

This research was supported by the Netherlands Foundation for the Advancement of Tropical Research (WOTRO 93-467), the Netherlands Organization for Scientific Research (NWO 050-60-810), the Netherlands Genomics Initiative/Netherlands Organisation for Scientific Research (NGI/NWO 911-03-016) and the EU funded Network of Excellence Lifespan (FP6 036894).

PART 5

Life history plasticity

Thomas Flatt and Andreas Heyland

Phenotypic plasticity, the ability of a single genotype to produce multiple phenotypes across different environments, is a concept of major importance in life history evolution (Stearns 1992, Roff 1992; see also Chapters 1 and 2). If there is genetic variation in the extent of life history plasticity among genotypes, selection can shape an optimal response ("reaction norm") to changes in the environment, and this adaptive plasticity enables organisms to optimize the expression of life history traits in an environment-dependent way that maximizes fitness (e.g., Stearns and Koella 1986). Despite the importance of plasticity in life history evolution, however, relatively little is understood about the proximate basis of plasticity and how it evolves. The chapters in this part of the book aim to give a mechanistic perspective of life history plasticity; for in-depth treatments of phenotypic plasticity and reaction norms we refer the interested reader to four books: Schlichting and Pigliucci (1998), Pigliucci (2001), West-Eberhard (2003), and DeWitt and Scheiner (2004).

In Chapter 17 Miner reviews the mechanisms underlying plasticity in feeding structures in echinoderm larvae. While these feeding structures represent morphological rather than life history traits, they are major determinants of fitness. As Miner discusses, feeding larvae from several classes of echinoderms alter the size of their feeding structures in response to food concentrations, so that when food is scarce larvae produce longer arms and smaller stomachs. The author reviews the mechanisms of this response, whereby larvae perceive exogenous cues from algae and alter the production of skeleton forming cells, possibly by changes in transcription factors that appear to play a role in skeleton formation.

In Chapter 18 Schmidt presents another example of life history plasticity, namely reproductive diapause in insects, with an emphasis on *Drosophila*. As the author discusses, insects that inhabit seasonal environments often express a diapause syndrome that affects a suite of traits, including energy allocation, resistance to environmental stress, patterns of reproduction, and age-specific mortality rates, and recent work has begun to elucidate the genetics and physiology of insect diapause in multiple organisms. Since states of arrested development and reproductive quiescence have widespread effects on life history traits in many taxa (see also Chapters 13 and 22), the research reviewed by Schmidt has the potential to outline common pathways and mechanisms of seasonal adaptations as well as the molecular basis of variation in the expression of such pleiotropic life history syndromes.

Chapter 19 by Brakefield and Zwaan examines what we have learned about polyphenisms, discrete phenotypes produced by a single genotype in response to changing environments. Such polyphenisms, most prominently in butterflies and moths, have been a constant source of inspiration to entomologists since well before Darwin's time. While traditionally the often strikingly divergent morphologies among alternative phenotypes or seasonal forms, for example in wing patterns, have attracted a lot of interest, polyphenisms are now proving extremely useful in unraveling the evolution of alternative suites of life history traits, including rates of aging and reproductive schedules. In their chapter, Brakefield and Zwaan discuss how polyphenisms can provide exceptional material to investigate how molecular, cellular, and ecological processes have become coordinated in evolution to yield extreme examples of developmental plasticity in which a single genome can yield two or more discrete life history phenotypes matched to their specific environments.

Finally, in Chapter 20, Rascón and colleagues review evolutionary and mechanistic insights into the plastic regulation of life history traits in workers of the honey bee (*Apis mellifera*), from early development and well into senescence. Worker life history progression is highly flexible and plastic, and is affected by social environment and genotype, and nutritional and sensory signaling cascades are known to mediate numerous behavioral, physiological, and metabolic changes in workers that translate into a largely plastic pattern of aging. The authors contrast these findings to life history control in the solitary insect *Drosophila* and outline how aging plasticity can be better understood through an experimental synthesis of two model systems: honey bees and fruit flies.

Many other examples of life history plasticity and its mechanistic basis are discussed throughout this book; various chapters explore the role of phenotypic plasticity in trade-offs and resource allocation, body size regulation, dauer formation, flowering time, and transitions between different reproductive modes. Together, these examples illustrate the importance of phenotypic plasticity for our understanding of life history evolution.

As many of these chapters make clear, plastic responses at the level of whole-organism life history traits require a complex sequence of events, from sensory detection of environmental signals or cues and neuronal processing down to physiological and metabolic integration and tissue-specific effects. As is the case for life history trade-offs (see Part 6), endocrine mechanisms seem to play a key role in mediating such plasticity: hormones transduce and integrate environmental signals via signaling cascades that affect gene transcription in target tissues and can thereby coordinate the expression of multiple phenotypes with effects on life history. Hormones thus provide a mechanistic link between the environment, genes, and whole-organism traits. Consequently, because of their major importance in the plastic regulation of suites of life history traits, it is likely that evolutionary changes in hormonal signaling play a key role in the evolution of life history plasticity (e.g., Finch and Rose 1995, Flatt *et al.* 2005, Heyland *et al.* 2005).

CHAPTER 17

Mechanisms underlying feeding-structure plasticity in echinoderm larvae

Benjamin G. Miner

17.1 Introduction

As research on the molecular mechanisms of life history evolution in marine organisms expands, we need to also expand the breadth of research. Researchers commonly study the transition between distinct life history stages (see also Chapter 3). These studies are necessary and important, but alone do not provide a complete understanding of how natural selection has molded the life histories of marine organisms. An understanding of underlying molecular mechanisms for traits that ensure life history stages survive and reproduce is also needed. In many cases, there is good evidence for which traits are critical for certain stages. These traits present an opportunity to understand which genes are expressed and how these genes are regulated in traits that influence Darwinian fitness and how they have likely influenced the evolution of particular life history stages.

For marine invertebrates with complex life histories, mothers either provision their offspring with more than enough energy and materials in the egg to metamorphose, or too little. In the latter case, larvae must acquire energy and materials from food that they capture. The evolution of complex life histories in many marine species is therefore influenced by structures that larvae use to capture food. In this chapter, I review research on or related to molecular mechanisms that underlie feeding-structure plasticity in marine invertebrate larvae. I restrict this review to echinoderms because there has been very little research on other phyla. My goal is to inspire more research on the molecular mechanisms of larval traits that strongly affect

fitness to broaden and improve our understanding of life history evolution in marine animals.

17.2 Plasticity of feeding structures

All five extant classes within the phylum Echinodermata have species with microscopic larvae—direct development in which there is no larval stage is rare (McEdward and Miner 2001). Larvae are bilaterally symmetrical, but metamorphose into pentamerously symmetrical juveniles that are similar in morphology to adults. In addition to the striking difference in morphology between individuals before and after metamorphosis, the ecology of larvae and post-larvae is also very different. Most species spawn gametes into the water, where fertilization occurs, and embryos develop while in the plankton. Individuals can spend days to weeks in the plankton before metamorphosing and beginning their benthic lives attached or crawling on the substratum.

There are two general forms of echinoderm larvae: feeding and non-feeding. Biologists have observed only non-feeding larvae in crinoids (McEdward and Miner 2001, Nakano *et al.* 2003). By contrast, both feeding and non-feeding larvae occur in the remaining four classes (McEdward and Miner 2001)—Asteroidea, Echinoidea, Ophiuroidea, and Holothuroidea—which make up the clade Eleutherozoa. Similarities among feeding larvae within the Eleutherozoa, and vestigial traits in some species with non-feeding larvae and phylogenetic reconstructions of ancestral states suggest that feeding larvae are homologous in the Eleutherozoa, and that non-feeding larvae have evolved multiple times (Strathmann 1978, Emlet 1995, Wray 1996).

Like many larvae of marine invertebrates, feeding larvae of the Eleutherozoa use a ciliary band to capture particles of food. However, unlike larvae of other phyla (with the exception of the tornaria larva of the hemichordates), feeding larvae of echinoderms capture particles in a parcel of water—not by maintaining direct contact with a particle (Strathmann 1971, 1975, 2007, Hart 1991). When a larva detects a particle near its ciliary band it reverses the effective stroke of cilia within a short region of the ciliary band near the particle. The parcel of water and particle are retained upstream, and then transported to the mouth along the ciliary band via ciliary reversals at adjacent regions of the ciliary band. Larvae select particles to ingest by capturing and retaining particles with their ciliary band and selectively ingesting captured particles at their mouth (Appelmans 1994).

Larvae of several species of echinoderms alter their feeding structures in response to different food concentrations (Table 17-1). When food is scarce, larvae produce longer ciliary bands and smaller stomachs relative to their size than when food is abundant. In all four classes with feeding larvae—Asteroidea, Echinoidea, Holothuroidea, and Ophiuroidea—biologists have observed feeding-structure plasticity (Table 17-1).

17.3   Evidence for adaptive plasticity

Two lines of evidence suggest that feeding-structure plasticity in echinoderm larvae is adaptive. From a functional point of view, a longer ciliary band can capture more food than a shorter band (Hart and Strathmann 1994). However, a longer ciliary band might come at a cost. Strathmann *et al.* (1992) suggested that larvae "pay" for a longer ciliary band by diverting energy from the developing juvenile (i.e., the rudiment). By doing this, development should be slowed and larvae will spend more time in the plankton. This additional time in the plankton will reduce the probability a larva survives because of high probability of mortality.

Although plausible, this hypothesis is difficult to test because long-armed larvae are fed less food than short-armed larvae, and food concentrations

Table 17-1 Studies demonstrating feeding-structure plasticity in echinoderms.

Species	Reference
Asteroidea	
Luidia foliolata	George 1994
Pisaster ochraceus	George 1999
Echinoidea	
Clypeaster subdepressus	Reitzel and Heyland 2007
Dendraster excentricus	Boidron-Metairon 1988, Hart and Strathmann 1994, Miner 2007
Evechinus chloroticus	Sewell *et al.* 2004
Heliocidaris tuberculata	Soars *et al.* 2009
Lytechinus variegatus	Boidron-Metairon 1988, McEdward and Herrera 1999, Miner and Vonesh 2004
Melitta tenuis	Reitzel and Heyland 2007
Paracentrotus lividus	Strathmann *et al.* 1992, Fenaux *et al.* 1994
Strongylocentrotus droechachiensis	Bertram and Strathmann 1998
Strongylocentrotus franciscanus	Miner 2005, McAlister 2007
Strongylocentrotus purpuratus	Miner 2005, McAlister 2007
Tripneustes gratilla	Byrne *et al.* 2008
Holothoroidea	
Australostichopus mollis	Morgan 2008
Ophiuroidea	
Macrophiothrix longipeda	Podolsky and McAlister 2005
Macrophiothrix koehleri	Podolsky and McAlister 2005

and morphology are confounded. Miner (2005) suggested that larvae "pay" for a longer ciliary band by diverting energy from the stomach, and suggested that the trade-off is between capturing food and processing it. When food is scarce, larvae might divert food to structures that capture food, but when food is abundant they invest in processing over capturing food. Consistent with this hypothesis, larvae morphologically respond to food concentrations before they can ingest particles, and produce a long ciliary band and small stomach when food is scarce and the opposite when food is abundant (Miner 2005). Because the response is pre-feeding, larvae are not altering their morphologies due to the direct effects of energy and material assimilated from exogenous particulate food—although any effects of dissolved organic matter remain. The trade-offs between food-capturing structures and the developing juvenile, or food-capturing structures and stomach size, are not mutually exclusive, and both are viable hypotheses for why larvae have evolved plasticity over constitutive defenses.

There is also comparative evidence that feeding-structure plasticity is adaptive. Larvae that develop from well-provisioned eggs are less dependent on exogenous food to complete larval development (Levitan 2000). This leads to a hypothesis that energy, or other vital components in an egg, and feeding-structure plasticity should be negatively correlated. There is evidence for this relationship in both echinoids and ophiuroids (Podolsky and McAlister 2005, Reitzel and Heyland 2007). However, there is also evidence that contradicts this hypothesis and supports an alternative, namely that larvae that develop from larger eggs have a greater capacity for feeding-structure plasticity (Bertram and Strathmann 1998, McAlister 2007, 2008, Bertram et al. 2009). Despite the apparent conflicting evidence, these hypotheses are not mutually exclusive and can be combined (Fig. 17-1). For example, the selective pressures for gaining exogenous food and constraints on the capacity for plasticity might both occur, but more strongly affect different egg sizes. If so, we might predict a humped relationship between feeding-structure plasticity and egg size. Larvae that develop from small sizes of egg are heavily dependent on exogenous food but do not have the capacity to produce large feeding structures when food is scarce. Larvae that develop from large eggs have the capacity for plasticity, but are not dependent on exogenous food. However, larvae that develop from intermediate-sized eggs have the capacity for the plasticity *and* a dependence on exogenous food. This hypothesis could be tested by combining meta-analyses and evolutionary contrasts to estimate the shape of the relationship between feeding-structure plasticity and egg size.

Larval feeding-structure plasticity is not unique to echinoids and occurs in the phylum Mollusca (Strathmann et al. 1993, Klinzing and Pechenik 2000). In two species of mollusc, larvae produce a larger ciliary band when food is scarce than when food is abundant. The distant phylogenetic relationship between molluscs and echinoderms and the different larval feeding mechanisms used by these two phyla together suggest that the similar response to food concentrations is convergent.

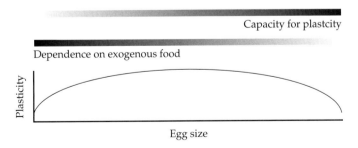

Figure 17-1 Hypothesized humped-shaped relationship between feeding-structure plasticity and egg size in marine invertebrate larvae. Opposing gradients for the capacity for plasticity and dependence of exogenous food are predicted to drive the shape of the relationship.

17.4 Developmental regulation

Echinoderm larvae provide a superb system for understanding the molecular mechanisms of feeding-structure plasticity. Sea urchin larvae are a model system for development, and the genome for a sea urchin (*Strongylocentrotus purpuratus*) was recently sequenced (Sodergren *et al.* 2006). In addition, larval feeding-structure plasticity is well studied in echinoderms, and there is a diversity of species with different degrees of larval feeding-structure plasticity (Table 17-1). Surprisingly, researchers have yet to directly study how this plasticity occurs at a sub-organismal level. Despite this, our knowledge of development in echinoderms can help us quickly develop and test hypotheses, and rapidly improve our understanding of the mechanisms by which larvae morphologically respond to algal concentrations.

There are two functional steps that must occur for larvae to adjust their feeding structures in response to food concentrations. Larvae must first perceive information about food concentrations, and then adjust the length of their ciliary band and stomach size. The mechanisms of these two functional steps are described below.

17.5 Mechanisms of perception

Gaining accurate information about future conditions in the surrounding environment is necessary to evolve plasticity (Harvell 1990). Unfortunately, there is little direct evidence to explain how larvae detect food concentrations. Here, based on past research, I develop hypotheses of how larvae perceive algal concentrations.

Larvae might detect algal concentrations at several locations on their bodies. Currently there is some evidence that larvae perceive algal concentrations with their epithelium. Sand dollar and sea urchin (echinoid) larvae adjust the size of their feeding structures before they can ingest algae, which suggests that larvae are detecting algal concentrations with their epithelium and not with cells in the mouth or digestive tract (Miner 2005). Additionally, larvae of echinoids and molluscs respond to algal concentration and not algal qual-

ity (Klinzing and Pechenik 2000, Miner 2007), suggesting that larvae are not using information from the amount of energy or material that is ingested or assimilated.

If larvae are detecting algal concentrations with their epithelium, then one hypothesis is that larvae use the same mechanism to detect algal cells for consumption and morphological plasticity. There is evidence that indicates that the mechanism that larvae use for perceiving cells near the ciliary band for feeding is different to the one used for morphological plasticity. Echinoid larvae capture unflavored plastic spheres (Appelmans 1994) but do not alter the size of their feeding structures in response to them (Miner 2007). However, larvae might use a similar mechanism to detect cells for ingestion and to control morphological plasticity. Echinoid larvae ingested algal-flavored plastic spheres at a greater rate than unflavored spheres, and the site of selection was near the mouth (Appelmans 1994). This suggests that receptors near the mouth inform larvae about which particles are edible, and could be used to assess algal concentration for morphological plasticity.

Ecological studies point to a general class of receptors that larvae might use to detect algal concentrations. Echinoid larvae do not adjust their morphology in response to different concentrations of plastic micro-spheres (Miner 2007). This suggests that larvae do not use the physical presence of algae alone to detect the concentration of algae. Larvae do appear to use chemicals emitted from algae to detect food concentrations, but different species use different types of information to detect algal concentration (Miner 2007). The larva of a species of sand dollar changes the size of its feeding structures in response to chemicals emitted from algae, whereas larvae of a species of sea urchin respond only to the presence of the intact algal cells. Furthermore, echinoid larvae consume more plastic micro-spheres if these are soaked in water with algae (Rassoulzadegan *et al.* 1984, Appelmans 1994). Apparently, some species use a chemical signal to assess algal concentrations, while others use a combination of physical and chemical cues.

Thyroid hormone, used by eukaryotes as a signaling molecule, is a candidate chemical that echinoderm larva might use to detect algal concentrations (Hart and Strathmann 1994). Thyroid hormone is used by eukaryotes as a signaling molecule (see also Chapter 7). Some species of unicellular algae produce thyroid hormones (Chino *et al.* 1994), although the role these hormones play in algae is unclear. Thyroid hormones also affect the development of larval and juvenile structures in echinoderms. In echinoderm larvae, thyroid hormones accelerate the development of juvenile structures (Chino *et al.* 1994, Hodin *et al.* 2001, Heyland and Hodin 2004, Heyland *et al.* 2004, 2006b), possibly at the expense of larval structures. Feeding larvae of echinoids treated with thyroid hormone can resemble larvae reared with abundant food (Heyland and Hodin 2004, Heyland *et al.* 2006b).

Once information is gained from epithelial receptors, larvae might use their nervous system to transmit information through their bodies. In all four classes of echinoderms with feeding larvae (i.e., the Eleutherozoa), the ciliary band is well innervated and the nervous system is well developed early in larval development before they begin to feed (Burke 1978, 1983a, Bisgrove and Burke 1986, 1987, Burke *et al.* 1986, Nakajima 1986, Moss *et al.* 1994, Chee and Byrne 1999a,b, Byrne *et al.* 2001, Cisternas *et al.* 2001, Cisternas and Byrne 2003, Nakajima *et al.* 2004, Nakano *et al.* 2006, Hirokawa *et al.* 2008, Dupont *et al.* 2009). Despite these numerous studies, the role played by the nervous system early in development is unknown.

The genes Orthodenticle (*Otx*) and Distal-less (*Dlx*) may play a role in specifying cells for the larval nervous system. *Otx* and *Dlx* encode for homeodomain transcription factors. *Otx* is expressed in the larval ciliary bands of holothuroids but not echinoids, whereas *Dlx* is expressed in the larval ciliary bands of both echinoids and holothuroids (Shoguchi *et al.* 2000, Lowe *et al.* 2002). However, expression patterns of *Otx* are not uniform along the band and the timing of expression, typically after the band has been specified, which suggests that it might be downstream in a regulatory pathway that specifies the larval nervous system. *Dlx* may play a role in specifying neurons, at least in echinoids (Lowe *et al.* 2002).

17.6 Mechanisms of morphological response

Researchers studying feeding-structure plasticity in echinoderm larvae have focused on the length of skeletal rods. This is because researchers typically investigate echinoid larvae that have skeletons (Table 17-1), and the lengths of skeletal rods correlate with the length of the ciliary band and are easy to measure (McEdward and Herrera 1999). However, mechanisms that regulate the length of the skeleton likely evolved secondarily when echinoids and ophiuroids (brittle stars) evolved larval skeletons—probably independently (Hendler 1978, Hotchkiss 1995). The initial appearance of larval feeding-structure plasticity was probably in an ancestor with a skeleton-less larva. Research should therefore focus on:

- the mechanisms that regulate the length of the ciliary band, to understand the evolution of feeding-structure plasticity in echinoderms
- the mechanisms that regulate the length of skeletal rods to understand subsequent evolutionary modification.

Currently there is much more data on how larvae produce their skeleton than their ciliary band.

In eleutherozoans, the ciliary band arises near the end of embryogenesis. During early embryogensis, cell interactions between the animal and vegetal regions of the embryo specify the location of the mouth (Hörstadius 1939, 1973). By the time the blastula is formed, four general embryological territories (endoderm, mesoderm, and oral and aboral ectoderm) are specified by maternal factors and cell–cell interactions (reviewed by Davidson *et al.* 1998). The ciliary band forms from oral ectodermal cells in a circum-oral ring (Czihak 1962). The initial ciliary band forms from ectodermal cells at the oral–aboral boundary, and interactions between oral and aboral ectoderm appear to determine the location of the ciliary band (Davidson *et al.* 1998, Su *et al.* 2009). The ciliary band then expands to take the characteristic location for a species.

Researchers have recently begun to focus on the molecular mechanisms that specify the cells of the oral and aboral ectoderm and the location of the ciliary band. Su *et al.* (2009) described the genetic regulatory network for the ectodermal cell types in a sea urchin. The products of three genes localize to the ciliary band: forkhead g (Foxg) (Tu et al. 2006), hepatocyte nuclear factor 6 (Hnf6) (Optim, et al. 2004), and Otxβ1/2. The localized expression of these genes along the ciliary band make these genes good candidates to study the control of ciliary band length. In addition, unlike Foxg, Hnf6 and Otxβ1/2 are not regulated by the common genes that establish the oral–aboral axis. Hnf6 and Otxβ1/2 are not regulated by Nodal, which regulates many of the genes involved in the oral–aboral axis (Duboc et al. 2004), whereas Foxg is up-regulated by Nodal and Lim1 and down-regulated by Bra and Gsc (Su et al. 2009). This could allow Hnf6 and Otxβ1/2 to regulate specific details of the ciliary band without disrupting the general morphological development of the larvae.

Although Foxg, Hnf6, and Otxβ1/2 may be involved in specifying cells of the ciliary band, the location of the band might be regulated by genes that specify oral and aboral ectoderm because the ciliary band forms at the boundary between these two cell types. The genetic regulatory network for the ectoderm (Su et al. 2009) provides many candidate genes for specifying the location of the ciliary band. In addition, there are a few genes missing from this network that are specific to the aboral ectoderm, which might be involved, for example cytoskeletal actin IIIa (CyIIIa) and calcium bind protein (Spec1 and Spec2a) (Davidson et al. 1998). There are also a few intercellular signaling molecules that might be important for specifying ectodermal cell types. The presence of platelet-derived growth factor and epidermal growth factor are necessary for development of aboral ectoderm (Govindarajan et al. 1995, Ramachandran et al. 1997). Although understanding early development of the ciliary band is a first step, it is unlikely to lead to a complete understanding of how larvae adjust the length of the ciliary band. We ultimately need to understand how larvae control the number of cells in the ciliary band later in development, when larvae are adjusting the length of their ciliary band.

In contrast to our understanding of development of the ciliary band, there is a large body of literature relevant to the formation of the skeleton in echinoids (but not ophiuroids). The skeleton is formed from cells that ingress into the blastocoel from the vegetal plate. These cells, called primary mesenchyme cells (PMCs), aggregate and form an extracellular matrix of proteins in which calcium carbonate is precipitated (Killian and Wilt 2008). The skeletal rods are lengthened by the PMCs into the characteristic shape of a species.

During skeletogenesis the PMCs interact with the ectoderm, and these interactions influence the size of the skeletal rods. Primary mesenchyme cells of different ages transplanted into blastulae produce similar skeletons that are appropriate for the stage of the blastula (Ettensohn and McClay 1986). Larvae that develop from quartered, halved, or doubled embryos produce proportionally sized skeletons (Hörstadius 1957, 1975, McEdward 1996). More focused experiments have shown that PMCs transplanted into a blastula treated with $NiCl_2$, which disrupts the specification of the ectoderm (Hardin et al. 1992), altered skeletogenesis, whereas $NiCl_2$-treated PMCs transplanted in an untreated blastula resulted in a normal skeleton (Armstrong et al. 1993). Furthermore, PMCs removed from a blastula produce larger skeletons than PMCs in vivo, and increasing the number of PMCs in a blastula did not change the size of the skeleton (Armstrong et al. 1993). Guss and Ettensohn (1997) provided further support for the hypothesis that signals from the ectoderm influence the PMCs and moderate the size of the skeleton. Recently, two signals from the epithelium that target PMCs have been reported. Misexpression of vascular endothelial growth factor (VEGF) and fibroblast growth factor (FGF) results in aberrant skeletons (Duloquin et al. 2007, Röttinger et al. 2008). Interestingly, the type of skeleton a larva produces is either specified very early in development or is genetically fixed. For example, Armstrong and McClay (1994) transplanted PMCs into blastulae of different species, and the type of skeleton was determined by the PMCs and not the blastula.

The genetic regulatory network of the primary mesenchyme cell of early development is well understood in sea urchins (see Bio Tapestry at www.biotapestry.org). There are at least 70 genes involved in the network (Ettensohn 2009), and our understanding of how the skeleton is formed is rapidly improving (Killian and Wilt 2008). Despite this complexity, there are a few genes that warrant particular consideration. Spicule matrix protein (SM50) and transmembrane protein (P16) are necessary for spicules to elongate, and P16 may play a role in PMCs receiving extracellular signals (Peled-Kramar *et al.* 2002, Cheers and Ettensohn 2005). SM50 is also expressed most strongly at the end of the skeletal rods, where plasticity in skeletal rods likely occurs, and is up-regulated by VEGF (Duloquin *et al.* 2007). Orthopedia (Otp) is expressed in a few cells of the oral ectoderm close to where the PMCs are beginning to produce the skeletal rods in the mid gastrula. Altering the expression of the Otp results in an abnormal skeleton (Di Bernardo *et al.* 1999, Cavalieri *et al.* 2003, 2007). This suggests that *Otp* is produced in the oral ectoderm and directs the PMCs as to where to produce the skeleton. Mesenchyme specific protein (*MSP* 130) is a cell-surface protein that is expressed in mesenchyme cells during skeletogenesis (Leaf *et al.* 1987), and appears to regulate calcium deposition (Carson *et al.* 1985). Love *et al.* (2007) demonstrated that three genes are expressed later in skeletogenesis, when the skeletal arm rods are elongating. Tetraspanin is expressed at the tip of the growing arms in the ectoderm. Advillin and carbonic anhydrase are expressed in the leading edge of the primary mesenchyme cells in the growing arms. The timing and expression pattern of these genes suggest they are involved in the arm-length plasticity in echinoids.

The mechanisms underlying the plasticity of the ciliary band likely differ in each of the four classes of the eleutherozoans. The holothuroids and asteroids have a skeleton-less feeding larva that is probably similar to the ancestral dipleurula larva of echinoderms. By contrast, the echinoids and ophiuroids have feeding larvae with skeletons, and both the length of the ciliary band and skeleton must change in concert. However, the skeleton of each class is likely convergent, which suggests that different mechanisms coordinate the ciliary band and skeletal response in these two classes, although a similar mechanism might still control the length of the ciliary band.

There is not much information about genes that regulate the size of the stomach. The genetic regulatory network for endoderm is well understood, but only for early development. As with the ciliary band and skeleton, information on later development is needed.

17.7 Integrative response

Larvae need to coordinate a response among different cell types. When food is scarce larvae increase the length of their ciliary band and decrease the size of the stomach. For asteroids, larvae must coordinate the response between the ectodermal cells of the ciliary band and the endodermal cells of the stomach. In echinoids and ophiuroids, larvae must additionally coordinate the mesodermal cells of the skeleton. Larvae almost certainly coordinate among cells in different structures with both electrical and chemical signals.

There is some evidence that suggests information is quickly passed among different structures in plutei. Starved larvae capture particles rapidly for several minutes and then capture particles less rapidly as the stomach fills (Strathmann 1971). The timescale suggests that the nervous system is involved in sending information between the stomach and ciliary band. Both structures are well innervated early in development. In addition, this behavior suggests that larvae can perceive information in the gut. The nervous system is probably partially involved in coordinating feeding-structure plasticity.

Hormones are also probably involved in coordinating a response to algal concentrations. *VEGF*, *FGF*, and *Otp* are involved in cell–cell communications between the epithelium and the PMCs (Cavalieri *et al.* 2007, Duloquin *et al.* 2007, Röttinger *et al.* 2008). With these molecules, larvae can pass information gathered at the epithelium to the PMCs. Another candidate for coordinating responses between stomach cells and PMCs is partitioning-defective (*Par*) protein, which is concentrated in both locations (Shiomi and Yamaguchi 2008). In contrast to the internal regulation of the above hormones, thyroid hormones from algae might allow larvae to coordinate a response. If thyroid hormone receptors are present on ciliary band cells, PMCs, and stomach cells, then exogenous thyroid hormones from algae could target all tissue types.

Lastly, energy limitations might coordinate structures that are negatively correlated. The ciliary band and stomach, and ciliary band and developing juvenile are the two candidates. If growth of these structures is regulated by the amount of energy available, then shunting energy to certain structures could regulate the coordinated responses. Larvae get energy to fuel development from lipid from their mothers. The pre-feeding plasticity in plutei suggests that regulating lipid metabolism could regulate a coordinated response. Genes that up- or down-regulate the metabolism of lipids are therefore good candidates to investigate.

17.8   Future directions

We are poised to understand the molecular mechanisms of feeding-structure plasticity in marine larvae, and echinoderms provide an excellent system to quickly advance our knowledge. There are a few obvious lines of research that should be fruitful. A first step is to use echinoid larvae and the plethora of molecular tools now available to gain a detailed understanding of how these larvae morphologically respond to food concentrations. Our current understanding of genetic regulatory networks and biomineralization will allow researchers to begin this step immediately. The genes and signaling molecules mentioned above are excellent candidates to begin with. A second step is to compare gene-expression patterns in echinoids and other echinoderm classes. These data will allow researchers to determine what aspects of feeding-structure plasticity are similar or different among different echinoderm classes. For example, is the regulation of larval skeletal rods convergent between echinoids and ophiuroids? A third step is to determine how larvae perceive algal concentrations. Identifying receptors would be especially helpful because researchers could then determine how and where information is transmitted throughout larvae. Furthermore, it might allow researchers to develop ways to experimentally "trick" larvae into producing different morphologies after larvae begin to feed, and remove the effects of energy and materials. This should allow researchers to identify downstream targets and the mechanisms of cell–cell communications more easily.

Non-developmental research is also needed. Researchers need to test more species for feeding-structure plasticity from different classes. In addition, larvae of hemichordates are thought to be homologous with larvae of echinoderms. Because they share a similar feeding mechanism and a common ancestor, it would be interesting if hemichordate larvae responded similarly to algal concentrations. The more species that are examined, the better our comparative tool kit becomes. These ecological studies can then guide developmental research, allowing researchers to determine what genes are modified and when plasticity evolved (e.g., Zhou *et al.* 2003).

17.9   Summary

1. Feeding larvae of echinoderms, as well as molluscs, adjust their feeding structures in response to different concentrations of algae.
2. There is functional and comparative evidence that this plasticity is adaptive.
3. Although researchers have yet to directly study the underlying mechanisms of feeding-structure plasticity in marine larvae, three fields of research are very relevant: development of the nervous system, specification of embryonic cell types, and biomineralization.
4. Given that feeding-structure plasticity probably evolved in an ancestor without a skeleton, researchers should investigate how the length and location of the ciliary band is modified by algal concentrations.
5. Future research is needed on the types of receptors that are located on the epithelia. Receptors for thyroid hormones are of particular interest.
6. Genes that specify the oral and aboral ectoderm (e.g., CyIIIa, Spec1, and Spec2a), and ciliary band (e.g., Foxg, Hnf6, and Otxβ1/2), provide excellent candidate genes to investigate.
7. Among the structures involved in feeding-structure plasticity in larvae, the larval skeleton in sea urchins is currently the best understood at the

molecular level. Studies that focus on the plasticity of the skeleton will likely advance the quickest.

8. Very little is known about the mechanisms that might be involved in adjusting stomach size. The well-understood genetic regulatory network of early development provides a good starting point to understand development of the stomach at later stages.

9. Comparative studies provide exceptional opportunities to understand the evolution of feeding-structure plasticity, especially across classes of echinoderms.

17.10 Acknowledgments

I would like to thank D. Leaf and K. Fielman for discussions, G. Wray and L. McEdward for motivating me to care about the mechanisms of life history evolution, and two reviewers for their hard work and thoughtful comments.

CHAPTER 18

Evolution and mechanisms of insect reproductive diapause: A plastic and pleiotropic life history syndrome

Paul S. Schmidt

18.1   Introduction

Life history traits are complex, polygenic, and prone to developmental and environmental lability. Age-specificity adds another layer to the genotype–phenotype map: although early-life and late-life characters can be correlated, any independence generates the potential for additional genetic and environmental contributing factors. Despite the apparent potential for widespread genetic and environmental variance, life history traits often exhibit relatively low estimated heritabilities. This is generally associated with the erosion of genetic variance due to strong directional selection (Roff 2002). There is an apparent paradox: on the one hand, life histories should exhibit reduced genetic variance as a function of intense selection; on the other, life history traits are often highly variable within and among natural populations. In many respects this is analogous to the classical discussion of genetic load and maintenance of heterozygosity (Crow 1992). Life history traits are polygenic and plastic to the extent that multiple multilocus genotypes may produce functionally equivalent phenotypes, and polymorphism is maintained despite directional selection. For example, heterozygosity is maintained across the genome despite intense artificial selection for hundreds of generations (Teotónio *et al.* 2009).

The concept of life history trade-offs (Roff 1992, Stearns 1992; also Part 6 of this volume) has been widely used to explain observed variance in natural populations, and posits that the relative fitness of life history phenotypes varies with environmental parameters that vary in space and/or time (Zera and Harshman 2001, Roff 2007b; also Chapter 24). Molecular polymorphism must form the basis for adaptive life history differentiation, but the identification of specific, functionally significant polymorphisms underlying adaptive life history evolution remains elusive (Flatt and Schmidt 2009). While genetically based life history trade-offs may be pervasive in nature, their molecular basis may be idiosyncratic to particular sets of populations and taxa. In addition to elucidating the causative nucleotide changes that underlie life history adaptation, a major challenge is to examine whether life history adaptation utilizes homologous pathways across species (Chapter 13).

Here, the phenotype of reproductive diapause in *Drosophila melanogaster* will be presented as a useful model for a molecular, mechanistic dissection of life history adaptation in natural populations. Diapause is in many respects the ultimate life history phenotype in insects, as it is a critical fitness component and has widespread effects on physiology, reproduction, and survivorship. While the ecology and natural history of *D. melanogaster* remain poorly understood, the genetic resources available are sufficient to identify molecular polymorphism of functional significance. The development of genetic and genomic resources will allow for novel investigations in ecological models; at present, however, evaluating the genetic models in their natural populations provides the best opportunity to elucidate the genetic architecture and molecular basis of life history variation (Mitchell-Olds and Schmitt 2006, Fitzpatrick *et al.* 2007, Tauber *et al.* 2007).

18.2   Advances and methods

The most important advance in the analysis of life history adaptation will be in the determination of mechanism. The study of life history evolution at the phenotypic level can proceed independently of the molecular underpinnings for the observed phenotypic variation and the pathways involved. However, this information is vital to a comprehensive understanding of:

- how molecular polymorphism underlies adaptive change in life histories
- what pathways are commonly associated with variation in suites of life history traits
- the mediation of variation to phenotype through physiology and networks
- how epistasis affects life history adaptation (Roff 2007b; also Chapter 2).

These central themes have been examined in a variety of ways. There is a history of using artificial selection to drive life histories in distinct directions (e.g., Zwaan *et al.* 1995, Zera and Harshman 2009). This has advantages, such as the ease of comparisons between selected and progenitor/control populations. Particularly with the use of whole-genome comparisons, this approach could identify those portions of the genome that have responded to a well-defined selection regime (Teotónio *et al.* 2009). One question, however, relates to spurious artifacts that may be driven by the source populations used. Artificially selected lines are also adapted to a specific (usually laboratory) environment; depending on context, this can be both an advantage and a disadvantage.

Similarly, inbred lines are routinely used in quantitative genetics. Investigations utilizing inbred lines have been essential to the development and testing of life history theory (Roff 2002). Adaptation to laboratory culture proceeds rapidly (Sgrò and Partridge 2000, Hoffmann *et al.* 2001, Linnen *et al.* 2001), and the inbreeding process itself will change life history profiles. Inbreeding depression results in intense selection for heterozygosity, and there exists a variety of mechanisms by which heterozygosity can be maintained despite inbreeding (e.g., a balanced lethal system). The somewhat random fixation of alleles at different loci can also result in

epistatic interactions and genetic correlations that may not be representative of dynamics in nature.

A classical yet complementary approach is the use of natural populations for the comprehensive dissection of the molecular basis and mechanisms of adaptive life history evolution. Working with natural populations has been likened to jumping into a moving car, but the relative advantage of entering a vehicle that is going nowhere may be dubious. Recent genomic comparisons among selected and control populations have shown that allelic diversity persists despite intense directional selection (Teotónio *et al.* 2009); this suggests that even laboratory-adapted, artificially selected populations are not at population genetic equilibrium. Experimental evolution in bacteria has also shown repeatedly that a variety of unpredictable changes take place in populations over various timescales (Blount *et al.* 2008), and that the only equilibrium may be lack of long-term stasis.

There are a variety of mechanisms by which variation for life history traits may be maintained within and/or among populations. When focusing on specific molecular polymorphisms, this effectively translates to some form of balancing selection. As such, population genetic models may be applicable, but the polygenic nature of life histories introduces a variety of complications. Epistasis, for example, may result in a particular nucleotide variant, contributing to phenotype in one genetic background (population) but not in another. Epistatic interactions may therefore result in the functional neutrality of polymorphic sites, which may be under selection in other circumstances. In such cases, the effectively neutral variant will proceed to fixation or loss dependent on its frequency in the population.

18.2.1   Clines

Of the various modes of selection, spatial variation in selection regimes may be of widespread importance in generating variable life history strategies. The spatial scale of analysis may be local, at mesoscales, or across regions. Latitudinal clines in life history phenotypes are widespread in a variety of organisms. Not only does the occurrence of a cline identify a phenotype of potential adaptive significance, it may also provide a point of departure for the identification of its molecular basis.

Latitudinal clines for life history traits are widespread in *Drosophila melanogaster*, which has been underutilized in ecological genetics despite a wealth of research on life history evolution. Clines have been observed for traits related to size (Gockel *et al.* 2001), central metabolism and metabolic pools (Sezgin *et al.* 2004), fecundity and longevity (Schmidt and Paaby 2008), and stress tolerance (Hoffmann *et al.* 2001), among many others. As *D. melanogaster* is presumed to have no significant genetic structure among populations, clines are often interpreted as the result of selection. This inference is strengthened when clines are replicated across independent samples (Paaby *et al.* 2010), but additional analyses are required to demonstrate selection directly on the phenotype of interest.

The true utility of clines is to identify the traits that vary across environments in a potentially adaptive way. In turn, such phenotypes can then be used as a system in which to identify the genes and pathways that form the molecular basis of adaptive life history variation in nature. These analyses can proceed in two general ways: quantitative trait locus-mapping to examine the association between genetic markers and phenotype, and candidate gene approaches. The quantitative trait locus (QTL) analyses of lifespan in *Drosophila* provide an excellent example of the former approach. Lifespan and rates of aging are highly variable among *Drosophila* strains, and the genetic architecture of longevity is quite complex. Mackay *et al.* (2006) examined the genetic architecture of longevity using laboratory parental strains and identified *catecholamines up* (*catsup*) as a potential gene affecting lifespan in *Drosophila*. Subsequently, Carbone *et al.* (2006) extended this work to a set of alleles derived from natural populations, and identified specific nucleotide polymorphisms within *catsup* that affected longevity and associated traits. Within the gene, various polymorphisms appear to be independent and contribute differentially to phenotype. Two quantitative trait nucleotides (QTNs) were identified for locomotor behavior and one for longevity; these are distinct polymorphisms located in different domains of the functional protein. This suggests that genes may affect more than one trait by means

of distinct segregating sites that are effectively independent.

The results of Carbone *et al.* (2006) illustrate that a comprehensive analysis can establish a causative link between nucleotide polymorphism and a trait of interest; a separate issue is whether or not QTNs can be identified in association studies using natural populations. Generally the power to detect genotype–phenotype associations is small, requiring large samples and extensive phenotype data. However, the use of natural populations and associative studies can be particularly effective if a candidate gene approach is also used. The insulin receptor (*InR*) in *D. melanogaster* provides an example. Many of the traits that exhibit latitudinal clines among populations in *D. melanogaster* are affected by various genes in the insulin/insulin-like signaling pathway (IIS): body size, lipid content, lifespan, reproduction, and stress tolerance. Mutations in both the insulin receptor (*InR*) and its substrate (*chico*) were shown to have robust and predictable effects on life history traits (Clancy *et al.* 2001, Tatar *et al.* 2001b). Thus the IIS pathway represents a logical candidate for the analysis of molecular variation that forms the basis for adaptive life history divergence. Paaby *et al.* (2010) screened both the *InR* and *chico* loci for patterns of geographic variation, nucleotide polymorphism, and divergence. Whereas *chico* evidenced patterns consistent with neutrality, *InR* did not. In particular, an amino acid repeat polymorphism showed an allele frequency cline in both North America and Australia. Subsequent experimental work further established the association between the polymorphism and divergent life history profiles that characterize temperate and neotropical populations in North America (Schmidt *et al.* 2005a, Schmidt and Paaby 2008). This suggests that the *InR* gene may contribute to adaptive life history differentiation across habitats, whereas molecular variation at *chico* may be of limited functional significance. It remains to be determined whether other components of the IIS pathway, such as the forkhead transcription factor *dFOXO*, also harbor variation that is associated with phenotypic variance for life history traits.

The analysis of spatial patterns of phenotypic variation, such as a latitudinal cline, may therefore represent a point of departure for comprehensive analysis of life history adaptation at the molecular level. The life history adaptation of primary interest to organismal biologists takes place in natural populations, not in a laboratory environment in which the actual selection regime may be ambiguous and/or unnatural. Natural populations describe how life history traits are evolving, not how they could potentially evolve following the pronounced adaptation to laboratory culture, which proceeds quickly (Hoffmann *et al.* 2001).

18.2.2   Temporal variation and seasonality

Similar to the mechanisms by which environmental variation over space may structure life history variation, selection regimes may also vary over distinct temporal scales. Seasonality is a pronounced part of the environment for many organisms. In many temperate insect taxa, individuals respond to a series of environmental cues and express forms of dormancy, in many respects similarly to vertebrate hibernation, dauer formation in nematodes, and vernalization in plants. Insect diapause is a genetically determined, neuroendocrine-mediated phenotype that is cued by token environmental signals such as photoperiod and/or temperature. Diapause is a very complex syndrome that is highly idiosyncratic among taxa; common themes include altered energy acquisition and storage, reproductive quiescence, elevated stress tolerance, and extended longevity (Tauber *et al.* 1986, Danks 1987). Thus, diapause can be viewed as a programmed and reversible switch between alternative life history profiles: the nondiapause portion of the life cycle is characterized by development, acquisition, and reproduction; the diapause portion by somatic persistence during periods of environmental challenge. In this sense diapause mediates plasticity in almost all aspects of life histories. The general dynamics of insect diapause have been extensively reviewed (e.g., Tauber *et al.* 1986, Denlinger 2002, Williams *et al.* 2009b). Although referred to as an "escape in time," recent data suggest that diapause may be a more actively regulated state (Denlinger 2002).

The timing of the switch between reproduction and dormancy is critical to fitness in seasonal environments: entering diapause too early results in a substantial cost in terms of reproduction, and entering diapause too late significantly elevates extrinsic risk of mortality. As might be predicted, the cues that elicit diapause vary among populations that occupy distinct environments. For example, latitudinal clines in the critical photoperiod required for diapause expression are commonly observed (e.g., Bradshaw 1976, Lumme and Lakovaara 1983, Kimura 1988).

18.3   Diapause as a model system for life history evolution

The phenotype of diapause, physiological and/or reproductive, integrates spatial and temporal variation in fundamental aspects of life history. Diapause is intimately associated with both reproductive- and survivorship-related fitness parameters, and aspects of its expression vary substantially within populations, among populations, and are divergent among closely related taxa. As such, the genetic analysis of diapause can provide an outstanding system for the molecular, mechanistic dissection of adaptive life history change in natural populations. Despite the apparent advantages associated with diapause expression, this appears to come at a significant cost. Natural populations, when brought into the laboratory environment, can lose responsiveness to environmental cues or the ability to express dormancy altogether (Oikarinen and Lumme 1979). Similarly, tropical species that evolve from temperate ancestors tend to lose the diapause response (e.g., Kimura 1988). Within populations, genotypes that have a high propensity to express reproductive diapause can be phenotypically distinct for a variety of fitness traits from genotypes with a low propensity to express diapause (e.g., Schmidt *et al.* 2005a,b). The genes that underlie the switch between direct development (nondiapause) and diapause are extensively pleiotropic, and must also be segregating for molecular variants that have differential effects on suites of correlated life history traits.

In the genetic analysis of diapause, three main goals can be proposed:

• the identification of genes that are responsible for induction of diapause under specific environmental conditions—what genes lie in between the sensing

of light and temperature and the resulting neuroendocrine response
• identification of those genes that are differentially regulated by the genes comprising the seasonal timer
• the elucidation of pathways and networks that mediate established physiological connections.

The genes that determine the seasonal response have proven difficult to identify, even in extremely well-characterized systems. One trait that has been the subject of QTL analysis is critical photoperiod. At a given temperature, lines derived from populations at different latitudes often enter diapause under different amounts of daylight (Danks 1987); critical photoperiod describes the light regime that prompts the switch to dormancy expression. Classical backcross analyses have been used to isolate factors associated with critical photoperiod to chromosomes (Lumme and Keranen 1978). Simple marker-based analyses have extended this localization to linkage groups within chromosomes (Lumme 1981), but further dissection may be precluded by the presence of paracentric inversions.

Recently, the first robust QTL map of animal photoperiodism was published in the pitcher plant mosquito *Wyeomyia smithii* (Mathias *et al.* 2007). This species undergoes a photoperiodic diapause in larval instars; both the critical photoperiod and the larval stage in which diapause is expressed vary predictably with latitude (Bradshaw 1976). Mathias *et al.* (2007) used an F2 population from lines recently derived from a northern and a southern population to construct an amplified fragment polymorphism (AFLP)-based map of critical photoperiod and stage of diapause. A number of QTL were observed for each trait, with one overlapping peak between the phenotypes. Interestingly, circadian genes did not appear to be directly involved, again demonstrating potential independence between genes involved in the regulation of circadian rhythms and those involved in the regulation of seasonal rhythms (Bradshaw *et al.* 2006, Emerson *et al.* 2009a). This landmark study provides a solid foundation for the identification and analysis of the genes that actually comprise the photoperiodic timer for seasonality. Unfortunately, the genome of *W. smithii* is not available and each QTL peak

harbors a nontrivial number of genes whose identities are unknown.

A complementary approach to identifying the genes involved in the regulation of diapause begins with candidates established by first principles. For example, ecdysteroids are intimately involved in the regulation of diapause, and one component of the heterodimeric ecdysone receptor changes in expression as a function of diapause (Rinehart *et al.* 2001). As diapause is associated with elevated tolerance to environmental stressors, genes involved in stress response are likely candidates for having central roles in the diapause program. A variety of genes encoding molecular chaperones change expression and/or protein abundance as a function of diapause (Kayukawa *et al.* 2005, Rinehart *et al.* 2007), as do genes involved in cryoprotection (Kostál *et al.* 2008). Similarly, insights into the genetic basis of diapause have been generated by screens for proteins and genes differentially expressed in diapausing versus nondiapausing individuals (Flannagan *et al.* 1998, Robich *et al.* 2007).

However, these direct comparative approaches cannot by themselves identify the actual causative nucleotide changes that underlie differences in critical photoperiod, thermoperiod, or between genotypes that express dormancy and those that do not. A given gene may be directly involved as a regulator, or it may lie downstream and be a component of a cascade. Showing that knockout of a gene changes phenotype merely demonstrates that it is in the pathway: a gene-to-phenotype connection does not necessarily translate to the identification of a gene that is segregating the molecular variation that underlies phenotypic variance. Similarly, associations between environmental cues and subsequent transcript or protein abundance demonstrate association, not causality. Use of natural or induced mutants that are nonphotoperiodic or nondiapausing is also useful but limited. Any complex phenotype with intersecting genetic networks that contribute to correlated phenotypes can be affected by a number of mutational events that can be direct or indirect. For example, a mutation that produces an achlorophyllous plant need not lie in a gene that underlies variance in chlorophyll density in leaves. The relevant variation must be mapped and subsequently analysed with respect to function.

18.4 Identifying genes for seasonality

The unambiguous identification of genes that regulate dormancy is a challenging task. Simple manipulations, such as an RNAi-mediated knockout, can demonstrate that a gene is involved in a pathway that underlies a phenotype, but do not necessarily pinpoint the gene that drives observed variance in a trait such as critical photoperiod. Similarly, QTL studies can be limited by the idiosyncrasies of the lines used to map, the recombinational landscape, genomic platform or lack thereof, and marker density. A given QTL peak may contain hundreds of genes (Mackay *et al.* 2006). Perhaps the most effective approach would be:

- phenotypic and quantitative genetic characterizations of natural populations, with the identification of what traits vary adaptively
- mapping of those traits by a combination of approaches, depending on the strengths and limitations of a particular organismal system
- broad scale sequencing, assembly, and annotation to ultimately provide an indication of what genes and open reading frames (ORFs) lie under QTL peaks
- targeted analysis of likely candidates by complementation, RNAi, deletions, or similar approaches
- identification of the QTNs that explain the observed variance
- functional and ecological analysis of the candidate polymorphisms in laboratory and field-based experiments.

At present, there are few systems in which this is feasible, but this will likely change in the near future. The continued development and decreasing cost of DNA sequencing is allowing genome-scale analyses in ecologically interesting yet less genetically tractable species. The assembly and annotation of such information is not a trivial enterprise, however, and most systems will never provide the methodologies and resolution of the genetic models. There are concrete reasons, aside from historical inertia, why *D. melanogaster* was incorporated as, and continues to be, a genetic model.

In the comprehensive study of the genetics of diapause, there are several keystones that have been neither shaped nor put in place. First is the identification of those genes that are causally responsible for the variation in timing, onset, and termination of dormancy that is routinely observed in nature. Given that the genetic basis of diapause expression has been well established, the variable diapause response of individual genotypes, populations at different latitudes/altitudes, and of closely related species demonstrates that diapause genes are segregating for nucleotide polymorphisms that are of immense functional significance.

The second keystone is the elucidation of the pathways or networks that regulate seasonal phenotypes. The hypothesis that seasonal rhythms are measured by the same mechanisms as circadian rhythms (Bünning's hypothesis) is unsupported in some systems (Bradshaw *et al.* 2006) but remains viable in others (e.g., Danks 1991, Stoleru *et al.* 2007; Stéhlík *et al.* 2008). Again, it is not possible to establish a direct causal connection between a candidate gene and variance in diapause phenotype by associations and classical mutant analyses alone. In the absence of sophisticated genetic manipulations, phenotypic variance must be mapped in addition to gene-by-gene and associative analyses.

The third keystone is to generate genome-wide expression patterns for appropriate stages of the life cycle under dormancy restrictive and permissive environmental conditions; this should also be extended to genotypes that vary in dormancy phenotype (e.g., diapause versus nondiapause, northern versus southern strains with different critical photoperiods), across age classes, and throughout the circadian cycle. This will allow for a comprehensive determination of what genes are differentially regulated specifically as a function of dormancy. It is worth emphasizing the critical distinction between the genes that explain the variance in dormancy expression phenotype and those involved in its onset, progression, and termination.

The fourth keystone is the comparative work and establishment of context. Dormancy syndromes are notoriously divergent among taxa and almost certainly have evolved multiple times independently (Danks 1994, 2002). The question is whether particular genes and pathways have been utilized multiple times independently, that is, the evolution of dormancy is more likely to involve particular mechanisms and proceed along certain avenues with probabilistic determination. A general prediction is that the genes that explain variance in onset of diapause may be more idiosyncratic, but that the pathways regulating dormancy may be more highly conserved. This stems from the observation that dormancy is regulated in fairly conserved fashion by components of the endocrine system, and particular signals have generally predictable effects on dormancy response (Tauber *et al.* 1986). Thus anything causing a differential input into these networks may be sufficient to create variation in dormancy phenotype given a set of cues. In contrast, association studies have shown widespread variation in which genes are differentially regulated during diapause (Williams *et al.* 2009b). As with any complex trait, dormancy is likely to contain elements that are homologous across taxa and elements that are unique to a particular group. The establishment of the genetic basis of dormancy in at least one organism would provide a useful construct for testing homology versus analogy in other systems with different relative strengths.

It has been hypothesized that dormancy in insects (diapause) and *C. elegans* (dauer) may be homologous phenotypes regulated by common pathways (Tatar and Yin 2001). The genes of the dauer formation pathway (*daf*) have widespread and predictable effects on life histories (see Chapter 22). The insect homologs, at least in *Drosophila*, have also been shown to determine age-specific patterns of reproduction and survivorship, as well as correlated characters (Clancy *et al.* 2001, Hwangbo *et al.* 2004, Flatt *et al.* 2005). This provides a concrete hypothesis that insect diapause is mediated, at least in part, by insulin signaling. This hypothesis is supported by the observations that genes in the IIS pathway control fundamental aspects of female reproductive development (LaFever and Drummond-Barbosa 2005, Hsu *et al.* 2008), as well as the identified role of IIS genes in dormancy in dipterans (Williams *et al.* 2006a, Sim and Denlinger 2008). Nevertheless, there has been no systematic investigation of the role of the IIS pathway in insect diapause.

18.4.1 Dormancy in *D. melanogaster*

An adult reproductive diapause was first described in *D. melanogaster* 20 years ago, and demonstrated many aspects of classical insect diapause (Saunders *et al.* 1989). The photoperiodic response was, however, limited to a narrow range of temperatures, suggesting a facultative diapause perhaps better suited to its ancestral habitat and natural history. Additionally, manipulations of the candidate circadian gene *period* did not alter the seasonal phenotype, measured as extent of reproductive quiescence. In hindsight, given recent discussion of the independence of circadian genes and seasonal phenotypes, this may not be particularly surprising (Bradshaw *et al.* 2006, Emerson *et al.* 2009a).

As with other *Drosophila* species (Lumme and Lakovaara 1983), *D. melanogaster* overwinters as an adult, resulting in temporal population continuity. Reproductive diapause is regulated by the neuroendocrine system (Saunders *et al.* 1990, Richard *et al.* 2001) and appears to involve time measurement (Saunders *et al.* 1990). Expression of diapause results in reproductive quiescence, increased longevity, elevated stress tolerance, and reduced rates of age-specific mortality (Tatar *et al.* 2001a). Furthermore, the propensity to express diapause is highly variable within and among populations. Incidence of diapause varies predictably with latitude (Schmidt *et al.* 2005a) and season (Schmidt and Conde 2006); these dynamics reflect the differential fitness of high- versus low-diapause propensity genotypes in distinct environments. The genetic variance for diapause has pleiotropic effects on a variety of other traits, including lifespan, age-specific mortality rates, fecundity profiles, multiple forms of stress tolerance, lipid content, and development time (Schmidt *et al.* 2005b). In natural populations of *D. melanogaster* in eastern North America, diapause genotype accounts for a significant amount of the observed variance for life histories that exhibit latitudinal clines (Schmidt and Paaby 2008).

One potential stumbling block in the utility of *D. melanogaster* as a system for a genetic dissection of diapause is the conflicting evidence regarding photoperiodism. Adult reproductive diapause can be photoperiodic in temperate endemic *Drosophila* (e.g., Lumme and Lakovaara 1983, Kimura 1988), and the strain Canton-S did appear to be photoperiodic at some temperatures (Saunders *et al.* 1989). Furthermore, the mechanism for translation between circadian and seasonal timing appears to be present in *D. melanogaster* (Stoleru *et al.* 2007).

Similar characterizations of European populations also indicate photoperiodism (Tauber *et al.* 2007), but North American populations appear to express reproductive quiescence based solely on temperature cues (Tatar *et al.* 2001a, Emerson *et al.* 2009b).

This observation raises the intriguing possibility that the use of photoperiodic versus thermal cues in *D. melanogaster* populations may depend on the colonization history and length of exposure to temperate environments. Diapause appears absent or at low frequency in ancestral populations from east and west Africa; in the standard assay described by Saunders *et al.* (1989), no isofemale lines exhibited a reproductive quiescence (Zimbabwe, $n = 19$; Kenya, $n = 15$; Gabon, $n = 14$; Malawi, $n = 8$). Similarly, a screen of isofemale lines of *D. simulans* from Arizona ($n = 96$) and Pennsylvania ($n > 200$) has not yielded any line that exhibits reproductive diapause at 11°C and 10L : 14D (P. Schmidt, unpublished data). This also corresponds to differences between *D. melanogaster* and *D. simulans* in their distributional and abundance patterns in northern temperate habitats. Collections made in the last week of September 2009 yielded the following, given as ratio of *D. simulans: D. melanogaster*:

- Shoreham, VT: 0 : 163
- Bowdoin, ME: 0 : 113
- Harvard, MA: 35 : 116
- Middlefield, CT: 7 : 77
- Churchville, MD: 93 : 4

Pennsylvania collections (Media, PA) are initially heavily skewed towards *D. melanogaster* (9 July 2009: 1 : 106) but increase the proportion of *D. simulans* over the reproductive season (11 August 2009, 46 : 39; 14 September 2009, 39 : 21; 15 October 2009, 159 : 57; 10 November 2009, 257 : 39). These patterns are consistent with the hypothesis that *D. melanogaster* is more persistent than *D. simulans* in temperate habitats due to expression of reproductive diapause.

The low incidence or absence of diapause in ancestral African populations and in *D. simulans* provides support for the hypothesis that dormancy in *D. melanogaster* is of recent origin and in its early stages of evolution (Saunders *et al.* 1989). The dormancy phenotype was originally described as oligopause, a facultative dormancy that used both photoperiodic and thermal cues. In the evolution of dormancy, it might be expected that the use of temperature cues would precede the ability to use changes in day length as a predictor of environment. It seems more straightforward, in a mechanistic sense, to arrest function by temperature thresholds than to switch between life history modes by means of light-mediated, anticipatory cues.

Tropical insects also offer a window into the mechanisms and evolution of dormancy, as seasonal phenotypes are widespread, yet there may be insufficient variation in light-based cues to measure seasonal time (Denlinger 1986). In this sense *D. melanogaster* represents a natural experiment, being a tropical endemic that has subsequently expanded its distribution to the temperate zones on multiple continents (David and Capy 1988). Evidence suggests widespread adaptation to novel temperate environments (Sezgin *et al.* 2004). Given the pleiotropic nature of reproductive diapause and its effect on fitness, it is likely that selection on reproductive diapause and correlated phenotypes drives some portion of the observed adaptive response to climatic heterogeneity. At present there is no consensus on what type of dormancy is actually present in *D. melanogaster* (Emerson *et al.* 2009b); the answer to this question may determine what types of genes are candidates for regulating the seasonal response.

Kostál (2006) articulated the distinctions previously made between two distinct types of invertebrate dormancy (e.g., Tauber *et al.* 1986, Danks 1987). Dormancy is simply defined as a generic phenotype in which there is a developmental arrest that is associated with a subsequent increase in organismal fitness; while such a trait may be of utility, it is not necessarily adaptive in the strict sense. Quiescence is the form of dormancy in which the response to the environment is immediate and facultative; again, this need not be adaptive; it may simply be physiologically inevitable. For example, if temperature drops below a particular threshold, reproductive development may be arrested for reasons that have nothing to do with overwintering. The cessation of reproduction may cause increases in longevity and stress tolerance due to underlying genetic and phenotypic correlations. In contrast, diapause is a specific type of dormancy that by definition is an adaptation to seasonality. In diapause there is profound physiological change, central regulation, and programmed times of initiation and termination. Thus, the diapause program can be divided into distinct phases: prediapause, involving preparation and accumulation of metabolic reserves; diapause, in which endogenous neuroendocrine regulation arrests development; postdiapause, in which exogenous factors such as temperature maintain the organism in a state of quiescence; and finally, the end of the entire diapause program and the return to reproduction and development (Kostál 2006).

The functional relationship between quiescence and diapause is unresolved. It is tempting to arrange them in series, with dormancy initiating a direct facultative response to exogenous factors (e.g., temperature, moisture), which subsequently becomes an anticipatory response, utilizing additional cues (e.g., photoperiod). This could be tested by mapping dormancy phenotype onto multiple independent, well-resolved phylogenies. Although quiescence and diapause may be considered quite distinct from an insect physiological perspective, both may share underlying neuroendocrine regulation. The major distinction may be one of degree and the anticipatory versus facultative nature of the life history plasticity. Regardless of whether dormancy in *D. melanogaster* may be quiescence or diapause, this system offers an unparalleled opportunity to dissect the genetic basis of dormancy and dormancy-mediated life history evolution.

18.4.2   Genes for diapause in *Drosophila*

Through both mapping and candidate gene approaches, three diapause genes have been identified in *D. melanogaster*:

- the *phosphoinositide 3-kinase (PI3K, aka Dp110)*, which is involved in insulin signaling
- *timeless (tim)*, a light responsive component of the circadian clock
- *couch potato (cpo)* an ecdysteroid-responsive RNA binding protein that is expressed in the peripheral nervous system and endocrine structures in both larvae and adults.

Williams *et al.* (2006a) demonstrated a robust association between natural variation at *Dp110* and diapause; transgenics and complementation analyses confirmed that a reduction in *Dp110* function and insulin signaling results in elevated diapause expression. While the causative nucleotide polymorphism(s) at *Dp110* that underlie diapause variation have not been elucidated, it is likely to reside in cis-regulatory regions (Williams *et al.* 2009b). It will be of great interest to determine whether the effects of *Dp110* are directly due to its function in IIS, as would be expected based on the potential homology between insect dormancy and nematode dauer (Tatar and Yin 2001).

As diapause is dependent on light and day length, circadian genes have long been proposed as candidates for diapause and photoperiodism (Danks 2005). The circadian clock is highly conserved in animals and regulates a variety of daily rhythms in many taxa, including insects (Saunders 2002). While the connection between the circadian clock and seasonal photoperiodism is controversial (Danks 2005, Emerson *et al.* 2009a), the clock gene *tim* was shown to have a pronounced effect on diapause expression (Tauber *et al.* 2007). A newly derived allele, *ls-tim*, produces both a long (L-TIM) and short (S-TIM) protein, whereas the ancestral *s-tim* allele produces only the short S-TIM protein. The S-TIM and L-TIM proteins are functionally distinct, differing in effective binding to cryptochrome and ultimately resulting in variable clock sensitivity to light (Sandrelli *et al.* 2007). The *ls-tim* allele confers increased diapause expression and is selectively favored in European populations (Tauber *et al.* 2007).

Schmidt *et al.* (2008) used a combination of direct mapping approaches and complementation analyses to identify the RNA binding protein encoding gene *couch potato (cpo)* as a major factor determining diapause in North American populations. Linkage association and complementation studies subsequently demonstrated that the variance for diapause phenotype is determined by a single nucleotide substitution that replaces isoleucine with lysine at amino acid 462 (Schmidt *et al.* 2008). This polymorphism is in a portion of a *cpo* exo (where diapause maps) that is only present in one of the *cpo* transcripts, suggesting that the effects of *cpo* on diapause may be transcript specific. Manipulations with mutant allelic series, P-element excisions that restore wild-type, *cpo* copy number, and quantitative PCR all suggest that *cpo* dosage determines diapause phenotype: increasing expression results in a low diapause phenotype whereas reducing expression causes the high diapause phenotype. The 462 polymorphism is present in only the *cpo*-RH transcript; for other transcripts and polypeptides, this region is spliced out. Interestingly, this region of the *cpo* locus is highly variable among the 12 *Drosophila* genomes; *D. simulans*, which does not exhibit a high incidence of diapause, appears to lack the *cpo*-RH transcript entirely.

18.5   Pathway and genomic analyses

Considering that all three *D. melanogaster* diapause genes (*Dp110, tim, cpo*) have been shown to directly affect diapause, these genes may functionally interact in a defined network that regulates diapause.

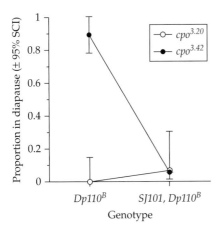

Figure 18-1   Diapause incidence as a function of multilocus genotype at *cpo* and *Dp110*. The *Dp110*^B allele is a deletion of the functional *Dp110* locus as well as neighboring genes; the *SJ101* allele is the transgenic rescue in the *Dp110B* background (Weinkove *et al.* 1999). The *cpo*^{3.42} genotype is a deletion of a small genomic region containing the *cpo* gene, and *cpo*^{3.20} is a duplication that effects rescue (Schmidt *et al.* 2008). The *cpo*^{3.20} allele is homozygous viable and maintained as such, whereas the other alleles were maintained over TM balancers. Crosses were made between stocks to create the four genotypes depicted. All crosses were replicated 10 times in both directions and resulting progeny were collected within 2 h of eclosion and placed at 11°C, 10L : 14D for a period of four weeks. The results demonstrate complementation of *cpo* and *Dp110* with respect to diapause. Diapause is high when the *cpo* and *Dp110* deletions are combined. Both a single copy of the *cpo* duplication and the *Dp110* transgene rescue the low diapause phenotype.

Alternatively, multiple pathways may be involved and these genes may exhibit failure to complement in standard analyses. For both *tim* and *cpo*, the QTNs for diapause have been identified and are point mutations resulting in change at the level of the protein; they may also affect expression and/or stability. Similarly, for both *cpo* and *Dp110* a decrease in expression level is associated with an increase in diapause incidence. Whether or not these genes interact is easily addressed by standard complementation analyses using mutant alleles and transgenic rescue strains. Data indicate that *cpo* and *tim* complement with respect to stress-related phenotypes but not for diapause (unpublished manuscript); *cpo* and *Dp110* complement for diapause and a series of correlated traits (see Fig. 18-1). This provides additional support for the hypothesis that diapause may be ultimately regulated by insulin signaling.

18.5.1   Diapause and insulin signaling

The general roles of juvenile hormone (JH) and ecdysteroids on patterns of insect reproduction, survivorship, and dormancy have been well described (Nijhout 1994, Tatar and Yin 2001, Flatt *et al.* 2005; see Chapter 13). As in other systems, *Drosophila* diapause is governed by JH and ecdysteroids (Richard *et al.* 2001) and results from a block to JH-release by the corpus allatum (Saunders *et al.* 1990). Whole-body ecdysteroid (20-hydroxyecdysone, 20E) titers are also distinct between *D. melanogaster* lines that differ in diapause incidence, with low diapause lines characterized by an increased level of 20E during diapause-inducing conditions (K.J. Min, T. Flatt and P. Schmidt, unpublished data). Manipulations using topical application of JH analogs and ecdysteroid injections change diapause and patterns of vitellogenesis in the developing ovaries (Saunders *et al.* 1990, Soller *et al.* 1999, Tatar *et al.* 2001a). The secretion of JH by cells in the corpus allatum is induced by the release of *Drosophila* insulin-like peptides (DILPs) from two clusters of neurosecretory cells (NSCs) in the midbrain (e.g., Broughton *et al.* 2005). Insulin signaling also regulates JH synthesis (Tu *et al.* 2005), and DILPs secreted from the pars intercerebralis can stimulate the production of ecdysteroids and juvenile hormone, which affects development of the ovary.

D. melanogaster females that express diapause arrest reproductive development at stage 7 (pre-vitellogenic), whereas flies that do not express diapause undergo vitellogenesis (stages 8–14 of oogenesis) and normal reproductive development. Vitellogenesis and this ovarian checkpoint at stages 7–8 are also regulated by the DILPs produced in the NSCs of the adult brain, whereas follicle cell development is regulated by the germ line (LaFever and Drummond-Barbosa 2005). Subsequently it was shown that DILPs control the G2 phase of the germ-line stem cell cycle through *PI3K* (*Dp*110) and dFOXO, the IIS downstream transcription factor (Hsu *et al.* 2008). The identified associations between the endocrine control of diapause, patterns of ovarian development, and the DILPs suggest that diapause is regulated by insulin signaling. Furthermore, the connections between diapause and IIS are not restricted to *Drosophila*. In the mosquito *Culex pipiens,* dsRNAi manipulations of the insulin receptor (*InR*) or the downstream forkhead transcription factor *foxo* disrupted the normal photoperiodic physiological responses of the ovaries (Sim and Denlinger 2008).

18.5.2 Expression analyses

In addition to analysing candidate genes and pathways, whole-genome approaches are readily performed in *D. melanogaster*. It would be straightforward to assess changes in the metabolome as a function of either diapause genotype or diapause expression, as has been done with classical mutants (Kamleh *et al.* 2007) and cold-related phenotypes (Overgaard *et al.* 2007). Ideally, genome-wide analysis would also include transcriptomics, proteomics, and arrays to examine genome-wide differentiation among populations (Turner *et al.* 2008). Whole-genome allele-specific expression analyses could also be used to test whether differentiation between populations for diapause phenotype is driven primarily by altered expression (Landry *et al.* 2005); such assays could be expanded across the insect life cycle as well as between restrictive and permissive environmental conditions for diapause expression.

While direct RNA sequencing generates more comprehensive data (Ozsolak *et al.* 2009), cDNA-based transcriptional profiling may be the simplest and most widely available method by which to generate preliminary data on the impact of diapause expression at a genome level. This could be performed in an untargeted way, in which global expression profiles are generated:

- for the same genotypes before, during, and after diapause
- over various temporal scales, such as daily rhythms and over the life cycle
- between genotypes under control and diapause conditions.

Alternatively, such profiling can be targeted with respect to a candidate gene or pathway, that is, a gene is manipulated in a variety of ways and the effects on global expression profiles are examined under a variety of conditions.

We performed such a targeted expression-profile analysis based on manipulations of *cpo*. In a common genetic background, two genotypes were constructed: *cpo* was manipulated to produce both a high-diapause genotype and a low-diapause genotype. These genotypes were exposed to two treatments and then analysed on the standard *Drosophila* Affymetrix v.2 microarray platform. The genotypes and treatments were designed in a simple 2 × 2 format as an initial pass to identify genes that are differentially expressed as a function of diapause expression, not temperature or genetic background. The results of this simple experiment were quite intriguing: of the transcripts identified, the majority (27 out of 42) were uncharacterized accessions (Table 18-1). Of the remainder, most of the ascribed functions are based on sequence similarity and prediction; very few of the genes given in Table 18-1 have been well studied. One interpretation of these data is that *cpo* regulates or participates in a pathway that has not been characterized. Clearly, there is a pronounced response when diapause is expressed. For example, gene *CG14456* is more than 90-fold up-regulated in the high-diapause genotype, compared to the low-diapause genotype, during diapause.

Although these genes may be of unknown function, there exist many methods by which to exam-

Table 18-1 Transcriptional profiling for cpo genotypes under control (25°C, 12L : 12D) and diapause-inducing (11°C, 10L : 14D) conditions. The cpo genotypes were w; 6326;6326/cpo3.20 (low diapause) and w; 6326;6326/cpo3.42 (high diapause). The cpo3.42 is the deletion, cpo3.20 is the duplication. All genotypes were constructed in both directions in 25 replicates. Progeny were cultured at low density and collected within 2 h of eclosion. Flies were then exposed to the appropriate treatment for a period of 7 days, collected from 12:00–1:00 pm, then frozen in liquid nitrogen and stored at −80°C until analysis. mRNA was hybridized to Drosophila Affymetrix v.2 chips with four replicates per treatment combination (two genotypes × two environmental conditions). A sequential filtering was used to identify transcripts that exhibited a significant genotype × environment interaction based on stepwise p values; and the interaction was due to differences between genotypes under diapause, not control, conditions; and had a G-fold change greater than 3.0 when the high diapause genotype was compared to the low diapause genotype under the diapause-inducing conditions. Thus, the filtering was designed as an initial screen to identify those genes that are differentially regulated as a function of diapause expression, not simply low temperature. The comparison between the high and low diapause genotypes served as a biological control in this regard.

Gene	Gene title	Gene ontology biological process	GFoldChange
CG14456	CG14456	—	90.285
CG9897	CG9897	Proteolysis	72.705
CG7194	CG7194	Gonad development	54.956
CG12038	CG12038	—	52.165
CG11619	CG11619	Glycerol metabolic process, signal transduction	37.836
CG8942	CG8942	Wnt receptor signaling pathway	26.017
CG40485	CG40485	Metabolic process	20.492
Cyp313b1	Cyp313b1	Electron transport, steroid metabolic process	15.234
CG17127	CG17127	—	11.913
CG3165	CG3165	—	10.369
Cyp49a1	Cyp49a1	Electron transport, steroid metabolic process	8.911
tko	mt ribosomal protein S12	Sensory perception, courtship behavior, mechanosensory behavior	8.575
Lcp65Ag3	Larval cuticle protein	—	5.923
CG5381	CG5381	—	5.092
Nmdar1	NMDA receptor	Potassium ion transport, nerve–nerve synaptic transmission	5.049
CG13192	CG13192	—	4.879
CG12486	CG12486	—	3.443
CG31826	CG31826	Transport	3.375
NaPi-T	Na[+]-dependant inorganic phosphate cotransporter	Phosphate metabolic process//cation transport	3.301
CG13826	CG13826	—	−3.002
fj	Four-jointed	Notch signaling pathway, cell–cell signaling	−3.058
CG8083	CG8083	Cation transport	−3.075
CG9642	CG9642	—	−3.185
CG8563	CG8563	Proteolysis	−3.332
Asph	Aspartyl β-hydroxylase	Proteolysis, transmembrane receptor protein tyrosine kinase signaling pathway	−3.416
α-Est2	Fragment B	—	−3.424
CG31772	CG31772	—	−3.750
skd	Blind spot	Embryonic development//cell fate commitment//sex comb development	−3.914
CG7912	CG7912	Sulfur metabolic process//anion transport//extracellular transport	−4.341
CG9505	CG9505	Proteolysis	−4.421
CG15556	CG15556	G-protein coupled receptor protein signaling pathway	−5.075
Osi15	Osiris	—	−5.299
CG13144	CG13144	—	−5.429
bnl	Fibroblast growth factor	Chemotaxis//open tracheal system development	−6.435
CG30488	CG30488	Defense response	−6.551
CG4835	CG4835	Chitin metabolic process	−7.235
spn-E	Spindle-E (homeless)	Oocyte maturation//germarium-derived oocyte fate determination	−8.282
CG14460	CG14460	—	−8.823
Vanin-like	Vanin-like	Coenzyme metabolic process//signal transduction	−9.348
CG30272	CG30272	Carbohydrate metabolic process//cation transport	−14.044
MtnB	Metallothionein B	Cellular metal ion homeostasis	−20.523
CG32564	CG32564	—	−38.638

ine how they affect diapause and life histories. A powerful method for dissecting the pathways and systems of interacting genes is to screen for strong epistatic interactions in partial and full knockouts. In *D. melanogaster* many deletion sets are publicly available and specific regional deletions can also be created by FRT-FLP facilitated recombination as was previously done with *cpo* (Schmidt *et al.* 2008).

18.6 Summary

1. Insect diapause offers an outstanding system for the comprehensive analysis of life history adaptation; it is a critical trait for organismal fitness that has widespread effects on survivorship, reproduction, stress tolerance, and general physiology.

2. Aspects of diapause are polymorphic within and divergent among taxa; diapause incidence and thresholds for diapause induction vary predictably with geography and season. This phenotypic variance is, in part, genetically determined and can be successfully mapped in both model and non-model systems.

3. *Drosophila melanogaster* expresses a reproductive diapause when cultured under short days and low temperature. This allows for a detailed genetic and mechanistic dissection of a phenotype that has been the subject of intensive physiological research for the past 60 years. The genetic and genomic resources in *Drosophila* can be directed at outlining those genes and pathways that cause the switch between direct development and dormancy and those that are differentially regulated to cause the observed phenotypic cascade of increased longevity, reduced sensescence, elevated stress tolerance, altered metabolism, and reproductive quiescence.

4. The genetic networks and genomic architecture elucidated in *Drosophila* would be directly applicable to other, better understood ecological systems and could generate invaluable insights into the molecular basis of adaptive life histories in natural populations.

18.7 Acknowledgments

This work was supported by grants DEB-0542859 and DEB-0921307 from the US National Science Foundation. The author wishes to acknowledge the generosity of M. Sokolowski in sharing strains, C. Kyriacou and K. Williams for discussion, and D. Denlinger and an anonymous reviewer for comments on the manuscript.

CHAPTER 19

Seasonal polyphenisms and environmentally induced plasticity in the Lepidoptera: The coordinated evolution of many traits on multiple levels

Paul M. Brakefield and Bas J. Zwaan

19.1 Introduction

Polyphenisms in butterflies and moths have been a continuous source of inspiration to entomologists since well before Darwin's time. In several well-known cases, including the European map butterfly, *Araschnia levana* (L.), the alternative adult forms associated with the phenomenon are so different visually that they were originally named as separate species. The forms may differ with respect to wing morphology in their color, patterning, size, and shape. They may be associated with different seasons, in which case the term "seasonal polyphenism" is applied to them (Shapiro 1976). Polyphenisms are, however, not only associated with the adult stage, since many fascinating examples are also found in pupae and in larvae (e.g., Greene 1996, Hazel 2002).

In polyphenic species in which the alternative forms have adaptive traits specific to the environment in which they occur, all eggs laid are able to develop into each of the forms; the developmental pathway that a particular individual follows is dependent on an environmental cue perceived during growth and which acts to predict a forthcoming environment in which natural selection occurs. Such predictive adaptive responses (Gluckman *et al.* 2005; Saastamoinen *et al.* 2010) involving discrete phenotypes provide exceptional material to investigate how a range of molecular, cellular, behavioral, and ecological processes have become coordinated in evolution to yield extreme examples of adaptive developmental plasticity in which a single genome can yield two or more discrete phenotypes matched to their specific environments. They may also prove to be very useful in unraveling how epigenetic phenomena contribute, alongside genetic mechanisms, to the evolution of developmental plasticity (see Gilbert and Epel 2009, West-Eberhard 2003). Evidence of the adaptive nature of seasonal polyphenisms has now been obtained for several species by demonstrating that seasonal forms have their highest relative fitness in the environment in which they typically live (e.g., Kingsolver 1996, Brakefield and Frankino 2009).

19.2 Frameworks for dissecting the evolution of polyphenisms

Studies on seasonal polyphenisms have, until recently, focused on the difference in morphology between the alternative phenotypes. In this chapter we will discuss recent work that is revealing how the seasonal forms of butterflies differ, not only in wing pattern, but in many other traits, from the metabolic and physiological to the behavioral and life history. Much of this research is on the dry- and wet-season forms of the African butterfly, *Bicyclus anynana* (Brakefield *et al.* 2009). The phenomenon of seasonal polyphenism clearly reflects the evolution of alternative forms involving suites of traits that are components of the complete functional pheno-type in each environment. Whether all differences in individual traits between forms are adaptive in nature requires further investigation, since some differences may involve the occurrence of unavoid-able "scarring" due to development in an environment with stress, for example with respect to nutrition. Seasons in both temperate and tropical regions of the world typically differ in the extent to which they are favorable for development, survival, and reproduction, and in many cases one season can be considered favorable for growth and reproduction whereas the other reflects a generally more stressful environment. In the latter type of season, the adult stage is typically prolonged until reproduction can occur. This distinction has made some seasonal polyphenisms attractive systems with which to explore the evolution of life histories and, in particular, of traits associated with differences in lifespan (see Brakefield *et al.* 1996, 2005, 2007).

In addition to developmental plasticity, processes of acclimation in response to changing environments within a life stage can be important components of the overall adaptive responses to environmental change. Once evolution of an axis of plasticity has occurred in a particular lineage, it may be flexible to evolutionary tinkering, and even to loss and gain, but resistant to any fundamental change in the directionality of the relationship between phenotypic range and environmental change (Brakefield *et al.* 2007).

Polyphenisms have been especially well studied in several species of Lepidoptera, where many such examples involve temporal heterogeneity in the environment (Shapiro 1976, Brakefield and Larsen 1984, Brakefield *et al.* 2007, Brakefield and Frankino 2009). However, numerous case studies, especially in other insect groups, are not in the first instance associated with seasonal environments and changes in time, but rather with variation in space, including in opportunities for dispersal (e.g., wing polyphenism in crickets and water striders), or in conditions of density (e.g., solitary versus gregarious forms in locusts). Others, for example the castes in social insects and males with divergent horn-morphologies in dung beetles, involve differences in nutrition during development (Emlen *et al.* 2007, Keller and Gordon 2009, Moczek 2009). However,

there are common features among these groups since they all involve hormonal regulation to coordinate the development of sets of different traits. In the case of adaptive polyphenisms, some environmental cue must occur during development that predicts effectively a forthcoming environment in which a specific mode of natural selection is expected. The insect can then respond to this cue to regulate development of the particular phenotype that maximizes reproductive success in the prospective environment (or its contribution to colony success in the case of worker social insects). In comparison to genetic polymorphism, in which segregating genotypes yield two or more discrete phenotypes, developmental plasticity can minimize any genetic load that is associated with the occurrence of phenotypes that are not matched in terms of functional design to the environment in which they typically live.

A second crucial requirement for the evolution of developmental phenotypic plasticity and polyphenism is a matching, in time and space, of the life cycle and dispersal behaviour of the organism to the scale and dynamics of the heterogeneity in the environment (see Nylin and Gotthard 1998). Given such a match and an accurate environmental cue providing effective prediction of the forthcoming environment in which selection happens, the evolution of seasonal polyphenism can, in theory, yield a highly efficient process of adaptation to environmental variation. The issue of matching of phenotype to environment is also highly relevant when considering the consequences of a change in the environmental conditions experienced by a species. If a particular phenotype with respect to plasticity was favored in one of the original environments in which it evolved but now occurs in, and experiences natural selection in, a novel environment, this may lead to a mis-match and a negative impact on fitness. The so-called Barker hypothesis, applied to human populations in developed societies, suggests that a metabolic syndrome associated with earlier experience of nutritional stress *in utero* can have major consequences on events in later life in an environment of plenty with excess nutrition (see Brakefield *et al.* 2005).

A further concept that is proving useful when considering the evolution of seasonal polyphenisms is that of the thrifty phenotype. Many polyphenisms, and perhaps the majority of those associated with seasonal cycles, can be viewed as involving one form that lives in a favorable environment for growth, survival, and reproduction, and an alternative form—a thrifty phenotype—which spends much of its life in a more stressful environment in which growth is often slower, adult lifespan is extended, and reproduction is delayed. Such forms often include a stage of diapause or reproductive dormancy in their life cycle, as well as a lipid-based metabolism. In the examples of polyphenisms associated with alternations of wet–dry seasons in the tropics, the dry-season form appears in many respects to represent such a thrifty phenotype (Brakefield *et al.* 2005, 2007). We will consider these frameworks in the present chapter.

19.3   Case studies on the adaptive nature of seasonal polyphenisms

Joel Kingsolver performed a series of experiments in North America that revealed much about how different patterns of natural selection in spring and summer favour the respective seasonal forms of the western white butterfly, *Pontia occidentalis* (Kingsolver 1995a, b, 1996). In this species, larvae developing under a short day length in the spring metamorphose into adults with rather darker pigmented regions of the wings. These same areas are paler when larvae develop in long-day conditions to yield the summer form. Earlier work by Watt (1968) and Kingsolver (1987) suggested that the increased melanization and specific basking behaviors of the spring form are likely to lead to a more effective thermoregulation and higher activity in the cooler conditions it experiences as an adult. In contrast, the paler wings of the summer form were predicted to reduce solar radiative heating and decrease the risk of overheating.

Butterflies of each seasonal form can typically be reared in the laboratory to emerge in a synchronized fashion, enabling cohort analyses to be made in the field to compare their survival curves under different conditioning of matching between phenotype and environment. Kingsolver's mark-release-recapture (MRR) experiments performed in each season compared the survival of cohorts of *Pontia*

butterflies involving various combinations of the natural forms and the variation they exhibit, together with cohorts of butterflies with painted-on, manipulated wing patterns. The latter can act as powerful controls for possible differences in physiology or behaviour between the forms. The results showed expected effects of variation in wing color patterns on survival, and supported an involvement of the proposed differences in thermal properties and theromoregulatory behaviors in the mechanisms which account for a changing pattern of natural selection with the seasons. Overall, Kingsolver's results provide a persuasive argument for the adaptive nature of this polyphenism (and see also the results on Colias butterflies, Ellers and Boggs 2004).

The adaptive nature of the seasonal polyphenism of the African squinting brown butterfly, *Bicyclus anynana*, has also been explored using MRR experiments with comparable results. This species has a dry-season form (DSF) with uniform brown wings and a wet-season form (WSF) with marginal eyespots and a white medial band across the wings. Brakefield and Larsen proposed an adaptive explanation for this polyphenism, together with those of other species of brown butterflies living in the wet–dry seasonal environments of the tropics that are characterized by a similar difference in wing pattern (Brakefield and Larsen 1984). They suggested that the DSF survives a long dry season to reproduce at the start of the following rains through being inactive and having an effective crypsis when at rest on dead leaf litter. In contrast, the WSF reproduces rapidly and is active on a background of green herbage. It has wing eyespots that may deflect at least some bird and lizard attacks away from the vulnerable body in such a way that some individuals can escape, albeit having lost part of their wing margins (which tear away extremely easily). MRR experiments, performed in Malawi by N. Reitsma, and involving the release of the WSF or the DSF butterflies with painted-on eyespots in the dry season environment, strongly supported the first part of this hypothesis about selection: butterflies with conspicuous eyespots have a substantially higher mortality than the DSF in the dry season (Brakefield and Frankino 2009). The second part of the hypothesis has received some support from experiments with captive birds, which suggested that the feeding attacks of naïve individuals are sometimes misdirected to the wing margins by eyespot-markings (Lyytinen *et al.* 2004). MRR experiments in the field, however, indicate only a very small advantage in terms of survival for the WSF in the wet season, in comparison to their very strong disadvantage in the dry season.

Some recent studies also suggest that, in the absence of any strong natural selection favoring crypsis, ventral eyespots may be subject to sexual selection, with larger eyespots being advantageous because of mate choice (Breuker and Brakefield 2002, Robertson and Monteiro 2005, Costanzo and Monteiro 2007). Selection on the seasonal forms of butterflies may often involve complex interactions between natural selection and sexual selection, reminiscent of selection on variation in the color pattern of male guppy fish in the streams of Trinidad with differing populations of predators (Endler 1995). Increasing numbers of studies are revealing an important role of butterfly wing color patterns in mate choice (e.g., Jiggins *et al.* 2001, Lukhtanov *et al.* 2005, Chamberlain *et al.* 2009), and it will therefore be interesting to determine the extent to which mate choice behaviour shows plasticity across seasonal forms of species with dramatic examples of polyphenism, including *Arashnia levana*, *Precis octavia*, and *Junonia coenia*.

Whilst the two case studies of *Pontia* and *Bicyclus* suggest that seasonal polyphenisms are likely to have an adaptive explanation, there are many other examples, such as those of *A. levana* and *P. octavia*, where we scarcely even have a working hypothesis about how natural selection works, let alone any experimental test of such a scenario. There is therefore substantial scope for careful experimental studies that can make an important contribution to understanding how natural selection works in the wild.

19.4 Environmental cues and the physiological regulation of plasticity

Experimental studies of the environmental regulation of seasonal polyphenisms in the Lepidoptera indicate that, whereas photoperiod is typically the critical variable in temperate regions, temperature is usually more important in the tropics and subtropics. However, there are many examples that have yet to be studied in any detail, and others in which the mechanism remains unclear (e.g., *Melanitis leda* and *Precis octavia*). Even in well-studied examples in the laboratory, it is not always clear that the results can account fully for the seasonal cycling of the forms in the field. For example, the alternative wing pattern phenotypes of the seasonal forms of *B. anynana* are readily generated in the laboratory by rearing the final two instars of the larva either at a low temperature (DSF) or at a high temperature (WSF; Kooi and Brakefield 1999). Analysis of daily trap captures and climatic variation for several species of *Bicyclus*, including *B. anynana*, from a single locality in Malawi also neatly demonstrated that temperature at this stage of the life cycle is closely correlated with cycling in wing pattern (Windig *et al.* 1994). However, the cooler temperatures in the field are not only associated with a DSF lacking marginal eyespots, but also with a tendency for larger adult size. This feature is, however, not generally reflected in laboratory results using low rearing temperatures. This variable probably does not act alone in the wild, but rather is one component of the environmental heterogeneity to which the larvae respond. In this context, we know that any variable that results in a longer development is associated with a tendency to develop a more DSF-like adult phenotype (e.g., Brakefield *et al.* 1998), and at the time larvae of the DSF are approaching pupation in the field, their grass food plants are tending to desiccate and die back (Brakefield and Reitsma 1991).

Remarkably little is known about the details of how the larvae of polyphenic butterflies sense the environmental cues, although critical thresholds have sometimes been examined for the response, for example with respect to photoperiod in *A. levana* (e.g., Koch 1992). Perhaps one of the best case studies is the fascinating account of the diet-induced larval polyphenism in the emerald moth, *Nemoria arizonaria*. In this species, one form of larva is a wonderfully close mimic of the catkins on which it feeds, whereas the other form is a typical twig mimic, as found in numerous species of geometrid moths (Greene 1996). Even here though, the precise dietary component that cues the switch from a twig- to a catkin-matching form is unclear. Similarly, color polyphenisms in the pupae of several species of Pierid and Papilio butterfly have been shown to involve background color or texture; in the latter case smooth leaf surfaces in the field are associated with a green color and rough surfaces such as bark with brown. Colour-matching of the pupae to their background has been demonstrated to enhance survival in several experimental studies. It would be interesting to take some of these examples and investigate how exactly the sensory system operates, in a way parallel to the recent work on locust, which reveals how stimulation of specific tergi under conditions of high density induces development of the *gregaria*, rather than the *solitaria* form (Anstey *et al.* 2009).

The initial response once the environmental cues have been sensed and in some way registered in the relevant developmental stage is a physiological one (see also Chapters 4 and 5). The ecdysteroid hormones mediate the development of the seasonal forms of *B. anynana* (Koch *et al.* 1996), as well as those of some other species, including *A. levana*. The increase in ecdysteroid titer following pupation occurs later in pupae of the DSF of *B. anynana* than in those of the WSF. If animals reared in an environment designed to produce the DSF are micro-injected as young pupae with a low dose of ecdysteroid hormone, they subsequently show a shift in wing pattern determination towards the larger ventral eyespots (and white medial band) characteristic of the adult WSF (Brakefield *et al.* 1998). Thus, the genetic pathway of eyespot pattern determination is up-regulated by early ecdysteroid release in WSF pupae to yield development of larger ventral wing eyespots (Brakefield *et al.* 1996). It is noteworthy that an uncoupling exists with the specification of the eyespots on the dorsal wing surfaces, as these are unaffected by ecdysteroid titers and are present in both seasonal forms (Brakefield *et al.* 1998).

19.5 Genetics of the evolution of the seasonal polyphenism in wing pattern

Seasonal polyphenisms in butterflies are considered to typically involve discrete phenotypes in the field—the different seasonal forms. Similarly, textbook accounts tend to use the terminology of a switch mechanism in the induction of alternative developmental pathways. However, in some cases intermediate phenotypes are not uncommon in the field (e.g., Windig *et al.* 1994), and in many species that have been investigated in detail, such phenotypes can be generated rather readily in the laboratory by using an appropriate intermediate rearing environment. Such an environment may be very rare in the field, at least at the time when the life stage that is sensitive to the environmental cue is present. In *B. anynana*, a broad gradient in rearing temperature in the laboratory generates the full range of wing pattern phenotypes, from one extreme form to the other, and a more or less continuous norm of reaction mapping wing phenotype on to temperature. New research is examining the degree to which the mediation of other traits is also described by a continuous reaction norm or one with a more stepped function (Oostra *et al.* 2010).

Artificial selection has been applied on quantitative variation in eyespot size for cohorts of *B. anynana* reared under standard temperature conditions to explore the availability of genetic variation for evolution of the developmental plasticity. Rapid responses to selection occur with respect to the elevation of the reaction norm describing the relationship at a population level between eyespot phenotype and rearing temperature (e.g., Brakefield *et al.* 1996). Eventually, a high line produces the WSF across all rearing temperatures, whereas a low line yields only butterflies lacking eyespots (DSF). Interestingly, high line butterflies retained strong plasticity, with larger eyespots developing in higher temperatures. Indeed, more recent selection experiments have been largely unsuccessful in changing the reaction norm to make it shallower or steeper (Wijngaarden *et al.* 2002, but see Brakefield and Frankino 2009). The overall results indicate high additive genetic variance for reaction norm elevation, but in combination with high genetic covariances across environments. One consequence of the latter may be that short-term responses to changes in the degree of seasonality of the environment that favour a change in reaction norm shape may be limiting.

The ecdysteroid hormones which mediate the seasonal wing patterns of *B. anynana* are also involved in the coordination of metamorphosis and developmental time. We had also noted in several studies how, in a particular cohort of butterflies, those eclosing earlier with faster development tended to have larger eyespots and a more wet-season-like phenotype. Therefore, more recent experiments have examined the genetic interactions between these traits by applying either an antagonistic or a concerted pattern of artificial selection on the traits of ventral eyespot size and developmental time (Zijlstra *et al.* 2003, 2004). Even in these short-term experiments, it proved possible to uncouple the traits, suggesting that there is at least some potential for their independent evolution. In addition, we showed that the hormone titers were associated with the direction of selection for developmental time but not for ventral eyespot size. Individuals from lines selected for slow development had lower levels of ecdysteroids early on in the pupal stage, and were less sensitive to hormone injection compared to the lines selected for fast development, irrespective of the selection direction on eyespot size. Based on these combined results, we concluded that there are at least two sources of genetic variation for eyespot size: one associated with hormone dynamics and one with the developmental pathway of ventral eyespot morphogenesis. When selecting only on eyespot size, both sources contribute to the selection response. However, when selecting for both developmental time and eyespot size, responses to selection for eyespot size could only be achieved in certain combinations of directions from the source of genetic variance in the developmental pathways because of the pleiotropic effects of the hormone physiology on developmental time and eyespot size.

19.6 Life history evolution in polyphenic butterflies

Phenotypic plasticity for life history traits has been studied in *Bicyclus anynana* with a particular emphasis on egg size, starvation resistance, and lifespan. Whereas the ecdysteroid hormones appear to play the major role in the physiological regulation of the wing patterning, juvenile hormone and vitellogenin are major players in modulating reproductive traits (Steigenga *et al.* 2006, Geister *et al.* 2008a,b). Since the process of adaptation to (changing) environments

can both involve direct genetic adaptation (genetic tracking) and phenotypic plasticity, these life history traits have been examined through both environmental and genetic manipulation. This will eventually enable us to address the question to what extent the genetic and phenotypic responses share the same molecular genetic and physiological mechanisms.

To survive the dry season, both a long lifespan and high resistance to starvation are required. The cooler environmental temperature regime in this season will ensure a longer lifespan and a higher starvation resistance because both these processes are related to metabolic rate in ectothermic species such as insects. Indeed, the starvation resistance of *Bicyclus* in the laboratory is much higher at dry season temperatures compared to wet season temperatures, and this can be largely explained by a decrease in metabolic rate (Pijpe *et al.* 2007). The effects of the developmental temperature of the butterflies was also investigated, with the expectation that butterflies reared at the dry season temperature would have a higher starvation resistance than wet season butterflies, thus adding to the adult temperature effect. However, the starvation resistance of the dry season butterflies was lower than that of the wet season butterflies, independent of the adult temperature environment (Pijpe *et al.* 2007). Again, this effect was best explained by the increase in resting metabolic rate (RMR) of the dry season compared to the wet season butterflies (Pijpe *et al.* 2007), a result that is consistent across all experiments (P.M. Brakefield *et al.*, unpublished data; Oostra *et al.* 2010). Although developmental temperature significantly affects starvation resistance and RMR, the magnitude of the effect is much smaller than the adult environmental effect. In other words, the realized starvation resistance of dry season butterflies in the dry season is much higher than the starvation resistance of wet season butterflies in the wet season. Our interpretation of these results is that the selection for an increased RMR (and as a consequence, a lower starvation resistance) was driven either by the necessity to counteract the effect of lower developmental temperature on growth rate and survival of the larvae and pupae, and/or to counteract the effect of lower adult temperature on RMR and the related flight ability necessary to find adult food sources, i.e., fruit falls, that are patchily distributed in the dry season. These hypotheses are part of our ongoing research.

The interplay between the effects of developmental and adult temperatures is also demonstrated by work on egg size variation in *Bicyclus anynana*. In the laboratory, the reproductive output in terms of number of eggs is significantly and substantially lower for DSF than WSF adults at temperatures that resemble the reproductive wet season (Brakefield *et al.* 2007). At lower temperatures resembling those of the dry season, reproduction is much lower for both seasonal forms. It is important to note that reproduction in the field in the dry season will be effectively zero for most of the time because of the lack of suitable larval host plants (plants that are provided in laboratory experiments).

Interestingly, not only egg number but also egg size effects are components of the phenotypic plasticity. Egg size in *Bicyclus anynana* responds to both developmental and adult temperature. When females are transferred shortly after eclosion to cooler temperature conditions they lay significantly larger eggs after about five days than those kept continuously at warmer temperatures (Fischer *et al.* 2003b). Intriguingly, this acclimation effect can be reversed, even at late ages or after a long period at either cool or warm temperatures, as was demonstrated in cross-transfer experiments (Fischer *et al.* 2003b). Moreover, females that developed at the cool dry season temperature lay much larger eggs after eclosion and onset of reproduction than females that developed at the warm wet season temperature (Fischer *et al.* 2003c). This effect persists throughout the lifespan of the females when the females are kept at the same temperature as that experienced in pre-adult development. As expected from earlier results, when adult females are cross-transferred, egg size increases at the cool temperature and decreases at the warm temperature, irrespective of the developmental temperature (Fischer *et al.* 2003c). There is thus no interaction between developmental plasticity and adult acclimation, and the effects of developmental temperature are still visible and significant when comparing the egg size of females at the same adult temperature and after adult acclimation (Fischer *et al.* 2003c). Therefore, contrary to the patterns for starvation

resistance and RMR described above, the effects of the developmental and adult temperatures operate in the same direction for egg size.

Relevant to the role of plasticity for life history adaptation is the question of what the functional implications and fitness consequences of these phenotypic effects are in the seasonal environments in nature. Larger eggs produced at a lower temperature in the laboratory have a higher hatching success than eggs from females at a warmer temperature (Fischer *et al.* 2003a,b) and a shorter developmental time (Fischer *et al.* 2003b). The higher fitness of larger eggs was also found in experiments where egg size was increased after several generations of artificial selection (Fischer *et al.* 2006). Importantly, a lower rearing temperature was found to be more detrimental for smaller eggs than larger eggs (Fischer *et al.* 2003a). When the effects of developmental and adult temperatures on egg size, egg number, egg survival, and larval survival are combined, the highest fitness is apparently reached for females laying larger but fewer offspring at cooler temperatures, and smaller but more eggs at warmer temperatures (Fischer *et al.* 2003b). All these results, when considered together, support the notion that developmental and adult plasticity (acclimation) for egg size do reflect an adaptation to the seasonal environments that *Bicyclus anynana* encounters in the field.

As mentioned above, starvation resistance is a key trait for the survival of adults of the DSF until the females can oviposit with the coming of the rains and growth of larval food plants (grasses). Therefore, we have investigated the mechanism of starvation resistance and its correlation with other life history traits using artificial selection. Selection for increased starvation resistance was highly effective, with the time until death when having access only to water following adult eclosion roughly doubling by generation 14 in the selected lines (Pijpe *et al.* 2008). As has also been found in other organisms (e.g., in *Drosophila*: Zwaan *et al.* 1995, Baldal *et al.* 2006), starvation resistance and lifespan were positively correlated: butterflies from lines selected for increased starvation resistance had a superior lifespan under *ad libitum* food conditions. Sex-specific trade-offs were observed. Thus, females reached a higher starvation resistance through an increase in body size and a reallocation of reproductive effort to an increase in egg size, at the cost of a decrease in egg number. Males increased their starvation resistance by lowering their metabolic rate (Pijpe *et al.* 2008). When in competition with unselected males, starvation-resistant males showed a lower mating success under *ad libitum* food conditions, but a superior one when all males were starved for three days prior to the experiment (an exposure which has no affect on mortality; J. Pijpe *et al.*, unpublished results). These effects may have been a consequence of a lower RMR and a better physiological condition compared to the unselected control, respectively.

There is an interesting similarity between the genetic and environmental effects for the traits discussed above. The starvation-selected females have a phenotype that is very similar to that of the temperature-induced DSF: they have a larger, more durable body and they lay fewer, but larger eggs. In contrast, the genetically more starvation-resistant males have a lower RMR than control males, an effect that is opposite to that observed for dry-season males (Brakefield *et al.* 2007). Currently our research focuses on uncovering the genetic mechanism of these two responses using various genomics tools (Fig. 19-1).

One other avenue of research that we are pursuing is comparing populations of *Bicyclus anynana* along latitudinal gradients. *B. anynana* populations range extensively throughout east Africa, from Ethiopia to South Africa. Thus, the predictability of seasonal change, as well as the range of temperature variation, differs substantially over the species distribution (Roskam and Brakefield 1999). Recently we have shown that populations near the equator in Malawi and South Africa differ significantly in their response to developmental temperature variation for the wing pattern traits, but not for life history traits such as developmental time, body size, starvation resistance, and RMR (de Jong *et al.* 2010). This information suggests that developmental plasticity may be more important for traits that are fixed during development than for those that can still be fine-tuned via adult acclimation. Currently, we are analysing gene sequence data obtained for a number of populations, looking for footprints of past selection for candidate genes involved in the relevant wing pattern and life history traits. This type of approach is essential to bridge the gap between laboratory experiments and the actual process of adaptation in the field.

Some unpublished data on *B. anynana* also indicate an involvement of behavioral plasticity in the seasonal polyphenism of this species. Butterflies of the WSF, when presented in laboratory flight cages with a mosaic of patches of green or brown leaves on which to rest, show a strong preference for green. When DSF butterflies, which spend most of their adult lifespan in an environment without any green foliage, were given the same choice, they showed no bias in their resting behaviour. This may follow from the uniform nature of their dry season environment or it could be related to their

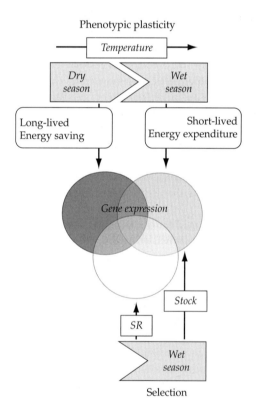

Figure 19-1 Interplay between the genetic mechanisms underpinning the plastic response to temperature and the genetic response to artificial selection in *Bicyclus anynana*. By using gene expression and other genomics tools future research will be able to uncover the mechanisms of plasticity and determine whether they involve separate pathways or whether the plasticity effects are reached via modulation of specific genetic pathways. These latter pathways may also be involved in the genetic response to selection. SR, starvation resistant.

having to switch to a reproductive lifestyle in the early rains.

19.7 Perspectives: Suites of adaptive traits in combination with an ability to acclimate

The most recent research on *B. anynana* has indicated that whole suites of traits are involved in the ability of adults to survive and reproduce in two different environments. Wing pattern evolution is only one component of a phenomenon which extends to a range of physiological, metabolic, and life history traits (for *Pararge aegeria* see also Gotthard 2008). More detailed work on other species is likely to reveal that this is a ubiquitous property of seasonal polyphenisms in the Lepidoptera.

Developmental phenotypic plasticity regulated via insect hormones provides the potential to produce alternative functional phenotypes. Adults of the DSF of *Bicyclus* emerge in the early dry season and must then survive by effective crypsis, opportunistic feeding, and a lipid-based physiology for up to six or seven months before, with the advent of rains and regeneration of larval food plants, they can reproduce. Thus, each cohort of the DSF is confronted in the early rains with a rapid change in both abiotic and biotic conditions. The wing pattern cannot change but other aspects of the phenotype, from metabolism and physiology to behaviour, are dramatically transformed. In east Africa an increase in temperature precedes the onset of rains by up to several weeks. Field observations indicate that this is the cue for *Bicyclus* butterflies to switch their metabolism and behaviour: males become sexually mature and court females, and females mate and develop eggs (Brakefield and Reitsma 1991). Unfortunately we have not yet mirrored this particular part of the life cycle in the laboratory and so these observations remain largely untested with respect to underlying mechanisms.

The most rigorous demonstration of acclimation for *B. anynana* in laboratory conditions involves the larger eggs laid by butterflies of the DSF when maintained at cool temperatures (Fischer *et al.* 2006, Geister *et al.* 2009). Combinations of developmental plasticity and the ability of some fitness-related phenotypic traits to acclimate to a change in envi-

ronment within a particular life stage is probably typical of many seasonal forms and alternative phenotypes in polyphenic systems (see Nylin 1992, Brakefield *et al.* 2007). This is likely to be particularly important in those adult forms which spend most of their life in an unfavorable environment for reproduction, but which must then reproduce quickly once the environment switches to one which favors reproduction.

Closely similar phenomena appear to occur for species of *Mycalesis* inhabiting wet–dry seasonal environments in northern Australia (Braby and Jones 1994; Braby 2002). These butterflies are members from an independent evolutionary radiation within the same subtribe of Mycalesina as the genus, *Bicyclus*. There are also indications of parallel phenomena in temperate species of the subfamily Satyrinae, which may indicate that there is some fundamental axis of plasticity involving temperature common to the whole clade (Brakefield *et al.* 2007, Brakefield and Frankino 2009). Plasticity in life history is especially striking in *Pararge aegeria*. The overwintering generation of this species, which produces the spring or early summer cohort of adult butterflies in bivoltine populations, can overwinter in diapause in either the larval or pupal stage, and there are clearly many interactions among life history, behavioral, and morphological traits that underpin responses to both spatial and temporal variation in its environment (see e.g., Wiklund *et al.* 1983, Nylin *et al.* 1989, Gotthard *et al.* 1994, Van Dyck and Wiklund 2002, Berger *et al.* 2008, Berwaerts *et al.* 2008). There are thus exceptionally complex interactions between plasticity at a developmental level, processes of acclimation, and variation in the timing of winter diapauses. The coordination of these processes presumably results in a close matching of the timing of development and reproduction of different cohorts to changing environments through time. Populations in different climates exhibit differences in the frequencies of the alternative life cycle profiles.

Seasonal polyphenisms may tend to be gained or lost comparatively readily within lineages (see Brakefield and Frankino 2009), rather than evolve to yield novel directions of relationships amongst the traits concerned, and between these traits and the underlying environmental heterogeneity

(Roskam and Brakefield 1999, Zijlstra *et al.* 2003, 2004). For example, evolution from a seasonal to a non-seasonal environment may be readily accompanied by loss of polyphenism and one of the alternative forms, but retention of the seasonal polyphenism when the relationship between the selection regime and the abiotic environment is in a fundamentally different direction may be highly unlikely (Brakefield *et al.* 2007). Future prospects are rich for comparative studies that seek to expand from experimental work on single model species in the field and laboratory to more comparative approaches across whole lineages of species. It is an intriguing possibility (see e.g., Nylin and Wahlberg 2008) that the evolution of plasticity could drive successful colonizations that take advantage of novel ecological opportunities and lead to radiations involving new ecological adaptations and the formation of new species. In such situations it appears especially likely that the plasticity will involve the evolution and coordination of suites of traits involved in functional phenotypes adapted to the novel environments.

19.8 Summary

1. Visually striking examples of polyphenisms in the Lepidoptera probably always have an explanation in terms of natural selection, with respect to either present or, occasionally, past environments. However, their adaptive nature has only been tested rigorously in a few case studies.

2. Polyphenisms typically involve differences in phenotype among alternative forms that extend across suites of traits. Thus, they not only involve variation in color patterns and morphology, but also in physiology, metabolism, behaviour, and life histories.

3. Some traits involved in polyphenisms may be inflexible to change once development and morphogenesis has occurred. In contrast, others, especially those involving metabolism and behaviour, may show an additional level of flexibility via an ability to acclimate to environmental change, for example within or across seasons.

4. The alternative forms characteristic of polyphenisms reflect the evolution of coordinated responses to environmental heterogeneity in time or space.

5. As such, they represent fascinating systems for understanding how a single genome can evolve the ability to yield, via genetic, epigenetic, physiological, and developmental mechanisms, the expression of alternative phenotypes and a predictive adaptive response.

6. Future studies which track the evolution of polyphenisms within lineages of the Lepidoptera in a multidisciplinary framework will be able to open up these mechanisms and, in particular, reveal how hormone dynamics evolve together with the sensitivity of their diverse downstream target tissues.

7. The evolution of developmental phenotypic plasticity and the mechanisms of acclimation for coordinated suites of traits may have played a crucial role in radiations in some lineages of Lepidoptera that have involved expansions into seasonal environments with new ecological opportunities.

19.9 Acknowledgments

We are most grateful to the community of researchers working with *Bicyclus anynana*, and especially to Klaus Fischer who very kindly commented on the manuscript for us. We also wish to thank the EU funded Network of Excellence LifeSpan (FP6 036894) for financial support.

CHAPTER 20

Honey bee life history plasticity: Development, behavior, and aging

Brenda Rascón, Navdeep S. Mutti, Christina Tolfsen, and Gro V. Amdam

20.1   Introduction

Honey bees exhibit a complex pattern of social organization that is embodied in their division of labor, making them some of the most ecologically successful insects. Recently, Hölldobler and Wilson (2008) resurrected the early 20th century metaphor of the insect society as a "superorganism" with physiological, reproductive, communication, and information-processing properties not unlike that of the single individual. The metaphor works well at the phenomenological level of the colony, but it is not always applicable to the study of the development, behavior, and aging of individuals in a society. Individual social insects display different biases in the kinds of behavioral tasks they perform, and these are often associated with changes in physiology that are correlated with age and adult morphological differences. A single "social genome" that is responsible for the ontogeny of development, and on which natural selection can act, does not exist. Instead, each individual is a product of development derived from its own genome. A challenge for scientists will be to understand how the regulation of development, behavior, and aging is achieved in such an advanced social group.

20.2   Development

The honey bee, *A. mellifera*, is characterized by complete metamorphosis (holometabolism). This developmental process is demarcated by four distinct stages, egg, larva, pupa, and adult, and is controlled by the endocrine regulators juvenile hormone (JH) and ecdysone (Winston 1987). Honey bees have a haploid–diploid sex determination system in which a fertilized egg develops into a female and an unfertilized egg develops into a drone (male bee) (Winston 1987). The embryo grows for three days and hatches into a larva. Honey bee larvae develop rapidly and proceed through five larval instars in about 5–6 days. At the end of the fifth larval instar, feeding ceases and pupation begins. During the pupal stage, which lasts about 14 days, the larval structures are broken down and adult anatomical features are formed. Thereafter, the bee emerges and metamorphosis is complete. The duration of development is caste and sex-specific, and ranges from 16 days for a queen to 21 days for a worker and 24 days for a drone (male honey bee).

20.2.1   Female caste morphology: Physiology, function, and reproduction

Honey bee females can develop into two castes: reproductive queens or essentially sterile workers. The behavioral and functional distinctions between queens and workers are primarily shaped during larval life. This is achieved through differential nutrition received by larvae that are largely genetically identical. Caste fate is determined by adult nurse bees, which control the amount and type of food provisioned to the larvae (Fig. 20-1). In honey bee society, queens are solely responsible for egg-laying. The queen has a well-developed reproductive system with more than 150 ovarioles (ovary filaments that produce eggs) per ovary and can lay

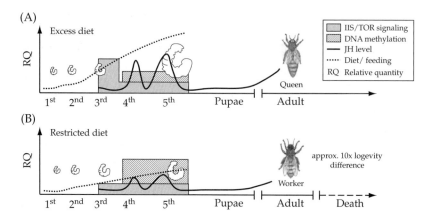

Figure 20-1   Female caste fate and longevity in the honey bee is determined by larval feeding. Molecular changes are depicted throughout larval development (five instars) and into adulthood. (A) Larvae fed a nutrient-rich diet (queen jelly, dotted line) early in life develop into reproductively active queens. Queen-destined larvae up-regulate IIS/TOR genes (relative quantities (RQ)), leading to enhanced IIS/TOR signaling followed by decreased DNA methylation in the fourth and fifth larval instar and a concomitant surge in JH titers. (B) Larvae fed a less nutrient-rich diet (worker jelly, dotted line) develop into workers and exhibit lowered IIS/TOR signaling (RQ) in the third and fourth larval instars relative to queen-destined larvae. These larvae show a higher degree of DNA methylation, and a less dramatic rise in JH titers in comparison to queen-destined larvae. The differences in IIS/TOR signaling cascade in the two female castes ensure more rapid growth in queen-destined larvae than in worker-destined larvae, which may underlie the differences in stress resistance and longevity in the adult stage. Interestingly, despite increased IIS/TOR signaling during development and high rates of reproduction, the queen can live markedly longer than her sibling worker bees.

up to 2000 eggs per day following a single or a few mating flights, which usually take place during the first weeks of her life (Winston 1987). In contrast, workers have reduced reproductive systems (only 2–20 ovarioles per ovary) and are functionally sterile. Instead of participating in direct reproduction, worker honey bees carry out necessary colony maintenance such as brood rearing and foraging for food resources, and take part in the reproductive swarm activities that are essential for colony-level reproduction. While worker honey bees have the potential to lay viable eggs, they will generally not do so under normal circumstances, as worker ovary development and egg-laying behavior are suppressed by pheromones secreted by the queen and brood (Ratnieks 1993). In the absence of the queen and young larvae, however, worker honey bees can lay unfertilized eggs that develop into haploid male drones. However, a colony with only laying worker bees is not sustainable and generally collapses within two months (Winston 1987).

As adults, queens, and workers are highly specialized in terms of morphology, physiology, and behavior. For instance, workers possess slim abdomens, corbiculae (a structure for carrying pollen) on their

hind legs, and a well-developed proboscis (long tongue) for feeding, cleaning, and food collection. In contrast, the corpulent, full-bodied queen bee is fed and groomed by workers, but does not possess corbiculae or a long proboscis, nor does she take part in colony nourishment, construction, or maintenance activities (Winston 1987). Moreover, workers have hypopharyngeal head glands that synthesize nutritious brood food (jelly). Workers and queens both possess stingers, but they use them for different behaviors. Worker honey bees will use their barbed (unretractable) stingers to attack intruders as part of their suicidal aggressive response during colony defense. On the contrary, the queen does not engage in colony defense and seldom uses her smooth and retractable stinger except in cases of supersedure (to attack, kill, and supersede a competitor) (Winston 1987).

20.2.2   An integrative molecular model for caste development: Differential nutrition during larval development triggers caste differentiation

A major molecular player in this phenotypic switch is JH. JH is a major systemic lipophilic hormone that is sensitive to ambient and social environment, nutrition, and physiology, and is an important transcriptional regulator in insects (Hartfelder and Engels 1998). The level of circulating JH is dynamic in both queen and worker-destined larvae throughout development. In fourth to fifth instar female larvae, JH levels in both whole-body extracts and in hemolymph are higher in queen-destined individuals than in worker-destined larvae of the same age (Rachinsky et al. 1990). Also, the application of synthetic JH causes worker-destined larvae to develop queen-like traits (Rembold et al. 1974, Barchuk et al. 2007). However, JH is only one of many factors that play a role during caste development.

Vertebrate studies show that environmental factors like food availability (affecting nutrient uptake) can influence gene expression by acting on transcription factors and the epigenome (Jaenisch and Bird 2003, Burdge et al. 2007). Because early-life social environment and nutrition are critical for the reliable segregation of honey bee castes, it has been postulated that caste differentiation may involve changes in the epigenome (Kucharski et al. 2008). In honey bees, a full complement of functional DNA cytosine-5-methyltransferases, similar to that of vertebrates, has been identified (Wang et al. 2006b). Interestingly, Kucharski et al. (2008) showed that cytosine–phosphate–guanosine (CpG) methylation by DNA methyltransferase 3 can be lower in queen-destined larvae than in developing worker-destined bees, supporting the hypothesis that DNA methylation may play an important role in caste development.

Large-scale transcript studies have identified hundreds of genes that are differentially expressed in queen- and worker-destined larvae (Evans and Wheeler 1999, Barchuk et al. 2007). Queen-destined larvae show up-regulation of genes involved in metabolism and nutrient sensing (Barchuk et al. 2007), including key components of the insulin/insulin-like signaling (IIS) and target of rapamycin (TOR) pathways (Wheeler et al. 2006). Recently, it was shown that decreasing the expression of TOR kinase (Patel et al. 2007) and insulin receptor substrate (IRS, a member of IIS) via RNA interference (RNAi) in young larvae, causes queen-destined individuals to develop worker-like traits (Wolschin et al. 2011). Moreover, Mutti and colleagues observed that suppression of IRS, TOR, and queen fate is

accompanied by decreased JH titers and increased DNA methylation levels, consistent with the results that elevated JH and reduced DNA methylation are associated with normal queen development (Mutti et al. submitted). Taken together, this suggests that the honey bee caste-differentiation cascade may be organized with IIS and TOR as the upstream regulators of both DNA methylation and endocrine effectors like JH. These findings can be summarized in our model, which illustrates how nutritional input signal variation in genetically identical sisters can be canalized to produce two distinct female phenotypes (Fig. 20-1) (see also Chapter 25).

20.3 Behavioral maturation and specialization

In honey bees the division of labor is characterized by temporal polyethism, a maturational schedule in which worker bees move through an age-correlated series of tasks (Winston 1987). Young bees initially perform within-colony activities, such as nursing of brood, cleaning, and taking care of the queen. At two to three weeks of age, worker bees transition to more risky outdoor foraging tasks, which they usually carry out for the remainder of their lives. During this behavioral change, the physiology of the young bee is remodeled for foraging. Some of the gross physiological changes that take place include an overall drop in body weight of 40%, reduced innate immunity, reduced stress resistance, and altered hormonal and molecular profiles (Page et al. 2006, Amdam et al. 2009b, Whitfield et al. 2006).

JH and vitellogenin (Vg) have been proposed as major endocrine regulators of behavioral maturation. Vg, an egg yolk-precursor and phospholipoglycoprotein, serves non-reproductive functions in worker honey bees, and elicits a positive influence on immunity, oxidative stress resistance, and longevity (see Seehuus et al. 2006a,b). Vg, which is mainly produced by young nurse bees before foraging initiation, is also an important transportable and transferable nutrient reserve for the colony as it serves as a source of protein for the brood and is distributed to other colony members by mouth (see Amdam et al. 2009b and references therein).

Vg synthesis occurs in the abdominal fat body (the functional homolog of vertebrate liver and white fat)

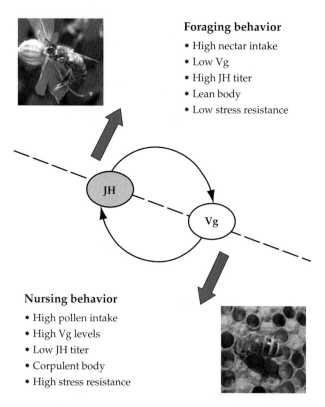

Foraging behavior
- High nectar intake
- Low Vg
- High JH titer
- Lean body
- Low stress resistance

Nursing behavior
- High pollen intake
- High Vg levels
- Low JH titer
- Corpulent body
- High stress resistance

Figure 20-2   Behavioral maturation in worker bees is regulated by a negative feedback loop between Vg and JH. In worker bees, haemolymph Vg titres rise at emergence and remain high throughout the nursing developmental period. The transition to foraging is accompanied by a drop in Vg and an increase in JH levels. In bees, Vg may act as a free-radical scavenger and as a rich source of amino acids, lipids, and carbohydrate. The Vg–JH axis consequently also modulates stress sensitivity and nutrient status in the adult worker. Thus, while the bee is performing nursing tasks, her Vg levels and oxidative stress resistance are elevated. After transitioning to foraging, her Vg levels drop and JH titers rise. At foraging onset, the worker bee becomes more susceptible to oxidative stress. In the negative feedback loop, high Vg levels may block JH synthesis and delay foraging onset. JH may reciprocally inhibit Vg synthesis and induce early foraging onset.

and is released into the hemolymph. The production of Vg is initiated immediately prior to adult emergence and is detectable in the hemolymph of bees older than three days (Pinto *et al.* 2000). From this age on, Vg steadily increases and reaches a maximum level during the nursing stage of behavioral development. JH follows an inverse pattern: when Vg levels are high, JH titers are low. Prior to the foraging transition, Vg levels decline and JH increases.

At the molecular level, the temporal division of labor among worker bees appears to be orchestrated by a mutually antagonistic feedback loop between Vg and JH (Amdam and Omholt 2003). In the double repressor network proposed by Amdam and Omholt (2003), Vg suppresses JH to delay foraging in nurses (Fig. 20-2). The mutually antagonistic

relationship between Vg and JH has been verified by RNAi-mediated knockdown of Vg, which causes an elevation of JH titer (Guidugli *et al.* 2005) and accelerates the transition to foraging (Nelson *et al.* 2007). Furthermore, the treatment of young bees with JH analogues such as methoprene or pyriproxyfen induces precocious foraging and lowers Vg levels (Pinto *et al.* 2000, Schulz *et al.* 2002 and references therein). These findings support the hypothesis of a feedback relationship in which high JH levels suppress Vg synthesis and/or accelerate its degradation (Amdam and Omholt 2003).

A mechanistic description for the causation route between Vg, JH, and onset of foraging remains elusive, but a putative Vg receptor has been localized to the head, fat body, and ovaries (Guidugli-Lazzarini

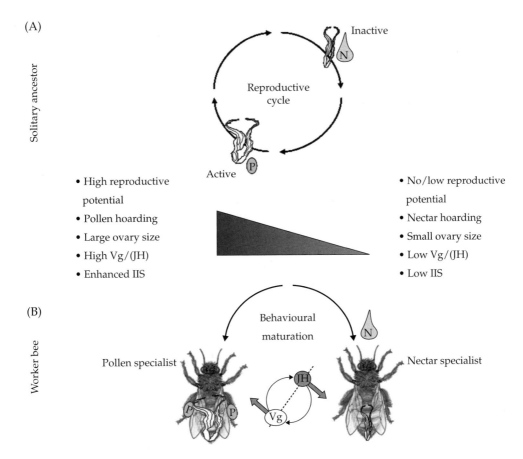

Figure 20-3 The modulation of foraging specialization by the negative feedback loop between Vg–JH as explained though the Reproductive Ground Plan Hypothesis of Amdam and colleagues (2004). This hypothesis posits that the reproductive physiology and reproductive genes of solitary ancestors were co-opted with foraging behavior during the social evolution of the honey bee. Pollen—a source of proteins, lipids, and vitamins—is hoarded by reproductively active individuals and required by brood for proper development. (A) In solitary insects, the reproductive state is characterized by large ovaries and many ovarioles, high Vg and JH titres, and a predisposition for pollen hoarding. In the non-reproductive state, ovary sizes are reduced, Vg and JH levels are low, and the solitary insect primarily forages for nectar. (B) Analogous to the solitary ancestor, high pollen-hoarding (high strain) honey bees forage earlier, possess larger ovaries (more, non-activated ovarioles), high Vg titres, and preferentially forage for pollen. On the other hand, low pollen-hoarding (low strain) bees forage later, possess smaller ovaries with fewer ovarioles, lower levels of Vg, and bias their foraging collection towards nectar. In worker bees, slow behavioral maturation and late foraging onset is typically associated with a longer life. Vg affects IIS signaling, which in turn, signals back to Vg via JH.

et al. 2008). This finding suggests that Vg could influence molecular networks in a variety of tissues, including those in which it is not expressed. Recently, Vg was also detected in brain tissue, and dynamic regulation has been localized to the central brain (Muench, Ihle and Amdam unpublished data).

20.3.1 Specialization of foraging behavior

A forager bee is capable of collecting both pollen and nectar in a single foraging trip, but she may bias her collection of food resources towards either nectar or pollen (Winston 1987). An explanatory framework for the evolution of foraging specialization is provided by the Reproductive Ground Plan Hypothesis (RGPH) of Amdam and colleagues (Amdam *et al.* 2004), which posits that gene networks that coordinated foraging behavior with reproductive physiology in ancestral solitary insects were co-opted to serve as a basis for behavioral specialization during the social evolution of the honey bee (Fig. 20-3).

The RGPH predicts that maternal reproductive traits such as Vg titer and ovary size, which are not normally geared toward actual reproduction in workers, are components of a suite of traits that influence foraging behavior. Support for the RGPH is evident in high- and low-pollen-hoarding honey bee strains that are bi-directionally selected for different foraging behavior toward pollen. Artificial selection resulted in bee colonies that collect and store low quantities (low sub-lines) versus higher (high sub-lines) quantities of pollen. These strains differ in several physiological traits that are generally associated with reproduction in insects (Page and Amdam 2007). Overall, the physiology of workers that collect pollen resembles the maternal or active reproductive stage of solitary insects (pollen hoarding is a specific maternal behavior in many solitary bees), including higher hemolymph levels of Vg, larger ovaries, and increased sensory sensitivity and motor activity (Page et al. 2006). The association between pollen hoarding and maternal behavior in solitary bees led to the proposition that honey bee foraging behavior is governed by the same ancestral molecular networks that have their roots in maternal care.

The link between reproductive traits and social behavior in workers has been corroborated by the mapping of major quantitative trait loci (QTL) for foraging behavior. The QTL architecture confirms that foraging behavior is influenced by a pleiotropic gene network, and that these genome regions show an over-abundance of IIS genes, which are central to nutritional regulation, reproduction, and food-related behavior in animals (Amdam et al. 2009a) (see Section 12.2). Ovarian factors may influence IIS (Flatt et al. 2008b), providing a potential explanation for the link between worker ovary size and behavior. These relationships are currently under investigation, but in the meantime the brain is generally regarded as the more autonomous pacemaker for behavior.

20.3.2   Central nervous system changes during behavioral maturation

Nurse bees mainly navigate in the darkness of the colony, where communication depends on odor- and mechanosensory perception (Winston 1987). Foraging, on the other hand, requires the processing of visual and olfactory stimuli for the learning and memorization of food sources and landmarks. In accordance with this, the most notable CNS changes during honey bee behavioral maturation occur within the olfactory glomeruli and the mushroom body (MB), higher order centers for olfactory perception and learning, respectively. The MB exhibits morphological plasticity through an increase in volume and outgrowth of brain neuropiles (Fahrbach et al. 1998), which cannot be explained by neurogenesis (Fahrbach et al. 1995) nor can these changes be correlated with the JH upregulation that is characteristic of the forager transition (Fahrbach et al. 1998).

Forager honey bee brains that show MB neuropil growth have higher expression levels of the transcription factor Krüppel homolog 1 (Kr-h1) (Fussnecker and Grozinger 2008). Kr-h1 expression is induced by cGMP, and recent work on Kr-h1 in Drosophila melanogaster implicates this transcription factor in ecdysone-mediated developmental MB plasticity (Hewes 2008). In honey bees, this relationship has not yet been confirmed, but it is possible that Kr-h1 plays a role comparable to the one in Drosophila.

During behavioral development, honey bees also show an up-regulation of two candidate genes for foraging behavior: malvolio (Ammvel) and foraging (Amfor) (Ben-Shahar et al. 2002, 2004). Malvolio encodes a putative manganese transport protein that is associated with increased sucrose responsiveness in honey bees; foraging is an ortholog of the D. melanogaster foraging (for) gene that encodes a cGMP-dependent protein kinase (PKG) (Ben-Shahar et al. 2002). In bees, the manipulation of PKG levels causes precocious foraging. Foragers with increased levels of foraging navigate towards light, which suggests that foraging influences honey bee foraging behavior by stimulating phototaxis (Ben-Shahar et al. 2003). Since Kr-h1 expression correlates with PKG activation and the Krh-1 promoter contains a putative cGMP-response element, foraging/PKG is believed to be a master regulator of a gene network for foraging behavior that includes Kr-h1 (Fussnecker and Grozinger 2008).

Microarray studies suggest that there are additional independent molecular pathways that are correlated with honey bee behavioral maturation (Whitfield *et al.* 2006). Past molecular studies conducted on the honey bee brain show that the mRNA levels of many genes differ between nurse bees and foragers (Whitfield *et al.* 2003). Genes with alleged roles in signal transduction, glutamate biosynthesis, and chemical homeostasis are increased in foragers, whereas genes with presumed roles in structural development are up-regulated in nurse bees. Several genes involved in translation are up-regulated in foragers, while others of the same category are up-regulated in nurse bees (Whitfield *et al.* 2006).

Additional independent molecular pathways have been correlated with honey bee behavioral maturation. Biogenic amines, several protein kinases, and second messengers are all part of an intricate network that modulates sensory sensitivity, motor function, and learning in response to behavioral task and foraging specialization (Page *et al.* 2006, Amdam *et al.* 2009a). Biogenic amines such as dopamine, serotonin, octopamine, and tyramine modulate aspects of gustatory, olfactory, and visual sensitivity in honey bees. Foragers show elevated levels of dopamine, octopamine, and serotonin (Schulz and Robinson 1999). Specifically, octopamine has been implicated in the nurse bee to foraging transition and recently a study showed that this biogenic amine can increase the likelihood of waggle dancing, a behavioral display that signals food resource quality (Barron *et al.* 2007).

Further work is needed to decipher the events that take place in the CNS during behavioral maturation, but it is clear that multiple pathways may act in conjunction with one another to elicit behavioral changes in the honey bee.

20.3.3 Metabolic changes during behavioral maturation

The behavioral transition of the honey bee from within-colony labor to foraging duties is marked by metabolic changes that remodel the physiology of the bee to alter oxidative requirements during flight and foraging behavior. Honey bee foragers engage in extensive food hoarding that involves frequent and long flights, as well as other energy-demanding and complex behaviors such as navigation, recruitment dances, and associative learning, which are used for communication and the memorization of foraging sources (Winston 1987). As a result, the physiological demands of foraging depend on increased oxidative capacity and altered nutrient processing.

Flight induces a dramatic change in the basal metabolic rate of the honey bee. Overall, foragers have mass-specific oxygen consumption rates that are 50% higher than those of nurse bees (Harrison 1986). During behavioral ontogeny, honey bee flight metabolic rate is 10-fold higher than the resting rate and is paralleled by a 10-fold increase in flight muscle cytochrome and a significant rise in glycolytic and antioxidant protein levels (Harrison and Fewell 2002, Roberts and Elekonich 2005, Wolschin and Amdam 2007a). To fuel their flights, foragers utilize carbohydrates (Winston 1987) and have thoracic glycogen stores double that of nurse bees (Harrison 1986). Indicative of an elevated metabolism, foragers typically exhibit higher overall protein turnover levels than younger nurse bees (Crailsheim 1986).

Foragers also display changes in IIS, as evidenced by heightened levels of insulin-like peptide 1 (ilp1) in the head, and insulin receptor 1 and 2 (InR1 and InR2) in the abdomen (Ament *et al.* 2008). Evidence from Vg RNAi and protein injection experiments in honey bees suggests that Ilp1 and Ilp2 are part of two separate paracrine systems that control fat body metabolism and govern somatic resource allocation during behavioral maturation. The Ilp2–JH axis likely modulates fat body lipid/carbohydrate resources and Ilp1 regulates protein synthesis and storage (Nilsen *et al.*, submitted). Recently, Wang and co-workers demonstrated the effects of reduced peripheral IIS on honey bee foraging behavior by down-regulating IRS in the abdominal fat body. They showed that IRS knockdowns biased their foraging efforts towards the collection of pollen (protein source) rather than nectar (carbohydrate source) (Wang *et al.* 2010).

Collectively, these findings indicate that there is an ontogenetic shift in total metabolism and nutrient processing during behavioral maturation that can be mediated by IIS signaling.

20.4 Worker aging

In the past, many senescence-driven studies have utilized the ubiquitous *D. melanogaster* and the roundworm, *Caenorhabditis elegans*, due to their short, tractable lifespans and widely used genetic tools available for these species. Recently, the honey bee has emerged as a promising new model system for senescence research due to its remarkable aging plasticity and socio-behavioral repertoire (Munch *et al.* 2008). However, due to the high degree of complexity associated with the elastic, ontogenetic specialization of tasks and the compartmentalization of alloparental functions within the honey bee colony, many of the well-known theories of aging are not always applicable to the life history of the honey bee.

Life history theory postulates that the pressure of natural selection on survival, which favors fitness during the reproductive phase of life, decreases after reproductive capacity has been exhausted. However, because the worker bee is functionally sterile and behaviorally moves through a series of stage-dependent tasks (see Sections 20.2 and 20.3), established theories of aging may not adequately explain the honey bee pattern of senescence (Amdam and Page 2005).

Classic evolutionary theories of aging, such as Medawar's mutation accumulation theory (1952), Williams' antagonistic pleiotropy theory (1957), and Kirkwood's (1977) disposable soma theory, all attempt to explain why rising mortality rates accompany old age. All theories rest on the concept of extrinsic mortality, such that when hazard (risk of dying) is high, then natural selection will not favor further investment of resources into the soma. In contrast, if extrinsic mortality is low, then selection favors somatic maintenance and would act more weakly to reduce mortality rates at older ages. Although these theories are generally regarded as the dominant explanatory paradigms for the evolution of aging, the sole focus on reproduction limits the application of these ideas to sterile worker honey bees, which act as alloparental caregivers that experience low mortality risk for part of their lives (during nursing) before moving to more hazardous tasks outside the hive (foraging) (Winston 1987). However, a postulate that integrates social resource transfers (e.g., brood care) with classical evolutionary thinking on aging (Lee 2003) would center on parental investment and resource transfers between individuals of a multitude of ages. This approach could prove fruitful in describing the aging characteristics of honey bee workers, which, depending on age and environment, engage in different forms of social resource transfers throughout their lives. However, empirical evaluation of this theory will require measurement of a myriad of behaviors, such as guarding, food exchange, fanning, warming, and foraging (Amdam and Page 2005).

20.4.1 Plasticity of aging

Honey bee senescence appears to differ in certain aspects from the aging patterns of the solitary model organisms traditionally used in aging research, which are characterized by progressive and irreversible aging. In contrast, honey bee aging is largely related to social task performance, rather than to chronological age (Fig. 20-4). Bees do not exhibit the characteristics commonly associated with aging during the first 30 days of the nursing period, while aging accelerates after transition to foraging. Thus, the shift from nursing to foraging is the most crucial determinant of the overall lifespan expectancy for a honey bee (Rueppell *et al.* 2008, Amdam *et al.* 2009b).

The number of days honey bees spend within the colony performing nest tasks can vary depending on the season, and strongly influences lifespan. Nurse bees can survive for more than 130 days, while diutinus or "winter" bees, which develop in the absence of brood and nursing activity, can survive for more than 280 days (negligible senescence) before they segregate into nurse bees (slow aging) and foragers (rapid aging) (Seehuus *et al.* 2006a). Thus, senescence in honey bees can be remarkably plastic. This plasticity is tested by social environmental manipulations, where removal of nurse bees can cause forager bees to behaviorally and physiologically revert to nursing tasks (see Section 20.3 and Fig. 20-4 for details). This role-reversal alters many important biomarkers of senescence. Reverted nurse bees (former foragers) undergo a reversal of immunosenescence and exhibit some of

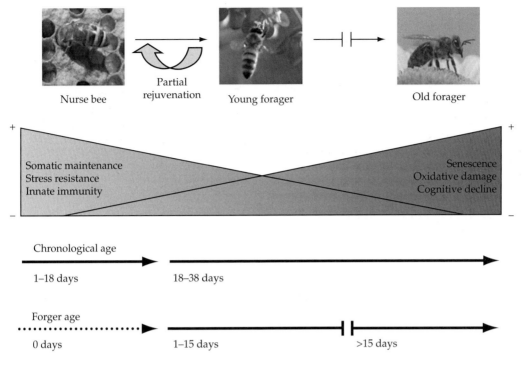

Figure 20-4 Honey bee worker senescence is a function of the behavioral task they perform within the colony. Senescence in workers is a plastic process that may be manipulated by changes in colony demography and/or environmental conditions. Nurse bees have high levels of Vg and are more resistant to oxidative and environmental stressors. During foraging onset, the somatic maintenance of the bee drops, and the worker bee experiences a dramatic decline in immune function. Young foragers are capable of a partial rejuvenation if they switch back to nursing tasks. As the forager bee becomes older, the cell repair and maintenance machinery becomes overwhelmed and the bee gradually accumulates protein oxidative damage in the brain. After approximately two active weeks of foraging, she experiences cognitive decline. While both nurses and young foragers perform well in associative learning, forager bees aged more than 15 days typically perform poorly. In the bee, as in many species, the accumulation of oxidatively modified proteins and a reduction in cognitive performance are unequivocal signs of senescence.

the biological hallmarks of the younger nurse bees they replace, for example elevation of Vg, and a suppression of JH (Amdam *et al.* 2005). However, reverted worker bees are not identical to normal nurse bees, but rather exhibit a mixed physiology that includes characteristics of the former forager state. Wolschin and Amdam (2007b) found that reverted worker bees have proteomic profiles that resemble both nurse bees and foragers. Moreover, there appear to be limits to aging reversal. Some aged foragers that have surpassed a "point of no return" appear unable to behaviorally revert and continue to progressively age.

In summary, although counterintuitive to our conventionally accepted axiom of aging, the revelation that chronological age can be decoupled from social task is intriguing, particularly for those interested in the reversal of the aging process.

20.4.2 Oxidative stress

One of the mechanisms that may underlie senescence in honey bees is oxidative stress. The free radical theory of aging proposes that molecular oxygen can serve as a source of reactive oxygen species (ROS), which can induce cumulative macromolecular damage and ultimately cause aging (Harman 1956). ROS are involved in normal cell respiration as by-products of aerobic mitochondrial metabolism, but they can also inflict damage on proteins, lipids, and DNA if not successfully scavenged by cellular antioxidants. The modulation of oxygen tension in invertebrates has become an essential tool for manipulating oxidative stress *in vivo*. Increased oxygen levels are known to augment the rate of ROS production, reduce the lifespan of *C. elegans* and *D. melanogaster*, and induce physiological changes that are of relevance to senescence (von Zglinicki and Sitte 2003).

Aside from its previously described functions during behavioral maturation (see Section 20.3), Vg can also act as an antioxidant (Seehuus *et al.* 2006b). Using RNAi-mediated knockdown of Vg, Seehuus *et al.* (2006b) showed that Vg protects worker honey bees from oxidative stress. Moreover, a subsequent RNAi-mediated Vg knockdown study confirmed that Vg can extend life (Nelson *et al.* 2007). Collectively, these studies further reinforce the observation that foragers display a greater vulnerability to senescence and also highlight that Vg may have acquired new functions in the honey bee.

Honey bee foragers must expand oxidative capacity to accommodate the energetic demands of flight (see Section 20.3). The elevated oxygen consumption rates associated with flight presumably also augment ROS production and accelerate senescence. A honey bee study focused on examining metabolically active tissues—flight muscle and brain—found that flight behavior induces an up-regulation of antioxidants in young forager flight muscles in comparison to older foragers (Williams *et al.* 2008). In the same study it was reported that as foragers grew older, their antioxidant capacity diminished within the span of a day. The authors speculated that the reduction of antioxidant defenses in the bees likely led to an acceleration of senescence via an excessive production of ROS. Interestingly, the observed changes in antioxidant defense were only evident in the thorax and not in the heads, which suggests that senescence-related damage may be tissue-specific. However, it should be noted that Corona *et al.* (2005) did not observe a clear correlation between age and antioxidant mRNA in the multiple profiled tissues (abdominal, neural, and thoracic) of worker honey bees. This could be attributed, at least in part, to the fact that the authors did not control for behavioral task, but rather used chronological age as their main metric of senescence despite evidence that illustrates the importance of social role in honey bee aging

(Behrends *et al.* 2007). Also, measurements of antioxidant activity or protein levels are often better indicators of cellular active state than transcripts. This may have also contributed to the lack of a correlation between antioxidant status and age.

The evidence for protein carbonylation, a marker of oxidative stress commonly used in aging experiments, is inconsistent and appears complex in honey bees. Although Seehuus *et al.* (2006b) showed that Vg in the brain of old workers is carbonylated in response to paraquat injections, a proxy for elevated ROS (Seehuus *et al.* 2006b), Williams *et al.* (2008) did not detect protein carbonylation differences in the thoraces or heads of honey bees of varying ages or activity levels. An explanation for this apparent discrepancy is that different stressors intended to mimic aging may preferentially target specific proteins, as shown in *D. melanogaster* (Das *et al.* 2001), or other macromolecules such as lipids. Also, oxidative damage may target proteins that are small and escape detection via common experimental methods. Collectively, the results of Seehuus *et al.* (2006b) and Williams *et al.* (2008) warrant further refinement of the current methods for detecting oxidative stress in honey bees.

Alternative strategies involving other molecular targets like lipids, DNA, and mitochondria should also be explored (Barja 2002, Kaneko 2003, Haddad *et al.* 2007). The measurement of proteasomal activity may also hold promise for oxidative stress detection in honey bees, particularly since it has been shown to change with age in *D. melanogaster* (Vernace *et al.* 2007).

20.4.3 Metabolic patterns of senescence

Global transcript profiling studies have shown that the mRNA levels of antimicrobial proteins and heat shock proteins, the latter of which are often involved in protein folding, are up-regulated during aging (reviewed by Munch *et al.* 2008). Interestingly, the accumulation of unfolded or misfolded proteins (proteotoxicity), and the subsequent lack of clearance by the proteasome, is thought to play a role in senescence and age-related neurodegeneration (Gray *et al.* 2003). Thus, the up-regulation of heat shock proteins could be a mechanism to counteract these processes. In contrast, mRNA for reproductive proteins (e.g., yolk and fatty acid binding proteins) and ATP synthesis proteins are down-regulated with increasing age, which points to the often observed decrease of motility and fecundity that accompanies old age.

Tissue-specific profiling of *D. melanogaster* has revealed that distinct tissues of the fly have different propensities for aging (Zhan *et al.* 2007). In this study, the transcript reveals changes in important functional categories like energy metabolism, protein degradation, stress resistance, immunity, and neurotransmitter release. In honey bees, a recent transcriptional study of the abdominal, neural, and thoracic tissue of queens by Corona *et al.* (2005) revealed changes in antioxidant proteins with age. The researchers found decreased levels of mRNAs with presumed roles in longevity during aging. In contrast, worker honey bees, which generally live much shorter lives than queens, exhibited no clear pattern of decline in antioxidant protein mRNA. It is possible that the chronological ages that were chosen for the queen versus worker comparison in this study (one-month worker versus one-year queen) are not indicative of true physiological age as senescence is not a simple function of chronological age in honey bees (Seehuus *et al.* 2006a). Moreover, reports on queen longevity generally place "old" queens in the range of three to four years (Winston 1987). Corona *et al.* (2005) also reported no marked mRNA expression differences between queens and workers, despite vast lifespan dissimilarities. Overall, the results illustrate that antioxidant enzymes measured at the transcript level are not necessarily correlated with organismal longevity responses.

Proteomic and metabolomic data are regarded as more representative of current cellular conditions than transcript information, but few proteomic studies of aging focus on invertebrates. Those studies that do use invertebrates center on *D. melanogaster* and *A. mellifera* (Sowell *et al.* 2007, Wolschin and Amdam 2007a,b, Wolschin *et al.* 2009). Whole-body studies utilizing both proteomic and transcript methods have shown an up-regulation of proteins and peptides with antimicrobial properties with increased age (reviewed by Munch *et al.* 2008). Moreover, during aging, a down-regulation of prophenoloxidase, an enzyme implicated in immune defense, was also detected in fruit fly brains (Sowell *et al.* 2007). Sowell and colleagues found that, as indicated above, the affected proteins in the aged *Drosophila* brain not only segregate to categories of immunity, but have functions in reproductive, developmental, metabolic, and cellular defense networks of the cell.

In honey bees, the behavioral maturational shift, a determinant of overall lifespan, is accompanied by changes in the abundance of proteins with roles in glycolysis, ATP synthesis, and free-radical defense (Schippers *et al.* 2006, Wolschin and Amdam 2007a) (see Section 20.3). Because honey bee senescence is tied to social task rather than chronological age, Wolschin and Amdam (2007b) used nest bees and foragers, before and after behavioral reversion, to profile their respective proteomes (see Section 20.4.1). The use of such a social–environmental technique enabled the researchers to study the proteome of bees of differing ages that performed identical tasks. This study showed significant alterations in the proteome due to age and behavioral task. Moreover, the study illustrated that worker bees of distinct life histories (nest bees versus foragers) and task-matched worker bees (nest bees and foragers) of differing ages show distinct protein expression profiles (Wolschin and Amdam 2007b). The proteomic profile of reverted nest bees resembled that of nest bees prior to the behavioral reversion, supporting the idea that behavioral tasks have a particular proteomic signature. Two proteins with putative roles in lipid and cholesterol metabolism, two α-glucosidases, and a malate dehydrogenase-like protein were associated with behavioral tasks, while others, including an odorant-binding protein, showed a clear age-dependent abundance.

Transcriptome and proteome studies can impart correlative information about the trajectory of aging and can provide valuable insight into the search for viable candidates for further research. Such large-scale profiling has spawned intervention into the aging process and has started to offer insight into the prolongation of cognitive span.

20.4.4 Cognitive senescence

Aging intervention studies have shown that lifespan extension is possible in vertebrates and invertebrate models, but the quality of life (health span) question still remains unanswered. How do we achieve enhanced longevity that is not coupled to diminished mental capacity? This is inarguably one of the main questions of our time. The honey bee system boasts a rich marriage of naturally occurring quantifiable behaviors and a history as a neuroscience model. Added to this, its aging plasticity makes it a particularly compelling organism for the study of cognitive senescence.

In honey bees, some cognitive functions, but not all, decline in foragers with extended foraging experience (Behrends et al. 2007, Rueppell et al. 2007). More specifically, a study of nurses and foragers of identical chronological ages revealed that bees that foraged for more than 15 days, in comparison to those that foraged for 6–13 days, performed more poorly in olfactory learning trials. However, the ability to more accurately discern between two odors is enhanced in foragers of long foraging duration relative to younger foragers (Behrends et al. 2007), indicative of greater learning acuity. These findings highlight the importance of considering life history and workload when studying the impacts of aging on cognitive performance.

To explore whether differences in foraging duration are actually linked to biochemical and structural changes in the honey bee brain, Wolschin et al. (2009) used proteomics and immunocytochemistry to examine the central brain and MB calyces of bees with varying foraging experience. The calyx is usually regarded as the input region for the MB, the site of memory formation, whereas the central brain is thought to be the higher order integration center. The central brain of foragers with more than 15 days of foraging experience showed a down-regulation of kinases and synaptic/neuronal growth-related proteins in comparison to workers with less foraging experience. The calyx appeared to stay intact regardless of foraging duration (Wolschin et al. 2009). These findings point to a complex interaction between the central brain and the MB calyx, and suggest that regions other than the calyx are responsible for foraging-dependent decline. Senescence patterns that are characteristic of some but not all brain regions appear to be common in aging, since they have also been reported in chimpanzees (Fraser et al. 2005).

20.4.5 Impact of nutrition sensing pathways on lifespan

The conserved nutrient sensing IIS and TOR pathways that play a role in honey bee caste differentiation (see Section 20.2) have been connected to aging in *C. elegans*, *D. melanogaster*, and *M. musculus*. (Piper et al. 2005, Kaeberlein and Shamieh 2010). In model invertebrates such as *D. melanogaster* and *C. elegans*, decreased IIS and TOR signaling lengthen lifespan (Tatar et al. 2001b, Piper et al. 2005, Kaeberlein and Shamieh 2010). In honey bees, the same pathways have also been associated with major adult life history transitions. The switch from nest tasks to foraging can be accompanied by increased IIS (Ament et al. 2008) and causes increased mortality independent of predation (Neukirch 1982). Ament et al. (2008) showed that foragers and nurses differ in ilp1, InR1, and InR2 levels, integral components of the IIS pathway (see Section 20.3). This same study demonstrated that the onset of foraging could be delayed in worker bees fed rapamycin, a TOR inhibitor. Thus, IIS and TOR signaling are likely involved in the behavioral transition, which plays a preeminent role in the determination of lifespan in worker honey bees.

The IIS/TOR pathways influence the course of life history in both queens and workers during development and may govern some of the adult longevity differences seen between the two castes, which also vary greatly in fecundity. In general, the relationship between the IIS/TOR cascades and lifespan in honey bees parallels findings in *D. melanogaster*, except when fecundity is taken into account.

In a comparison between queens and workers, Corona et al. observed lower levels of InR expression in long-lived queens (2007), consistent with observations in *D. melanogaster* and *C. elegans* in which a downregulation of nutrient sensing genes or gene partners such as InR, IRS-1, TOR, FOXO (forkhead box, sub-group O), and others extend lifespan (see Chapters 11, 13, and 22). Low fecundity is known to confer longevity in classic models of aging (Flatt and Kawecki 2007). However, the highly fecund honey bee queen can live about ten times longer than essentially sterile short-lived workers, yet shows higher levels of IIS expression as 3rd instar larvae (Wheeler *et al.* 2006, de Azevedo and Hartfelder 2008). Thus, queens depend on increased IIS during development to achieve an adult morphology geared toward reproduction, including large body size and ovaries, whereas adulthood lifespan *per se* may be extended via reduced IIS. This regulatory ontogeny has been mimicked in *C. elegans*, which confirms that IIS can be decoupled between life stages (Dillin *et al.* 2002).

The IIS and TOR signaling pathways are also thought to underlie the increased longevity response seen with caloric restriction, which is known to have a conserved effect on the reduction of aging rate (Bishop and Guarente 2007a). Two protein deacetylases, Sir2 (Silent information regulator 2) and Rpd3 (a transcriptional regulator), are thought to mediate the life-lengthening effect of caloric restriction and have spawned a variety of dietary mimetics and medical intervention therapies (Partridge *et al.* 2005b). For instance, resveratrol, which activates a class of conserved proteins (sirtuins) that includes Sir2, extends the lifespan of *C. elegans* and *D. melanogaster*, without sacrificing fecundity (Wood *et al.* 2004). It has been proposed that resveratrol's mechanism of action mimics that of caloric restriction and that Sir2 underlies this response (Wood *et al.* 2004), but its specific modes of interaction remain to be discovered.

Caloric restriction research remains highly promising for understanding the mechanistic underpinnings of aging, but such research remains unexplored in honey bees. In the future, honey bees could potentially add to this area of research because of their dynamic social nature, food provisioning strategies, and lifespan plasticity (see Chapter 14).

20.5 Concluding remarks

A unifying theory of life history progression and aging does not presently exist, but our current understanding warrants a holistic approach. Historically, aging has been depicted as a deregulated and unavoidable death sentence. This axiom gave rise to simple deterioration and preprogrammed theories that made the process of aging seem static and perhaps not investigation worthy to some scientists. Today, however, many studies of lifespan extension, genetic intervention, and caloric restriction have revealed that life history progression and aging are plastic, regulated processes that are governed by physiological and evolutionary trade-offs. These dynamics are broadly shared between taxa but some of the traits can have species-specific features. In advanced social insects, many species-specific traits result from colony-level selection. This context provides challenges, but it also provides unique opportunities for progress towards the goal of understanding how life history plasticity evolves.

20.6 Summary

1. Honey bees are social and possess a complex behavioral repertoire along with compensatory mechanisms that allow them to respond to environmental variation.

2. Female honey bees develop into two castes: long-lived reproductive queens or shorter-lived functionally sterile worker bees. Caste differentiation depends on nutritional variation, and is mediated by the action of large metabolic networks like IIS and TOR, hormonal titers (Vg and JH), and DNA methylation.

3. In the adult worker bee, the Vg–JH axis regulates behavioral maturation, foraging specialization, and longevity in a manner that suggests that foraging behaviors derived from an ancient solitary ancestor were co-opted during the social evolution of the honey bee.

4. Behavioral maturation in honey bee workers involves physiological, anatomical, and biochemical changes that prepare the bee for metabolically-demanding processes like flight and foraging. These behaviors require the expansion of oxidative capacity, and the formation and maintenance of complex spatial memories.

5. The honey bee is a new and promising model system for aging research because of its aging plasticity and socio-behavioral repertoire. Worker bees show classic features of aging, but these are partially reversible since senescence is tied to social task rather than chronological age.

20.7   Acknowledgments

We thank two anonymous reviewers for their constructive feedback and apologize to those colleagues whose work we could not cite due to space considerations. Many thanks to Svea Hohensee for her contributions to an early version of the manuscript, Bente Smedal and Adam Siegel for photographic material, and Jim Hunt, Øyvind Halskau, Heli Havukainen, and Florian Wolschin for their critical review of the manuscript. Financial support to GVA was provided by the Norwegian Research Council (#175413, 180504, 185306) and the PEW Foundation.

PART 6

Life history integration and trade-offs

Thomas Flatt and Andreas Heyland

The authors in this part of the book focus on trade-offs: how they functionally work and how they integrate and constrain life history traits and their evolution. Organisms have to work as integrated wholes if they are to survive and reproduce (Stearns 1992). If organisms were a collection of unconnected modules that did not interact with each other, they could not function, and adaptation by natural selection would not be possible. Life history traits are thus bound together and integrated by numerous relationships provided by the mechanisms of genetics, physiology, and development, and one major aspect of this integration of life history traits is trade-offs.

A trade-off between two (or more) life history traits exists when an increase in one life history trait that improves fitness is coupled to a decrease in another life history trait that reduces fitness. Such trade-offs are ubiquitous; their existence can be established through negative phenotypic and genetic correlations, phenotypic manipulations, or negative correlated responses to artificial selection. Evolutionarily relevant trade-offs are those that are reflected in negative genetic correlations between life history traits or correlated responses to selection; they are of evolutionary interest because they imply that traits cannot evolve independently from each other. Thus, trade-offs constrain the simultaneous evolutionary optimization of correlated suites of life history traits with respect to reproductive success (Stearns 1989, 1992, Roff 1992, 2007b; see also Chapters 1, 2, 27, and 28).

Well-known examples of trade-offs include the classical trade-offs between survival and reproduc-

tion, number and size of offspring, or current versus future reproduction (Stearns 1992, Roff 1992). For example, the evidence for genetic correlations and trade-offs between life history traits from selection experiments in *Drosophila* can be summarized as follows (Stearns and Partridge 2001): adult survival and either lifetime or early fecundity are negatively correlated; development time and body size are positively correlated; there exists often but not always a trade-off between early and late fecundity; there is no positive correlation between development time and lifespan; and changes in life history traits are consistent with trade-offs among physiological traits so that, for example, long-lived flies or flies with a higher fecundity late in life have higher fat and glycogen content, higher starvation, desiccation, and alcohol resistance, and can fly longer than flies with shorter lifespans or lower fecundity late in life. However, while it is clear that life history traits are connected to each other and cannot respond to selection without changing some other trait that also contributes to fitness, we still have very little understanding of the mechanisms that cause them (e.g., Stearns 2000, Leroi 2001, Barnes and Partridge 2003, Roff 2007b; see also Chapters 1, 2, 27, and 28).

The major aim of the research reviewed in the chapters throughout this part of the book is to illuminate the proximate basis of life history trade-offs. Elucidating the mechanistic underpinnings of trade-offs is important for our understanding of:

- whether constraints upon life history evolution have to be understood as a problem resource

allocation among competing life history functions, or whether trade-offs exist for functional and structural reasons that have nothing to do with differential energy investment
• the dependency of trade-offs upon the environmental and physiological context, i.e., the conditions under which trade-offs can be decoupled or not
• the antagonistic pleiotropy (or "trade-off") theory for the evolution of aging.

Chapter 21 by Nedelcu and Michod describes the significance of life history trade-offs during evolutionary transitions in individuality (such as the transitions from unicellular to multicellular individuals), using volvocalean green algae as an example. In particular, the authors discuss the mechanisms whereby individual cells that are part of an integrated group are released from the survival–reproduction trade-off that constrains their unicellular relatives, so that for a group of cells these two fitness components can be maximized independently and simultaneously.

In Chapter 22 Gerisch and Antebi discuss the molecular and genetic basis of life history integration and trade-offs in *C. elegans* and relate these findings to related events in other model organisms. Recent work in this model system reveals that insulin/IGF, TGF-β, serotonergic, and steroid hormone signaling pathways interact and converge to regulate many aspects of the plastic trade-off between reproduction and survival. As Gerisch and Antebi point out, these and similar integratory hormonal pathways suggest a common, evolutionarily conserved molecular basis for the regulation of phenotypic plasticity and life history trade-offs in metazoans.

Chapter 23 by McKean and Lazzaro assesses our current understanding of the mechanisms that mediate trade-offs with immune function, the so-called "costs of immunity." In their review, the authors discuss and evaluate trade-offs between reproduction and immunity and their resource-allocation basis; the costs associated with the induction of an immune response and how the mechanisms regulating immunity have likely evolved to limit the costs of immunological deployment; and evidence for trade-offs between different components of defense.

Zera and Harshman in Chapter 24 focus on metabolic aspects of trade-offs, by reviewing recent studies in *Drosophia* and *Gryllus* crickets that demonstrate how changes in intermediary metabolism are correlated with, and likely contribute to, evolutionary changes in individual life history traits and trade-offs. Importantly, the authors discuss how the evolution of life histories can evolve through coordinated regulatory changes in many enzymes, leading to altered flux of single pathways and integrated flux changes in multiple pathways.

Chapter 25 by Lancaster and Sinervo reviews and conceptualizes the contribution of social interactions to life history trade-offs. While traditional life history theory ignores trade-offs due to social interactions, social systems can significantly expand the set of possible trade-offs by introducing asymmetric interactions between sexes, age classes, and the invasion of alternative strategies. In their chapter the authors outline the mechanisms underlying gene epistasis caused by signaler–receiver dynamics and gene interactions among individuals and how such mechanisms impact life history trade-offs, with a particular emphasis on the endocrine regulation of trade-offs.

In the last chapter of this section, Chapter 26, Hau and Wingfield review the role of hormones as internal signals that mediate life history trade-offs. By focusing on sex steroids in vertebrates such as testosterone, the authors outline possible endocrine pathways by which such hormones could mediate life history trade-offs and discuss evolutionary scenarios that may account for interspecific variation in hormonally-regulated life history trade-offs.

In summary, some key insights about trade-offs discussed in these chapters (see also Chapters 27 and 28) are that:

• life history trade-offs are not necessarily caused by competitive resource allocation and can be mediated by signaling pathways independent of energy investment (see also Chapters 11 and 13)
• hormonal pathways, including the insulin/IGF-1 signaling (IIS) pathway, are major physio-

logical mediators and modulators of life history trade-offs

• trade-offs traditionally thought to be obligatory, for example between reproduction and survival, can be decoupled

• trade-offs that are caused by a module performing two competing functions can be broken by duplication and specialization of modules

• social interactions and dynamics are a very important but neglected factor in shaping life history trade-offs.

Together, these chapters demonstrate that examining the mechanistic basis of trade-offs is of major importance for our understanding of the evolution of life histories.

CHAPTER 21

Molecular mechanisms of life history trade-offs and the evolution of multicellular complexity in volvocalean green algae

Aurora M. Nedelcu and Richard E. Michod

21.1 Introduction

Although life history trade-offs are recognized as central to life history evolution, the mechanisms underlying trade-offs are not well understood (e.g., Roff 2007b, Monaghan *et al.* 2009). Both artificial selection and experimental evolution have demonstrated the presence of genetically-based trade-offs, but relatively few studies have been able to pinpoint the underlying molecular mechanisms (see Roff 2007b and Chapters 1 and 2 for discussion). Often, the functional basis of a trade-off is understood in terms of competition for limited resources among competing traits—such as reproduction, somatic growth, and maintenance—within an organism (also known as adaptive resource allocation) (Zera and Harshman 2001, Chapter 24). Searches for genes that affect, in opposite ways, two life history traits (i.e., "trade-off" genes) suggest that life history trade-offs could be the result of adaptive resource allocation at the organismal level (Bochdanovits and de Jong 2004, St-Cyr *et al.* 2008), and that this differential allocation might be caused by a trade-off between protein biosynthesis (growth) and energy metabolism (survival)—likely mediated by signal transduction pathways at the level of cellular metabolism (Bochdanovits and de Jong 2004). However, several studies propose that trade-offs could, in fact, be the result of signaling genes or pathways that simultaneously regulate two life history traits in opposite directions, independent of

resource allocation (Leroi 2001, Chapter 11). Furthermore, trade-offs can also be produced as a consequence of the performance of one activity generating negative consequences for other traits; for instance, aerobic metabolism generates reactive oxygen species that, if not fully neutralized, can be damaging to biological molecules and negatively affect other activities (Monaghan *et al.* 2009; also see Chapter 15).

Because the loci involved in life history trade-offs are expected to show antagonistic pleiotropy (Stearns 1992), life history trade-offs are generally thought to limit the set of possible trait combinations, and thus restrict the range of possible evolutionary trajectories and end-points (Roff 2007b). Life history trade-offs gain unique significance during evolutionary transitions in individuality (Box 21-1), such as the transitions from unicellular to multicellular individuals and solitary individuals to eusocial societies. Specifically, it has been suggested that trade-offs constraining the evolutionary trajectories of solitary individuals can be uncoupled in the context of a group through the evolution of specialized cells in multicellular individuals (Michod 2006, Michod *et al.* 2006) and castes in eusocial insects (Roux *et al.* 2009). In this way, traits (and fitness components) negatively correlated in previously solitary individuals can be optimized independently and simultaneously in a group, and new levels of fitness at the group level can emerge. The ability to break life history trade-offs through

---

Box 21-1 Glossary

Chlamydomonas

A polyphyletic genus that includes many unicellular species of bi-flagellated green algae. The most well-known species is *Chlamydomonas reinhardtii* (Fig. 21-1A), which is also a close relative of multicellular volvocine algae.

Eudorina

Spherical colonial forms comprising 16–32 undifferentiated cells separated by a considerable amount of extracellular matrix (Fig. 21-1C).

Evolutionary transitions in individuality

An evolutionary process whereby a group of previously independent individuals become stably integrated into a new functional, physiological, and reproductively autonomous and indivisible evolutionary unit; that is, a new individual. Such transitions took place during the evolution of the eukaryotic cell, the evolution of multicellularity, and the evolution of eusociality.

Flagellation constraint

Stems from the fact that in flagellated cells a single structure—known as the microtubule organizing center (MTOC)—has to perform two distinct functions: during the growth phase, MTOCs act as flagellar basal bodies (and organize the flagellar microtubules), whereas during cell division MTOCs act as centrioles (and direct the formation of the mitotic spindle). This constraint was proposed to have been at the origin of differentiated multicellularity in early metazoans (Margulis 1981). Notably, the flagellation constraint in volvocalean algae (Koufopanou 1994) has a slightly different structural basis than the one invoked in the origin of metazoans. Specifically, in contrast to other protists, in most green flagellates MTOCs can move laterally while attached to the flagella, and can act simultaneously as basal bodies and centrioles during cell division. However, in volvocalean algae, due to a coherent rigid cell wall the position of flagella is fixed and, thus, the basal bodies cannot move laterally and take the position expected for centrioles during cell division while remaining attached to the flagella.

Gonium

Colonial forms comprising 8–16 undifferentiated cells organized as flat or slightly curved sheets in a single layer (Fig. 21-1B).

Palintomy

The process during which a giant parental cell undergoes a rapid sequence of repeated divisions, without intervening growth, to produce numerous small cells.

Pleodorina

Spherical colonial forms consisting of 64–128 cells, including up to 50% differentiated somatic cells at the anterior pole (Fig. 21-1D).

Volvocalean algae

A group of green algae (order Volvocales) in the class Chlorophyceae, comprising both unicellular species (such as those in the genus *Chlamydomonas*) and multicellular species with various levels of morphological and developmental complexity (spanning many genera, including *Gonium*, *Pandorina*, *Eudorina*, *Pleodorina*, *Volvox*; Fig. 21-1). Most genera are poly- or paraphyletic, as many characters used to define specific genera have evolved independently several times (Herron and Michod 2008).

Volvox

A polyphyletic genus that comprises large spherical colonies with a high ratio of somatic to reproductive cells. *Volvox* species differ in many developmental traits, including the timing of germ-line segregation (early versus late in development) and the presence/absence of cytoplasmatic bridges in the adult. The most studied species is *Volvox carteri* (Fig. 21-1E), which is a species with complete and early separation between somatic and reproductive cells, and loss of cytoplasmatic bridges by the end of the embryonic development (Kirk 1998).

the evolution of specialization and division of labor is likely to have contributed to the evolutionary success of multicellularity and sociality.

During transitions in individuality, fitness becomes reorganized. For instance, in unicellular individuals, the same cell contributes to both fitness components, typically these contributions being separated in time. In multicellular groups, however, cells can specialize in either component, and this leads to the differentiation of reproductive cells (germ) and survival-enhancing cells (soma). Thus during the evolution of multicellularity, new levels of fitness can emerge at the higher level of organization (i.e., the multicellular individual) following the reorganization of survival- and reproduction-related functions between somatic and reproductive cells (Michod and Nedelcu 2003, Michod 2006, Michod *et al.* 2006). From a mechanistic point of view, we have argued that the evolution of germ–soma separation in multicellular individuals involved the co-option of molecular mechanisms underlying life history trade-offs in unicellular lineages, by changing their expression from a temporal into a spatial context (Nedelcu and Michod 2006, Nedelcu 2009).

Here, we discuss further this proposal using the volvocalean green algal group as a model system. First, we introduce the volvocalean algae and discuss the aspects of their life history that are relevant to the evolution of multicellularity in this lineage. Then, we briefly review what is known about the mechanistic basis of acclimation—a specific adaptive response to environmental changes that magnifies the survival–reproduction trade-off in these algae. We suggest that the molecular basis for this trade-off is independent of resource allocation. Instead, this survival–reproduction trade-off is mediated through photosynthesis, whose down-regulation has a pleiotropic effect on the two fitness components; that is, it promotes survival (through avoiding the production of damaging reactive oxygen species (ROS) at a cost to immediate growth and reproduction. Lastly, we propose a hypothesis for the evolution of somatic cells in which, by simulating the general acclimation signal (i.e., a change in the redox status of the cell) in a spatial rather than temporal context, a life history trade-off gene can be co-opted into a "specialization" gene. In particular,

we suggest that the gene responsible for the differentiation of sterile somatic cells in the multicellular alga, *Volvox carteri*, evolved from a gene involved in the down-regulation of photosynthesis during acclimation to stressful environmental conditions, as a means to ensuring survival at a cost to immediate reproduction.

21.2 The volvocalean green algal group

21.2.1 Overview

The volvocalean group comprises photosynthetic bi-flagellated green algae with discrete generations and a single reproductive episode that marks the end of the generation. The so-called "volvocine lineage" contains closely related unicellular (*Chlamydomonas*-like) and multicellular species that show an increase in cell number, volume of extracellular matrix per cell, division of labor, and ratio between somatic and reproductive cells (Larson *et al.* 1992; Fig. 21-1). Cell specialization evolved multiple times in this group, and the different levels of complexity among volvocalean species are thought to represent alternative stable states (among which evolutionary transitions have occurred several times during the evolutionary history of the group), rather than a monophyletic progression in organizational and developmental complexity (Larson *et al.* 1992, Herron and Michod 2008).

The observed morphological and developmental diversity among volvocine algae appears to result from the interaction of conflicting structural and functional constraints and strong selective pressures. All volvocalean algae share the so-called "flagellation constraint" (Koufopanou 1994; Box 21-1); that is, due to a rigid cell wall, the flagellar basal bodies cannot move laterally and take the position expected for centrioles during cell division while remaining attached to the flagella. Therefore, cell division and motility can take place simultaneously only for as long as flagella can beat without having the basal bodies attached (i.e., up to five cell divisions).

The presence of a cell wall is coupled with the second conserved feature among volvocalean algae, which is their unique way of reproduction—namely, palintomy (Box 21-1). Volvocalean cells do not

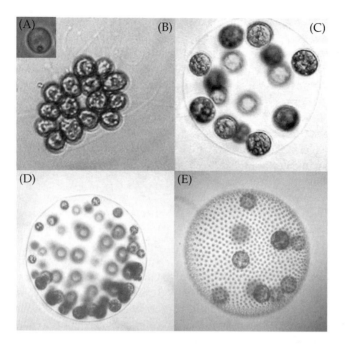

Figure 21-1   Unicellular (A, *Chlamydomonas reinhardtii*) and multicellular volvocalean algae without (B, *Gonium pectorale* and C, *Eudorina elegans*) and with (D, *Pleodorina californica* and E, *Volvox carteri*) cell differentiation (smaller cells are somatic bi-flagellated cells; larger cells are nonmotile reproductive cells).

double in size and then undergo binary fission. Rather, each cell grows about 2^n-fold in volume, then a rapid and synchronous series of n divisions (multiple fission) without intervening growth is initiated (under the mother cell wall) to produce 2^n small cells. Because clusters, rather than individual cells, are produced in this way, this type of reproduction is believed to have been an important precondition facilitating the evolution of multicellularity in this group (Kirk 1998). Indeed, while in the unicellular members of the group—such as *Chlamydomonas* (Fig. 21-1A; Box 21-1)—the daughter cells (2–2^4) separate from each other after division; in the multicellular species, the cluster of 2^n cells does not disintegrate, and coenobial forms (a type of multicellular organization in which the number of cells in the adult is determined by the number of cleavage divisions during embryogenesis; i.e., there is no increase in number of cells in the adult; Kirk 1998) are produced. For instance, in *Gonium*, the resulting cells (2^2–2^5) stay together and form a convex discoidal colony (Fig. 21-1B; Box 21-1), whereas in *Eudorina*,

Pleodorina and *Volvox* (Fig. 21-1C, D and E; Box 21-1), the cells (2^4–2^6, 2^6–2^7, and 2^9–2^{16}, respectively) form spherical colonies up to 3 mm in size.

The two selective pressures that are thought to have contributed to the increase in complexity in volvocine algae are the advantages of a large size (potentially to escape predators, achieve faster motility, homeostasis, or better exploit eutrophic conditions) and the need for flagellar activity (to access to the euphotic/photosynthetic zone and nutrients, and to achieve better mixing of the surrounding environment) (Bell *et al.* 1985, Sommer and Gliwicz 1986, Solari *et al.* 2006a,b). Interestingly, given the background imposed by the volvocalean type of organization presented above, namely the flagellar constraint and the palintomic mode of reproduction, it is difficult to achieve the two selective advantages—larger size and motility—simultaneously. Indeed, the larger the colonies (and the number of cells), the larger the mother cell and the number of cell divisions (up to 15–16 in some *Volvox* species). Consequently, the motility (and thus survival) of a colony whose reproduction requires more than five cell divisions is negatively impacted during the growth and reproductive phase. In other words, the larger the size of the colonies, the more acute the trade-off between reproduction and viability.

The negative impact of the flagellation constraint can be overcome by cellular specialization/division of labor: some cells become involved mostly in motility, while the rest of the cells specialize for reproduction. The proportion of cells that remain motile throughout most or all of the life cycle is directly correlated with the number of cells in a colony: from none in *Gonium* and *Eudorina*, to up to one-half in *Pleodorina* and more than 99% in *Volvox* (Larson *et al.* 1992) (Fig. 21-1). In *Volvox carteri*, the division of labor is complete (Fig. 21-1 E): the reproductive cells (gonidia)—which give rise to new daughter colonies—are set apart early during development; the somatic cells are terminally differentiated and undergo cellular senescence and death once the progeny is released from the parental colony (Kirk 1998). Overall, the present diversity in morphological and developmental complexity in the volvocalean algae reflects distinct strategies and solutions to the same set of constraints, selective pressures, and life history trade-offs (discussed further below).

21.2.2   Life history trade-offs and the evolution of multicellularity in volvocalean algae

Volvocalean algae are constrained by life history trade-offs that are similar to those in other lineages (e.g., the trade-off between reproduction and survival mediated through body size and developmental time). These trade-offs are also rather dynamic; they can change during development, in response to the environment, and can evolve. Notably, in this group, some of the common life history trade-offs, especially in relation to survival and reproduction, are amplified by developmental traits (i.e., palintomy), constraints (i.e., the flagellar constraint), and selective pressures specific to volvocalean algae (i.e., the need to maintain flagellar activity throughout their entire life-cycle).

As in other lineages in which fecundity increases with body size (e.g., most invertebrates, plants, and some small mammals), in unicellular volvocalean algae, fecundity is directly correlated to (and can be predicted from) body size at the end of the growth phase; that is, the number of offspring is dependent on the maternal body size, specifically on how many times the mother cell increases in volume. However, as attaining a large body size requires a longer time to maturity, survival rates may decrease as the time to reproduction increases. This trade-off between reproduction and survival mediated through body size and developmental time may therefore limit the increase in fecundity (Fig. 21-2A). In environments in which predation imposes a strong selective pressure, a large body size can also be beneficial in terms of survival (Morgan *et al.* 1980). In such cases, the survival benefits of a large body size can offset the survival cost incurred due to longer developmental times. Nevertheless, the decrease in the cell surface-to-volume ratio (which affects metabolic exchanges) puts a limit on the maximum size a unicellular organism can reach. Further survival benefits associated with escaping predators can only be achieved if the increase in body size is attained by forming multicellular groups—such as through the failure of the offspring (daughter cells) to separate at the end of cleavage (Fig. 21-2B). Indeed, it has been shown experimentally that predation by a phagotroph selects for eight-celled colonies in the green alga, *Chlorella vulgaris* (Boraas *et al.* 1998).

Such a scenario is postulated to have happened during the evolution of multicellularity in this group (Kirk 1998), and is reflected in the presence of colonial forms developed from a single mother cell—such as in the genera *Gonium* and *Eudorina*, comprised of up to 32 cells (Fig. 21-1; Box 21-1). However, an additional increase in size (and a potential increase in survival through avoiding a different class of predators) will be constrained by new trade-offs set up by the cost of maintaining, moving, and reproducing a large colony. For instance, as the flagellation constraint puts a limit on the number of cell divisions (i.e., five) that can take place without negatively affecting the survival of the colony during the reproductive phase, an increase in the number of cell divisions above five

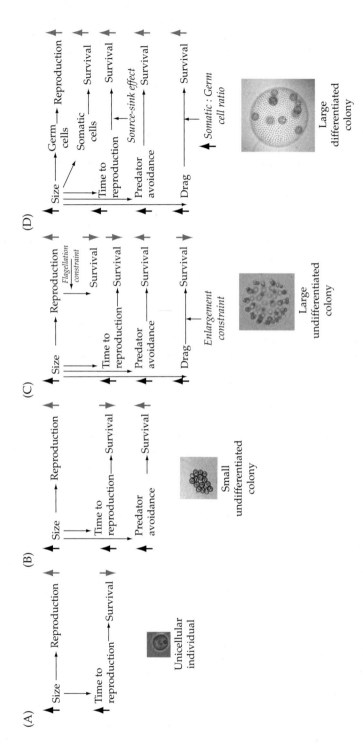

Figure 21-2 Life history trade-offs during the evolution of multicellularity in the volvocine lineage (see text for discussion). Solid black upward arrows denote an increase in the value of the life history trait or cost/benefit indicated on the right; gray upward and downward arrows, respectively, indicate positive and negative effects on the two components of fitness. A, Trade-offs in unicellular individuals. Basic survival–reproduction trade-off mediated through body size and time to reproduction in unicellular volvocalean algae such as *Chlamydomonas*. B, Trade-offs in small undifferentiated multicellular colonies. The survival benefits of size in terms of predator avoidance can offset the survival cost of reproduction in terms of increased time to reproduction in small colonial volvocine (8–16 cells) algae such as *Gonium*. C, Trade-offs in larger undifferentiated colonies. Increased colony size in terms of number of cells increases the survival costs of reproduction via the flagellation and enlargement constraints in volvocine colonial species with more than 32 cells, such as *Pleodorina*. D, Release from trade-offs in species with differentiated somatic and germ cells. In large multicellular *Volvox* species, the survival costs of reproduction imposed by the flagellation and enlargement constraints and by the increased time to reproduction are alleviated through the specialization of somatic and germ cells, an increased somatic-to-germ cell ratio, and the source-sink effect.

(and an increase in size of the offspring above 32 cells) will result in the mother colony losing motility during the reproductive phase. Furthermore, due to their specific way of reproduction (autocolony, i.e., each cell in a colony will produce a daughter colony), each cell in the colony needs to grow 2^n-fold in volume in order to produce a daughter colony of 2^n cells. This massive increase in volume of all cells in the colony will impose a high cost on motility (and survival) as the mass (and the drag) of the colony will increase. This results in the motility of the colony being negatively affected even before the division phase (the enlargement constraint; Solari et al. 2006a,b). Thus, the survival cost of moving a large body size and the survival cost during the reproduction phase (imposed by the flagellation constraint) will likely offset any survival benefits of large size (such as escaping predators) (Fig. 21-2C). The trade-off between reproduction and viability becomes more and more acute as body size increases and at some point it becomes untenable to have further increases in body size, unless the trade-off can be broken and the two components of fitness (reproduction and survival) maximized independently.

The evolution of specialized sterile somatic cells, in colonies comprising more than 32 cells, released multicellular volvocalean algae from the constraints imposed by the survival–reproduction trade-off specific to this group (Fig. 21-2D). By not growing and by not reproducing, the somatic cells alleviate the survival costs of moving and reproducing a large colony, thus allowing the benefits of a large body size to be realized further. The benefits of soma specialization are multiple and include:

- colony motility while reproducing (overcoming the flagellation constraint)
- motility while large (overcoming the enlargement constraint)
- increased resource uptake due to the "source-sink" effect (in which somatic cells serve as the source and the germ cells act as the sink) (Bell et al. 1985, Koufopanou and Bell 1993)
- enhanced uptake of resources and removal of waste by flagellar beating (Solari et al. 2006a,b).

Germ specialization can also provide additional benefits, such as decreased generation time, increased productivity by specialization at photosynthesis, and hydrodynamic advantages stemming from the location of germ in the interior of the colony (Solari et al. 2006b). The specialization of germ and soma also imposes some direct costs, as germ specialization reduces the number of cells available for vegetative functions (survival cost), and soma specialization reduces the number of reproducing cells (fecundity cost). However, these costs are overcome by the benefits of germ–soma separation, and are mediated by the ratio between the number of germ and somatic cells in a colony. The observed increase in the soma-to-germ ratio among species with increasing numbers of cells is thought to be a reflection of the trade-off between survival and fecundity imposed by the cost of reproduction as body size (in terms of number of cells) increases.

21.3 Mechanisms of life history trade-offs and the evolution of multicellularity in volvocalean algae

21.3.1 Overview

Trade-offs can occur between physiological traits (whether expressed at the same or at different times in the life cycle), and result from variations in genetic factors (pleiotropy), environmental factors, or combinations of these two types of factors (Zera and Harshman 2001, Chapter 24). As in other systems, little is known about the causal mechanisms underlying life history trade-offs (and how trade-offs are manifested or modulated at the genetic level) in volvocalean algae. A special case is the trade-off between survival and reproduction during the reproductive phase, which is mediated through the flagellation constraint. In this case, the functional basis of the trade-off between motility and cell division is understood in terms of competition for the same structure, as the microtubule-organizing centers act both as flagellar basal bodies and centrioles (Box 21-1). In other cases, trade-offs between survival and reproduction can be thought of in terms of adaptive allocation of limited internal resources. This might be especially true in cases where somatic and reproductive cells are connected through cytoplasmatic bridges (such as in some species of *Volvox*, e.g., *Volvox aureus*) and resources are believed to be transferred from somatic cells into gonidia (Kirk 1998). Differential allocation of limited internal resources to survival and reproduction can also be invoked during embryonic development in volvocalean species in which cell size at the end of embryonic cleavage is indicative of cell fate (e.g., *V. carteri*); that is, small cells are destined to become somatic cells while larger cells develop into gonidia (Kirk 1995). This strategy can be interpreted as a type of maternal control whereby internal resources are differentially allocated (during embryonic cell division) among cells that will specialize for either survival- or reproduction-related functions.

Reduced nutrient availability is known to substantially magnify, while increased nutrient availability can diminish (or obviate), an apparent trade-off; such plastic responses are thought to be determined by priority rules that govern the relative allocation of resources to organismal processes as a function of nutrient input (Zera and Harshman 2001). In some lineages, laboratory and field experiments showed that, under nutrient-poor or stressful conditions, allocation to maintenance or storage take precedence over allocation to reproduction (see Zera and Harshman 2001). For instance, in yeast, high nutrient conditions favor high levels of cellular cAMP and cell division, while nutrient-deprived environments trigger a drop in the concentration of cAMP and cells cease to divide (Eraso and Gancedo 1985). Yeast strains that cannot properly regulate cAMP level continue to divide during nutrient deprivation, and ultimately die due to starvation (Wilson and Tatchell 1988).

21.3.2 Acclimation and life history trade-offs in *Chlamydomonas*

In volvocalean algae—as in other photosynthetic organisms—nutrient-poor or stressful environments trigger a series of metabolic alterations collectively known as acclimation, which favor survival when the potential for cell growth and division is reduced (Grossman 2000). One of the consequences of this complex series of responses is a temporary inhibition of cell division (and thus reproduction), to ensure long-term survival. However, the mecha-

nistic basis underlying this trade-off is very different from that described above in yeast.

Acclimation involves both specific responses (e.g., scavenging for a specific nutrient) and general responses. The general responses include: a decline in the rate of photosynthetic activities, the accumulation of starch (diverting energy and fixed carbon from cell growth), a general metabolic slowdown, and cessation of cell division (Grossman 2000, Wykoff *et al.* 1998). Photosynthetic organisms use light energy to generate chemical energy (ATP) and reductants (NADPH) that are subsequently used to fix carbon dioxide (which will regenerate ADP and NADP+). This coupling renders photosynthesis and its efficiency highly dependent on environmental conditions; changes in various abiotic factors, including light, temperature, water, and nutrient availability, have an immediate impact on photosynthetic activities and subsequently on other metabolic processes (Pfannschmidt *et al.* 2009).

The down-regulation of photosynthesis is critical for sustaining cell viability under conditions of nutrient deprivation (Davies *et al.* 1996, Wykoff *et al.* 1998). The lack of nutrients in the environment blocks cell growth and limits the consumption of NADPH and ATP generated via photosynthesis. Consequently, the photosynthetic electron transport becomes reduced and the redox potential of the cell increases (Wykoff *et al.* 1998, Grossman 2000). Furthermore, because NADPH is not rapidly recycled (due to the slow-down of anabolic processes and the decreased demand for reductant in nutrient-poor environments), excited chlorophyll molecules and high potential electrons will accumulate and could interact with oxygen to create ROS. ROS are a series of partially reduced and highly reactive forms of oxygen, including the superoxide anion (O_2^-), the hydroxyl radical (OH·), and hydrogen peroxide (H_2O_2). Although ROS are by-products of normal metabolism and can act as secondary messengers in various signal transduction pathways (e.g., see Van Breusegem *et al.* 2001, Mittler 2002, Eberhard *et al.* 2008; also see Chapter 15), increased intracellular levels of ROS (i.e., oxidative stress) can alter cellular functions and damage many biological structures, most importantly DNA (e.g., Marnett and Plastaras 2001).

Consequently, the regulation of the photosynthetic electron transport is an important hallmark of the general response to nutrient deprivation in *Chlamydomonas*. A series of processes, including reduced photosynthetic electron transport and the redirection of energy absorbed from photosystem II to photosystem I, can decrease NADPH production, favor ATP production through cyclic electron transport, and allow a more effective dissipation of the excess absorbed excitation energy. Altogether these changes decrease the potential toxic effect of excess light energy (and thus serve to increase survival) and help coordinate cellular metabolism and cell division with the growth potential of the cell (Grossman 2000, Chang *et al.* 2005).

Several mutants that affect general acclimation responses have been isolated in *Chlamydomonas reinhardtii* (Davies *et al.* 1996, Chang *et al.* 2005, Moseley *et al.* 2006). These mutants are unable to down-regulate photosynthetic electron transport and die sooner than the wild-type. Notably, their death appears to be the consequence of the accumulation of photodamage. Consistent with this suggestion is the fact that an electron-transport inhibitor that can induce the down-regulation of photosynthesis can rescue these mutants (Davies *et al.* 1996, Moseley *et al.* 2006). Moreover, if maintained in the dark, these mutants can survive nutrient deprivation as well as the wild-type strains do (Davies *et al.* 1996).

Overall, the available data suggest that, in *Chlamydomonas*, survival under nutrient limitation is primarily determined by the ability to down-regulate the photosynthetic electron transport to avoid light-induced oxidative damage. How is this accomplished? Changes in the status of the photosynthetic apparatus during acclimation to nutrient limitation result in changes in the redox potential of the chloroplast. Photosynthetic redox signals (including photosynthetically generated ROS) are then transduced in the chloroplast or cytosol and affect the expression of nuclear genes coding for various proteins involved in acclimation, including chloroplast light-harvesting proteins (Moseley *et al.* 2006, Pfannschmidt *et al.* 2009). Since, in *Chlamydomonas*, cell division is dependent on cell size (Umen and Goodenough 2001), and cell growth is dependent on photosynthesis, the down-regulation of photosynthesis in order to ensure survival will also suppress cell growth and thus reproduction. In other words, the down-regulation of photosynthesis has a pleiotropic effect on two life history traits: it promotes survival at a cost to immediate reproduction. Therefore, the observed trade-off between survival and reproduction under nutrient limitation does not appear to be a direct consequence of resource allocation (although, as a side-effect, the inhibition of growth and reproduction might release resources that could, in theory, be re-allocated to survival-related functions).

Recently, we have identified in *C. reinhardtii* a gene—currently known as *rls1* (Duncan *et al.* 2007), which is induced under specific nutrient limitation (including phosphorus and sulfur-deprivation), during the stationary phase, as well as under light-deprivation (Nedelcu and Michod 2006, Nedelcu 2009). Furthermore, we showed that the induction of *rls1* coincides with the down-regulation of a nuclear-encoded light-harvesting protein (Nedelcu and Michod 2006) and with the decline in the reproduction potential of the population under limiting conditions (Nedelcu 2009). The fact that *rls1* is expressed under multiple environmental stresses, and its induction corresponds with a decline in reproduction, suggests that *rls1* is part of the general acclimation response and might function as a regulator of acclimation in *C. reinhardtii*. Supporting this suggestion is the finding that an inhibitor of the photosynthetic electron flow that triggers general acclimation-like responses (Wykoff *et al.* 1998) also induces the expression of *rls1* (Nedelcu 2009).

Thus, altogether, *rls1* appears to be specifically induced under environmental conditions where the temporary down-regulation of photosynthesis is beneficial in terms of survival though costly in terms of immediate reproduction. In other words, *rls1* might act as a life history trade-off gene (i.e., a gene that affects, in opposite ways, two life history traits, Bochdanovits and de Jong 2004).

21.3.3 The genetic basis for cell differentiation in *Volvox carteri*

V. carteri consists of approximately 2000 permanently biflagellated somatic cells and up to 16 nonflagellated reproductive cells. Terminal differentiation of somatic cells in *V. carteri* involves the expression of *regA*, a master regulatory gene that encodes a transcriptional repressor (Kirk *et al.* 1999), thought to suppress nuclear genes coding for chloroplast proteins (Meissner *et al.* 1999). Consequently, the cell growth (dependent on photosynthesis) and division (dependent on cell growth) of somatic cells are suppressed. Interestingly, the closest homolog of *regA* in *C. reinhardtii* is *rls1* (Nedelcu and Michod 2006, Duncan *et al.* 2007). *RegA*, like *rls1*, contains a SAND domain, which is found in a number of nuclear proteins, many of which function in chromatin-dependent or DNA-specific transcriptional control. Proteins containing a SAND domain have been reported in both animal and land plants; one such protein, ULTRAPETALA1, acts as a key negative regulator of cell accumulation in *Arabidopsis* shoot and floral meristems (Carles *et al.* 2005).

Mutations in *regA* result in the somatic cells *regA*ining reproductive abilities, which in turn results in them losing their flagellar capabilities (e.g., Starr 1970, Kirk *et al.* 1987). As flagellar activities are very important for these algae, the survival and reproduction of *V. carteri* individuals in which such mutant somatic cells occur is negatively affected (Solari *et al.* 2006b). Although *regA* belongs to a gene family that comprises 14 members in *V. carteri* (Duncan *et al.* 2007), *regA* is currently known as the only locus that can mutate to yield Reg mutants (Kirk *et al.* 1999).

The expression of *regA* is strictly determined by the size of cells at the end of embryogenesis; cells below a threshold size develop into somatic cells (Kirk *et al.* 1993). Which cells express *regA* and differentiate into somatic cells is determined during development through a series of symmetric and asymmetric cell divisions. The asymmetric divisions ensure that some cells (i.e., the germ line precursors) remain above the threshold cell size associated with the expression of *regA* (Kirk 1995). *RegA* is induced in very young somatic cells immediately after the end of embryogenesis, but is never expressed in gonidia (Kirk *et al.* 1999). The mechanism underlying the differential expression of *regA* (i.e., ON in the somatic cells and OFF in the gonidia) is not known; it has been postulated that specific transcription factors bind to the cis-regulatory elements identified in three of the introns (i.e., two enhancers and one silencer) and act in concert to either silence or induce *regA* expression (Stark *et al.* 2001).

21.4 Co-opting mechanisms underlying environmentally induced life history trade-offs for cell differentiation

The evolution of specialized cells in multicellular volvocalean algae can be understood in terms of the need to break survival–reproduction trade-offs, such that survival and reproduction can be maximized independently and simultaneously. We have previously suggested that the evolution of soma in multicellular lineages involved the co-option of life history trade-off genes whose expression in their unicellular ancestors was conditioned on environmental cues (as an adaptive strategy to enhance survival at an immediate cost to reproduction). This, we suggested, happened through shifting their expression from a temporal (environmentally induced) into a spatial (developmental) context (Nedelcu and Michod 2004, 2006; Fig. 21-3). Furthermore, we have reported that the closest homolog of *V. carteri regA* in its unicellular relative, *C. reinhardtii*, is *rls1*—a life history gene that is involved in the general acclimation response to various environmental stresses (Nedelcu and Michod 2006, Nedelcu 2009).

How can general acclimation responses in unicellular organisms be co-opted for cell differentiation in multicellular groups? As we discussed above, in photosynthetic organisms, the flux of electrons through the electron-transport system (ETS) has to be balanced with the rate of ATP and NADPH consumption; imbalances between these processes can result in the generation of toxic ROS (e.g., Wykoff *et al.* 1998). When a nutrient (e.g., sulfur, phosphorus) becomes limiting in the environment, ATP and NADPH consumption declines. This results in an excess of excitation energy and a subsequent change in the redox state of the photosynthetic apparatus, which will trigger a suite of short- and long-term acclimation responses (e.g., Wykoff *et al.* 1998, Pfannschmidt *et al.* 2009; Fig. 21-4). Other environmental factors (e.g., cold, water stress) are also known to result in changes

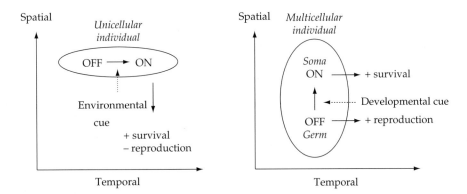

Figure 21-3 Schematic representation of the change in expression pattern from a temporal context (environmentally-induced) into a spatial context (developmentally-induced) of a life-history trade-off gene in a unicellular individual as it becomes a "specialization" gene in a multicellular individual (adapted from Nedelcu and Michod 2006).

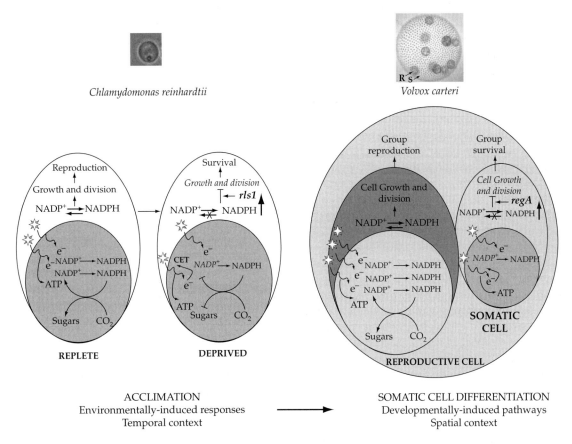

Figure 21-4 A model for the co-option of acclimation responses into somatic cell differentiation in *Volvox carteri* (R and S denote reproductive and somatic cells, respectively; see text for discussion). Although many components are involved, for simplicity, changes in redox status are symbolized by the over-reduction of the NADP+ pool due to either decreased NADPH consumption—in nutrient-deprived *Chlamydomonas*, or excess of excitation energy (owing to a higher surface/volume ratio)—in *Volvox* somatic cells. The switch to cyclic electron transport (CET), which can maintain ATP synthesis (and thus vital processes) in acclimated *Chlamydomonas* cells (Eberhard *et al.* 2008) and possibly in *Volvox* somatic cells, is also indicated (adapted from Nedelcu 2009).

in the cellular redox status and trigger similar acclimation responses (e.g., Eberhard *et al.* 2008). Thus, in principle, any factor that can elicit a similar redox change could prompt acclimation-like responses, and ultimately induce cessation of cell division. In a group context, if such a change is restricted to a subset of cells, and if the suppression of reproduction in this subset of cells is beneficial to the group, sterile somatic cells can evolve.

In *V. carteri*, the expression of *regA* is restricted (by an unknown mechanism) to cells whose size at the end of embryonic divisions falls below 8 μm (Kirk *et al.* 1993). As cell surface area and volume change at different rates, we proposed that in these small cells the ratio between membrane-bound proteins (including ETS and ETS-associated components) and soluble factors (including $NADP^+$ and ADP) changes relative to the ratio in larger cells. Specifically, there will be an excess of membrane-bound proteins (Nedelcu 2009; Fig. 21-4). Consequently, these small cells could experience an imbalance between the flux of electrons and the availability of final acceptors, which would result in a change in the intra-cellular redox status and the induction of acclimation-like responses, culminating in the suppression of division (Fig. 21-4). Supporting this scenario is the fact that cytodifferentiation in *V. carteri* is light-dependent (Stark and Schmitt 2002).

Hence, by simulating the general acclimation signal (i.e., a change in the redox status of the cell) in a spatial rather than temporal context, an environmentally induced trade-off gene can be differentially expressed between cell types, allowing for the two components of fitness to be maximized independently and simultaneously. This hypothesis also predicts that somatic cell differentiation is more likely to evolve in lineages with enhanced acclimation mechanisms, or, more generally, in lineages that can trade-off reproduction for survival in stressful environments. Because environments that vary in time, such as those in which volvocalean algae live (Kirk 1998), will select for enhanced and efficient acclimation responses (note that temporally varying environments have been shown to select for phenotypic plasticity, i.e., generalists in *C. reinhardtii*; Reboud and Bell 1997), such environments are likely to be more conducive to the evolution of somatic cell dif-

ferentiation. In this context, it is noteworthy that cast differentiation in social wasps is also thought to have evolved in variable environments, and specific adaptations to seasonal environments that control sequential shifts between life-cycle phases (such as diapause; Chapter 18) have been proposed as prerequisites to the evolution of sociality in this group (Hunt and Amdam 2005).

21.5   Conclusion

Volvocalean algae are an excellent model system with which to study life history trade-offs from both evolutionary and mechanistic perspectives. They exhibit a diverse range of morphological, developmental, and life history traits, inhabit different habitats (soil, temporal water bodies, permanent lakes), and are subjected to various ecological pressures. In addition, the genomes of *C. reinhardtii* (Merchant *et al.* 2007) and *V. carteri* (Prochnik *et al.* 2010) have been recently sequenced, which will allow the integration of life history theory with genomics, facilitate the development of mechanistic models of life history trade-offs, and address the extent to which evolutionary trajectories are deterministic versus stochastic (Chapter 2). Perhaps the unique feature of this system is the possibility to address the evolution of life history traits and trade-offs during evolutionary transitions in individuality. Although we focused here on the transition from unicellular to multicellular life, we argue that similar principles are likely to apply to other evolutionary transitions as well.

21.6   Summary

1. Although life history trade-offs are generally thought to limit the possible set of trait combinations and can constrain the evolutionary trajectory of a lineage, during evolutionary transitions in individuality such trade-offs can be uncoupled by differentiation or specialization, and thus can promote evolution at the new level of organization.

2. Understanding the molecular mechanisms of life history trade-offs in solitary individuals is important for understanding evolutionary transitions in individuality and the evolution of new traits at the higher level (such as sterile somatic cells during the evolution of multicellularity and sterile castes

during the evolution of eusociality), in both proximate and ultimate terms.

3. Life history trade-offs can be caused by factors other than resource allocation. In volvocalean algae, the trade-off between survival and reproduction in nutrient-limited environments is mediated through photosynthesis, whose down-regulation has a pleiotropic effect on the two fitness components, that is, it promotes survival (by avoiding the overproduction of damaging reactive oxygen species) at a cost to immediate reproduction.

4. We propose that during the transition to multicellularity and the evolution of germ–soma separation, life history trade-off genes associated with ensuring long-term survival at a cost to immediate reproduction have been co-opted into "specializa-tion" genes by changing their expression patterns from a temporal into a spatial context. In particular, we suggest that the gene responsible for the differentiation of sterile somatic cells in the multicellular alga, *Volvox carteri*, evolved from a gene involved in the down-regulation of photosynthesis during acclimation to stressful environmental conditions.

21.7 Acknowledgments

We thank Deborah Shelton, Matt Herron, Armin Rashidi, and Pierre Durand for discussion and comments on an earlier draft of the manuscript, and the editors and two reviewers for their suggestions. This work was supported by an NSERC grant to AMN and an NSF grant to REM and AMN.

CHAPTER 22

Molecular basis of life history regulation in *C. elegans* and other organisms

Birgit Gerisch and Adam Antebi

22.1 Introduction

All animals evaluate their environment and adjust their metabolism, development, and reproduction according to circumstances. Nutrient quality, day length, population density, and temperature are major environmental cues that affect plastic life history traits (Finch 1990). Phenotypic plasticity, the ability of a given genotype to produce different phenotypes in response to various environmental conditions (Roff 1992, Stearns 1992) can lead to presumably adaptive, regulatory adjustments in all aspects of the life cycle—from development, growth, metabolism, and reproduction, to longevity—which are geared to maximize reproductive success. In some cases, cues can induce developmental alternatives such as diapause, a state of arrested development, and, relatedly, torpor and hibernation, all of which are specialized to allow animals to outlast adversity until conditions improve.

The diversity and evolution of life history strategies have been intensely studied through comparative and field studies, and have contributed greatly towards our understanding of nature (Finch 1990). Another successful approach has arisen from the genetic dissection of life history traits in invertebrate model organisms. Model organisms typically have simple, well-described development, anatomy, and life cycle. The powerful molecular and genetic methods available in model organisms can identify genes, pathways, and causal mechanisms that regulate life history traits. Because such animals themselves have been selected for laboratory cultivation,

however, it remains to be seen how far these findings apply to natural populations.

Studies in the nematode *Caenorhabditis elegans* have contributed enormously to our understanding of the molecular basis of life history regulation. In particular, the study of the dauer diapause has illuminated the relationship between environment and basic strategies geared towards reproduction (fast life history) or survival and longevity (slow life history). Studies in the worm have revealed a fundamental paradigm whereby environmental and physiological cues are sensed by the neurosensory apparatus and internal signaling, and are processed and integrated into hormonal signals. Hormones, working through their cognate receptors, coordinate metabolism, growth, reproduction, and homeostasis, and probably mediate trade-offs between reproduction and survival. Molecular analysis has led to the realization that evolutionarily conserved hormonal pathways, including serotonin, steroids, Transforming growth factor beta (TGF-beta), insulin, and other peptide hormones, govern this process (Fielenbach and Antebi 2008).

This review focuses on how *C. elegans* dauer formation has helped to elucidate the molecular basis of life history plasticity and longevity. It also touches upon other pathways governing developmental timing, stage structure, diet, reproduction, and longevity. At several points we draw parallels to other species where similarities are apparent. Our goal is to provide a resource for framing life history traits in molecular and mechanistic terms.

22.2   *C. elegans* life history

C. elegans is a free-living hermaphroditic species that dwells in soil and humus, but is easily cultivated in the laboratory on petri plates containing lawns of *E. coli* supplemented with salts and cholesterol. Like many species, *C. elegans* development and lifespan are highly dependent on environmental conditions.

In favorable environments supporting reproduction—moderate temperatures (e.g., 20°C), abundant food, and low population density—*C. elegans* undergoes continuous development from embryo, through four larval stages (L1–L4), to reproductive adults in 3.5 days. During the reproductive period, adults produce broods of 250–300 progeny over 4–5 days, and live another 2–3 weeks. If, however, *C. elegans* encounters harsh environmental conditions of high temperature (27°C), dwindling food, or overcrowding during L1 or early L2 stages, they will enter a developmentally arrested quiescent third larval stage, called the dauer diapause (Fielenbach and Antebi 2008). Dauer larvae suppress growth and cell division, and exhibit a number of adaptations geared towards extended survival. They are highly resistant to oxidative and heat stress, and can live for months without food. In preparation for diapause, they store fats and carbohydrates and spend down these reserves within the dauer stage itself, as they are non-feeding and sealed to the environment. During dauer, respiration is suppressed in favor of glycolysis. Dauer larvae also have a number of morphological changes, including remodeled neural architecture, slim body, and a specialized thickened cuticle to withstand desiccation. Yet when returned to favorable environments, worms will resume normal reproductive development and have a normal lifespan, revealing remarkable plasticity. Dauer diapause can be thought of as a developmental checkpoint, analogous to reproductive diapause in overwintering insects (see Chapters 13 and 18), or hibernation/torpor in mammals, as discussed in more detail below.

22.3   Genetics of dauer formation

Dauer formation mutants fall into two general classes: Daf-c (dauer formation constitutive) mutants always go into dauer, while Daf-d (dauer formation defective) mutants never go into the dauer stage, regardless of environmental conditions (Fielenbach and Antebi 2008). These loci have been placed into genetic pathways based on experiments of epistasis and synergy. Molecular cloning has identified components involved in pheromone production, sensory cell structure, and function, as well as endocrine pathways of serotonin, TGF-beta, insulin/IGF, and steroid hormone signaling, discussed below (Fig. 22-1A).

22.4   Dauer pheromone

Pheromones are small signaling molecules that mediate intraspecies communication, conveying information about population density, mating receptivity, location of food sources, or nesting grounds. A potent trigger of *C. elegans* dauer formation, dauer pheromone is an excreted indicator of population density (Golden and Riddle 1984). It is a mixture of at least four ascarosides (ascr1–4), consisting of a 3,6-dideoxyhexose ascarylose sugar backbone conjugated to fatty acids of varying length and saturation (Butcher *et al.* 2007). At high concentrations, these molecules work synergistically to promote dauer formation, while at lower concentrations they act as an attractive mating pheromone for males (Srinivasan *et al.* 2008). Evidently, alternate phenotypic plastic outcomes arise (developmental arrest, mating) in response to a common set of signal molecules depending on concentration.

Pheromone biosynthesis is carried out through peroxisomal beta-oxidation of long-chain fatty acids, namely by DAF-22, a sterol carrier protein homolog to the human SCPx thiolase, and DHS-28, a homolog of the D-bifunctional protein that works just upstream of SCPx (Butcher *et al.* 2009). This involvement of beta-oxidation products may indicate that animals can couple their nutritional and metabolic status to pheromone production.

Although by definition, pheromones are species specific, their study can also illuminate conserved signaling pathways and physiological responses. The identification of the dauer pheromone is particularly important because it potently induces a state of somatic endurance and extended survival through downstream conserved pathways. In

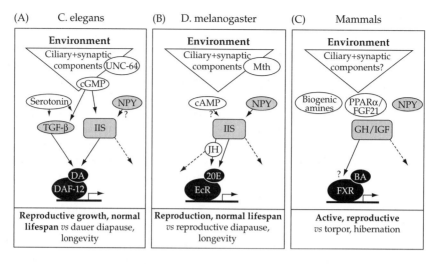

Figure 22-1   Comparative models of diapause and torpor. A, *C. elegans* dauer diapause. B, *D. melanogaster* reproductive diapause. C, Mammalian topor. In A and B, environmental signals are sensed and integrated by neurons using ciliary and synaptic components, such as *C. elegans* UNC-64 and the *D. melanogaster Methuselah* (*Mth*) and cGMP. Signals are further transduced through downstream endocrine signaling, including serotonin, neuropeptide Y-like peptide (NPY), TGF-beta, and insulin/IGF signaling (IIS). In worm and fly, IIS modulates steroid hormone synthesis (worm: dafachronic acids (DA), fly: 20-hydroxy-ecdysone, (20E)). IIS, TGF-beta, and juvenile hormone (JH) may also have other functions (dotted lines). In favorable environments, steroid-bound nuclear hormone receptors DAF-12 (worm) and ecdysone receptor (EcR, fly) promote programs for reproductive growth and normal lifespan, whereas in unfavorable environments no ligand is produced and programs for diapause and dauer longevity result. In the mammalian torpor model, C, in fed states, peroxisome proliferator-activated receptor/fibroblast growth factor 21 (PPAR/FGF21) signaling and NPY are inactive. In starvation states, PPAR/FGF21 signaling and NPY promote torpor, and may inhibit growth hormone/insulin-like growth factor (GH/IGF) signaling. Speculatively, bile acids (BA) and the farnesoid X receptor (FXR) could play a role in torpor comparable to DA/DAF-12 in regulating dauer diapause. For details see text.

Drosophila, long-chain fatty acids found in the cuticle also serve as mating pheromones (Ferveur 2005). Conceivably, they too could serve to signal population density and metabolic state.

22.5   Neurosensory signaling and processing

Environmental cues are perceived and integrated by the nervous system, which enables animals to assess environmental quality. *C. elegans* has a simple nervous system consisting of 302 neurons, including numerous sensory neurons that mediate modalities of taste, smell, touch, and thermosensation. By using laser ablation microsurgery to kill individual cells, chemosensory amphid neurons ASI, ADF, and ASG have been shown to prevent dauer, while ASJ promotes dauer entry and recovery (Bargmann 2006). The thermosensory interneuron AIY promotes dauer at higher temperatures and prevents it at lower ones. Moreover, CHE-1 mutants, which

specifically alter ASE fate and function, modulate dauer formation, conceivably through changes in perception of soluble food signals (Reiner *et al.* 2008).

Many of the identified endocrine pathways described below also reside within these and other neurons, showing a direct link between sensory signaling and neuroendocrine activity. For example, the *daf-7*/TGF-beta peptide is expressed in ASI, *daf-28*/insulin-like peptide in ASI and ASJ, and *tph-1*, which is involved in serotonin biosynthesis in ADF (Bargmann 2006). The NPY/RFamide peptide hormone *flp-18* is expressed in AIY (Cohen *et al.* 2009), and the *scd-2*/ALK-peptide ligand in AIY and ASE (Reiner *et al.* 2008). Furthermore, the XXX neuroendocrine cells are important sites of steroid production (Fielenbach and Antebi 2008). Finally, IL2 neurons express a Notch ligand *lag-2*/DSL, involved in dauer recovery (Ouellet *et al.* 2008). How sensory information is integrated and converted into congruent endocrine signals is not well understood, but is clearly an important area of research.

As might be expected, components of neuro-sensory architecture, signal transduction, and integration impact dauer formation (Bargmann 2006). Sensory neurons have a highly ordered ciliary structure within the dendrites that project from cell bodies in the nerve ring (the worm brain) to the nose tip, where sensory information is detected. Mutants that perturb the ciliary structure of sensory neurons result in defects in chemotaxis and dauer formation. Such "senseless" mutants generally are Daf-d at lower temperatures, but Daf-c at higher temperatures, revealing that neurosensory signaling both prevents and promotes dauer formation. Molecularly identified genes include a set of highly conserved proteins that make up the structural components of the cilia themselves or regulate intraflagellar transport, assembly, and biogenesis, such as the Bardet–Bietl syndrome proteins implicated in human disease (Bargmann 2006).

Sensory signal transduction also regulates dauer formation and includes components of the cGMP pathway, such as G-proteins, guanylyl cyclases, cGMP dependent kinases, and ion channel subunits, whose signaling is organized through the cilium (Bargmann 2006). In particular, the transmembrane guanylyl cyclase DAF-11 evidently regulates dauer through neuronal cGMP levels: low levels promote dauer, while high levels signal non-dauer development (Birnby *et al.* 2000). Finally, highly conserved proteins, functioning in synaptic vesicle trafficking and release, influence dauer formation. Loss-of-function mutations in *unc-64*/syntaxin or *unc-31*/CAPS (calcium-dependant activator protein for secretion) result in Daf-c phenotypes (Ailion *et al.* 1999). These complexes are thought to regulate the exocytosis of insulin-like and other peptide hormones (see below).

22.6   Sensory signal transduction and longevity

Surprisingly, many mutations that perturb neuro-sensory cell structure and function also impact lifespan. Mutants with sensory deficits or reduced cGMP signaling are long lived and are thought to

work by reducing insulin signal transduction (Apfeld and Kenyon 1999, Hahm *et al.* 2009). Similarly, laser ablation of certain sensory neurons can positively or negatively influence longevity (Alcedo and Kenyon 2004, Lee and Kenyon 2009). Conceivably, the absence of favorable environmental cues triggers downstream physiological responses that induce states of somatic endurance similar to dauer. Remarkably, this feature is evolutionarily conserved. *Drosophila Or83b* odorant receptor mutants, which are globally deficient in olfaction, are long lived, and food-derived odors partly suppress longevity due to dietary restriction (Libert *et al.* 2007). Similarly, a reduction of function in *Drosophila* Methuselah, a G-protein coupled receptor that affects synaptic function (Song and Liao 2000), results in extended life (McGarrigle and Huang 2007). Altogether, these studies reveal that sensory perception and neuronal cell function can have dramatic effects on animal life history traits and longevity.

22.6.1   Insulin/IGF-1 signal transduction

Insulin/IGF-1 signaling (IIS) is well known for its physiological function regulating animal growth and metabolism. In recent years it has emerged as the key endocrine pathway regulating the major life history parameters of reproduction, survival, and longevity. A modest reduction of IIS in worms, flies, mice, and perhaps even humans results in exceptional longevity, demonstrating that this physiological function is remarkably conserved (Fielenbach and Antebi 2008, Willcox *et al.* 2008, Barzilai and Bartke 2009). In *C. elegans*, a strong down-regulation of the pathway during larval development promotes dauer formation, while a modest down-regulation in adults result in greater stress resistance and a doubling of adult lifespan, with surprisingly little consequence for fertility, revealing that longevity and reproduction can be independently regulated. Depending on the allele and the maintenance temperature of the *daf-2* mutant animals, additional phenotypes can include slower development, decreased mobility, increased fat deposition, and extended reproductive period, but most of these phenotypes can also be uncoupled from longevity (Gems *et al.* 1998).

Molecular dissection of IIS has led to important insights into the mechanistic basis of organismal plasticity. Perhaps most importantly, the forkhead transcription factor DAF-16/FOXO is a "master regulator" of longevity, normally inhibited by IIS. In particular, the balance between reproduction and dauer development is visible in the nuclear activity of FOXO and associated molecules. Molecular genetic studies suggest a model whereby, in favorable environments, various insulin-like peptides are synthesized primarily in sensory neurons and gut, which activate the DAF-2/insulin/IGF receptor in diverse target tissues (Fielenbach and Antebi 2008). This stimulates a PI3 lipid kinase, which in turn activates the PDK1/AKT/SGK kinase cascade, resulting in the phosphorylation of DAF-16/FOXO. When phosphorylated, FOXO is retained in the cytoplasm, allowing the expression of reproductive programs. In unfavorable environments or under stress conditions, down-regulation of the IIS pathway results in FOXO nuclear translocation and activation of genes involved in dauer programming, stress resistance (heat, oxidative, UV, pathogen), somatic homeostasis, autophagy, thrifty metabolism, and longevity (Melendez et al. 2003, Murphy et al. 2003, McElwee et al. 2004). FOXO inhibits TOR signaling to down-regulate somatic growth (Jia et al. 2004). It also regulates secondary hormones, including the insulin and steroid hormones that propagate signals between tissues (Fielenbach and Antebi 2008). A number of nuclear factors, including heat shock factor (Hsu et al. 2003), beta-catenin (Essers et al. 2005), and sirtuins (Berdichevsky et al. 2006, Wang and Levy 2006), modulate expression of FOXO targets. In addition, a number of kinases (JNK; Oh et al. 2005, MST; Lehtinen et al. 2006, MAP; Nanji et al. 2005, Troemel et al. 2006, AMP; Apfeld et al. 2004, Greer et al. 2007) and phosphatases (SMK-1; Wolff et al. 2006, PPTR-1; Padmanabhan et al. 2009) modulate the phosphorylation state and activity of FOXO, presumably integrating inputs from various stress-response pathways.

IIS is visibly connected to environmental and physiological cues in multiple ways. Extrinsic cues such as starvation and dauer pheromone suppress expression of the *daf-28*/insulin gene in sensory neurons. Similarly, FOXO translocation is induced by starvation, oxidative stress, and heat stress, as well as by mutants defective in sensory signaling (Henderson and Johnson 2001, Lee and Ambros

2001, Lin et al. 2001). In accord with this, the longevity of sensory mutants is largely FOXO-dependent. Intrinsic cues also impact this signaling pathway. For example, laser removal of germ-line stem cells, which results in extended lifespan (see Section 22.8), correlates with translocation of FOXO into the nuclei of intestinal cells (Berman and Kenyon 2006), demonstrating the tissue–tissue communication that underlies this process. These observations give visible molecular evidence of inputs into the endocrine signaling underlying life history plasticity.

22.6.2 TGF-beta signaling

TGF-beta signaling comprises another major endocrine system that impacts various *C. elegans* life history traits (Savage-Dunn 2005). Similar to IIS, a reduction in TGF-beta signaling promotes fat deposition and constitutive dauer formation during larval development and longevity in adults. In addition, reduced TGF-beta signaling also modifies feeding behavior and reproductive biology. Animals clump together at the edge of bacterial lawns, termed social feeding, which is thought to maximize access to food. They also retain eggs and often larvae hatch inside, feeding off the mother's carcass, allowing the young access to nutrients even under harsh environments (Chen and Caswell-Chen 2004).

Elegant molecular studies have elucidated the core components of TGF-beta signaling. The peptide hormone DAF-7/TGF-beta is regulated within the ASI neurosensory cells in response to environmental signals (Savage-Dunn 2005). In favorable environments, the hormone is expressed, and it binds to DAF-1/4 TGF-beta type I/II receptor kinases, which are found in most tissues throughout the body. Receptor activation results in the stimulation of the transcription factors DAF-8 and DAF-14 SMAD, which override a repressive transcriptional complex of DAF-3/Co-SMAD and DAF-5/SNO/SKI, allowing continuous development. In unfavorable environments, *daf-7*/TGF-beta is not expressed and consequently DAF-3/Co-SMAD and DAF-5/SNO/SKI promote dauer formation and slow life history traits. Notably, DAF-3 and DAF-5 are required for all the major life history alterations induced by reduction of

TGF-beta signaling. In addition, there is substantial crosstalk to FOXO. For example, the longevity of *daf*-7 mutants ultimately depends on DAF-16/FOXO (Shaw *et al.* 2007).

A number of modulators of the TGF-beta pathway have been uncovered, some of which influence the actual signal transduction pathway or impact TGF-beta production itself. Notable among them are components of ALK (nematode anaplastic lymphoma kinase) signaling, which form a coherent sub-pathway influencing TGF-beta signaling (Reiner *et al.* 2008). Interestingly, this pathway was discovered through a convergence of conventional genetic screening and studies on wild polymorphic *C. elegans* variants. By studying isolates from desert habits, a strain was isolated that was resistant to high-temperature-induced dauer formation. The Daf-d phenotype was caused by a mutation in the *scd*-2 gene, which encodes a homolog of the anaplastic lymphoma protooncogene receptor, tyrosine kinase. The other known components of this signaling pathway were also required for dauer formation, suggesting a simple model. *hen*-1 encodes the ALK peptide ligand, and is produced in ASE and AIY neurons. HEN-1 then binds to SCD-2/ALK receptor, distributed in various tissues throughout the body. Signal transduction relies on the SOC-1/RTK adaptor protein and the SMA-5/MAP kinase. Genetic epistasis experiments and molecular studies suggest that these components stimulate the transcriptional repressive activity of DAF-3/co-SMAD or DAF-5/sno-ski. The expression of the *HEN-1* ligand in the ASE is notable because this neuron is nearly exclusively responsible for detecting soluble chemoattractants (Bargmann 2006). Moreover, the ZnF protein CHE-1 specifies the ASE fate and mutants have Daf-d phenotypes similar to *scd*-2. It is therefore tempting to speculate that the ALK signaling pathway mostly mediates soluble food cues into the dauer decision.

22.6.3   Biogenic amine signaling

Behavioral changes as well as metabolic adaptation to survive under reduced food availability are postulated to involve biogenic amines such as dopamine and serotonin. Serotonin works as a neurohormone governing behavior, metabolism, development, and immunity. *C. elegans* mutants deficient for the tryptophan hydroxylase homolog *tph*-1 have reduced serotonin production and display phenotypes consistent with a slow life history mode. For example, mutants are prone to constitutive dauer formation, have altered movement and feeding behavior, show increased sensitivity to starvation cues, store fat, retain eggs, and have an extended reproductive period (Sze *et al.* 2000). Moreover, they are more resistant to heat stress and pathogen challenge, and promote nuclear localization of DAF-16 (Liang *et al.* 2006). *tph*-1 Daf-c phenotypes are suppressed by *daf*-16 and *daf*-3 loss-of-function mutations, suggesting *tph*-1 works early in the dauer pathways (Sze *et al.* 2000). *tph*-1 mutant animals have a normal lifespan, but mutants in serotonin receptors *ser*-1 and *ser*-4 are long-lived and short-lived respectively, suggesting that the receptors signal antagonistically to modulate longevity (Murakami and Murakami 2007).

22.6.4   Neuropeptide-Y-like signaling

Neuropeptide Y and its receptors regulate feeding behavior and metabolism throughout the animal kingdom. *C. elegans* harbors 12 neuropeptide-Y-like receptors (nprs), several of which affect feeding behavior, fat metabolism, and dauer formation. Natural variation in the *npr*-1 locus influences social versus solitary foraging mutants (de Bono and Bargmann 1998, Cohen *et al.* 2009). Variants with reduced NPR-1 are more prone to social feeding. Mutants of the NPY/RF-amide peptide hormone FLP-18 accumulate excess fat and have decreased oxygen consumption. Loss-of *flp*-18 enhances the Daf-c phenotypes of *daf*-7/TGF beta mutants, but not that of *daf*-2/InR mutants (Cohen *et al.* 2009), suggesting NPY-like signaling could work through IIS. FLP-18 is proposed to bind NPR-4 and NPR-5 receptors based on biochemical and genetic studies (Cohen *et al.* 2009). The implication of NPY-like signaling in worms, flies, and mammals (below) suggests an ancient mechanism of coupling diet to regulation of behavior, metabolism, and life history traits.

22.6.5   Steroid hormone signaling

Nuclear hormone receptors (NHR) are transcription factors that respond to fat-soluble hormones, such as steroids and retinoids, to directly regulate gene expression. They are well suited to coordinate animal metabolism, development, reproduction, and homeostasis. Steroid receptors, such as the ecdysone receptor of insects and the estrogen and androgen receptors of mammals, trigger stage transitions, such as metamorphosis and puberty, respectively. NHR signaling is conserved in *C. elegans*, where DAF-12, a homolog of vertebrate LXR, FXR, and vitamin D receptors, constitutes a major endocrine pathway that governs dauer formation, fat metabolism, reproduction, and lifespan (Fielenbach and Antebi 2008).

Molecular genetic studies reveal that IIS and TGF-beta signaling converge on DAF-12, which then works as a hormone-regulated switch to control dauer and continuous development (Motola *et al.* 2006, Fielenbach and Antebi 2008) (Fig. 22-1A). Molecular genetic studies suggest a model whereby in favorable environments, TGF-beta and insulin-like peptides are synthesized predominately in sensory neurons. Working through their respective signaling pathways, they stimulate hormone biosynthetic genes in steroidogenic endocrine tissues to produce the DAF-12 ligands. These ligands are derived from dietary cholesterol and modified through a series of steps to the active hormone, the bile acid-like steroids, called the dafachronic acids (DA). In tissues throughout the body, liganded DAF-12 results in bypass of dauer and promotion of continuous development to reproductive maturity. In unfavorable environments, TGF-beta and insulin endocrines are down-regulated, stimulating the activity of DAF-3/5 and DAF-16 transcriptional complexes, respectively. Consequently, hormone biosynthetic genes are thought to be down-regulated, DA production is suppressed, and the unliganded DAF-12 together with the co-repressor DIN-1, a homolog of mammalian SHARP, repress reproductive gene expression and specify dauer formation. Notably, biosynthetic mutants completely deficient in DA production, such as *daf-9/* CYP450, constitutively enter dauer and are also stress-resistant and long lived as adults at lower temperatures (Fielenbach and Antebi 2008).

It should be noted that DAF-12's activity during dauer formation is absolutely essential. In this light, DA effectively prevents dauer by disabling DAF-12's repressive activity. By contrast, DAF-12's activity for reproductive development is non-essential, and null mutants reach reproductive maturity with minor heterochronic defects in the epidermis and other tissues (see below) (Antebi *et al.* 1998). Although TGF-beta and IIS signaling converge on DAF-12 for dauer formation, these pathways clearly have independent outputs. For example, IIS regulates adult longevity somewhat interdependently with NHR signaling, while TGF-beta regulates foraging and egg laying (Thomas *et al.* 1993, Gems *et al.* 1998).

At least two endogenous DAF-12 ligands have been found: delta-4 and delta-7 DA. These stimulate DAF-12 transcriptional activity and rescue dauer formation in the nanomolar range (Motola *et al.* 2006). Whether or not these ligands have distinct physiological roles is unknown. A third ligand, cholestenoic acid, works at higher concentrations and also stimulates the vertebrate LXR (Held *et al.* 2006). A number of enzymes involved in the manufacture of DA have been identified genetically and suggest that a branched pathway catalyses ligand synthesis. DAF-36, a Rieske oxygenase, carries out the first step in the delta-7 branch, the conversion of cholesterol to 7-dehydrocholesterol (Fielenbach and Antebi 2008). The *Drosophila* homolog, *neverland*, works at a similar step in the synthesis of ecdysteroids. HSD-1, a 3-hydroxy steroid dehydrogenase homolog, is proposed to work in the delta-4 branch, making delta-4 keto steroids (Patel *et al.* 2008). These pathways ultimately converge on DAF-9/CYP450, which carries out the last step, oxidizing the cholesterol side chain to the bile acid moiety (Motola *et al.* 2006), a reaction that is biochemically orthologous to mammalian CYP27A1, a key enzyme in bile acid synthesis (Russell 2003). Additional biosynthetic enzymes in the worm are predicted to exist based on the proposed intermediates in the pathway. In addition, Niemann-Pick homologs NCR-1 and 2 are thought to mediate trafficking of sterol intermediates and work upstream of DAF-12, similar to counterparts in flies and mammals (Fielenbach and Antebi 2008).

Steroidogenic tissues in the worm have been identified based on the expression patterns of the hormone biosynthetic genes (Fielenbach and Antebi 2008). DAF-36 is expressed primarily in the intestine, while DAF-9 is found in neuroendocrine-like cells in the head called the XXX cells, as well as the hypodermis and spermatheca. HSD-1 is expressed exclusively in the XXX cells. NCR-1 is widely expressed, while NCR-2 is expressed in the XXX and somatic gonad. By inference, the XXX cells are thought to be critical to steroid production, integrating TGF-beta and IIS as well as other inputs. Accordingly, laser ablation of these cells promotes transient dauer formation. However, the hypodermis also plays a key role. In fact, the most dramatic regulation of *daf*-9/CYP450 is seen in the hypodermis, which is proposed to mediate feedback control. *daf*-9 regulation in this tissue gives clear visible evidence of molecular endocrine regulation as the basis for the dauer decision. Typically, *daf*-9 hypodermal up-regulation is a sign of commitment to reproductive growth; conversely its down-regulation (e.g., by loss of IIS or TGF-beta signaling) is a sign of dauer formation (Gerisch and Antebi 2004). Conceivably, tonic levels of hormone are made by the XXX cells, while the hypodermis fine-tunes regulation to ensure an all-or-none decision. Distributed hormone biosynthesis may be one way in which the various tissues can vote on the dauer decision.

In vertebrates, bile acids were previously thought to play a simple role in emulsifying lipids. In recent years, however, it has become clear that they are also signaling molecules, which regulate glucose and fat metabolism, cholesterol homeostasis, and thermogenesis, working predominately through FXR nuclear receptor and G-protein coupled receptor signaling (Hylemon *et al.* 2009). A simple notion is that bile acids serve as a satiety signal, not unlike DA. Interestingly, a number of long-lived mouse mutants such as the Ames Dwarf, have altered bile acid levels, which are thought to stimulate xenobiotic metabolism and chemical defense (Amador-Noguez *et al.* 2007). Finally, polymorphisms in the DAF-12 homolog LXR, which binds oxysterols and bile acids, are associated with human longevity (Mooijaart *et al.* 2007). Clearly, further studies on the role of bile acids as potential signaling molecules influencing energy homeostasis, reproduction, and longevity should be fertile areas of future research.

22.7 Developmental timing and life history specification

Dauer formation and longevity are life history traits that are plastic and responsive to the environment. They represent natural stage alternatives or physiological states that enhance survival. When considering life history traits and their evolution, genetic determinants of developmental timing, life stages, and life plan must also come into play. During development, each cell in the body acquires both positional and temporal identity. Transformations in temporal fate, termed heterochrony, can have dramatic consequences on the relative timing of tissue differentiation, maturation, or the temporal organization of life stages themselves. Groundbreaking studies in *C. elegans* reveal that a regulatory hierarchy, called the heterochronic circuit, controls biological time, determining the temporal identity of stage-specific programs, the number of life stages, and coordination of soma and gonadal development (Moss 2007). These loci may give important insight into the stage structure of life histories, from the cellular to organismal level.

With only 959 somatic cells, the entire *C. elegans* cellular development is known from fertilization to adult, and each stage has discernible and invariant stage-specific patterns of cell division, migration, differentiation, and death. For example, during each larval molt, epidermal stem cells, called seam cells, undergo asymmetric cell division to generate a terminally differentiated hypodermal cell and a seam stem cell, but with stage-specific variations (Moss 2007). Mutations in the heterochronic circuit cause temporal transformations, such that earlier patterns are expressed at later larval stages (retarded) or later patterns are expressed at earlier larval stages (precocious). Interestingly, precocious mutants stop molting one stage early, while retarded mutants often undergo additional molts. Conceivably, different life stage structures could be generated by such mechanisms operating in evolution.

Molecular analysis reveals that the heterochronic components work as genetic switches that turn on programs of the next larval stage, while turning off those of the previous stage. Identified loci work in a stage- and tissue-specific manner, and encode various transcriptional, translational, and post-translational regulators.

Among them, the microRNAs, small 20–26 nucleotide RNAs, play a critical role in this process by complementary binding to target mRNAs and

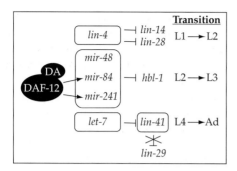

Figure 22-2 Heterochronic genes regulate developmental timing of seam cells. *lin-4* miRNA down-regulates LIN-14/nuclear protein and LIN-28/RNA binding protein, resulting in the transition from L1 to L2. DAF-12/NHR-dependent up-regulation of miRNAs of the *let-7* family, such as *mir-48, mir-84,* and *mir-241* results in down-regulation of *hunchback-like* 1, *hbl-*1, and transition from L2 to L3. L4-to-adult transition is induced by miRNA *let-7* expression, which results in a down-regulation of *lineage-abnormal* 41 (*lin-41*/TRIM71). *lin-41* and other heterochronic genes (not shown) prevent the expression of the zinc-finger transcription factor *lin-29*.

hindering gene expression within cells. For example, the microRNA *lin-4* regulates L1/L2 transitions by down-regulating the nuclear protein LIN-14 (Moss 2007; Fig. 22-2). Remarkably, *lin-4* null mutants repeat L1 programs again and again at later larval stages; animals are still fertile because gonadal maturation is unaffected. Similarly, *mir-48, mir-84,* and *mir-241* (which comprise a family of microRNAs related to *let-7*) regulate L2/L3 transitions by down-regulation of the Zn finger protein, HBL-1/hunchback (Moss 2007). *let-7* itself regulates L4/ault transitions by down-regulating *lin-41/*TRIM71, which derepresses the LIN-29/Zinc finger protein (Moss 2007). Importantly, homologs of these microRNAs are conserved from worms to humans, and many of them also control differentiation and stem cell progression.

A number of other heterochronic loci regulate production and maturation of microRNAs. For example, the LIN-28 RNA binding protein inhibits maturation of the precursor form of the *let-7* microRNA (Viswanathan *et al.* 2008). Remarkably, the human LIN28 works in concert with pluripotency factors OCT4, NANOG, and SOX2 to re-specify somatic cells to induced pluripotent stem cells (Yu *et al.* 2007a), and probably does so by inhibiting microRNAs. Taken together, these data show a

fundamental link between heterochronic loci, developmental progression, and stem cell biology.

Among the heterochronic loci the nuclear hormone receptor DAF-12 has a unique role, converting global hormonal signals into expression of cell-intrinsic programs. DAF-12 is widely expressed throughout the body. In response to DA, DAF-12 coordinates programs in soma and gonad during L2/L3 transitions; mutants exhibit retarded heterochronic phenotypes in which they repeat L2 patterns of seam cell division and gonadal outgrowth during the L3 stage. Heterochronic defects are mild in *daf*-12 null mutants, but are more severe in ligand binding domain mutants (Antebi *et al.* 1998). Because the phenotype of *daf*-12 null mutants is impenetrant, it is likely that other factors are involved in this transition. Indeed, additional inactivation of other loci, such as the DRE-1/FBXO11 in gonad or microRNAs in epidermis, reveal that redundant pathways specify reproductive development in various tissues (Abbott *et al.* 2005, Fielenbach *et al.* 2007).

Recently, DAF-12 has been shown to directly regulate the transcription of *mir-241* and *mir-84* in various tissues, thus revealing a molecular link between hormonal and cell intrinsic signaling (Bethke *et al.* 2009). It is proposed that, in the presence of DA, the nuclear receptor activates the expression of these microRNAs, which themselves help down-regulate the target mRNA of *hbl-1/*hunchback to effect L2/L3 transitions. In the absence of its ligand, DAF-12 specifies developmental arrest at the dauer diapause, and strongly represses microRNA expression in most tissues. Together, the nuclear receptor and microRNA comprise a hormone-coupled switch that turns off earlier programs and allows for later ones. Conceivably, similar mechanisms are at work during hormone-dependent progression in mammalian stem cells, differentiation, and tumorigenesis.

Importantly, by virtue of working at the convergence of the dauer and heterochronic pathways, DAF-12 conveys environmental information into the heterochronic timer, shutting it down under adverse conditions and advancing it under favorable ones. Such a mechanism could allow developmental timing to be closely coupled to nutrient and other environmental signals.

Indeed, an intimate connection between nutrient conditions and reproductive transitions are seen in other animals. For example, *Drosophila* larval growth, metamorphosis, and developmental timing are governed by interaction of nutrients, IIS, and ecdysteroid signaling (McBrayer *et al.* 2007, Walkiewicz and Stern 2009). Moreover, ecdysteroids working through the ecdysone receptor drive microRNA expression (Sempere *et al.* 2003), insect metamorphosis, and developmental progression in various tissues. In mammals, time to maturity and fertility are regulated similarly by nutrients, IIS, and steroids. Conceivably, developmental progression in other species is embedded in a rich circuitry not unlike that of the *C. elegans* dauer and heterochronic pathways.

22.8   Reproduction and longevity

The disposable soma theory postulates that aging arises as a result of investment in reproduction at the cost of somatic maintenance, suggesting trade-offs between reproduction and survival (see Chapters 11 and 24). In classical experiments, long-lived lines of *Drosophila* were found by selecting for delayed reproduction (Rose and Charlesworth 1981). Conversely, strains selected for rapid reproduction often showed shorter total lifespan, leading to the notion that there exists a trade-off between reproduction and survival. Despite this important observation, numerous studies indicate that there is not always an obligate coupling between increased longevity and lower fecundity. For example, many long-lived *daf*-2 mutants have broods similar to wild type (Gems *et al.* 1998). Queen bees produce thousands of eggs yet have much longer lifespans than isogenic worker bees (Finch 1990). Conversely, many sterile mutants are not necessarily long lived.

The reason they can be uncoupled is because the energetic costs of reproduction likely represent only a part of a coordinated regulatory response. Indeed, elegant studies in *C. elegans* reveal that regulatory signaling from gonad to soma may underlie the relationship of survival and reproduction (Hsin and Kenyon 1999). At hatch, the *C. elegans* gonadal primordium consists of two germ-line precursor cells (Z2, Z3) and two somatic gonadal precursors

(Z1, Z4). When these germ-line precursors are removed by laser microsurgery, animals become sterile adults that live 50% longer than mock-treated animals. Surprisingly, when somatic gonadal precursors are further removed, animals have normal lifespans. Thus the energetic costs of germ-line production cannot account for increased longevity. Furthermore, genetic removal of germ-line stem cells, but not oocytes or sperm, is required for extended survival (Arantes-Oliveira *et al.* 2002). Conversely, mutants that have tumorous germ-lines with increased numbers of stem cells have shortened lifespans. Taken together, these studies suggest that life-shortening signals from germ-line stem cells and life-lengthening signals from the somatic gonad impact animal longevity. Consistent with intercellular signaling, germ-line ablation results in the nuclear localization of DAF-16/FOXO in intestinal cells (Berman and Kenyon 2006). FOXO translocation is visibly dependent on DAF-12 and DA (Gerisch *et al.* 2007). Moreover, germ-line longevity requires DAF-16/FOXO, as well as DAF-12/ nuclear receptor signaling (Hsin and Kenyon 1999, Gerisch *et al.* 2007). These molecular events strongly indicate that tissue communication and hormonal regulation underlie germ-line longevity, and that diversion of resources to the germ-line may not adequately describe why reproductively active animals are relatively shorter lived.

Recently, germ-line longevity has been tied to organismal fat metabolism (Wang *et al.* 2008). Germlineless *glp*-1 animals have excess fat, as measured by oil red O staining (O'Rourke *et al.* 2009). This change may be critical to longevity since knockdown of a specific lipid/cholesterol lipase abolishes longevity, while overexpression results in increased lifespan (Wang *et al.* 2008). Thus fat metabolism, either through changes in energy balance or hormonal signaling, impinges upon enhanced survival.

Importantly, germ-line longevity is also evolutionarily conserved. *Drosophila* mutant strains in which germ-line stem cells have been depleted live longer than controls (Flatt *et al.* 2008b). Paradoxically, such animals have increased expression of FOXO target genes, yet have increased levels of insulin-like peptides (ILPs) and insulin-like binding protein, mimicking a state of insulin resistance. Conceivably the insulin-like binding proteins sequester ILPs and impede IIS, but the mechanism uncoupling insulin production from downstream events has to be further analysed.

What is the physiological significance of germline ablation induced longevity? Germ-line longevity may represent an aspect of reproductive diapause in which germ cells arrest due to adverse circumstances. Interestingly, a reproductive diapause has been recently discovered in *C. elegans*. Here, animals trim back their germ-lines and induce longevity in response to overcrowding and starvation (Angelo and Van Gilst 2009). It remains to be seen what molecules are responsible for this process.

22.9   Dietary restriction

Dietary restriction (DR), a reduction in nutrient intake without malnutrition, can extend health span and lifespan in species as diverse as worms, flies, and rodents, suggesting evolutionarily ancient mechanisms. DR links nutrient conditions to other key life history changes such as slowed developmental timing and maturation, reduced fecundity, altered fat metabolism, and reduced growth. In *C. elegans*, DR is induced in a variety of ways, including bacterial dilution in liquid, solid, or semisolid culture, as well as reduced ingestion in the pumping deficient mutant *eat*-2 (Mair and Dillin 2008). These regimens may identify overlapping yet distinct processes.

Despite its importance, the molecular basis of DR is only now coming to light. In the worm, the handful of molecules and processes identified indicate that the response to DR is regulated, and not simply due to passive changes. For example, DR-induced longevity via bacterial dilution in liquid depends on homologs of the transcription factors SKN-1/NRF-2 as well as PHA-4/FOXA1 but not DAF-16/FOXO (Bishop and Guarente 2007b, Panowski *et al.* 2007). Instead, *daf*-16/FOXO is required for DR induced by alternate day feeding or bacterial dilution on plates (Greer and Brunet 2009, Honjoh *et al.* 2009). Interestingly, SKN-1 is required in the ASI neurons to provoke DR-induced longevity, implying that DR is coordinated via a hormonal mechanism (Bishop and Guarente 2007b). The nature of these hormones and their

receptors should be areas of future interesting research. Within cells, DR induces states of thrifty metabolism such as increased autophagy and respiration, as manipulations that reduce these processes abrogate DR-induced longevity (Bishop and Guarente 2007b, Hansen *et al.* 2008). Moreover, a reduction of protein synthesis has been shown to extend lifespan in worms—depletion of ribosomal subunits, translational regulators, and the nutrient sensor TOR kinase all extend nematode lifespan (Hansen *et al.* 2007, Pan *et al.* 2007). As DR itself reduces protein synthesis, some of the beneficial effects may accrue from this.

22.10   Diapause in other nematode strains

Most free-living nematodes have a dauer-like stage. Likewise, the infective forms of parasitic larvae resemble dauers, having similar morphology and metabolic characteristics. Accordingly, some of the dauer signaling pathways identified in *C. elegans* also occur in other nematodes. For example, recovery from the infective stage of the parasitic hookworm *Ancylostoma caninum* is stimulated by 8-bromo-cGMP and muscarinic agonists (Tissenbaum *et al.* 2000, Hawdon and Datu 2003), implicating cGMP and cholinergic signaling. The PI3 kinase inhibitor LY294002, which down-regulates insulin/IGF signaling, prevents feeding-associated recovery (Brand and Hawdon 2004). By contrast, a role for TGF-beta signaling is less apparent.

Recent evidence suggests that the most critical conserved module in dauer formation and infective parasitism is the DAF-12/NHR signaling pathway (Ogawa *et al.* 2009, Wang *et al.* 2009c). Mutations in *daf*-12 or exogenous treatment with the DAF-12 ligand Δ7-dafachronic acid prevent dauer formation in the free-living nematode, *Pristionchus pacificus*, which diverged from *C. elegans* over 120 million years ago (Ogawa *et al.* 2009). Δ7-dafachronic acid also stimulates exit from the infective stages of several important parasitic nematodes (Ogawa *et al.* 2009, Wang *et al.* 2009c), further demonstrating ancient origins for steroid control of diapause. These findings could be useful for the development of unique therapeutic targets against parasitic nematodes.

22.11 *D. melanogaster* reproductive diapause

A comparison of *C. elegans* dauer diapause with the *D. melanogaster* reproductive diapause (see Chapter 18) reveals similar mechanisms for regulating fast and slow life histories. In temperate climates, *D. melanogaster* has evolved the ability to overwinter through reproductive diapause. Diapausing females are defined by arrested ovarian development, indicated by the absence of vitellogenin in the eggs (Tatar and Yin 2001, Flatt *et al.* 2005). They also show increased lipid deposition, stress resistance, and longevity, similar to *C. elegans* diapause (Clancy *et al.* 2001, Tatar *et al.* 2001a). At the molecular level, animals in diapause induce genes involved in somatic endurance, heat shock proteins, cryoprotectants, and the innate immune response, which all help survival in inhospitable conditions (Tatar and Yin 2001). Many of the same gene families are also up-regulated in dauer.

Environmental cues triggering reproductive diapause include shortening day length and lower temperatures. Indeed one of the loci identified as critical to the regulation of diapause among *Drosophila* populations is *timeless*, first implicated in circadian rhythmnicity and photoperiodism (Tauber *et al.* 2007). However, it is thought that *timeless* can influence diapause through pathways independent of photoperiodicity, since the allelic variant, *ls*, promotes greater entry into diapause than allelic variant *s* at all day lengths (Tauber *et al.* 2007). Nutrient conditions strongly influence the fecundity, metabolism, and longevity of laboratory fly stocks (Skorupa *et al.* 2008), although it is unknown exactly how nutrients influence reproductive diapause in the wild. Presumably IIS and TOR mediate nutrient sensing and impact reproductive diapause, as described below. Finally, it is unknown whether or not pheromones affect reproductive diapause.

Reproductive diapause is associated with various hormonal changes, including suppression of IIS, juvenile hormone (JH) (see Chapter 13), and ecdysone production (Fig. 22-1B). Based on studies in a variety of insects, the insulin-producing neurosecretory cells (IPCs) are thought to integrate inputs from sensory neurons and those mediating photoperiodicity. The IPCs project directly or signal humorally onto the corpora allata (CA, the site of JH production) and the ovaries (Tatar and Yin 2001, Flatt *et al.* 2005). It is thought that when insulin signaling is down, JH drops, resulting in a failure to synthesize ecdysone in the ovary and vitellogenins in the fat body, and thus an arrest of oocytes at pre-vitellogenic stages.

In *Drosophila*, reduced IIS results in a number of changes reminiscent of reproductive diapause, including suppressed ovarian development and vitellogenesis, increased fat deposition, stress resistance, and longevity (Clancy *et al.* 2001, Tatar *et al.* 2001b). Ablation of IPCs of the brain, where three of the seven ILPs are produced, induce similar diapause-like phenotypes (Broughton *et al.* 2005). In addition, polymorphic deletions in wild isolates that increase the frequency of individuals entering diapause may be linked to the *Dp110* locus (Williams *et al.* 2006a), a homolog of the PI3 kinase subunit, or to *couch potato* (Schmidt *et al.* 2008), an RNA binding protein whose *C. elegans* counterpart, *mec-8*, is implicated in neurosensory architecture, IIS, and longevity (Apfeld and Kenyon 1999). Recent studies of IIS in the mosquito, *Culex pipiens*, also support the hypothesis that IIS is involved in insect reproductive diapause: insulin receptor RNAi triggered a follicular arrest, while FOXO RNAi blocked fat deposition phenotypes associated with diapause (Sim and Denlinger 2008). The production of ILPs themselves is regulated by neuropeptide F, a homolog of mammalian NPY, as well as by cAMP and ERK signal transduction (Lee *et al.* 2008c, Walkiewicz and Stern 2009), providing a potential link to nematode diapause and mammalian torpor (Fig. 22-1B).

Juvenile hormone is a sesquiterpenoid that regulates a number of life history traits, mediating trade-offs between reproduction and survival, and it is considered a primary regulator of reproductive diapause in *D. melanogaster* and other insects (Tatar and Yin 2001, Flatt *et al.* 2005) (see Chapter 13). In adults the main endocrine source of the JH is the CA, where a shutdown of JH synthesis is associated with reproductive diapause. A causative role for JH is evident, since removal of the CA triggers reproductive diapause, whereas supplementation with juvenile hormone analogs stimulates diapause recovery (Tatar and Yin 2001, Flatt *et al.* 2005). Despite a number of candidate receptors, JH's molecular signaling pathway remains elusive.

Moreover, peptide hormones such as allatostatins and allatotropins also modulate JH production, and are candidates for influencing reproductive diapause.

How is JH linked to IIS? Mutations in InR decrease the JH level as well as ecdysteroids (Tatar and Yin 2001, Flatt *et al.* 2005), and treatment of a long-lived and sterile InR mutant with a JH analog, methoprene, initiates vitellogenesis and restores normal lifespan. Consistent with a direct role in regulation, InR and JH are both located in the CA. Taken together, these results suggest that JH works downstream of IIS in an endocrine cascade.

The steroid hormone 20-hydroxy-ecdysone (20E) works through the ecdysone nuclear hormone receptor transcription factor to regulate diverse processes during insect development and adulthood. During *Drosophila* larval development, metamorphosis is triggered by ecdysteroids produced in the prothoracic gland. Growth and hormonal signals that regulate larval ecdysteroid production include IIS and PTTH (prothoracicotropic hormone) signaling, which thereby impact the timing of metamorphosis and body size (McBrayer *et al.* 2007, Walkiewicz and Stern 2009). By adulthood, the prothoracic gland degenerates, and ecdysone production is concentrated in the ovarian follicles. Here, regulation is not as well understood, but seems to be under control of IIS, JH, and perhaps other inputs (Tatar and Yin 2001, Flatt *et al.* 2005). In particular, during reproductive diapause, ecdysteroid production is shut down at the ovary. As a consequence of suppressed JH and ecdysteroid production, vitellogenin synthesis in the fat body and uptake into the oocyte comes to a halt.

Interestingly, a modest reduction in ecdysteroid signaling by reduced EcR dosage results in extended lifespan but, surprisingly, increased fecundity (Simon *et al.* 2003). Presumably, more severe reductions negatively impact fertility. Conceivably long-lived IIS and EcR mutants recapitulate various aspects of reproductive diapause. As mentioned above, ablation of germ-line stem cells in *C. elegans* and *Drosophila* results in long-lived adults. In principle, reproductive diapause and germ-line longevity may represent overlapping cellular and molecular pathways. Along these lines, *C. elegans* DAF-12 may be analogous to the EcR in regulating various aspects of developmental timing, lifespan, and life history in response to steroids.

22.12   Torpor/hibernation of mammals

To survive food and water shortage or other hostile environmental conditions, some mammalian species save energy in torpor, a state characterized by low body temperature and reduced metabolic rates (Geiser 2004). Torpor usually lasts only for hours or days, and is typically triggered by fasting or by dietary restriction. Hibernation is prolonged torpor that is mostly seasonal, lasting for the duration of winter. Hibernators rely on fat stores, allowing animals to survive months without feeding. Like overwintering insects, shorter photoperiods and low temperature are environmental cues that induce the response. Torpor/hibernation are states of somatic endurance that enable animals to survive stress. However, the relationship between hibernation and longevity has not been fully explored, although some studies suggest that hibernating species can have reduced IIS and can be longer lived (Storey 2010).

To date, the molecular and physiological basis of torpor is only partly understood. Notably, different long-lived mouse models such as the Ames dwarf show torpor-like reduction of body temperature (Hunter *et al.* 1999). In fact, mutants are often housed with wild-type animals to stay warm. Ames dwarf mice lack the hypothalamic transcription factor, Prop-1, and are deficient for growth hormone (GH), IGF-1, prolactin, and thyroid-stimulating hormone (Bartke 2008b). Snell dwarf mice lacking the transcription factor Pit-1 have similar hormone deficiencies, and reduced body temperature and metabolic rates (Hunter *et al.* 1999). Mice engineered to have lowered core body temperatures themselves have extended lifespan (Bartke 2008b), revealing that this is a determinant of longevity. Alternately, longevity in dwarf mouse models may stem largely from reduced GH–IGF signaling. More generally, reduced insulin/IGF signaling in mammals decreases fecundity and extends life, perhaps in analogy to dauer and reproductive diapause (Bartke 2008b).

The nuclear hormone receptor PPAR-alpha plays a central role in fatty acid oxidation, ketogenesis, glucose metabolism, and torpor induction. Elegant molecular genetic studies in the mouse suggest that, in response to fasting, PPARalpha directly up-regulates FGF21 in the liver (Inagaki *et al.* 2007). FGF21 works as a hormone to coordinate the fasting response in different tissues, stimulating genes involved in lipolysis in white adipose tissue, fatty acid oxidation, and ketogenesis in the liver. PPAR agonists or FGF21 overexpressing transgenics also induce an accompanying reduction in body temperature and physical activity. Thus, PPAR and FGF21 play a critical role in the adaptive response to fasting (Fig. 22-1C).

How might PPAR/FGF21 signaling cascades relate to IIS? Notably, FGF21 transgenics are small and have reduced GH/IGF signaling through STAT5 regulation (Inagaki *et al.* 2008). Transgenics also stimulate PGC-1alpha, a coactivator that works together with FOXO, PPARs, and sirtuins to promote thrifty metabolism (Potthoff *et al.* 2009). A simple hypothesis is that PPAR/FGF21 cascades may induce a metabolic state similar to reduced IIS, with accompanying traits of increased somatic endurance and survival. Interestingly, the long-lived Ames dwarf mice, with reduced GH/IGF signaling, have increased bile acid levels, which are thought to stimulate xenobiotic metabolism and chemical defense through FXR (Amador-Noguez *et al.* 2007). Additionally, FXR mutants are more prone to go into torpor (Cariou *et al.* 2007). The relationship between FXR and IIS is complex in mice, but may parallel regulation of dauer in worms by the bile acid nuclear receptor DAF-12 and IIS (Fig. 22-1C).

Torpor in mammals is also promoted by neuropeptide Y signaling and catecholamines, and opposed by leptin. Injection of NPY into rats or Siberian hamsters induces a reduction in body temperature similar to torpor (Paul *et al.* 2005b). By contrast, loss of NPY-like signaling in worms induces dauer formation, and in flies reduces insulin gene expression (Lee *et al.* 2008c, Cohen *et al.* 2009). Thus, superficially, NPY signaling appears to work in opposite ways in mammals and invertebrates, but it remains possible that specific agonists and antagonists, as well as feedback regulation, can shift the activity of these signaling pathways.

The biogenic amines, epinephrine, and norepinephrine, signaling through the beta-adrenergic receptors, also stimulate lipolysis and torpor (Swoap and Weinshenker 2008). Dopamine beta-hydroxylase mutants (DBH) are deficient in these catecholamines and resistant to fasting-induced torpor. This resistance may stem from elevated leptin, since ob/ob mice lacking leptin signaling restore torpor sensitivity to dbh mice. Recently it was found that norepinephrin inhibits the secretion of GH in different species (Gahete *et al.* 2009). As mentioned above, biogenic amines regulate feeding behavior in worms, and reduction of serotonin signaling promotes dauer formation and stress resistance. Thus biogenic amines may influence feeding behaviors, metabolism, and slow life history traits across taxa.

22.13   Prospectus

The dissection of *C. elegans* dauer formation has led to a wealth of information on the molecular basis of life history regulation. Many of the pathways are conserved in evolution and their counterparts in other organisms may play homologous roles. A comparative approach may help to further elucidate conserved molecular mechanisms and physiology, which couple environmental and physiologic cues to life history traits. It also remains to be seen how many of the discoveries in the laboratory setting will apply to animals in the wild. In future it will be important to understand how the various signaling pathways interact to generate a coherent and consistent output. Furthermore, the molecular mechanisms that constitute proximal components of the output need to be more fully understood.

22.14   Summary

The study of *C. elegans* dauer formation has provided fundamental insights into the phenotypic plasticity of life history traits. A genetic dissection of dauer formation has helped to elucidate the molecular basis of life history plasticity and longevity, and also illuminated intertwined pathways governing developmental timing, stage structure, metabolism, reproduction, and lifespan.

1. A fundamental paradigm has emerged, whereby environmental, dietary, and physiological cues are sensed by the neurosensory apparatus, and integrated and processed into hormonal signals within endocrine cells. In target tissues, cognate receptors coordinate metabolism, growth, reproduction, homeostasis, and stress responses to ensure maximal reproductive success.

2. Molecular analysis of worms, flies, and mice has led to the realization that evolutionarily conserved hormonal pathways, including biogenic amines, steroid, TGF-beta, insulin, and other hormones, govern these processes.

3. Although the interacting pathways are incredibly complex, deeper forays into these areas in the various model systems are bound to turn up new and interesting parallels at physiological and molecular levels.

22.15 Acknowledgments

We would like to acknowledge members of the Antebi laboratory for comments on the manuscript. We apologize to those authors whom we could not cite directly due to space limitations.

CHAPTER 23

The costs of immunity and the evolution of immunological defense mechanisms

Kurt A. McKean and Brian P. Lazzaro

23.1 Introduction

Why we get sick is a complicated question (Nesse and Williams 1996, Stearns and Koella 2008). Immunologists, pathologists, parasitologists, and other practitioners of what have historically been described as the medical sciences have focused almost exclusively on mechanisms of host immune defense and the virulence mechanisms pathogens use to overcome those defenses. For the evolutionary biologist, the focus on proximate mechanism has been supplanted by broader questions concerning why pathogens may evolve to be more or less virulent, or how susceptibility to infectious disease is maintained in populations over time. In many populations successful defense against parasites and pathogens is an important determinant of fitness, and the selection pressure imposed by pathogens is often very strong. The presence of genetic variation for defense in populations experiencing strong directional selection pressure poses an evolutionary problem that we call the"susceptibility paradox": despite the ubiquitous selective pressure that would seem to favor improved host defense, why is there persistent susceptibility to diseases that have been encountered by host populations for generations?

Solutions to the susceptibility paradox involve understanding how evolutionary forces contribute to and shape genetic variation for traits affecting the host–pathogen interaction. While inbreeding and mutational processes may certainly play a role in

generating susceptible individuals, the broader maintenance of susceptibility in large outbred populations, where selection for resistance should be efficient, suggests that natural selection also acts to maintain disease susceptibility (Wakelin and Blackwell 1988). One possibility is that susceptibility is maintained as a consequence of balancing selection, due either to overdominant selection or from negative frequency-dependent coevolutionary interactions between host and pathogen (Jaenike 1978, Hamilton 1980, May and Anderson 1990, Frank 1991, Lambrechts et al. 2006). While examples of balancing selection have been documented (Lively and Dybdahl 2000, Garrigan and Hedrick 2003, Fumagalli et al. 2009), the conditions necessary for maintaining balanced polymorphisms are quite stringent and balancing selection seems unlikely to be the sole factor contributing to the maintenance of susceptibility.

More recently, ecological immunologists have begun to explore more nuanced selective hypotheses, adopting a whole-organism perspective when examining the genetic and environmental factors affecting pathogen defense phenotypes (Rolff and Siva-Jothy 2003, Schmid-Hempel 2003, Siva-Jothy et al. 2005, Lazzaro and Little 2009). Explicit in this approach is the recognition that pathogen defense is costly and that further improvements in defense may not be adaptive because they correlate negatively, or "trade off" with other components of fitness. While there are obvious advantages of pathogen defense, resistance to infection is certainly not the sole determinant of fitness. Organisms must also develop properly and in a timely manner and, as adults, find mates and successfully reproduce. In the absence of genetic correlation among these traits, pathogen defense, rates of development, and reproduction would be predicted to evolve to limits determined solely by physical constraints. Empirical observations showing that populations can respond to artificial selection for resistance (Kraaijeveld and Godfray 1997, Fellowes *et al.* 1998, Ye *et al.* 2009), development time (Nunney 2007), and patterns of reproduction (Richardson and Kojima 1965, Holland and Rice 1999, McKean and Nunney 2008) suggest that organisms are limited not by physical constraints, but rather by the patterns of correlation among these fitness-associated traits.

The assumption that organisms are selected to optimally allocate limited resources among fitness-associated traits is a basic tenet of life history theory and the presence of trade-offs may act as an important limit to adaptation (Stearns 1992, Roff 2002). Trade-offs involving mechanisms of pathogen defense are referred to as the costs of immunity (Kraaijeveld *et al.* 2002, Rolff and Siva-Jothy 2003, Schmid-Hempel 2003, 2005, Siva-Jothy *et al.* 2005), of which there are three types: maintenance costs, deployment costs, and multiple-fronts costs (see Table 23-1). Maintenance costs are the expense of resource allocation toward developing the infrastructure of an immune system and to constitutively expressed (non-induced) immunological mechanisms. Deployment costs are associated with induction of immune mechanisms. Induced immune mechanisms are initiated through the recognition of potential disease-causing agents and result in the production of factors important in mediating defense. Mechanistically, deployment costs could arise

Table 23-1 The evolutionary costs of immunity. Costs can be measured as negative phenotypic or genetic correlation between traits. Evolutionary costs represent genetic correlation and act as potential limits to adaptation. Shown are methods of measuring the evolutionary costs of immunity and representative examples of the measurement of such costs in the literature.

Type of cost	Measurement	Example
Maintenance cost	Correlated responses to selection on defense in fitness-associated traits measured in the absence of infection	Selection for increased resistance to parasitoid wasps saw a correlated decrease in larval competitive ability (Kraaijeveld and Godfray 1997, Fellowes *et al.* 1998)
	Negative genetic correlation between defense and other fitness-associated traits (measured in the absence of infection)	A negative genetic correlation between resistance to a bacterial infection and fecundity of uninfected individuals in an environment in which food was limiting (McKean *et al.* 2008)
Deployment cost	Correlated change in the magnitude of the physiological costs of immune induction following selection for increased defense	Direct measurement of changes in deployment costs in selection lines have not been measured; however, changes in patterns of gene expression following selection for increased defense to *Pseudomonas aeruginosa* are suggestive (Ye *et al.* 2009)
	Genotype-by-immune status (induced versus uninduced) interaction for fitness-associated traits	Reductions in starvation and desiccation resistance associated with successful parasitoid defense varied among iso-female lines of *Drosophila* (Hoang 2001)
Multiple-fronts cost	Selection for improved defense against one pathogen or pathogen strain has a correlated decrease in defense against another	Selection for increased resistance to the parasitoid wasp *Leptopilina boulardi* saw a correlated *increase* in resistance to *Asobara tabida*, a result contrary to the prediction under multiple-fronts costs (Fellowes *et al.* 1999a); significant local adaptation between pathogens and hosts does indicate that divergence in defense may come at a cost of increased susceptibility to other pathogen strains (Lively *et al.* 2004)
	Negative genetic correlation in defense against different pathogens or host-genotype-by-pathogen-genotype interaction for defense against an array of pathogens	Host-genotype-by-pathogen-clone interaction for *Daphnia magna* infected with the bacteria *Pasteuria ramosa* (Carius *et al.* 2001)

either as a consequence of the reallocation of resources during the induced response or from immunopathology.

The last category of the costs of immunity, multiple-fronts costs, is relatively little studied (Table 23-1). Multiple-fronts costs are trade-offs between the efficacy of defense against diverse pathogens or pathogen strains, such that an increase in resistance to pathogen A results in a concomitant decrease in resistance to pathogen B. Thus, multiple-fronts costs are conceptually distinct from defense being pleiotropically linked to fitness traits, such as reproduction, that are not directly involved in defense. Multiple-fronts costs are required for negative frequency-dependent coevolution between pathogen and host. However, even in the absence of negative frequency-dependent coevolutionary interactions, multiple-fronts costs could act to maintain susceptibility if host populations encounter a heterogeneous microbial community containing species or strains to which mechanisms of defense show such antagonistic effects.

The presence of the costs of immunity ensures that high-fitness genotypes are not necessarily those possessing the greatest pathogen defense, but instead are those making the best compromise between the competing needs for whole-organism lifetime reproductive success. The evidence of such costs, especially those associated with immunological maintenance and deployment, has been thoroughly reviewed elsewhere (Kraaijeveld *et al.* 2002, Rolff and Siva-Jothy 2003, Schmid-Hempel 2003, 2005, Siva-Jothy *et al.* 2005). It is not our goal to simply restate this evidence. Instead we will focus on the growing understanding of the molecular mechanisms mediating such costs.

While much is being learned of costs of immunity in birds and mammals, including the molecular mechanisms mediating these costs (see, for example, Martin *et al.* 2008), we emphasize here research carried out in insects, especially the *D. melanogaster* model system. Our focus on *D. melanogaster* is not merely one of convenience. In the past 30 years, *Drosophila* has emerged as a model for studies of both the mechanisms and evolution of immune function and pathogen defense (Brennan and Anderson 2004, Royet and Dziarski 2007, Ferrandon *et al.* 2007, Lemaitre and Hoffmann 2007, Sackton

et al. 2007, Dionne and Schneider 2008, Lazzaro 2008, Imler and Eleftherianos 2009) as well as for the study of life history evolution (see the many examples cited in Roff 2002).

In this chapter we assess our current understanding of the underlying mechanisms mediating the various costs of immunity. First, we evaluate evidence of trade-offs between reproduction and immunity, and the potential mechanisms mediating the allocation of energy and resources between these competing needs. Next we examine costs associated with the induction of an immune response, and how mechanisms of immune regulation have likely evolved to limit the costs of immunological deployment. Lastly we examine evidence for trade-offs between different components of defense, including trade-offs between resistance and tolerance as well as potential trade-offs in defense against different pathogens. We conclude by offering our view on the future of efforts to elucidate the molecular mechanisms of the costs of immunity.

23.2 Innate immune defense in *Drosophila*

Mechanisms of defense act to limit pathogen-associated fitness costs through two separate but interconnected mechanisms: "resistance" to pathogen establishment and proliferation and "tolerance" of established infections. Operationally, resistance is measured as the inverse of the pathogen load and has been the traditional focus in studies of pathogen defense in animals (Schneider and Ayres 2008). Tolerance limits the fitness impact of a given pathogen load, and is measured as the slope of the reaction norm describing the relationship between variation in pathogen load and fitness, or "health" (Boots 2008, Råberg *et al.* 2009, Schneider and Ayres 2008). In other words, genotypes that suffer greater health consequences of a given pathogen load are less tolerant than genotypes suffering less from a similar load. The importance of tolerance mechanisms has long been understood in the plant pathology literature (reviewed in Råberg *et al.* 2009), but only recently have animal pathologists begun to experimentally examine the importance of tolerance as a mechanism of host defense (Råberg *et al.*

2007, 2009, Ayres and Schneider 2008, Schneider and Ayres 2008).

Both tolerance and resistance mechanisms will act to reduce the fitness impacts of infections capable of successful colonization. However, resistance mechanisms reducing the per-contact rate of pathogenic infection are also vital components of overall defense. In the septic environment inhabited by *Drosophila* there is likely a constant interaction between the host and a host-associated microbial community of natural pathogens, opportunistic pathogens, commensals, and mutualists. Indeed, the first line of defense, acting to limit the per-contact rate of infection, may be niche occupation by native gut microbiota acting to prevent colonization of potential pathogens (Dillon and Dillon 2004). Barrier epithelial tissues are a second line of defense limiting the per-contact rate of infection. These barrier defense mechanisms include physical and physiological mechanisms as well as immune mechanisms constitutively expressed in various epithelial tissues. For example, the production of lysozyme, antimicrobial peptides (AMPs), and reactive oxygen species in the *Drosophila* gut creates a hostile environment that is not conducive for the survival of many microbes (Daffre *et al.* 1994, Hultmark 1996, Ryu *et al.* 2006, 2008). There is both constitutive and induced expression of AMPs in epithelial tissues, including the salivary glands, gut, and male and female reproductive tissues (Ferrandon *et al.* 1998, Tzou *et al.* 2000). The expression of immune mechanisms at barrier epithelia may interact with the native microbiota in providing defense. For example, disregulation of antimicrobial peptide expression in the *Drosophila* gut can cause changes in the microbiome, resulting in pathology (Ryu *et al.* 2008).

Various active defense mechanisms form the next line of defense against an incidental or pathological breach of barrier defenses. The mechanisms of immunological defense include encapsulation, melanization, phagocytosis, coagulation, the inducible production of antimicrobial peptides, RNAi, and nutrient sequestration, such as the production of iron-chelating transferrin that may act to limit microbial growth. These defense mechanisms have been reviewed elsewhere (Yoshiga *et al.* 1999, Brennan and Anderson 2004, Dunkov and Georgieva

2006, Ferrandon *et al.* 2007, Lemaitre and Hoffmann 2007, Royet and Dziarski 2007, Dionne and Schneider 2008, Stuart and Ezekowitz 2008, Imler and Eleftherianos 2009). To date, research has focused almost exclusively on resistance-promoting defense mechanisms; only recently have researchers begun to examine defense mechanisms mediating tolerance (Boots 2008, Schneider and Ayres 2008, Råberg *et al.* 2009). Perhaps even less appreciated is the role native microbial symbionts may play in defense to more systemic infections. For example, the secondary symbionts *Hamiltonella defensa* and *Serratia symbiotica* aid in resistance of pea aphids to the parasitoid *Aphidius ervi* (Ferrari *et al.* 2004, Oliver *et al.* 2003, 2005) and another, *Regiella insecticola*, provides increased resistance to the fungus *Pandora neoaphidis* (Scarborough *et al.* 2005).

Innate immune mechanisms are phylogenetically ancient and highly effective in their defense against a broad array of viruses, bacteria, fungi, trypanosomes, nematodes, and parasitoids, indicative of a long evolutionary history of interaction with potential pathogens (Danilova 2006). Furthermore, evolutionary genetic analysis has revealed genes involved in immune function to be some of the most rapidly adaptively genes in the genomes of insects and vertebrates alike (Schlenke and Begun 2003, Nielsen *et al.* 2005, Evans *et al.* 2006, Sackton *et al.* 2007, Waterhouse *et al.* 2007, Lazzaro 2008). Yet genetic variation for disease susceptibility within populations remains, presumably due to costs of immunity that act to slow or prevent the evolution of even greater defense (reviewed in Kraaijeveld *et al.* 2002, Zuk and Stoehr 2002, Schmid-Hempel 2003, 2005, Siva-Jothy *et al.* 2005). We now turn our attention to recent research examining the potential mechanisms mediating such costs and how the costs of immunity have likely played an important role in the evolution of immune system function.

23.3 Trade-offs between reproduction and immunity

Perhaps the trade-off most critically important to host fitness is that between immunity and reproduction. It has become increasingly well established, particularly in insect systems, that mating and reproduction negatively impact pathogen defense through a variety of mechanisms. The mechanisms and costs may vary between sexes, due to differences in mating activity, physiological construction, and interactions with progeny.

Male *D. melanogaster* suffer reduced ability to clear an infection of non-pathogenic *Eschericia coli* after engaging in extensive courting of and copulation with females (McKean and Nunney 2001). This effect can be attributed to energetic expenditure and a simple decrease in the amount of time spent feeding relative to less vigorously reproductive males (McKean and Nunney 2001, 2005). Intense energetic expenditure has also been shown to transiently decrease immune capacity in crickets and other insects (Adamo and Parsons 2006), as well as in humans (Nieman 1999).

Reproduction and immunity also trade-off in female *D. melanogaster*, although through different mechanisms. Despite the fact that mating weakly induces expression of some defense genes (Lawniczak and Begun 2004, Mack *et al.* 2006, McGraw *et al.* 2004, Peng *et al.* 2005, Domanitskaya *et al.* 2007), mated females actually experience a transiently reduced resistance to infection (Fedorka *et al.* 2007, Short and Lazzaro 2010). This post-mating susceptibility does not occur in *D. melanogaster* mutant females that lack germ lines, and depends on the transfer of sperm and seminal fluid proteins by males (Short and Lazzaro 2010). This indicates that the effect is not just due to the act of copulation, but is instead a function of female reproductive physiology and communication with somatic tissue. Similarly, infertile *Caenorhabditis elegans* exhibit an increased resistance to pathogenic bacteria that depends on immunosuppressive signals from the germ line and developing embryos (Miyata *et al.* 2008, Alper *et al.* 2010). Although the current literature is not definitive, there are several lines of evidence supporting the generalized hypothesis that reproductively induced susceptibility to infection is a consequence of the trade-off between energy allocation to reproduction and the need to maintain homeostasis (see Chapter 24 for a broader discussion of metabolically driven trade-offs).

Both immune defense and reproduction are energetically demanding and subject to nutritional availability (see also Chapter 11). In humans, malnutrition and susceptibility to infection are closely intertwined (Katona and Katona-Apte 2008). Intense athletic activity increases susceptibility to infection in humans, although this effect can be mitigated through carbohydrate ingestion (Nieman 1999, 2008). Energetic expenditure and nutritional deprivation also decreases resistance to infection in *D. melanogaster* and other insects (McKean and Nunney 2001, 2005, Adamo and Parsons 2006), suggesting that energy accumulation and storage are important for resistance. Negative genetic correlation between resistance and fecundity (measured prior to immune challenge) has been observed in *D. melanogaster*, demonstrating the potential for evolutionary costs of immune maintenance. These costs, however, were apparent only in an environment in which adult food availability was limiting. In a nutritionally superior environment the maintenance costs were erased (McKean *et al.* 2008), suggesting that food availability may affect the expression of genetic variation for these traits.

Reciprocally, infection status can impact metabolic state. Severe and chronic infection causes substantial metabolic change in humans (Powanda and Beisel 2003) and insects (Dionne *et al.* 2006, Schilder and Marden 2006). In insects, metabolic phenotypes of infection can include elevation of hemolymph (blood) sugar levels and loss of glycogen and triglyceride stores. This infection-mediated metabolic syndrome is similar to that of mammalian diabetes and is regulated by insulin-like signaling (Dionne *et al.* 2006, Schilder and Marden 2006). In *D. melanogaster*, immune stimulation results in the marked down-regulation of a large number of genes involved in basal metabolism (De Gregorio *et al.* 2001). There is also a complex interaction between immunity and lipid metabolism in vertebrates (Kyriazakis *et al.* 1998, van den Elzen *et al.* 2005) and insects (Mullen *et al.* 2004, Adamo and Parsons 2006). The complete mechanistic basis for the interaction between dietary lipid content and immune performance is not known, but in at least some insects high dietary lipid levels can reduce resistance to infection through depletion of a hemolymph lipid carrier that pleiotropically functions in immune defense (Adamo and Parsons 2006, Adamo *et al.* 2007).

Although the effects of egg production on female metabolic state have not been thoroughly studied in *D. melanogaster*, there is considerable indirect evidence attesting to its substantial energetic requirements. Mating significantly alters expression of metabolic (as well as immune-related) genes in *D. melanogaster* females relative to virgin controls (Innocenti and Morrow 2009). *D. melanogaster* females cease egg production under starvation conditions, and may even apoptose and resorb incompletely developed eggs (Bownes *et al.* 1988, Terashima and Bownes 2004). The fat body, which is the organ primarily responsible both for systemic humoral immunity in response to challenge and for the nutritional provisioning of developing eggs, may be the tissue most acutely involved in mediating the competing demands between reproduction and immunity.

The insulin/insulin-like growth factor (IGF) and juvenile hormone (JH) endocrine signals are prime candidates for mediating a physiological trade-off between immunity in fecundity in insects (see Chapter 13 for further discussion of the role of endocrine signaling in life history trade-offs). As mentioned above, chronic infection in insects can result in an insulin-mediated metabolic syndrome (Dionne *et al.* 2006, Schilder and Marden 2006). IGF signaling also directly influences egg production by regulating JH levels, which are reduced 80% in insulin insensitive mutants (Tatar *et al.* 2001b). JH, which is activated in females following mating by the male seminal protein, sex peptide (Moshitzky *et al.* 1996, Innocenti and Morrow 2009), initiates yolk protein production and vitellogenesis in the fat body (Bownes 1982). Ectopic provision of JH can stimulate yolk protein production and egg maturation under what would otherwise be starvation conditions (Bownes *et al.* 1988). At the same time, JH is itself a strong inhibitor of insect immune systems under standard nutritional conditions (Rolff and Siva-Jothy 2002, Flatt *et al.* 2008a). In an exciting development, the expression of AMP genes was shown to be directly induced by FOXO under starvation conditions and in the absence of infection (Becker *et al.* 2010). This regulation is independent of the Toll and Imd signaling pathways, which regulate canonical inducible AMP expression, and provides one potential mechanism for metabolic modulation of immune activity.

Although it has not been experimentally demonstrated, it is in principle possible that genetic polymorphism in insulin-like signaling or reproductive endocrinology could mediate evolutionary trade-offs between immunity and reproduction. *D. melanogaster* artificially selected for longevity and late-life fecundity show correlated responses of decreased metabolism and increased energy storage through fat accumulation (Djawdan *et al.* 1998), very similar to the phenotype observed in mutant flies with decreased insulin signaling (Broughton *et al.* 2005, Giannakou *et al.* 2007). Common genetic variation in pleiotropic regulators of immunity and other fitness traits, including reproduction and metabolism, could provide an evolutionary basis for life history trade-offs involving immune defense.

23.4 Deployment costs, tolerance, and the evolution of immune regulation

The very fact that some defense mechanisms are inducible suggests that their constitutive expression would be costly (Harvell 1990, Frost 1999). A number of studies have now demonstrated significant costs associated with the induction of immune defense mechanisms in both vertebrate and invertebrate systems (reviewed in: Kraaijeveld *et al.* 2002, Zuk and Stoehr 2002, Schmid-Hempel 2003, 2005, Siva-Jothy *et al.* 2005). In this section we discuss two important issues concerning the evolution of the costs of immunological deployment and the mechanisms mediating those costs. First, we argue that the evolution of mechanisms for actively downregulating immune responses has come about as a consequence of natural selection favoring genotypes with reduced deployment costs. Therefore, understanding the mechanism of immune regulation provides insight into the mechanisms specifically evolved to mitigate the costs of immunological deployment. Second, we discuss the need to demonstrate that populations exhibit genetic variation in the physiological cost of immunological deployment in order to understand whether such costs may affect the evolution of defense.

For many of the same reasons that immune responses have evolved to be inducible as opposed to constitutive, we predict that selection will favor the eventual termination of the induced state. The null hypothesis, which does not require the action of selection favoring termination, is that there is a passive diminution of the induced state arising simply as a consequence of the gradual elimination of immune-inducing agents. Alternatively, the diminution of the induced state could arise as a consequence of active down-regulation (Schneider 2007). We predict active down-regulation will evolve if:

• future encounters with the pathogen are variable and unpredictable
• the continued expression of the defense is costly
• more rapid termination of the response, compared to passive diminution, reduces deployment costs without drastically affecting the efficacy of defense.

There is now abundant evidence that active negative regulation of immune responses have evolved in both vertebrates and invertebrates (Schneider 2007, Serhan and Savill 2005). Negative regulators bring about the resolution of induction of both the Toll and Imd pathways (Aggarwal and Silverman 2008). For example, wntD is a gene in the Wnt gene family that has been found to inhibit the translocation of the NF-κB transcription factor, Dorsal, into the nucleus (Gordon *et al.* 2005). Knockout mutants for wntD show dramatically increased constitutive and induced expression of the antimicrobial peptide, diptericin. Paradoxically, while these mutants show increased antimicrobial peptide expression they appear more susceptible to infection by both *Micorococcus luteus* and *Listeria monocytogenes* (Gordon *et al.* 2005, 2008). The wntD mutants also show increased expression of the immune-induced gene, Edin, a gene of unknown function but whose expression is required for resistance to *L. monocytogenes* infections. Interestingly, over-expression of Edin using the UAS/Gal4 system also results in reduced survival following infection, suggesting that this gene is at least partially responsible for the observed cost of immune-induced pathology (Gordon *et al.* 2008). The precise cause of the observed pathology, due either to disruption of energy reallocation during the immune response or perhaps the immune defense attacking host tissues, is unknown.

The transcription factors AP-1 and STAT appear to be inhibitors of the Imd-pathway NF-κB transcription factor Relish, acting to form complexes that replace Relish at promoter regions and preventing further transcription (Kim *et al.* 2005, 2007). Mutation in these genes results in increased expression of antimicrobial peptide genes such as Attacin A. Similar to wntD mutants, the over-expression of Relish target genes resulted in an increased rate of clearance of *E. coli* but a more rapid death of flies following infection (Kim *et al.* 2007). Again, whether this immune-induced pathology results from the disruption of changes in energy balance during the immune response or due to immunopathology is not known. Whether mutations in these negative regulators of induced immune responses affect deployment costs in other fitness-associated traits, such as fecundity, has to date not been investigated.

The discovery of these and other negative regulators of induced immunity suggest that a significant portion of the evolution of immune systems and immune responses may be toward the mitigation of costs associated with the activation and deployment of the mechanisms of defense. This would include not only the active regulation of the transcriptional response discussed here but also in the localized containment of inflammation to limit tissue damage (e.g., Sadd and Siva-Jothy 2006), and perhaps even the evolution of the various mechanisms of immunological memory, which would act to reduce deployment costs upon encountering the same pathogen.

Our view of the importance of deployment costs in shaping the evolution of immune function has important implications for discussion of the mechanisms mediating tolerance to infection in *Drosophila* and other organisms. Schneider and Ayres (2008) have defined three classes of tolerance mechanisms based on the effect the tolerance mechanism has on resistance. Class I mechanisms are typified by effector molecules causing immunopathology. The constitutive or induced expression of such molecules may act to increase resistance, but will decrease tolerance compared to molecules better able to target non-self entities without collateral damage to self. Class II mechanisms are typified by regulators with pleiotropic effects on both resistance and tolerance.

Such regulators enhance resistance by increasing the strength or duration of an immune response but may act to decrease tolerance to infection if these immune responses have immunopathologic consequences or greatly disrupt energy balance. The examples we have provided of the negative regulation of immune responses in *Drosophila*, by molecules such as WntD, AP-1, and STAT, all fall into this second category, and indeed research has indicated that mutations in these genes act to increase resistance but negatively affect the fly because of decreased tolerance. In our view, these first two classes of tolerance mechanism simply reflect the costs of immunity.

The third class of tolerance mechanisms described by Schneider and Ayers (2008) are those that can be separated from resistance mechanisms. Examples include detoxifying enzymes that "clean up" toxins produced by the host or pathogen during infection, genes involved in energy reallocation during the immune response, mechanisms preventing physiological damage, and repair mechanisms. Certainly the reallocation of energy as a consequence of immune system induction is at the heart of how we think about the costs of immunological deployment. Other mechanisms promoting tolerance in a resistance-independent manner may contribute to the costs of defense, but these costs are not necessarily a direct consequence of the expression of resistance mechanisms, as is the case for class I and class II tolerance mechanisms.

Our hypothesis that active down-regulation of immune responses is adaptive assumes that ancestral populations possessed genetic variation in immunological cost of deployment upon which natural selection could act. Whether contemporary populations possess such variation is a little-studied question, but one of vital importance if we are to understand the potential for future evolutionary change in deployment costs and the contribution of such costs to the maintenance of disease susceptibility. The evolutionary costs of immunological deployment may be evaluated in two ways. First, using standard quantitative genetic designs, a significant genotype-by-immune status interaction, comparing naïve and immune-induced individuals, would indicate that the population possesses genetic variation in the physiological cost of deployment upon

which selection could act. An alternative would be to evaluate whether there has been evolutionary divergence in the physiological cost of deployment following experimental evolution for increased defense. In both types of experiment, it is necessary to evaluate the physiological cost of deployment using non-infectious immune inducers so that the costs of deployment can be distinguished from those costs caused by the pathogen itself.

Because of the relative ease of experimental induction of the immune system, there are many examples, in both vertebrates and invertebrates, of physiological costs of deployment (reviewed in: Kraaijeveld *et al.* 2002, Zuk and Stoehr 2002, Schmid-Hempel 2003, 2005, Siva-Jothy *et al.* 2005). However, there is a great paucity of studies examining evolutionary costs of deployment, which evaluate whether genotypes vary in the physiological cost experienced. Flies having successfully defended themselves against the parasitoid wasp *Asobara tabida* are smaller and have reduced fecundity, indicating a significant physiological cost of immune deployment (Fellowes *et al.* 1999b). In a study examining whether deployment costs vary among genotypes, successful defense against attack resulted in reduced desiccation and starvation resistance, and the magnitude of this cost varied among iso-female lines (Hoang 2001). It should be noted, however, that in this type of experiment it is impossible to distinguish the costs of deployment from costs arising from the activities of the wasp itself. We recently examined the evolutionary costs of deployment in response to heat-killed bacteria, which would therefore control for confounding costs associated with pathogenicity, and found more equivocal results (McKean *et al.* 2008). None of the cited studies have attempted to elucidate potential underlying genetic mechanisms. Nonetheless, where evolutionary costs of immunological deployment are found, the logical next experimental step will be to evaluate whether polymorphisms in pathways responsible for regulation of immune responses are associated with variation in the physiological costs of deployment experienced by different genotypes.

In addition to active negative regulation of immune responses, there may be coordinated changes in other physiological systems during an immune response that can reveal mechanisms of deployment costs. The use of microarray technology, allowing researchers to examine genome-wide changes in patterns of gene expression, is likely to be an important tool in understanding the mechanisms underlying deployment costs. This is well illustrated in a recent study conducted by Ye *et al.* (2009).

Ye *et al.* (2009) artificially selected for increased resistance to the bacterial pathogen, *Pseudomonas aeruginosa*. In the three replicate populations, defense increased from ~15% survival after infection to ~70% survival in 10 generations of selection. Correlated decreases in female longevity and egg viability were observed in uninfected flies from the selection lines, compared to controls, indicating evolutionary costs of immunological maintenance. The researchers then examined evolved changes in patterns of gene expression following infection using whole-genome microarrays. The experimental design did not include a comparison of expression patterns in uninfected control and selection lines, which means that it is impossible to determine if the observed differences are due to evolved changes in the induced transcription of genes following the challenge, or if the changes represent evolved changes in the constitutive expression (or elements of both). Nonetheless, a number of interesting patterns do emerge, including the fact that genes in both the Toll and Imd pathways, as well as genes involved in phagocytosis, evolved significantly greater expression in the selection lines (Ye *et al.* 2009).

There are, however, a number of genes that evolved decreased expression in the selection lines. Overall there were 110 genes showing significant reductions in expression in at least two of the selection lines and 67 which evolved reduced expression evolved in all three lines. We obtained this list of genes from the authors in order to determine whether the function of these significantly downregulated genes may provide clues associated with potential costs of immunity. First, we averaged the p-values of the three comparisons of control and selection lines and rank-ordered the 67 genes showing evolved decreases in expression. The gene showing the most consistent evolved reduction in expression was Chorion protein 15 (average p-value = 8.52×10^{-7}). This gene is important for eggshell

formation and is expressed late during the process of oogenesis (Cavaliere *et al.* 2008). A previous microarray analysis of induced expression following infection with *P. aeruginosa* did not indicate chorion protein 15 was down-regulated following infection (Apidianakis *et al.* 2005), raising the interesting possibility that evolved reductions in the constitutive expression of this gene may contribute to the observed maintenance cost of reduced egg viability in the selected lines.

We also examined whether the 110 genes that had evolved significant reductions in expression could be assigned to specific biological processes based on gene ontology terms that were over-represented (enriched), using the functional annotation clustering tool in the bioinformatic resource DAVID, the database for annotation, visualization, and integrated discovery (Dennis *et al.* 2003, Huang *et al.* 2007, 2009). Using the highest stringency for the DAVID fuzzy clustering, we found genes involved in chitin metabolism to be over-represented in this gene list (enrichment score = 3.5, where scores greater than 1.3 are considered to be "interesting"). We found similar results when the list of 67 genes that had evolved reduced expression in all three selection replicates was submitted to DAVID. The biological significance of decreased chitin metabolism is unknown, as is whether or not the evolutionary change was in the constitutive or induced expression of the genes. Regardless, the combination of experimental evolution and microarray analysis will undoubtedly prove to be a rich source of hypotheses for further research concerning the molecular mechanisms mediating evolutionary trade-offs.

23.5 Multiple-fronts costs of immunity

Multiple-fronts costs of immunity arise when increased defense against one type of pathogen or pathogen strain has a correlated decrease in defense to another. Such costs are very little studied, and in *D. melanogaster* the two studies examining such costs have failed to find supporting evidence. In the first study, selection for increased resistance to the parasitoid wasp *L. boulardi* had a correlated increase in resistance to *A. tabida*, while selection for increased resistance to *A. tabida* saw no change in resistance to

L. boulardi (Fellowes *et al.* 1999a; see Table 23-1). In the other, pairwise correlation coefficients for resistance to four different bacterial pathogens (*S. marcescens*, *P. burhodogranaria*, *E. faecalis*, and *L. lactis*) tended to be nonsignificantly positives across a set of 95 inbred lines (Lazzaro *et al.* 2006). There is mechanistic evidence that suggests that multiple-fronts costs may exist in insects, but that these are based in antagonisms between different components of the immune system. For instance, in the Egyptian cotton leafworm *Spodoptera littoralis*, humoral antibiotic activity is slightly negatively correlated with the number of circulating hemocytes, which may be involved in phagocytosis and whose abundance is positively correlated with phenoloxidase activity (Cotter *et al.* 2004). Hemolymph antibacterial activity and phenoloxidase activity have also been negatively correlated in bumblebees (Moret and Schmid-Hempel 2001, Wilfert *et al.* 2007) and the cabbage looper *Trichoplusia ni* (Freitak *et al.* 2007). As different arms of the immune system are differentially important in resistance to distinct pathogens, trade-offs among immune system arms can be taken to suggest that multiple-fronts costs probably exist.

Intriguingly, there are multiple instances of loss-of-function mutations in immune-related genes that have conflicting effects on defense against different microbial pathogens (Dionne and Schneider 2008). For example, loss-of-function mutation in the TNF family ligand, eiger, increases *D. melanogaster* defense against *S. typhimurium* and *M. marinum* but results in reduced defense against *S. aureus*, *E. faecalis*, *S. pneumonia*, *B. bassiana*, and *B. cepacia* (Brandt *et al.* 2004, Schneider *et al.* 2007). Similarly, inactivation of insulin signaling in chico mutants increases *D. melanogaster* defense against *E. faecalis* and *P. aeruginosa* (Libert *et al.* 2008), but such inactivation reduces defensive capabilities in infections with *M. marinum* (Dionne *et al.* 2006). Lastly, loss-of-function mutation in serine protease 7 (CG3066) has complex effects on the defense of *D. melanogaster* against a panel of pathogens, where the mutation increases resistance to *S. pneumoniae*, decreases resistance to *L. monocytogenes* and *S. typhimurium*, increases tolerance but decreases resistance to *E. faecalis*, decreases tolerance but increases resistance to *B. cepacia*, and lastly had no effect on resistance or tol-

erance to infection with *E. coli* (Ayres and Schneider 2008). The examination of pathogenic outcomes across a broad array of pathogens and genetic mutants demonstrates the complexity of innate immune function and indicates potential molecular mechanisms mediating trade-offs in defense against different pathogens. It remains untested but fully expected that naturally occurring genetic variation has similar pleiotropic effects on the outcome of infections with these different pathogens.

23.6 Future directions

Studies of life history evolution, including the costs of immunity, have proceeded largely without understanding of the physiological and genetic mechanisms mediating trait variation and trade-offs (Roff 2007b). Indeed, our understanding of the costs of immunological maintenance and deployment has advanced almost exclusively without reference to underlying mechanisms. This stands in stark contrast to our functional understanding of immunological defense, where an incredible amount is known concerning mechanism (Brennan and Anderson 2004, Ferrandon *et al.* 2007, Royet and Dziarski 2007, Dionne and Schneider 2008, Imler and Eleftherianos 2009). The detailed understanding of the molecular mechanisms of immunological defense is a testament to the power of the reductionist research paradigm of the biomedical sciences, but this reductionist approach has itself come at a cost, neglecting the whole-organism perspective that can provide insight into how the competing demands of development, reproduction, and pathogen defense may contribute to the maintenance of disease susceptibility. Recognizing the importance of the costs of immunity in determining patterns of immune function and the pathogenic outcomes of infection requires a reappraisal of what is meant by effective defense and an evaluation of the heuristic value of exclusively reductionistic approaches (Little *et al.* 2008).

We have identified a number of potential molecular mediators of the various costs of immunity, including hormonal mechanisms, insulin signaling, negative regulators of immune induction, changes in global patterns of gene expression, and genes showing antagonistic pleiotropic effects on patterns of defense against different pathogens. What is missing is an explicit test of the hypothesis that polymorphism in these various mechanisms promotes the observation of evolutionary trade-offs (i.e., negative genetic correlation) that could act to slow the adaptive evolution of greater immune defense. In addition to analysis of how polymorphism in these particular mechanisms affects the observation of the costs of immunity, we suggest three areas of research that we feel will be important as the field moves forward.

First, it is becoming clear that genotype-by-environment interactions may affect the magnitude or even appearance of the costs of immunity. For example, the maintenance cost of reduced larval competitive ability associated with selection for resistance to parasitoid wasps was only observed in a high-density larval environment (Kraaijeveld and Godfray 1997, Fellowes *et al.* 1998). Likewise, an evolutionary trade-off between fecundity and resistance to a bacterial pathogen was only observed in an environment in which food was limiting (McKean *et al.* 2008). Even host-genotype-by-pathogen-genotype interactions may vary in different environments, although such three-way interactions have been little studied (Vale and Little 2009). Such environmental heterogeneity may itself act to maintain disease susceptibility (Lazzaro and Little 2009). Moving life history analysis forward, it seems that researchers should be very aware of the environmental conditions under which they are carrying out their experiments and report them, including the density, temperature, sex of the study organism, whether individuals were kept in same-sex or mixed sex conditions, the time of day immune challenges were carried out, the infective dose of immune challenge, and the food availability. Ideally, the impact of variation in these and other environmental variables on the costs of immunity and the pathogen-host interaction will continue to be investigated. One sobering lesson is that the strong effect of environmental variation on the observation of the costs of immunity may make it difficult to infer the importance of such costs in nature from experiments carried out in a laboratory setting.

Second, we know very little of the totality of interactions between hosts and their associated microbial community. As already mentioned, this community is composed of natural pathogens, opportunistic pathogens, commensals, and mutualists, each of which poses unique problems influencing the evolution of immune function and potential costs associated with that evolution. For example, if the vast majority of interactions between hosts and the microbial community are not with natural pathogens capable of causing disease, then we may expect immune responses to be tightly regulated so that potential threats are dealt with quickly, reducing potential deployment costs. Furthermore, the presence of microbial mutualists promoting defense through either niche occupation or other means likely pose a significant problem to immune defense and its evolution. It is conceivable that evolutionary increases in the expression of immune defense mechanisms against various natural pathogens may come at a cost of adversely affecting the community of mutualists and commensals that themselves provide some protection. Indeed, under normal conditions there appears to be a delicate balance between NF-κB mediated transcription of antimicrobial peptides and the repression of expression by the homeobox transcription factor Caudal. Knocking down Caudal expression by RNAi in D. melanogaster in increased antimicrobial peptide production and a pronounced change in the community of bacteria found in the gut, causing pathology in the flies (Ryu *et al.* 2008). In other words, the costs of immunity could be manifest in part through the effect of evolutionary changes in patterns of defense on the community of mutualists and commensals associated with the host.

Lastly, the analysis of the molecular mechanisms mediating the costs of immunity have for the most part ignored developmental changes that could lead to the observation of trade-offs. It seems highly likely that, especially for maintenance costs, decisions concerning the allocation of resources toward the competing needs of developing an immune system and the development of other structures that affect fitness may be made early on. For example, the increased resistance to the parasitoid wasp *Asobara tabida* appears to be due to a two-fold increase in the number of circulating hemocytes in the larvae (Kraaijeveld *et al.* 2001). Kraaijeveld *et al.* (2001) have hypothesized that the mechanism promoting the observed cost of reduced larval competitive ability in the selected lines was due to an evolutionary change in the allocation of resources toward hemocytes versus toward jaw muscles, an allocation decision that would have affected developmental trajectories very early in development. The developmental signals mediating these allocation decisions, and therefore leading to the observation of maintenance costs, are completely unexplored.

23.7   Summary

1. Given the presumed benefits of improved pathogen defense, the continued maintenance of disease susceptibility poses an evolutionary problem that we call the "susceptibility paradox."

2. The costs of immunity are manifest as trade-offs between defense and other components of fitness and include maintenance costs, deployment costs, and multiple-fronts costs (Table 23-1). The presence of such costs may impose limits on immune defense evolution and thus act to maintain disease susceptibility and provide a solution to the susceptibility paradox.

3. Trade-offs between defense and reproduction may result from the dependence of both processes on nutritional and metabolic stores. This suggests that metabolic genes or metabolism-related endocrine signals may pleiotropically link reproduction and immunity, and that variation in these genes may underlie observed trade-offs.

4. Deployment costs may arise from energy reallocation or immunopathology. We argue that such costs are intimately associated with mechanisms of immune regulation and that negative regulators of immunity have evolved specifically to reduce deployment costs.

5. Very little is known of the potential evolutionary costs of deployment. Should, however, genotypes vary in the physiological cost of deployment experienced, then polymorphism in negative regulators of immune induction become obvious candidates mediating these evolutionary trade-offs.

6. Multiple-fronts costs are manifest as trade-offs in defense against different pathogens and have been little examined. Some studies have found negative genetic correlation between different components of the immune system, suggestive of such costs. Recent genetic studies have identified loss-of-function mutations with antagonistic pleiotropic effects on defense. It is unknown if naturally occurring genetic variation is similarly shaped by such antagonistic effects.

7. The field of ecological immunology and the study of trade-offs associated with the costs of immunity is a relatively recent development in evolutionary ecology. There is still a tremendous amount to be learned concerning the molecular mechanisms mediating the costs of immunity and how such costs affect the evolution of immune defense.

CHAPTER 24

Intermediary metabolism and the biochemical-molecular basis of life history variation and trade-offs in two insect models

Anthony J. Zera and Lawrence G. Harshman

24.1 Introduction

For decades, the physiological causes of life history evolution have been a central focus of research in evolutionary biology (Townsend and Calow 1981, Stearns 1989, Rose and Bradley 1998, Zera and Harshman 2001, 2009, Flatt *et al.* 2005, Harshman and Zera 2007, Roff 2007b, Boggs 2009; also see Chapters 2, 11, and 26). Physiology is used here in a broad sense to denote organismal function (e.g., metabolism, endocrine regulation) at all levels of biological organization (e.g., whole-organism, biochemical, and molecular). A central issue has been the nature of evolutionary changes in internal resource allocation to key organismal processes, such as growth, maintenance (respiration), energy storage, and reproduction. The ultimate goal has been to identify how these physiological changes have contributed to the evolution of individual life history traits (e.g., increased early-age fecundity), and trade-offs between traits (i.e., negative associations between traits such as reproductive effort and somatic investment or longevity).

Until recently, most physiological studies have been undertaken at the level of whole-organism physiology and considerable progress has been made in this area at the interspecific and, more recently, the intraspecific levels (Rose and Bradley 1998, Zera and Harshman 2001, 2009, Prasad and Joshi 2003, Harshman and Zera 2007). For example, as discussed in more detail below, numerous labo-

ratory selection studies have identified strong positive genetic associations between somatic energy reserves (e.g., lipid, glycogen) and specific life history adaptations, such as resistance to starvation, extended longevity, and increased locomotion (e.g., dispersal). Conversely, somatic reserves are often reduced in individuals from populations selected for elevated reproductive effort, a trait that often trades off with the life history traits mentioned above. These studies collectively point to a central role played by evolutionary changes in relative nutrient allocation to somatic reserves versus reproduction in the evolution of life histories. However, until recently, the biochemical and molecular mechanisms underlying these evolutionary changes in physiology have been under-studied, and thus have largely remained a black box.

During the past decade, there has been increasing focus on the biochemical and molecular bases of resource allocation to life history traits, at the level of intermediary metabolism (Arking *et al.* 2000, Zera and Harshman 2001, 2009, Zera 2005). Intermediary metabolism essentially comprises the ensemble of biochemical pathways involved in the production, conversion, and utilization of key classes of molecules, such as lipid (triglyceride, fatty acid), carbohydrate (glycogen and glucose), protein (amino acids), and various antioxidants. The major impetus for undertaking these studies has been the growing appreciation that important constituents of life histories, such as yolk protein for eggs, triglyceride and glycogen for somatic functions, energy for growth and maintenance, and antioxidants to combat oxidative stress, are products of specific metabolic pathways. Thus, a deep understanding of many important mechanistic aspects of life history evolution, such as the nature of life history adaptations and the causes of life history trade-offs, requires detailed information on biochemical and molecular modifications of specific pathways of metabolism that produce molecular components of life histories.

Because of space limitation we will mainly focus on two experimental models that have figured prominently in recent studies of intermediary metabolism and life history microevolution: the wing-polymorphic cricket *Gryllus firmus* and the fruitfly *Drosophila melanogaster*. Studies of these two insect models complement each other. Wing-polymorphic crickets have been especially useful in illuminating the mechanisms by which variation in enzyme activity gives rise to variation in flux through pathways of lipid metabolism and accumulation of lipids important for different life history functions (Zhao and Zera 2002, Zera 2005, reviewed in Zera and Harshman 2009). These types of biochemical data are importance because they provide the functional link between molecular variation, such as variation in DNA sequence or gene expression (e.g., transcript abundance), and variation in whole-organism physiology. These biochemical data are especially important given the increasing number of studies reporting on transcriptomic data relating to life histories (e.g., Pletcher *et al.* 2002, McCarroll *et al.* 2004, Sorensen *et al.* 2007, St-Cyr *et al.* 2008). As pointed out by Feder and Walser (2005), transcriptome data are often not reliable indicators of corresponding variation in protein activity or fitness.

Drosophila melanogaster has been an important genetic model in the study of life history trade-offs (For reviews see Rose and Bradley 1998, Zera and Harshman 2001, 2009, Prasad and Joshi 2003, Flatt *et al.* 2005, Harshman and Zera 2007), and a number of investigations have focused on biochemical and molecular correlates of life history variation in the laboratory. There is a wealth of genetic and molecular tools available for this species, which makes it an extremely powerful model for functional studies of

life history evolution. In addition, *D. melanogaster* has been extensively studied since the 1970s with regard to the population genetics of enzymes of intermediary metabolism (Eanes 1999, Sezgin *et al.* 2004, Flowers *et al.* 2007). Recent life history studies in natural populations of this species (e.g., Schmidt and Paaby 2008) allow an integration of the extensive data on enzyme polymorphism with variation in life history traits.

24.2 *Gryllus firmus*: Biochemical and molecular studies of trade-offs in lipid metabolism and life histories

24.2.1 Background on life history variation in *Gryllus* and methodological perspective

During the past decade, dispersal polymorphism in *Gryllus* has developed into a prominent experimental model for the investigation of the proximate mechanisms underlying life history variation and trade-offs (for reviews see Zera and Harshman 2001, 2009, Roff and Fairbairn 2007b, Zera 2009). The polymorphism involves morphs (discrete phenotypes) that differ substantially in numerous morphological, biochemical, endocrine, and behavioral traits related to dispersal ability and reproduction. These differences can result from variation in genetic and/or environmental (e.g., photoperiod, temperature, density) factors, although the majority of physiological studies have focused on genetic differences (i.e., between artificially-selected populations; see below). The collective result of these differences is a short-winged (SW) morph that cannot fly but which has considerably greater (about 200–400%) egg production compared with the flight-capable morph—long-winged with functional flight muscles = LW(f)—even in the absence of flight (see Figure 1 of Zera and Harshman 2009 or Zera 2009 for morphological and reproductive characteristics of morphs). In other words, reproductive effort is reduced in the LW(f) morph as a consequence of the cost of flight *capability*. Wing polymorphism is the most dramatic example of the trade-off between dispersal (dispersal capability) and reproduction, a prominent trade-off in animals.

There are several noteworthy advantages of dispersal polymorphism in *Gryllus* as an experimental model to study life history physiology. Most importantly, as mentioned above and discussed in detail below, dispersing and flightless/reproductive morphs exhibit very large differences in life history traits and their underlying physiological correlates, and occur commonly in natural populations. The existence of these large-magnitude differences considerably simplifies experimental analyses at every level. Individuals of this species are large (0.5–1 g adults), which also allows experimental analysis of many aspects of physiology, such as measurement of organ-specific enzyme activity and energy reserves, measurement of blood hormone levels, and organ transplants, which are not possible or are much more difficult to undertake in smaller model organisms such as *Drosophila*. Finally, there is currently extensive background information on numerous physiological aspects of the dispersal–reproduction trade-off in various *Gryllus* species (nutrient input, systemic endocrinology, energy reserves), which provides an important functional context for studies of the biochemical and molecular causes of life history variation and trade-offs. On the other hand, a major limitation of an organism such as *Gryllus*, which is not a traditional genetic model organism, is the lack of genomic information and the lack of sophisticated genetic and molecular tools that are readily available in other genetic models such as *D. melanogaster*, *Caenorhabditis elegans*, and *Mus musculus*.

A hallmark of the physiological studies of life history specialization in *Gryllus* is their integrative (both horizontal and vertical) nature (Fig. 24-1). Horizontal studies have investigated multiple contributors to a particular level in the functional hierarchy, most notably activities of various enzymes that contribute to flux through the lipogenic pathway. Such studies are useful in pinpointing the identity and extent to which adaptive variation at a particular hierarchical level is due to variation in one or multiple factors (e.g., one or many enzymes in a particular pathway). Such studies constitute the core of "systems" analyses (e.g., Fell 2003, Flowers *et al.* 2007). Vertical studies of *Gryllus*, by contrast, have focused on the chain of causality through various levels of biological organization, for example from the expression of an individual gene product (e.g., *Nadp⁺-Idh* transcript abundance; discussed below), through aspects of intermediary

metabolism (enzyme specific activity, lipogenic flux), to the terminal whole-organism feature, triglyceride concentration. Vertical studies are important because they identify the extent to which adaptive variation at one level of the biological hierarchy is transformed into adaptive variation at another level (e.g., transcript abundance to enzyme activity, or enzyme activity to pathway flux). Because of the often highly non-linear relationships between variation at different hierarchical levels (Fell 2003, Dykhuizen and Dean 2009), the extent to which such transformations occur is an open issue that needs to be investigated empirically. However, such studies are rare and our understanding of the mapping of genotype onto phenotype remains minimal (Houle 2010). To our knowledge, integrative studies of *Gryllus* currently comprise the most comprehensive analyses of the relationship between variation in a particular aspect of intermediary metabolism and life history variation and trade-offs found in outbred populations.

24.2.2   Lipid reserves: The physiological context of biochemical studies of life history trade-offs

As mentioned previously, there is extensive information on morph-specific differences in the concentration of various macromolecules, most notably various types of lipid. This provides important context for biochemical and molecular studies of life history trade-offs in *G. firmus*. The two main components of lipid are triglyceride, which comprises about 85% of total lipid, and phospholipid, which comprises most of the remaining 15%. Triglyceride is energetically expensive to biosynthesize and is the main flight fuel in *G. firmus*, as is the case for many other insects (grasshoppers, cockroaches), and it typically occurs in high concentration in these species (Beenakkers *et al.* 1985, Downer 1985, both cited in Zera 2005; note that because of space limitation, some references will be cited in other references rather than being listed in the References section). Because triglyceride accumulation is expected to be a significant somatic cost of flight capability that trades-off with early age fecundity, triglyceride accumulation has been a primary focus of physiological studies of life history specializations in *Gryllus*.

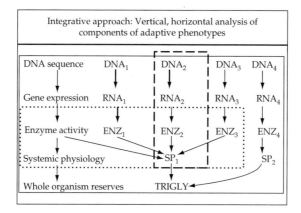

Figure 24-1 Vertical and horizontal integrative studies of the functional causes of life history variation in *Gryllus*. Vertical studies (within dashed-lined box) focus on the chain of causality of a single factor through several levels of the biological hierarchy. In this case, variation in the sequence/expression of a particular gene results in associated variation in enzyme activity, flux through the pathway in which the enzyme functions, and, finally, variation in a whole-organism trait. Horizontal studies (within stippled-lined box) focus on multiple components of variation at a particular level of the biological hierarchy. In this case, variation in multiple enzymes gives rise to variation in flux through the pathway in which the enzymes function. Subscripts refer to different genes, products of the genes, etc. In this example, the focal aspect of systemic (whole-organism) physiology (SP_1) is flux through the pathway of lipid biosynthesis, which contributes significantly to the standing whole-organism triglyceride concentration (TRIGLY). Other aspects of systemic physiology (SP_2; e.g., rate of fatty-acid oxidation) also influence (reduce) the standing level of triglyceride.

Various classes of lipid are used for a variety of reproductive as well as somatic functions, a fact that has often been ignored in studies of life history physiology (Zera 2005). For example, triglyceride, while typically being considered a key somatic energy reserve, also is a key energetic component of eggs, providing energy during embryonic development. Furthermore, phospholipid is in especially high concentration in vitellogenin (yolk protein), eggs, and ovaries, (Beenakkers *et al.* 1985 as cited in Zera 2005) where it is used for extensive membrane construction during embryogenesis. Thus background studies in *Gryllus* have focused on accumulation of phospholipid as well as triglyceride. Given the variety of somatic and reproductive roles of these two lipid classes, we expected that alterations of pathways of lipid metabolism would be an important aspect of the evolution of morph specialization for dispersal versus egg production. Specifically, we expected morph-specific trade-offs in the relative accumulation, production, utilization, and organ allocation of triglyceride and phospholipid. Furthermore, because pathways of intermediary metabolism strongly interact, we also expected to see interactions between aspects of lipid

metabolism and other aspects of intermediary metabolism, such as amino acid and protein metabolism. This topic is discussed near the end of Section 24.2.6.

Because our primary focus has been life history microevolution, physiological–genetic studies of life history traits were conducted in replicated (three pairs of LW(f) and SW) artificially selected lines of *Gryllus firmus* initiated from field-collected individuals and raised under constant environmental conditions. The lines used were close to being pure-breeding, with the frequency of the selected morph being greater than 90%. Studies focused primarily on females because of the dramatic and easily quantified differences in reproductive effort (ovarian mass). Background studies (Zera and Larsen 2001) demonstrated substantially elevated whole-body concentrations of total lipid and triglyceride, but a lower level of phospholipid in LW(f) versus SW selected lines by the end of the first week of adulthood. Importantly, because triglyceride and phospholipid concentrations were low and did not differ between the morphs on the day of adult emergence, these differences are produced during the first week of adulthood, precisely when ovarian growth occurs to a much greater degree in SW versus LW(f) females. The strong negative genetic associations between triglyceride and phospholipid accumulation or triglyceride accumulation and ovarian growth suggests a potential direct causal relationship (functional trade-off) between these factors.

Only minor differences in nutrient acquisition (consumption and absorption) occur between the morphs during the first week of adulthood, and the morphs do not differentially assimilate nutrients from their food (Zera and Brink 2000, A. J. Zera unpublished data). Thus, the differences between morphs in accumulation of various lipid classes and ovarian growth, are almost exclusively produced by internal processes rather than by differential nutrient input, which is very useful for investigations of the biochemical basis of allocation trade-offs. It is critically important to assess nutrient input when investigating the physiological causes of life history variation and trade-offs, although this has not often been done with any degree of rigor (Zera and Harshman 2001, 2009). The impact of variation in nutrient input on allocation and life history trade-offs and the importance of assessing nutrient input in studies of life history physiology is discussed further in Section 24.3.

24.2.3   Morph-specific differences in flux through pathways of lipid biosynthesis and oxidation

A widely held assumption in life history physiology is that variation in internal nutrient allocation, which contributes significantly to life history trade-offs, results from the differential flow of nutrients through bifurcating pathways of metabolism. However, direct empirical information bearing on this widely held assumption is minimal. We directly measured relative flux through the pathway of fatty-acid biosynthesis and the triglyceride versus phospholipid arms of glyceride biosynthesis by injecting radiolabeled lipid precursors (^{14}C-acetate or ^{14}C-palmitic acid) into LW(f) or SW adults, and quantifying incorporation into triglyceride and phospholipid end products (Fig. 24-2). We also trapped CO_2 to identify morph differences in the oxidation of fatty acid for energy production. The results of these studies showed that the greater accumulation of triglyceride and lesser accumulation of phospholipid in the LW(f), compared with the SW selected lines, resulted from the combined effects of three flux trade-offs:

- greater rate of incorporation of precursors into fatty-acids (greater overall rate of *de novo* fatty acid biosynthesis)
- reduced rate of oxidation of fatty acid (greater sparing of fatty acids for incorporation into total glycerides—triglyceride and phospholipid)
- greater diversion of biosynthesized fatty acid into subsequent biosynthesis of triglyceride as opposed to phospholipid.

See Figure 24-3 and Table 24-1; for details and more extensive discussion of these studies see Zhao and Zera 2002, Zera 2005, and references therein.

In addition to these whole-organism trade-offs, radiotracer studies identified several important organ-allocation trade-offs. For example, a greater proportion of biosynthesized triglyceride was diverted to the soma in the LW(f) morph, while a greater diversion of triglycerides into the ovaries was observed in the SW morph. Because of this differential organ allocation, the energetic trade-off between the soma and reproductive organs is even greater than indicated by whole-organism measures of lipid differences between morphs (Zhao and Zera 2002, Zera 2005). This finding demonstrates the importance of measuring organ-specific aspects of physiology in life history studies. Yet in many cases, most notably in studies of small species such as *Drososphila* (discussed in Section 24.3), organ-specific allocation has typically not been measured. Demonstrating multiple, genetically-based flux trade-offs in lipid metabolism underlying a life history trade-off is one of the most important contributions of studies of *Gryllus* to life history physiology. These lipogenic flux differences between LW(f) and SW *G. firmus* provide the key link between variation in whole-organism energy reserves and variation in the expression or characteristics of gene products (e.g., enzymes), which is discussed next.

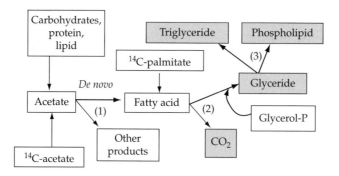

Figure 24-2 Trade-offs in flux through various pathways of lipid biosynthesis and oxidation indentified using radiotracers (^{14}C-acetate and ^{14}C-palmitate). As described in the text, in Figure 24-3, and more fully in Zhao and Zera (2002), Zera and Zhao (2003), Zera (2005), and Zera and Harshman (2009), three metabolic trade-offs have been identified (1, 2, and 3), in which there is morph-specific differential flux through bifurcating pathways of lipid metabolism. Relative flux values for trade-off 3 are given in Table 24-1.

24.2.4 Enzymatic basis of flux trade-offs: Digging deeper into the functional hierarchy of life history trade-offs

Activities of lipogenic enzymes were quantified to identify the specific points in metabolic pathways that are responsible for morph-differences in flux through the pathway of lipid biosynthesis. How are the morph-specific flux differences in lipid metabolism produced? By alternating activities of a few, many, or most enzymes in the pathway? Thus far the focus has been on enzymes of the *de-novo* pathway of fatty acid biosynthesis (production of new fatty acid from acetyl CoA) in the fat body of *G. firmus*, the main organ of lipid biosynthesis in insects (Fig. 24-4). Specific activities of representative enzymes of each of the three contributors to overall fatty-acid biosynthesis were studied:

- *de-novo* fatty acid biosynthetic pathway *per se* (fatty-acid synthase and ATP-citrate lyase)
- pentose shunt, which produces a significant proportion of NADPH required for fatty acid biosynthesis (glucose-6-phosphate dehydrogenase)

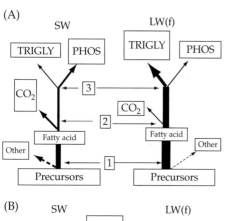

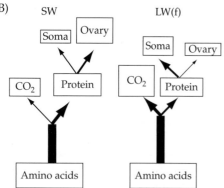

Figure 24-3 Biochemical trade-offs between dispersing and flightless morphs of *G. firmus*, which differ in life history. LW(f), long-winged morph with functional flight muscles; SW, short-winged morph. The width of the lines denotes relative flux differences between morphs through pathways of lipid (panel A, top) and amino acid (panel B bottom) metabolism determined by radiotracer studies. For example, in the top panel, relative to the SW morph, the LW morph diverts a greater amount of lipid precursors into production of fatty acid. Also, less fatty acid is oxidized to CO_2 and a greater amount is diverted to triglyceride and less to phospholipid in the LW morph. With regard to amino acid metabolism, the LW morph oxidizes a greater amount of amino acid, converts less amino acid to total protein, and allocates a lesser amount of biosynthesized protein to the ovaries. Data are from Zera (2005) and Zera and Harshman (2009). See text for additional discussion.

Table 24-1 Relative biosynthesis of triglyceride (% of total lipid biosynthesis) in LW(f) and SW morphs of *G. firmus* (trade-off 3 in Figures 24-2 and 24-3)

Line	Block-1	Block-2	Block-3	Paired *t*-test
LW(f)	69.7 ± 1.7**	64.5 ± 2.4**	67.7 ± 3.3**	$t_{(2)} = 4.90$
SW	45.3 ± 2.0	46.9 ± 2.2	51.6 ± 2.4	$P < 0.04$

- other cytoplasmic NADPH-producing enzymes, which also contribute significantly to the NADPH pool for lipognesis (NADP⁺-isocitrate dehydrogenase and NADP⁺-malate dehydrogenase).

The activities of all studied enzymes were substantially elevated in LW(f) compared with SW *G. firmus* (Zera and Zhao 2003, Zera 2005), indicating a global alteration of activities of enzymes involved in lipogenesis. Thus increased flux through the *de novo* pathway of fatty acid biosynthesis appears to have evolved via modulation of numerous enzymes of the pathway (indeed, every enzyme studied to date). These results support the theoretical predictions and experimental findings of "systems analyses" of flux control of metabolic pathways (e.g., metabolic control analysis, Fell 2003): in general, large changes in flux through a pathway are thought to require modulation of multiple enzymes of the pathway. For example, an increase/decrease in the rate of gluconeogenesis, during starvation or after refeeding, occurs by the up- or down-regulation of numerous enzymes of the gluconeogenic pathway (Graner and Pilkis 1990, cited in Zera 2005). A similar coordinate alteration of multiple enzymes underlies the increased lipogenesis in the *obese* strain of mice (discussed in Fell 2003), and in strains of mice artificially-selected for increased fat content (Asante et al 1989, Hastings and Hill 1990, both cited in Zera and Zhao 2003). An important exception to this rule appears to be enzymes that occur at branch points of metabolic pathways, which appear to exert strong control of flux (LaPorte and Koshland 1984, cited in Flowers *et al.* 2007). An excellent example of this point are the enzymes at the glucose-6-phosphate branchpoint studied by Eanes and co-workers (e.g., Flowers 2007) and discussed in detail below.

Intriguingly, specific activities of all studied lipogenic enzymes were elevated by about the same magnitude in the LW(f)- vs SW-selected lines (Zera and Zhao 2003). Furthermore, when LW(f) and SW lines were crossed and backcrossed, the activities of enzymes differed between LW(f) and SW F₂ individuals by about the same magnitude as between LW(f) and SW parentals, and activities were very highly correlated with each other and with wing morph (Zera and Zhao 2003 and A.J. Zera, unpublished data). This result strongly implicates polymorphic regulator(s) as the explanation for the coordinate expression of multiple enzyme activities and their correlation with wing morph. A similar explanation has been proposed for the coordinate up-regulation of numerous enzymes of lipogenesis in the *obese* strain of mice (discussed in Fell 2003), and the co-ordinate response to selection of transcript abundance of numerous metabolic proteins in lines of yeast selected in a low nutrient environment (Ferea *et al.* 1999).

An elevated level of, or increased tissue sensitivity to, juvenile hormone (JH) has long been suspected of coordinating the expression of the many components of the trade-off between flight capability and fecundity in *Gryllus* (Zera and Harshman 2001, 2009, Zera 2009), and many life history trade-offs in insects in general (Zera *et al.* 2007; see also Chapter 13). When the effective JH titer in LW(f) females was artificially increased by topical application of a JH analogue, methoprene, the LW(f) morph was converted into a remarkable SW phenocopy: activities of numerous lipogenic enzymes were reduced, triglyceride biosynthesis was reduced, and fatty acid oxidation was increased, as was ovarian growth (Zera and Zhao 2004). The results of this JH study further implicate genetically variable endocrine regulator(s) of intermediary metabolism as an important cause of the enzyme activity differences between LW(f)- and SW-selected lines, which, in turn, give rise to flux differences in lipogenesis and, subsequently, differences in whole-organism triglyceride reserves.

(A)

De novo pathway of fatty-acid biosynthesis

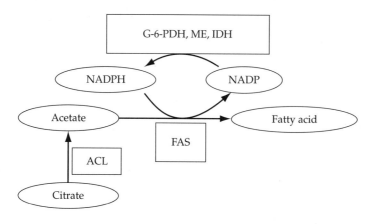

(B)

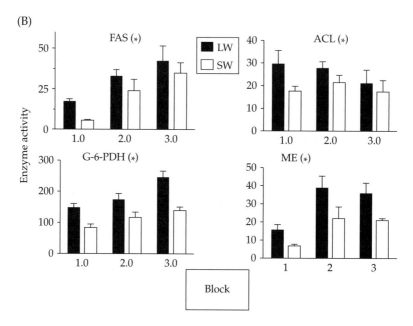

Figure 24-4 A, Enzymes of the *de novo* pathway of fatty acid biosynthesis investigated in *G. firmus*. G-6-PDH, glucose-6-phosphate dehydrogenase; ME, malic enzyme; IDH, NADP⁺-isocitrate dehydrogenase; ACL, ATP-citrate lyase, FAS, fatty acid synthase. B, Differences between LW- and SW-selected lines in specific activities of representative enzymes. (Note that in this chapter the LW morph is sysnonymous with the LW(f) morph; see text). Numbers on the *x*-axis refer to block (independent selection trial; see text). Asterisks in parentheses (*$P < 0.05$; 2 df) refer to the results of paired t-tests of LW versus SW lines across the three blocks. Data are from Zera (2005) and Zera and Harshman (2009).

Although it is tempting to conclude from the hormone application experiment that JH itself is the polymorphic regulator that coordinates the expression of morph differences in lipid metabolism and other traits, such an inference would not be justified. The gross-level nature of hormone manipulation experiments typically precludes inferring the role of a specific hormone or its mode of action, unless other data are available (Zera *et al.* 2007). Although a number of studies have implicated various insect hormones (e.g., juvenile hormones, ecdysteorids, insulin-like peptides) as regulators of metabolism and life histories, most of these inferences are questionable, given the limited or problematic data on which they were based (for a critical assessment of this topic see Zera *et al.* 2007, Zera and Harshman, 2009). The endocrine causes of variation in intermediary metabolism (and other life history traits) in *outbred* populations remains one of the most important but currently one of the least understood aspects of allocation and life history trade-offs (Harshman and Zera 2007, Zera and Harshman 2009).

24.2.5 Enzymological and molecular causes of differences in enzyme activities between morphs

Ongoing work (R. Schilder and A. J. Zera, manuscripts in preparation) is focusing on the causes of

the enzymatic-specific activity differences between LW(f) and SW morphs. This issue, by itself, is very complex because there can be numerous causes of genetic differences in enzyme activities (Fig. 24-5, Panel A). Our approach has been to focus on three representative enzymes, each of which functions in one of the three main components of lipogenesis:

- ATP-citrate lyase, which functions in the *de novo* pathway of fatty acid biosynthesis
- NADP$^+$-isocitrate dehydrogenase (NADP$^+$-IDH), an NADPH producer in the cytoplasm
- 6-phosphogluconate dehydrogenase (6-PGDH), a pentose shunt enzyme that also is an important NADPH producer.

Comprehensive enzymological and molecular data obtained for each of these enzymes in *G. firmus*

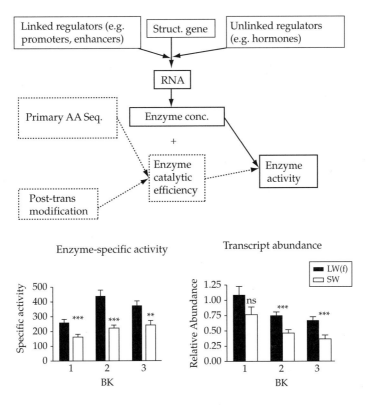

Figure 24-5   Top, potential causes of variation in enzyme specific activity. Variation in enzyme specific activity can either be due to variation in catalytic properties of the enzyme (either pre or posttranslationally), or due to enzyme concentration (due to changes in gene expression). Bottom, specific activity and transcript abundance for NADP$^+$-isocitrate dehydrogenase (NADP$^+$-IDH) in LW(f) (note that LW(f) and LW are synonymous in this chapter) and SW lines of *G. firmus*. Values are line means ± SEM. Note the consistently higher enzyme activity and transcript abundance in LW(f) lines (which also exhibit higher concentration of NADP$^+$-IDH enzyme protein; data not shown). (R. Schilder and A. J. Zera, unpublished data).

has allowed an assessment of the relative importance of morph-specific differences in gene expression (transcript abundance), enzyme concentration, enzyme kinetics, and DNA sequence to morph differences in enzyme-specific activity.

For each of the three cases mentioned above:

• the enzyme was purified to homogeneity and polyclonal antibodies were produced to be subsequently used to quantify enzyme protein
• the purified enzyme, as well as unpurified enzyme, were characterized with respect to various catalytic parameters (e.g., the Michaelis constant (K_M), or turnover number (k_{cat}), two key contributors to the rate of catalysis, Zera *et al.* 1985, Eanes 1999).

Furthermore, a portion or all of the sequence of the gene encoding the mature enzyme protein was amplified and used to measure transcript abundance by real-time PCR. All lines were also characterized for electrophoretic mobility for 6-PGDH and NADP⁺-IDH, and, for the latter enzyme, the entire coding region of its gene also was sequenced in several individuals from each of the three LW(f) and each of the three SW selected lines (18 sequences). Verrelli and Eanes (2001a) have identified functionally important, but cryptic amino acid variation at the *Pgm* locus in *D. melanogaster,* which results in no observable electrophoretic mobility differences among genotypes. This study illustrates the importance of obtaining DNA sequence (and kinetic) information on lines subjected to biochemical and genetic studies. Here we will discuss preliminary results for one of the enzymes, NADP⁺-isocitrate dehydrogenase (NADP⁺-IDH) (R. Schilder and A. J. Zera, manuscript in preparation).

Cytoplasmic NADP⁺-IDH is an important contributor to the NADPH pool in insects such as *D. melanogaster,* and in many other organisms (Zera *et al.* 2011). The activity of this enzyme is high in the fat body of *G. firmus,* as expected for an enzyme involved in lipogenesis. All but one of 18 *Nadp⁺-Idh* sequences, obtained from each of the three LW-selected and each of the three SW-selected lines, exhibited identical amino-acid sequences. Furthermore, no difference in any kinetic parameter (e.g., K_M, k_{cat}) was observed between any of the three pairs of LW(f) or SW lines. By contrast, the LW(f)

morph exhibited significantly elevated *Nadp⁺-Idh* transcript abundance relative to the SW morph, which paralleled morph differences in NADP+-IDH enzyme protein concentration measured immunologically (R. Schilder and A. J. Zera, manuscript in preparation), and differences in specific activity (Fig. 24-5, Panel B). Thus, NADP⁺-IDH-specific activity differences between LW(f) and SW lines are not due to kinetic differences resulting from either changes in DNA sequence or post-translational modification. Rather, they appear to be due largely, if not exclusively, to morph differences in gene expression. which in turn give rise to differences in enzyme protein concentration. Current data for NADP⁺-IDH together with results of endocrine manipulation (Zera and Zhao 2004, see above) collectively suggest that morph-specific differences in endocrine regulation lead to differences in gene expression, enzyme activity, lipogenic flux, and lipid accumulation.

Other than NADP⁺-IDH in *G. firmus* discussed above, we know of no other case in which a comprehensive analysis of the causes of specific enzyme activity variation has been undertaken for an enzyme whose activity strongly covaries genetically with flux through a pathway of metabolism that, in turn, is strongly correlated genetically with a life history trade-off. Because activities of numerous enzymes differ between the LW(f) and SW morphs, the precise contribution of morph differences in NADP+-IDH activity to morph differences in flux through lipogenesis cannot be assessed at present. A key experiment for the future would be to experimentally reduce NADP⁺-IDH-specific activity using RNAi and to assess the impact of this alteration on the NADPH pool and rate of lipogenesis.

24.2.6 Amino acid metabolism and life history trade-offs in *Gryllus*

Like lipids, amino acids can be used for a variety of important reproductive and somatic functions. They can be oxidized for energy, used to biosynthesize a variety of important somatic (e.g., flight muscle) or reproductive (yolk) proteins, or can be converted into other macromolecules such as fatty acids or glucose. Because of its contributions to numerous somatic and reproductive functions, amino acid metabolism is a prime candidate as a key biochemical aspect of intermediary metabolism whose modification contributes significantly to life history evolution (Zera and Zhao 2006). To investigate the existence of morph-specific differences in amino acid metabolism in *G. firmus* we followed the metabolism of injected ^{14}C-glycine into somatic or ovarian protein, somatic, or ovarian lipid (triglyceride or phospholipid), or oxidation to CO_2 for energy. As was the case for lipid, dramatic morph-specific genetic trade-offs were observed for many aspects of amino acid metabolism (Zera and Zhao 2006, Fig. 24-3 panel B, Table 24-2). Basically, the SW morph preferentially converted ^{14}C-glycine into reproductive protein, at the expense of oxidation or conversion to lipid, while the LW(f) morph oxidized this compound for energy or converted it into somatic lipid to a much greater degree. These differences in *in vivo* pathway flux identified using radiotracers are consistent with morph-differences in enzyme-specific activities. For example, the higher specific activities of the catabolic enzymes aspartate amino transferase and alanine amino transferase, in LW(f) versus SW-selected lines, also indicate higher rates of amino acid oxidation in LW(f) lines (Zera 2005). By contrast, the activity of carnitine palmitoyl tranferase, a key enzyme in fatty acid oxidation, is higher in SW versus LW(f) fat body (Zhao and Zera

2001 as cited in Zera 2005). These studies of amino acid metabolism underscore the remarkable morph-specific global remodeling of intermediary metabolism discussed above with regard to lipid metabolism.

24.3   *Drosophila melanogaster*

The biochemical and molecular aspects of life history evolution in *Drosophila* have been most directly investigated in laboratory populations that have diverged in various life history traits (e.g., longevity) and associated aspects of whole organism physiology (e.g., resistance to oxidative damage, energy reserves) as a consequence of laboratory selection (Luckinbill *et al.* 1984, Rose *et al.* 1984, 1996 as cited in Zera and Harshman 2009, Prasad and Joshi 2003). The explicit goal of these biochemical and molecular analyses has been to identify the functional causes of observed evolutionary changes in life history. Another important line of research has been molecular and biochemical characterization of genetically variable enzymes (allozymes) in natural populations of *D. melanogaster* (studies by Eanes and co-workers; see below). Because various life history traits have recently been measured in these field populations, the relationship between population divergence in enzyme activity and life history traits in the context of climatic adaptation can be

Table 24-2   Summary of differences in flux through various pathways of intermediary metabolism between the flight-capable and the flightless/reproductive morphs of *G. firmus* differing in life histories

Aspect of metabolism	LW(f) relative to SW	References
Biosynthetic rate of total lipid	Higher	A, B
Absolute biosynthetic rate of triglyceride	Higher	A
Absolute biosynthetic rate of phospholipid	Lower	A
Relative rates of triglyceride/phospholipid biosynthesis	Higher	A
Rate of fatty acid oxidation	Lower	C
Relative utilization rate of fatty acid for biosynthesis versus oxidation	Higher	C
Conversion rate of amino acids into lipid	Higher	D
Rate of amino acid oxidation	Higher	D
Biosynthetic rate of ovarian protein from amino acids	Lower	D
Allocation of biosynthesized lipid to soma versus ovaries: Triglyceride	Higher	A
Allocation of biosynthesized lipid to soma versus ovaries: Phospholipid	Lower	A

See text, Zera (2005), and Zera and Zhao (2006) for detailed descriptions of morphs and experimental conditions. LW(f), flight-capable morph; SW, flightless/reproductive morph. Sources: A, Zhao and Zera, 2002; B, Zhao and Zera, 2001; C, Zera and Zhao, 2003; D, Zera and Zhao, 2006. Table is modified from Table 2 of Zera and Zhao (2006).

assessed. Finally, a number of other laboratory studies have investigated various aspects of biochemistry and intermediary metabolism in the context of life history. These studies are briefly discussed at the end of the *Drosophila* section.

24.3.1   Laboratory selection on life history

Laboratory selection experiments, conducted over several decades on *Drosophila*, have identified aspects of whole-organism physiology that have consistently evolved in concert with individual, or correlated sets of, life history traits (Luckinbill *et al.* 1984, Service 1987, Djwadan *et al.* 1996, 1998; all cited in Zera and Harshman 2009). For example, lines selected for extended longevity typically exhibit decreased early age-fecundity, resistance to oxidative stress, and (sometimes) increased lipid reserves. Conversely, lines selected for starvation resistance exhibit increased lipid stores and increased duration of juvenile development. These results imply that modifications in pathways of intermediary metabolism involved in the production of energy reserves, antioxidants, and detoxication of reactive oxygen species are important contributors of life history evolution in *Drosophila*.

One notable set of studies focused on gene expression and activities of enzymes involved in resistance to oxidative damage in laboratory populations of *D. melanogaster* selected by Luckinbill *et al.* (1984) for extended longevity. Extensive data, obtained since the 1950s, strongly implicate oxidative damage as a significant cause of aging. Elevated mRNA levels of antioxidant enzymes such as superoxide dismutase (SOD), catalase (CAT), xanthine dehydrogenase, and glutathione transferase were found in long-lived lines (Dudas and Arking 1995, Force *et al.* 1995). The selected lines also had higher levels of SOD and CAT enzyme activity. Reverse selection of the selected lines resulted in a relative decrease in mRNA for antioxidant enzymes and a decrease in the age at which oxidative damage was manifest (Arking *et al.* 2000, Fig. 24-6). In this experiment, there was no decrease in the mRNA corresponding to enzymes involved in general metabolism, indicating that the decrease in antioxidant mRNAs was not an artifact of an overall reduction in transcrip-

tion. These studies clearly implicate changes in the expression of genes encoding antioxidant enzymes as an important factor in the evolution of oxidative stress resistance. Because oxidative stress resistance might be a general basis of extended longevity in laboratory-selected populations of *Drosophila* (Harshman and Haberer 2000), this topic is worthy of extensive investigation in the future.

A number of studies have also compared the activities of a variety of enzymes, such as NADPH-producing dehydrogenases and enzymes that potentially have detoxification activity, in the context of life history trait variation. The cofactor NADPH is important in a variety of metabolic contexts, such as fatty-acid and triglyceride biosynthesis (see *Gryllus* section, Zera *et al.* 1985, Eanes 1999), as well as in the production of antioxidant molecules such as glutathione. Luckinbill *et al.* (2001) found that long-lived selected lines exhibit a high frequency of a high-activity glucose-6-phosphate dehydrogenase (G-6-PDH) allozyme, and an elevated level of G-6-PDH enzyme activity. Activities of NADPH-producing enzymes such as malic enzyme and glucose-6 phosphate dehydrogenase were elevated in *Drosophila* lines selected for starvation resistance and which have increased lipid stores (Harshman *et al.* 1999). Finally, flies selected to survive on decomposing lemon, which presumably has secreted toxins from the many bacteria present, exhibited slower rates of development, which was the life history trait measured. Activities of detoxication enzymes, specifically multiple epoxide hydrolases and glutathione hydrolases, were also measured in these lines (Harshman *et al.* 1991). An intriguing finding of this study was that the activity of one of the glutathione transferases was markedly elevated when the selected line flies were first held on lemon before the assay, but not when held on standard *Drosophila* food before the assay. Thus, the selected lines evolved an inducible response to fresh lemon, which has a cue that triggers increased activity of one specific form of an enzyme. This study illustrates how a specific and potentially functionally important type of phenotypic plasticity can evolve in the laboratory, with general implications for evolution in natural populations. Additional studies on the biochemical basis of life history evolution, but not involving laborato-

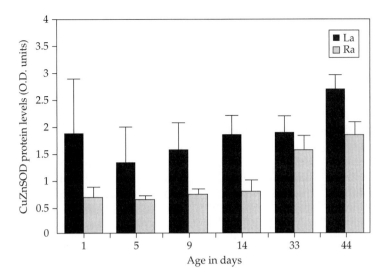

Figure 24-6 *D. melanogaster* from a line selected for extended life span (La) differed in superoxide dismutase protein abundance (CuZnSOD) at mid-life relative to control line flies (Ra). Figure from Arking *et al.* (2000), with permission from Elsevier. Error bars are standard deviations.

ry-selected lines, are dealt with at the end of the *Drosophila* section.

24.3.2 Clinal variation in intermediary metabolism and life history traits in the field

Life history traits in *D. melanogaster* exhibit latitudinal clines on most continents (De Jong and Bochdanovits 2003). Clines in lifespan and reproduction are negatively correlated in *D. melanogaster* along the east coast of North America (Schmidt and Paaby 2008), with increased longevity and decreased reproduction occurring in northern populations (Schmidt *et al.* 2005a,b as cited in Schmidt and Paaby 2008). *Drosphila melanogaster*, originally a native of tropical Africa, has colonized Australia and North America in the recent past (Schmidt and Paaby 2008). Latitudinal clines in life history traits such as longevity and lifetime fecundity have been interpreted as evolved adaptations of topically derived *D. melanogaster* to temperate environments. Schmidt and Paaby (2008) have shown that latitudinal variation in life history traits is largely explained by variation in reproductive diapause, a potentially important but understudied phenomenon in *D. melanogaster*,which appears to allow its northern populations to overwinter. Reproductive diapause

is a physiological syndrome involving many life history attributes, most notably depression of reproduction and lifespan extension. Diapause typically results in a substantial modulation of carbohydrate and lipid metabolism to produce, among other things, the increased energy reserves (glycogen, triglyceride) necessary for survival during an extended non-feeding period (Denlinger *et al.* 2004), Thus, latitudinal variation in the metabolic pathways involved in the biosynthesis of triglyceride and glycogen is likely to be an important biochemical component of these life history clines.

The physiological and biochemical mechanisms that underlie the life history clines have not been a major direct focus of study to date. However, independent of the life history investigations, clinal variation in polymorphic enzymes of intermediary metabolism (allozymes) have been extensively studied during the past three decades in the context of the population genetics of climatic adaptation (Eanes 1999, Sezgin *et al.* 2004, Flowers *et al.* 2007). These data provide insight into the likely biochemical basis of some aspects of the latitudinal life history cline in *D. melanogaster*.

A major focus of allozyme studies has been the enzymes of glycolysis and various branchpoints associated with glycolysis, in particular the enzymes associated with the glucose-6-phosphate branch-point (e.g., G-6-PDH and, to a lesser degree, phosphoglucomutase (PGM); Fig. 24-7). In addition, alcohol dehydrogenase (ADH), which detoxifies ethanol and converts it to lipid, an important energy reserve associated with several life history traits (discussed above), has also been extensively studied. Central biochemical issues in allozyme studies have been:

• the extent and underlying causes of variation in the activity of enzyme variants
• the degree to which differences in enzyme activity result in differences in flux through the pathways in which they function
• the extent to which these flux differences affect fitness via production of energy reserves such as glycogen and triglyceride (Eanes 1999).

This approach essentially amounts to the "vertical analysis" (Fig. 24-1) discussed above in the *Gryllus* section. More recently, the focus in *D. mela-*

nogaster is shifting from studies of individual enzymes to systemic properties of variation throughout glycolysis and associated branch points (Flowers *et al.* 2007, Zera *et al.* 2011), essentially the "horizontal" approach discussed previously (Fig. 24-1). Because of space limitations we will focus mainly on three well-studied polymorphisms that potentially influence life history variation: G-6-PDH, PGM, and ADH.

The G-6-PDH polymorphism in *D. melanogaster* has been intensively studied from both functional and evolutionary perspectives (Eanes 1999). The polymorphism mainly consists of two allozymes that differ by a single amino acid substitution: leucine versus proline at residue 382. This amino acid substitution results in a polymorphism for quaternary structure: the allele encoding the electrophoretically fast allozyme (A) is a dimer while the other allozyme (B) is a tetramer. African populations are almost entirely composed of the B allele while European populations are mostly characterized by

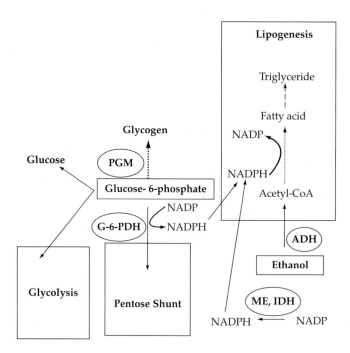

Figure 24-7 Metabolic pathways illustrating the various potential metabolic fates of glucose-6-phosphate, the various enzymes that contribute to the NADPH pool used for lipid biosynthesis, and the route of metabolism of ethanol to lipid in *Drosophila melanogaster*. PGM, phosphoglucomutase; G-6-PDH, glucose-6-phosphate dehydrogenase; ME, NADP⁺-malate dehydrogenase (malic enzyme); IDH, NADP⁺-isocitrate dehydrogenase. Dashed lines indicate multiple steps in the pathway.

a high frequency of the A allele, and the A allozyme increases clinally with latitude in North America and Australia. Thus, as is the case for many of the other allozyme clines in North America in *D. melanogaster* (Sezgin *et al.* 2004), and the cline in diapause incidence (Schmidt and Paaby 2008), the derived phenotype (the "A" allele in the case of the G-6-PDH polymorphism) is the phenotype which increases in northern populations. The A allozyme is less active and has a higher K_M for glucose-6-phosphate than the B allozyme. Kinetic studies indicate a 40% difference in catalytic efficiency between the allozymes *in vitro*, and a 20–40% difference in activity between the allozymes *in vivo*. Several studies have demonstrated that the low activity allele is associated with reduced flux through the pentose shunt (e.g., Cavener and Clegg 1981, Labate and Eanes 1992, both cited in Eanes 1999).

Molecular evolution studies show a pattern of silent substitution that indicates the polymorphism is ancient, similar to the situation for the *Adh* polymorphism. Somewhat paradoxically, natural selection appears to favor decreased pentose shunt flux as a function of increasing latitude (low activity "A" allele is more common in northern populations). Because lipid content increases with latitude, and the pentose shunt is thought to provide a significant proportion of NADPH for lipid biosynthesis, one might expect the B allozyme to increase in frequency in northern latitudes. Interestingly, the pentose shunt also appears to be reduced in larvae that contain a high-activity ADH allele in which there is increased flux from ethanol into lipid (Eanes, 1999 and see below).

Verrelli and Eanes (2001a,b) provided an answer to this paradox by noting that glucose-6-phosphate (G-6-P), the substrate of G-6-PDH, occurs at the branchpoint of glycogen metabolism, and glycogen occurs in higher concentration in diapausing individuals. G-6-P can be oxidized for energy through glycolysis, diverted through the pentose shunt, or converted into glycogen (Fig. 24-7). Thus, an important function of reduced pentose shunt activity, mediated by the A allozyme of G-6-PDH, may actually be to divert G-6-P away from lipid biosynthesis and into glycogen biosynthesis (Verrelli and Eanes 2001a). According to this hypothesis, other

NADPH-producing enzymes, such as NADP+-isocitrate dehydrogenase and NADP+-malate dehydrogenase, would play a more important role in providing NADPH for enhanced lipid biosynthesis in diapausing individuals in northern populations (Merritt *et al.* 2006). Additional support for this interesting idea is the higher frequency, in northern populations, of a high-activity allozyme of PGM, the enzyme involved in the diversion of G-6-P to glycogen. This high activity allozyme is associated with higher glycogen content in flies. Thus, reduced G-6-PDH and pentose shunt activity, coupled with elevated PGM activity, appear to both divert G-6-P away from lipid and into glycogen (Verrelli and Eanes 2001b).

The alcohol dehydrogenase (ADH) polymorphism is probably the most extensively studied enzyme polymorphism in *D. melanogaster*. The two major ADH allozymes differ by only one amino acid substitution, which results in electrophoretic variation (Eanes 1999). Electrophotetically fast and slow allozymes differ in k_{cat} and differ twofold in the level of protein, although there are no appreciable differences in transcript abundance. Latitudinal clines in the northern and southern hemispheres run in the same direction, with the fast (high activity) allele increasing from lower to higher latitudes. There is convincing evidence that the nucleotide site responsible for the ADH amino acid substitution, or a closely linked regulatory site, is the direct target of selection.

Originally, alcohol tolerance/detoxification has been the focus of selection acting on the ADH polymorphism. However, subsequent studies suggested that increased utilization of ethanol for lipid biosynthesis in individuals bearing the Fast allozyme might be important. Triglyceride content increases almost 40% from southern to northern US populations of *D. melanogaster*, in parallel with the increase in diapause discussed above. The polymorphic enzyme α-glycerophosphate dehydrogenase (α-GPDH), which provides the glycerol backbone for triglyceride biosynthesis, exhibits a cline in allele frequencies in North America that parallels the ADH cline. The more active α-GPDH allele increases in frequency with latitude, like the more active ADH allele, and the activity of both of these enzymes are induced by alcohol in the diet.

In larvae and adults, flux studies indicate that ethanol is used extensively for lipid synthesis. ADH has a high flux coefficient from ethanol into lipid in larvae, and ADH allozyme activity differences exert a strong differential influence on flux from ethanol to lipid at this life stage (Freriksen *et al.* 1994 as cited in Eanes 1999). As is the case for the G6PDH polymorphism, a potential functional interaction between clinal ADH and clinal life history polymorphisms can be envisioned via the effect of ADH on the biosynthesis of lipid, an energy reserve that often plays an important role in reproductive diapause and overwintering.

In summary, the careful, detailed studies of Eanes and colleagues have identified the major pieces of the puzzle regarding the relationship between clinal variation in the activities of enzymes of intermediary metabolism, energy reserves, life history, and climatic adaptation. The unique contribution of these studies for life history evolution is that they constitute a detailed dissection of enzymes and pathways of metabolism (as described above for *Gryllus*), but undertaken in field populations, in the context of geographical variation in life history. All other major physiological analyses of life history variation have focused on within-population variation. This experimental system shows immense promise for identifying how the microevolution of intermediary metabolism contributes to the microevolution of life histories.

24.4   Other studies and issues relevant to *Drosophila*

24.4.1   Additional biochemical and molecular studies

In addition to the *Drosophila* laboratory selection experiments described above, a large number of other studies have investigated associations between enzyme activities and various life history traits in lines that were not used in laboratory selection experiments. Because of space limitations, only a few examples can be presented here. Transgenic over-expression of G-6-PDH, in the whole body or neuronal tissue, resulted in lifespan extension in *D. melanogaster*, without affecting fecundity or metabolic rate (Legan *et al.* 2008). Flies over-expressing

G6PDH exhibited elevated *in vitro* G-6-PDH activity, elevated *in vivo* levels of reduced cofactors (NADPH and NADH), and an elevated *in vivo* ratio of reduced/oxidized glutathione (reduced glutathione is an important antioxidant). Also indicating a connection between longevity and G6PDH, flies held under dietary restriction conditions that extend lifespan exhibited an elevated level of G-6-PDH mRNA (Pletcher *et al.* 2002). Experimental elevation of glutamate-cystine-ligase, an important enzyme in the biosynthesis of glutathione, also extended lifespan in *D. melanogaster* (Orr *et al.* 2005 as cited in Legan *et al.* 2008). These studies indicate that enzymes involved in redox balance can increase lifespan, perhaps markedly.

There are other studies indicating a relationship between metabolism and aging with implications for life history trade-offs. For example, various microarray studies show that mitochondrial function, the citric acid cycle, and glycolysis markedly decrease as a function of age in *Drosophila* (Pletcher *et al.* 2002). This down-regulation is evolutionarily conserved (McCarroll *et al.* 2004). There may be a general metabolic response to age that affects life history traits and influences how life histories evolve.

24.4.2   Influence of changes in allocation versus nutrient input on life history evolution

Both theoretical and experimental studies indicate that the magnitude and variance of nutrient acquisition can strongly affect allocation trade-offs and life history trade-offs dependent upon allocation (van Noordwijk and de Jong 1986, De Jong and van Noordwijk 1992, Zera and Harshman 2001, Boggs 2009). Increased amount or variation in input reduces the magnitude of trade-offs, while reduced amount or variance has the opposite effect. Thus, an important issue in life history physiology is the relative contributions of changes in nutrient input and changes in internal allocation to evolutionary changes in life history trade-offs. In the *Gryllus* studies described above, this issue could be explicitly addressed because nutrient input was monitored for the flight-capable and flightless morphs during the first week of adulthood when the experiments were conducted. No large difference in acquisition was found between the morphs (less than 10% difference during the first week of adulthood), and thus the enormous morph-specific differences in production of various lipids, proteins, etc., were due almost exclusively due to differential conversion and allocation of a similar internal resource base by the LW(f) and SW morphs (Zera and Brink 2000, Zera and Harshman 2009, A. J. Zera, unpublished data). By contrast, until very recently (e.g., Lee *et al.* 2008b), nutrient input has been monitored much less rigorously in studies of *Drosophila* life history physiology. Thus, the importance of changes in allocation versus acquisition in life history evolution is much more difficult, if impossible, to assess quantiatively in earlier studies. Interestingly, Foley and Luckinbill (2001) found that direct selection on feeding rate resulted in changes in caloric intake, energy reserves, and various life history traits, suggesting that evolutionary modification of nutrient input may have been an important contributor to life history evolution in the laboratory selection studies described above. Rose and Bradley (1998) came to a similar conclusion regarding the importance of nutrient acquisition, largely based on indirect evidence. Most studies of caloric restriction in *Drosophila*, which have also played a prominent role in life history physiology, also are compromised because of problems in quantifying food intake. Recent well-controlled studies of nutrient input on life history traits and trade-offs in *Drosophila* (e.g., Lee *et al.* 2008b) provide experimental protocols and a new theoretical perspective to rigorously address this issue.

24.4.3 Quantitative-genetic variative in enzyme activities and fitness

In a pioneering quantitative-genetic investigation of the microevolution of intermediary metabolism, Clark and coworkers (Clark 1990, Clark *et al.* 1990, and references therein) reported extensive standing additive genetic variation for activities of each of 10 enzymes of carbohydrate and lipid metabolism and concentrations of two key energy reserves (glycogen, triglyceride) in *D. melanogaster*. In addition, a number of functionally significant positive or negative genetic correlations between various enzyme activities and concentrations of energy reserves was observed. For example, similar to the situation observed in *Gryllus* (see above), activities of the lipogenic enzymes G-6-PDH and fatty acid synthase were positively correlated with triglyceride content. In some cases, biochemical phenotype covaried with components of fitness. For example, triglyceride or glycogen content was positively correlated with viability or fecundity. The significance of the studies by Clark and coworkers is that they were among the first to demonstrate widespread genetic variation and covariation for enzyme activities and energy reserves in intermediary metabolism impacting fitness and life history.

24.5   Summary

1. Altered flux through core pathways of intermediary metabolism is expected to be a key contributor to life history evolution. However, only during the past decade have studies begun to directly investigate this topic in detail.

2. This phenomenon has been best studied in insects in which laboratory populations have been selected for divergent life histories, field populations that differ clinally in life history, or in naturally-occurring life history (dispersal) polymorphism. These cases of life history evolution often involve correlated changes in energy reserves or resistance to oxidative damage, providing the opportunity to investigate how evolutionary changes in metabolic pathways involved in these physiological traits contribute to life history evolution.

3. Studies in the wing-polymorphic cricket *Gryllus firmus* have provided the first direct demonstration that differential flux through pathways of intermediate (lipid and amino acid) metabolism are correlated with and contribute significantly to variation in individual life history traits and trade-offs between traits. This is a key assumption of the "Y" model of nutrient allocation. These flux differences appear to be produced by global changes in numerous enzymes of the pathways of lipogenesis and amino acid metabolism. Morph-specific expression of these global changes are coordinated by altered hormonal regulation, pointing to the key but understudied role of endocrine modifications in life history evolution.

4. Biochemical studies of laboratory-selected populations of *Droosphila melanogaster* have identified specific enzymes involved in the production of lipid energy reserves and antioxidants, and deactivation of reactive oxygen species. Enhanced expression of antioxidant enzymes may be an important general contributor to the evolution of extended longevity, at least in *Drosophila*.

5. Molecular, biochemical, and genetic studies of populations of *Drosophila melanogaster* along the east coast of the USA have identified large-scale functional changes in allozymes involved in the production of triglyceride and glycogen, and parallel clines in life history (reproductive diapause) dependent upon increased production of these reserves. This model shows exceptional promise for identifying the biochemical–molecular bases of recent life history evolution as an adaptation to colder climate in northern latitudes.

24.6 Acknowledgments

A. J. Zera acknowledges grants from the National Science Foundation (IOS-0516973 and IBN-0212486). L. Harshman acknowledges grants from the National Institutes of Health (RO1-DK074136) and Army Research Office (W911NF-07–1–0307).

CHAPTER 25

Epistatic social and endocrine networks and the evolution of life history trade-offs and plasticity

Lesley T. Lancaster and Barry Sinervo

25.1 Introduction

Life history theory attributes trade-offs to the action of pleiotropy, which tethers resource allocation to two or more (life history) traits by single genes. Here, we consider trade-offs to arise from negative genetic correlations between traits that contribute to fitness (see Chapter 2), but we also view sources of social selection as capable of generating such negative genetic correlations (Sinervo *et al.* 2008). We review epistatic and pleiotropic endocrine source(s) of trade-offs in resource allocation, including genetically based social traits that are often only considered tangentially in life history analysis. In particular, we explore the roles of social genes in generating far more complex trade-offs than can be explained by pleiotropy alone. We suggest that many trade-offs arise from a logical extension of the epistatic effects of multiple genes on single traits to the epistatic regulation of two or more traits. Epistasis results from gene interaction within individuals, but this idea can be extended to interactions between maternal–progeny genotypes (Wolf and Hager 2006), between maternal and paternal genotypes (Haig and Westoby 1989, Haig 1996), and among interacting genotypes in the case of density or frequency-dependent regulation (Sinervo and Calsbeek 2006, Sinervo *et al.* 2006, 2008).

Most trade-offs are invoked along two-dimensions of life history traits, with the caveat that pairs of life history traits interact with at least one dimension involving a fitness component, such as progeny or adult survival (Fig. 25-1: Lack 1947, Williams 1966b), growth (Gadgil and Bossert 1970) or time to maturity (Cole 1954). Notice that even simple trade-offs require ordination in three dimensions (Fig. 25-1A,B). It is not easy to visualize such spaces, but dimensions can be flattened by layering multiple survival components on composite axes (e.g., survival to maturity, during maturation, or between reproductive events, Fig. 25-1A–C). We depict steps in the analysis of constituent pairwise interactions (Fig. 25-1A–C) and a five-axis space (Fig. 25-1D). Phillips and Arnold (1989) provide visualizations of three- and four-dimensional surfaces for selection, but the important point for life history theory is that we have no general methods for predicting the outcome of complex selection on optimal life history allocation, except in the case of two and three dimensions (Sinervo and Clobert 2008), largely because the statistical assessment of selection on life history components requires large sample sizes. Moreover, fitness surfaces for life history trade-offs in nature may generate single fitness optima, but such surfaces are often highly non-linear and involve multiple optima generated by social games (Sinervo and Svensson 2002, Sinervo *et al.* 2008). Our thesis is that trade-offs are likely to involve many more dimensions than three, and these added dimensions arise from social system interactions (Sinervo *et al.* 2008).

Consider surfaces that govern optimal offspring size in the case of density regulation. At high density survival is strongly related to progeny size, and

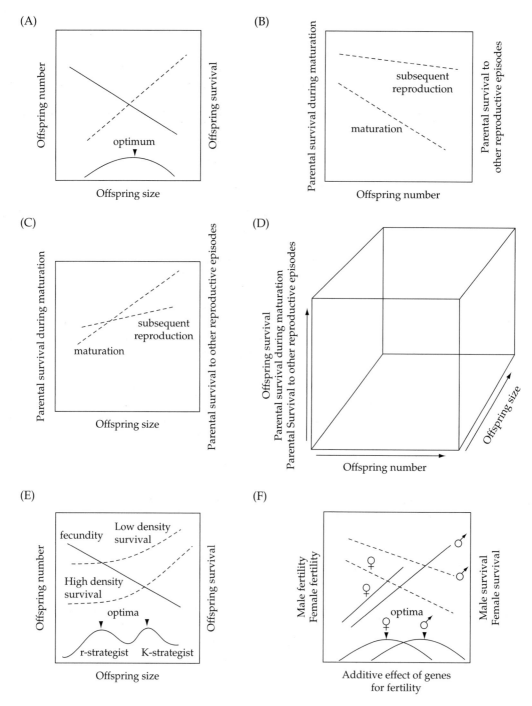

Figure 25-1 The dimensions of life history trade-offs in sexual organisms. In all panels, dashed lines depict survival components of fitness, solid lines depict reproductive component of fitness, and heavy lines the product of survival and reproductive effects on optima. A, The trade-off between offspring size, offspring number, and offspring survival along with optimal allocation. B, The trade-off between offspring number on survival of the female parent during maturation and parental survival after reproduction to subsequent reproductive episodes. C, The trade-off between offspring size on parental survival during maturation and survival after reproduction to subsequent reproductive episodes. D, The multidimensional space required to compute optima for the trade-offs depicted in A–C, reduced to a three-dimensional space, with three axis labels indicating the three survival functions involved in the complex trade-off. E, Under density competition, survival functions are expected to be steeper or even accelerate at high compared to low density (fecundity is assumed to be static), and these can generate two distinct optima for an r-strategist versus a K-strategist. F, Genes for fertility are shared between the sexes and arise from the genes of the HP-gonadal axis, but impacts of selection on these genes can vary between the sexes generating conflicting optima (survival and reproductive components for each sex are shown), which are referred to as antagonistic selection or ontogenetic conflict.

may even exhibit an accelerating slope (Fig. 25-1E), but at low density survival is only weakly related to progeny size. These two functions, when multiplied by the size-number trade-off, can generate two optima, and multiple life history strategies will evolve to match the optima and be maintained through density cycles. Recent theory has begun to analyse the complex life history adaptations for density recruitment, size, survival, and reproduction (Falster *et al.* 2008), but most theory still ignores multiple optima.

The problem of visualizing such complex adaptive surfaces is further dimensionalized if we add the differences that two sexes add to the life history problem. Females tend to be shaped by natural selection, but males tend to be shaped by both natural and sexual selection. The sexes differ considerably in the form of natural versus sexual selection that shapes endocrine function and yet they share most genes (Pishcedda and Chippindale 2006, Foerster *et al.* 2007, Sinervo and McAdam 2008), except for genes that govern sex determination in species with sex chromosomes. Different life history optima between the sexes (Fig. 25-1F) generate a trade-off that operates on the endocrine networks governing life history traits expressed by males and females (Sinervo and Miles 2010). The presence of alternative strategies within one or both sexes adds further dimensions to the life history problem (Sinervo and Calsbeek 2003, Sinervo and Clobert 2008).

25.2 Endocrine networks and life history trade-offs

Endocrine networks generate homeostasis via regulation of energy flow and metabolism among competing life history functions, but also respond to abiotic and social stressors. The structure of an endocrine network that evolves to regulate optimal life history allocation, under requirements of homeostasis, is a positive–negative regulatory feedback loop, which is instantiated in response to exogenous triggers like photoperiod (Fig. 25-2). For example, clutch size is positively regulated by the gonadotropon follicle stimulating hormone (FSH), and egg size is positively regulated by estrogen (E), but E negatively regulates FSH production by the ante-

rior pituitary (Fig. 25-2). The basic feedback loop in males is only different in that negative regulation of gonadotropins is due to testosterone (T), which is secreted by the testes in response to luteinizing hormone (LH) and FSH (Mills *et al.* 2008). In females, LH positively regulates maintenance of the corpora lutea, which secrete progesterone. This is yet another feedback loop controlling reproductive cycle length (Guillette 1993).

Networks of positive and negative regulation are often modulated by gated switches (Milo *et al.* 2002), which are on–off switches for energy allocation to life history functions. The on–off state of the switch is controlled by different endocrine systems than the one involved in the feedback loop, and the action of the switch responds to environmental stressors (Fig. 25-2B).

In vertebrates, the hypothalamic–pituitary–adrenal (HPA) axis forms network circuitry for key gated switches. In response to abiotic and/or biotic stressors, the central nervous system triggers the HPA to secrete corticotropin releasing hormone, which stimulates release of adrenocorticotropin hormone, which travels via the blood stream to the adrenals where it stimulates release of cortisol or corticosterone (CORT) (Wingfield and Moore 1987). Bound to corticosterone binding globulins (CBG), CORT travels to diverse cellular and tissue targets. Released from CBG, CORT binds to mineralocorticoid and glucocorticoid receptors in the HP to alter levels of FSH or to components of the CNS to effectuate alterations in behavior (Korte *et al.* 2005) and energy flow (Fig. 25-2).

We suggest that the physiological architecture of network circuits be the focus of life history analysis for the following reasons. The routes by which the HPA affects reproduction are forms of pure life history plasticity (Fig. 25-2A). However, we are also interested in life history adaptations. Genetic variants in a population provide an opportunity to understand the genetic bases of life history trade-offs. Genetic variants can arise anywhere in such networks. In Figure 25-2, we depict a hypothetical situation in which two types exist in one population: an obligate type that always keeps immune function primed, and a facultative type that down-regulates immune function and only upregulates in response to disease stressors. These two genetic

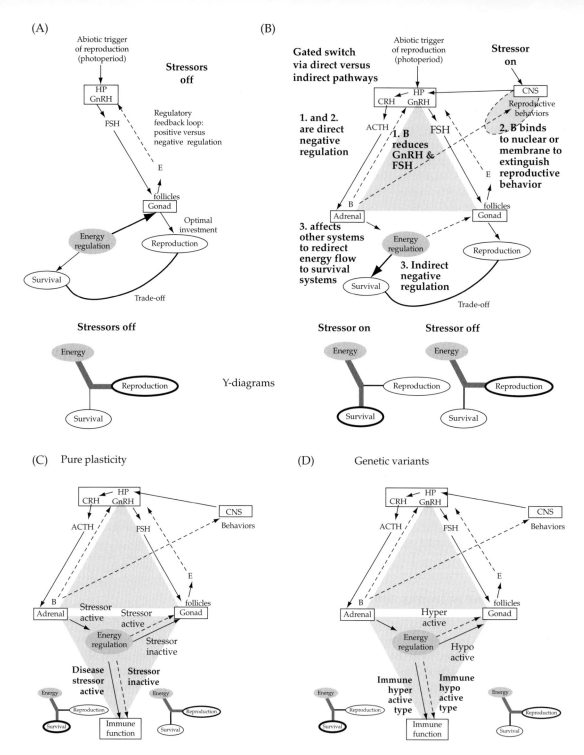

Figure 25-2 The structure of endocrine networks and impacts on Y-diagrams of energy allocation (Zera and Harshman 2001). A, A negative (dashed lines)/ positive (solid black lines) feedback loop regulating optimal allocation to reproduction (Sinervo and Licht 1991), triggered by abiotic cues like photoperiod (Phillips *et al.* 1987). B, The same network modified to include effects of adrenal corticosteroids, corticosterone (labeled CORT) or cortisol, which is triggered in response to a stressor that stimulates corticotropin-releasing hormone (CRH) and adrenocorticotropin hormone (ACTH). Corticosteroids can act as a gated switch at two points: to shut off FSH production in the HP or to directly bind to receptors in the CNS to alter behaviors. Corticosterone (CORT) can redirect energy from the metabolic pool for immune function at the expense of reproduction. C, The gated switch action of CORT can be viewed as pure plasticity in the presence or absence of a stressor. D, In principle, genetic variants can arise at multiple control points in networks governing homeostasis and cascading effects on endocrine networks generate norms of reactions. For example, the action of CORT can also be viewed as mediating the life history trade-offs generated by hypothetical genetic variants that either have immune function chronically elevated (as in a hyperactive immune type) or respond facultatively (in a hypoactive immune type). Mutations leading to alternative types can, in principle, occur anywhere in endocrine networks.

variants might be maintained in the context of a cycling disease agent, and yet the pathways they invoke are the same circuits as those invoked by the plasticity of the HP–adrenal axis. Trade-offs are the result of both forms of network circuitry, yet their effects on selection are dramatically different. Plasticity attenuates selection on traits (although selection on the underlying mechanism promoting plasticity may be quite strong), whereas additive genetic variants exacerbate exposure of traits to selection. Because selection can be chronic under density regulation, plasticity is often favored.

25.3   An example of a gated switch in developmental life history trade-offs

Sinervo and Clobert (2008) extensively review endocrine networks governing trade-offs of amphibian development (Wilbur 1977). Cues used to induce early time to metamorphosis, but smaller size, arise from the HPA axis, a gated switch that responds to social crowding and density-dependent regulation (a biotic cue) as well as by pond drying (an abiotic cue) (Hayes 1997, Denver 2000). Endocrine plasticity involving density regulation might also involve prey abundance. For example, spadefoot toads exhibit plasticity with respect to a prey item, the fairy shrimp, which is either present or absent in their natal pond (Pfennig 1990, 1992a,b). Alternative tadpole strategies of omnivory versus carnivory are induced in spadefoot toads by the HP-thyroid axis (Fig. 25-3). Fairy shrimp naturally contain thyroxine (T_3). When tadpoles consume shrimp they ingest thyroxine (Pfennig 1992b), which triggers development of carnivores that metamorphose rapidly, but at small size. Larval strategies of spadefoot toads are related to antisocial behaviors of cannibalism versus social behaviors of schooling. Carnivores adopt cannibalistic behavior, owing to their having strong jaw muscles and keratinized beak. When compared to solitary carnivores, omnivore schools churn up and feed efficiently on detritus. Even though carnivores eat conspecifics, kin altruism and gene relatedness confers protection for larvae from their cannibalistic kin (Pfennig et al. 1993), thereby invoking social system trade-offs (Hamilton 1964, Sinervo et al. 2008) in what would otherwise be simple trade-offs. Below, we show that endocrine plasticity often mediates life history trade-offs due to social causes and density-dependent selection. In the absence of a plastic switch, selection on tadpoles in ponds that vary in shrimp density would be much more severe. The shrimp provide a local cue that ameliorates selection on tadpole survival.

25.4   Social networks and life history trade-offs

It is apparent from the spadefoot toad example that social system trade-offs can be synthesized with endocrine system trade-offs because social interactions commonly affect endocrine interactions within individuals and vice versa (i.e., hormones affect social behavior, and social cues affect hormones). As in endocrine networks (Fig. 25-2), social networks (Fig. 25-4) can be depicted as positive and negative interactions that generate system cycling or stability in the numbers of each strategy type (analogous to homeostasis in physiological networks). Consider a population composed of two types such as an r-strategist, which produces large clutches of small eggs, and a K-strategist, which produces small clutches of large eggs (Sinervo et al. 2000). Systems of r- and K-strategists can generate oscillatory density cycles, stabilize on a population size with both types present, fix on r-strategists, or fix on K-strategists (Slatkin 1981). Under the oscillatory scenario, when the population exceeds carrying capacity, strong negative regulation of r-strategists on self-types allows the K-strategist to invade during the population crash. After the population has crashed, the K-strategist replaces the r-strategist. As the population recovers from low density, the r-strategist can outcompete the K-strategist by virtue of a release of self-regulation. The corresponding social regulatory network is depicted in Figure 25-5A.

For over two decades, we have investigated the ensuing life history trade-offs invoked by density regulation in the side-blotched lizard, *Uta stansburiana*. We have studied trade-offs that involve regulation of clutch and egg size (HP-gonadal, Fig. 25-6), and which affect survival of both offspring and parents (Sinervo and DeNardo 1996, Sinervo 1999, Sinervo et al. 1992, 2000), and trade-offs that are not typically considered in life history theory such as

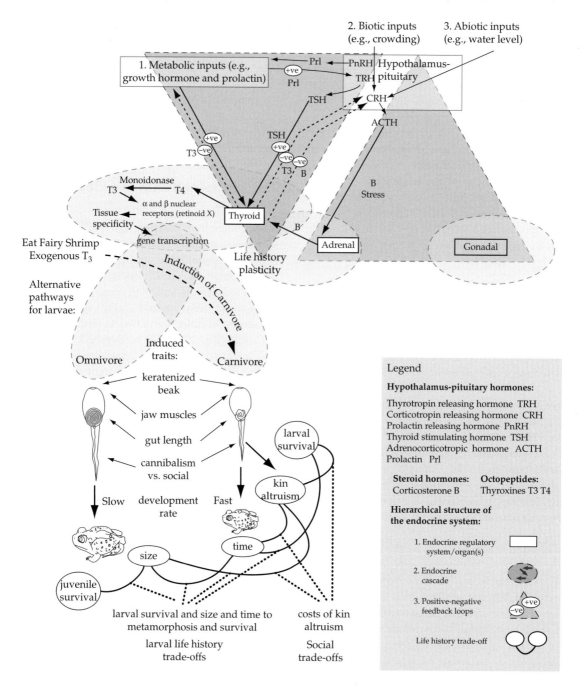

Figure 25-3 Regulatory endocrine network and the trade-off between survival, size, and time to metamorphosis in spadefoot toad larval development (Pfennig 1992b). In amphibians the hormone prolactin (Prl) plays a role that is similar to growth hormones in other vertebrates, in that Prl regulates larval growth and interacts with the HP–thyroid axis to achieve metamorphic climax. Control arises from the hypothalamic–pituitary axis (HP) owing to gene regulation of peptide hormones such as prolactin-releasing hormone (PnRH) and thyroxine-releasing hormone (TRH), which respond to metabolic inputs (growth) and abiotic inputs (e.g., temperature, pond water level). Adrenal axis. Depending on either abiotic or biotic stressors (Hayes 1997, Denver 2000), the hormone CRH triggers ACTH production by the pituitary, which in turns stimulates the adrenal to produce corticosterone (labeled CORT). CORT can induce earlier metamorphosis by acting on levels of PnRH, TRH, or the sensitivity of the thyroid to TSH. Under such stressful pond conditions, metamorphosis is accelerated and survival is enhanced, reflecting adaptive plasticity. Life history plasticity in the larval development of a generic amphibian, and the environmental cues that generate plasticity, are shown (social crowding, pond drying). In the case of spadefoot toads, presence versus absence of fairy shrimp prey serves as an additional switch involving thyroxine. Social trade-offs. Spadefoot toad larvae are subject to higher dimension trade-offs because of social interactions. The basic trade-offs of amphibian development involve the time and size of metamorphosis, balanced against larval survival and juvenile survival after metamorphosis. Because carnivore morphs are more predisposed to cannibalism, the carnivores are selected to detect and avoid eating kin, but still eat unrelated tadpoles. Not eating kin constitutes a kin altruistic cost (of reduced growth and thus slower time to metamorphosis or smaller size) that benefits a genetically related tadpole. Thus, the alternative strategies of carnivore and omnivore invoke additional social trade-offs.

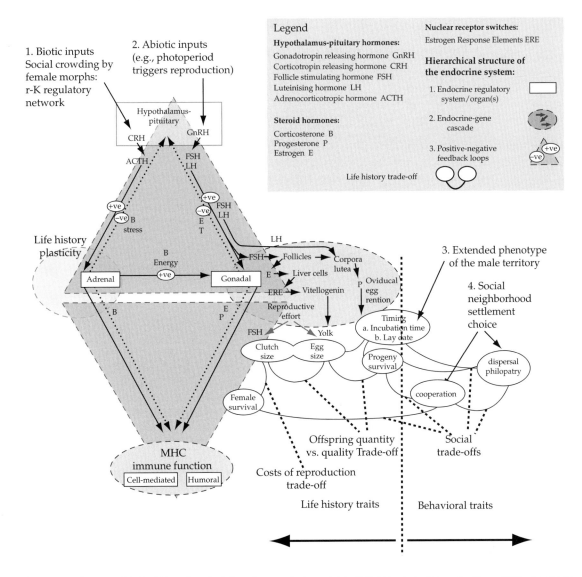

Figure 25-4 The HP-gonadal axis, which regulates clutch size and progeny size, is associated with life history trade-offs in reptiles (Sinervo 1999) and mammals (Oksanen *et al.* 2002), based on manipulations of the endocrine system. An example of an endocrine cascade for egg size and egg number in lizards is depicted. (GnRH triggers FSH, which induces follicle proliferation. The theca of resulting follicles then produce E. E travels to the liver and initiates gene transcription via an ERE, an estrogen-response element, to effectuate vitellogenin transcription and translation, which is packaged in many small ova or a few large ova by ovaries.) A single cascade causes trade-offs in egg and clutch size, forming a pleiotropy. Additional endocrine cascades effectuate levels of progesterone (P) via stimulation of corpora lutea by luteinizing hormone (LH). However, clutch size and thus FSH determine corpora lutea number, which secrete progesterone. Progesterone affects the degree of egg retention in reptiles and incubation time. Endocrine pleiotropy affects two trade-offs: progeny survival and survival costs of reproduction for females (Sinervo 1999). Even though many genes are involved, all pleiotropic effects can be traced back to FSH through one endocrine cascade.

However, more complex life history theories implicate immune function in reproductive costs through effects on resource allocation and metabolic pathways (Lochmiller and Dabbert 1993). For example, in the side-blotched lizard, we have observed a strong trade-off between immune function versus survival in r-versus K-selected color morphs of females. A single gene of major effect called the OBY locus, named for the colors orange, blue, and yellow, governs female reproductive strategy (r- versus K-selected). The strong negative genetic correlation (Svensson *et al.* 2001a,b, 2009) between immune function and survival costs of reproduction (Svensson *et al.* 2002), and survival before maturity (Svensson *et al.* 2009) are due to interactions among physiological pathways: 1) the adrenal-immune system-survival, path interacts with 2) the adrenal-gonadal-clutch size versus egg mass path) (Lancaster *et al.* 2008).

In summary, while the offspring size–number trade-off appears to arise from the pleiotropic effects of a single endocrine cascade, the costs of reproduction trade-off arise from two or more endocrine cascades, or physiological epistasis. In addition, one of these cascades involves the HPA of reproductive females. As noted in Fig. 25-3, the HPA axis modulates life history plasticity in the context of social stressors of female density and female morph frequency (Comendant *et al.* 2003). Even more complex life history trade-offs affect egg size, and lay date, incubation time, and density regulation (Svensson and Sinervo 2000) via the choice of social territory by the female (Calsbeek and Sinervo 2004) or choice of mate (discussed in text). Male progeny also experience very complex social trade-offs of altruism, mutualism, and competition that involve offspring size and offspring settlement strategies (reviewed in Sinervo *et al.* 2008). Standard life history theory does not allow for assessment of such social trade-offs but extensions of the Price (1970) Equation can be used to partition trade-offs due to pleiotropy versus epistatic social interactions (see Sinervo *et al.* 2008).

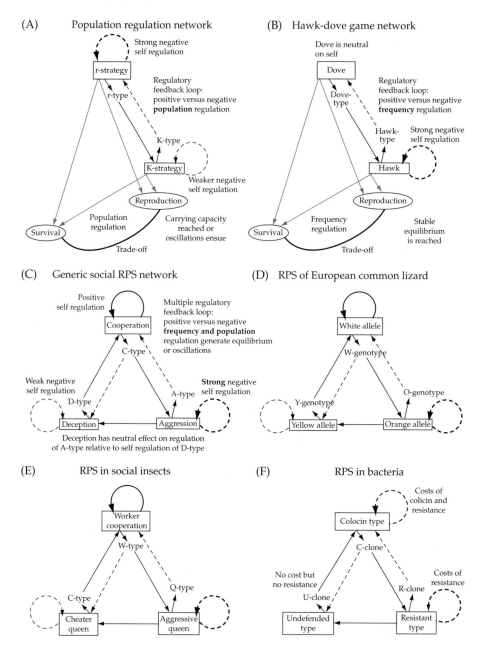

Figure 25-5 As in the case of endocrine networks, social networks of organisms can be depicted as networks of positive regulation (solid black lines) and negative regulation between non-self types or negative or positive regulation with self-types (re-curved arrows and dashed lines). Social interactions influence allocation to survival and reproduction (Y-diagrams, grey arrows) and thus outcome of life history trade-offs (bottom of panel A and B, but ignored in C–E). A, Network diagram for density genotypes of r- and K-strategists like the side-blotched lizard is depicted. B, Network diagram for a system of Hawks and Doves is depicted in which the Hawk is strongly self-regulating but benefits from a high frequency of Dove. C, Network diagram for a system of three density types: a cooperative type that exhibits positive self-regulation, an aggressive type that exhibits strong negative self-regulation, and a deceptive type that exhibits weaker self-regulation. This rock-paper-scissors (RPS) system is observed in European common lizards (shown) and side-blotched lizards (not shown). This network would have also been present in primordial eusocial insects (shown). The network is also constantly invading eusociality. For example, in the stingless bee *Schwarziana quadripunctata* (E), some individuals reared in worker cells avoid a worker fate by developing into fully functional dwarf queens (Wenseleers *et al.* 2005). F, The generic RPS social network also governs density regulation in the bacterium *E. coli* (Kerr *et al.* 2002), involving trade-offs among the production of offensive (colicin) and defensive genes (colicin detoxification), and the intrinsic rate of increase (replication).

antagonistic selection between the sexes (Sinervo and McAdam 2008), social neighborhood choice and its impacts on female allocation (Calsbeek and Sinervo 2004), and social and life history trade-offs involving cooperative male strategies (Sinervo and Clobert 2003, Sinervo *et al.* 2006). Group-structured versions of the Price (1970) equation are critical in unraveling such complex social/life history trade-offs and we refer interested readers to Sinervo *et al.* (2008). Our goal here remains focused on links between endocrine systems and social systems. In females, the stress hormone CORT responds to social stress and differential crowding by r- and K-strategists (discussed below). CORT has cascading affects on costs of reproduction and size–number trade-offs. Thus, the full regulatory network involves social and endocrine networks (Fig. 25-4,5). Endocrine networks of vertebrates (Licht *et al.* 1977), exemplified by *Uta*, are generally conserved across vertebrates (Oksanen *et al.* 2002, Mappes *et al.* 2008).

With minor modifications, the network structure of r/K strategists can be applied to systems of aggressive and non-aggressive personality types involved in Hawk–Dove dynamics (see Fig. 25-5; Maynard Smith 1982). Korte *et al.* (2005) review endocrine, behavioral, and metabolic networks governed by the HPA, which regulate Hawk and Dove strategies. Hawk exhibits strong negative self-regulation, but frequency of Hawk increases when Doves are abundant. Dove, on the other hand, is neutral to self, but contracts in frequency when subjected to more Hawks. This system will equilibrate with Hawk and Dove if the negative effects between Hawks are great enough to allow for invasion of Dove, otherwise it fixes on Hawk (Maynard Smith 1982).

Hawk–Dove interactions generalize to systems with three strategies (Sinervo *et al.* 2007), with the addition of a cooperative type, which often evolves in social systems. The cooperative type generates positive self-regulation—more cooperators (in a group) enhance recruitment of self. However, cooperation is vulnerable to invasion by Aggression (i.e., Hawk-like type), which is self-limiting and generates negative self-regulation. Aggression is vulnerable to invasion by Deception (e.g., Dove-like) that is immune to or avoids Aggressive types and by virtue of lower levels of negative self-regulation can invade Aggressive types. Cooperative types, which thwart Deception via cooperation, benefit from invasion of Deceptive types. A rock–paper–scissors (RPS) social system ensues from Hawk–Dove–Cooperator social interactions. RPS systems can either stabilize with all three types preserved or it can cycle in frequency (and/or density) of all three types (Sinervo *et al.* 2007). RPS networks are common across organisms (Fig. 25-5C–E). For example, European common lizards and side-blotched lizards exhibit similar RPS density regulation. In general, any social organism demonstrating cooperation can generate similar density regulation and complex social trade-offs. Selection in such systems will operate on endocrine responses within and among individuals of each type. For example, Deception is likely to be under facultative or plastic control, being submissive when dominants are around but aggressive when confronting self-types (Sinervo and Miles (2010) discuss the reptilian HPA).

Social trade-offs arise directly from social competition (density regulation) and cooperation that abounds in natural systems (Sinervo *et al.* 2007, 2008). Social networks can also provide exogenous cues that trigger endogenous endocrine networks of organisms. In many cases, individuals in social networks actually produce potent hormones/pheromones that induce gated switch effects in the internal endocrine networks of conspecifics (see below).

Within most populations of organisms, intraspecific interactions such as competition and mating are critical to fitness components and ensuing trade-offs. Local density of conspecifics is one of the main reasons why organisms might experience trade-offs based on limited resources, because individuals are expected to compete most closely with members of their own species, and density is a primary trigger for altering resource allocation in response to limitation. Local conspecific density also has secondary effects on the evolution of trade-offs: the evolution of alternative strategies. Both theoretical models and empirical evidence have demonstrated that average fitness of competition-regulated populations is optimized if individuals vary in their life history/trade-off allocation phenotypes (Skulason and Smith 1995, Gray and McKinnon 2007). The presence of multiple strategies within populations reduces intraspecific competition, and ensuing niche partitioning

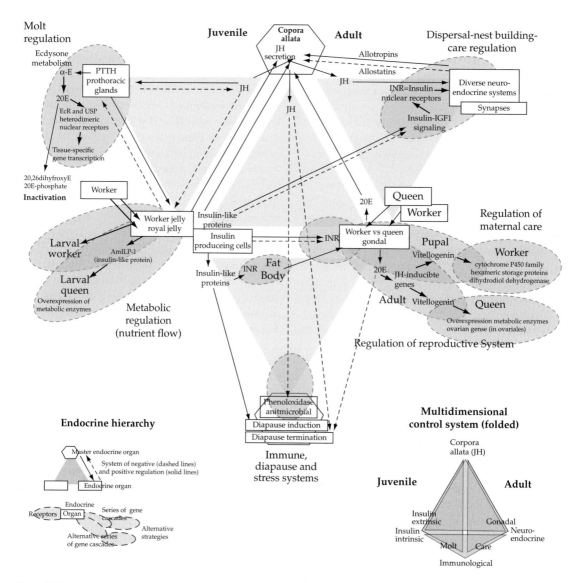

Figure 25-6 Mini-review of endocrine network diagrams for life history trade-offs for a generic insect in which resources are allocated to competing functions f growth, metabolic regulation, immune system regulation, reproduction, and other more complex life history behaviors (for the specific case of honeybees) like parental care and dispersal/migration that directly interact with reproduction. This network is a composite from reviews of *Drosophila* (Flatt *et al.* 2005), social hymenoptera (Wheeler *et al.* 2006; references in the text), and other insects (Nijhout 1994, Truman and Riddiford 2002). Hormones that largely affect juvenile life history are clustered on the left, and adult life history on the right. The immune system, generic to both stages, is at the bottom. *JH and ecdysone regulation.* The developmental hormones in insects are hydroxylated steroid hormones, the ecdysones, and the family of sesquiterpene hormones, the juvenile hormones (Nijhout 1994). Ecdysones are molting hormones, with pulses promoting cuticular development, whereas JHs maintain the insect in its current form as the insect responds to a molting surge of ecdysone. Thus, they have a negative feedback relationship similar to vertebrate HP–thyroxine axis (Fig. 25-3). In immature insects, ecdysones are typically produced by the prothoracic glands. The secretory source of JH is the corpus allatum, a major secretory structure on the insect brain. The endocrine system of insects can be represented by a multidimensional tetrahedron, shown folded at lower right. *Nuclear receptors for ecdysone and JH.* The ecdysone receptor complex consists of ligand-binding ecdysone receptor (EcR) (Truman and Riddiford 2002). The heterodimeric partner of EcR (heterodimeric partners have two protein parts) is ultraspiracle (USP), which is conserved in the arthropods. USP is an ancient nuclear receptor whose ortholog, RXR, is found in vertebrates in the form of the thyroid hormone and retinoic acid receptors (see Figs 25-3 and 25-7). *Drosophila* USP binds JH with low affinity, but recent analysis shows that neither JH nor methoprene (a JH analogue) has very high affinity in its ligand-binding pocket (Truman and Riddiford 2002). JH acid (a JH precursor with some hormonal activity) could fit into this pocket. There is still no clear consensus about the JH receptor. USP is a potential candidate, however, it does not bind with high-affinity, which is

increases niche breadth. Once polymorphic life history strategies evolve in response to local competition, fitness of each life history strategy in the population will tend to be negatively frequency-dependent: a rare strategy, experiencing decreased competition, will flourish at the expense of common strategies (Sinervo and Calsbeek 2006).

25.5   Social dimensions to trade-offs: Endocrine mediation and fitness consequences

In Sections 25.6–25.9, we turn to examples of how life history phenotypes evolve in response to density and frequency cues. The evolution of plasticity in trade-off allocation decisions, in response to conspecifics, is theoretically adaptive (Schlichting and Pigliucci 1998, Wolf *et al.* 1998), but few studies have reported on the entire process. This requires an understanding of:

(i)   perception of social cues
(ii) physiological transduction of social cues into internal (usually endocrine) signaling molecules
(iii) effects of these hormones on plastic adjustment of life history strategy
(iv) concomitant fitness effects.

Here we review a few study systems in which at least two physiological/ecological sub-processes are known. We also suggest further research that may fill in gaps and allow generalizations about how evolved social plasticity leads to greater complexity in life history and trade-off evolution (see Sinervo and Clobert 2008) than is currently predicted by simple "Y" models (Zera and Harshman

2001) and other density- and frequency-independent models. Ultimately, more studies should be applied to a fifth component of socially mediated, evolved life histories, which is the effect of fitness (item (iv) above) on frequency in the next generation (which comprises future social cues (item (i) above)) (see Alonzo and Sinervo 2001, 2007 for a mate choice example). This would allow us to close the feedback cycle and to understand how the evolution of socially-mediated phenotypic plasticity and frequency/density-dependent selection influence each other over evolutionary timescales.

25.6   Corticosterone, egg size, and the trade-off between aspects of offspring quality in a lizard

Traditional life history theory considers offspring quality to be synonymous with offspring size (as in the classic offspring size-number trade-off; Smith and Fretwell 1974). This is because offspring size is under maternal control and is a direct function of her resource allocation to reproduction. However, the size/number model does not take into account genetic variation among offspring, which could render maternal resource investment more or less beneficial, depending on progeny genotype. The Trivers–Willard model (1973) expanded on this early theory and explicitly incorporated differences in resource allocation to each progeny sex, owing to differential forces of sexual selection in sons versus natural selection in daughters. This is a form of parent–offspring conflict. When considered from the offspring's perspective, parent–offspring conflict is known as ontogenetic conflict (Rice and Chippindale 2001). An intuitive example of this is intersexual

considered a hallmark of a hormone receptor. The gene product of *Methoprene tolerant* (*Met*), isolated from *Drosophila* (Flatt *et al.* 2005), is another candidate suggested for the JH receptor. *Social interaction.* In social hymenoptera such as honeybees, signaling pathways for royal jelly (RJ) determine the queen (Wheeler *et al.* 2006). Recent molecular work (Wheeler *et al.* 2006) suggests that RJ triggers an insulin-like protein called *AmILP*-1 (Fig. 16-13). A molecule in RJ, fed by workers to eggs destined to be queens, triggers the cascade of changes in JH titer, which alters gene expression and in particular upregulates metabolic pathways. *Metabolic regulation in social insects.* Elevated expression of metabolic enzymes by queen-destined larvae reflects the enhanced growth rate of queens during late larval development. Many differentially expressed genes are tied to metabolism and cellular responses of JH, a result consistent with physiological differences between queen and worker larvae. JH and ecdysone, which govern expression of ovarian genes (Nijhout 1994), are reduced in larvae fated to become workers relative to those fated to become queens. Vitellogenin acts as a novel hormone during pupal development that triggers maternal care in "mature" workers (see text). Queens down-regulate many genes expressed by bipotential larvae and turn on a distinct set of caste-related genes. Queens overexpress metabolic enzymes. Workers showed elevated expression of a gene in the cytochrome P450 family (implicated in detoxification mechanisms), hexameric storage proteins, and dihydrodiol dehydrogenase (Wheeler *et al.* 2006). Young larvae overexpress two heat-shock proteins (70 and 90 kDa), and several proteins related to RNA processing.

ontogenetic conflict, in which traits that are beneficial when inherited by one sex are detrimental when inherited by the other sex (Fig. 25-1F) (Rice and Chippindale 2001, see Sinervo and McAdam 2008 for a clutch size example). However, ontogenetic conflict can occur with respect to any genotype that varies among offspring, and is not limited to a genotype by sex interaction.

In side-blotched lizards, the value of offspring size depends on its inherited genotype for the social strategy it exhibits, indicated by alternative throat colors. Here we consider conflict from the mother's viewpoint, and how her egg size allocation strategy has evolved in the context of a polymorphic social system that generates ontogenetic conflict among offspring for the benefit of large egg size. Side-blotched lizards can be yellow-, orange-, or blue-throated. In males, color reflects alternative mating strategies. Orange males are territorial and aggressive, blue males are mate-guarders that cooperate with other blue males, and yellow males are sneakers (Sinervo and Lively 1996, Zamudio and Sinervo 2000, Sinervo et al. 2006). In females, alternative throat colors correlate to alternative egg-laying strategies. Orange females lay larger clutches of small eggs. Yellow and blue females lay small clutches of large eggs (Sinervo et al. 2000). Throat color and life history traits are heritable (Sinervo and Svensson 2002).

In yellow and blue females, but not orange females, increased egg size (which leads to increased progeny size) leads to enhanced progeny survival, reflecting an egg size–genotype interaction (Lancaster et al. 2008). One reason for this is that large versus small progeny engage in different escape behaviors, and escape behaviors exhibited by smaller progeny benefit orange progeny (Lancaster et al. 2010). Specifically, hatchling morphology is directly affected by egg size, with hatchlings from small eggs exhibiting a leaner body shape relative to hatchlings from large eggs, which are stockier at birth. Body shape is repeatable from birth to adulthood. Body shape, in turn, affects escape behavior, with stockier hatchlings more likely to engage in cryptic behaviors and leaner hatchlings more likely to jump up and away from predators. To summarize, yellow-throated females lay large eggs, resulting in stocky hatchlings that

perform cryptic behaviors; these behaviors are consistent with the "sneaker" strategy employed by yellow-throated males. Similarly, orange females lay small eggs, resulting in lean hatchlings, which perform leaping behaviors consistent with territorial behaviors of orange-throated males that defend rock piles with numerous protected, elevated crevasses (Lancaster et al. 2010).

An individual's throat color genotype is inherited equally from both parents, and color expression is additive, so we hypothesized that in addition to evolving an adaptive genetic correlation between egg-laying strategy and throat color in dams (Sinervo et al. 2000, 2001, Sinervo and McAdam 2008), in response to selection for differing egg sizes by offspring genotype, females should also plastically modulate their egg size in response to social cues that predict the sire's contribution to offspring throat color. In general, maternal-effect plasticity is any contribution of maternal phenotype to offspring phenotype in response to external cues. Traditionally, such maternal-effect plasticity has been shown to be adaptive when environments fluctuate, and maternal environments accurately predict offspring environments (Mousseau and Fox 1998). We extend this prediction and posit that when environments contain genes (i.e., when females can detect genotypes of social neighbors), then local social environments could represent an adaptive cue, allowing mothers to predict which genes progeny are likely to inherit. In other words, females have cues not only about future environments that offspring will likely *inhabit*, but also about genes at high frequency in potential sires and near neighbors that offspring will likely *inherit*—because genes found at high frequency in potential sires are more likely to be inherited by their progeny than genes at low frequencies in the local environment (Lancaster et al. 2007, 2010). If females can incorporate social cues into maternal-effect allocations, they can potentially mold some offspring traits (such as body shape, see above) to match other offspring traits that are not amenable to maternal-effect manipulation (such as throat color genotypes).

The local frequency of throat colors in the social environment plastically affects maternal egg-size allocation strategy via impacts of social cues on the HPA (Fig. 25-4). Specifically, orange females experience decreased CORT in response to orange-throated neighbors in nature (Comendant *et al.* 2003). Conversely, yellow and blue females experience increased CORT when crowded by yellow- and blue-throated individuals (Comendant *et al.* 2003). Chronically elevated CORT induces larger egg-mass of females, without a concomitant decrease in clutch size (Lancaster *et al.* 2008). This likely occurs because of CORT's effect on increasing appetite, so females experiencing chronically elevated CORT are more likely to feed more during oogenesis, resulting in a net energy gain that can be translated into both survival and increased gamete size (Jonsson 1997). This mechanism allows each maternal morph to appropriately adjust offspring size in response to social cues, which can inform her of her offspring's likely genotype. Small size benefits orange hatchlings, and orange females lay small eggs. Combining the results of Comendant *et al.* (2003) and Lancaster *et al.* (2008), we suggest that orange females further decrease egg size via reducing plasma CORT when local frequency of orange throat color is high (indicating that offspring will likely inherit that orange color through the sire as well as the dam), while yellow and blue females further increase egg size when the social environment predicts that offspring are also likely to inherit yellow and blue throat color from sires. We performed a second experiment assigning females to controlled mating in the laboratory. As expected, females of all throat colors mated to yellow-throated sires (but not sires of other throat colors) responded by increasing average egg size (Lancaster *et al.* 2010).

HPA-mediated allocation plasticity is an adaptive mechanism by which females respond to the trade-off among different aspects of offspring quality. Females use social cues to predict offspring inherited traits (Alonzo and Sinervo 2001, 2007), and then use that information, via endocrine modulation, to alter reproductive allocation. This trade-off among different aspects of offspring quality exists because the population was *already* polymorphic for social strategies. This pre-existing polymorphism created selection pressure favoring evolution of a secondary polymorphism (in egg size plasticity), in which yellow and orange females differ in whether they increase or decrease egg size in

response to crowding by a similar phenotype. This supports the hypothesis that once a population or species becomes polymorphic along one axis, numerous other traits also evolve polymorphism as an adaptive response to changes in the selective environment caused by the initial polymorphic trait. Populations polymorphic for mating strategy or antipredator phenotypes often display polymorphic life history strategies (Ahnesjo and Forsman 2003, Badyaev and Vleck 2007). It is generally unknown which polymorphic axis (life history versus mating or antipredator strategy) arose first in a species' evolutionary history, but phylogenetic analysis of *Uta* (Corl *et al.* 2009, 2010) suggests that changes in social strategy alter life history traits of females and sexually-selected traits of males.

25.7 Juvenile hormone, vitellogenin, and reproductive trade-offs in eusocial honeybees

Eusocial honeybees (*Apis mellifera*) are one of the best-studied systems for how social interactions affect individual phenotype and (colony) fitness (see also Chapter 20). Queens and workers have evolved pheromones and other chemical signals to regulate hive activity (Seeley 1989, Page and Fondrk 1995), and these optimize overall hive productivity. Recently, researchers have begun to investigate individual variation in worker behavior within hives, leading to an emerging hypothesis that reproductive trade-offs in solitary ancestors were critical to the evolution of eusociality and division of labor (Amdam *et al.* 2004). Specifically, individual worker bees vary in their propensity to gather nectar versus pollen during foraging bouts. Artificial selection experiments and QTL studies (Ruppell *et al.* 2004, Hunt *et al.* 2007) provide compelling evidence of genetic control in individual foragers, revealing substantial genetic variation (polymorphism) among workers in their tendency to gather nectar versus pollen. However, foragers also plastically switch from foraging on nectar to foraging on pollen when pollen reserves are experimentally removed from colonies (Fewell and Winston 1992), suggesting that the abundance or lack of previously stored pollen in hives represents a social cue that prompts individual bees to switch their foraging strategies. Cross-fostering studies indicate that foraging phenotype is significantly affected by an interaction among forager genotype (high versus low pollen hoarding) and colony type (high versus low pollen hoarding). Similar norms of reaction were uncovered in high- versus low-selected colonies among resource abundance treatments (Fewell and Page 1993). Thus, propensity to forage on nectar versus pollen is regulated by genes and environment.

Females of many solitary insects feed predominantly on nectar during non-reproductive times and feed on pollen during oogenesis. This plastic strategy reflects a self-maintenance versus reproduction trade-off. Nectar, which is high in sucrose, supports the female's metabolism, while pollen provides protein to be invested into offspring (Page and Amdam 2007). As discussed above, interactions with conspecifics are predicted to commonly result in polymorphic life history strategies. In the case of honeybee ancestors, increasing social interactions among solitary females may have led to the trade-off between self-maintenance and reproduction.

In comparison to an ancestral female insect, which would have performed reproductive, foraging, and parental care behaviors, female bees have resolved these trade-offs by dividing the task among castes (polymorphism). Queens primarily perform the reproductive function, while workers perform maternal care early in life, and forage later in life (after about day 10–14 of their adult life). Evidence from hormonal data suggests that these two distinct kinds of honeybee polymorphisms (the queen/worker reproductive polymorphism reflecting the foraging versus reproduction trade-off, and the pollen versus nectar preference polymorphism among workers reflecting an ancestral self-maintenance versus reproduction trade-off) may arise from similar ancestral endocrine machinery. Specifically, these evolved worker–queen and worker–worker polymorphisms appear to have been achieved via heterochronic shifts in the development of bees' endocrine systems (Fig. 25-6). The usual pre-reproductive adult phase expressed in solitary insects is bypassed in honeybees, and timing of hormonal events associated with reproductive maturity has been shifted from adult to pupal and pre-pupal stages (Page *et al.*

2006). Juvenile hormone (JH) determines ovariole development during the fifth larval instar in both queens and workers (Capella and Hartfelder 2002), leading to production of vitellogenin (Vg, a yolk protein that has evolved novel endocrine function that promotes maternal care behaviors) during pupal development. JH and Vg are both critical in organizing both queen versus worker castes, and pollen versus nectar hoarding preferences among workers. This early difference in JH between queens and workers, which in part determines caste differentiation, is mediated by a gated switch that involves insulin-like proteins and nutrition (Fig. 25-6), analogous to gated switches of prolactin/thyroxine in anuran development (Fig. 25-3) and the vertebrate HPA (Fig. 25-2). Experiments demonstrate that increased larval nutrition increases ovary size and ovariole number in both worker and queen castes. Workers feed larval queens royal jelly for longer durations than larval workers, and larval queens consume sugar and protein in different ratios. Furthermore, larvae destined to become queens are fed more food than are workers (Winston 1987). Early nutritional differences result in differing levels of JH, and therefore Vg, between queens and workers.

Furthermore, high JH has positive effects on ovary development and Vg production, which in workers, but not queens, promotes maternal care behavior expressed heterochronically at birth instead of later—after dispersal, nest building, and reproduction (as expected in non-eusocial females). Elevated Vg also primes worker bees for pollen preference (versus nectar preference), when they commence their foraging phase of life.

In lines of bees selected for high pollen preference, many correlated changes in non-selected traits were observed, including increased titers of Vg and JH expressed during pupal and early adult phases (Amdam *et al.* 2004). Selection for pollen preference enhanced development rate, and frequency of worker reproduction in the absence of a queen (Page *et al.* 2006), which resulted in overexpression of genes for ovarian development and insulin-like signaling components (Hunt *et al.* 2007). Therefore, selection for high pollen preference of workers also increased expression of reproductive and maternal phenotypes among workers, mediated by increases in insulin/insulin-like signaling, JH, and Vg. Manipulation of worker nutrition and Vg and JH titers corroborate the selection experiments. Experimental application of JH during the fifth instar decreased apoptosis within ovarioles, thus rescuing bees fated to become workers (and restoring reproductive function). Furthermore, inhibition of Vg production via RNA interference results in bees who forage precocially, skip the maternal care portion of their lives, and preferentially feed on nectar (Amdam *et al.* 2007).

Results from high-pollen selected lines suggest a potential mechanism for evolution of eusociality (Amdam *et al.* 2004). Prior to evolution of eusociality, cooperative breeding would have resulted in some females obtaining more nutrients and also suppressing their nest mates' nutrient intake. Dominant females would translate increased nutrition into enhanced and earlier expression of JH, and thus increased ovary development and increased Vg titers. Increased Vg in reproductive females leads to pollen preference because proteinaceous pollen enhances offspring production. Nutrient-deprived females would have decreased JH and Vg, and thus be more likely to forage on nectar (i.e., self-maintenance). Over time, selection for gene regulation of alternative strategies could have fine-tuned the two roles, resulting in evolved heterochronic shifts in insulin, JH, and Vg among queens and workers (Amdam *et al.* 2004).

This polymorphic endocrine mechanism follows our and others' predictions that social interactions and competition promotes polymorphism in the evolution of life history strategies in response to trade-offs. Furthermore, within this evolved social context, pollen-hoarding and nectar-hoarding strategies have evolved to respond plastically to local frequencies of other such strategies within the hive (whereas an ancestral solitary female's life history plasticity was likely more closely attuned to resources in the environment and/or abiotic conditions). In cooperatively breeding vertebrates relative to ancestral social species, evolved changes in HPA responsiveness of dominants relative to subordinates appears to modulate the gated-switch of individuals that are breeders versus helpers (See Fig. 2 in Creel 2001 and Chapter 26 in this volume).

25.8 Testosterone, growth hormone, social dominance, and smolting in Atlantic salmon

Atlantic salmon, *Salmo salar*, are well characterized for plasticity in and polymorphic expression of male life history strategies (Thorpe *et al.* 1998). Male *S. salar* undergo two complementary developmental conversions between juvenile and adult phases: sexual maturation and smolting, which involves growth and changes in coloration, osmoregulation, behavior, and metabolism associated with transition from freshwater to marine environments (Hoar 1976). Due to potential trade-offs between sexual maturation and metamorphosis (Thorpe 1987), these two developmental events are only loosely physiologically coupled. Specifically, eggs hatch in spring and juveniles grow during summer, transforming from fry to parr. Around mid-summer, some parr commit to smolt after one year (Thorpe 1986). Individuals that smolt within one year exhibit increased appetite, enhanced growth, and undergo metamorphic changes that prime the next spring's smolt. Individuals not undergoing physiological preparations to smolt by late summer are unable to smolt next spring. Priming before smolting is necessary (Pickering *et al.* 1987). Individuals smolting in year one travel to the ocean, where they grow and accumulate reserves for one to two years prior to migration and reproduction. Individuals not smolting after one year remain parr for up to seven years (Metcalfe *et al.* 1989, Metcalfe and Thorpe 1990). All individuals are capable of smolting in one year, thus smolt timing is based on environmental factors (Thorpe *et al.* 1998), although genetic differentiation in smolt propensity is present among populations, likely reflecting differences in the threshold response to environmental cues (Horton *et al.* 2009).

Overall fitness is higher in males that undergo smolting, both in terms of survivorship (Myers 1984) and reproductive success (Hutchings and Myers 1988). The delay in smolting increases mortality as parr. However, some males that delay smolting can mature as parr and sire offspring during their second autumn, when all one-year smolts are still out at sea. Spawning success of male parr is always much less than that of a smolt during a single fertilization episode, but parr gain paternity as small, inconspicuous sneaker males. An individual parr will only gain 10% paternity, and this gain is traded-off against earlier breeding. The parr paternity fraction goes down as number of parr at nests increases (Hutchings and Myers 1988), while large territorial males sire ~70%. Furthermore, increased human predation on smolts may increasingly favor parr maturation in nature (Myers 1984, Hard et al. 2008).

Within cohorts, variation in the smolt decision results in a bimodal distribution of parr size by autumn (Thorpe 1977). The smolt switch appears to be mediated by local density and conspecific competition (Horton et al. 2009). Individuals that hatched earlier and/or are more aggressive earlier may obtain more resources and larger size by the time the smolting decision is made in their first year of life (Metcalfe et al. 1989). Furthermore, feedback via aggressive social interactions directly mediates life history plasticity in this system (Metcalfe and Thorpe 1990), but it is not yet known whether such social feedback involves the HPA-axis. Smolting exhibits complex endocrine regulation in S. salar and molecular mechanisms are not completely understood (Fig. 25-7). Testosterone (T), cortisol, thyroid hormones, and growth hormone (GH) are key hormones that rise during the parr–smolt transformation (Yamada et al. 1993, Ebbesson et al. 2008), suggesting that these hormones may have a role in implementing developmental conversion. One action of GH is to increase aggression and feeding rates, and GH is thought to be primarily responsible for rapid growth during transformation. Thyroid hormones (i.e., T_3) are also thought to influence growth, color change, metabolism, lipid mobilization, downstream migration, and olfactory imprinting during metamorphosis (Ebbesson et al. 2008).

While molecular mechanisms responsible for initiation of smolting in response to social cues are unknown, the process may involve T. T may increase as a by-product of elevated GH (see above). In Atlantic salmon, gonadal maturation coincides with onset of feeding (Thorpe 1994a,b), the same developmental time point when social status influences smolting (Metcalfe and Thorpe 1990). As T controls aggression and gonadal development, it is possible that differential expression of T during gonadal development is responsible for setting dominance

hierarchies at a very early age in S. salar. Following gonadal development, T wanes but remains throughout the parr stage. However, relative social status, once established, is stable in salmonid fish (Jenkins 1969), and may not rely on high T titers in dominant parr. Once dominance is established, low-ranking individuals are excluded from resources, which partially determines their decision not to smolt (Metcalfe and Thorpe 1990), perhaps via HPA suppression of GnRH release.

Atlantic salmon face trade-offs of smolt timing and early sexual maturation as parr. Local density limits the ability of parr to garner enough resources to make smolt transitions and acquire competitive dominance, generating alternative life history strategies. Maturation without smolting is likely ancestral, but increasing competition for resources within natal streams may have been the agent of selection favoring the evolution of smolting, in which some individuals delay reproduction to migrate elsewhere in search of resources (Thorpe 1994a,b). As in honeybees, the transition between alternative life histories is phenotypically plastic, and is regulated by social interactions with conspecifics, in addition to having a genetic basis.

25.9 Gibberellins, auxin, ethylene, and reproductive allocation in monoecious plants

Plants and other hermaphroditic organisms commonly experience a resource allocation trade-off between male versus female reproductive function. This trade-off is specifically predicted to be frequency-dependent, with a rare-sex advantage (Fisher 1930, Hamilton 1967, Trivers and Willard 1973). In hermaphrodites, gender allocation is environmentally determined at some point after birth, and the trade-off between male versus female function is manifest at the individual level. Thus, individual sex allocation can plastically respond to local conditions (Charnov and Bull 1977). Hermaphrodites are, like unisexuals in gonochoristic/dioecious species, predicted to favor either maleness or femaleness over a more equitable sex allocation strategy. This is because differences in plant quality and/or local environments will differentially affect reproductive success as males versus females (Charnov and Bull

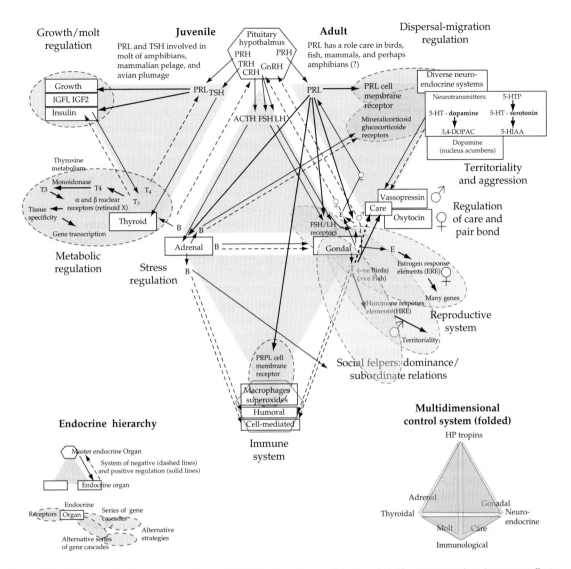

Figure 25-7 Mini-review of endocrine network diagrams for life history trade-offs due to behavior, unfolded for a generic vertebrate (components affecting various vertebrates are labeled), in which is depicted: resource allocation to competing functions of growth regulation, metabolic regulation (homeostasis), immune system regulation, reproduction, and other more complex life history behaviors like parental care and dispersal/migration that directly interact with reproduction. The tetrahedron structure of the endocrine network is shown in the lower right and analogous to that of insects (Fig. 25-6). Hormones that largely affect juveniles are clustered on the left, and adults on the right. The immune system, generic to both stages, is at the bottom. Detailed discussion of the hierarchical structure of the endocrine system can be found in Sinervo and Clobert (2008) and Sinervo and Calsbeek (2004), but most interactions are tripartite (grey triangles) involving at least three endocrine secretory organs (gated switches that control positive/negative regulation). *Hypothalmic and pituitary control.* The master endocrine regulator of life history function is found in the hypothalamus and pituitary, which produces the various releasing hormones (RH), such as prolactin RH (PnRH), thyroid RH (TRH), corticotropin RH (CRH), and gonadotropin RH (GnRH), among others. In vertebrates, regulation of growth (and metamorphosis if present) is achieved by the thyroxine, the metabolically active T3 and its precursor T4 produced by the thyroid, in interaction with brain prolactin (Prl) pathways and/or other growth hormones.

Prl is an important but as yet understudied hormone from the perspective of life history trade-offs (Sinervo and Miles 2010). Prl interacts with diverse life history trade-offs, including metamorphosis (in amphibians, refs in Fig. 25-3), and molt cycles of amphibians, birds, and mammals (Bole-Feysot *et al.* 1998), and regulates dispersal and/or migration behaviors and expression of parental care in both sexes in all vertebrates (Schradin and Anzenberger 1999). While Prl plays a role in regulating progeny care in vertebrates, neuroendocrine hormones Vassopression and argeninevassopressin (AVP) control care in male vertebrates while the octopeptide oxytocin controls care in female vertebrates. The HPA affects diverse CNS targets (reviewed by Sinervo and Miles 2010), including strategies of social helpers in communal breeders (see Creel 2001). Behaviors can be modulated at many control points, including the monoamine system of neurotransmitters (dopamine and serotonin). For example, pair-bonding behaviors (in the context of care) arise from dopamine activity in the nucleus accumbens.

1977), and also because the fitness function for male versus female allocation may be nonlinear, i.e., a unit increase in female fitness may accelerate as investment in maleness decreases, and vice versa (Charnov *et al.* 1976). This is theoretically expected to occur whenever equal investment in male and female structures decreases overall reproductive output, in comparison with investment in predominantly sperm or predominantly egg production, and male versus female behavior.

Here we review effects of hormones and social environment on adaptive allocation in monoecious plants. Monoecious plants are those in which individual plants are bisexual; however their sexual function is segregated by flower, such that on each plant, some flowers are staminate (male), some flowers are carpellate (female), and no flowers express both male and female function. In monoecious species, therefore, relative investment of each individual into male versus female function, and environmental and physiological conditions leading them to favor one sex over the other, are relatively easy to measure. This makes them excellent model systems for studying evolved plasticity in reproductive trade-off allocation.

In a study of the monoecious plant *Begonia gracilis*, experimental manipulation of the social environment resulted in shifts in resource allocation towards male or female flowers (Lopez and Dominguez 2003). Specifically, in high-pollen load treatments, plants shifted allocation towards an increased percentage of female flowers, while in low-pollination treatments, plants up-regulated male flower production. These results support an adaptive sex-allocation mechanism explanation, in which individuals alter gender in response to operational sex ratios of local populations (Werren and Charnov 1978).

In contrast to pollen-mediated social regulation of gender expression in *Begonia*, individuals of the monoecious, wind-pollinated ragweed, *Ambrosia artemisiifolia*, have an alternative social regulation of gender expression due to the individual's height relative to heights of conspecific neighbors. The proportion of male flowers is increased in relatively taller individuals, and decreased in relatively shorter individuals. Furthermore, all individuals that produced only female flowers inhabited the dense clumps of conspecifics in the understory (Lundholm and Aarssen 1994). Experiments revealed that in this case light (irradiance) was the proximate cue promoting development of male, rather than female flowers (Lundholm and Aarssen 1994). In *Ambrosia* (in contrast to *Begonia*), wind pollination means that an individual's male versus female fitness is dependent on relative heights of other individuals in the population (lateral wind transfers pollen down), in addition to local sex ratio (Lundholm and Aarssen 1994).

Evidence from exogenous hormone application in plants of the Cucurbitaceae suggests an evolutionarily-labile hormonal basis for gender determination in monoecious plants (Yamasaki *et al.* 2005). In cultivated monoecious bitter melon (*Momordica charantia*), exogenous application of gibberellic acid (GA; a plant hormone most commonly known for its effects of promoting cell elongation and seed germination) increased relative production of female flowers to 26%, from 15% in the absence of treatment (Thomas 2008). However, in cucumbers (*Cucumis sativus*), GA increased male flowers, while ethylene (a hormone with a broad range of tissue-specific effects) increased female flowers (Yin and Quinn 1995). In cucumbers, exogenous application of IBA (indole-3-butyric acid, an auxin) increased female flowers in one cultivated strain, while it decreased female flowers in a second cultivated strain (Diola *et al.* 2008). In cucumber, melon, and zucchini, exogenous application of brassinosteroids (a class of plant steroid integral to pollen tube formation, along with other tissue-specific effects) increased ethylene production and the proportion of female flowers produced by individual plants (Papadopoulou and Grumet 2005). In monoecious conifers, spraying with GA resulted in male-biased floral production in summer, but female-biased floral development in autumn, because proteins that bind to GA differ by season in conifers (Khryanin 2007). Therefore, gender expression in monoecious plants is hormonally regulated, with specific mechanisms being evolutionarily labile and/or mediated by redundant endocrine pathways. In the future, it would be interesting to examine correlations among endogenous hormones and male:female flower ratio (Papadopoulou *et al.* 2005), and to examine a phylogenetically more diverse array of monoecious species.

Plant hormonal levels can be influenced by both light levels and by pollen germination on stigmas. Exposure to UV-B decreased auxin levels and increased GA levels in leaves and pistils of tomato (Yang *et al.* 2004). Alternatively, auxin is also involved in shade avoidance responses in many plants, and auxin production is stimulated by low ratios of red:far red light (such low ratios signal shade) (Alabadi and Blazquez 2009). One of these phytochrome- or UV-mediated hormone-synthesis pathways is likely involved in *Ambrosia*'s socially- and hormonally-mediated gender allocation. In *Begonia*, pollen germination on stigmas likely induces hormonal changes. In tobacco, endogenous auxin increased in stigmas in response to pollen germination, perhaps via pollen tube growth, which hastens floral senescence (Chen and Zhao 2008), and also from a downstream increase in ethylene production (De Martinis *et al.* 2002). Effects of shading or receipt of outcrossed pollen modulate a plant's "perception" of their social environment to downstream effects on gender expression, and thus reflect gated switches for plant reallocation.

Social regulation of gender allocation in gonochoristic animals and monoecious plants reflect complementary phenomena. In both cases, individuals experiencing social regulation of gender allocation usually deviate significantly within individuals from a 50:50 ratio of male to female function (Fisher 1930). As in the honeybee, tadpole, lizard, and salmon examples, this supports our predictions that socially mediated plasticity in life history trade-off evolution leads to polymorphic life history allocation strategies. In the case of sex allocation, any perturbation of one individual from its optimal sexual allocation will alter optimal investment strategy by each of its neighbors. Such social feedback results in increased variance in reproductive allocation strategies, and ultimately ameliorates intensity of mate competition.

25.10 Conclusions and future directions

The structure of all endocrine networks that mediate life history trade-offs are likely to involve systems of positive and negative regulatory feedback loops modulated by gated switches that turn off one allocation pathway (e.g., reproduction) and open energy flow to another pathway (e.g., survival). From our review and synthesis of the structure of invertebrate (Fig. 25-6) and vertebrate endocrine systems (Fig. 25-7) it is clear that regulatory structures of the two groups are remarkably similar given the hundreds of millions of years of evolution that separates the groups. This similarity can be explained either by convergence or common ancestry. We suggest the similarities in endocrine regulation among vertebrates and invertebrates (and potentially, plants) are due to a fundamental similarity in the topology of social networks that is related to density regulation and how social cues can be used to invoke life history plasticity (Figs 25-3 and 25-5). The negative and positive regulatory loops generated by density and frequency regulation impose selection on endocrine plasticity to generate remarkably similar endocrine network structures.

Future work should focus on further understanding the general features of life history endocrine regulation in a social context as we have done in summary points 1–4 below. Finally, further work is needed in learning how endocrine plasticity affects selective environments in which social organisms operate. Understanding the role of plastic phenotypes in shaping social environment and therefore local selection pressures will lead to vast improvement in our understanding of the dynamics of life history evolution.

25.11 Summary

The preceding five examples on taxonomically diverse organisms (vertebrates, insects, plants) illustrate how social dimensions contribute to evolution and expression of life history strategies. Despite this, certain generalities in the social and endocrine regulation of trade-offs are apparent (Fig. 25-5). Using the preceding five examples, we have summarized below four general features of endocrine mechanisms in life history trade-off evolution in a social context.

1. In all cases, social interactions promote polymorphism in life strategies (spadefoot toad polyphenism, lizard egg size/escape behavior, division of labor and trade-offs in social insects, alternative male strategies in salmon, male versus female flowers).

2. Evolution of life history trade-offs (networks of negative/positive regulation) in social contexts lead to socially mediated plasticity (gated-switches that alter negative/positive regulation) in life history traits, often involving gene-by-environment interactions in which individuals vary in their thresholds of response to social cues from conspecifics.

3. Life history plasticity in response to social cues involves translation of external (social) signals into endocrine responses, which are responsible for implementing life history polymorphism. These endocrine mechanisms are thought to have originated from ancestral, monomorphic states, in which these same endocrine mechanisms would have regulated life history switches in response to timing or other external (but not social) cues.

4. Fitness consequences of having evolved socially-responsive, polymorphic life history strategies are usually positive at the population level (i.e., average fitness increases via enhanced niche breadth or more efficient population regulation).

Socially mediated plastic life history polymorphisms experience both frequency dependent expression and frequency dependent fitness; they are thus likely to produce complex population and evolutionary dynamics. Future modeling of social and endocrine networks and empirical studies involving experimental deviations of phenotypic frequencies will be integral to beginning to understand the generalized impacts of the social dimension of life history evolution.

CHAPTER 26

Hormonally-regulated trade-offs: Evolutionary variability and phenotypic plasticity in testosterone signaling pathways

Michaela Hau and John C. Wingfield

26.1 Introduction

A major part of life history theory revolves around the existence of trade-offs (e.g., Stearns 1992, Roff 2002; see also Chapter 1). A trade-off exists when a change in one trait has either positive or negative consequences on another trait, in both an evolutionary and a functional sense (Stearns 1989). Trade-offs manifest themselves at various organismal levels, such as among life history traits, behaviors, and physiological processes. For example, female birds with experimentally enlarged clutch sizes (an addition of eggs) in the first breeding attempt show a reduced clutch size in a subsequent breeding attempt in that same year and/or a reduced survival in the following non-breeding season compared with controls (e.g., Dijkstra *et al.* 1990). This illustrates the existence of a life history trade-off between current and future reproduction, with high investment into current reproductive effort impairing survival and/or future fecundity (see also Chapter 11). Trade-offs can occur among behaviors, as when male rodents treated with exogenous testosterone display increased sexual behavior but at the same time show reduced paternal behavior (e.g., Clark and Galef 1999). Trade-offs can also occur among physiological processes, such as when the investment of energy and nutrients into reproduction comes at the expense of growth in fish species (e.g., Warner 1984). Because many life history, behavioral, and physiological trade-offs involve

endocrine control mechanisms, hormones may constitute an important part of the physiological machinery that underlies vertebrate life history strategies (Ketterson and Nolan 1992, Finch and Rose 1995, Sinervo and Svensson 1998, Zera and Harshman 2001; see also Chapter 24).

One major question in the field of evolutionary endocrinology is to what extent hormonally-regulated trade-offs represent adaptive physiological mechanisms that can readily evolve, and to what degree they represent constraints that limit optimal responses of individuals, populations, and species (Ketterson and Nolan 1999, Wingfield *et al.* 2001, Ricklefs and Wikelski 2002, Hau 2007, Adkins-Regan 2008, Lessells 2008, McGlothlin and Ketterson 2008, Ketterson *et al.* 2009).

To some extent this relates, on a physiological level, to the long-standing question about the processes that create trade-offs. One possibility is that trade-offs are caused by the limited availability of critical resources, such as energy, nutrients, and time, requiring differential allocation into competing processes that simultaneously require such resources (functional or intrinsic trade-offs; e.g., Lessells 2008; see also Chapter 11). If so, hormonal regulation of traits may have been favored by selection to mediate/signal allocation decisions, and quantitative variations in hormone signaling could lead to changes in the resolution of the trade-off. Such quantitative variations in hormone signaling conceivably could occur over short evolutionary time scales or even within individuals, as during different reproductive stages. Alternatively, trade-offs may be caused by a linkage of traits on a molecular level, for example from past selection for a correlation of certain traits, from linkage disequilibrium or other processes (molecular or genetic trade-offs, e.g., Roff and Fairbairn 2007b). Such linkages could require longer evolutionary times to change or break, and also might not be possible for an individual to change within its lifetime. In this latter case hormones may be part of the mechanism causing the trade-off because they could be molecules contributing to those links (e.g., Leroi 2001). Conceivably, such a molecular linkage of traits could be the result of past selection for allocation signaling by hormones but, due to evolutionary inertia, persists without there necessarily being a functional (e.g., resource-based) trade-off at present. Some processes, for example strong past selection on the correlation of traits or the functioning of complex physiological systems (see below) could also have dramatically reduced the amount of standing genetic variation that selection could act on, thereby greatly slowing down the rate of evolutionary change in the resolution of trade-offs (see also Heideman 2004, Heideman and Pittman 2009). Such evolutionary inertia could potentially provide a physiological explanation for the remarkable one-dimensionality of vertebrate life histories, i.e., the finding that organisms vary in life history strategies primarily along a slow or fast pace-of-life axis (e.g., Ricklefs and Wikelski 2002).

Here we will discuss evolutionary implications of the hormonal regulation of vertebrate life history, behavioral, and physiological trade-offs, focusing on one well-known endocrine system, the hypothalamic–pituitary–gonad axis (HPG) and the steroid hormone testosterone. We will approach this question primarily from a physiological standpoint by describing known patterns of conservation and variation. We will first consider the potential for adaptations in testosterone-regulated trade-offs via evolutionary change by analyzing interspecific patterns (comparing populations and species). Such analyses will be important to understand which components in the complex HPG axis and testosterone signaling cascade can evolve, for example when a different resolution of trade-offs

would be optimal in populations experiencing major changes in ecological conditions, such as after invading new habitats. Second, we will discuss the degree of plasticity in the hormonal regulation of traits to environmental conditions on an intraspecific level (comparing individuals within a population). Knowing phenotypic plasticity in hormonally-regulated traits can help assess how well individuals can adjust the resolution of trade-offs to changes in circumstances within their lifetime, for example when the environment fluctuates seasonally or social conditions change (e.g., group size, density etc.). Our focus will be on birds because of their importance as study subjects in natural and laboratory environments, and particularly on male birds because of the availability of data. Even though we will concentrate our discussion here on testosterone, vertebrate life history trade-offs are likely regulated by various, and probably interacting, hormonal systems, including the gonadotropins (e.g., Mills *et al.* 2008), the hypothalamo–pituitary–adrenal system (Bókony *et al.* 2009, Hau *et al.* 2010), and others (e.g., insulin-like growth factor-1, Sparkman *et al.* 2009; thyroid hormones, e.g., Chapters 6 and 7).

26.2 Testosterone and trade-offs

Testosterone is a molecule whose main functions concern the regulation of reproductive function in male vertebrates (Knobil and Neill 1988, Wingfield 2006). Testosterone synthesis and release, as in most other hormones, results from the activation of a complex endocrine cascade. For testosterone, this cascade is the HPG axis, which comprises several hierarchical steps and feedback loops (Fig. 26-1). The activity of the HPG cascade is triggered by environmental information, transduced from sensory receptor systems to the hypothalamus where gonadotropin-releasing hormone (GnRH) release is regulated (Fig. 26-1). Several types of GnRH are known to influence the release into the circulation of the two gonadotropins, luteinizing hormone (LH) and follicle-stimulating hormone (FSH) from the anterior pituitary. LH and FSH release lead to a subsequent stimulation of testosterone production from the testes, and production of negative feedback systems, which in turn regulate GnRH and

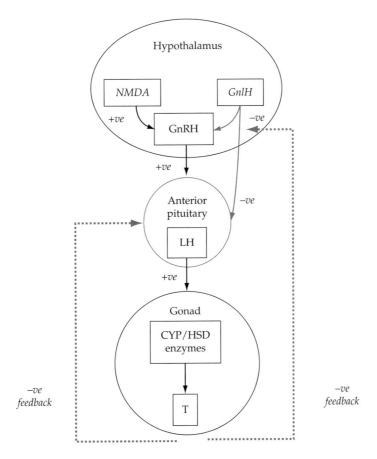

Figure 26-1 Classical schematic of the hypothalamo–pituitary–gonad axis regulating the secretion of testosterone in reproductive mature vertebrate individuals. Note that gonadotropin-inhibitory hormone (GnIH) can inhibit gonadotropin-releasing hormone (GnRH) centrally or hypophysiotropically to inhibit GnRH-stimulated release of luteinizing hormone (LH) at the anterior pituitary (e.g. Tsutsui *et al.* 2009). Glutamate can also stimulate GnRH and LH release as shown by injections of its agonist NMDA (N-methyl-D-aspartate; e.g., Meddle *et al.* 1999). CYP, cytochrome P450 enzymes; HSD, hydroxysteroid-dehydrogenase. Dotted lines, feedback loops (positive or negative). In general, this neuroendocrine and endocrine cascade leading to release of testosterone into blood is highly conserved across vertebrates. However, the interactions of GnRH, GnIH, glutamate, and other hypothalamic neurotransmitters and neuropeptides can provide great diversity of pathways by which environmental information can be transduced into testosterone release.

gonadotropin release (Nelson 2005, Norris 2007). Recently, a new peptide has been discovered that can inhibit gonadotropin release (gonadotropin-inhibitory hormone, GnIH; Tsutsui *et al.* 2009), thus modulating the activity of the HPG axis. Testosterone is usually synthesized within the Leydig cells of the testes from cholesterol in various enzymatic steps (Fig. 26-2).

A brief summary of the known signal transduction pathways of testosterone (Nelson 2005, Fig. 26-3) is as follows. Being a small, lipophilic molecule, testosterone can diffuse across cell walls and passively enter the circulation (unless bound to large binding proteins, see below). Once in the blood, testosterone circulates to target tissues where it enters cells and interacts (usually) with intracellular receptors that then become gene transcription factors regulating many morphological, physiological, and behavioral traits. In blood, testosterone is often carried bound to sex-hormone binding-globulins or, as in birds, to corticosteroid binding globulin (e.g., Breuner and Orchinik 2002, Fig. 26-3). One major function of these binding globulins might be to "buffer" the actions of testosterone, although

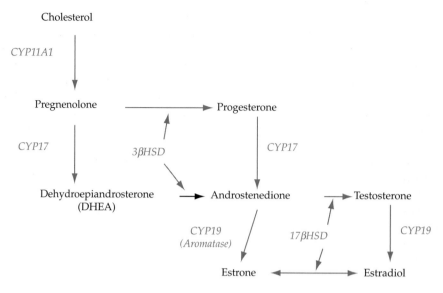

Figure 26-2 Simplified schematic of testosterone-production pathways. CYP, cytochrome P450 enzymes; HSD, hydroxysteroid-dehydrogenase. Modified from Soma (2006).

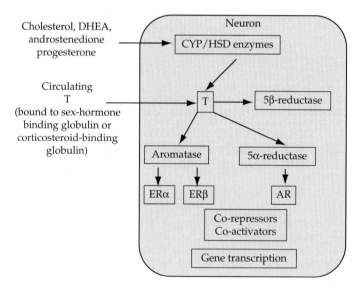

Figure 26-3 Potential fates of circulating testosterone (T) after it enters a target cell such as a brain neuron. Steroid metabolizing enzymes play a key role in determining which steroid hormone genomic receptor will be activated and thus which set of genes is expressed. AR, androgen receptor; ER, estrogen receptor (α or β). Their activation can also be modified by co-activators and co-repressors which are also potential foci of diversity of regulation. Note also that in at least some neurons, steroid metabolizing enzymes are expressed that can synthesize T or estradiol *de novo* from cholesterol, or sequester circulating androgen precursors such as androstenedione, progesterone, or dehydroepiandrosterone (DHEA). CYP, cytochrome P450 enzymes; HSD, hydroxysteroid-dehydrogenase. This basic "response" system is generally conserved across vertebrates, but the ways in which specific components may be emphasized or de-emphasized can be highly variable, providing diversity of mechanistic pathways. Compiled from Schlinger and London (2006), Soma (2006), Wingfield *et al.* (2001).

alternative functions of aiding in testosterone trafficking across the membrane can at this point not yet be excluded (e.g., Breuner and Orchinik 2002).

In general, unbound testosterone enters target cells, where steroid-metabolizing enzymes may modify it, resulting in profound influences on its mechanisms of action (Fig. 26-3). Basically there are four major fates of testosterone when it enters a target cell, for example a neuron in the brain (e.g., Zakon 1998). First, it can directly bind to the androgen receptor (AR) or, second, it can be aromatized to estradiol, which then binds to estrogen receptor alpha or beta (ERα, ERβ). Activation of these receptors then results in a completely different set of genes being transcribed. Third, testosterone can be converted to 5α-dihydrotestosterone, which also binds to AR but cannot be aromatized to estradiol, thus potentiating the AR pathway. Fourth, conversion of testosterone to 5α-dihydrotestosterone results in de-activation of testosterone because this metabolite does not bind to AR and also cannot be aromatized to estradiol.

Testosterone's effects on male reproduction include the regulation of reproductive behavior (courtship, copulation), physiology (sperm maturation), morphology (muscle hypertrophy), and many other processes (Fig. 26-4). In addition to reproduction, testosterone exerts actions on several other processes that are major components of the life history strategy of an organism: metabolism, immune function, and many more (e.g., Ketterson and Nolan 1992, Wingfield et al. 2001). Some of these traits are antagonistically affected by testosterone and one important function of this hormone might therefore be the physiological regulation of trade-offs (Ketterson and Nolan 1992). Indeed, intraspecific studies, such as a long-term study on dark-eyed juncos, *Junco hyemalis*, have shown that individuals with naturally or experimentally increased circulating testosterone concentrations have increased reproductive output, larger home range sizes, and higher extra-pair fertilization rates (Reed et al. 2006, McGlothlin and Ketterson 2008). Conversely, male juncos with higher testosterone concentrations often suffer from decreased parental care, lower survival rates, impaired immune function, and suppressed molt, indicating that testosterone mediates life history, and behavioral and physiological trade-offs. Similarly, large-scale comparative analyses in bird species have shown that absolute testosterone concentrations of males during the breeding season are

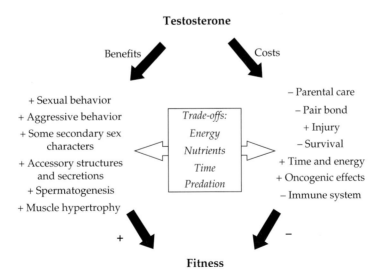

Figure 26-4 Effects of testosterone on various traits that either enhance fitness (benefits) or that decrease fitness (costs). Potential currencies of trade-offs are indicated in the middle box. + and − indicate the direction of the effect of an increase in testosterone concentrations on a certain trait. Redrawn from Wingfield (2006).

positively related to reproductive intensity, as they are higher in species with shorter breeding seasons (which usually lay a larger numbers of eggs in each breeding attempt and thus show a higher reproductive effort per unit time) (Goymann *et al.* 2004, Garamszegi *et al.* 2008, Hau *et al.* in revision). Antagonistic effects of testosterone on traits, corresponding to those shown in intraspecific studies, have long been hypothesized, for example a suppression of immune function (Folstad and Karter 1992), which could impair survival rate. Indeed there is evidence that testosterone can suppress immune function in some vertebrate species, but whether this is a general pattern and indeed mediated by testosterone is not yet fully understood (e.g., Roberts *et al.* 2004).

Hence, there is ample evidence both from intra- and interspecific studies that testosterone is involved in regulating life history trade-offs. Does that limit individuals and populations in their ability to optimally resolve trade-offs when environmental conditions vary?

26.3   Conservation versus variation in testosterone-regulated traits on an interspecific level

"It is impossible for any population of organisms to evolve optimal solutions to all selective changes at once" (quoted from Freeman and Herron 2007, p. 383). Here, we will ask whether the regulation of traits by testosterone potentially creates impediments (i.e., constraints) to evolutionary variation in the resolution of trade-offs or whether it promotes evolutionary adaptations in trade-offs. Essentially, we ask: does a male that has increased testosterone concentrations during the breeding season to promote traits that increase reproductive success, such as the display of aggression to establish a territory and courtship to attract a female, always suffer from negative effects of testosterone on other traits. Such effects could be, for example, on parental behavior or immune responses. This could be the case despite ecological conditions in its population of origin rendering male contributions to raising offspring critical to reproductive success (e.g., low overall food availability) or favor a strong investment in immune function (e.g., high pathogen pressure)?

We will approach this question by describing known patterns of conservation in the testosterone signaling cascade, which might indicate the presence of evolutionary impediments, as well as known patterns of variation that indicate the existence of evolutionary variability in that endocrine system. It is unlikely that "hard" constraints exist that prevent the evolutionary variation of hormonally-regulated trade-offs; rather there might exist processes that slow down the rate of evolutionary change, especially over short time scales (Adkins-Regan 2008, McGlothlin and Ketterson 2008, Ketterson *et al.* 2009). The pleiotropic linkage of traits via hormones is one factor that has been suggested to impede the rate of evolutionary variation in trade-offs, but there is also the interesting possibility that this linkage of traits by hormones might speed up evolutionary adaptations (Ketterson *et al.* 2009). Also, the remarkable conservation in the endocrine system of vertebrates, in which a limited number of vertebrate hormones exists, and probably more than 80% of the vertebrate endocrine system has been conserved for around 200–300 million years (Wingfield 2006, Norris 2007) could be taken as evidence for constraints, as it implies that it is highly unlikely that a population evolves a new hormone to resolve a particular trade-off. However, an alternative interpretation for the observed conservation is that it is the result of a physiological organization favored by selection (see also Heideman 2004). Hence, from such large-scale patterns it is yet not possible to distinguish the contributions of evolutionary impediments to the patterns that we observe in natural systems. Below we will therefore discuss in more detail known conservation and variation in signal production (i.e., synthesis pathways and control of secretion mechanisms for testosterone release; Figs 26-1 and 26-2) and signal transduction pathways (i.e., transport to target tissues and mechanisms of action of testosterone; Fig. 26-3).

26.4   Signal production pathway

The main components of the HPG axis are conserved from fish to mammals (e.g., Wingfield 2006, Norris 2007, Fig. 26-1). One possible reason for this conservation could stem from the complexity of the HPG cascade (e.g., Adkins-Regan 2008), with selection having minimized genetic variation in its components to prevent the collapse of the entire system should one component malfunction as a result of alteration. Selection against genetic variation in components of endocrine axes might be particularly strong for processes at the top of the cascade because variation in such upstream processes will have ramifications for all processes downstream (see also Heideman and Pittman 2009). Conservation in the HPG axis could also be due to linkage disequilibrium, for example as a result of multilocus selection on functional components of the HPG axis (e.g., Sinervo and Svensson 1998). However, strong selection pressures can be expected to break such linkages (e.g., Roff and Fairbairn 2007b) and therefore such conservation is more likely the result of a physiological organization that has been optimized by selection.

Even if certain components of the HPG axis and their interactions were constrained in their variation to some extent, it would still be questionable whether this system would present an impediment to the diversification of life history, or behavioral or physiological trade-offs. Indeed, Adkins-Regan (2008) has argued that a crucial component to the functioning of the HPG axis lies in its neural regulation, which in turn is evolutionarily highly variable in vertebrates. The neural input into the HPG axis includes sensory systems that gather environmental information as well as the pathways that lead to neuroendocrine secretions (Wingfield 2008). Furthermore, GnIH, which can act both centrally and hypophysiotropically (Tsutsui *et al.* 2009), could play a potentially vital role in generating variation.

There exists well-known interspecific variation in the sensitivity to and reliance on photoperiod as the primary environmental cue for regulating the activity of the HPG axis (Bronson 1985, Wingfield *et al.* 1992). Most species of temperate zone vertebrates rely strongly on photoperiod for regulating HPG axis activity. But even among temperate zone species, there exists considerable variation among species and populations in the importance of other cues such as temperature, food, and social stimuli (Bronson 1985, Hahn *et al.* 1997). Furthermore, in species living at lower latitudes, food cues and social stimuli appear to be equally or even more important than photoperiod (e.g., Hau *et al.* 2008,

Schoech and Hahn 2008). Hence, even though the backbone of the HPG axis is highly conserved in vertebrates, there appears to exist ample variability in the neural input pathways into this axis, which could be selected for and result in altered resolutions of trade-offs.

Although it has not yet been comprehensively analysed, there is likely also interspecific variation in the stimulatory effects of the GnRHs on the gonadotropins LH and FSH, and their respective effects on testosterone-producing enzymes. For example, across avian species, concentrations of LH and testosterone in males do not occur at the same ratio. Two closely related sparrow species in North America (the white-crowned sparrow, *Zonotrichia leucophrys*, and the song sparrow, *Melospiza melodia*) show a ratio of LH/testosterone of about 1.5 at the beginning of the breeding season (Wingfield and Farner 1993). By contrast, neotropical spotted antbirds (*Hylophylax n.naevioides*) have an LH/testosterone ratio of 4.5 (Wikelski *et al.* 2000), while afrotropical stonechats (*Saxicola torquata axillaris*) show a ratio of 0.7 (Dittami and Gwinner 1985, Goymann *et al.* 2006). Although absolute concentrations of LH might vary between these studies due to differences in hormone assays, the data from several studies were analysed in the same laboratory and patterns are tantalizing in suggesting that species may differ in their sensitivity to LH for stimulating testosterone production. This could be tested further in studies determining direct responses to LH and characterizing/quantifying receptors for LH. The role of GnIH in this system is only just beginning to be appreciated and could provide further valuable insight.

There also exists interspecific variation in avian species in the relative circulating concentrations of testosterone and some of its precursors and metabolites (Fig. 26-2), such as dehydroepiandrosterone (DHEA), androstenedione, dihydrotestosterone (DHT), and others. For example, the ratio of circulating testosterone to DHT concentrations in males at the beginning of the breeding season is about 12:1 in song sparrows, 8:1 in white-crowned sparrows (Wingfield and Farner 1993), but 4:1 in Wilson's phalarope (*Phalaropus tricolor*, Fivizzani *et al.* 1986). It is therefore likely that the relative densities and/or activities of steroidogenic enzymes differ across species, even though their role in the endocrine cascade is the same.

Lastly, there may also exist interspecific variation in the sites of testosterone production. There is ample evidence now that many target tissues (particularly in the brain) can utilize cholesterol or androgen precursors, such as progesterone, DHEA, androstenedione, and possibly others, and convert them to testosterone (Schlinger and London 2006, Soma 2006). All the enzymes necessary to synthesize testosterone are expressed in these cells, at least in some avian species. Since this is an important emerging field and comparative data are still sparse, at present it is hard to estimate how much interspecific variation exists. However, if sites of testosterone production varied across species, such extragonadal production of testosterone could contribute to circulating concentrations or conversely provide a mechanism for highly localized effects independent of the functioning of the HPG axis and circulating testosterone.

26.5 Signal transduction pathway

Testosterone exerts its biological actions via transduction pathways that equal or even exceed the complexity of the production cascade. Because of processes similar to the ones discussed above, in the signal production section such complexity could constrain interspecific variation. However, being further downstream in the endocrine signaling cascade, one could expect transduction pathways generally to show more genetic variability. Indeed, evolutionary adaptations at the tissue sensitivity level (e.g., hormone receptor) rather than the signaling level (e.g., circulating hormone concentrations) have been suggested to be more likely to occur in natural populations (Adkins-Regan 2005). However, a realistic scenario could also be that an interaction of both variation at the signal production level (i.e., secretion of testosterone via HPG activation) and sensitivity of the responding tissues (i.e., receptors and other factors) is what varies among populations and species (see also Ketterson et al. 2009). For example, within individuals there is ample crosstalk between the signal and its effector systems. Long-term environmental cues such as photoperiod ("initial predictive cues"; Wingfield et al. 1992) usually affect signal production, and the hormonal signal then activates hormone-sensitive tissues. Then other, more short-term environmental cues, such as food availability and social factors ("supplementary cues"), may influence tissue-response systems further to customize responses of individuals in time and space. Thus it is likely that any evolutionary change in one component of this interaction would affect the other one as well.

There is indeed substantial interspecific variation in many of the components of the testosterone transduction cascade. For reasons of brevity, here we will outline just a few select examples. First, interspecific variation in circulating concentrations, temporal dynamics, and binding affinities of CBG has been documented (Breuner et al. 2006), which could affect the bioavailability of testosterone for tissues in various ways. Second, avian species that have been looked at so far possess only one nuclear androgen receptor (Wingfield et al. 2001, Fig. 26-3), but in fish a second form has been described (Sperry and Thomas 1999). Third, even though androgen receptor forms appear to be strongly conserved, their spatial distribution within the avian brain shows considerable interspecific variation (e.g., Gahr 2001). Finally, there exists large variation in the distribution and activity of the enzyme aromatase (or CYP19), which converts testosterone into estradiol, which in turn acts via estrogen and not androgen receptors (Fig. 26-2; e.g., Schlinger 1997, Silverin et al. 2000, Soma 2006).

Taken together, there is some evidence for conservation of components of the HPG axis and their function, seemingly most prominently in signal production pathways but also to some degree in signal transduction. Conversely, there is much evidence for interspecific variation in neural inputs into the HPG axis and in signal transduction pathways. Increasing our understanding of role of GnIH is a particularly interesting challenge, as it might modulate neural input, behavioral responses, and signal production processes.

26.6 Plasticity of testosterone-regulated trade-offs on an individual level

To what extent do hormonally-regulated trade-offs impede, versus provide plasticity to, individuals in achieving optimal resolutions of trade-offs under varying environmental or social contexts? For example, would an individual in breeding condition, which shows elevated testosterone concentrations to combat male competitors or attract potential mates, always suffer from a suppressed immune system or show decreased parental care, even if there was plenty of food in a given year allowing for the simultaneous investment of resources into processes supporting reproduction and survival? Evolutionarily, the mechanisms of selection on hormonally-regulated trade-offs are expected be similar on the inter- and the intraspecific levels, since the individual phenotype will be the target of selection. However, when analysing potential limitations to an individual imposed by the regulation of trade-offs through testosterone it seems important to focus on the degree of plasticity within its lifetime. Below we will discuss three aspects of hormonally-regulated trade-offs that might limit individual plasticity. First, there might be significant costs associated with the neuroendocrine system that have a negative impact on fitness, irrespective of the phenotype. Second, individual phenotypic plasticity may be limited by the specifics of endocrine reaction norms. Third, developmental processes might create potent limits to individuals by permanently altering their phenotype and possibly their phenotypic plasticity.

26.7   Costs

Costs are fitness detriments incurred by the expression of a trait, even when an optimal phenotype is expressed (Pigliucci 2005). Endocrine signaling could be associated with various costs, a topic excellently discussed by Lessells (2008). For example, an obvious energetic cost could arise at the signal production level if either the molecule itself was expensive to produce or the signal-producing machinery was expensive to maintain. However, as pointed out by Lessells (2008), testosterone as a molecule is likely rather cheap to produce because it requires neither limited nutritional resources, such as rare amino acids or elements, nor energetically expensive compounds like proteins. Hence, unless one key enzyme in the biosynthetic pathway is particularly "expensive" to produce and maintain, it is unlikely that testosterone as a molecule is costly to produce. At least

for males, the metabolic costs of maintaining active testes and the added weight of enlarged testes could impose energetic challenges (for estimates in birds see, e.g., Ricklefs 1974). However, the part of the testis that produces testosterone (Leydig cells; e.g., Nelson 2005) is a miniscule part of the gonad as a whole. Thus, proposed costs of maintaining a testis likely refer almost completely to producing sperm and not to testosterone. Toxicity of testosterone could represent another major cost as it would have immediate negative fitness consequences for an individual (Wingfield *et al.* 2001, Lessells 2008). However, most of the known toxic effects of hormones have been observed at pharmacological doses, which also holds for the proposed oncogenic effects of steroid hormones. Taken together, there do not appear to be significant costs associated with producing or having testosterone in the circulation *per se*.

26.8   Phenotypic plasticity and reaction norms

Perhaps the most important determinant of limitations to individuals in their ability to achieve an optimal resolution of hormonally-regulated trade-offs within their lifespan is their degree of endocrine phenotypic plasticity. For our purposes, a useful definition for phenotypic plasticity is an "environmentally-based change in the phenotype" (Via *et al.* 1995). A reaction norm can be defined as the "genotype-specific environment–phenotype function" (Pigliucci 2005) and describes the way in which a trait changes as a function of environmental variation (i.e., it specifies the elevation, slope, and shape of the relationship between the trait and the environmental variable). Phenotypic plasticity can likely be found in many components of the testosterone system, i.e., both in signal production/release pathways (sensory perception, neural transmission, and production/release pathways) as well as in signal transduction pathways (cellular sensitivity and responsiveness) to testosterone.

When applied to the hormonal control of traits and trade-offs, reaction norms can help visualize and analyze aspects of phenotypic plasticity of individuals (Fig. 26-5, Nussey *et al.* 2007). The intercept (elevation) of a reaction norm would show, for example, how much testosterone is required for the regulation of a trait under certain conditions, how sensitive a certain tissue is to the effects of testosterone, the duration of testosterone secretion in response to stimulation, and many other aspects. To understand limitations to individuals, knowing the intercept of a reaction norm in testosterone-regulated trade-offs is likely less informative than the steepness of slope (genotype I in Fig. 26-5). The steepness of the slope would indicate, for example, how much plasticity exists in an individual across environments or social contexts in circulating testosterone concentrations, tissue sensitivity, or duration of testosterone secretion. A shallow slope (genotype II in Fig. 26-5) would indicate little plasticity, whereas a steep slope (genotype I in Fig. 26-5) would indicate great plasticity. The shape of the reaction norm (not illustrated in Fig. 26-5, for example, curve versus straight line) would determine the degree of linearity in the relationship between testosterone and the trait. Thus, if an individual shows a rather shallow reaction norm, its ability to plastically vary the resolution of hormonally-regulated trade-offs likely is quite limited. But if an individual possesses a reaction norm with a steep slope, it might readily be able to adjust to variations in environmental or social contexts by means of phenotypic plasticity.

There appear to exist variations in how much plasticity individuals from certain populations or species show in their responsiveness of circulating testosterone concentrations to social stimulation (Wingfield *et al.* 1990, Goymann 2009). For example, Arctic-breeding bird species such as white-crowned sparrows (*Zonotrichia leucophrys gambelli*) display high circulating concentrations of testosterone throughout their short breeding season, but cannot further increase plasma testosterone in response to social stimuli, such as a territorial intrusion by a conspecific male (Wingfield *et al.* 2007). Hence, Arctic sparrows likely have a shallow reaction norm in testosterone responses across a range of social stimulation. By contrast, individuals from a subspecies of white-crowned sparrow living at temperate-zone latitudes (*Zonotrichia leucophrys pugetensis*), which enjoy much longer breeding seasons, can show large social modulations of plasma testosterone concentrations during the breeding season, suggesting the existence of a steeper reaction norm to the intensity of social stimuli (Wingfield *et al.* 2007). If one moves to tropical latitudes, individuals might have even greater plasticity in their social modulation of testosterone (e.g., Neotropical spotted antbirds, Wikelski *et al.* 1999). Similar variations in plasticity among individuals from populations

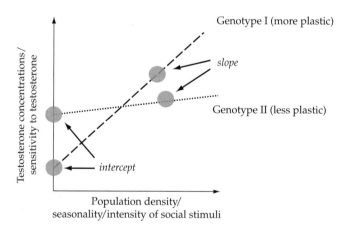

Figure 26-5 Schematic representations of divergent reaction norms. *y*-axis denotes endocrine traits, either in testosterone-production or transduction pathways, *x*-axis denotes environmental or social variables. Two divergent genotypes are illustrated. Genotype I has a low intercept value (for example low testosterone concentrations might be sufficient for the regulation of a trait) but its elevation increases across varying environmental or social circumstances (i.e., increasingly more testosterone is required). The reaction norm of genotype I has a steep slope indicating great phenotypic plasticity. By contrast, genotype II starts out with a higher intercept value (indicating a requirement of generally higher testosterone concentrations), but shows a very shallow slope. The latter indicates little plasticity across different environmental or social contexts.

with different social systems or life history strategies exist in other vertebrate taxa (Hirschenhauser and Oliveira 2006) and likely also in humans (Archer 2006). Furthermore, elegant studies in *Peromyscus* mice have shown individual, and genetically heritable, phenotypic plasticity in responses to environmental cues such as photoperiod and food stimuli (reviewed in Heideman and Pittman 2009).

There is also remarkable plasticity in the brain, for example in the activity of the enzyme aromatase (Fig. 26-3). Although, in light of technical limitations, it is currently not possible to determine individual phenotypic plasticity for most neural processes, in birds we know that the activity of aromatase can change both over very rapid (within minutes) and slower (hours to days) timescales in response to social stimuli (Balthazart *et al.* 2003). Individuals within a population or from different populations/species could vary in their amplitude of changes in aromatase activity (generally, or at specific sites) in response to social stimuli, with individuals with steeper reaction norms being able to more plastically adjust the resolution of trade-offs to variations in social environment. Similarly, there is known seasonal and social plasticity in androgen receptor and aromatase (and other steroid-metabolizing enzymes) expression in various avian species (e.g., Soma *et al.* 1999, 2003, Canoine *et al.* 2007) that could contribute to individual plasticity in trade-off resolution in similar ways.

If it is beneficial to adjust trade-offs to environmental or social contexts, why do not all individuals show maximal plasticity? The most plausible explanation is that plasticity itself is costly, potentially due to various processes (reviewed in deWitt *et al.* 1998). Just as one example, costs of plasticity of hormonally-regulated trade-offs could arise from the need to maintain both the sensory and regulatory machineries to support plastic responses. A number of different sensory systems would have to be involved, for example in the perception of social information such as visual, auditory, and olfactory senses. To process and integrate this information, brain areas would need to be maintained so that signals to regulate components of the HPG axis would be sent out when appropriate. However, costs might also arise from the production of plasticity, developmental instability, and many other processes (deWitt *et al.* 1998, West-Eberhard 2003). At present, we have a very limited understanding of the specific costs that hormonal plasticity might generate.

26.9 Development

Endocrine processes during development such as hormonal, environmental, and experiential effects can permanently determine the embryonic, postnatal, and adult phenotype of an individual. Since this subject has recently been excellently summarized (Groothuis *et al.* 2005, Adkins-Regan 2008, Monaghan 2008), here we will only mention a few select examples. In principle, such permanent differentiation could be considered to limit individuals in their plasticity—at least when compared with the undifferentiated state. A good example could be sexual differentiation, where hormones (among other processes) permanently determine the morphology, physiology, and behavior in individuals according to their genetic sex; those changes cannot be reversed in adulthood. However, it is also conceivable that some developmental effects increase phenotypic plasticity later in life. For example, maternal effects transmitted via hormones to developing offspring can have potent effects on the adult phenotype of the offspring (Mousseau and Fox 1998). In birds, increased yolk androgen concentrations in developing eggs affect offspring begging intensity, food competitiveness, nestling growth, and immune function (among many other effects, e.g., Groothuis *et al.* 2005). If such traits were expressed in a context-dependent manner more strongly in chicks hatched from eggs containing higher yolk androgen concentrations compared to chicks with lower androgen exposure, these maternal effects could serve to enhance phenotypic plasticity. Likewise, in birds, yolk hormone concentrations can affect the "personality" (consistent differences between individuals in their behavior across time and contexts) of the offspring (Groothuis *et al.* 2005), and different personalities may show different levels of phenotypic plasticity (Dingemanse *et al.* 2010a). Hence, it will be an exciting field for the future to study whether effects during early development (or "programming," reviewed in Monaghan 2008) increase or limit individual plasticity.

26.10 Future directions of evolutionary endocrinology

We have discussed several aspects relating to the evolution of the testosterone signaling cascade, but it is important to note that similar principles likely apply to other hormonal systems as well. From our short overview it has become apparent that there is conservation in certain parts and ample variation in other parts of the testosterone-signaling cascade. However, to understand the evolutionary processes that generate the observed patterns, more work on the evolutionary endocrinology of vertebrates is needed (for excellent summaries of some recent advances in this nascent field see Adkins-Regan 2005, Williams 2008, Heideman and Pittman 2009, Ketterson *et al.* 2009). To better understand the degree and rate of evolutionary variation in hormonally mediated trade-offs, detailed work on different populations of one species of vertebrate or on different species that face divergent trade-offs (e.g., due to differences in environment or social systems, in ecologies, life histories etc.) will be important. Latitudinal comparisons could be a fruitful approach (e.g., Goymann *et al.* 2004, Wingfield *et al.* 2007, Hau *et al.*, 2010) because they can utilize the well-known diversity in life history strategies that exists across latitudes. Such types of studies would ideally use a comprehensive approach, in which hormonal traits as well as their linkage to behaviors and life history traits were assessed (e.g., McGlothlin *et al.* 2007, 2008). Experimental approaches, for example administration of hormones and pharmacological blockers of hormone actions to determine the resulting effects on traits (e.g., Hau *et al.* 2000, Hau 2007), could be used to confirm causal relationships between hormones and traits. Furthermore, interactions between different hormonal systems should be included (such as between the HPG and the hypothalamo–pituitary–adrenal axes, e.g., Hau *et al.* 2010), as they often regulate trade-offs in concert. Likewise, the hormonal regulation of other physiological systems, such as the immune system, the metabolic system, and many others should be studied in more detail, as they are part of the proximate mechanisms that underlie life history strategies (e.g., Ricklefs and Wikelski 2002, Wiersma *et al.* 2007, Lee *et al.* 2008d).

To determine the rate of evolutionary change in the hormonal regulation of traits two additional approaches will likely be useful. First, directional selection experiments could reveal how fast and which components of hormone-signaling cascades can evolve (e.g., Satterlee and Johnson 1988, Evans *et al.* 2006). Second, quantitative genetic approaches could be used to determine heritabilities, rate of evolutionary change, and the strength of selection on hormonally-mediated trade-offs in both natural and experimental populations (e.g., Zera *et al.* 2007, McGlothlin and Ketterson 2008).

For individual-level studies of limitations to an optimal resolution of trade-offs, a reaction-norm approach appears important. Analyses of hormonal data from a reaction-norm perspective are just beginning to be integrated in the field of evolutionary and ecological endocrinology (e.g., Crews 2003, Heideman and Pittman 2009, Dingemanse *et al.* 2010b). However, data could be extracted from longitudinal studies on individuals that span different seasons (e.g., Wikelski *et al.* 1999), social situations (e.g., Kralj-Fiser *et al.* 2007), or gradients in other contexts. Indeed, some studies already seem to provide such data and would only need to be revisited, for example studies that determined variation in individuals sampled across different breeding stages in their testosterone responses to injections of GnRH (e.g., Jawor *et al.* 2006) or in their behavioral sensitivity at different stages of the breeding cycle to the administration of exogenous hormones (e.g., Lynn *et al.* 2005). Likewise we need to determine whether costs exist that limit the hormonally mediated expression of plasticity in certain populations or individuals.

Knowing evolutionary trajectories in endocrine-regulated trade-offs, as well as limitations to individuals in ecological times, will enable us to better predict, for example, which individuals, populations, or species might be threatened by alterations in their environment due to global change, or gauge the potential colonization success of invasive species. More generally, such data will provide basic knowledge for an improved understanding of evolution of the physiological processes that are the bases for adaptations of organisms to environmental and social circumstances.

26.11   Summary

1. One major question in evolutionary endocrinology is how, at what rate, and where in complex endocrine cascades do links between hormones and traits involved in life history, and behavioral and physiological trade-offs evolve? Does the hormonal regulation of traits limit or promote the adaptation of individuals or populations to environmental and social conditions? Here we approach this question from a physiological standpoint, describing known patterns of conservation and variation in testosterone signaling pathways. We focus our analysis on the regulation of life history trade-offs by testosterone in male vertebrates, specifically in birds. First, we will consider adaptations in trade-offs via evolutionary change (comparing populations and/or species), and second, plastic responses to environmental conditions (comparing individuals).

2. The main components of the testosterone production pathway—the hypothalamic–pituitary–gonad (HPG) axis—are remarkably conserved across vertebrates. However, neural input mechanisms into the HPG axis as well as sites and activities of testosterone-synthesizing enzymes show considerable interspecific variation. Likewise, there appears to exist ample interspecific variation in the testosterone transduction pathways, specifically in the way in which testosterone is metabolized and exerts its biological effects. Hence, ample evolutionary variation appears to occur predominantly at the neural input level as well as the signal transduction pathway.

3. Adjustments of trade-offs to ecological conditions in individuals might depend greatly on their degree of phenotypic plasticity, i.e., the specifics of their reaction norms in the regulation of trade-offs by testosterone. Individuals with great phenotypic plasticity likely are less constrained than individuals with less plasticity in their ability to alter trade-offs depending on circumstances. This plasticity could be achieved at various levels of the testosterone-signaling cascade, but has not yet been studied in detail in vertebrates.

4. To increase our understanding of evolutionary dynamics in endocrine systems we will need detailed comparisons among populations or species, complemented by quantitative genetics approaches and directional selection studies. Furthermore, it will be crucial to study phenotypic plasticity of individuals from a reaction-norm perspective and also determine whether such plasticity incurs costs.

26.12   Acknowledgments

MH would like to acknowledge highly valuable discussions with Ellen Ketterson, Jesko Partecke, Dan Nussey, Tom Hahn, Becca Safran, and Tim Greives. Two anonymous reviewers provided valuable comments on an earlier version of the manuscript. JCW acknowledges support from the USA National Science Foundation (IOS-0750540).

PART 7

Concluding remarks

Does impressive progress on understanding mechanisms advance life history theory?

Stephen C. Stearns

27.1 Introduction

In this chapter I will summarize this book in two ways. The first assesses the progress reported here on the intra-organismal mechanisms of all sorts that mediate life history evolution broadly construed. On this issue, the glass is much more than half full. The second asks whether the understanding we have gained from mechanisms advances life history theory. On that issue, the glass is at least half empty.

This synthesis chapter therefore has two central parts. In the first part, in section 27.2, I highlight some striking contributions that this book makes to our understanding of the mechanistic basis of life histories (see Section 27.2). I emphasize three conceptual advances:

- trade-offs can be mediated by switches in signaling pathways, independent of the allocation of energy and materials to competing functions
- at least one of those signaling pathways—IIS—appears to be ancient, conserved, and widely shared
- if a trade-off results from constraints on a module that must perform two functions, it can be broken by duplicating the module, dividing the labor, and allowing the daughter modules to specialize.

In the second part I address three issues (see Section 27.3):

- Do these data force the theory to change?
- Is a general picture of intermediate structure in the genotype-phenotype map emerging?

- Are the empirical problems with the theory being addressed?

I conclude with some reflections on the striking differences between the theoretical-synthetic approach that characterized the early decades of research on life history evolution and the analytical–reductive approach that characterizes this volume (cf. Stearns 1992, Flatt and Schmidt 2009,).

It may help to put the development of these ideas in context. After the initial consolidation of life history theory two decades ago (Roff 1992, Stearns 1992, Charnov 1993, Charlesworth 1994) the field continued to develop in at least four directions.

- First, the theory was used to help organize research in neighboring fields, chief among them biological anthropology (e.g., Hill and Hurtado 1996, Boesch and Boesch-Achermann 2000, Howell 2000; see also Chapter 12) and behavioral ecology (e.g., Clutton-Brock 2009; see also Chapter 25).
- Second, the theory was criticized to put it on a sounder footing and extended to increase its range of application, a path taken by those who have been developing adaptive dynamics (e.g., Dunlop *et al.* 2009), working on the implications of density, frequency, and state dependence (e.g., Mangel and Clark 1988, Houston and McNamara 1999, Sinervo and Calsbeek 2006, Bonsall and Mangel 2009), exploring eco-evolutionary feedbacks (e.g., Mougi and Nishimura 2006, Metcalf *et al.* 2008, Jones *et al.* 2009), and developing the theory of life history invariants (e.g., Charnov 2009).

- Third, there have been further direct tests of the theory, mostly in the laboratory, often using fruit flies or bacteria (e.g., Stearns 2000, Ackermann *et al.* 2007), and sometimes in the field, often using fish (e.g., Reznick *et al.* 2004).
- Fourth, a group accepted that trade-offs are pervasive, constituting the central intrinsic constraint on life history evolution, and tried to uncover the mechanisms that produce them. That effort expanded into a general attempt to understand the molecular mechanisms that mediate life history evolution broadly construed, which is the subject of this volume.

None of these efforts were driven by a sense that the theory had failed. In fact, its predictions were successful, as far as they went, but the theory was incomplete where the consequences of alternative assumptions had not been explored. There was plenty left to be done at the level of general theory uninformed by intra-organismal mechanisms (e.g., Abrams 1993), and there still is. Nor was it the case that testing the theory exposed empirical crises, for most of the qualitative tests resulted in confirmations (but see Bryant and Reznick 2004). However, there were problems encountered both in making quantitative predictions and in testing them, problems that originated in our lack of quantitative understanding of trade-offs and constraints. Thus one motivation to explore mechanisms was to see whether we could find there a quantitative understanding of trade-offs that was missing from a theory that was qualitatively but not quantitatively successful. Another was curiosity about mechanisms, irrespective of the state of theory. While both motivations have figured in producing the work summarized here, it appears that the second has dominated.

27.2 How research on mechanisms is changing views on life history evolution

27.2.1 The nature of trade-offs: Signals, allocation, or both?

The most important general insight that research into mechanisms has contributed to life history evolution is that trade-offs can be caused by signaling networks as well as by energy allocation. This idea explains how reproduction can, in some cases, be uncoupled from lifespan. The evidence is summarized by Edward and Chapman (Chapter 11) as follows: "There are three main lines of evidence to suggest that the literal application of 'Y' models may not fully explain the proximate mechanisms underlying the relationships between life history traits. Firstly, the elimination of reproduction does not necessarily extend life span....Secondly, although mutations that extend lifespan generally cause reduced fecundity, some apparently also increase fecundity....Thirdly, there are marked sex differences in the response of males and females to interventions that increase longevity....A proximate explanation for this challenge to traditional life history theory is that trade-offs are mediated by molecular signals....The putative molecular signals represent arbitrary connections between life history traits and may be independent of resource availability" (cf. Barnes and Partridge 2003, Leroi 2001).

Based on their elegant dissection of signaling pathways in *Caenorhabdites elegans*, Gerisch and Antebi (Chapter 22) describe the mechanisms as follows: "A fundamental paradigm has emerged, whereby environmental, dietary, and physiological cues are sensed by the neurosensory apparatus, and integrated and processed into hormonal signals within endocrine cells. In target tissues, cognate receptors coordinate metabolism, growth, reproduction, homeostasis, and stress responses to ensure maximal reproductive success." And based on their work on lizards, Sinervo and Lancaster (Chapter 25) arrive at a similar picture, one more informed by social interactions: "Endocrine networks generate homeostasis via regulation of energy flow and metabolism among competing life history functions, but also respond to abiotic and social stressors. The structure of an endocrine network that evolves to regulate optimal life history allocation, under requirements of homeostasis, is a positive–negative regulatory feedback loop, which is instantiated in response to exogenous triggers like photoperiod."

However, the fact that some trade-offs are mediated by molecular signals does not necessarily mean that other trade-offs are not mediated by the allocation of energy and materials. Zera and Harshman (Chapter 24) report studies that track the movement of amino and fatty acids among tissues in a cricket, providing direct evidence that Y-allocations of resources do occur. Measuring the contributions of acquisition and allocation to trade-offs is a major empirical challenge on which some progress is being made (Djawdan *et al.* 1996, O'Brien *et al.* 2008). Both switches in signaling pathways and changes in allocation of energy and materials are probably involved in most trade-offs.

Does the insight that some trade-offs are mediated by signals without accompanying Y-allocations actually change life history theory? If it were generally true that reproduction is completely decoupled from survival—that variation in reproduction had no consequences at all for variation in lifespan, either within a single generation for physiological reasons or over many generations for evolutionary reasons—then a central assumption of the evolutionary theory of aging (Williams 1957) would fall, and we would have to start anew. Interestingly, often when no direct trade-off between survival and reproduction is found, survival is found to trade-off with another fitness trait (Flatt and Schmidt 2009). On balance, the weight of the evidence gathered at the phenotypic level, much of it correlated responses in selection experiments, suggests that reproduction usually does trade-off with survival (Roff 1992, Stearns 1992). For example, while *Drosophila* is one of the model systems in which reproduction can be decoupled from lifespan, in that case by careful adjustment of the amino acids in the diet (Grandison *et al.* 2009), the classical coupling between reproduction and lifespan is well confirmed by correlated responses in selection experiments (Stearns and Partridge 2001). This suggests that when trade-offs are not found in molecular studies, it may be because they are not expressed in the environments in which the experiments are being done.

That contrast between physiological and selection experiments does emphasize that we do not know much about the links between proximate and ultimate causes and more specifically about the mechanisms mediating negative genetic correlations detected in selection experiments. Such knowledge will be one key to developing a theory that is quantitatively successful. Those genetic correlations could be mediated by switches in signaling pathways, by shifts in Y-allocations of energy and materials, by some combination of both, or by mechanisms not yet discovered. We do not yet have a procedure for translating genetic correlations, critical in evolutionary theory, into intra-organismal mechanisms

of the sort reviewed here. What mechanisms, of the many present, might be the main targets of selection? Do we have theoretical expectations about which they might be, given the architecture of a control network? Shingleton (Chapter 4) suggests one answer: "In general, we might expect that changes in those…regulatory mechanisms that have the fewest pleiotropic effects will occur first, since pleiotropy is thought to constrain evolutionary change (Hansen and Houle 2004). For example, changes in cell size should have fewer pleiotropic effects (on body size) than changes in cell number because the former can, in principle, occur at the very end of development, while the latter require alterations in the rate and duration of cell proliferation during development."

This resembles a common argument in evo-devo, which claims that cis-regulatory evolution is the predominant mode of morphological evolution because changes in modular enhancers restrict pleiotropy. Remarkably, much of life history evolution seems to involve pervasive pleiotropy. Why do we see all this pleiotropy and all these trade-offs? Why does the evolution of life histories not proceed through cis-regulatory changes that have minimal or no pleiotropic effects? This important empirical problem in life history evolution will require mechanistic approaches (T. Flatt, personal communication).

27.2.2 Ancient, conserved, broadly shared mechanisms?

The key elements of a central signaling pathway that mediates life history trade-offs in a broad range of species has been identified. The insulin/insulin-like signaling pathway (IIS) affects growth rate, developmental time, aging, and fertility in flies, worms, and mice, in all three of which a reduction in IIS causes an increase in longevity (Chapters 4, 13, and 14; see also Tatar *et al.* 2003). Why might a signaling system be ancient, widely shared, and deeply conserved? Once a signaling system gets co-opted into coordinating multiple responses, it is subjected to evolutionary embedding, for its interactions in a network of traits imply that the continued successful functioning of the organism depends upon the signaling system remaining in the same state. It is therefore conserved (cf. Stearns 1994).

Another signaling pathway with effects on life history transitions in a broad range of deuterostomes is organized around the thyroid hormones, with excellent work reported here on organisms ranging from sea urchins (Chapter 3), lampreys and salmon (Chapter 6), to amphibians (Chapter 7). There is certainly ample evidence that thyroid hormones coordinate changes in many tissues during life history transitions. Denver (personal communication) has suggested that they may also participate in the transition from the uterine to the air-breathing environment in mammals, which would imply an interesting homology of parturition and amphibian metamorphosis.

These remarkable findings raise the hope that the study of ancient, conserved mechanisms might bring to the study of life history traits the sorts of insights that the study of homeobox and other developmental control genes have brought to evo-devo, for example the notion that the mechanisms causing trade-offs may be as ancient and widely shared as those causing developmental patterns. A good example of the latter is provided by Banta and Purugganan (Chapter 9), who point out that "many aspects of the genetic network controlling flowering" in *Arabidopsis* and rice are the same in both lineages.

That hope is moderated by two observations mentioned in this volume. First, Kuningas and Westendorp (Chapter 16) summarize the status of research that aims to see whether processes that extend lifespan in flies, worms, and mice also do so in humans. Reducing IIS signaling to increase lifespan in humans has to be mild to be beneficial; otherwise the side effects are serious (cf. Austad and Finch 2008). Certain alleles of superoxide dismutase increase disease risk and effects on lifespan are unknown. Evidence on the impact of variation in heat shock proteins on lifespan is mixed. Forkhead transcription factor variants have clear effects on health, and evidence of their impact on lifespan is accumulating (Flachsbart *et al.* 2009). The evidence for sirtuins, which are stimulated by resveratrol, is mixed and conclusions are unclear. As for dietary restriction, evidence from a prospective cohort study suggests that an optimal body mass index rather than low body weight improves lifespan. One could react to this mixed picture by saying that clinical research is not as rigorous as basic research and that

the effects found in model systems will eventually be found in humans. Or one could react by saying that only in humans is pathology well enough understood, and clinical investigation of the entire individual is thorough enough to document side effects, which are probably there in the worms, flies, and mice, but invisible to us because those model organisms cannot tell us about their discomfort. While there is probably some truth in both views, I lean towards the latter, for the chapters by Kuningas and Westendorp (Chapter 16) and Muehlenbein and Flinn (Chapter 12) remind us that some of our best evidence on mechanisms comes from humans, where our inability to do manipulation experiments is more than compensated for by the extensive details that the medical research community has learned about physiology and metabolism.

The second sobering observation concerns honey bees (Chapter 20), where "the highly fecund . . . queen can live about ten times longer than essentially sterile short-lived workers, yet shows higher levels of IIS as larvae." If the relationship between reproduction and lifespan is mediated by several mechanisms, and if demographic conditions favor long life in queens, there are probably enough options among the different mechanisms for selection eventually to find one that will allow long life despite high reproductive effort, particularly in this case, where queens have been released from most of the other reproductive trade-offs, such as those associated with predation risk.

Our hope that a few pathways will be shown to mediate life history variation in most animals should also be tempered by the realization that the causes of variation in traits such as growth rate, fertility, and survival are often more complex than the causation of traits such as digit number, limb components, and body axes. Life history traits are strongly influenced by environmental factors, chief among them temperature and diet. That does not mean, for example, that IIS is not important in mediating trade-offs among plastic traits in a complex and changing environment. It very probably plays a key role. But it does mean that dissecting the causes influencing such traits is a major experimental challenge.

Let me illustrate that point with a result from an experiment that used P-element inserts to create mutational covariance, and which uncovered a trade-off between reproduction and survival in *Drosophila melanogaster* (Stearns and Kaiser 1996). The difference in length and position of treatment inserts interacted with genetic backgrounds and temperature to elicit a trade-off between lifetime fecundity and lifespan that was only detected as a *three-way* interaction effect. It would have been missed by a simpler experimental design. That result, which is silent on mechanism, suggests the sorts of designs that might be necessary to evaluate the contribution of variation in IIS signaling to overall phenotypic variation in fecundity and survival.

27.2.3 Decoupling functions by duplicating modules

If a module must perform more than one function and is constrained by over-commitment, then the resulting trade-offs can be uncoupled by duplicating the module, dividing the labor, and allowing the daughter modules to specialize. Nedelcu and Michod (Chapter 21) note that such uncoupling occurred in at least two major evolutionary transitions, from single-celled to multicellular and from solitary to eusocial organisms. The macroevolutionary context of those examples might suggest that the process is not common in microevolution, but in yeast there is a good example of a protein that originally functioned both as a signaling molecule and as an enzyme, its dual functions preventing it from being optimized for either. When the gene for the protein was duplicated, each copy could specialize so that one produced a protein that became more efficient at signaling and the other a protein that became a more efficient enzyme (Hittinger and Carroll 2007). The concept is general and should apply to any level of the biological hierarchy at which modules can be identified.

27.3 Is work on mechanisms changing theory?

27.3.1 Are these data forcing the theory to change?

When attempting to show that a new body of work makes a difference, it is often wise to select a classical example that plays a central role in an established field and show that by incorporating new insights, some important prediction is changed or some outstanding puzzle is solved. The tension between theoretical generality and mechanistic detail has been present since the study of life history evolution began; people have been asking what difference the mechanistic detail might make right from the start. Those concentrating on developing the theory thought that there would be no reason to go into mechanisms unless it could be shown that such detail makes a difference. While it has always been clear that mechanisms must exist that underpin notions like trade-offs and constraints, experience suggests that discovering them is one thing and showing that they make any difference to the general structure of the theory is another.

I did not find a single case in this volume in which the incorporation of mechanistic detail forced significant change in a specific model of life history theory, nor, with some notable exceptions, did the authors in this volume try to bring their data into intimate contact with the theory. In most cases, they were pursuing a different agenda, for most of the science described here is not so much concerned with testing theory as with describing mechanisms. For them, life history theory provided a general motivational structure, not a set of specific predictions to be tested. The theory said, in effect, that understanding the causes of trade-offs and the mechanistic basis of major life history transitions is important simply because they were unknown, not because it was clear that knowing them would make a difference to the theory.

In this book there is at least one apparent exception: the work of Sinervo and his colleagues on syndromes of morphological, behavioral, and life history traits in lizards. They expanded the theory by including density, frequency, and state dependence to explain cycles of syndromes. However, that expansion was not driven by experimental results reporting details of intra-organismal mechanisms. It was driven by recognition of the logical inadequacies of an existing theory that could not yet explain the cycling polymorphism observed at the level of whole organisms, and it was then followed up with work on the mechanisms underlying the syndromes.

Thus the answer to the question, "Is knowledge of mechanisms changing the theory?" is, "Not yet."

One reason is that we have not yet taken the theory seriously enough. If we demanded of it not just qualitative but precise quantitative agreement with the evidence, then we would be forced to acknowledge that it in fact fails to explain important details (see *Caulobacter* example below), and that acknowledgement would motivate a serious effort to specify the intermediate structure—the intra-organismal features that constrain the evolution of phenotypic traits—whose incorporation into the theory would permit quantitative prediction.

27.3.2 Are we identifying general features of intermediate structure in the genotype–phenotype map?

Evolutionary biology is faced with a puzzle. Understanding the general features of the genotype–phenotype map is widely accepted as one of its major current challenges, the key to connecting genetic evolution to trait evolution. Where the traits are morphological, the standard approach is now through evo-devo. But where the traits are plastic and quantitative, as they are in life history evolution and behavioral ecology, the approach is the one taken here, through studying the integration of the organism by hormones and other molecular signals that change physiological state. The result is a body of knowledge in which greater and greater detail threatens to obscure the simplifying features needed to construct a general theory of the organism.

Those whose priority is description focus on detail; those whose priority is prediction focus on simplicity. In this volume description, and therefore detail, dominates. Once the lens is focused on the mechanistic details, their interaction with the theory is often set aside, or even forgotten, in the search for more details. This is characteristic of a field driven by an experimental rather than a theoretical paradigm. Such a field does not try to simplify systems to extract general properties. It accumulates more and more detail, in the belief that if enough detail is accumulated, generality and simplicity will emerge.

If simplification had been a higher priority, we might have found here a clearer statement of intermediate structure (Stearns 1986) that condensed the complex connections between genotype and phenotype by making an artful chunking of reality at an appropriate level. We do have part of that picture, but it is not yet connected to the whole-organism traits that constitute a life history in such a way that we can see whether it makes any difference to a prediction like this: if the extrinsic mortality rates of adults increase, then age and size at maturity should decrease, reproductive effort should increase, and lifespan should decrease. Despite the very interesting results on alternative ways to view trade-offs in worms and flies presented here, I see no reason to think that that prediction would not still hold.

While a general model of the organism might not yet have been achieved, some conceptual issues have been clarified. For example, in Chapter 2 Roff asks whether different genetic architectures can yield the same developmental pathways and whether different developmental pathways can yield the same phenotype. If such options exist at several levels in the genotype–phenotype map among closely related species, as appears to be the case for wing dimorphism in crickets (Chapter 2) and sex determination in flies (Saccone *et al.* 2002), and even among populations, as is the case for wing size in *Drosophila subobscura* in North and South America (Calboli *et al.* 2003), then a general theory of the organism will be difficult to construct and perhaps will only be possible at a level too abstract to be helpful. Whether this issue is serious for traits such as fecundity, maturation, and survival remains to be seen.

Whatever general patterns finally emerge from mechanistic studies, they have at least emphasized the recognition that abstract terms like antagonistic pleiotropy, negative mutational covariance, negative genetic correlations, Y-allocations, and trade-offs are operationalized in different contexts, covering a diverse array of different types of molecular linkages and not necessarily mapping directly onto each other. Edward and Chapman (Chapter 11) put it well: "…it is not yet clear whether physiological and evolutionary trade-offs occur via the same underlying mechanisms. It would be interesting to know, for example, whether individuals selected for early or late age reproduction retain equal capacity to express physiological trade-offs; that is, whether the effects underlying these different kinds of trade-offs are additive. It would also be useful to know whether nutrient signaling evolves during artificial selection for early- and late-age reproduction." In other words, we do not know how the genetic changes involved in correlated responses are connected to physiological allocations.

To do so will require solving some serious puzzles in genomics, transcriptomics, proteomics, and metabolomics, and in the validation of laboratory results in natural populations. The authors of these chapters are well aware of the challenge. As Schmidt (Chapter 18) points out, "Showing that knockout of a gene changes phenotype merely demonstrates that it is in the pathway: a gene-to-phenotype connection does not necessarily translate to the identification of a gene that is segregating molecular variation that underlies phenotypic variance." Kuningas and Westendorp (Chapter 16) note, "…in most cases laboratory conditions poorly mimic the evolutionary niche in which these genes come to expression. It is largely unknown to what extent mutations that influence lifespan in laboratory conditions affect fitness in natural environments." And Zera and Harshman (Chapter 24) state, "The endocrine causes of variation in intermediary metabolism (and other life history traits) in *outbred* populations remains one of the most important but currently one of the least understood aspects of allocation and life history trade-offs."

27.3.3 Are the empirical problems with the theory being addressed?

There are at last three problems in life history theory whose solution might be found in the study of mechanisms but that are not addressed in this volume.

The first concerns the shape of trade-offs. Whether a trade-off is linear, concave upward, or concave downward has been an issue since Gadgil and Bossert (1970) discovered that it was a distinction that made a difference. After 40 years of hard work by many people we still do not know the general shape of trade-offs. That this problem has not been taken up by the community of scientists working on mechanisms could be because they are not aware of

the issue, or it could be because it is hard to establish at the level of mechanism whether marginal effects on trade-offs are linear, increasing, or decreasing. It was certainly an issue that theory suggested in the early 1970s would be important to resolve (Stearns 1976); its lack of resolution by a large group of people working on trade-offs remains striking.

The second concerns the co-evolution of life histories in hosts and pathogens, where the evolution of virulence has become a central issue in the last 20 years. Here, a key assumption is that transmission and virulence trade off. This assumption was initially suggested by the classic study of co-evolution between myxomatosis and rabbits in Australia (Fenner 1983), and was later confirmed by a comparative pattern in nematodes and fig wasps (Herre 1993). The existence of the virulence–transmission trade-off has been supported by several experimental studies conducted at the level of the whole organism (e.g., Ebert and Mangin 1997, Turner *et al.* 1998, Cooper *et al.* 2002, Jensen *et al.* 2006, Berenos *et al.* 2009), but not by all such studies (Alizon *et al.* 2009). Sometimes the trade-off is found; sometimes it is not. This pattern suggests that an investigation of mechanisms could explain the pattern of occurrence. Because these are not systems in which mechanisms are easy to investigate, for the processes mediating trade-offs in the pathogen are occurring in part inside a host, it is quite understandable that progress has been slow on this front. Nonetheless it is worth pointing out that a breakthrough engineered by attention to mechanisms in this particular case would be extremely important.

The third case is notable because it indicates how theory might more often drive experiments; it concerns a quantitative discrepancy in the responses measured in experimental evolution that could be explained by intra-organismal mechanisms. Ackermann *et al.* (2007) tested the evolutionary theory of aging with *Caulobacter crescentus*, a bacterium that divides asymmetrically. They evolved three populations for 2000 generations under conditions where selection was strong on traits expressed early in life but very weak on traits expressed later in life. While all three populations evolved faster growth rates, mostly by decreasing the age at first division (Fig. 27-1), the decrease in

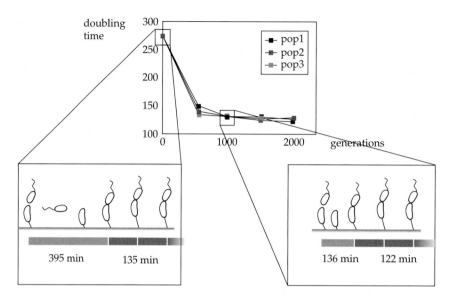

Figure 27-1 The results of an experiment in which *Caulobacter crescentus* was subjected to 2000 generations of experimental evolution with strong selection to improve reproductive performance early in life. As expected, both age at first division and interbirth interval (stalked cell cycle length) decreased. Figure courtesy of Martin Ackermann and Urs Jenal.

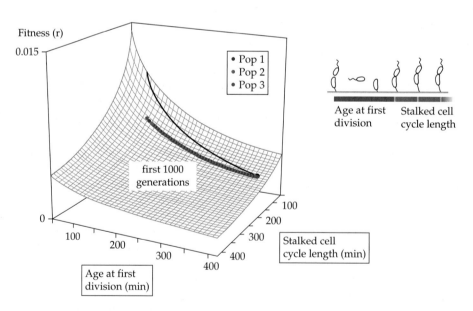

Figure 27-2 Fitness in *C. crescentus* as a function of age at first division and stalked cell cycle length. Changes in interbirth interval (stalked cell cycle length) were smaller than expected. Theory predicts that the populations should move upward on the fitness surface in the dimensions corresponding to both age at first division and stalked cell cycle length, but they did so with large changes in the first dimension but only small changes in the second. That strongly suggests that stalked cell cycle length is constrained, but the constraints are not yet known. Figure courtesy of Martin Ackermann and Urs Jenal.

interbirth interval (stalked cell cycle length) was much less than expected (Fig. 27-2).

While one could speculate that interbirth interval was constrained by the series of processes that must be completed before cell division can take place, no attempt has yet been made to expose those processes, link them together, time each link, and figure out why each link could not be completed faster. This example emphasizes a quantitative issue highlighted by a strong test in which theory failed to account for important details—precisely the sort of place where an understanding of the mechanisms mediating life history evolution could be expected to help.

27.4   Conclusion

Of the many barriers to the conceptual unification of knowledge, some of the most important must be differences in research paradigms. Reading this volume has brought home to me how those working on life history theory and on intra-organismal mechanisms differ in their definition of the research enterprise. The theory seeks general predictions based on simple assumptions tested with comparative patterns and experimental evolution. Work on intra-organismal mechanisms is not driven by evolutionary theory and does not seek to test it. It does have an implicit theory—a theory of networks of communication and control, of signal transduction and cascade—but that theory is not expressed as predictions to be tested. Instead, the ideas that are tested are local, usually based on the results of the last experiment, and progress is made in small steps. This is a picture very familiar to anyone who has worked in reductionist biology, where the method has brought great success, if success is measured by the level of detail achieved in a description rather than by the identification of general, simple, unifying principles that explain variation among individual organisms.

This issue is well understood by McKean and Lazzaro (Chapter 23): "The detailed understanding of the molecular mechanisms of immunological defense is a testament to the power of the reductionist research paradigm of the biomedical sciences, but this reductionist approach has itself come at a cost, neglecting the whole-organism perspec-

tive that can provide insight into how the competing demands of development, reproduction, and pathogen defense may contribute to the maintenance of disease susceptibility." Their chapter on the costs of immunity and resistance versus tolerance should be required reading for anyone interested in evolutionary medicine.

While impressive progress is here reported on our understanding of the mechanistic basis of life history variation, this volume should be taken as an intermediate progress report, for if the goal is to connect the data to the theory in such a way that the theory is forced to change, there is still a long way to go.

27.5   Summary

1. Trade-offs can be mediated by signaling pathways as well as by allocations of energy and materials, and in some cases traits thought to be generally involved in trade-offs, e.g., reproduction and survival, can be decoupled.

2. The insulin/insulin-like signaling pathway appears to mediate growth, reproduction, and lifespan in a range of species spanning flies, worms, and mice. Deeply conserved signaling pathways may play a role in research on life history traits similar to the role that the homeobox genes have played in evo-devo. Results from honeybees qualify this claim.

3. If a module performs two functions and is constrained from optimizing performance for both, then duplicating it and allowing the daughter modules to specialize can break the trade-off and improve performance. This concept applies anywhere in the biological hierarchy where modules are so constrained.

4. Data of the sort reported here are not yet forcing the theory to change because we are not yet requiring of them that they explain the inability of the theory to make quantitative predictions.

5. It is proving difficult to construct a general picture of the key features of organisms that govern expression, variation, and co-variation in life history traits.

6. We should take both the theory and the evidence more seriously, demanding of the theory that it make quantitative predictions and of the evidence that it change the theory so that it can do so.

7. Although we now know much more about the intra-organismal mechanisms that mediate expression and variation in life history traits, that knowledge has not yet changed the theory. Where the theory has been expanded, the expansion was not motivated by results on intra-organismal mechanisms.

8. Some important problems in life history evolution—among them the shape of trade-offs, the nature of the virulence-transmission trade-off, and the inability of the theory to deal with quantitative discrepancies between prediction and observation—have not yet been addressed directly by mechanistic studies.

CHAPTER 28

What mechanistic insights can or cannot contribute to life history evolution: An exchange between Stearns, Heyland, and Flatt

Thomas Flatt, Andreas Heyland, and Stephen C. Stearns

When discussing Steve Stearns' chapter (Chapter 27) and the rationale of this book, the three of us realized that Steve and we (the editors) differ in how we see the contribution mechanistic insights can make to life history evolution. Steve suggested that life might be more interesting for our readers if we wrote a rebuttal to some of Steve's claims and if he then wrote an answer. This postscript summarizes our discussion.

28.1 Why mechanisms are important for life history theory: A response by Flatt and Heyland

Life history theory seeks to explain, from first principles, the evolution of the main features of life cycles, in particular the design of organisms with regard to reproductive success (Stearns 1976, 1992, Roff 1992, Charlesworth 1994). The theory claims that we need to understand two things in order to understand the evolution of life histories (Stearns 2000):

• the environmental factors that affect the age- or stage-specific schedules of survival and reproduction
• the connections (trade-offs) between life history traits and the internal constraints these connections (and other factors) impose upon how life history traits can vary.

By building on these simplifying claims, life history theory has focused on general principles and deliberately ignored complicating details, such as the mechanisms of genetics, development, and physiology. For example, a keystone of life history theory is the Euler–Lotka equation, which relates demographic traits such as age at maturity, survival rates, and rates of reproduction to fitness. By using this or related mathematical frameworks, and by assuming constraining trade-offs as boundary conditions, one can derive theoretical predictions about life history evolution, most—if not all—of which have been confirmed empirically with tremendous success. This large body of work has been summarized and reviewed in several excellent books by Stearns (1992) and Roff (1992, 2002). In contrast to classical life history theory, the present book has a different aim. By focusing on explicit genetic, developmental, and physiological mechanisms, it attempts to provide a description of the "molecular natural history" of life history evolution. Given the clear-cut success of classical life history theory, what can knowledge about mechanistic details contribute to our understanding of life history evolution?

In his chapter, Stearns argues that, while mechanistic studies have resulted in several striking conceptual advances, none of these insights is forcing the theory to change, at least not yet. Currently, most mechanistic descriptions of life history traits and their evolution neither directly test theoretical predictions, nor require the existing theory to be revised. While the predictive theory is deliberately void of explicit mechanism (as is common in evolutionary biology), the study of mechanisms is motivated by a reductionist description and analysis of function (as is typical in molecular biology). As Stearns points out, there is still a long way to go before mechanistic insights and the theory can cross-fertilize each other. While we agree with Stearns that, despite impressive progress, the study of molecular mechanisms that mediate life history evolution is still in its infancy and has not yet made sufficient contact with the theory, there are in our opinion several important issues in life history evolution that can be illuminated by taking a mechanistic approach. Here we discuss four of them:

- trade-offs
- antagonistic pleiotropy and the evolution of aging
- hormones as a feature of the genotype-phenotype map
- the evolution of life history transitions.

Some but not all of these issues contact the classical theory; for others, mechanistic insights are likely to improve our general understanding of life history evolution and might lead to extensions of the classical theory in the future.

The first issue is that of trade-offs. The existence and importance of trade-offs in life history evolution is undisputed (Roff 1992, Stearns 1992), but as has been pointed out by Stearns (e.g., Stearns 2000, 2005, Stearns and Magwene 2003) and many others, we lack fundamental knowledge about their genetic, developmental, and physiological causes. In most cases, trade-offs are simply assumed to exist, for example as constraining boundary conditions in models of life history evolution, and even when they are measured (through genetic correlations or correlated responses to selection), their underlying mechanisms remain typically unknown (Stearns 1989, 2000, Stearns and Magwene 2003). This lack of mechanistic insight into the nature and structure of trade-offs is problematic for at least three reasons.

First, while the theory is in most cases silent on the causes of trade-offs (S. C. Stearns, personal communication; Stearns 1992), many life history models assume that trade-offs are caused by competitive resource allocation (Kirkwood 1977, van Noordwijk and de Jong 1986, Houle 1991, de Jong and van Noordwijk 1992, Shanley and Kirkwood 2000, Worley *et al.* 2003). However, as work in this volume

illustrates, this might not always necessarily be the case. Trade-offs can occur for structural reasons or be caused by signaling processes independent of resource allocation (Tatar and Carey 1995; also see Chapters 11 and 13). Moreover, when reading the recent experimental life history literature it becomes obvious that most empirical studies of trade-offs uncritically assume that such trade-offs *must be* caused by competitive resource allocation (Barnes and Partridge 2003, Harshman and Zera 2007, Flatt and Schmidt 2009; also see Chapter 24). However, only very few studies have actually attempted to demonstrate the resource basis of trade-offs by quantifying acquisition or allocation of resources, i.e., energetic investment (Stearns 1992; see also Chapter 24). Thus, uncovering the causes of trade-offs might lead to us reconsider, at least in some instances, one of our most commonly made assumptions about the nature of trade-offs, an issue that directly touches upon aspects of the theory. While many life history models and predictions will be unaffected by this knowledge, others that rely on the assumption of resource allocation might need to be revisited. In any case, as good scientists, we should probe and question our assumptions and find out when they hold or not. Moreover, where resource allocation trade-offs do exist, different species might vary in the way resources are acquired and allocated to survival and reproduction, suggesting that unknown internal (developmental and physiological) mechanisms interact in unknown ways with extrinsic factors that affect survival and reproduction: without mechanistic insight we cannot develop a theory to explain such variation (Stearns 2000).

Second, life history theory distinguishes between physiological and microevolutionary trade-offs (Stearns 1989, 1992). A physiological trade-off is caused by competition among two or more processes for limited resources (or by mutually exclusive signaling processes independent of allocation) within a single individual, whereas a microevolutionary trade-off is defined at the population level as a change in a trait that increases fitness and which is linked to a change in another trait that decreases fitness—thus, there must be a negative genetic correlation in the population between the two fitness components in question. What is the relationship between physiological and microevolutionary trade-offs? For example, a physiological (intra-individual) trade-off can be genetically fixed in the population, and all individuals will physiologically respond in the same way, without the physiological trade-off contributing to the microevolutionary trade-off. Alternatively, physiological trade-offs might be genetically variable and contribute to, or modulate, evolutionary trade-offs among individuals in the population and can thus contribute to the response to selection (Stearns 1989, 1992). Without detailed knowledge about the physiological nature of trade-offs, it will be difficult to understand or predict when and how physiological trade-offs contribute to, or modulate, evolutionary trade-offs or not (for an empirical example see Flatt and Kawecki 2007).

Third, development and physiology (for example, hormonal signaling, as amply demonstrated in this volume) can modulate the expression of genetic (microevolutionary) trade-offs in a way that depends on the environment (Stearns 1989, 1992). This is illustrated by the fact that the sign of a genetic correlation can change across different environments. For example, a genetic correlation might be negative in one environment but positive in another. Similarly, a trade-off might be present in one environment but absent in an alternative environment, suggesting that trade-offs can be "broken" or "uncoupled." In addition to this environmental dependency, trade-offs might or might not change under selection. In none of these cases do we really understand the causes that underlie this dynamic and flexible behavior of trade-offs (Stearns 1989). As several chapters in this volume suggest, mechanistic insights might help us to understand this issue better. Even though recent insights into the mechanistic nature of trade-offs might not yet directly change the way we construct life history models, we agree with Stearns (2000) that "the explanation of variation in life histories will not be complete until those mechanisms are understood."

The second issue relates to the importance of trade-offs in the evolutionary theory of aging, which can be considered a part of life history theory (Stearns 1992, Charlesworth 1994). While the evolutionary theory of aging has not been given much attention in this volume (for overviews see Rose 1991, Stearns 1992, Stearns and Partridge 2001, Flatt and Schmidt 2009), work on aging serves as a good

example for how mechanistic insights can touch upon the theory. In short, two major models have been proposed to explain the evolution of aging (Medawar 1952, Williams 1957). One of them posits that early fitness components are negatively connected to fitness components late in life by alleles with antagonistic pleiotropic effects; under this model, the evolution of aging can be seen as a byproduct of selection for increased fitness early in life, with weak selection at advanced ages being unable to oppose the pleiotropic and negative effects late in life (Williams 1957). Thus, Williams' postulate predicts that early and late fitness components are coupled by trade-offs that are due to the existence of alleles with antagonistic pleiotropic effects. This is clearly a mechanistic prediction of an evolutionary model that is part of life history theory broadly construed (Stearns 1992, Charlesworth 1994).

While selection experiments and phenotypic manipulations have lent much support to Williams' hypothesis (Stearns and Partridge 2001, Flatt and Promislow 2007, Flatt and Schmidt 2009), formally testing the "antagonistic pleiotropy" model requires us to identify genes and to functionally characterize them. Without this knowledge we cannot know whether the observed trade-off pattern is caused by antagonistic pleiotropy, linkage disequilibrium, or a correlation with a third unobserved trait (Clark 1987, Charlesworth 1990, Houle 1991). This is an important mechanistic issue regarding trade-offs: it not only concerns trade-offs between early and late fitness traits that are relevant in the evolution of aging, but in fact relates to trade-offs between fitness components in general. While knowledge about the genetic nature of trade-offs might not affect most classical life history models, where trade-offs are simply assumed as boundary conditions, it is relevant for our understanding of the evolution of aging. So what have genetic insights contributed to this problem?

Recent research in molecular genetics of aging now tells us that Williams probably got it right and that many mutations (at least in the laboratory) exhibit the kind of life history pleiotropy he envisaged. With the advent of improved methods for genetic mapping, evolutionary geneticists have also recently begun to successfully identify single loci and quantitative trait nucleotides with major pleiotropic effects on life history traits in natural populations, including polymorphisms in insulin/IGF-1 signaling and other pathways (Carbone *et al.* 2006, Schmidt *et al.* 2008, Paaby *et al.* 2010; also see Chapter 18). We consider this an example where genetic insights have confirmed a postulate that can be considered part of life history theory broadly construed (Flatt and Promislow 2007, Flatt and Schmidt 2009). Moreover, as discussed by Stearns and other authors in this book, this work has led to the realization that the insulin/IGF-1 signaling pathway might play a major evolutionarily conserved role in pleiotropically affecting lifespan and other life history traits connected with it. This finding has come as somewhat of a surprise to evolutionary biologists working on aging who, starting with Williams (1957), had intuitively assumed that it would be hard to identify major homologous genetic effects upon lifespan (Flatt and Schmidt 2009).

The third issue concerns hormones as intermediate components of the genotype–phenotype map. Since hormones are critical physiological regulators of decisions about growth, storage, maturation, and reproduction, life history theorists postulated that they might play a major role in life history evolution, for example in terms of acting as mediators of trade-offs (Stearns 1989, Finch and Rose 1995). Indeed, over the last 15 years or so we have accumulated a wealth of data that seem to confirm this intuition. As many chapters in this book showcase, we have now many excellent examples that suggest that hormones are key players in physiologically mediating life history transitions, trade-offs, and phenotypic plasticity (see many chapters in this book and reviews by Zera and Harshman 2001, Flatt *et al.* 2005, Heyland *et al.* 2005, Toivonen and Partridge 2009). Again, one might say that insights into hormonal mechanisms do not force the theory to change, but it is also amply clear now that changes in endocrine function are central to the coordination, integration, and modulation of life history traits in response to changes in the environment. One (perhaps naive) prediction would then be that genes involved in endocrine signaling might be "hot spots" for the evolution of life histories. If so, we would have learned something important about the functional architecture of life history traits.

A fourth issue is the evolution of life history transitions, such as metamorphosis and transitions between alternative life history strategies. Although the classical theory does not make specific predictions about these phenomena (Stearns 1992, 2000; also see foreword to this volume), an improved knowledge about the biology of life history transitions is clearly important for our understanding of life history evolution. We believe that mechanistic insights will be particularly helpful for extending the classical theory into this uncharted area. Indeed, many chapters in this book provide good examples of how mechanistic studies are beginning to uncover important commonalities and differences in developmental aspects of life history transitions (see Chapters 3, 5, 6, 7, and 8.).

More generally, this body of work holds great promise for expanding the scope of classical work on life histories by conceptually integrating it with recent advances in evolutionary developmental biology (evo-devo). One major current and successful avenue is to integrate evo-devo with molecular population genetics and phylogenomics (Carroll 2005, Barrett *et al.* 2008, Rebeiz *et al.*, 2009, Stern 2010). We predict that in the future life history evolution will significantly benefit from cross-talk with these disciplines; such advances will allow us, for example, to test for signatures of selection at the DNA level and, in a comparative and phylogenetic context, to determine whether particular molecular changes that affect life histories are homologous or caused by convergence.

There is no doubt that classical life history models have derived their impressive predictive power from the fact that they are phenotypic models that are silent on mechanism. Yet we are convinced that in order for us to arrive at a more complete understanding of life history evolution we need to learn more about the mechanisms that affect life history traits and their evolution. Regarding the classical tension between organismal and molecular biologists Stearns (2000) notes: "We will see more cooperation between these fields, for it is in the interest of both to use each other's methods and to ask each other's questions, and there are solid grounds for mutual respect. People with one foot solidly in each camp will do exciting research." One of the principal motivations for editing this volume was to bring these two camps together more closely. If this book has helped this integration a bit, then it has fulfilled its purpose.

28.2 Reply by Stearns: Mechanisms do not yet force the theory to change

Thomas and Andreas have done a good job of defending their project, and I find it hard to disagree with a statement that cites my work 28 times in 7 pages. We do not have any fundamental differences. Here I only want to state explicitly where we agree and reiterate some distinctions that I tried to draw in my chapter.

We agree that studies of mechanism have illuminated the nature of trade-offs. Uncovering those mechanisms has revealed that energy allocation is not all there is to trade-offs, and some of the mechanisms mediating aging appear to be ancient and widely shared. These are significant advances, but because life history theory was silent about the mechanisms that produce trade-offs, uncovering those mechanisms was not going to change that *kind* of theory.

Some of our disagreement arises from what we consider to be life history theory. I consider it to be the part of the study of life history evolution that can be expressed in mathematical models. They consider it to include, in addition to that, assumptions about the mechanisms that cause trade-offs. Where those mechanisms make a difference to the predictions of theory, they should be included in the theory. My point is, that has not yet been done, and doing so does not appear to have been on the agenda of most of those who authored chapters in this book. That is not a criticism, it is simply a comment about what future research might address. I do not expect it to be a top priority for those working on mechanisms, to whose strengths it does not play, but it might attract those with a gift for theory.

References

Abbasi, A. A., Prasad, A. S., Rabbani, P. & Dumouchelle, E. (1980) Experimental zinc deficiency in man. Effect on testicular function. *Journal of Laboratory and Clinical Medicine*, 96, 544–550.

Abbott, R. J. & Gomes, M. F. (1989) Population genetic structure and outcrossing rate of *Arabidopsis thaliana* (L.) Heynh. *Heredity*, 62, 411–418.

Abbott, A. L., Alvarez-Saavedra, E., Miska, E. A., Lau, N. C., Bartel, D. P., Horvitz, H. R. & Ambros, V. (2005) The *let-7* MicroRNA family members *mir-48*, *mir-84*, and *mir-241* function together to regulate developmental timing in *Caenorhabditis elegans*. *Developmental Cell*, 9, 403–414.

Abe, T., Suzuki, T., Unno, M., Tokui, T. & Ito, S. (2002) Thyroid hormone transporters: recent advances. *Trends in Endocrinology & Metabolism: TEM*, 13, 215–220.

Abouheif, E. & Wray, G. A. (2002) Evolution of the gene network underlying wing polyphenism in ants. *Science*, 297, 249–252.

Abrams, P. (1993) Does increased mortality favor the evolution of more rapid senescence *Evolution*, 47, 877–887.

Ackermann, M., Schauerte, A., Stearns, S. C. & Jenal, U. (2007) Experimental evolution of aging in a bacterium. *BMC Evolutionary Biology*, 7, 126.

Adamo, S. A. & Parsons, N. M. (2006) The emergency life history stage and immunity in the cricket, *Gryllus texensis*. *Animal Behaviour*, 72, 235–244.

Adamo, S. A., Fidler, T. L. & Forestell, C. A. (2007) Illness-induced anorexia and its possible function in the caterpillar, *Manduca sexta*. *Brain, Behavior, and Immunity*, 21, 292–300.

Adkins-Regan, E. (2005) *Hormones and Animal Social Behavior*. Princeton, Princeton University Press.

Adkins-Regan, E. (2008) Do hormonal control systems produce evolutionary inertia? *Philosophical Transactions of the Royal Society of London B*, 262, 1599–1609.

Adler, L. & Jarms, G. (2009) New insights into reproductive traits of scyphozoans: special methods of propagation in *Sanderia malayensis* GOETTE, 1886 (Pelagiidae, Semaeostomeae) enable establishing a new classification of asexual reproduction in the class Scyphozoa. *Marine Biology*, 156, 1411–1420.

Adolphs, R. (2003) Cognitive neuroscience of human social behavior. *Nature Reviews Neuroscience*, 4 165–178.

Aggarwal, K. & Silverman, N. (2008) Positive and negative regulation of the *Drosophila* immune response. *BMB Reports*, 41, 267–77.

Ahnesjo, J. & Forsman, A. (2003) Correlated evolution of colour pattern and body size in polymorphic pygmy grasshoppers, *Tetrix undulata*. *Journal of Evolutionary Biology*, 16, 1308–1318.

Aiello, L. C. & Wheeler, P. (1995) The expensive-tissue hypothesis: The brain and the digestive system in human and primate evolution. *Current Anthropology*, 36, 199–221.

Ailion, M., Inoue, T., Weaver, C. I., Holdcraft, R. W. & Thomas, J. H. (1999) Neurosecretory control of aging in *Caenorhabditis elegans*. *Proceedings of the National Academy of Sciences of the United States of America*, 96, 7394–7397.

Alabadi, D. & Blazquez, M. (2009) Molecular interactions between light and hormone signaling to control plant growth. *Plant Molecular Biology*, 69, 409–417.

Alcedo, J. & Kenyon, C. (2004) Regulation of *C. elegans* longevity by specific gustatory and olfactory neurons. *Neuron*, 41, 45–55.

Alcendor, R. R., Gao, S., Zhai, P., Zablocki, D., Holle, E., Yu, X., Tian, B., Wagner, T., Vatner, S. F. & Sadoshima, J. (2007) Sirt1 regulates aging and resistance to oxidative stress in the heart. *Circulation Research*, 100, 1512–1521.

Alemseged, Z., Spoor, F., Kimbe, W. H., Bobe, R., Geraads, D., Reed, D. & Wynn, J. G. (2006) A juvenile early hominin skeleton from Dikika, Ethiopia. *Nature*, 443, 296–301.

Alexander, R. D. (1974) The evolution of social behavior. *Annual Review of Ecology and Systematics*, 5, 352–383.

Alexander, R. D. (1979) *Darwinism and Human Affairs*. Seattle, University of Washington Press.

Alexander, R. D. (1989) Evolution of the human psyche. In Mellars, P. & Stringer, C. (Eds.) *The Human Revolution*. Chicago, University of Chicago Press.

Alexander, R. D. (1990) *How humans evolved: Reflections on the uniquely unique species. Museum of Zoology (Special Publication No. 1)*. Ann Arbor, The University of Michigan.

Alexander, R. D. (2005) Evolutionary selection and the nature of humanity. In Hosle, V. & Illies, C. (Eds.) *Darwinism and Philosophy*. South Bend, University of Notre Dame Press.

Alexander, R. D. & Noonan, K. M. (1979) Concealment of ovulation, parental care, and human social evolution. In Chagnon, N. & Irons, W. (Eds.) *Evolutionary Biology and Human Social Behavior*. North Scituate, Duxbury Press.

Alizon, S., Hurford, A., Mideo, N. & Van Baalen, M. (2009) Virulence evolution and the trade-off hypothesis: history, current state of affairs and the future. *Journal of Evolutionary Biology*, 22, 245–259.

Allard, J. S., Heilbronn, L. K., Smith, C., Hunt, N. D., Ingram, D. K., Ravussin, E. & De Cabo, R. (2008) In vitro cellular adaptations of indicators of longevity in response to treatment with serum collected from humans on calorie restricted diets. *PLoS One*, 3, e3211.

Allman, J., Rosin, A., Kumar, R. & Hasenstaub, A. (1998) Parenting and survival in anthropoid primates: Caretakers live longer. *Proceedings of the National Academy of Sciences of the United States of America*, 95, 6866–6869.

Alonso-Alvarez, C., Bertrand, S., Devevey, G., Prost, J., Faivre, B. & Sorci, G. (2004) Increased susceptibility to oxidative stress as a proximate cost of reproduction. *Ecology Letters*, 7, 363–368.

Alonso-Alvarez, C., Bertrand, S., Faivre, B., Chastel, O. & Sorci, G. (2007) Testosterone and oxidative stress: the oxidation handicap hypothesis. *Proceedings of the Royal Society of London B*, 274, 819–825.

Alonso-Blanco, C., Aarts, M. G. M., Bentsink, L., Keurentjes, J. J. B., Reymond, M., Vreugdenhil, D. & Koornneef, M. (2009) What has natural variation taught us about plant development, physiology, and adaptation? *Plant Cell*, 21, 1877–1896.

Alonso-Blanco, C., El-Assal, S. E. D., Coupland, G. & Koornneef, M. (1998) Analysis of natural allelic variation at flowering time loci in the Landsberg erecta and Cape Verde islands ecotypes of *Arabidopsis thaliana*. *Genetics*, 149, 749–764.

Alonzo, S. & Sinervo, B. (2001) Mate choice games, context-dependent good genes, and genetic cycles in the side-blotched lizard *Uta stansburiana*. *Behavioral Ecology and Sociobiology*, 49, 176–186.

Alonzo, S. & Sinervo, B. (2007) The effect of sexually antagonistic selection on adaptive sex ratio allocation. *Evolutionary Ecology Research*, 9, 1–21.

Alper, S., McElwee, M. K., Apfeld, J., Lackford, B., Freedman, J. H. & Schwartz, D. A. (2010) The *Caenorhabditis elegans* germ line regulates distinct signaling pathways to control lifespan and innate immunity. *Journal of Biological Chemistry*, 285, 1822–1828.

Al-Regaiey, K. A., Masternak, M. M., Bonkowski, M. S., Panici, J. A., Kopchick, J. J. & Bartke, A. (2007) Effects of caloric restriction and growth hormone resistance on insulin-related intermediates in the skeletal muscle. *Journal of Gerontology A*, 62, 18–26.

Amador-Cano, G., Carpizo-Ituarte, E. & Cristino-Jorge, D. (2006) Role of protein kinase C, G-protein coupled receptors, and calcium flux during metamorphosis of the sea urchin *Strongylocentrotus purpuratus*. *Biological Bulletin*, 210, 121–131.

Amador-Noguez, D., Dean, A., Huang, W., Setchell, K., Moore, D. & Darlington, G. (2007) Alterations in xenobiotic metabolism in the long-lived Little mice. *Aging Cell*, 6, 453–470.

Amdam, G. V. & Omholt, S. W. (2003) The hive bee to forager transition in honeybee colonies: the double repressor hypothesis. *Journal of Theoretical Biology*, 223, 451–464.

Amdam, G. V. & Page, R. E. J. (2005) Intergenerational transfers may have decoupled physiological and chronological age in a eusocial insect. *Aging Research Reviews*, 4, 398–408.

Amdam, G. V., Norberg, K., Fondrk, M. K. & Page, R. E. J. (2004) Reproductive ground plan may mediate colony-level selection effects on individual foraging behavior in honey bees. *Proceedings of the National Academy of Sciences of the United States of America*, 101, 11350–11355.

Amdam, G. V., Aase, A. L., Seehuus, S. C., Fondrk, M. K., Norberg, K. & Hartfelder, K. (2005) Social reversal of immunosenescence in honey bee workers. *Experimental Gerontology*, 40, 939–947.

Amdam, G. V., Norberg, K., Page, R. E. J., Erber, J. & Scheiner, R. (2006) Downregulation of vitellogenin gene activity increases the gustatory responsiveness of honey bee workers (*Apis mellifera*). *Behavioural Brain Research*, 169, 201–205.

Amdam, G. V., Nilsen, K. A., Norberg, K., Fondrk, M. K. & Hartfelder, K. (2007) Variation in endocrine signaling underlies variation in social life history. *The American Naturalist*, 170, 37–46.

Amdam, G. V., Ihle, K. E. & Page, R. E. J. (2009a) Regulation of honey bee (*Apis mellifera*) life histories by vitellogenin. In Pfaff, D., Arnold, A., Etgen, A., Fahrbach, S. E. & Rubin, R. (Eds.) *Hormones, Brains and Behavior*. 2 ed. San Diego, CA, Elsevier Academic Press.

Amdam, G. V., Rueppell, O., Fondrk, M. K., Page, R. E. J. & Nelson, C. M. (2009b) The nurse's load: early-life exposure to brood-rearing affects behavior and lifespan in honey bees (*Apis mellifera*). *Experimental Gerontology*, 44, 467–471.

Ament, S. A., Corona, M., Pollock, H. S. & Robinson, G. E. (2008) Insulin signaling is involved in the regulation of worker division of labor in honey bee colonies. *Proceedings of the National Academy of Sciences of the United States of America*, 105, 4226–4231.

Andersen, C. H., Jensen, C. S. & Petersen, K. (2004) Similar genetic switch systems might integrate the floral inductive pathways in dicots and monocots. *Trends in Plant Science*, 9, 105–107.

Anderson, K., Rosner, W., Khan, M., New, M., Pang, S., Wissel, P., & Kappas, A. (1987) Diet-hormone interactions: protein/carbohydrate ratio alters reciprocally the plasma levels of testosterone and cortisol and their respective binding globulins in man. *Life Sciences*, 40, 1761–1768.

Andreelli, F., Hanaire-Broutin, H., Laville, M., Tauber, J. P., Riou, J. P. & Thivolet, C. (2000) Normal reproductive function in leptin-deficient patients with lipoatropic diabetes. *Journal of Clinical Endocrinology and Metabolism* 85, 715–719.

Andrews, Z. B. & Horvath, T. L. (2009) Uncoupling protein-2 regulates lifespan in mice. *American Journal of Physiology. Endocrinology and Metabolism*, 296, E621–E627.

Andries, J. C. (1979) Effect of exogenous JHI on imaginal determination in *Aeshna cyanea*. *Journal of Insect Physiology*, 25, 261–267.

Angelo, G. & Van Gilst, M. R. (2009) Starvation protects germline stem cells and extends reproductive longevity in *C. elegans*. *Science*, 326, 954–958.

Angilletta, M. J., Steury, T. D. & Sears, M. W. (2004) Temperature, growth rate, and body size in ectotherms: Fitting pieces of a life history puzzle. *Integrative and Comparative Biology*, 44, 498–509.

Anstey, M. L., Rogers, S. M., Ott, S. R., Burrows, M. & Simpson, S. J. (2009) Serotonin mediates behavioral gregarization underlying swarm formation in desert locusts. *Science*, 323, 627–630.

Antebi, A., Culotti, J. G. & Hedgecock, E. M.(1998) daf-12 regulates developmental age and the dauer alternative in *C. elegans*. *Development*, 125, 1191–1205.

Antebi, A. (2006) Nuclear hormone receptors in *C. elegans*. In The *C. elegans* Research Community (Ed.), *WormBook*, doi/10.1895/wormbook.1.7.1, http://www.wormbook.org.

Apfeld, J. & Kenyon, C. (1999) Regulation of lifespan by sensory perception in *Caenorhabditis elegans*. *Nature*, 402, 804–809.

Apfeld, J., O'Connor, G., McDonagh, T., Distefano, P. S. & Curtis, R. (2004) The AMP-activated protein kinase AAK-2 links energy levels and insulin-like signals to lifespan in *C. elegans*. *Genes & Development*, 18, 3004–3009.

Apidianakis, Y., Mindrinos, M. N., Xiao, W., Lau, G. W., Baldini, R. L., Davis, R. W. & Rahme, L. G. (2005) Profiling early infection responses: *Pseudomonas aeruginosa* eludes host defenses by suppressing antimicrobial peptide gene expression. *Proceedings of the National Academy of Sciences of the United States of America*, 102, 2573–2578.

Appelmans, N. (1994) Sites of particle selection determined from observations of individual feeding larvae of the sand dollar *Dendraster excentricus*. *Limnology and Oceanography*, 39, 404–411.

Apter, D., Raisanen, I., Ylostalo, P. & Vihko, R. (1987) Follicular growth in relation to serum hormonal patterns in adolescents compared with adult menstrual cycles. *Fertility and Sterility*, 47, 82–88.

Aragona, B. J., Liu, Y., Curtis, J. T., Stephan, F. K. & Wang, Z. (2003) A critical role for nucleus accumbens dopamine in partner-preference formation in male prairie voles. *Journal of Neuroscience* 23, 3483–3490.

Arai, M. N. (1997) *A Functional Biology of Scyphozoa*. London, Chapman and Hall.

Arantes-Oliveira, N., Apfeld, J., Dillin, A. & Kenyon, C. (2002) Regulation of life-span by germ-line stem cells in *Caenorhabditis elegans*. *Science*, 295, 502–505.

Archer, J. (2006) Testosterone and human aggression: an evaluation of the challenge hypothesis. *Neuroscience and Biobehavioral Reviews*, 30, 319–345.

Archer, M. A., Phelan, J. P., Beckman, K. A. & Rose, M. R. (2003) Breakdown in correlations during laboratory evolution. II. Selection on stress resistance in *Drosophila* populations. *Evolution*, 57, 536–543.

Arking, R. & Giroux, C. (2001) Antioxidant genes, hormesis, and demographic longevity. *Journal of Anti-Aging Medicine*, 4, 125–136.

Arking, R. A., Burde, V., Graves, K., Hari, R., Feldmand, E., Zeevi, A., Soliman, S., Araiya, A., S., B., Vettraino, J., Sathrasala, K., Wehr, N. & Levine, R. L. (2000) Forward and reverse selection for longevity in *Drosophila* is characterized by alteration of antioxidant gene expression and oxidative damage pattern. *Experimental Gerontology*, 35, 167–185.

Arlotti, J. P., Cottrell, B. H., Lee, S. H. & Curtin, J. J. (1998) Breastfeeding among low-income women with and without peer support. *Journal of Community Health Nursing*, 15, 163–178.

Arlt, W., Martens, J. W., Song, M., Wang, J. T., Auchus, R. J. & Miller, W. L. (2002) Molecular evolution of adrenarche: structural and functional analysis of p450c17 from four primate species. *Endocrinology*, 143, 4665–4672.

Armstrong, N. & McClay, D. R. (1994) Skeletal pattern is specified autonomously by the primary mesenchyme cells in sea urchin embryos. *Developmental Biology*, 162, 329–338.

Armstrong, N., Hardin, J. & McClay, D. R. (1993) Cell–cell interactions regulate skeleton formation in the sea urchin embryo. *Development*, 119, 833–840.

Arnold, S. J. (1981) Behavioral variation in natual-populations II. The inheritance of feeding response in crosses between geographic races of the garter snake, *Thamnophis elegans*. *Evolution*, 35, 510–515.

Arquier, N., Geminard, C., Bourouis, M., Jarretou, G., Honegger, B., Paix, A. & Leopold, P. (2008) *Drosophila* ALS regulates growth and metabolism through functional interaction with insulin-like peptides *Cell Metabolism*, 7, 333–338.

Arum, O. & Johnson, T. E. (2007) Reduced expression of the *Caenorhabditis elegans* p53 ortholog cep-1 results in increased longevity. *Journal of Gerontology A* 62, 951–959.

Ashizawa, K. & Cheng, S. Y. (1992) Regulation of thyroid hormone receptor-mediated transcription by a cytosol protein. *Proceedings of the National Academy of Sciences of the United States of America*, 89, 9277–9281.

Ashizawa, K., McPhie, P., Lin, K. H. & Cheng, S. Y. (1991) An *in vitro* novel mechanism of regulating the activity of pyruvate kinase M2 by thyroid hormone and fructose 1, 6-bisphosphate. *Biochemistry*, 30, 7105–7111.

Ashok, M., Turner, C. & Wilson, T. G. (1998) Insect juvenile hormone resistance gene homology with the bHLH-PAS family of transcriptional regulators. *Proceedings of the National Academy of Sciences of the United States of America*, 95, 2761–2766.

Atkinson, D. (1994) Temperature and organism size – a biological law for Ectotherms. *Advances in Ecological Research*, 25, 1–58.

Attardo, G. M., Hansen, I. A. & Raikhel, A. S. (2005) Nutritional regulation of vitellogenesis in mosquitoes: Implications for anautogeny. *Insect Biochemistry & Molecular Biology*, 35, 661–675.

Atwell, S., Huang, Y. S., Vilhjalmsson, B. J., Willems, G., Horton, M., Li, Y., Meng, D., Platt, A., Tarone, A. M., Hu, T. T., Jiang, R., Muliyati, N. W., Zhang, X., Amer, M. A., Baxter, I., Brachi, B., Chory, J., Dean, C., Debieu, M., De Meaux, J., Ecker, J. R., Faure, N., Kniskern, J. M., Jones, J. D., Michael, T., Nemri, A., Roux, F., Salt, D. E., Tang, C., Todesco, M., Traw, M. B., Weigel, D., Marjoram, P., Borevitz, J. O., Bergelson, J. & Nordborg, M. (2010) Genome-wide association study of 107 phenotypes in *Arabidopsis thaliana* inbred lines. *Nature*, 465, 627–31.

Aubert, F. & Shine, R. (2009) Genetic assimilation and the postcolonization erosion of phenotypic plasticity in island tiger snakes. *Current Biology*, 19, 1932–1936.

Aukerman, M. J., Hirschfeld, M., Wester, L., Weaver, M., Clack, T., Amasino, R. M. & Sharrock, R. A. (1997) A deletion in the PHYD gene of the *Arabidopsis* Wassilewskija ecotype defines a role for phytochrome D in red/far-red light sensing. *Plant Cell*, 9, 1317–1326.

Austad, S. N. (1989) Life extension by dietary restriction in the bowl and doily spider, *Frontinella pyramitela*. *Experimental Gerontology*, 24, 83–92.

Austad, S. N. (1997) Comparative aging and life histories in mammals. *Experimental Gerontology*, 32, 23–38.

Austad, S. N. (1999) *Why we age: what science is discovering about the body's journey through life*. New York, Wiley.

Austad, S. N. & Finch, C. E. (2008) The evolutionary context of human aging and degenerative disease. In Stearns, S. C. & Koella, J. C. (Eds.) *Evolution in Health and Disease*. Oxford, Oxford University Press.

Austad, S. N. & Fischer, K. E. (1992) Primate longevity: its place in the mammalian scheme. *American Journal of Primatology*, 28, 251–261.

Averof, M. & Patel, N. H. (1997) Crustacean appendage evolution associated with changes in Hox gene expression. *Nature*, 388, 682–686.

Ayre, D. J. (1983) The effects of asexual reproduction and inter-genotypic aggression on the genotypic structure of populations of the sea anemone *Actinia tenebrosa*. *Oecologia*, 57, 158–165.

Ayres, J. S. & Schneider, D. S. (2008) A signaling protease required for melanization in *Drosophila* affects resistance and tolerance of infections. *PLoS Biology*, 6, 2764–2773.

Ayson, F. G. & Lam, T. J. (1993) Thyroxine injection of female rabbitfish (*Siganus guttatus*) broodstock: Changes in thyroid hormone levels in plasma, eggs, and yolk-sac larvae, and its effect on larval growth and survival. *Aquaculture*, 109, 83–93.

Babin, P. J. (1992) Binding of thyroxine and 3,5,3′-triiodothyronine to trout plasma lipoproteins. *American Journal of Physiology. Cell Physiology*, 262, E712–E720.

Badyaev, A. V. & Vleck, C. M. (2007) Context-dependent development of sexual ornamentation: implications for a trade-off between current and future breeding efforts. *Journal of Evolutionary Biology*, 20, 1277–1287.

Bagatell, C. J. & Bremner, W. J. (1990) Sperm counts and reproductive hormones in male marathoners and lean controls. *Fertility and Sterility*, 53, 688–692.

Baker, B. S. & Tata, J. R. (1990) Accumulation of proto-oncogene c-erb-A related transcripts during *Xenopus* development: association with early acquisition of response to thyroid hormone and estrogen. *EMBO Journal*, 9, 879–885.

Baker, B. S. & Tata, J. R. (1992) Prolactin prevents the autoinduction of thyroid hormone receptor mRNAs during amphibian metamorphosis. *Developmental Biology*, 149, 463–467.

Baker, J., Liu, J. P., Robertson, E. J. & Efstratiadis, A. (1993) Role of insulin-like growth factors in embryonic and postnatal growth. *Cell*, 75, 73–82.

Bakker, W. J., Harris, I. S. & Mak, T. (2007) FOXO3a is activated in response to hypoxic stress and inhibits HIF1-induced apoptosis via regulation of CITED2. *Molecular Cell*, 28, 941–953.

Balaban, R. S., Nemoto, S. & Finkel, T. (2005) Mitochondria, oxidants, and aging. *Cell*, 120, 483–495.

Balasubramanian, S., Sureshkumar, S., Agrawal, M., Michael, T. P., Wessinger, C., Maloof, J. N., Clark, R., Warthmann, N., Chory, J. & Weigel, D. (2006) The *PHYTOCHROME C* photoreceptor gene mediates natural variation in flowering and growth responses of *Arabidopsis thaliana*. *Nature Genetics*, 38, 711–715.

Baldal, E. A., Baktawar, W., Brakefield, P. M. & Zwaan, B. J. (2006) Methuselah life history in a variety of conditions, implications for the use of mutants in longevity research. *Experimental Gerontology*, 41, 1126–1135.

Bales, K. L., Kim, A. J., Lewis-Reese, A. D. & Carter, C. S. (2004) Both oxytocin and vasopressin may influence alloparental behavior in male prairie voles. *Hormones and Behavior*, 45, 354–361.

Ballard, P. L. (1979) Glucocorticoids and differentiation. In Baxter, J. D. & Rousseau, G. G. (Eds.) *Glucocorticoid Hormone Action*. New York, Springer.

Balon, E. K. (1999) Alternative ways to become a juvenile or a definitive phenotype (and on some persisting linguistic offenses). *Environmental Biology of Fishes*, 56, 17–38.

Balthazart, J., Baillien, M., Charlier, T. D., Cornil, C. A. & Ball, G. F. (2003) Multiple mechanisms control brain aromatase activity at the genomic and non-genomic level. *Journal of Steroid Biochemistry & Molecular Biology*, 86, 367–379.

Barbieri, M., Bonafe, M., Franceschi, C. & Paolisso, G. (2003) Insulin/IGF-1-signaling pathway: an evolutionarily conserved mechanism of longevity from yeast to humans. *American Journal of Physiology – Endocrinology and Metabolism*, 285, E1064–E1071.

Barchuk, A. R., Cristino, A. S., Kucharski, R., Costa, L. F., Simoes, Z. L. & Maleszka, R. (2007) Molecular determinants of caste differentiation in the highly eusocial honeybee *Apis mellifera*. *BMC Developmental Biology*, 7, 70.

Barger, J. L., Kayo, T., Vann, J. M., Arias, E. B., Wang, J., Hacker, T. A., Wang, Y., Raederstorff, D., Morrow, J. D., Leeuwenburgh, C., Allison, D. B., Saupe, K. W., Cartee, G. D., Weindruch, R. & Prolla, T. A. (2008) A low dose of dietary resveratrol partially mimics caloric restriction and retards aging parameters in mice. *PLoS One*, 3, e2264.

Bargmann, C.I. (2006) Chemosensation in *C. elegans*. In The *C. elegans* Research Community (Ed.), *WormBook*, doi/10.1895/wormbook.1.7.1, http://www.wormbook.org.

Barja, G. (2002) Endogenous oxidative stress: relationship to aging, longevity and caloric restriction. *Aging Research Reviews*, 1, 397–411.

Barja, G. (2004) Aging in vertebrates, and the effect of caloric restriction: a mitochondrial free radical production-DNA damage mechanism? *Biological Reviews of the Cambridge Philosophical Society*, 79, 235–251.

Barnes, A. I. & Partridge, L. (2003) Costing reproduction. *Animal Behaviour*, 66, 199–204.

Barnes, A. I., Boone, J. M., Jacobson, J., Partridge, L. & Chapman, T. (2006) No extension of lifespan by ablation of germ line in *Drosophila*. *Proceedings of the Royal Society of London B* 273, 939–947.

Barnes, A. I., Wigby, S., Boone, J. M., Partridge, L. & Chapman, T. (2008) Feeding, fecundity and lifespan in female *Drosophila melanogaster*. *Proceedings of the Royal Society of London B*, 275, 1675–1683.

Barrett, R. D. H., Rogers, S. M. & Schluter, D. (2008) Natural selection on a major armor gene in threespine stickleback. *Science*, 322, 255–257.

Barron, A. B., Maleszka, R., Vander Meer, R. K. & Robinson, G. E. (2007) Octopamine modulates honey bee dance behavior. *Proceedings of the National Academy of Sciences of the United States of America*, 104, 1703–1707.

Bartels, A. & Zeki, S. (2004) The neural correlates of maternal and romantic love. NeuroImage *NeuroImage*, 21, 1155–1166.

Bartke, A. (2008a) Impact of reduced insulin-like growth factor-1/insulin signaling on aging in mammals: novel findings. *Aging Cell*, 7, 285–290.

Bartke, A. (2008b) New findings in gene knockout, mutant and transgenic mice. *Experimental Gerontology*, 43, 11–14.

Bartke, A. (2008c) Insulin and aging. *Cell Cycle*, 7, 3338–3343.

Bartke, A., Wright, J. C., Mattison, J. A., Ingram, D. K., Miller, R. A. & Roth, G. S. (2001) Extending the lifespan of long-lived mice. *Nature*, 414, 412.

Bartke, A., Bonkowski, M. & Masternak, M. (2008) How diet interacts with longevity genes. *Hormones*, 7, 17–23.

Barzilai, N. & Bartke, A. (2009) Biological approaches to mechanistically understand the healthy lifespan extension achieved by calorie restriction and modulation of hormones. *Journal of Gerontology A* 64, 187–191.

Bass, T. M., Weinkove, D., Houthoofd, K., Gems, D. & Partridge, L. (2007) Effects of resveratrol on lifespan in *Drosophila melanogaster* and *Caenorhabditis elegans*. *Mechanisms of Ageing and Development*, 128, 546–552.

Bauer, J. H. & Helfand, S. L. (2006) New tricks of an old molecule: lifespan regulation by p53. *Aging Cell*, 5, 437–440.

Bauer, J. H., Poon, P. C., Glatt-Deeley, H., Abrams, J. M. & Helfand, S. L. (2005) Neuronal expression of p53 dominant-negative proteins in adult *Drosophila melanogaster* extends life span. *Current Biology*, 15, 2063–2068.

Bauer, J. H., Chang, C., Morris, S. N., Hozier, S., Andersen, S., Waitzman, J. S. & Helfand, S. L. (2007) Expression of dominant-negative Dmp53 in the adult fly brain inhibits insulin signaling. *Proceedings of the National Academy of Sciences of the United States of America*, 104, 13355–13360.

Bauer, J. H., Morris, S. N., Chang, C., Flatt, T., Wood, J. G. & Helfand, S. L. (2009) dSir2 and Dmp53 interact to mediate aspects of CR-dependent, lifespan extension in *D. melanogaster*. *Aging*, 1, 38–48.

Bauer, J. H., Chang, C., Bae, G., Morris, S. N., & Helfand, S. L. (2010). Dominant-negative Dmp53 extends life span through the dTOR pathway in *D. melanogaster*. *Mechanisms of Ageing and Development*, 131, 193–201.

Baumeister, R. F. (2005) *The Cultural Animal: Human Nature, Meaning, and Social Life*. New York, Oxford University Press.

Baur, J. A. & Sinclair, D. A. (2008) What is xenohormesis? *American Journal of Pharmacology and Toxicology*, 3, 152–159.

Baur, J. A., Pearson, K. J., Price, N. L., Jamieson, H. A., Lerin, C., Kalra, A., Prabhu, V. V., Allard, J. S., Lopez-Lluch, G., Lewis, K., Pistell, P. J., Poosala, S., Becker, K. G., Boss, O., Gwinn, D., Wang, M., Ramaswamy, S., Fishbein, K. W., Spencer, R. G., Lakatta, E. G., Le Couteur, D., Shaw, R. J., Navas, P., Puigserver, P., Ingram, D. K., De Cabo, R. & Sinclair, D. A. (2006) Resveratrol improves health and survival of mice on a high-calorie diet. *Nature*, 444, 337–342.

Baxter, G. & Morse, D. E. (1987) G-Protein and diacylglycerol regulate metamorphosis of planktonic Molluscan larvae. *Proceedings of the National Academy of Sciences of the United States of America*, 84, 1867–1870.

Bayer, C. A., Holley, B., and Fristrom, J.W. (1996) A switch in broad-complex Zinc-finger isoform expression is regulated post-transcriptionally during the metamorphosis of *Drosophila* imaginal discs. *Developmental Biology*, 177(1), 1–14.

Beadle, G., Tatum, E. & Clancy, C. (1938) Food level in relation to rate of development and eye pigmentation in *Drosophila melanogaster*. *Biological Bulletin*, 75, 447–462.

Beavis, W. D. (1994) The power and deceit of QTL experiments: lessons from comparative QTL studies. In Wilkison D. B. (Ed.) *49th Annual Corn and Sorghum Industry Research Conference*. American Seed Trade Association, Chicago.

Becker, K. B., Stephens, K. C., Davey, J. C., Schneider, M. J. & Galton, V. A. (1997) The type 2 and type 3 iodothyronine deiodinases play important roles in coordinating development in *Rana catesbeiana* tadpoles. *Endocrinology*, 138, 2989–2997.

Becker, T., Loch, G., Beyer, M., Zinke, I., Aschenbrenner, A. C., Carrera, P., Inhester, T., Schultze, J. L. & Hoch, M. (2010) FOXO-dependent regulation of innate immune homeostasis. *Nature*, 463, 369–373.

Behrends, A., Scheiner, R., Baker, N. & Amdam, G. V. (2007) Cognitive aging is linked to social role in honey bees (*Apis mellifera*). *Experimental Gerontology*, 42, 1146–1153.

Bell, G. (1985) The origin and early evolution of germ cells as illustrated by the Volvocales. In Halvorson, H. O. & Monroy A. (Eds.) *The Origin and Evolution of Sex*. Alan R. Liss, New York, pp. 221–256.

Bell, G. & Koufopanou, V. (1986) The cost of reproduction. In Dawkins, R. & Ridley, M. (Eds.) *Oxford Surveys in Evolutionary Biology*. Oxford, Oxford University Press.

Bell, R., Hubbard, A., Chettier, R., Chen, D., Miller, J. P., Kapahi, P., Tarnopolsky, M., Sahasrabuhde, S., Melov, S. & Hughes, R. E. (2009) A human protein interaction network shows conservation of aging processes between human and invertebrate species. *PLoS Genetics*, 5, e1000414.

Bellen, H. J., Vaessin, H., Bier, E., Kolodkin, A., Develyn, D., Kooyer, S. & Jan, Y. N. (1992) The *Drosophila couch potato* gene: an essential gene required for normal adult behavior. *Genetics*, 131, 365–375.

Belles, X. (2010) Beyond *Drosophila*: RNAi *in vivo* and functional genomics in insects. *Annual Review of Entomology*, 55, 111–128.

Bellizzi, D., Rose, G., Cavalcante, P., Covello, G., Dato, S., De Rango, F., Greco, V., Maggiolini, M., Feraco, E., Mari, V., Franceschi, C., Passarino, G. & De Benedictis, G. (2005) A novel VNTR enhancer within the SIRT3 gene, a human homologue of SIR2, is associated with survival at oldest ages. *Genomics*, 85, 258–263.

Belsky, J. (1997) Attachment, mating, and parenting: An evolutionary interpretation. *Human Nature*, 8, 361–381.

Belsky, J. (2005) Differential susceptibility to rearing influence: An evolutionary hypothesis and some evidence. In Ellis, B. J. & Bjorklund, D. F. (Eds.) *Origins of the Social Mind: Evolutionary Psychology and Child Development*. New York, Guilford Press.

Bely, A. E. & Nyberg, K. G. (2010) Evolution of animal regeneration: re-emergence of a field. *Trends in Ecology & Evolution*, 25, 161–170.

Belyaeva, E. S., Aizenzon, M. G., Semeshin, V. F., Kiss, I. I., Koczka, K., Baritcheva, E. M., Gorelova, T. D. and Zhimulev, I. F. (1980) Cytogenic analysis of the 2B3-4-2B11 region of the X chromosome of *Drosophila melanogaster*. I. Cytology of the region and mutant complementation groups. *Chromosoma*, 81, 281–306.

Bensaad, K., Tsuruta, A., Selak, M. A., Vidal, M. N., Nakano, K., Bartrons, R., Gottlieb, E. & Vousden, K. H.

(2006) TIGAR, a p53-inducible regulator of glycolysis and apoptosis. *Cell*, 126, 107–120.

Ben-Shahar, Y., Robichon, A., Sokolowski, M. B. & Robinson, G. E. (2002) Influence of gene action across different time scales on behavior. *Science*, 296, 741–744.

Ben-Shahar, Y., Leung, H. T., Pak, W. L., Sokolowski, M. B. & Robinson, G. E. (2003) cGMP-dependent changes in phototaxis: a possible role for the foraging gene in honey bee division of labor. *Journal of Experimental Biology*, 206, 2507–2515.

Ben-Shahar, Y., Dudek, N. L. & Robinson, G. E. (2004) Phenotypic deconstruction reveals involvement of manganese transporter malvolio in honey bee division of labor. *Journal of Experimental Biology*, 207, 3281–3288.

Bentley, G. R., Harrigan, A. M., Campbell, B. & Ellison, P. T. (1993) Seasonal effects on salivary testosterone levels among Lese males of the Ituri Forest, Zaire. *American Journal of Human Biology*, 5, 711–717.

Berdichevsky, A., Viswanathan, M., Horvitz, H. R. & Guarente, L. (2006) *C. elegans* SIR-2.1 interacts with 14-3-3 proteins to activate DAF-16 and extend life span. *Cell*, 125, 1165–1177.

Berenos, C., Schmid-Hempel, P. & Wegner, K. M. (2009) Evolution of host resistance and trade-offs between virulence and transmission potential in an obligately killing parasite. *Journal of Evolutionary Biology*, 22, 2049–2056.

Berg, B. N. & Simms, H. S. (1960) Nutrition and longevity in the rat. II. Longevity and onset of disease with different levels of food intake. *Journal of Nutrition*, 71, 255–263.

Berger, D., Walters, R. & Gotthard, K. (2008) What limits insect fecundity? Body size- and temperature-dependent egg maturation and oviposition in a butterfly. *Functional Ecology*, 22, 523–529.

Bergland, A. O., Agotsch, M., Mathias, D., Bradshaw, W. E. & Holzapfel, C. M. (2005) Factors influencing the seasonal life history of the pitcher-plant mosquito, *Wyeomyia smithii*. *Ecological Entomology*, 30, 129–137.

Bergland, A. O., Genissel, A., Nuzhdin, S. V. & Tatar, M. (2008) Quantitative trait loci affecting phenotypic plasticity and the allometric relationship of ovariole number and thorax length in *Drosophila melanogaster*. *Genetics*, 180, 567–582.

Bergot, B. J., Baker, F. C., Cerf, D. C., Jamieson, G. & Schooley, D. A. (1981) Qualitative and quantitative aspects of juvenile hormone titers in developing embryos of several insect species: discovery of a new Jh-like substance extracted from eggs of *Manduca sexta*. In Pratt, G. E. & Brooks, G. T. (Eds.) *Juvenile Hormone Biochemistry*. Amsterdam, Elsevier.

Berking, S., Czech, N., Gerharz, M., Herrmann, K., Hoffmann, U., Raifer, H., Sekul, G., Siefker, B., Sommerei,

A. & Vedder, F. (2005) A newly discovered oxidant defence system and its involvement in the development of *Aurelia aurita* (Scyphozoa, Cnidaria): reactive oxygen species and elemental iodine control medusa formation. *International Journal of Developmental Biology*, 49, 969–976.

Berman, J. R. & Kenyon, C. (2006) Germ-cell loss extends *C. elegans* lifespan through regulation of DAF-16 by *kri-1* and lipophilic-hormone signaling. *Cell*, 124, 1055–1068.

Berreur, P., Porcheron, P., Berreur-Bonnenfant, J. & Simpson, P. (1979) Ecdysteroid levels and pupariation in *Drosophila melanogaster*. *Journal of Experimental Zoology*, 210, 347–352.

Bertram, D. F. & Strathmann, R. R. (1998) Effects of larval and maternal nutrition on growth and form of planktotrophic larvae. *Ecology*, 79, 315–327.

Bertram, D. F., Phillips, N. E. & Strathmann, R. R. (2009) Evolutionary and experimental change in egg volume, heterochrony of larval body and juvenile rudiment, and evolutionary reversibility in pluteus form. *Evolution & Development*, 11, 728–739.

Bertrand, J. F. & Woollacott, R. M. (2003) G protein-linked receptors and induction of metamorphosis in *Bugula stolonifera* (Bryozoa). *Invertebrate Biology*, 122, 380–385.

Berwaerts, K., Matthysen, E. & Van Dyck, H. (2008) Take-off flight performance in the butterfly *Pararge aegeria* relative to sex and morphology: a quantitative genetic assessment. *Evolution*, 62, 2525–2533.

Bethke, A., Fielenbach, N., Wang, Z., Mangelsdorf, D. J. & Antebi, A. (2009) Nuclear hormone receptor regulation of microRNAs controls developmental progression. *Science*, 324, 95–98.

Bhaskaran, G., Sparagana, S.P., Barrera, P. and Dahm, K.H. (1986) Change in *corpus allatum* function during metamorphosis of the tobacco hornworm, *Manduca sexta*. *Archives of Insect Biochemistry and Physiology* 3, 321–338.

Bianco, A. C., Salvatore, D., Gereben, B., Berry, M. J. & Larsen, P. R. (2002) Biochemistry, cellular and molecular biology, and physiological roles of the iodothyronine selenodeiodinases. *Endocrine Reviews*, 23, 38–89.

Bickler, P. E. & Buck, L. T. (2007) Hypoxia tolerance in reptiles, amphibians, and fishes: Life with variable oxygen availability. *Annual Review of Physiology*, 69, 145–170.

Biggers, W. J. & Laufer, H. (1999) Settlement and metamorphosis of Capitella larvae induced by juvenile hormone-active compounds is mediated by protein kinase C and ion channels. *Biological Bulletin*, 196, 187–198.

Bimbaum, K. D. & Sanchez Alvarado, A. (2008) Slicing across kingdoms: Regeneration in plants and animals. *Cell*, 132, 697–710.

Birnby, D. A., Link, E. M., Vowels, J. J., Tian, H., Colacurcio, P. L. & Thomas, J. H. (2000) A transmembrane guanylyl cyclase (DAF-11) and Hsp90 (DAF-21) regulate a common set of chemosensory behaviors in *C. elegans*. *Genetics*, 155, 85–104.

Bisgrove, B. W. & Burke, R. D. (1986) Development of Serotonergic neurons in embryos of the sea urchin, *Strongylocentrotus purpuratus*. *Development, Growth & Differentiation*, 28, 569–574.

Bisgrove, B. W. & Burke, R. D. (1987) Development of the nervous system of the pluteus larva of *Strongylocentrotus droebachiensis*. *Cell and Tissue Research*, 248, 335–343.

Bishop, N. A. & Guarente, L. (2007a) Genetic links between diet and lifespan: shared mechanisms from yeast to humans. *Nature Reviews Genetics*, 8, 835–844.

Bishop, N. A. & Guarente, L. (2007b) Two neurons mediate diet-restriction-induced longevity in *C. elegans*. *Nature*, 447, 545–549.

Bishop, C. D. & Hall, B. K. (2009) Sniffing out new data and hypotheses on the form, function, and evolution of the Echinopluteus post-oral vibratile lobe. *Biological Bulletin*, 216, 307–321.

Bishop, C. D., Erezyilmaz, D. F., Flatt, T., Georgiou, C. D., Hadfield, M. G., Heyland, A., Hodin, J., Jacobs, M. W., Maslakova, S. A., Pires, A., Reitzel, A. M., Santagata, S., Tanaka, K. & Youson, J. H. (2006a) What is metamorphosis? *Integrative and Comparative Biology*, 46, 655–661.

Bishop, C. D., Huggett, M., Heyland, A., Hodin, J. & Brandhorst, B. P. (2006b) Interspecific variation in metamorphic competence in marine invertebrates: the significance for comparative investigations of regulatory systems. *Integrative and Comparative Biology*, 46, 662–682.

Bjedov, I., Toivonen, J. M., Kerr, F., Slack, C., Jacobson, J., Foley, A. & Partridge, L. (2010) Mechanisms of lifespan extension by rapamycin in the fruit fly *Drosophila melanogaster*. *Cell Metabolism*, 11, 35–46.

Bjorklund, D. F. & Pellegrini, A. D. (2002) *The Origins of Human Nature: Evolutionary developmental psychology*. Washington, APA Press.

Black, R. E. & Bloom, L. (1984) Heat shock proteins in *Aurelia* (Cnidaria, Scyphozoa). *Journal of Experimental Zoology*, 230, 303–307.

Blackstone, N. W. (1999) Redox control in development and evolution: Evidence from colonial hydroids. *Journal of Experimental Biology*, 202, 3541–3553.

Blackstone, N. W. (2001) Redox state, reactive oxygen species and adaptive growth in colonial hydroids. *Journal of Experimental Biology*, 204, 1845–1853.

Blackstone, N. W. (2003) Redox signaling in the growth and development of colonial hydroids. *Journal of Experimental Biology*, 206, 651–658.

Blackstone, N. W. (2006) Multicellular redox regulation: integrating organismal biology and redox chemistry. *BioEssays*, 28, 72–77.

Blackstone, N. W. (2008) Metabolic gradients: A new system for old questions. *Current Biology*, 18, R351–R353.

Blackstone, N. W. & Bridge, D. M. (2005) Model systems for environmental signaling. *Integrative and Comparative Biology*, 45, 605–614.

Blackstone, N. W. & Jasker, B. D. (2003) Phylogenetic considerations of clonality, coloniality, and mode of germline development in animals. *Journal of Experimental Zoology B* 297B, 35–47.

Blackstone, N. W., Cherry, K. S. & Glockling, S. L. (2004a) Structure and signaling in polyps of a colonial hydroid. *Invertebrate Biology*, 123, 43–53.

Blackstone, N. W., Cherry, K. S. & Van Winkle, D. H. (2004b) The role of polyp-stolon junctions in the redox signaling of colonial hydroids, *Hydrobiologia*, 530, 291–298 (8).

Blakemore, S.-J., Winston, J. & Frith, U. (2004) Social cognitive neuroscience: Where are we heading? *Trends in Cognitive Neurosciences*, 8, 216–222.

Blanton, M. L. & Specker, J. L. (2007) The hypothalamic–pituitary–thyroid (HPT) axis in fish and its role in fish development and reproduction. *Critical Reviews in Toxicology*, 37, 97–115.

Blount, Z. D., Borland, C. Z. & Lenski, R. E. (2008) Historical contingency and the evolution of a key innovation in an experimental population of *Escherichia coli*. *Proceedings of the National Academy of Sciences of the United States of America*, 105, 7899–7906.

Bluher, M., Kahn, B. B. & Kahn, C. R. (2003) Extended longevity in mice lacking the insulin receptor in adipose tissue. *Science*, 299, 572–574.

Bochdanovits, Z. & De Jong, G. (2003) Experimental evolution in *Drosophila melanogaster*: interaction of temperature and food quality selection regimes. *Evolution*, 57, 1829–1836.

Bochdanovits, Z. & De Jong, G. (2004) Antagonistic pleiotropy for life history traits at the gene expression level. *Proceedings of the Royal Society of London B*, 271, S75–S78.

Bode, H. R. (2003) Head regeneration in *Hydra*. *Developmental Dynamics*, 226, 225–236.

Bode, H. R. (2009) Axial patterning in *Hydra*. *Cold Spring Harbor Perspectives in Biology*, 1, a000463.

Bodkin, N. L., Alexander, T. M., Ortmeyer, H. K., Johnson, E. & Hansen, B. C. (2003) Mortality and morbidity in laboratory-maintained Rhesus monkeys and effects of long-term dietary restriction. *Journal of Gerontology A*, 58, 212–219.

Boesch, C. & H. Boesch-Achermann (2000) *The Chimpanzees of the Taï Forest: Behavioural ecology and evolution*. Oxford, Oxford University Press.

Boggs, C. L. (2009) Understanding insect life histories and senescence through a resource allocation lens. *Functional Ecology*, 23, 27–37.

Bogin, B. (1994) Adolescence in evolutionary perspective. *Acta Paediatrica*, 406, 29–35.

Bogin, B. (1999) Evolutionary perspective on human growth. *Annual Review of Anthropology*, 28, 109–153.

Boidron-Metairon, I. F. (1988) Morphological plasticity in laboratory-reared echinoplutei of *Dendraster excentricus* (Eschscholtz) and *Lytechinus variegatus* (Lamarck) in response to food conditions. *Journal of Experimental Marine Biology and Ecology*, 119, 31–41.

Boily, G., Seifert, E. L., Bevilacqua, L., He, X. H., Sabourin, G., Estey, C., Moffat, C., Crawford, S., Saliba, S., Jardine, K., Xuan, J., Evans, M., Harper, M. E. & McBurney, M. W. (2008) SirT1 regulates energy metabolism and response to caloric restriction in mice. *PLoS One*, 3, e1759.

Bókony, V., Lendvai, Á. Z., Liker, A., Angelier, F., Wingfield, J. C. & Chastel, O. (2009) Stress response and the value of reproduction: are birds prudent parents? *The American Naturalist*, 173, 589–598.

Bole-Feysot, C., Fgroffin, V., Edery, M., Binart, N. & Kelley, P. A. (1998) Prolactin (PRL) and its receptor: Actions, signal transduction pathways and phenotypes observed in PRL receptor knockout mice. *Endocrine Reviews*, 3, 225–268.

Bolker, J. A. (2000) Modularity in development and why it matters to evo-devo. *American Zoologist*, 40, 770–776.

Bollback, J. P. & Huelsenbeck, P. (2009) Parallel genetic evolution within and between bacteriophage species of varying degrees of divergence. *Genetics*, 181, 225–234.

Bolton, J. P., Collie, N. L., Kawauchi, H. & Hirano, T. (1987) Osmoregulatory actions of growth hormone in rainbow trout (*Salmo gairdneri*). *Journal of Endocrinology*, 112, 63–68.

Bonafe, M. & Olivieri, F. (2009) Genetic polymorphism in long-lived people: cues for the presence of an insulin/IGF-pathway-dependent network affecting human longevity. *Molecular and Cellular Endocrinology*, 299, 118–123.

Bonafe, M., Barbieri, M., Marchegiani, F., Olivieri, F., Ragno, E., Giampieri, C., Mugianesi, E., Centurelli, M., Franceschi, C. & Paolisso, G. (2003) Polymorphic variants of insulin-like growth factor I (IGF-1) receptor and phosphoinositide 3-kinase genes affect IGF-1 plasma levels and human longevity: cues for an evolutionarily conserved mechanism of lifespan control. *Journal of Clinical Endocrinology and Metabolism*, 88, 3299–3304.

Bonett, R. M. & Chippindale, P. T. (2004) Speciation, phylogeography and evolution of life history and morphology in plethodontid salamanders of the *Eurycea multiplicata* complex. *Molecular Ecology*, 13, 1189–1203.

Bonkowski, M. S., Rocha, J. S., Masternak, M. M., Al Regaiey, K. A. & Bartke, A. (2006) Targeted disruption of growth hormone receptor interferes with the beneficial actions of calorie restriction. *Proceedings of the National Academy of Sciences of the United States of America*, 103, 7901–7905.

Bonsall, M. B. & Mangel, M. (2009) Density dependence, lifespan and the evolutionary dynamics of longevity. *Theoretical Population Biology*, 75, 46–55.

Booth, A. G., Shelley, A., Mazur, G., Tharp, G. & Kittock, R. (1989) Testosterone and winning and losing in human competition. *Hormones and Behavior* 23, 556–571.

Boots, M. (2008) Fight or learn to live with the consequences? *Trends in Ecology & Evolution*, 23, 248–250.

Boraas, M. E., Seale, D. B. & Boxhorn, J. E. (1998) Phagotrophy by a flagellate selects for colonial prey: A possible origin of multicellularity. *Evolutionary Ecology*, 12, 153–164.

Bordone, L., Motta, M. C., Picard, F., Robinson, A., Jhala, U. S., Apfeld, J., McDonagh, T., Lemieux, M., McBurney, M., Szilvasi, A., Easlon, E. J., Lin, S. J. & Guarente, L. (2006) Sirt1 regulates insulin secretion by repressing UCP2 in pancreatic beta cells. *PLoS Biology*, 4, e31.

Bordone, L., Cohen, D., Robinson, A., Motta, M. C., Van Veen, E., Czopik, A., Steele, A. D., Crowe, H., Marmor, S., Luo, J., Gu, W. & Guarente, L. (2007) SIRT1 transgenic mice show phenotypes resembling calorie restriction. *Aging Cell*, 6, 759–767.

Bosch, T. C. G. (2003) Ancient signals: peptides and the interpretation of positional information in ancestral metazoans. *Comparative Biochemistry and Physiology B*, 136, 185–196.

Bosch, T. C. G., Krylow, S. M., Bode, H. R. & Steele, R. E. (1988) Thermotolerance and synthesis of heat shock proteins: these responses are present in *Hydra attenuata* and Absent in *Hydra oligactis*. *Proceedings of the National Academy of Sciences of the United States of America*, 85, 7927–7931.

Boulétreau-Merle, J., Allemand, R., Cohet, Y. & David, J. R. (1982) Reproductive strategy in *Drosophila melanogaster* – significance of a genetic-divergence between temperate and tropical populations. *Oecologia*, 53, 323–329.

Bourke, A. F. G. & Franks, N. F. (1995) *Social Evolution in Ants*. Princeton, NJ, Princeton University Press.

Bower, F. O. 1908. The Origin of a Land Flora. MacMillan & Co, London.

Bownes, M. (1982) NJ Hormonal and genetic regulation of vitellogenesis in *Drosophila*. *Quarterly Review of Biology*, 57, 247–274.

Bownes, M., Scott, A. & Shirras, A. (1988) Dietary components modulate yolk protein gene transcription in *Drosophila melanogaster*. *Development*, 103, 119–128.

Bownes, M., Ronaldson, E. & Mauchline, D. (1996) 20-hydroxyecdysone, but not juvenile hormone, regulation of yolk protein gene expression can be mapped to cis-acting DNA sequences. *Developmental Biology*, 173, 475–489.

Braby, M. F. (2002) Life history strategies and habitat templets of tropical butterflies in north-eastern Australia. *Evolutionary Ecology*, 16, 399–413.

Braby, M. F. & Jones, R. E. (1994) Effect of temperature and hostplants on survival, development and body-size in 3 tropical satyrine butterflies from North-Eastern Australia. *Australian Journal of Zoology*, 42, 195–213.

Bradshaw, W. E. (1976) Geography of photoperiodic response in a diapausing mosquito. *Nature*, 262, 384–386.

Bradshaw, W. E., Holzapfel, C. M. & Mathias, D. (2006) Circadian rhythmicity and photoperioidism in the pitcher-plant mosquito: can the seasonal timer evolve independently of the circadian clock? *The American Naturalist* 167, 601–605.

Brakefield, P. M. (2005) Bringing Evo Devo to Life. PLoS Biology, 3, e340.

Brakefield, P. M. & Frankino, W. A. (2009) Polyphenisms in Lepidoptera: Multidisciplinary approaches to studies of evolution. In Ananthakrishnan, T. N. & Whitman, D. W. (Eds.) *Phenotypic Plasticity in Insects: Mechanisms and Consequences*. Plymouth, Science Publishers, Inc.

Brakefield, P. M. & Larsen, T. B. (1984) The evolutionary significance of dry and wet season forms in some tropical butterflies. *Biological Journal of the Linnean Society*, 22, 1–12.

Brakefield, P. M. & Reitsma, N. (1991) Phenotypic plasticity, seasonal climate and the population biology of *Bicyclus* butterflies (Satyridae) in Malawi. *Ecological Entomology*, 16, 291–303.

Brakefield, P. M., Gates, J., Keys, D., Kesbeke, F., Wijngaarden, P. J. & Al, E. (1996) Development, plasticity and evolution of butterfly eyespot patterns. *Nature*, 384, 236–242.

Brakefield, P. M., Kesbeke, F. & Koch, P. B. (1998) The regulation of phenotypic plasticity of eyespots in the butterfly *Bicyclus anynana*. *The American Naturalist*, 152, 853–860.

Brakefield, P. M., Gems, D., Cowen, T., Christensen, K., Grubeck-Loebenstein, B. & Al, E. (2005) What are the effects of maternal and pre-adult environments on aging in humans, and are there lessons from animal models?. *Mechanisms of Ageing and Development*, 126, 431–438.

Brakefield, P. M., Pijpe, J. & Zwaan, B. J. (2007) Developmental plasticity and acclimation both contribute to adaptive responses to alternating seasons of plenty and of stress in *Bicyclus* butterflies. *Journal of Biosciences*, 32, 465–475.

Brakefield, P. M., Beldade, P. & Zwaan, B. J. (2009) The African butterfly *Bicyclus anynana*: evolutionary genetics and evo-devo. In Behringer, R. R., Johnson, A. D. &

Krumlauf, R. E. (Eds.) *Emerging Model Organisms: A Laboratory Manual*. New York, Cold Spring Harbor Laboratory Press.

Brand, M. D. (2000) Uncoupling to survive? The role of mitochondrial inefficiency in aging. *Experimental Gerontology*, 35, 811–820.

Brand, A. & Hawdon, J. M. (2004) Phosphoinositide-3-Oh-kinase inhibitor LY294002 prevents activation of *Ancylostoma caninum* and *Ancylostoma ceylanicum* third-stage infective larvae. *International Journal for Parasitology*, 34, 909–914.

Brand, M. D., Couture, P., Else, P. L., Withers, K. W. & Hulbert, A. J. (1991) Evolution of energy metabolism: Proton permeability of the inner membrane of the liver mitochondria is greater in mammal than in reptile. *Biochemistry Journal*, 275, 81–86.

Brandt, S. M., Dionne, M. S., Khush, R. S., Pham, L. N., Vigdal, T. J. & Schneider, D. S. (2004) Secreted bacterial effectors and host-produced Eiger/TNF drive death in a *Salmonella*-infected fruit fly. *PLoS Biology*, 2, e418.

Bremner, W. J., Vitiello, M. V. & Prinz, P. N. (1983) Loss of circadian rhythmicity in blood testosterone levels with aging in normal men. *Journal of Clinical Endocrinology and Metabolism*, 56, 1278–1281.

Brennan, C. A. & Anderson, K. V. (2004) *Drosophila*: the genetics of innate immune recognition and response. *Annual Review of Immunology*, 22, 457–483.

Brent, G. A. (2000) Tissue-specific actions of thyroid hormone: insights from animal models. *Reviews in Endocrine & Metabolic Disorders*, 1, 27–33.

Bretman, A., Fricke, C. & Chapman, T. (2009) Plastic responses of male *Drosophila melanogaster* to the level of sperm competition increase male reproductive fitness. *Proceedings of the Royal Society of London B*, 276, 1705–1711.

Breuker, C. J. & Brakefield, P. M. (2002) Female choice depends on size but not symmetry of dorsal eyespots in the butterfly *Bicyclus anynana*. *Proceedings of the Royal Society of London B*, 269, 1233–1239.

Breuner, C. W. & Orchinik, M. (2002) Plasma binding proteins as mediators of corticosteroid action in vertebrates. *Journal of Endocrinology*, 175, 99–112.

Breuner, C. W., Lynn, S. E., Julian, G. E., Cornelius, J. M., Heidinger, B. J., Love, O. P., Sprague, R. S., Wade, H. & Whitman, B. A. (2006) Plasma-binding globulins and acute stress response. *Hormone and Metabolic Research*, 38, 260–268.

Bribiescas, R. G. (1996) Testosterone levels among Aché hunter/gatherer men: a function interpretation of population variation among adult males. *Human Nature*, 7, 163–188.

Bribiescas, R. G. (2001) Reproductive ecology and life history of the human male. *Yearbook of Physical Anthropology*, 44, 148–176.

Bribiescas, R. G. (2006) On the evolution, life history, and proximate mechanisms of human male reproductive senescence. *Evolutionary Anthropology*, 15, 132–141.

Bribiescas, R. G. & Ellison, P. T. (2008) How hormones mediate trade-offs in human health and disease. In Stearns, S. C. & Koella, J. (Eds.) *Evolution in Health and Disease*, 2nd edn. New York, Oxford University Press.

Bridge, D., Cunningham, C. W., Schierwater, B., Desalle, R. & Buss, L. W. (1992) Class-level relationships in the phylum Cnidaria: evidence from mitochondrial genome structure. *Proceedings of the National Academy of Sciences of the United States of America*, 89, 8750–8753.

Bridges, R. S. (2008) *Neurobiology of the Parental Brain*. Maryland Heights, Academic Press.

Britton, J. R., Britton, H. L. & Gronwaldt, V. (2006) Breastfeeding, sensitivity, and attachment. *Pediatrics*, 118, 1436–1443.

Brogiolo, W., Stocker, H., Ikeya, T., Rintelen, F., Fernandez, R. & Hafen, E. (2001) An evolutionarily conserved function of the *Drosophila* insulin receptor and insulin-like peptides in growth control. *Current Biology*, 11, 213–221.

Brommer, J. E., Merila, J. & Kokko, H. (2002) Reproductive timing and individual fitness. *Ecology Letters*, 5, 802–810.

Bronikowski, A. M. (2000) Experimental evidence for the adaptive evolution of growth rate in the garter snake *Thamnophis elegans*. *Evolution*, 54, 1760–1767.

Bronikowski, A. M. (2008) The evolution of aging phenotypes in snakes: a review and synthesis with new data. *Age*, 30, 169–176.

Bronikowski, A. M. & Arnold, S. J. (1999) The evolutionary ecology of life history variation in the garter snake *Thamnophis elegans*. *Ecology*, 80, 2314–2325.

Bronikowski, A. M. & Arnold, S. J. (2001) Cytochrome b phylogeny does not match subspecific classification in the Western terrestrial garter snake, *Thamnophis elegans*. *Copeia*, 2001, 508–513.

Bronikowski, A. M. & Promislow, D. E. L. (2005) Testing evolutionary theories of aging in wild populations. *Trends in Ecology & Evolution*, 20, 271–273.

Bronikowski, A. M. and D. Vleck (2010). Metabolism, body size and life span: A case study in evolutionarily divergent populations of the garter snake (*Thamnophis elegans*) Integrative and Comparative Biology 50: 880–887.

Bronson, F. H. (1985) Mammalian reproduction: an ecological perspective. *Biology of Reproduction*, 32, 1–26.

Brookes, P. S., Buckingham, J. A., Tenreiro, A. M., Hulbert, A. J. & Brand, M. D. (1998) The proton permeability of the inner membrane of liver mitochondria from ectothermic and endothermic vertebrates and from obese rats: correlations with standard metabolic rate and phospholipid fatty acid composition. *Comparative Biochemistry and Physiology B, Biochemistry & Molecular Biology*, 119, 325–334.

Bross, T. G., Rogina, B. & Helfand, S. L. (2005) Behavioral, physical, and demographic changes in *Drosophila* populations through dietary restriction. *Aging Cell*, 4, 309–317.

Broughton, S. J., Piper, M. D., Ikeya, T., Bass, T. M., Jacobson, J., Driege, Y., Martinez, P., Hafen, E., Withers, D. J., Leevers, S. J. & Partridge, L. (2005) Longer lifespan, altered metabolism, and stress resistance in *Drosophila* from ablation of cells making insulin-like ligands. *Proceedings of the National Academy of Sciences of the United States of America*, 102, 3105–3110.

Broughton, S., Alic, N., Slack, C., Bass, T., Ikeya, T., Vinti, G., Tommasi, A. M., Driege, Y., Hafen, E. & Partridge, L. (2008) Reduction of DILP2 in *Drosophila* triages a metabolic phenotype from lifespan revealing redundancy and compensation among DILPs. *PLoS One*, 3, e3721.

Browder, M. H., D'Amico, L. J. & Nijhout, H. F. (2001) The role of low levels of juvenile hormone esterase in the metamorphosis of *Manduca sexta*. *Journal of Insect Science*, 1–11.

Brown, D. D. (1997) The role of thyroid hormone in zebrafish and axolotl development. *Proceedings of the National Academy of Sciences of the United States of America*, 94, 13011–13016.

Brown, D. D. (2005) The role of deiodinases in amphibian metamorphosis. *Thyroid*, 15, 815–821.

Brown, D. D. & Cai, L. (2007) Amphibian metamorphosis. *Developmental Biology*, 306, 20–33.

Brown, R. P. & Griffin, S. (2005) Lower selected body temperatures after food deprivation in the lizard *Anolis carolinensis*. *Journal of Thermal Biology*, 30, 79–83.

Brown, C. L., Doroshov, S. I., Nunez, J. M., Hadley, C., Vaneenennaam, J., Nishioka, R. S. & Bern, H. A. (1988) Maternal triiodothyronine injections cause increases in swimbladder inflation and survival rates in larval striped bass, *Moronev saxatilis*. *Journal of Experimental Zoology*, 248, 168–176.

Brown, D. D., Cai, L., Das, B., Marsh-Armstrong, N., Schreiber, A. M. & Juste, R. (2005) Thyroid hormone controls multiple independent programs required for limb development in *Xenopus laevis* metamorphosis. *Proceedings of the National Academy of Sciences of the United States of America*, 102, 12455–12458.

Brunet, A., Sweeney, L. B., Sturgill, J. F., Chua, K. F., Greer, P. L., Lin, Y., Tran, H., Ross, S. E., Mostoslavsky, R., Cohen, H. Y., Hu, L. S., Cheng, H. L., Jedrychowski, M. P., Gygi, S. P., Sinclair, D. A., Alt, F. W. & Greenberg, M. E. (2004) Stress-dependent regulation of FOXO transcription factors by the SIRT1 deacetylase. *Science*, 303, 2011–2015.

Bruning, E., Saxer, A. & Lanzrein, B. (1985) Methyl far-nesoate and juvenile hormone III in the normal and pre-cocene treated embryos of the ovoviviparous cockroach *Nauphoeta cinerea*. *International Journal of Invertebrate Reproduction and Development*, 8, 269–278.

Bryant, M. & Reznick, D. (2004) Comparative studies of senescence in natural populations of guppies. *The American Naturalist*, 163, 55–68.

Bubliy, O. A. & Loeschcke, V. (2005) Correlated responses to selection for stress resistance and longevity in a labo-ratory population of *Drosophila melanogaster*. *Journal of Evolutionary Biology*, 18, 789–803.

Buchan, J. C., Alberts, S. C., Silk, J. B. & Altmann, J. (2003) True paternal care in a multi-male primate society. *Nature*, 425, 179–181.

Buchholz, D. R. & Hayes, T. B. (2002) Evolutionary pat-terns of diversity in spadefoot toad metamorphosis (Anura: Pelobatidae). *Copeia*, 2002, 180–189.

Buchholz, D. R. & Hayes, T. B. (2005) Variation in thyroid hormone action and tissue content underlies species dif-ferences in the timing of metamorphosis in desert frogs. *Evolution & Development*, 7, 458–467.

Buchholz, D. R., Hsia, S. C., Fu, L. & Shi, Y. B. (2003) A dominant-negative thyroid hormone receptor blocks amphibian metamorphosis by retaining corepressors at target genes. *Molecular and Cellular Biology*, 23, 6750–6758.

Buchholz, D. R., Tomita, A., Fu, L., Paul, B. D. & Shi, Y. B. (2004) Transgenic analysis reveals that thyroid hormone receptor is sufficient to mediate the thyroid hormone signal in frog metamorphosis. *Molecular and Cellular Biology*, 24, 9026–9037.

Buchholz, D. R., Paul, B. D. & Shi, Y. B. (2005) Gene-specific changes in promoter occupancy by thyroid hormone receptor during frog metamorphosis. Implications for developmental gene regulation. *Journal of Biological Chemistry*, 280, 41222–41228.

Buchholz, D. R., Paul, B. D., Fu, L. & Shi, Y. B. (2006) Molecular and developmental analyses of thyroid hor-mone receptor function in *Xenopus laevis*, the African clawed frog. *General and Comparative Endocrinology*, 145, 1–19.

Buckbinder, L. & Brown, D. D. (1993) Expression of the *Xenopus laevis* prolactin and thyrotropin genes during metamorphosis. *Proceedings of the National Academy of Sciences of the United States of America*, 90, 3820–3824.

Buena, F., Swerdloff, R. S., Steiner, B. S., Lutchmansingh, P., Peterson, M. A., Pandian, M. R., Galmarini, M. & Bhasin, S. (1993) Sexual function does not change when serum testosterone levels are pharmacologically varied within the normal male range. *Fertility and Sterility*, 59, 1118–1123.

Bullen, B. A., Skrinar, G. S., Beitins, I. Z., Von Mering, G., Turnbull, B. A. & McArthur, J. W. (1985) Induction of menstrual disorders by strenuous exercise in untrained women. *New England Journal of Medicine*, 312, 1349–1353.

Bult, A. & Lynch, C. B. (1996) Multiple selection responses in house mice bidirectionally selected for thermoregula-tory nest-building behavior: crosses of replicate lines. *Behavior Genetics*, 26, 439–446.

Bult, A. & Lynch, C. B. (2000) Breaking through artificial selection limits of an adaptive behavior in mice and the consequences for correlated responses. *Behavior Genetics*, 30, 193–206.

Burdge, G. C., Hanson, M. A., Slater-Jefferies, J. L. & Lillycrop, K. A. (2007) Epigenetic regulation of transcrip-tion: a mechanism for inducing variations in phenotype (fetal programming) by differences in nutrition during early life? *British Journal of Nutrition*, 97, 1036–1046.

Burke, R. D. (1978) The structure of the nervous system of the pluteus larva of *Strongylocentrotus purpuratus*. *Cell and Tissue Research*, 191, 233–247.

Burke, R. D. (1980) Neural control of Echinoid metamor-phosis. *American Zoologist*, 20, 911.

Burke, R. D. (1983a) Development of the larval nervous system of the sand dollar, *Dendraster excentricus*. *Cell and Tissue Research*, 229, 145–154.

Burke, R. D. (1983b) The induction of metamorphosis of marine invertebrate larvae – stimulus and response. *Canadian Journal of Zoology–Revue Canadienne De Zoologie*, 61, 1701–1719.

Burke, R. D., Brand, D. G. & Bisgrove, B. W. (1986) Structure of the nervous system of the Auricularia larva of *Parasticopus californicus*. *Biological Bulletin*, 170, 450–460.

Burn, J. E., Smyth, D. R., Peacock, W. J. & Dennis, E. S. (1993) Genes conferring late flowering in *Arabidopsis thaliana*. *Genetica*, 90, 147–155.

Burton, P. & Finnerty, J. (2009) Conserved and novel gene expression between regeneration and asexual fission in *Nematostella vectensis*. *Development, Genes and Evolution*, 219, 79–87.

Buss, D. M. & Schmitt, D. P. (1993) Sexual strategies the-ory: an evolutionary perspective on human mating. *Psychological Review*, 100, 204–232.

Butcher, R. A., Fujita, M., Schroeder, F. C. & Clardy, J. (2007) Small-molecule pheromones that control dauer development in *Caenorhabditis elegans*. *Nature Chemical Biology*, 3, 420–422.

Butcher, R. A., Ragains, J. R., Li, W., Ruvkun, G., Clardy, J. & Mak, H. Y. (2009) Biosynthesis of the *Caenorhabditis elegans* dauer pheromone. *Proceedings of the National Academy of Sciences of the United States of America*, 106, 1875–1879.

Byrne, M., Cisternas, P. & Koop, D. (2001) Evolution of larval form in the sea star genus *Patiriella*: conservation and change in the larval nervous system. *Development, Growth & Differentiation*, 43, 459–468.

Byrne, M., Sewell, M. A. & Prowse, T. A. A. (2008) Nutritional ecology of sea urchin larvae: influence of endogenous and exogenous nutrition on echinopluteal growth and phenotypic plasticity in *Tripneustes gratilla*. *Functional Ecology*, 22, 643–648.

Cai, L. & Brown, D. D. (2004) Expression of type II iodothyronine deiodinase marks the time that a tissue responds to thyroid hormone-induced metamorphosis in *Xenopus laevis*. *Developmental Biology*, 266, 87–95.

Caicedo, A. L., Stinchcombe, J. R., Olsen, K. M., Schmitt, J. & Purugganan, M. D. (2004) Epistatic interaction between *Arabidopsis FRI* and *FLC* flowering time genes generates a latitudinal cline in a life history trait. *Proceedings of the National Academy of Sciences of the United States of America*, 101, 15670–15675.

Caicedo, A. L., Richards, C., Ehrenreich, I. M. & Purugganan, M. D. (2009) Complex rearrangements lead to novel chimeric gene fusion polymorphisms at the *Arabidopsis thaliana* MAF2-5 flowering time gene cluster. *Molecular Biology and Evolution*, 26, 699–711.

Calboli, F. C., Gilchrist, G. W. & Partridge, L. (2003) Different cell size and cell number contribution in two newly established and one ancient body size cline of *Drosophila subobscura*. *Evolution*, 57, 566–573.

Calder, W. A. (1984) *Size, Function and Life History*. Cambridge, MA, Harvard University Press.

Caldwell, P. E., Walkiewicz, M. & Stern, M. (2005) Ras activity in the *Drosophila* prothoracic gland regulates body size and developmental rate via ecdysone release. *Current Biology*, 15, 1785–1795.

Callaini, G. & Dallai, R. (1987) Cuticle formation during the embryonic development of the dipteran *Ceratitis capitata*. *Italian Journal of Zoology*, 54, 221–227.

Callery, E. M. & Elinson, R. P. (2000) Thyroid hormone-dependent metamorphosis in a direct developing frog. *Proceedings of the National Academy of Sciences of the United States of America*, 97, 2615–2620.

Callery, E. M., Fang, H. & Elinson, R. P. (2001) Frogs without polliwogs: evolution of anuran direct development. *BioEssays*, 23, 233–241.

Calsbeek, R. & Sinervo, B. (2004) Within clutch variation in offspring sex determined by differences in sire body size: cryptic mate choice in the wild. *Journal of Evolutionary Biology*, 17, 464–470.

Campbell, A. (2002) *A Mind of her Own: The Evolutionary Psychology of Women*. New York, Oxford University Press.

Campbell, B. C. (2006) Adrenarche and the evolution of human life history. *American Journal of Human Biology*, 18, 569–589.

Campbell, B. C., Gillett-Netting, R. & Meloy, M. (2004) Timing of reproductive maturation in rural versus urban Tonga boys, Zambia. *Annals of Human Biology*, 31, 213–227.

Campbell, B. C., Leslie, P. W., Little, M. A. & Campbell, K. L. (2005) Pubertal timing, hormones, and body composition among adolescent Turkana males. *American Journal of Physical Anthropology*, 128, 896–905.

Canoine, V., Fusani, L., Schlinger, B. & Hau, M. (2007) Low sex steroids, high steroid receptors: Increasing the sensitivity of the nonreproductive brain. *Developmental Neurobiology*, 67, 57–67.

Capella, I. C. S. & Hartfelder, K. (2002) Juvenile-hormone-dependent interaction of actin and spectrin is crucial for polymorphic differentiation of the larval honeybee ovary. *Cell and Tissue Research*, 307, 265–272.

Capy, P., Pla, E. & David, J. R. (1993) Phenotypic and genetic variability of morphometrical traits in natural populations of *Drosophila melanogaster* and *D. simulans*. I. Geographic variations. *Genetics, Selection, Evolution*, 25, 517–536.

Carani, C., Scuteri, A., Marrama, P. & Bancroft, J. (1990) The effects of testosterone administration and visual erotic stimuli on nocturnal penile tumescence in normal men. *Hormones and Behavior*, 24, 435–441.

Carbone, M. A., Jordan, K. W., Lyman, R. F., Harbison, S. T., Leips, J., Morgan, T. J., De Luca, M., Awadelia, P. & Mackay, T. F. C. (2006) Phenotypic variation and natural selection at *Catsup*, a pleiotropic quantitative trait gene in *Drosophila*. *Current Biology*, 16, 912–919.

Caretta, N., Palego, P., Roverato, A., Selice, R., Ferlin, A. & Foresta, C. (2006) Age-matched cavernous peak systolic velocity: a highly sensitive parameter in the diagnosis of arteriogenic erectile dysfunction. *International Journal of Impotence Research* 18, 306–310.

Cariou, B., Bouchaert, E., Abdelkarim, M., Dumont, J., Caron, S., Fruchart, J. C., Burcelin, R., Kuipers, F. & Staels, B. (2007) FXR-deficiency confers increased susceptibility to torpor. *FEBS Letters*, 581, 5191–5198.

Carius, H. J., Little, T. J. & Ebert, D. (2001) Genetic variation in a host-parasite association: potential for coevolution and frequency-dependent selection. *Evolution*, 55, 1136–1145.

Carles, C. C., Choffnes-Inada, D., Reville, K., Lertpiriyapong, K. & Fletcher, J. C. (2005) ULTRAPETALA1 encodes a SAND domain putative transcriptional regulator that controls shoot and floral meristem activity in *Arabidopsis*. *Development*, 132, 897–911.

Carney, G. E. & Bender, M. (2000) The *Drosophila ecdysone receptor* (*EcR*) gene is required maternally for normal oogenesis. *Genetics*, 154, 1203–1211.

Caro, P., Gomez, J., Lopez-Torres, M., Sanchez, I., Naudi, A., Jove, M., Pamplona, R. & Barja, G. (2008) Forty percent and eighty percent methionine restriction decrease mitochondrial ROS generation and oxidative stress in rat liver. *Biogerontology*, 9, 183–196.

Carr, B. R. (1998) Disorders of the ovaries and female reproductive tract. In Larsen, P. R., Kronenberg, H. M., Melmed, S. & Polonsky, K. S. (Eds.) *Williams Textbook of Endocrinology*. 9 ed. Philadelphia, Saunders.

Carroll, S. B. (2005) *Endless forms most beautiful: the new science of evo devo*. New York, London, W. W. Norton and Company.

Carroll, S. B. (2008) Evo-devo and an expanding evolutionary synthesis: a genetic theory of morphological evolution. *Cell*, 134, 25–36.

Carroll, S. B., Grenier, J. K. & Weatherbee, S. D. (2000) *From DNA to Diversity: Molecular Genetics and the Evolution of Animal Design*, Oxford, Wiley Blackwell.

Carson, D. D., Farach, M. C., Earles, D. S., Decker, G. L. & Lennartz, W. J. (1985) A monoclonal antibody inhibits calcium accumulation and skeleton formation in cultured embryonic cells of the sea urchin. *Cell*, 41, 639–648.

Carter, C. S. (2002) Neuroendocrine perspectives on social attachment and love. In Cacioppo, J. T., Berntson, G. G., Adolphs, R., Carter, C. S., Davidson, R. J., McClintock, M. K., Mcewen, B. S., Meaney, M. J., Schacter, D. L., Sternberg, McEwen., Suomi, S. S. & Taylor, S. E. (Eds.) *Foundations in Social Neuroscience*. Cambridge, MIT Press.

Carter, M. E. & Brunet, A. (2007) FOXO transcription factors. *Current Biology*, 17, R113–R114.

Carter, H. B., Pearson, J. D., Metter, E. J., Chan, D. W., Andres, R., Fozard, J. L., Rosner, W. & Walsh, P. C. (1995) Longitudinal evaluation of serum androgen levels in men with and without prostate cancer. *Prostate*, 27, 25–31.

Cartland-Shaw, L. K., Cree, A., Skeaff, C. M. & Grimmond, N. M. (1998) Differences in dietary and plasma fatty acids between wild and captive populations of a rare reptile, the tuatara (*Sphenodon punctatus*). *Journal of Comparative Physiology B*, 168, 569–580.

Cartwright, P. (2003) Developmental insights into the origin of complex colonial Hydrozoans. *Integrative and Comparative Biology*, 43, 82–86.

Cartwright, P., Schierwater, B. & Buss, L. W. (2006) Expression of a Gsx parahox gene, Cnox-2, in colony ontogeny in *Hydractinia* (Cnidaria: Hydrozoa). *Journal of Experimental Zoology B*, 306B, 460–469.

Carvalho, G. B., Kapahi, P. & Benzer, S. (2005) Compensatory ingestion upon dietary restriction in *Drosophila melanogaster*. *Nature Methods*, 2, 813–815.

Carvalho, G. B., Kapahi, P., Anderson, D. J. & Benzer, S. (2006) Allocrine modulation of feeding behavior by the sex peptide of *Drosophila*. *Current Biology*, 16, 692–696.

Casey, M. L. & MacDonald, P. C. (1998) Endocrine changes of pregnancy. In Larsen, P. R., Kronenberg, H. M., Melmed, S. & Polonsky, K. S. (Eds.) *Williams Textbook of Endocrinology*, 9th edn. Philadelphia, Saunders.

Castoe, T. A., De Koning, A. P. J., Kim, H. M., Gu, W., Noonan, B. P., Naylor, G., Jiang, Z. J., Parkinson, C. L. & Pollock, D. D. (2009) Evidence for an ancient adaptive episode of convergent molecular evolution. *Proceedings of the National Academy of Sciences of the United States of America*, 106, 8986–8991.

Caswell, H. (1981) The evolution of 'mixed' life histories in marine invertebrates and elsewhere. *The American Naturalist*, 117, 529–536.

Catoni, C., Peters, A. & Martin Schaefer, H. (2008) Life history trade-offs are influenced by the diversity, availability and interactions of dietary antioxidants. *Animal Behaviour*, 76, 1107–1119.

Cavalieri, V., Spinelli, G. & Di Bernardo, M. (2003) Impairing Otp homeodomain function in oral ectoderm cells affects skeletogenesis in sea urchin embryos. *Developmental Biology*, 262, 107–118.

Cavalieri, V., Di Bernardo, M. & Spinelli, G. (2007) Regulatory sequences driving expression of the sea urchin *Otp* homeobox gene in oral ectoderm cells. *Gene Expression Patterns*, 7, 124–130.

Cavaliere, V., Bernardi, F., Romani, P., Duchi, S. & Gargiulo, G. (2008) Building up the *Drosophila* eggshell: first of all the eggshell genes must be transcribed. *Developmental Dynamics*, 237, 2061–2072.

Centanin, L., Ratcliffe, P. J. & Wappner, P. (2005) Reversion of lethality and growth defects in Fatiga oxygen-sensor mutant flies by loss of hypoxia-inducible factor-alpha/Sima. *EMBO Reports*, 6, 1070–1075.

Chamberlain, N. L., Hill, R. I., Kapan, D. D., Gilbert, L. E. & Kronforst, M. R. (2009) Polymorphic butterfly reveals the missing link in ecological speciation. *Science*, 326, 847–850.

Champlin, D. T. & Truman, J. W. (1998a) Ecdysteroid control of cell proliferation during optic lobe neurogenesis in the moth *Manduca sexta*. *Development*, 125, 269–277.

Champlin, D. T. & Truman, J. W. (1998b) Ecdysteroids govern two phases of eye development during metamorphosis of the moth, *Manduca sexta*. *Development*, 125, 2009–2018.

Chandler, J., Wilson, A. & Dean, C. (1996) *Arabidopsis* mutants showing an altered response to vernalization. *Plant Journal*, 10, 637–644.

Chang, C. W., Moseley, J. L., Wykoff, D. & Grossman, A. R. (2005) The LPB1 gene is important for acclimation of *Chlamydomonas reinhardtii* to phosphorus and sulfur deprivation. *Plant Physiology*, 138, 319–329.

Chapais, B. (2008) *Primeval Kinship*. Cambridge, Harvard University Press.

Chapelle, G. & Peck, L. S. (1999) Polar gigantism dictated by oxygen availability. *Nature*, 399, 114–115.

Chapman, T. & Partridge, L. (1996) Female fitness in *Drosophila melanogaster*: an interaction between the effect of nutrition and of encounter rate with males. *Proceedings of the Royal Society of London B*, 263, 755–759.

Chapman, T., Hutchings, J. & Partridge, L. (1993) No reduction in the cost of mating for *Drosophila melanogaster* females mating with spermless males. *Proceedings of the Royal Society of London B*, 253, 211–217.

Chapman, T., Liddle, L. F., Kalb, J. M., Wolfner, M. F. & Partridge, L. (1995) Cost of mating in *Drosophila melanogaster* females is mediated by male accessory-gland products. *Nature*, 373, 241–244.

Chapman, T., Miyatake, T., Smith, H. K. & Partridge, L. (1998) Interactions of mating, egg production and death rates in females of the Mediterranean fruit fly, *Ceratitis capitata*. *Proceedings of the Royal Society of London B*, 265, 1879–1894.

Chapman, J. A., Kirkness, E. F., Simakov, O., Hampson, S. E., Mitros, T., Weinmaier, T., Rattei, T., Balasubramanian, P. G., Borman, J., Busam, D., Disbennett, K., Pfannkoch, C., Sumin, N., Sutton, G. G., Viswanathan, L. D., Walenz, B., Goodstein, D. M., Hellsten, U., Kawashima, T., Prochnik, S. E., Putnam, N. H., Shu, S., Blumberg, B., Dana, C. E., Gee, L., Kibler, D. F., Law, L., Lindgens, D., Martinez, D. E., Peng, J., Wigge, P. A., Bertulat, B., Guder, C., Nakamura, Y., Ozbek, S., Watanabe, H., Khalturin, K., Hemmrich, G., Franke, A., Augustin, R., Fraune, S., Hayakawa, E., Hayakawa, S., Hirose, M., Hwang, J. S., Ikeo, K., Nishimiya-Fujisawa, C., Ogura, A., Takahashi, T., Steinmetz, P. R. H., Zhang, X., Aufschnaiter, R., Eder, M. K., Gorny, A. K., Salvenmoser, W., Heimberg, A. M., Wheeler, B. M., Peterson, K. J., Bottger, A., Tischler, P., Wolf, A., Gojobori, T., Remington, K. A., Strausberg, R. L., Venter, J. C., Technau, U., Hobmayer, B., Bosch, T. C. G., Holstein, T. W., Fujisawa, T., Bode, H. R., David, C. N., Rokhsar, D. S. & Steele, R. E. (2010) The dynamic genome of Hydra. *Nature*, 464, 592–596.

Charlesworth, B. (1980) *Evolution in Age-structured Populations*. First edition Cambridge, Cambridge University Press.

Charlesworth, B. (1990) Optimization models, quantitative genetics, and mutation. *Evolution*, 44, 520–538.

Charlesworth, B. C. (1994) *Evolution in Age-structured Populations*. Second edition, Cambridge, Cambridge University Press.

Charlesworth, B. & Hughes, K. A. (2000) The maintenance of genetic variation in life history traits. In Singh, R. S. & Krimbas, C. B. (Eds.) *Evolutionary Genetics: From Molecules to Morphology*. Cambridge, U.K., Cambridge University Press.

Charnov, E. L. (1991) Evolution of life history variation among female mammals. *Proceedings of the National Academy of Sciences of the United States of America*, 88, 1134–1137.

Charnov, E. L. (1993) *Life History Invariants: Some Explorations of Symmetry in Evolutionary Ecology*. New York, Oxford University Press.

Charnov, E. L. (2009) Optimal (plastic) life histories in growing versus stable populations. *Evolutionary Ecology Research*, 11, 983–987.

Charnov, E. L. & Berrigan, D. (1992) Why do female primates have such long lifespans and so few babies? Or life in the slow lane. *Evolutionary Anthropology*, 2, 191–194.

Charnov, E. L. & Bull, J. (1977) When is sex environmentally determined? *Nature*, 266, 829–830.

Charnov, E. L., Smith, J. M. & Bull, J. J. (1976) Why be an hermaphrodite? *Nature*, 263, 125–126.

Chatterton, R. T., Vogelsong, K. M., Lu, Y. C. & Hudgens, G. A. (1997) Hormonal responses to psychological stress in men preparing for skydiving. *Journal of Clinical Endocrinology and Metabolism*, 82, 2503–2509.

Chee, F. & Byrne, M. (1999a) Development of the larval serotonergic nervous system in the sea star *Patiriella regularis* as revealed by confocal imaging. *Biological Bulletin*, 197, 123–131.

Chee, F. & Byrne, M. (1999b) Serotonin-like immunoreactivitiy in the brachiolaria larvae of *Patiriella regularis*. *Invertebrate Reproduction & Development*, 36, 111–115.

Cheers, M. S. & Ettensohn, C. A. (2005) P16 is an essential regulator of skeletogenesis in the sea urchin embryo. *Developmental Biology*, 283, 384–396.

Chen, J. & Caswell-Chen, E. P. (2004) Facultative vivipary is a life history trait in *Caenorhabditis elegans*. *Journal of Nematology*, 36, 107–113.

Chen, D. & Zhao, J. (2008) Free IAA in stigmas and styles during pollen germination and pollen tube growth of *Nicotiana tabacum*. *Physiologia Plantarum*, 134, 202–215.

Chen, C., Jack, J. & Garofalo, R. S. (1996) The *Drosophila* insulin receptor is required for normal growth. *Endocrinology*, 137, 846–856.

Chen, D., Steele, A. D., Lindquist, S. & Guarente, L. (2005) Increase in activity during calorie restriction requires Sirt1. *Science*, 310, 1641.

Chen, J. J., Senturk, D., Wang, J. L., Muller, H. G., Carey, J. R., Caswell, H. & Caswell-Chen, E. P. (2007) A demographic analysis of the fitness cost of extended longevity in *Caenorhabditis elegans*. *Journal of Gerontology A*, 62, 126–135.

Chen, D., Steele, A. D., Hutter, G., Bruno, J., Govindarajan, A., Easlon, E., Lin, S. J., Aguzzi, A., Lindquist, S. & Guarente, L. (2008) The role of calorie restriction and SIRT1 in prion-mediated neurodegeneration. *Experimental Gerontology*, 43, 1086–1093.

Chen, M., Payne, W. S., Dunn, J. R., Chang, S., Zhang, H. M., Hunt, H. D. & Dodgson, J. B. (2009) Retroviral delivery of RNA interference against Marek's disease virus *in vivo*. *Poultry Science*, 88, 1373–1380.

Chia, F. S. & Rice, M. E. (Eds.) (1978) *Settlement and Metamorphosis of Marine Invertebrate Larvae*. New York, Elsevier.

Chiang, A. S. & Schal, C. (1994) Cyclic volumetric changes in corpus allatum cells in relation to juvenile hormone biosynthesis during ovarian cycles in cockroaches. *Archives of Insect Biochemistry and Physiology*, 27, 53–64.

Chiang, A. S., Gadot, M., Burns, E. L., & Schal, C. (1991) Sexual differentiation of nymphal corpora allata and the effects of ovariectomy on adult gland morphometrics in *Blattella germanica*. *Experientia*, 47, 81–83.

Chiang, G. C. K., Barua, D., Kramer, E. M., Amasino, R. M. & Donohue, K. (2009) Major flowering time gene, *Flowering Locus C*, regulates seed germination in *Arabidopsis thaliana*. *Proceedings of the National Academy of Sciences of the United States of America*, 106, 11661–11666.

Chiappe, D. & MacDonald, K. (2005) The evolution of domain-general mechanisms in intelligence and learning. *Journal of General Psychology*, 132, 5–40.

Chihara, C. J., Fristrom, J. W., Petri, W. H. & King, D. S. (1972) The assay of ecdysones and juvenile hormones on *Drosophila* imaginal disks *in vitro*. *Journal of Insect Physiology*, 18, 1115–1123.

Chino, Y., Saito, M., Yamasu, K., Suyemitsu, T. & Ishihara, K. (1994) Formation of the adult rudiment of sea urchins is influenced by thyroid hormones. *Developmental Biology*, 161, 1–11.

Chiori, R., Jager, M., Denker, E., Wincker, P., Da Silva, C., Le Guyader, H., Manuel, M. & Queinnec, E. (2009) Are Hox genes ancestrally involved in axial patterning? Evidence from the hydrozoan *Clytia hemisphaerica* (Cnidaria). *PLoS One*, 4, e4231.

Chown, S. L. & Gaston, K. J. (2010) Body size variation in insects: a macroecological perspective. *Biological Reviews of the Cambridge Philosophical Society*, 85, 139–169.

Christiansen, F. B. & Fenchel, T. M. (1979) Evolution of marine invertebrate reproductive patterns. *Theoretical Population Biology*, 16, 267–282.

Chung, H., Bogwitz, M. R., McCart, C., Andrianopoulos, A., Ffrench-Constant, R. H., Batterham, P. & Daborn, P.

J. (2007) Cis-regulatory elements in the *Accord* retrotransposon result in tissue-specific expression of the *Drosophila melanogaster* insecticide resistance gene *Cyp6g1*. *Genetics*, 175, 1071–1077.

Cisternas, P. & Byrne, M. (2003) Peptidergic and serotonergic immunoreactivity in the metamorphosing ophiopluteus of *Ophiactis resiliens* (Echinodermata, Ophiuroidea). *Invertebrate Biology*, 122, 177–185.

Cisternas, P., Selvakumaraswamy, P. & Byrne, M. (2001) Localisation of the neuropeptide S1 in an ophiuroid larva. In Barker, M. (Ed.) *Echinoderms 2000*. Rotterdam, Swets and Seitlinger.

Clancy, D. J., Gems, D., Harshman, L. G., Oldham, S., Stocker, H., Hafen, E., Leevers, S. J. & Partridge, L. (2001) Extension of life-span by loss of CHICO, a *Drosophila* insulin receptor substrate protein. *Science*, 292, 104–106.

Clancy, D. J., Gems, D., Hafen, E., Leevers, S. J. & Partridge, L. (2002) Dietary restriction in long-lived dwarf flies. *Science*, 296, 319.

Clare, A. S. (1996a) Natural product antifoulants: Status and potential. *Biofouling*, 9, 211–229.

Clare, A. S. (1996b) Signal transduction in barnacle settlement: Calcium re-visited. *Biofouling*, 10, 141–159.

Clark, A. G. (1987) Senescence and the genetic-correlation hang-up. *The American Naturalist*, 129, 932–940.

Clark, A. G. (1990) Genetic components of variation in energy storage in *Drosophila melanogaster*. *Evolution*, 44, 637–650.

Clark, M. M. & Galef, B. G. (1999) A testosterone-mediated trade-off between parental and sexual effort in male mongolian gerbils (*Meriones unguiculatus*). *Journal of Comparative Psychology*, 113, 388–395.

Clark, R. D. & Hatfield, E. (1989) Gender differences in receptivity to sexual offers. *Journal of Psychology and Human Sexuality* 2, 39–55.

Clark, A. G., Suzumski, F. M., Bell, K. A., Keith, L. E., Houtz, S. & Merriwether, D. A. (1990) Direct and correlated responses to artificial selection on lipid and glycogen contents in *Drosophila melanogaster*. *Genetical Research*, 56, 49–56.

Clark, K. M., Castillo, M., Calatroni, A., Walter, T., Cayazzo, M., Pino, P. & Lozoff, B. (2006) Breast-feeding and mental and motor development at 5 1/2 years. *Ambulatory Pediatrics*, 6, 65–71.

Clayton, W. S. J. (1985) Pedal laceration by the anemone *Aiptasia pallida*. *Marine Ecology Progress Series*, 21, 75–80.

Clemons, G. K. & Nicoll, C. S. (1977) Development and preliminary application of a homologous radioimmunoassay for bullfrog prolactin. *General and Comparative Endocrinology*, 32, 531–535.

Clutton-Brock, T. H. (1988) *Reproductive Success: Studies of individual variation in contrasting breeding systems*. Chicago, University of Chicago Press.

Clutton-Brock, T. (2009) Structure and function in mammalian societies. *Philosophical Transactions of the Royal Society B*, 364, 3229–3242.

Clyne, J. D. & Miesenböck, G. (2009) Postcoital finesse. *Neuron*, 61, 491–493.

Cohen, H. Y., Miller, C., Bitterman, K. J., Wall, N. R., Hekking, B., Kessler, B., Howitz, K. T., Gorospe, M., De Cabo, R. & Sinclair, D. A. (2004) Calorie restriction promotes mammalian cell survival by inducing the SIRT1 deacetylase. *Science*, 305, 390–392.

Cohen, M., Reale, V., Olofsson, B., Knights, A., Evans, P. & De Bono, M. (2009) Coordinated regulation of foraging and metabolism in *C. elegans* by RFamide neuropeptide signaling. *Cell Metabolism*, 9, 375–385.

Cole, L. C. (1954) The population consequences of life history phenomena. *Quarterly Review of Biology*, 29, 103–137.

Colman, R. J., Anderson, R. M., Johnson, S. C., Kastman, E. K., Kosmatka, K. J., Beasley, T. M., Allison, D. B., Cruzen, C., Simmons, H. A., Kemnitz, J. W. & Weindruch, R. (2009) Caloric restriction delays disease onset and mortality in rhesus monkeys. *Science*, 325, 201–204.

Colombani, J., Bianchini, L., Layalle, S., Pondeville, E., Dauphin-Villemant, C., Antoniewski, C., Carré, C., Noselli, S. & Leopold, P. (2005) Antagonistic actions of ecdysone and insulins determine final size in *Drosophila*. *Science*, 310, 667–670.

Comendant, T., Sinervo, B., Svensson, E. I. & Wingfield, J. (2003) Social competition, corticosterone and survival in female lizard morphs. *Journal of Evolutionary Biology*, 16, 948–955.

Conlon, I. & Raff, M. (1999) Size control in animal development. *Cell*, 96, 235–244.

Cook, C. E., Yue, Q. & M.E., A. (2005) Mitochondrial genomes suggest that hexapods and crustaceans are mutually paraphyletic. *Proceedings of the Royal Society of London B*, 272, 1295–1304.

Cooper, V., Reiskind, M., Miller, J., Shelton, K., Walther, B., Elkinton, J. & Ewald, P. (2002) Timing of transmission and the evolution of virulence of an insect virus. *Proceedings of the Royal Society of London B*, 269, 1161–1165

Cooper, T. M., Mockett, R. J., Sohal, B. H., Sohal, R. S. & Orr, W. C. (2004) Effect of caloric restriction on lifespan of the housefly, *Musca domestica*. *FASEB Journal*, 18, 1591–1593.

Cooper, T. F., Remold, S. K., Lenski, R. E. & Schneider, D. (2008) Expression profiles reveal parallel evolution of epistatic interactions involving the CRP regulon in *Escherichia coli*. *PLoS Genetics*, 4, e35.

Corbo, R. M., Ulizzi, L., Piombo, L. & Scacchi, R. (2008) Study on a possible effect of four longevity candidate genes (ACE, PON1, Ppar-gamma, and APOE) on human fertility. *Biogerontology*, 9, 317–323.

Corl, A., Davis, A., Kuchta, S., Comendant, T. & Sinervo, B. (2009) Alternative mating strategies and the evolution of sexual size dimorphism in the side-blotched lizard, *Uta stansburiana*: a population-level comparative analysis. *Evolution*, 64, 79–96.

Corl, A., Davis, A., Kuchta, S. & Sinervo, B. (2010) Selective loss of polymorphic mating types is associated with rapid phenotypic evolution during morphic speciation. *Proceedings of the National Academy of Sciences of the United States of America*, 107, 4254–4259.

Cornwallis, C. K. & Uller, T. (2010) Towards an evolutionary ecology of sexual traits. *Trends in Ecology & Evolution* 25, 145–152.

Corona, M., Hughes, K. A., Weaver, D. B. & Robinson, G. E. (2005) Gene expression patterns associated with queen honey bee longevity. *Mechanisms of Ageing and Development*, 126, 1230–1238.

Corona, M., Velarde, R. A., Remolina, S., Moran-Lauter, A., Wang, Y., Hughes, K. A. & Robinson, G. E. (2007) Vitellogenin, juvenile hormone, insulin signaling, and queen honey bee longevity. *Proceedings of the National Academy of Sciences of the United States of America*, 104, 7128–7133.

Coschigano, K. T., Clemmons, D., Bellush, L. L. & Kopchick, J. J. (2000) Assessment of growth parameters and lifespan of GHR/BP gene-disrupted mice. *Endocrinology*, 141, 2608–26013.

Costanzo, K. & Monteiro, A. (2007) The use of chemical and visual cues in female choice in the butterfly *Bicyclus anynana*. *Proceedings of the Royal Society of London B*, 274, 845–851.

Costanzo, J. P., Baker, P. J. & Lee, R. E. (2006) Physiological responses to freezing in hatchlings of freeze-tolerant and-intolerant turtles. *Journal of Comparative Physiology. B*, 176, 697–707.

Cotter, S. C., Kruuk, L. E. B. & Wilson, K. (2004) Costs of resistance: genetic correlations and potential trade-offs in an insect immune system. *Journal of Evolutionary Biology*, 17, 421–429.

Couper, J. M. & Leise, E. M. (1996) Serotonin injections induce metamorphosis in larvae of the Gastropod mollusc *Ilyanassa obsoleta*. *Biological Bulletin*, 191, 178–186.

Cowan, D. B., Jones, M., Garcia, L. M., Noria, S., Del Nido, P. J. & McGowan, F. X. (2003) Hypoxia and stretch regulate intercellular communication in vascular smooth muscle cells through reactive oxygen species formation. *Arteriosclerosis, Thrombosis and Vascular Biology*, 23, 1754–1760.

Coyne, J. A. & Beecham, E. (1987) Heritability of two morphological characters within and among natural populations of *Drosophila melanogaster*. *Genetics*, 117, 727–737.

Craig, S. F., Slobodkin, L. B., Wray, G. A. & Biermann, C. H. (1997) The 'paradox' of polyembryony: a review of the cases and a hypothesis of its evolution. *Evolutionary Ecology*, 11, 127–143.

Crailsheim, K. (1986) Dependence of Protein-Metabolism on Age and Season in the Honeybee (*Apis mellifica carnica*). *Journal of Insect Physiology*, 32, 629–634.

Crain, D. A., Bolten, A. B., Bjorndal, K. A., Guillerte, L. J. & Gross, T. S. (1995) Size-dependent, sex-dependent, and seasonal changes in Insulin-like Growth Factor I in the loggerhead sea turtle (*Caretta caretta*). *General and Comparative Endocrinology*, 98, 219–226.

Crawford, D., Libina, N. & Kenyon, C. (2007) *Caenorhabditis elegans* integrates food and reproductive signals in lifespan determination. *Aging Cell*, 6, 715–721.

Creel, S. (2001) Social dominance and stress hormones. *Trends in Ecology & Evolution*, 16, 491–497.

Cresko, W. A., Amores, A., Wilson, C., Murphy, J., Currey, M., Phillips, P., Bell, M. A., Kimmel, C. B. & Postlethwait, J. H. (2004) Parallel genetic basis for repeated evolution of armor loss in Alaskan threespine stickleback populations. *Proceedings of the National Academy of Sciences of the United States of America*, 101, 6050–6055.

Crews, D. E. (2003) *Human Senescence: Evolutionary and biocultural perspectives*. New York, Cambridge University Press.

Criscuolo, F., Gonzalez-Barroso, M. D., Bouillaud, F., Ricquier, D., Miroux, B. & Sorci, G. (2005) Mitochondrial uncoupling proteins: new perspectives for evolutionary ecologists. *The American Naturalist*, 166, 686–699.

Crow, J. F. (1992) Twenty-five years ago in genetics: identical triplets. *Genetics*, 124, 395–398.

Cullen, C. F. & Milner, M. J. (1991) Parameters of growth in primary cultures and cell-lines established from *Drosophila* imaginal disks. *Tissue & Cell*, 23, 29–39.

Currie, D. A., Milner, M. J. & Evans, C. W. (1988) The growth and differentiation *in vitro* of leg and wing imaginal disk cells from *Drosophila melanogaster*. *Development*, 102, 805–814.

Curtis, T. J. & Wang, Z. (2003) The neurochemistry of pair bonding. *Current Directions in Psychological Science*, 12, 49–53.

Cutler, W. B., Garcia, C. R., Huggins, G. R. & Preti, G. (1986) Sexual behavior and steroid levels among gynecologically mature premenopausal women. *Fertility and Sterility*, 45, 496–502.

Czihak, G. (1962) Entwicklungphysiologie der Echinodermen. *Fortschritte der Zoologie*, 14, 238–267.

Dabbs, J. M. & Dabbs, M. G. (2000) *Heroes, Rogues, and Lovers: Testosterone and Behavior*. New York, McGraw-Hill.

Dabbs, J. M. & Mohammed, S. (1992) Male and female salivary testosterone concentrations before and after sexual activity. *Physiological Behavior*, 52, 195–197.

Daffre, S., Kylsten, P., Samakovlis, C. & Hultmark, D. (1994) The lysozyme locus in *Drosophila melanogaster*: an expanded gene family adapted for expression in the digestive tract. *Molecular and General Genetics*, 242, 152–162.

D'Agati, P. & Cammarata, M. (2006) Comparative analysis of thyroxine distribution in Ascidian larvae. *Cell and Tissue Research*, 323, 529–535.

Daitoku, H., Hatta, M., Matsuzaki, H., Aratani, S., Ohshima, T., Miyagishi, M., Nakajima, T. & Fukamizu, A. (2004) Silent information regulator 2 potentiates Foxo1-mediated transcription through its deacetylase activity. *Proceedings of the National Academy of Sciences of the United States of America*, 101, 10042–10047.

D'Amico, L. J., Davidowitz, G. & Nijhout, H. F. (2001) The developmental and physiological basis of body size evolution in an insect. *Proceedings of the Royal Society of London B*, 268, 1589–1593.

Danilova, N. (2006) The evolution of immune mechanisms. *Journal of Experimental Zoology B*, 306, 496–520.

Danks, H. V. (1987) *Insect Dormancy: An Ecological Perspective*. Ottawa, Biological Survey of Canada.

Danks, H.V. (1991) Winter habitats and ecological adaptations for winter survival. In Lee, R. E.& Delinger, D. L. (Eds.) *Insects at Low Temperature*. Chapman and Hall, New York and London. pp. 231–259

Danks, H.V. (1994) *Insect life-cycle polymorphism: theory, evolution and ecological consequences for seasonality and diapause control*. Dordrecht, Kluwer.

Danks, H.V. (2002) The range of insect dormancy responses. *European Journal of Entomology*, 99, 127–142.

Danks, H. V. (2005) How similar are daily and seasonal biological clocks? *Journal of Insect Physiology*, 51, 609–619.

Darwin, C. (1859) *On the Origin of Species by Means of Natural Selection*. London, J. Murray.

Das, N., Levine, R. L., Orr, W. C. & Sohal, R. S. (2001) Selectivity of protein oxidative damage during aging in *Drosophila melanogaster*. *Biochemical Journal*, 360, 209–216.

Das, B., Schreiber, A. M., Huang, H. & Brown, D. D. (2002) Multiple thyroid hormone-induced muscle growth and death programs during metamorphosis in *Xenopus laevis*. *Proceedings of the National Academy of Sciences of the United States of America*, 99, 12230–12235.

Das, R. M., Van Hateren, N. J., Howell, G. R., Farrell, E. R., Bangs, F. K., Porteous, V. C., Manning, E. M., McGrew, M. J., Ohyama, K., Sacco, M. A., Halley, P. A., Sang, H. M., Storey, K. G., Placzek, M., Tickle, C., Nair, V. K. & Wilson, S. A. (2006) A robust system for RNA interference in the chicken using a modified microRNA operon. *Developmental Biology*, 294, 554–563.

Davey, J. C., Becker, K. B., Schneider, M. J., St Germain, D. L. & Galton, V. A. (1995) Cloning of a cDNA for the type II iodothyronine deiodinase. *Journal of Biological Chemistry*, 270, 26786–26789.

David, J. R. (1970) Le nombre d'ovarioles ches la drosophile en relation avec la fecondite et la valeur adaptive. *Archives de Zoologie Experimentale et Generale*, 111, 357–370.

David, J. R. & Capy, P. (1988) Genetic variation of *Drosophila melanogaster* natural populations. *Trends in Genetics*, 4, 106–111.

Davidowitz, G. & Nijhout, H. F. (2004) The physiological basis of reaction norms: The interaction among growth rate, the duration of growth and body size. *Integrative and Comparative Biology*, 44, 443–449.

Davidowitz, G., D'Amico, L. J., Roff, D. A. & Nijhout, H. F. (2002) The physiological regulation of insect body size. *Integrative and Comparative Biology*, 42, 1217–1217.

Davidowitz, G., D'Amico, L. J. & Nijhout, H. F. (2004) The effects of environmental variation on a mechanism that controls insect body size. *Evolutionary Ecology Research*, 6, 49–62.

Davidowitz, G., Roff, D. A. & Nijhout, H. F. (2005) A physiological perspective on the response of body size and development time to simultaneous directional selection. *Integrative and Comparative Biology*, 45, 525–531.

Davidson, B. & Swalla, B. J. (2002) A molecular analysis of ascidian metamorphosis reveals activation of an innate immune response. *Development*, 129, 4739–4751.

Davidson, J. M., Camargo, C. A. & Smith, E. R. (1978) Effects of androgen on sexual behavior in hypogonadal men. *Journal of Clinical Endocrinology and Metabolism* 48, 955–958.

Davidson, E. H., Cameron, R. A. & Ransick, A. (1998) Specification of cell fate in the sea urchin embryo: summary and some proposed mechanisms. *Development*, 125, 3269–3290.

Davidson, B., Jacobs, M. & Swalla, B. J. (2002) *The Individual as a Module: Metazoan evolution and coloniality*. Chicago, University of Chicago Press.

Davies, J. P., Yildiz, F. H. & Grossman, A. (1996) Sac1, a putative regulator that is critical for survival of *Chlamydomonas reinhardtii* during sulfur deprivation. *EMBO Journal*, 15, 2150–2159.

Davis, P. J., Leonard, J. L. & Davis, F. B. (2008) Mechanisms of nongenomic actions of thyroid hormone. *Frontiers in Neuroendocrinology*, 29, 211–218.

Day, T. & Rowe, L. (2002) Developmental Thresholds and the Evolution of Reaction Norms for Age and Size at Life History Transitions. *The American Naturalist*, 159, 338–350.

Deacon, T. W. (1997) *The Symbolic Species: The Co-evolution of Language and the Brain*. New York, Norton.

Dean, A. M. & Thornton, J. W. (2007) Mechanistic approaches to the study of evolution: the functional synthesis. *Nature Reviews Genetics*, 8, 675–688.

Dean, C., Leakey, M. G., Reid, D., Schrenk, F., Schwartz, G. T., Stringer, C. & Walker, A. (2001) Growth processes in teeth distinguish modern humans from *Homo erectus* and earlier hominins. *Nature*, 414, 628–631.

de Azevedo, S. V. & Hartfelder, K. (2008) The insulin signaling pathway in honey bee (*Apis mellifera*) caste development - differential expression of insulin-like peptides and insulin receptors in queen and worker larvae. *Journal of Insect Physiology*, 54, 1064–1071.

De Bono, M. & Bargmann, C. I. (1998) Natural variation in a neuropeptide Y receptor homolog modifies social behavior and food response in *C. elegans*. *Cell*, 94, 679–689.

De Gregorio, E., Spellman, P. T., Rubin, G. M. & Lemaitre, B. (2001) Genome-wide analysis of the *Drosophila* immune response by using oligonucleotide microarrays. *Proceedings of the National Academy of Sciences of the United States of America*, 98, 12590–12595.

De Groef, B., Goris, N., Arckens, L., Kuhn, E. R. & Darras, V. M. (2003) Corticotropin-releasing hormone (CRH)-induced thyrotropin release is directly mediated through CRH receptor type 2 on thyrotropes. *Endocrinology*, 144, 5537–5544.

de Heinzelin, J., Clark, J. D., White, T., Hart, W., Renne, P., Wolde Gabriel, G., Beyene, Y. & Vrba, E. (1999) Environment and behavior of 2.5-million-year-old Bouri hominids. *Science*, 284, 625–629.

de Jesus, E. G. (1994) Thyroid hormone surges during milkfish metamorphosis. *Israeli Journal of Aquaculture-Bamidgeh*, 46, 59–63.

de Jesus, E. G., Toledo, J. D. & Simpas, M. S. (1998) Thyroid hormones promote early metamorphosis in grouper (*Epinephelus coioides*) larvae. *General and Comparative Endocrinology*, 112, 10–16.

de Jong, M. A., Kesbeke, F. M. N. H., Brakefield, P. M., and Zwaan, B. J. (2010) Geographic variation in thermal plasticity of life history and wing pattern in *Bicyclus anynana*. *Climate Research* 43: 91–102.

de Jong, G. & Bochdanovits, Z. (2003) Latitudinal clines in *Drosophila melanogaster*: body size, allozyme frequencies, inversion frequencies, and insulin-signalling pathway. *Journal of Genetics*, 82, 207–223.

De Jong, G. & Van Noordwijk, A. J. (1992) Acquisition and allocation of resources: Genetic (co)variances, selections, and life histories. *The American Naturalist*, 139, 749–770.

Dekker, T., Geier, M. & Carde, R. T. (2005) Carbon dioxide instantly sensitizes female yellow fever mosquitoes to human skin odours. *Journal of Experimental Biology*, 208, 2963–2972.

Delahunty, K. M., McKay, D. W., Noseworthy, D. E. & Storey, A. E. (2007) Prolactin responses to infant cues in men and women: effects of parental experience and recent infant contact. *Hormones and Behavior*, 51, 213–220.

de La Rochebrochard, E., De Mouzon, J., Thepot, F. & Thonneau, P. (2006) Fathers over 40 and increased failure to conceive: the lessons of in vitro fertilization in France. *Fertility and Sterility*, 85, 1420–1424.

Delmotte, F., Leterme, N., Bonhomme, J., Rispe, C. & Simon, J.-C. (2001) Multiple routes to asexuality in an aphid species. *Proceedings of the Royal Society of London B*, 268, 2291–2299.

De Luca, M., Roshina, N. V., Geiger-Thornsberry, G. L., Lyman, R. F., Pasyukova, E. G. & Mackay, T. F. (2003) Dopa decarboxylase (Ddc) affects variation in *Drosophila* longevity. *Nature Genetics*, 34, 429–433.

De Martinis, D., Cotti, G., Heker, S. T., Harren, F. J. M. & Mariani, C. (2002) Ethylene response to pollen tube growth in *Nicotiana tabacum* flowers. *Planta*, 214, 806–812.

Demas, G. E. (2004) The energetics of immunity: A neuroendocrine link between energy balance and immune function. *Hormones and Behavior*, 43, 75–80.

Denlinger, D. L. (1986) Dormancy in tropical insects. *Annual Review of Entomology*, 31, 239–264.

Denlinger, D. L. (2002) Regulation of diapause. *Annual Review of Entomology*, 47, 93–122.

Denlinger, D. L., Yocum, G. D. & Rinehart, J. P. (2004) Hormonal control of diapause. In Gilbert, L. I., Iatrou, K. & Gill, S. S. (Eds.) *Comprehensive Molecular Insect Science*. Amsterdam, Elsevier.

Dennis, G., Sherman, B. T., Hosack, D. A., Yang, J., Gao, W., Lane, H. C. & Lempicki, R. A. (2003) DAVID: Database for Annotation, Visualization, and Integrated Discovery. *Genome Biology*, 4, P3.

Dent, J. N., Etkin, W. & Gilbert, L. I. (1968) A Survey of Amphibian Metamorphosis. In Gilbert, L. I. & Frieden, E. (Eds.) *Metamorphosis: A Problem in Developmental Biology*. New York, Plenum Press.

Denver, R. J. (1988) Several hypothalamic peptides stimulate in vitro thyrotropin secretion by pituitaries of anuran amphibians. *General and Comparative Endocrinology*, 72, 383–393.

Denver, R. J. (1996) Neuroendocrine Control of Amphibian Metamorphosis. In Gilbert, L. I., Tata, J. R. & Atkinson, B. G. (Eds.) *Metamorphosis: Postembryonic Reprogramming of Gene Expression in Amphibian and Insect Cells*. San Diego, Academic Press.

Denver, R. J. (2000) Evolution of the corticotropin-releasing hormone signaling and its role in stress-induced developmental plasticity. *American Zoologist* 40, 995–996.

Denver, R. J. (2009) Stress hormones mediate environment-genotype interactions during amphibian development. *General and Comparative Endocrinology*, 164, 20–31.

Denver, R. J., Glennmeier, K. A., Boorse, G. C. (2002) Endocrinology of Complex Life Cycles: Amphibians. In Pfaff, D. W., Arnold, A. P., Etgen, A. M., Fahrbach, S. E. & Ruben, R. T. (Eds.) *Hormones, Brain and Behavior*. USA, Elsevier.

Denver, R. J. & Licht, P. (1989a) Neuropeptide stimulation of thyrotropin secretion in the larval bullfrog: evidence for a common neuroregulator of thyroid and interrenal activity in metamorphosis. *Journal of Experimental Zoology*, 252, 101–104.

Denver, R. J. & Licht, P. (1989b) Neuropeptides influencing *in vitro* pituitary hormone secretion in hatchling turtles. *Journal of Experimental Zoology*, 251, 306–315.

Derby, A. (1975) An *in vitro* quantitative analysis of the response of tadpole tissue to thyroxine. *Journal of Experimental Biology*, 168, 147–156.

Déry, M. A., Michaud, M. D. & Richard, D. E. (2005) Hypoxia-inducible factor 1: regulation by hypoxic and non-hypoxic activators. *International Journal of Biochemistry & Cell Biology*, 37, 535–540.

Desalvo, M. K., Voolstra, C. R., Sunagawa, S., Schwarz, J. A., Stillman, J. H., Coffroth, M. A., Szmant, A. M. & Medina, M. (2008) Differential gene expression during thermal stress and bleaching in the Caribbean coral. *Montastraea faveolata*. *Molecular Ecology*, 17, 3952–3971.

De Witt, T. J. & Scheiner, S. M. (eds.) 2004. *Phenotypic plasticity: functional and conceptual approaches*. Oxford, Oxford University Press.

De Witt, T. J., Sih, A. & Wilson, D. S. (1998) Costs and limits of phenotypic plasticity. *Trends in Ecology & Evolution*, 13, 77–81.

Dibello, P. L. R., Withers, D. A., Bayer, C. A., Fristrom, J. W. & Guild, G. M. (1991) The *Drosophila broad-complex* encodes a family of related proteins containing zinc fingers. *Genetics* 129, 385–397.

Di Bernardo, M., Castagnetti, S., Bellomonte, D., Oliveri, P., Melfi, R., Palla, F. & Spinelli, G. (1999) Spatially restricted expression of PlOtp, a *Paracentrotus lividus* orthopedia-related homeobox gene, is correlated with oral ectodermal patterning and skeletal morphogenesis in late-cleavage sea urchin embryos. *Development*, 126, 2171–2179.

Dijkstra, C., Bult, A., Bijlsma, S., Daan, S., Meijer, T. & Zijlstra, M. (1990) Brood size manipulations in the kestrel (*Falco tinnunculus*) - effects on offspring and parent survival. *Journal of Animal Ecology*, 59, 269–285.

Dill, A. & Sun, T. P. (2001) Synergistic derepression of gibberellin signaling by removing *RGA* and *GAI* function in *Arabidopsis thaliana*. *Genetics*, 159, 777–785.

Dillin, A., Crawford, D. K. & Kenyon, C. (2002) Timing requirements for insulin/ IGF-1 signaling in *C. elegans*. *Science*, 298, 830–834.

Dillon, R. J. & Dillon, V. M. (2004) The gut bacteria of insects: nonpathogenic interactions. *Annual Review of Entomology*, 49, 71–92.

Dingemanse, N. J., Edelaar, P. & Kempenaers, B. (2010b) Why is there variation in baseline glucocorticoid levels? *Trends in Ecology & Evolution*, 25, 261–262.

Dingemanse, N. J., Kazem, A. J. N., Reale, D. & Wright, J. C. (2010a) Behavioural reaction norms: animal personality meets individual plasticity. *Trends in Ecology & Evolution*, 25, 81–89.

Diola, V., Orth, A. I. & Guerra, M. P. (2008) Reproductive biology in monoecious and gynoecious cucumber cultivars as a result of IBA application. *Horticultura Brasileira*, 26, 30–34.

Dionne, M. S. & Schneider, D. S. (2008) Models of infectious diseases in the fruit fly *Drosophila melanogaster*. *Disease Models and Mechanisms*, 1, 43–49.

Dionne, M. S., Pham, L. N., Shirasu-Hiza, M. M. & Schneider, D. S. (2006) Akt and FOXO dysregulation contribute to infection-induced wasting in *Drosophila*. *Current Biology*, 16, 1977–1985.

Dittami, J. P. & Gwinner, E. (1985) Annual cycles in the African stonechat *Saxicola torquata axillaris* and their relationship to environmental factors. *Journal of Zoology*, 207, 357–370.

Djawdan, M., Sugiyama, T., Schlaeger, L., Bradley, T. & Rose, M. (1996) Metabolic aspects of the trade-off between fecundity and longevity in *Drosophila melanogaster*. *Physiological Zoology*, 69, 1176–1195.

Djawdan, M., Chippindale, A. K., Rose, M. R. & Bradley, T. J. (1998) Metabolic reserves and evolved stress resistance in *Drosophila melanogaster*. *Physiological Zoology*, 71, 584–594.

Doane, W. W. (1960) Developmental physiology of the mutant *female sterile (2) adipose* of *Drosophila melanogaster* I. Adult morphology, longevity, egg production, and egg lethality. *Journal of Experimental Zoology*, 145, 1–21.

Dobretsov, S., Teplitski, M. & Paul, V. J. (2009) Mini-review: quorum sensing in the marine environment and its relationship to biofouling. *Biofouling*, 25, 413–427.

Dodd, M. H. I., & Dodd, J. M. (1976) The biology of metamorphosis. In Lofts, B. (Ed.) *Physiology of the Amphibia* vol. III. New York, Academic Press.

Domanitskaya, E. V., Liu, H., Chen, S. & Kubli, E. (2007) The hydroxyproline motif of male sex peptide elicits the innate immune response in *Drosophila* females. *FEBS Journal*, 274, 5659–5668.

Dominici, F. P., Hauck, S., Argentino, D. P., Bartke, A. & Turyn, D. (2002) Increased insulin sensitivity and upregulation of insulin receptor, insulin receptor substrate (IRS)-1 and IRS-2 in liver of Ames dwarf mice. *Journal of Endocrinology*, 173, 81–94.

Dominick, O. S. & Truman, J. W. (1986a) The physiology of wandering behavior in *Manduca sexta*, 3. Organization of wandering behavior in the larval nervous-system. *Journal of Experimental Biology*, 121, 115–132.

Dominick, O. S. & Truman, J. W. (1986b) The physiology of wandering behavior in *Manduca sexta*, 4. Hormonal induction of wandering behavior from the isolated nervous system. *Journal of Experimental Biology*, 121, 133–151.

Donahue, S. P. & Phillips, L. S. (1989) Response of IGF-1 to nutritional support in malnourished hospital patients: a possible indicator of short-term changes in nutritional status. *American Journal of Clinical Nutrition*, 50, 962–969.

Donehower, L. A., Harvey, M., Slagle, B. L., McArthur, M. J., Montgomery, C. A. J., Butel, J. S. & Bradley, A. (1992) Mice deficient for p53 are developmentally normal but susceptible to spontaneous tumours. *Nature*, 356, 215–221.

Donohue, K., Dorn, L., Griffith, C., Kim, E., Aguilera, A., Polisetty, C. R., Schmitt, J. & Galloway, L. (2005) Environmental and genetic influences on the germination of *Arabidopsis thaliana* in the field. *Evolution*, 59, 740–757.

Donovan, B. T. & van der Werff ten Bosch, J. J. (1965) *Physiology of puberty*. Baltimore, Williams and Wilkins.

Dorn, A. (1983) Hormones during embryogenesis of the milkweed bug, *Oncopeltus fasciatus* (Heteroptera: Lygaeidae). *Entomologia Generalis* 8, 193–214.

Dover, G. (2000) How genomic and developmental dynamics affect evolutionary processes. *BioEssays*, 22, 1153–1159.

Dowling, D. K. & Simmons, L. W. (2009) Reactive oxygen species as universal constraints in life history evolution. *Proceedings of the Royal Society of London B*, 276, 1737–1745.

Doyle, M. R., Bizzell, C. M., Keller, M. R., Michaels, S. D., Song, J., Noh, Y. S., & Amasino, R. M. (2005) HUA2 is required for the expression of floral repressors in *Arabidopsis thaliana*. *Plant Journal*, 41, 376–385.

Draper, P. & Harpending, H. (1988) A sociobiological perspective on the development of human reproductive strategies. In MacDonald, K. B. (Ed.) *Sociobiological*

Perspectives on Human Development. New York, Springer.

Draper, I., Kurshan, P. T., McBride, E., Jackson, F. R. & Kopin, A. S. (2007) Locomotor activity is regulated lay D2-like receptors in *Drosophila*: An anatomic and functional analysis. *Developmental Neurobiology*, 67, 378–393.

Drnevich, J. M., Reedy, M. M., Ruedi, E. A., Rodriguez-Zas, S. & Hughes, K. A. (2004) Quantitative evolutionary genomics: differential gene expression and male reproductive success in *Drosophila melanogaster. Proceedings of the Royal Society of London B*, 271, 2267–2273.

Drummond, A. J., Ashton, B., Buxton, S., Cheung, M., Heled, J., Kearse, M., Thierer, T. & Wilson, A. (2010) *Geneious v4.8*, available from http://www.geneious.com.

Drummond-Barbosa, D. (2008) Stem cells, their niches and the systemic environment: An aging network. *Genetics*, 180, 1787–1797.

Drummond-Barbosa, D. & Spradling, A. C. (2001) Stem cells and their progeny respond to nutritional changes during *Drosophila* oogenesis. *Developmental Biology*, 231, 265–278.

Duboc, V., Röttinger, E. & Lepage, T. (2004) Nodal and BMP2/4 signaling organizes the oral-aboral axis of the sea urchin embryo. *Development Cell*, 6, 397–410.

Dudas, S. P. & Arking, R. A. (1995) A coordinate upregulation of antioxidant gene activity is associated with the delayed onset of senescence in a long-lived strain of *Drosophila. Journal of Gerontology A*, 50A, B117–B127.

Dudley, R. (1998) Atmospheric oxygen, giant Paleozoic insects and the evolution of aerial locomotor performance. *Journal of Experimental Biology*, 201, 1043–1050.

Duellman, W. E. & Trueb, L. (1994) *Biology of Amphibians.* Baltimore, Johns Hopkins University Press.

Dufour, S. & Rousseau, K. (2007) Neuroendocrinology of fish metamorphosis and puberty: Evolutionary and ecophysiological perspectives. *Journal of Marine Science and Technology*, 15, 55–68.

Duloquin, L., Lhomond, G. & Gache, C. (2007) Localized VEGF signaling from ectoderm to mesenchyme cells controls morphogenesis of the sea urchin embryo skeleton. *Development*, 134, 2293–2302.

Dunbar, R. I. M. (1998) The social brain hypothesis. *Evolutionary Anthropology*, 6, 178–190.

Duncan, L., Nishii, I., Harryman, A., Buckley, S., Howard, A., Friedman, N. R. & Miller, S. M. (2007) The VARL gene family and the evolutionary origins of the master cell-type regulatory gene, regA, in *Volvox carteri. Journal of Molecular Evolution*, 65, 1–11.

Dunkov, B. & Georgieva, T. (2006) Insect iron binding proteins: insights from the genomes. *Insect Biochemistry & Molecular Biology*, 36, 300–309.

Dunlap, J. C. (1999) Molecular bases for circadian clocks. *Cell*, 96, 271–290.

Dunlop, E. S., Heino, M. & Dieckmann, U. (2009) Eco-genetic modeling of contemporary life-history evolution. *Ecological Applications*, 19, 1815–1834.

Dupont, S., Wilson, K., Obst, M., Skold, H., Nakano, H. & Thorndyke, M. C. (2007) Marine ecological genomics: when genomics meets marine ecology. *Marine Ecology Progress Series*, 332, 257–273.

Dupont, S., Thorndyke, W., Thorndyke, M. C. & Burke, R. D. (2009) Neural development of the brittlestar *Amphiura filiformis. Development, Genes and Evolution*, 219, 159–166.

Dykhuizen, D. E. & Dean, A. M. (2009) Experimental evolution from the bottom up. In Garland, T. & Rose, M. R. (Eds.) *Experimental Evolution: Methods and Applications.* Berkeley, University of California Press.

Eales, J. G. (1997) Iodine metabolism and thyroid-related functions in organisms lacking thyroid follicles: Are thyroid hormones also vitamins? *Proceedings of the Society for Experimental Biology and Medicine*, 214, 302–317.

Eales, J. G. & Brown, S. B. (1993) Measurement and regulation of thyroidal status in teleost fish. *Reviews in Fish Biology and Fisheries*, 3, 299–347.

Eales, J. G., Holmes, J. A., McLeese, J. M. & Youson, J. H. (1997) Thyroid hormone deiodination in various tissues of larval and upstream-migrant sea lampreys, *Petromyzon marinus. General and Comparative Endocrinology*, 106, 202–210.

Eales, J. G., McLeese, J. M., Holmes, J. A. & Youson, J. H. (2000) Changes in intestinal and hepatic thyroid hormone deiodination during spontaneous metamorphosis of the sea Lamprey, *Petromyzon marinus. Journal of Experimental Zoology*, 286, 305–312.

Eanes, W. F. (1999) Analysis of selection on enzyme polymorphisms. *Annual Review of Ecology and Systematics*, 30, 301–326.

Ebbesson, L. O. E., Bjoernsson, B. T., Ekstroem, P. & Stefansson, S. O. (2008) Daily endocrine profiles in parr and smolt Atlantic salmon. *Comparative Biochemistry and Physiology A*, 151, 698–704.

Eberhard, S., Finazzi, G. & Wollman, F. A. (2008) The dynamics of photosynthesis. *Annual Review of Genetics*, 42, 463–515.

Ebert, D. & Mangin, K. (1997) The influence of host demography on the evolution of virulence of a microsporidian gut parasite. *Evolution*, 51, 1828–1837.

Edeline, E., Bardonnet, A., Bolliet, V., Dufour, S. & Pierre, E. (2005) Endocrine control of *Anguilla anguilla* glass eel dispersal: Effect of thyroid hormones on locomotor activity and rheotactic behavior. *Hormones and Behavior*, 48, 53–63.

Edgar, B. A. (1999) From small flies come big discoveries about size control. *Nature Cell Biology*, 1, E191–E193.

Edgar, B. A. (2006) How flies get their size: genetics meets physiology. *Nature Reviews Genetics*, 7, 907–916.

Edwards, J. S. & Chen, S.-W. (1979) Embryonic development of an insect sensory system, the abdominal cerci of *Acheta domesticus*. *Roux's Archives of Developmental Biology*, 186, 151–178.

Ehrenreich, I. M. & Purugganan, M. D. (2006) The molecular genetic basis of plant adaptation. *American Journal of Botany*, 93, 953–962.

Ehrenreich, I. M., Honzawa, Y., Chou, L., Roe, J. L., Kover, P. X. & Purugganan, M. D. (2009) Candidate gene association mapping of *Arabidopsis* flowering time. *Genetics*, 183, 325–335.

Eigenmann, J. E., Patterson, D. F. & Froesch, E. R. (1984a) Body size parallels insulin-like growth factor 1 levels but not growth hormone secretory capacity. *Acta Endocrinologica*, 106, 448–453.

Eigenmann, J. E., Patterson, D. F., Zapf, J. & Froesch, E. R. (1984b) Insulin-like growth factor 1 in the dog - a study in different dog breeds and in dogs with growth hormone elevation *Acta Endocrinologica*, 105, 294–301.

El-Assal, S. E. D., Alonso-Blanco, C., Peeters, A. J. M., Raz, V. & Koornneef, M. (2001) A QTL for flowering time in *Arabidopsis* reveals a novel allele of *CRY2*. *Nature Genetics*, 29, 435–440.

El-Assal, S. E. D., Alonso-Blanco, C., Peeters, A. J. M., Wagemaker, C., Weller, J. L. & Koornneef, M. (2004) The role of cryptochrome 2 in flowering in *Arabidopsis*. *Plant Physiology*, 134, 539–539.

Elia, M. (1992) Organ and tissue contribution to metabolic rate. In McKinney, J. M. & Tucker, H. N. (Eds.) *Energy Metabolism: Tissue Determinents and Cellular Corollaries*. New York, Raven Press.

Elia, L., Selvakumaraswamy, P. & Byrne, M. (2009) Nervous system development in feeding and nonfeeding asteroid larvae and the early juvenile. *Biological Bulletin*, 216, 322–334.

Elinson, R. P. (2001) Direct development: an alternative way to make a frog. *Genesis* 29, 91–95.

Ellers, J. & Boggs, C. L. (2004) Functional ecological implications of intraspecific differences in wing melanization in *Colias* butterflies. *Biological Journal of the Linnean Society*, 82, 79–87.

Ellis, B. J. & Garber, J. (2000) Psychosocial antecedents of variation in girls' pubertal timing: maternal depression, stepfather presence, and marital and family stress. *Child Development*, 71, 485–501.

Ellison, P. T. (1982) Skeletal growth, fatness, and menarcheal age: a comparison of two hypotheses. *Human Biology*, 54, 269–281.

Ellison, P. T. (1990) Human ovarian function and reproductive ecology: new hypotheses. *American Anthropologist*, 92, 933–952.

Ellison, P. T. (2001) *On fertile ground, a natural history of human reproduction*. Cambridge, Harvard University Press.

Ellison, P. T. (2003) Energetics and reproductive effort. *American Journal of Human Biology*, 15, 342–351.

Ellison, P. T. & Lager, C. (1986) Moderate recreational running is associated with lowered salivary progesterone profiles in women. *American Journal of Obstetrics and Gynecology*, 154, 1000–1003.

Ellison, P. T. & Panter-Brick, C. (1996) Salivary testosterone levels among Tamang and Kami males of central Nepal. *Human Biology*, 68, 955–965.

Ellison, P. T. & Valeggia, C. R. (2003) C-peptide levels and the duration of lactational amenorrhea. *Fertility and Sterility*, 80, 1279–1280.

Ellison, P. T., Bribiescas, R. G., Bentley, G. R., Campbell, B. C., Lipson, S. F., Panter-Brick, C. & Hill, K. (2002) Population variation in age-related decline in male salivary testosterone. *Human Reproduction*, 17, 3251–3253.

Elzinga, J. A., Atlan, A., Biere, A., Gigord, L., Weis, A. E. & Bernasconi, G. (2007) Time after time: flowering phenology and biotic interactions. *Trends in Ecology & Evolution*, 22, 432–439.

Emerson, K. J., Bradshaw, W. E. & Holzapfel, C. M. (2009a) Complications of complexity: integrating environmental, genetic and hormonal control of insect diapause. *Trends in Genetics*, 25, 217–225.

Emerson, K. J., Uyemra, A. M., McDaniel, K. L., Schmidt, P. S., Bradshaw, W. E. & Holzapfel, C. M. (2009b) Environmental control of ovarian dormacy in natural populations of *Drosophila melanogaster*. *Journal of Comparative Physiology A*, 196, 825–829.

Emlen, D. J. & Allen, C. E. (2003) Genotype to phenotype: Physiological control of trait size and scaling in insects. *Integrative and Comparative Biology*, 43, 617–634.

Emlen, D. J., Lavine, L. C. & Ewen-Campen, B. (2007) On the origin and evolutionary diversification of beetle horns. *Proceedings of the National Academy of Sciences of the United States of America*, 104, 8661–8668.

Emlet, R. B. (1986) Facultative planktotrophy in the tropical Echinoid *Clypeaster rosaceus* (Linnaeus) and a comparison with obligate planktotrophy in *Clypeaster subdepressus* (Gray) (Clypeasteroida, Echinoidea).

Journal of Experimental Marine Biology and Ecology, 95, 183–202.

Emlet, R. B. (1995) Larval spicules, cilia, and symmetry as remnants of indirect development in the direct developing sea urchin *Heliocidaris erythrogramma*. *Developmental Biology*, 167, 405–415.

Endler, J. A. (1986) *Natural selection in the wild*. Monographs in population biology. Princeton, Princeton University Press.

Endler, J. A. (1995) Multiple-trait coevolution and environmental gradients in guppies. *Trends in Ecology & Evolution*, 10, 22–29.

Engelmann, K. & Purugganan, M. (2006) The molecular evolutionary ecology of plant development: flowering time in *Arabidopsis thaliana*. *Advances in Botanical Research*, 44, 507–526.

Eraso, P. & Gancedo, J. M. (1985) Use of glucose analogs to study the mechanism of glucose-mediated cAMP increase in yeast. *FEBS Letters*, 191, 51–54.

Erezyilmaz, D. F. (2006) Imperfect eggs and oviform nymphs: a history of ideas about the origins of insect metamorphosis. *Integrative and Comparative Biology*, 46, 795–807.

Erezyilmaz, D. F., Riddiford, L. M. & Truman, J. W. (2004) Juvenile hormone acts at embryonic molts and induces the nymphal cuticle in the direct-developing cricket. *Development, Genes and Evolution*, 214, 313–323.

Erezyilmaz, D. F., Riddiford, L. M. & Truman, J. W. (2006) The pupal specifier broad directs progressive morphogenesis in a direct-developing insect. *Proceedings of the National Academy of Sciences of the United States of America*, 103, 6925–6930.

Erezyilmaz, D. F., Rynerson, M. R., Truman, J. W. and Riddiford, L. M. (2010). The Role of the Pupal Determinant Broad During Embryonic Development of a Direct-Developing Insect. *Development, Genes and Evolution*. 219 (535–544).

Eri, R., Arnold, J. M., Hinman, V. F., Green, K. M., Jones, M. K., Degnan, B. M. & Lavin, M. F. (1999) Hemps, a novel Egf-like protein, plays a central role in Ascidian metamorphosis. *Development*, 126, 5809–5818.

Ernande, B., Clobert, J., McCombie, H. & Boudry, P. (2003) Genetic polymorphism and trade-offs in the early life history strategy of the Pacific oyster, *Crassostrea gigas* (Thunberg, 1795): a quantitative genetic study. *Journal of Evolutionary Biology*, 16, 399–414.

Escriva, H., Manzon, L., Youson, J. & Laudet, V. (2002) Analysis of lamprey and hagfish genes reveals a complex history of gene duplications during early vertebrate evolution. *Molecular Biology and Evolution*, 19, 1440–1450.

Essers, M. A., De Vries-Smits, L. M., Barker, N., Polderman, P. E., Burgering, B. M. & Korswagen, H. C. (2005) Functional interaction between beta-catenin and FOXO in oxidative stress signaling. *Science*, 308, 1181–1184.

Essner, J. J., Breuer, J. J., Essner, R. D., Fahrenkrug, S. C. & Hackett, P. B. (1997) The zebrafish thyroid hormone receptor alpha 1 is expressed during early embryogenesis and can function in transcriptional repression. *Differentiation*, 62, 107–117.

Essner, J. J., Johnson, R. G. & Hackett, P. B. (1999) Overexpression of thyroid hormone receptor alpha 1 during zebrafish embryogenesis disrupts hindbrain patterning and implicates retinoic acid receptors in the control of hox gene expression. *Differentiation*, 65, 1–11.

Ettensohn, C. A. (2009) Lessons from a gene regulatory network: echinoderm skeletogenesis provides insights into evolution, plasticity and morphogenesis. *Development*, 136, 11–21.

Ettensohn, C. A. & McClay, D. R. (1986) The regulation of primary mesenchyme cell migration in the sea urchin embryo: transplantations of cells and latex beads. *Developmental Biology*, 117, 380–391.

Evans, J. D. & Wheeler, D. E. (1999) Differential gene expression between developing queens and workers in the honey bee, *Apis mellifera*. *Proceedings of the National Academy of Sciences of the United States of America*, 96, 5575–5580.

Evans, J. D., Aronstein, K., Chen, Y. P., Hetru, C., Imler, J. L., Jiang, H., Kanost, M., Thompson, G. J., Zou, Z. & Hultmark, D. (2006) Immune pathways and defence mechanisms in honey bees *Apis mellifera*. *Insect Molecular Biology*, 15, 645–656.

Everitt, A. V. & Le Couteur, D. G. (2007) Life extension by calorie restriction in humans. *Annals of the New York Academy of Sciences*, 1114, 428–433.

Evgen'ev, M. B., Garbuz, D. G., Shilova, V. Y. & Zatsepina, O. G. (2007) Molecular mechanisms underlying thermal adaptation of xeric animals. *Journal of Biosciences*, 32, 489–499.

Fahrbach, S. E., Strande, J. L. & Robinson, G. E. (1995) Neurogenesis is absent in the brains of adult honey bees and does not explain behavioral neuroplasticity. *Neuroscience Letters*, 197, 145–148.

Fahrbach, S. E., Moore, D., Capaldi, E. A., Farris, S. M. & Robinson, G. E. (1998) Experience-expectant plasticity in the mushroom bodies of the honeybee. *Learning & Memory*, 5, 115–123.

Falconer, D. S. (1981) *Introduction to Quantitative Genetics*. London, Longman.

Falconer, D. S. & Mackay, T. F. C. (1996) *Introduction to Quantitative Genetics*. Fourth Edition, Essex, Pearson Prentice Hall.

Falster, D. S., Moles, A. T. & Westoby, M. (2008) A general model for the scaling of offspring size and adult size. *The American Naturalist*, 172, 299–317.

Farley, J. (1982). Gametes and spores: ideas about sexual reproduction 1750–1914. Johns Hopkins Univ Press, *Baltimore* USA.

Faulks, S. C., Turner, N., Else, P. L. & Hulbert, A. J. (2006) Calorie restriction in mice: effects on body composition, daily activity, metabolic rate, mitochondrial reactive oxygen species production, and membrane fatty acid composition. *Journal of Gerontology A*, 61, 781–794.

Fautin, D. G. (2002) Reproduction of Cnidaria. *Canadian Journal of Zoology-Revue Canadienne de Zoologie*, 80, 1735–1754.

Feder, M. E. & Walser, J.-C. (2005) The biological limitations of transcriptomics in elucidating stress and stress responses. *Journal of Evolutionary Biology*, 18, 901–910.

Fedorka, K. M., Linder, J. E., Winterhalter, W. E. & Promislow, D. (2007) Post-mating disparity between potential and realized immune response in *Drosophila melanogaster*. *Proceedings of the Royal Society of London B*, 274, 1211–1217.

Feldman, H. A., Longcope, C., Derby, C. A., Johannes, C. B., Araujo, A. B., Coviello, A. D., Bremner, W. J. & McKinlay, J. B. (2002) Age trends in the level of serum testosterone and other hormones in middle-aged men: longitudinal results from the Massachusetts male aging study. *Journal of Clinical Endocrinology and Metabolism 87*, 587–598.

Fell, D. (2003) *Understanding the Control of Metabolism*. Portland, Portland Press.

Felley, J. (1980) Analysis of morphology and asymmetry in bluegill sunfish (*Lepomis macrochirus*) in the southeastern United States. *Copeia*, 1980, 18–29.

Fellowes, M. D. E., Kraaijeveld, A. R. & Godfray, H. C. J. (1998) Trade-off associated with selection for increased ability to resist parasitoid attack in *Drosophila melanogaster*. *Proceedings of the Royal Society of London B*, 265, 1553–1558.

Fellowes, M. D. E., Kraaijeveld, A. R. & Godfray, H. C. J. (1999a) Cross-resistance following artificial selection for increased defense against parasitoids in *Drosophila melanogaster*. *Evolution*, 53, 966–972.

Fellowes, M. D. E., Kraaijeveld, A. R. & Godfray, H. C. J. (1999b) The relative fitness of *Drosophila melanogaster* (Diptera, Drosophilidae) that have successfully defended themselves against the parasitoid *Asobara tabida* (Hymenoptera, Braconidae). *Journal of Evolutionary Biology*, 12, 123–128.

Fenaux, L., Strathmann, M. F. & Strathmann, R. R. (1994) Five tests of food-limited growth of larvae in coastal waters by comparisons of rates of development and form of echinoplutei. *Limnology and Oceanography*, 39, 84–98.

Fenner, F. (1983) The Florey Lecture, 1983 - biological-control, as exemplified by smallpox eradication and myxomatosis. *Proceedings of the Royal Society of London B*, 218, 259–285.

Ferea, T. L., Botstein, D., Brown, P. O. & Rosensweig, R. F. (1999) Systematic changes in gene expression and patterns following adaptive evolution in yeast. *Proceedings of the National Academy of Sciences of the United States of America*, 96, 9721–9726.

Ferrandon, D., Jung, A. C., Criqui, M., Lemaitre, B., Uttenweiler-Joseph, S., Michaut, L., Reichhart, J. M. & Hoffmann, J. A. (1998) A drosomycin-GFP reporter transgene reveals a local immune response in *Drosophila* that is not dependent on the Toll pathway. *EMBO Journal*, 17, 1217–1227.

Ferrandon, D., Imler, J. L., Hetru, C. & Hoffmann, J. A. (2007) The *Drosophila* systemic immune response: sensing and signalling during bacterial and fungal infections. *Nature Reviews Immunology*, 7, 862–874.

Ferrari, J., Darby, A. C., Daniell, T. J., Godfray, H. C. J. & Douglas, A. E. (2004) Linking the bacterial community in pea aphids with host-plant use and natural enemy resistance. *Ecological Entomology*, 29, 60–65.

Ferretti, P. & Geraudie, J. (2001) *Cellular and Molecular Basis of Regeneration: From Invertebrates to Humans*. New York, John Wiley and Sons.

Ferveur, J. F. (2005) Cuticular hydrocarbons: their evolution and roles in *Drosophila* pheromonal communication. *Behavioral Genetics*, 35, 279–295.

Festucci-Buselli, R. A., Carvalho-Dias, A. S., De Oliveira-Andrade, M., Caixeta-Nunes, C., Li, H. M., Stuart, J. J., Muir, W., Scharf, M. E. & Pittendrigh, B. R. (2005) Expression of Cyp6g1 and Cyp12d1 in DDT resistant and susceptible strains of *Drosophila melanogaster*. *Insect Molecular Biology*, 14(1), 69–77

Fewell, J. H. & Page, R. E. J. (1993) Genotypic variation in foraging responses to environmental stimuli by honey bees, *Apis mellifera*. *Experientia*, 49, 1106–1112.

Fewell, J. H. & Winston, M. L. (1992) Colony state and regulation of pollen foraging in the honey bee, *Apis mellifera* L. *Behavioral Ecology and Sociobiology*, 30, 387–393.

Ffrench-Constant, R. H. (2007) Which came first: insecticides or resistance? *Trends in Genetics*, 23, 1–4.

Fielenbach, N. & Antebi, A. (2008) *C. elegans* dauer formation and the molecular basis of plasticity. *Genes & Development*, 22, 2149–2165.

Fielenbach, N., Guardavaccaro, D., Neubert, K., Chan, T., Li, D., Feng, Q., Hutter, H., Pagano, M. & Antebi, A. (2007) DRE-1: an evolutionarily conserved F box protein that regulates *C. elegans* developmental age. *Developmental Cell*, 12, 443–455.

Finch, C. E. (1990) *Longevity, senescence and the genome*. Chicago, IL, University of Chicago Press.

Finch, C. E. & Rose, M. R. (1995) Hormones and the physiological architecture of life history evolution. *Quarterly Review of Biology*, 70, 1–52.

Finkel, T. & Holbrook, N. J. (2000) Oxidants, oxidative stress and the biology of aging. *Nature*, 408, 239–247.

Finnerty, J. R., Paulson, D., Burton, P., Pang, K. & Martindale, M. Q. (2003) Early evolution of a homeobox gene: the parahox gene *Gsx* in the Cnidaria and the Bilateria. *Evolution & Development*, 5, 331–345.

Finnerty, J. R., Pang, K., Burton, P., Paulson, D. & Martindale, M. Q. (2004) Origins of bilateral symmetry: Hox and dpp expression in a sea anemone. *Science*, 304, 1335–1337.

Fischer, K., Bot, A. N. M., Brakefield, P. M. & Zwaan, B. J. (2003a) Fitness consequences of temperature-mediated egg size plasticity in a butterfly. *Functional Ecology*, 17, 803–810.

Fischer, K., Brakefield, P. M. & Zwaan, B. J. (2003b) Plasticity in butterfly egg size: Why larger offspring at lower temperatures? *Ecology*, 84, 3138–3147.

Fischer, K., Eenhoorn, E., Bot, A. N. M., Brakefield, P. M. & Zwaan, B. J. (2003c) Cooler butterflies lay larger eggs: developmental plasticity versus acclimation. *Proceedings of the Royal Society of London B*, 270, 2051–2056.

Fischer, K., Bot, A. N. M., Brakefield, P. M. & Zwaan, B. J. (2006) Do mothers producing large offspring have to sacrifice fecundity? *Journal of Evolutionary Biology*, 19, 380–391.

Fisher, R. A. (1930) *The genetical theory of natural selection*. Oxford, Claredon Press.

Fisher, H. (2002) *Why we love: the nature and chemistry of romantic love*. New York, Henry Holt and Company.

Fisher, S. E. (2005) On genes, speech, and language. *New England Journal of Medicine*, 353, 1655–1657.

Fitzpatrick, M. J., Feder, E., Rowe, L. & Sokolowski, M. B. (2007) Maintaining a behaviour polymorphism by frequency-dependent selection on a single gene. *Nature*, 447, 210–212.

Fiumera, A. C., Dumont, B. L. & Clark, A. G. (2006) Natural variation in male-induced 'cost-of-mating' and allele-specific association with male reproductive genes in *Drosophila melanogaster*. *Philosophical Transactions of the Royal Society of London B*, 361, 355–361.

Fivizzani, A. J., Colwell, M. A. & Oring, L. W. (1986) Plasma steroid hormone levels in free-living Wilson's Phalaropes *Phalaropus tricolor*. *General and Comparative Endocrinology*, 62, 137–144.

Flachsbart, F., Caliebe, A., Kleindorp, R., Blanche, H., Von Eller-Eberstein, H., Nikolaus, S., Schreiber, S. & Nebel, A. (2009) Association of FOXO3A variation with human longevity confirmed in German centenarians. *Proceedings of the National Academy of Sciences of the United States of America*, 106, 2700–2705.

Flannagan, R. D., Tammariello, S. P., Joplin, K. H., Cikra-Ireland, R. A., Yocum, G. D. & Denlinger, D. L. (1998) Diapause-specific gene expression in pupae of the flesh fly *Sarcophaga crassipalpis*. *Proceedings of the National Academy of Sciences of the United States of America*, 95, 5616–5620.

Flatt, T. (2004) Assessing natural variation in genes affecting *Drosophila* lifespan. *Mechanisms of Ageing Development*, 125, 155–159.

Flatt, T. (2005) The evolutionary genetics of canalization. *Quarterly Review of Biology*, 80, 287–316.

Flatt, T. (2009) Aging: Diet and longevity in the balance. *Nature*, 462, 989–990.

Flatt, T. & Kawecki, T. J. (2007) Juvenile hormone as a regulator of the trade-off between reproduction and lifespan in *Drosophila melanogaster*. *Evolution*, 61, 1980–1991.

Flatt, T. & Promislow, D. E. (2007) Physiology: Still pondering an age-old question. *Science*, 318, 1255–1256.

Flatt, T. & Schmidt, P. S. (2009) Integrating evolutionary and molecular genetics of aging. *Biochimica et Biophysica Acta*, 1790, 951–962.

Flatt, T., Tu, M. P. & Tatar, M. (2005) Hormonal pleiotropy and the juvenile hormone regulation of *Drosophila* development and life history. *BioEssays*, 27, 999–1010.

Flatt, T., Heyland, A., Rus, F., Porpiglia, E., Sherlock, C., Yamamoto, R., Garbuzov, A., Palli, S. R., Tatar, M. & Silverman, N. (2008a) Hormonal regulation of the humoral innate immune response in *Drosophila melanogaster*. *Journal of Experimental Biology*, 211, 2712–2724.

Flatt, T., Min, K. J., D'Alterio, C., Villa-Cuesta, E., Cumbers, J., Lehmann, R., Jones, D. L. & Tatar, M. (2008b) *Drosophila*, germ-line modulation of insulin signaling and lifespan. *Proceedings of the National Academy of Sciences of the United States of America*, 105, 6368–6373.

Flatt, T., Moroz, L. L., Tatar, M., & Heyland, A. (2006). Comparing thyroid and insect hormone signaling. *Integrative & Comparative Biology*, 46: 777–794.

Fleming, A. S., Ruble, D., Krieger, H. & Wong, P. Y. (1997) Hormonal and experimental correlates of maternal responsiveness during pregnancy and puerperium in human mothers. *Hormones and Behavior*, 31, 145–158.

Fleming, A. S., O'Day, D. H. & Kraemer, G. W. (1999) Neurobiology of mother-infant interactions: Experience and central nervous system plasticity across development and generations. *Neuroscience and Biobehavioral Reviews*, 23, 673–685.

Flinn, M. V. (2006) Cross-cultural universals and variations: The evolutionary paradox of informational novelty. *Psychological Inquiry*, 17, 118–123.

Flinn, M. V. & Alexander, R. D. (2007) Runaway social selection. In Gangestad, S. W. & Simpson, J. A. (Eds.) *The Evolution of Mind*. New York, Guilford Press.

Flinn, M. V. & Coe, K. C. (2007) The linked red queens of human cognition, coalitions, and culture. In Gangestad, S. W. & Simpson, J. A. (Eds.) *The Evolution of Mind*. New York, Guilford Press.

Flinn, M. V. & Ward, C. V. (2005) Evolution of the social child. In Ellis, B. & Bjorklund, D. (Eds.) *Origins of the Social Mind: Evolutionary Psychology and Child Development*. London, Guilford Press.

Flinn, M. V., Geary, D. C. & Ward, C. V. (2005) Ecological dominance, social competition, and coalitionary arms races: Why humans evolved extraordinary intelligence. *Evolution and Human Behavior*, 26, 10–46.

Flowers, J. M., Sezgin, E., Kumagai, S., Duvernell, D. D., Matzkin, L. M., Schmidt, P. S. & Eanes, W. F. (2007) Adaptive evolution of metabolic pathways in *Drosophila*. *Molecular Biology and Evolution*, 24, 1347–1354.

Foerster, K., Coulson, T., Sheldon, B., Pemberton, J. M., Clutton-Brock, T. H. & Kruuk, L. E. B. (2007) Sexually antagonistic genetic variation in the red deer. *Nature*, 447, 1107–1111.

Foley, P. A. & Luckinbill, L. S. (2001) The effects of selection for larval behavior on adult life history features in *Drosophila melanogaster*. *Evolution*, 55, 2493–2502.

Folstad, I. & Karter, A. J. (1992) Parasites, bright males, and the immunocompetence handicap. *The American Naturalist*, 139, 603–622.

Fong, S. S., Joyce, A. R. & Palsson, B. O. (2005) Parallel adaptive evolution cultures of Escherichia coli lead to convergent growth phenotypes with different gene expression states. *Genome Research*, 15, 1365–1372.

Fontana, L., Weiss, E. P., Villareal, D. T., Klein, S. & Holloszy, J. O. (2008) Long-term effects of calorie or protein restriction on serum IGF-1 and IGFBP-3 concentration in humans. *Aging Cell*, 7, 681–687.

Force, A. G., Staples, T., Soliman, T. & Arking, R. A. (1995) A comparative biochemical and stress analysis of genetically selected *Drosophila* strains with different longevities. *Developmental Genetics*, 17, 340–351.

Foucher, F., Morin, J., Courtiade, J., Cadioux, S., Ellis, N., Banfield, M. J. & Rameau, C. (2003) *DETERMINATE* and *LATE FLOWERING* are two *TERMINAL FLOWER1/CENTRORADIALIS* homologs that control two distinct phases of flowering initiation and development in pea. *The Plant Cell*, 15, 2742–2754.

Fowler, K. & Partridge, L. (1989) A cost of mating in female fruit-flies. *Nature*, 338, 760–761.

Francis, L. (1979) Contrast between solitary and clonal lifestyles in the sea anemone *Anthopleura elegantissima*. *American Zoologist*, 19, 669–681.

Francis, L. (1988) Cloning and aggression among sea anemones (Coelenterata: Actiniaria) of the rocky shore. *Biological Bulletin*, 174, 241–253.

Frank, S. A. (1991) Ecological and genetic models of host-pathogen coevolution. *Heredity*, 67, 73–83.

Frankino, W. A., Emlen, D. & Shingleton, A. (2009) Experimental approaches to studing the evolution of animal form: The shape of things to come. In Garland, T. & Rose, M. R. (Eds.) *Experimental Evolution: Concepts, Methods and Applications of Selection Experiments*. Berkley and Los Angeles, University of California Press.

Fraser, S. E., Green, C. R., Bode, H. R. & Gilula, N. B. (1987) Selective disruption of gap junctional communication interferes with a patterning process in *Hydra*. *Science*, 237, 49–55.

Fraser, A. M., Brockert, J. E. & Ward, R. H. (1995) Association of young maternal age with adverse reproductive outcomes. *New England Journal of Medicine*, 332, 1113–1117.

Fraser, H. B., Khaitovich, P., Plotkin, J. B., Paabo, S. & Eisen, M. B. (2005) Aging and gene expression in the primate brain. *PLoS Biology*, 3, e274.

Frazier, M. R., Harrison, J. F., Kirkton, S. D. & Roberts, S. P. (2008) Cold rearing improves cold-flight performance in *Drosophila* via changes in wing morphology. *Journal of Experimental Biology*, 211, 2116–2122.

Freeman, S. & Herron, J. C. (2007) *Evolutionary Analysis*. Upper Saddle River, NJ Pearson Prentice Hall.

Freeman, G. & Ridgway, E. B. (1990) Cellular and intracellular pathways mediating the metamorphic stimulus in Hydrozoan planulae. *Roux's Archives of Developmental Biology*, 199, 63–79.

Freitak, D., Wheat, C. W., Heckel, D. G. & Vogel, H. (2007) Immune system responses and fitness costs associated with consumption of bacteria in larvae of *Trichoplusia ni*. *BMC Biology*, 5, 56.

Fricke, C., Perry, J., Chapman, T. & Rowe, L. (2009) The conditional economics of sexual conflict. *Biology Letters*, 5, 671–674.

Fricke, C., Bretman, A. & Chapman, T. (2010) Female nutritional status determines the magnitude and sign of responses to a male ejaculate signal in *Drosophila melanogaster*. *Journal of Evolutionary Biology*, 23, 157–165.

Friedman, D. B. & Johnson, T. E. (1988) A mutation in the *age-1* gene in *Caenorhabditis elegans* lengthens life and reduces hermaphrodite fertility. *Genetics*, 118, 75–86.

Friedman, W. E. & Williams, J. H. (2003) Modularity of the angiosperm female gametophyte and its bearing on the early evolution of endosperm in flowering plants. *Evolution*, 57, 216–230.

Friesema, E. C., Docter, R., Moerings, E. P., Verrey, F., Krenning, E. P., Hennemann, G. & Visser, T. J. (2001) Thyroid hormone transport by the heterodimeric human system L amino acid transporter. *Endocrinology*, 142, 4339–4348.

Frisch, R. E. & McArthur, J. W. (1974) Menstrual cycles: fatness as a determinant of minimum weight for height necessary for their maintenance or onset. *Science*, 185, 949–951.

Frisch, R. E. & Revelle, R. (1971) Height and weight at menarche and a hypothesis of menarche. *Archive of Disease in Childhood*, 46, 695–701.

Fritzenwanker, J. H., Genikhovich, G., Kraus, Y. & Technau, U. (2007) Early development and axis specification in the sea anemone *Nematostella vectensis*. *Developmental Biology*, 310, 264–279.

Frost, S. D. W. (1999) The immune system as an inducible defense. In Tollrian, R. & Harvel, C. D. (Eds.) *The Ecology and Evolution of Inducible Defense*. Princeton, NJ, Princeton University Press.

Fumagalli, M., Cagliani, R., Pozzoli, U., Riva, S., Comi, G. P., Menozzi, G., Bresolin, N. & Sironi, M. (2009) Widespread balancing selection and pathogen-driven selection at blood group antigen genes. *Genome Research*, 19, 199–212.

Furlow, J. D. & Neff, E. S. (2006) A developmental switch induced by thyroid hormone: *Xenopus laevis* metamorphosis. *Trends in Endocrinology & Metabolism*, 17, 40–47.

Furtado, O. V., Polcheira, C., Machado, D. P., Mourao, G. & Hermes-Lima, M. (2007) Selected oxidative stress markers in a South American crocodilian species. *Comparative Biochemistry and Physiology C*, 146, 241–254.

Fussnecker, B. & Grozinger, C. (2008) Dissecting the role of Kr-h1 brain gene expression in foraging behavior in honey bees (*Apis mellifera*). *Insect Biochemistry & Molecular Biology*, 17, 515–522.

Gadgil, M. & Bossert, W. H. (1970) Life historical consequences of natural selection. *The American Naturalist*, 104, 1–24.

Gadot, M., Chiang, A. S., Burns, E. L. & Schal, C. (1991) Cyclic juvenile hormone biosynthesis in the cockroach, *Blattella germanica*: effects of ovariectomy and corpus allatum denervation. *General and Comparative Endocrinology*, 82, 163–171.

Gahete, M. D., Duran-Prado, M., Luque, R. M., Martinez-Fuentes, A. J., Quintero, A., Gutierrez-Pascual, E., Cordoba-Chacon, J., Malagon, M. M., Gracia-Navarro, F. & Castano, J. P. (2009) Understanding the multifactorial control of growth hormone release by somatotropes: lessons from comparative endocrinology. *Annals of the New York Academy of Sciences*, 1163, 137–153.

Gahr, M. (2001) Distribution of sex steroid hormone receptors in the avian brain: Functional implications for neural sex differences and sexual behaviors. *Microscopy Research and Technique*, 55, 1–11.

Galton, V. A. (1990) Mechanisms underlying the acceleration of thyroid hormone-induced tadpole metamorphosis by corticosterone. *Endocrinology*, 127, 2997–3002.

Galton, V. A. (1992) Thyroid hormone receptors and iodothyronine deiodinases in the developing Mexican axolotl, *Ambystoma mexicanum*. *General and Comparative Endocrinology*, 85, 62–70.

Galton, V. A. (2005) The roles of the iodothyronine deiodinases in mammalian development. *Thyroid*, 15, 823–834.

Garamszegi, L. Z., Hirschenhauser, K., Bokony, V., Eens, M., Hurtrez-Bousses, S., Moller, A. P., Oliveira, R. F. & Wingfield, J. C. (2008) Latitudinal distribution, migration, and testosterone levels in birds. *The American Naturalist*, 172, 533–546.

Garigan, D., Hsu, A. L., Fraser, A. G., Kamath, R. S., Ahringer, J. & Kenyon, C. (2002) Genetic analysis of tissue aging in *Caenorhabditis elegans*: a role for heat-shock factor and bacterial proliferation. *Genetics*, 161, 1101–1112.

Garrigan, D. & Hedrick, P. W. (2003) Perspective: detecting adaptive molecular polymorphism: lessons from the MHC. *Evolution*, 57, 1707–1722.

Garstang, W. 1928. The origin and evolution of larval forms. Report of the British Association for the Advancement of Science (D) 77–98.

Gateño, D. & Rinkevich, B. (2003) Coral polyp budding is probably promoted by a canalized ratio of two morphometric fields. *Marine Biology*, 142, 971–973.

Gauchat, D., Mazet, F., Berney, C., Schummer, M., Kreger, S., Pawlowski, J. & Galliot, B. (2000) Evolution of Antp-class genes and differential expression of *Hydra* Hox/paraHox genes in anterior patterning. *Proceedings of the National Academy of Sciences of the United States of America*, 97, 4493–4498.

Gazzani, S., Gendall, A. R., Lister, C. & Dean, C. (2003) Analysis of the molecular basis of flowering time variation in *Arabidopsis* accessions. *Plant Physiology*, 132, 1107–1114.

Geary, D. C. (2005) *The Origin of Mind: Evolution of brain, cognition, and general intelligence*. Washington D.C., The American Psychological Association.

Geary, D. C. & Bjorklund, D. F. (2000) Evolutionary developmental psychology. *Child Development*, 71, 57–65.

Geary, D. C. & Flinn, M. V. (2001) Evolution of human parental behavior and the human family. *Parenting, Science and Practice*, 1, 5–61.

Geary, D. C. & Flinn, M. V. (2002) Sex differences in behavioral and hormonal response to social threat. *Psychological Review*, 109, 745–750.

Geary, D. C. & Huffman, K. J. (2002) Brain and cognitive evolution: forms of modularity and functions of mind. *Psychological Bulletin*, 128, 667–698.

Geiser, F. (2004) Metabolic rate and body temperature reduction during hibernation and daily torpor. *Annual Review of Physiology*, 66, 239–274.

Geiser, F. & Learmonth, R. P. (1994) Dietary fats, selected body temperature and tissue fatty acid composition of agamid lizards (*Amphibolurus nuchalis*). *Journal of Comparative Physiology B*, 164, 55–61.

Geiser, F., Firth, B. T. & Seymour, R. S. (1992) Polyunsaturated dietary lipids lower the selected body temperature of a lizard. *Journal of Comparative Physiology B*, 162, 1–4.

Geister, T. L., Lorenz, M. W., Hoffmann, K. H. & Fischer, K. (2008a) Effects of the NMDA receptor antagonist MK-801 on female reproduction and juvenile hormone biosynthesis in the cricket *Gryllus bimaculatus* and the butterfly *Bicyclus anynana*. *Journal of Experimental Biology*, 211, 1587–1593.

Geister, T. L., Lorenz, M. W., Meyering-Vos, M., Hoffmann, K. H. & Fischer, K. (2008b) Effects of temperature on reproductive output, egg provisioning, juvenile hormone and vitellogenin titres in the butterfly *Bicyclus anynana*. *Journal of Insect Physiology*, 54, 1253–1260.

Geister, T. L., Lorenz, M. W., Hoffmann, K. H. & Fischer, K. (2009) Energetics of embryonic development: effects of temperature on egg and hatchling composition in a butterfly. *Journal of Comparative Physiology B*, 179, 87–98.

Geller, J. B. & Walton, E. D. (2001) Breaking up and getting together: evolution of symbiosis and cloning by fission in sea anemones (Genus *Anthopleura*). *Evolution*, 55, 1781–1794.

Geller, J. B., Fitzgerald, L. J. & King, C. E. (2005) Fission in sea anemones: integrative studies of life cycle evolution. *Integrative and Comparative Biology*, 45, 615–622.

Gems, D., Pletcher, S. & Partridge, L. (2002) Interpreting interactions between treatments that slow aging. *Aging Cell*, 1, 1–9.

Gems, D., Sutton, A. J., Sundermeyer, M. L., Albert, P. S., King, K. V., Edgley, M. L., Larsen, P. L. & Riddle, D. L. (1998) Two pleiotropic classes of *daf-2* mutation affect larval arrest, adult behavior, reproduction and longevity in *C. elegans*. *Genetics*, 150, 129–155.

Genade, T., Benedetti, M., Terzibasi, E., Roncaglia, P., Valenzano, D. R., Cattaneo, A. & Cellerino, A. (2005) Annual fishes of the genus *Nothobranchius* as a model system for aging research. *Aging Cell*, 4, 223–233.

George, S. B. (1994) Phenotypic plasticity in the larvae of *Luidia foliolata* (Echinodermata: Asteroidea). In David, B., Guille, A., Féral, J. & Roux, M. (Eds.) *Echinoderms Through Time*. Rotterdam, Balkema.

George, S. B. (1999) Egg quality, larval growth and phenotypic plasticity in a forcipulate seastar. *Journal of Experimental Marine Biology and Ecology*, 237, 203–224.

Geraldo, N., Baurle, I., Kidou, S.-I., Hu, X. Y. & Dean, C. (2009) *FRIGIDA* delays flowering in *Arabidopsis* via a cotranscriptional mechanism involving direct interaction with the nuclear cap-binding complex. *Plant Physiology*, 150, 1611–1618.

Gerisch, B. & Antebi, A. (2004) Hormonal signals produced by DAF-9/cytochrome P450 regulate *C. elegans* dauer diapause in response to environmental cues. *Development*, 131, 1765–1776.

Gerisch, B., Rottiers, V., Li, D. L., Motola, D. L., Cummins, C. L., Lehrach, H., Mangelsdorf, D. J. & Antebi, A. (2007) A bile acid-like steroid modulates *Caenorhabditis elegans* lifespan through nuclear receptor signaling. *Proceedings of the National Academy of Sciences of the United States of America*, 104, 5014–5019.

Geven, E. J. W., Nguyen, N. K., Van Den Boogaart, M., Spanings, F. A. T., Flik, G. & Klaren, P. H. M. (2007) Comparative thyroidology: thyroid gland location and iodothyronine dynamics in Mozambique tilapia (*Oreochromis mossambicus* Peters) and common carp (*Cyprinus carpio* L.). *Journal of Experimental Biology*, 210, 4005–4015.

Ghazanfar, A. A. & Santos, L. R. (2004) Primate brains in the wild: The sensory bases for social interactions. *Nature Reviews Neuroscience*, 5, 603–616.

Ghazi, A., Henis-Korenblit, S. & Kenyon, C. (2009) A transcription elongation factor that links signals from the reproductive system to lifespan extension in *Caenorhabditis elegans*. *PLoS Genetics*, 5, e1000639.

Giannakou, M. E. & Partridge, L. (2007) Role of insulin-like signalling in *Drosophila* lifespan. *Trends in Biochemical Sciences*, 32, 180–188.

Giannakou, M. E., Goss, M., Junger, M. A., Hafen, E., Leevers, S. J. & Partridge, L. (2004) Long-lived *Drosophila* with overexpressed dFOXO in adult fat body. *Science*, 305, 361.

Giannakou, M. E., Goss, M., Jacobson, J., Vinti, G., Leevers, S. J. & Partridge, L. (2007) Dynamics of the action of dFOXO on adult mortality in *Drosophila*. *Aging Cell*, 6, 429–438.

Giannakou, M. E., Goss, M. & Partridge, L. (2008) Role of dFOXO in lifespan extension by dietary restriction in *Drosophila melanogaster*: not required, but its activity modulates the response. *Aging Cell*, 7, 187–198.

Gifondorwa, D. J., Robinson, M. B., Hayes, C. D., Taylor, A. R., Prevette, D. M., Oppenheim, R. W., Caress, J. & Milligan, C. E. (2007) Exogenous delivery of heat shock protein 70 increases lifespan in a mouse model of amyotrophic lateral sclerosis. *Journal of Neuroscience*, 27, 13173–13180.

Gilbert, S. F. & Epel, D. (2009) *Ecological developmental biology: integrating epigenetics, medicine, and evolution*. Sunderland, Sinauer Associates.

Gilchrist, A. S. & Partridge, L. (1999) A comparison of the genetic basis of wing size divergence in three parallel body size clines of *Drosophila melanogaster*. *Genetics*, 153, 1775–1787.

Gilchrist, G. W., Huey, R. B. & Serra, L. (2001) Rapid evolution of wing size clines in *Drosophila subobscura*. *Genetica*, 112, 273–286.

Gilchrist, G. W., Huey, R. B., Balanya, J., Pascual, M. & Serra, L. (2004) A time series of evolution in action: a latitudinal cline in wing size in South American *Drosophila subobscura*. *Evolution*, 58, 768–780.

Gillooly, J. F., Charnov, E. L., West, G. B., Savage, V. M. & Brown, J. H. (2002) Effects of size and temperature on developmental time. *Nature*, 417, 70–73.

Glennemeier, K. A. & Denver, R. J. (2002) Small changes in whole-body corticosterone content affect larval *Rana pipiens* fitness components. *General and Comparative Endocrinology*, 127, 16–25.

Gluckman, P. D., Hanson, M. A. & Spencer, H. G. (2005) Predictive adaptive responses and human evolution. *Trends in Ecology & Evolution*, 20, 527–533.

Gockel, J., Kennington, W. J., Hoffman, A. A., Goldstein, D. B. & Partridge, L. (2001) Non-clinality of molecular variation implicates selection in maintaining a morphological cline of *Drosophila melanogaster*. *Genetics*, 158, 319–323.

Golden, J. W. & Riddle, D. L. (1984) The *C. elegans* dauer larva: developmental effects of pheromone, food, and temperature. *Developmental Biology*, 102, 368–378.

Gomez-Merino, D., Drogou, C., Chennaoui, M., Tiolloer, E., Mathieu, J. & Guezennec, C. Y. (2005) Effects of combined stress during intense training on cellular immunity, hormones and respiratory infections. *Neuroimmunomodulation*, 12, 164–172.

Good, R. A. & Lorenz, E. (1992) Nutrition and cellular immunity. *International Journal of Immunopharmacology*, 14, 361–366.

Goodrick, C. L. (1978) Body weight increment and length of life: the effect of genetic constitution and dietary protein. *Journal of Gerontology*, 33, 184–190.

Gorbman, A. & Bern, H. A. (1962) *A textbook of comparative endocrinology*. New York, John Wiley and Sons, Inc.

Gordon, M. D., Dionne, M. S., Schneider, D. S. & Nusse, R. (2005) WntD is a feedback inhibitor of Dorsal/NF-kappaB in *Drosophila* development and immunity. *Nature*, 437, 746–749.

Gordon, M. D., Ayres, J. S., Schneider, D. S. & Nusse, R. (2008) Pathogenesis of *Listeria*-infected *Drosophila* wntD mutants is associated with elevated levels of the novel immunity gene *edin*. *PLoS Pathogens*, 4, e1000111.

Goto, S. G., Yoshida, K. M. & Kimura, M. T. (1998) Accumulation of Hsp70 mRNA under environmental stresses in diapausing and nondiapausing adults of *Drosophila triauraria*. *Journal of Insect Physiology*, 44, 1009–1015.

Gotthard, K. (2008) Adaptive growth decisions in butterflies. *Bioscience, Biotechnology, and Biochemistry*, 58, 222–230.

Gotthard, K., Nylin, S. & Wiklund, C. (1994) Adaptive variation in growth rate: life history costs and consequences in the speckled wood butterfly, *Pararge aegeria*. *Oecologia*, 99, 281–289.

Govindarajan, V., Ramachandran, R. K., George, J. M., Shakes, D. C. & Tomlinson, C. R. (1995) An ECM-bound, PDGF-like growth factor and a TGF-alpha-like growth factor are required for gastrulation and spiculogenesis in the Lytechinus embryo. *Developmental Biology*, 172, 541–551.

Goymann, W. (2009) Social modulation of androgens in male birds. *General and Comparative Endocrinology*, 163, 149–157.

Goymann, W., Moore, I. T., Scheuerlein, A., Hirschenhauser, K., Grafen, A. & Wingfield, J. C. (2004) Testosterone in tropical birds: effects of environmental and social factors. *The American Naturalist*, 164, 327–334.

Goymann, W., Geue, D., Schwabl, I., Flinks, H., Schmidl, D., Schwabl, H. & Gwinner, E. (2006) Testosterone and corticosterone during the breeding cycle of equatorial and European stonechats (*Saxicola torquata axillaris* and *S. t. rubicola*). *Hormones and Behavior*, 50, 779–785.

Gräff, J., Jemielity, S., Parker, J. D., Parker, K. M. & Keller, L. (2007) Differential gene expression between adult queens and workers in the ant *Lasius niger*. *Molecular Ecology*, 16, 675–683.

Grandison, R. C., Piper, M. D. W. & Partridge, L. (2009) Amino-acid imbalance explains extension of lifespan by dietary restriction in *Drosophila*. *Nature*, 462, 1061–1064.

Grasso, L., Maindonald, J., Rudd, S., Hayward, D., Saint, R., Miller, D. & Ball, E. (2008) Microarray analysis identifies candidate genes for key roles in coral development. *BMC Genomics*, 9, 540.

Gray, P. B. (2003) Marriage, parenting, and testosterone variation among Kenyan Swahili men. *American Journal of Physical Anthropology*, 122, 279–286.

Gray, P. B. & Campbell, B. C. (2009) Human male testosterone, pair bonding and fatherhood. In Ellison, P. T. & Gray, P. B. (Eds.) *Endocrinology of Social Relationships*. Cambridge, Harvard University Press.

Gray, S. M. & McKinnon, J. S. (2007) Linking color polymorphism maintenance and speciation. *Trends in Ecology & Evolution*, 22, 71–79.

Gray, A., Feldman, H. A., McKinlay, J. B. & Longcope, C. (1991) Age, disease, and changing sex hormone levels in middle-aged men: results of the Massachusetts Male Aging Study. *Journal of Clinical Endocrinology and Metabolism*, 73, 1016–1025.

Gray, P. B., Kahlenberg, S. M., Barrett, E. S., Lipson, S. F. & Ellison, P. T. (2002) Marriage and fatherhood are associated with lower testosterone in males. *Evolution and Human Behavior*, 23, 193–201.

Gray, D. A., Tsirigotis, M. & Woulfe, J. (2003) Ubiquitin, proteasomes, and the aging brain. *Science of Aging Knowledge Environment*, 34, p. rE6.

Gray, P. B., Yang, C. F. & Pope, H. G. J. (2006) Fathers have lower salivary testosterone levels than unmarried men and married non-fathers in Beijing, China. *Proceedings of the Royal Society of London B*, 273, 333–339.

Gray, P. B., Parkin, J. C. & Samms-Vaughan, M. E. (2007) Hormonal correlates of human paternal interactions: a hospital-based investigation in urban Jamaica. *Hormones and Behavior*, 52, 499–507.

Green, R. F. & Noakes, D. L. G. (1995) Is a little bit of sex as good as a lot? *Journal of Theoretical Biology*, 174, 87–96.

Greene, E. (1996) Effect of light quality and larval diet on morph induction in the polymorphic caterpillar *Nemoria arizonaria* (Lepidoptera: Geometridae). *Biological Journal of the Linnean Society*, 58, 277–285.

Greer, E. L. & Brunet, A. (2008a) FOXO transcription factors in aging and cancer. *Acta Physiologica*, 192, 19–28.

Greer, E. L. & Brunet, A. (2008b) Signaling networks in aging. *Journal of Cell Science*, 121, 407–412.

Greer, E. L. & Brunet, A. (2009) Different dietary restriction regimens extend lifespan by both independent and overlapping genetic pathways in *C. elegans*. *Aging Cell*, 8, 113–127.

Greer, E. L., Dowlatshahi, D., Banko, M. R., Villen, J., Hoang, K., Blanchard, D., Gygi, S. P. & Brunet, A. (2007) An AMPK-FOXO pathway mediates longevity induced by a novel method of dietary restriction in *C. elegans*. *Current Biology*, 17, 1646–1656.

Grimaldi, D. & Engel, M. S. (2005) *Evolution of the Insects*. Cambridge University Press, Cambridge, USA

Grönke, S., Clarke, D. F., Andrews, T. D., Broughton, S., Andrews, D. & Partridge, L. (2010) Molecular evolution and mutagenesis of *Drosophila* insulin-like peptides. *PLoS Genetics*, 6, e1000857.

Groothuis, T. G. G., Muller, W., Von Engelhardt, N., Carere, C. & Eising, C. (2005) Maternal hormones as a tool to adjust offspring phenotype in avian species. *Neuroscience and Biobehavioral Reviews*, 29, 329–352.

Gross, T. N. & Manzon, R. G. (2011) Sea lamprey (Petromyzon marinus) contain four developmentally regulated serum thyroid hormone distributor proteins. *General and Comparative Endocrinology*, 170, 640–649.

Grossman, A. (2000) Acclimation of *Chlamydomonas reinhardtii* to its nutrient environment. *Protist*, 151, 201–224.

Grumbach, M. M. & Styne, D. M. (1998) Puberty: ontogeny, neuroendocrinology, physiology and disorders. In Wilson, J. D., Foster, D. W., Kronenberg, H. M. & Larsen, P. R. (Eds.) *Williams Textbook of Endocrinology*, 9th edn. Philadelphia, W.B. Saunders.

Gruntenko, N. E. & Rauschenbach, I. Y. (2008) Interplay of JH, 20E and biogenic amines under normal and stress conditions and its effect on reproduction. *Journal of Insect Physiology*, 54, 902–908.

Guarente, L. (2000) Sir2 links chromatin silencing, metabolism, and aging. *Genes & Development*, 14, 1021–1026.

Guatelli-Steinberg, D., Reid, D. J., Bishop, T. A. & Larsen, C. S. (2005) Anterior tooth growth periods in Neandertals were comparable to those of modern humans. *Proceedings of the National Academy of Sciences of the United States of America*, 102, 14197–14202.

Gubernick, D. J. (1990) A maternal chemosignal maintains paternal behaviour in the biparental California mouse, *Peromyscus californicus*. *Animal Behaviour*, 39, 936–942.

Gudernatsch, J. F. (1912) Feeding Experiments on tadpoles. I. The influence of specific organs given as food on growth and differentiation. A contribution to the knowledge of organs with internal secretion. *Roux's Archives of Developmental Biology*, 35, 457–483.

Guidugli, K. R., Nascimento, A. M., Amdam, G. V., Barchuk, A. R., Omholt, S., Simoes, Z. L. & Hartfelder, K. (2005) Vitellogenin regulates hormonal dynamics in the worker caste of a eusocial insect. *FEBS Letters*, 579, 4961–4965.

Guidugli-Lazzarini, K. R., Do Nascimento, A. M., Tanaka, E. D., Piulachs, M. D., Hartfelder, K., Bitondi, M. G. & Simoes, Z. L. (2008) Expression analysis of putative vitellogenin and lipophorin receptors in honey bee (*Apis mellifera* L.) queens and workers. *Journal of Insect Physiology*, 54, 1138–1147.

Guillette, L. J. J. (1993) The evolution of vivparity in lizards. *BioScience*, 43, 742–751.

Guillette, L. J. J., Cox, M. C. & Crain, D. A. (1996) Plasma Insulin-like Growth Factor-I concentration during the reproductive cycle of the American alligator (*Alligator mississippiensis*). *General and Comparative Endocrinology*, 104, 116–122.

Guss, K. A. & Ettensohn, C. A. (1997) Skeletal morphogenesis in the sea urchin embryo: regulation of primary mesenchyme gene expression and skeletal rod growth by ectoderm-derived cues. *Development*, 124, 1899–1908.

Gustafsson, L. & Sutherland, W. J. (1988) The costs of reproduction in the collared flycatcher *Ficedula albicollis*. *Nature*, 335, 813–815.

Guzick, D. S., Overstreet, J. W., Factor-Litvak, P., Brazil, C. K., Nakajima, S. T., Coutifaris, C., Carson, S. A., Cisneros, P., Steinkampf, M. P., Hill, J. A., Xu, D. & Vogel, D. L. (2001) Sperm morphology, motility and concentration in fertile and infertile men. *New England Journal of Medicine*, 345, 1388–1393.

Gwadz, R. W. (1969) Regulation of blood meal size in the mosquito. *Journal of Insect Physiology*, 15, 2039–2044.

Haddad, L. S., Kelbert, L. & Hulbert, A. J. (2007) Extended longevity of queen honey bees compared to workers is associated with peroxidation-resistant membranes. *Experimental Gerontology*, 42, 601–609.

Hadfield, M. G. & Paul, J. V. (2001) Natural Chemical Cues for Settlement and Metamorphosis of Marine-Invertebrate Larvae. In McClintock, J. B. & Baker, B. J. (Eds.) *Marine Chemical Ecology*. Boca Raton, CRC Press.

Hadfield, M. G., Meleshkevitch, E. A. & Boudko, D. Y. (2000) The apical sensory organ of a gastropod veliger is a receptor for settlement cues. *Biological Bulletin*, 198, 67–76.

Hadfield, M. G., Carpizo-Ituarte, E. J., Del Carmen, K. & Nedved, B. T. (2001) Metamorphic competence, a major adaptive convergence in marine invertebrate larvae. *American Zoologist*, 41, 1123–1131.

Hahn, T. P., Boswell, T., Wingfield, J. C. & Ball, G. F. (1997) Temporal flexibility in avian reproduction. In Nolan JR., V., Ketterson, E. D. & Thompson, C. F. (Eds.) *Current Ornithology*. New York, Plenum Press.

Hahm, J. H., Kim, S. & Paik, Y. K. (2009) Endogenous cGMP regulates adult longevity via the insulin signaling pathway in *Caenorhabditis elegans*. *Aging Cell*, 8, 473–483.

Haig, D. (1996) Gestational drive and the green-bearded placenta. *Proceedings of the National Academy of Sciences of the United States of America*, 93, 6547–6551.

Haig, D. & Westoby, M. (1989) Parent-specific gene expression and the triploid endosperm. *The American Naturalist*, 134, 147–155.

Haigis, M. C. & Guarente, L. P. (2006) Mammalian sirtuins – emerging roles in physiology, aging, and calorie restriction. *Genes & Development*, 20, 2913–2921.

Halme, A., Cheng, M. & Hariharan, I. K. (2010) Retinoids regulate a developmental checkpoint for tissue regeneration in *Drosophila*. *Current Biology*, 20, 458–463.

Hamilton, W. D. (1964) The evolution of social behavior. *Journal of Theoretical Biology*, 7, 1–52.

Hamilton, W. D. (1966) The moulding of senescence by natural selection. *Journal of Theoretical Biology*, 12, 12–45.

Hamilton, W. D. (1967) Extraordinary sex ratios. *Science*, 156, 477–488.

Hamilton, W. D. (1980) Sex versus non-sex versus parasite. *Oikos*, 35, 282–290.

Hamilton, W. D., Axelrod, R. & Tanese, R. (1990) Sexual reproduction as an adaptation to resist parasites (a review). *Proceedings of the National Academy of Sciences of the United States of America*, 87, 3566–3573.

Hansen, T. F. (2003) Is modularity necessary for evolvability? Remarks on the relationship between pleiotropy and evolvability. *Biosystems*, 69, 83–94.

Hansen, T. F. & Houle, D. (2004) Evolvability, stabilizing selection, and the problem of stasis. In Pigliucci, M. & Preston, K. (Eds.) *The Evolutionary Biology of Complex Phenotypes*. Oxford, Oxford University Press.

Hansen, M., Hsu, A. L., Dillin, A. & Kenyon, C. (2005) New genes tied to endocrine, metabolic, and dietary regulation of lifespan from a *Caenorhabditis elegans* genomic RNAi screen. *PLoS Genetics*, 1, 119–128.

Hansen, M., Taubert, S., Crawford, D., Libina, N., Lee, S. J. & Kenyon, C. (2007) Lifespan extension by conditions that inhibit translation in *Caenorhabditis elegans*. *Aging Cell*, 6, 95–110.

Hansen, M., Chandra, A., Mitic, L. L., Onken, B., Driscoll, M. & Kenyon, C. (2008) A role for autophagy in the extension of lifespan by dietary restriction in *C. elegans*. *PLoS Genetics*, 4, e24.

Hard, J. J., Gross, M. R., Heino, M., Hilborn, R., Kope, R. G., Law, R. & Reynolds, J. D. (2008) Evolutionary consequences of fishing and their implications for salmon. *Evolutionary Applications*, 1, 388–408.

Hardin, J., Coffman, J. A., Black, S. D. & McClay, D. R. (1992) Commitment along the dorsoventral axis of the sea urchin embryo is latered in response to $NiCl_2$. *Development*, 116, 671–685.

Harman, D. (1956) Aging: a theory based on free radical and radiation chemistry. *Journal of Gerontology*, 11, 298–300.

Harman, S. M., Metter, E. J., Tobin, J. D., Pearson, J. D. & Blackman, M. R. (2001) Longitudinal effects of aging on serum total and free testosterone levels in healthy men, Baltimore Longitudinal Study of Aging. *Journal of Clinical Endocrinology and Metabolism*, 86, 724–731.

Harper, J. M., Leathers, C. W. & Austad, S. N. (2006) Does caloric restriction extend life in wild mice? *Aging Cell*, 5, 441–449.

Harrington, L. S., Findlay, G. M., Gray, A., Tolkacheva, T., Wigfield, S., Rebholz, H., Barnett, J., Leslie, N. R., Cheng, S., Shepherd, P. R., Gout, I., Downes, C. P. & Lamb, R. F. (2004) The TSC1-2 tumor suppressor controls insulin-PI3K signaling via regulation of IRS proteins. *Journal of Cell Biology*, 166, 213–223.

Harris, M. P., Fallon, J. F. & Prum, R. O. (2002) Shh-Bmp2 signaling module and the evolutionary origin and diversification of feathers. *Journal of Experimental Zoology*, 294, 160–176.

Harrison, J. M. (1986) Caste-specific changes in honeybee flight capacity. *Physiological Zoology*, 59, 175–187.

Harrison, J. F. & Fewell, J. H. (2002) Environmental and genetic influences on flight metabolic rate in the honey bee, *Apis mellifera*. *Comparative Biochemistry and Physiology A*, 133, 323–333.

Harrison, D. E., Strong, R., Sharp, Z. D., Nelson, J. F., Astle, C. M., Flurkey, K., Nadon, N. L., Wilkinson, J. E., Frenkel, K., Carter, C. S., Pahor, M., Javors, M. A., Fernandez, E.

& Miller, R. A. (2009) Rapamycin fed late in life extends lifespan in genetically heterogeneous mice. *Nature*, 460, 392–395.

Harshman, L. G. (1999) Investigation of the endocrine system in extended longevity lines of *Drosophila melanogaster*. *Experimental Gerontology*, 34, 997–1006.

Harshman, L. G. & Haberer, B. A. (2000) Oxidative stress resistance: a robust correlated response to selection in extended longevity lines of *Drosophila melanogaster*? *Journal of Gerontology A*, 55, B415–B417.

Harshman, L. G. & Zera, A. J. (2007) The cost of reproduction: the devil in the details. *Trends in Ecology & Evolution*, 22, 80–86.

Harshman, L. G., Ottea, J. A. & Hammock, B. D. (1991) Evolved environment-dependent expression of detoxification enzyme activity in *Drosophila melanogaster*. *Evolution*, 45, 791–795.

Harshman, L. G., Hoffmann, A. A. & Clark, A. G. (1999) Selection for starvation resistance in *Drosophila melanogaster*: physiological correlates, enzyme activities and multiple stress responses. *Journal of Evolutionary Biology*, 12, 370–379.

Hart, M. W. (1991) Particle captures and the method of suspension feeding by Echinoderm larvae. *Biological Bulletin*, 180, 12–27.

Hart, M. W. (1996) Evolutionary loss of larval feeding: Development, form and function in a facultatively feeding larva, *Brisaster latifrons*. *Evolution*, 50, 174–187.

Hart, M. W. & Strathmann, R. R. (1994) Functional consequences of phenotypic plasticity in echinoid larvae. *Biological Bulletin* 168, 291–299.

Hartfelder, K. & Engels, W. (1998) Social insect polymorphism: hormonal regulation of plasticity in development and reproduction in the honeybee. *Current Topics in Developmental Biology*, 40, 45–77.

Hartnoll, R. G. (2001) Growth in Crustacea – twenty years on. *Hydrobiologia*, 449, 111–122.

Harvell, C. D. (1990) The ecology and evolution of inducible defenses. *Quarterly Review of Biology*, 65, 323–340.

Harvey, W. (1651) *Excercitationes De Generatione Animalium* (Anatomical exercitations concerning the generation of living creatures). London, Pulleyn, 1651.

Harvey, P. H. & Clutton-Brock, T. H. (1985) Life history variation in primates. *Evolution*, 39, 559–581.

Harvie, P. D., Filippova, M. & Bryant, P. J. (1998) Genes expressed in the ring gland, the major endocrine organ of *Drosophila melanogaster*. *Genetics*, 149, 217–231.

Hashimoto, T. & Watanabe, S. (2005) Chronic food restriction enhances memory in mice - analysis with matched drive levels. *Neuroreport*, 16, 1129–1133.

Hasty, P., Campisi, J., Hoeijmakers, J., Van Steeg, H. & Vijg, J. (2003) Aging and genome maintenance: lessons from the mouse? *Science*, 299, 1355–1359.

Hasunuma, I., Yamamoto, K. & Kikuyama, S. (2004) Molecular cloning of bullfrog prolactin receptor cDNA: changes in prolactin receptor mRNA level during metamorphosis. *General and Comparative Endocrinology*, 138, 200–210.

Hatase, H., Sudo, R., Watanabe, K. K., Kasugai, T., Saito, T., Okamoto, H., Uchida, I. & Tsukamoto, K. (2008) Shorter telomere length with age in the loggerhead turtle: a new hope for live sea turtle age estimation. *Genes & Genetic Systems*, 83, 423–426.

Hau, M. (2007) Regulation of male traits by testosterone: implications for the evolution of vertebrate life histories. *BioEssays*, 29, 133–144.

Hau, M., Wikelski, M., Soma, K. K. & Wingfield, J. C. (2000) Testosterone and year-round territorial aggression in a tropical bird. *General and Comparative Endocrinology*, 117, 20–33.

Hau, M., Perfito, N. & Moore, I. T. (2008) Timing of breeding in tropical birds: mechanisms and evolutionary implications. *Ornitologia Neotropical*, 19(suppl), 39–59.

Hau, M., Ricklefs, R. E., Wikelski, M., Lee, K. A. & Brawn, J. D. (2010) Corticosterone, testosterone and life history strategies of birds. *Proceedings of the Royal Society London B*, 277, 3203–3212.

Haussmann, M. F., Winkler, D. W., Huntington, C. E., Nisbet, I. C. T. & Vleck, C. M. (2007) Telomerase activity is maintained throughout the lifespan of long-lived birds. *Experimental Gerontology*, 42, 610–618.

Havelock, J. C., Auchus, R. J. & Rainey, W. E. (2004) The rise in adrenal androgen biosynthesis: adrenarche. *Seminars in Reproductive Medicine*, 22, 337–347.

Havenhand, J. N. (1993) Egg to juvenile period, generation time, and the evolution of larval type in marine invertebrates. *Marine Ecology Progress Series*, 97, 247–260.

Hawdon, J. M. & Datu, B. (2003) The second messenger cyclic GMP mediates activation in *Ancylostoma caninum* infective larvae. *International Journal for Parasitology*, 33, 787–793.

Hawkes, K. (2003) Grandmothers and the evolution of human longevity. *American Journal of Human Biology*, 15, 380–400.

Hawkes, K., O'Connell, J. F. & Blurton Jones, N. G. (1989) Hardworking Hadza grandmothers. In Standen, V. & Foley, R. A. (Eds.) *Comparative Socioecology: The Behavioural Ecology of Humans and Other Mammals*. London, Basil Blackwell.

Hawkes, K., O'Connell, J. F., Blurton Jones, N. G., Alvarez, H. & Charnov, E. L. (1998) Grandmothering, menopause, and the evolution of human life histories. *Proceedings of the National Academy of Sciences of the United States of America*, 95, 1336–1339.

Hay, J. M., Subramanian, S., Millar, C. D., Mohandesan, E. & Lambert, D. M. (2008) Rapid molecular evolution in a living fossil. *Trends in Genetics*, 24, 106–109.

Hayashi, H., Yamaza, H., Komatsu, T., Park, S., Chiba, T., Higami, Y., Nagayasu, T. & Shimokawa, I. (2008) Calorie restriction minimizes activation of insulin signaling in response to glucose: potential involvement of the growth hormone-insulin-like growth factor 1 axis. *Experimental Gerontology*, 43, 827–32.

Hayes, T. B. (1995) Interdependence of corticosterone and thyroid hormones in larval toads (*Bufo boreas*). I. Thyroid hormone-dependent and independent effects of corticosterone on growth and development. *Journal of Experimental Zoology*, 271, 95–102.

Hayes, T. B. (1997) Steroids as potential modulators of thyroid hormone activity in anuran metamorphosis. *American Zoologist*, 37, 185–194.

Hazel, W. N. (2002) The environmental and genetic control of seasonal polyphenism in larval color and its adaptive significance in a swallowtail butterfly. *Evolution*, 56, 342–348.

Hedner, E., Sjogren, M., Frandberg, P. A., Johansson, T., Goransson, U., Dahlstrom, M., Jonsson, P., Nyberg, F. & Bohlin, L. (2006) Brominated cyclodipeptides from the marine sponge *Geodia barretti* as selective 5-HT ligands. *Journal of Natural Products*, 69, 1421–1424.

Hedrick, S. M. (2009) The cunning little vixen: Foxo and the cycle of life and death. *Nature Immunology*, 10, 1057–1063.

Heidel, A. J., Clarke, J. D., Antonovics, J. & Dong, X. N. (2004) Fitness costs of mutations affecting the systemic acquired resistance pathway in *Arabidopsis thaliana*. *Genetics*, 168, 2197–2206.

Heideman, P. D. (2004) Top-down approaches to the study of natural variation in complex physiological pathways using the white-footed mouse (*Peromyscus leucopus*) as a model. *ILAR Journal*, 45, 4–13.

Heideman, P. D. & Pittman, J. T. (2009) Microevolution of neuroendocrine mechanisms regulating reproductive timing in *Peromyscus leucopus*. *Integrative and Comparative Biology*, 49, 550–562.

Hekimi, S., Burgess, J., Bussiere, F., Meng, Y. & Benard, C. (2001) Genetics of lifespan in *C. elegans*: molecular diversity, physiological complexity, mechanistic simplicity. *Trends in Genetics*, 17, 712–718.

Held, J. M., White, M. P., Fisher, A. L., Gibson, B. W., Lithgow, G. J. & Gill, M. S. (2006) DAF-12-dependent rescue of dauer formation in *Caenorhabditis elegans* by (25S)-cholestenoic acid. *Aging Cell*, 5, 283–291.

Helle, S., Lummaa, V. & Jokela, J. (2002) Sons reduced maternal longevity in preindustrial humans. *Science*, 296, 1085.

Henderson, S. T. & Johnson, T. E. (2001) *daf-16* integrates developmental and environmental inputs to mediate aging in the nematode *C. elegans*. *Current Biology*, 11, 1975–1980.

Henderson, S. T., Bonafe, M. & Johnson, T. E. (2006) *daf-16* protects the nematode *Caenorhabditis elegans* during food deprivation. *Journal of Gerontology A*, 61, 444–460.

Hendler, G. (1978) Development of *Amphioplus abditus* (Verrill) (Echinodermata: Ophiuroidea). II. Description and discussion of ophiuroid skeletal ontogeny and homologies. *Biological Bulletin*, 154, 79–95.

Hendler, G. & Dojiri, M. (2009) The contrariwise life of a parasitic, pedomorphic copepod with a non-feeding adult: ontogenesis, ecology, and evolution. *Invertebrate Biology*, 128, 65–82.

Hennemann, G., Docter, R., Friesema, E. C., De Jong, M., Krenning, E. P. & Visser, T. J. (2001) Plasma membrane transport of thyroid hormones and its role in thyroid hormone metabolism and bioavailability. *Endocrine Reviews*, 22, 451–476.

Hentschel, B. T. & Emlet, R. B. (2000) Metamorphosis of barnacle nauplii: Effects of food variability and a comparison with amphibian models. *Ecology*, 81, 3495–3508.

Herlihy, J. T., Stacy, C. & Bertrand, H. A. (1992) Long-term calorie restriction enhances baroreflex responsiveness in Fischer 344 rats. *American Journal of Physiology. Heart and Circulatory Physiology*, 263, H1021–H1025.

Herman, W. S. & Tatar, M. (2001) Juvenile hormone regulation of longevity in the migratory monarch butterfly. *Proceedings of the Royal Society of London B*, 268, 2509–2514.

Herre, E. (1993) Population structure and the evolution of virulence in nematode parasites of fig wasps. *Science*, 259, 1442–1445.

Herron, M. D. & Michod, R. E. (2008) Evolution of complexity in the volvocine algae: Transitions in individuality through Darwin's eye. *Evolution*, 62, 436–451.

Herskind, A. M., McGue, M., Holm, N. V., Sorensen, T. I., Harvald, B. & Vaupel, J. W. (1996) The heritability of human longevity: a population-based study of 2872 Danish twin pairs born 1870–1900. *Human Genetics*, 97, 319–323.

Heuer, H. & Visser, T. J. (2009) Minireview: Pathophysiological importance of thyroid hormone transporters. *Endocrinology*, 150, 1078–1083.

Hewes, R. S. (2008) The buzz on fly neuronal remodeling. *Trends in Endocrinology & Metabolism*, 19, 317–323.

Hewlett, B. S. & Lamb, M. E. (2005) *Hunter-gatherer Childhoods: Evolutionary, developmental, and cultural perspectives*. New Brunswick, Aldine Transaction.

Heyland, A. & Hodin, J. (2004) Heterochronic developmental shift caused by thyroid hormone in larval sand dollars and its implications for phenotypic plasticity and the evolution of nonfeeding development. *Evolution*, 58, 524–538.

Heyland, A. & Moroz, L. L. (2005) Cross-kingdom hormonal signaling: An insight from thryoid hormone functions in marine larvae. *Journal of Experimental Biology*, 208, 4355–4361.

Heyland, A. & Moroz, L. L. (2006) Signaling mechanisms underlying metamorphic transitions in animals. *Integrative and Comparative Biology*, 46, 743–759.

Heyland, A., Reitzel, A. M. & Hodin, J. (2004) Thyroid hormones determine developmental mode in sand dollars (Echinodermata: Echinoidea). *Evolution & Development*, 6, 382–392.

Heyland, A., Hodin, J. & Reitzel, A. M. (2005) Hormone signaling in evolution and development: a non-model system approach. *BioEssays*, 27, 64–75.

Heyland, A., Price, D. A., Bodnarova-Buganova, M. & Moroz, L. L. (2006a) Thyroid hormone metabolism and peroxidase function in two non-chordate animals. *Journal of Experimental Zoology B*, 306, 551–566.

Heyland, A., Reitzel, A. M., Price, D. A. & Moroz, L. L. (2006b) Endogenous thyroid hormone synthesis in facultative planktotrophic larvae of the sand dollar *Clypeaster rosaceus*: implications for the evolutionary loss of larval feeding. *Evolution & Development*, 8, 568–579.

Heyland, A., Vue Z. Voolstra, C. A., Medina, M. & Moroz L. L., Developmental transcriptome of *Aplysia californica*. JEZ Part B (online article).

Hill, K. (1993) Life history theory and evolutionary anthropology. *Evolutionary Anthropology*, 2, 78–88.

Hill, K. & Hurtado, A. M. (1991) The evolution of reproductive senescence and menopause of human females. *Human Nature*, 2, 315–350.

Hill, K. & Hurtado, A. M. (1996) *Ache life history: the ecology and demography of a foraging people*. New York, Aldine de Gruyter.

Hill, R. C., De Carvalho, C. E., Salogiannis, J., Schlager, B., Pilgrim, D. & Haag, E. S. (2006) Genetic flexibility in the convergent evolution of hermaphroditism in *Caenorhabditis* nematodes. *Developmental Cell*, 10, 531–538.

Hillman, R. & Lesnik, L. H. (1970) Cuticle formation in the embryo of *Drosophila melanogaster*. *Journal of Morphology*, 131, 385–395.

Hirata, Y., Kurokura, H. & Kasahara, S. (1989) Effects of thyroxine and thiourea on the development of larval red sea bream *Pagrus major*. *Bulletin of the Japanese Society of Scientific Fisheries*, 55, 1189–1195.

Hirn, M., Hetru, C., Lagueux, M. & Hoffman, J. A. (1979) Prothoracic gland activity and blood titers of ecdysone and ecdysterone during the last larval instar of *Locusta migratoria*. *Journal of Insect Physiology*, 25, 255–262.

Hirokawa, T., Komatsu, M. & Nakajima, Y. (2008) Development of the nervous system in the brittle star *Amphipholis kochii*. *Development, Genes and Evolution*, 218, 15–21.

Hirschenhauser, K. & Oliveira, R. F. (2006) Social modulation of androgens in male vertebrates: meta-analyses of the challenge hypothesis. *Animal Behaviour*, 71, 265–277.

Hitchler, M. J., Wikainapakul, K., Yu, L., Powers, K., Attatippaholkun, W. & Domann, F. E. (2006) Epigenetic regulation of manganese superoxide dismutase expression in human breast cancer cells. *Epigenetics*, 1, 163–171.

Hittinger, C. T. & Carroll, S. B. (2007) Gene duplication and the adaptive evolution of a classic genetic switch. *Nature*, 449, 677–681.

Hittinger, C. T., Goncalves, P., Sampaio, J. P., Dover, J., Johnston, M. & Rokas, A. (2010) Remarkably ancient balanced polymorphisms in a multi-locus gene network. *Nature*, 464, 54–58.

Hjelmborg, J., Iachine, I., Skytthe, A., Vaupel, J. W., McGue, M., Koskenvuo, M., Kaprio, J., Pedersen, N. L. & Christensen, K. (2006) Genetic influence on human lifespan and longevity. *Human Genetics*, 119, 312–321.

Hoang, A. (2001) Immune response to parasitism reduces resistance of *Drosophila melanogaster* to desiccation and starvation. *Evolution*, 55, 2353–2358.

Hoar, W. S. (1976) Smoth transformation – evolution, behavior, and physiology. *Journal of the Fisheries Research Board of Canada*, 33, 1233–1252.

Hoar, W. S. & Randall, D. J. (1988) The physiology of smolting salmonids. *Fish Physiology: The Physiology of Developing Fish*. Toronto, Academic Press, Inc.

Hodek, I. (1983) Role of environmental factors and endogenous mechanisms in the seasonality of reproduction in insects diapausing as adults. In Brown, V. K., Hodek, I., & Brown, V. K. (Eds.) *Diapause and Life Cycle Strategies in Insects*. Berlin, Springer.

Hodin, J. (2006) Expanding networks: signaling components in and a hypothesis for the evolution of metamorphosis. *Integrative and Comparative Biology*, 46, 719–742.

Hodin, J. (2009) She shapes events as they come: Plasticity in insect reproduction. In Whitman, D. & Anathakrishnan, T. N. (Eds.) *Insects and Phenotypic Plasticity*. Enfield, Science Publishers.

Hodin, J. & Riddiford, L. M. (2000) Different mechanisms underlie phenotypic plasticity and interspecific variation for a reproductive character in drosophilids (Insecta: Diptera). *Evolution*, 54, 1638–1653.

Hodin, J., Hoffman, J., Miner, B. G. & Davidson, B. J. (2001) Thyroxine and the evolution of lecithotrophic development in echinoids. In Barker, M. (Ed.) *Echinoderms 2000.* Rotterdam, Swets and Zeitlinger.

Hodkova, M. (1976) Nervous inhibition of corpora allata by photoperiod in *Pyrrhocoris apterus. Nature,* 263, 521–523.

Hodkova, M. (1999) Regulation of diapause and reproduction in *Pyrrhocoris apterus (L.)* (Heteroptera)-neuroendocrine outputs (mini-review). *Entomological Science,* 2, 563–566.

Hodkova, M. (2008) Tissue signaling pathways in the regulation of life-span and reproduction in females of the linden bug, *Pyrrhocoris apterus. Journal of Insect Physiology,* 54, 508–517.

Hodkova, M., Okuda, T. & Wagner, R. M. (2001) Regulation of corpora allata in females of *Pyrrhocoris apterus* (Heteroptera) (a mini-review). *In Vitro Cellular & Developmental Biology.* 37, 560–563.

Hodkova, M., Berkova, P. & Zahradnickova, H. (2002) Photoperiodic regulation of the phospholipid molecular species composition in thoracic muscles and fat body of Pyrrhocoris apterus (Heteroptera) via an endocrine gland, corpus allatum. Journal of Insect Physiology, 48, 1009–1019.

Hoekstra, H. E. & Coyne, J. A. (2007) The locus of evolution: Evo devo and the genetics of adaptation. *Evolution,* 61, 995–1016.

Hoffmann, M. H. (2002) Biogeography of *Arabidopsis thaliana* (L.) Heynh. (Brassicaceae). *Journal of Biogeography,* 29, 125–134.

Hoffmann, A. A., Hallas, R., Sinclair, C. & Partridge, L. (2001) Rapid loss of stress resistance in *Drosophila melanogaster* under adaptation to laboratory culture. *Evolution,* 55, 436–438.

Hoffmann, M. H., Bremer, M., Schneider, K., Burger, F., Stolle, E. & Moritz, G. (2003) Flower visitors in a natural population of *Arabidopsis thaliana. Plant Biology,* 5, 491–494.

Hoheisel, G. & Sterba, G. (1963) Über die Wirkung von Kaliumperchlorat (KCIO4) auf Ammocyten von *Lampetra planeri* BLOCH. *Zeitschrift fur Mikroskopisch-Anatomische Forschung,* 70 490–516.

Holehan, A. M. & Merry, B. J. (1985) Lifetime breeding studies in fully fed and dietary restricted female CFY Sprague-Dawley rats. 1. Effect of age, housing conditions and diet on fecundity. *Mechanisms of Ageing and Development,* 33, 19–28.

Holland, B. & Rice, W. R. (1999) Experimental removal of sexual selection reverses intersexual antagonistic coevolution and removes a reproductive load. *Proceedings of the National Academy of Sciences of the United States of America,* 96, 5083–5088.

Holland, L. Z., Albalat, R., Azumi, K., Benito-Gutierrez, E., Blow, M. J., Bronner-Fraser, M., Brunet, F., Butts, T., Candiani, S., Dishaw, L. J., Ferrier, D. E. K., Garcia-Fernandez, J., Gibson-Brown, J. J., Gissi, C., Godzik, A.,

Hallbook, F., Hirose, D., Hosomichi, K., Ikuta, T., Inoko, H., Kasahara, M., Kasamatsu, J., Kawashima, T., Kimura, A., Kobayashi, M., Kozmik, Z., Kubokawa, K., Laudet, V., Litman, G. W., McHardy, A. C., Meulemans, D., Nonaka, M., Olinski, R. P., Pancer, Z., Pennacchio, L. A., Pestarino, M., Rast, J. P., Rigoutsos, I., Robinson-Rechavi, M., Roch, G., Saiga, H., Sasakura, Y., Satake, M., Satou, Y., Schubert, M., Sherwood, N., Shiina, T., Takatori, N., Tello, J., Vopalensky, P., Wada, S., Xu, A. L., Ye, Y. Z., Yoshida, K., Yoshizaki, F., Yu, J. K., Zhang, Q., Zmasek, C. M., De Jong, P. J., Osoegawa, K., Putnam, N. H., Rokhsar, D. S., Satoh, N. & Holland, P. W. H. (2008) The amphioxus genome illuminates vertebrate origins and cephalochordate biology. *Genome Research,* 18, 1100–1111.

Hölldobler, B. & Wilson, E. O. (2008) *The Superorganism: The Beauty, Elegance, and Strangeness of Insect Societies.* New York, W.W. Norton & Co.

Holliday, R. (1989) Food, reproduction and longevity: is the extended lifespan of calorie-restricted animals an evolutionary adaptation? *BioEssays,* 10, 125–127.

Holm, E. R., Nedved, B. T., Carpizo-Ituarte, E. & Hadfield, M. G. (1998) Metamorphic signal transduction in *Hydroides elegans* (Polychaeta: Serpulidae) is not mediated by a G protein. *Biological Bulletin,* 195, 21–29.

Holman, D. J. & Wood, J. W. (2001) Pregnancy loss and fecundability in women. In Ellison, P. T. (Ed.) *Reproductive Ecology and Human Evolution.* Hawthorne, Aldine de Gruyter.

Holmes, J. A., Chu, H., Khanam, S. A., Manzon, R. G. & Youson, J. H. (1999) Spontaneous and induced metamorphosis in the American brook lamprey, *Lampetra appendix. Canadian Journal of Zoology-Revue Canadienne de Zoologie,* 77, 959–971.

Holzenberger, M., Dupont, J., Ducos, B., Leneuve, P., Geloen, A., Even, P. C., Cervera, P. & Le Bouc, Y. (2003) IGF-1 receptor regulates lifespan and resistance to oxidative stress in mice. *Nature,* 421, 182–187.

Holzenberger, M., Kappeler, L. & De Magalhaes Filho, C. (2004) IGF-1 signaling and aging. *Experimental Gerontology,* 39, 1761–1764.

Hone, D. W. & Benton, M. J. (2005) The evolution of large size: how does Cope's Rule work? *Trends in Ecology & Evolution* 20, 4–6.

Honegger, B., Galic, M., Koehler, K., Wittwer, F., Brogiolo, W., Hafen, E. & Stocker, H. (2008) Imp-L2, a putative homolog of vertebrate IGF-binding protein 7, counteracts insulin signaling in *Drosophila* and is essential for starvation resistance. *Journal of Biology,* 7, 10.

Honek, A. (1993) Intraspecific variation in body size and fecundity in insects – a general relationship. *Oikos,* 66, 483–492.

Honjoh, S., Yamamoto, T., Uno, M. & Nishida, E. (2009) Signalling through RHEB-1 mediates intermittent fasting-induced longevity in *C. elegans*. *Nature*, 457, 726–730.

Hörstadius, S. (1939) The mechanics of sea urchin development, studied by operative methods. *Biological Reviews of the Cambridge Philosophical Society*, 14, 132–179.

Hörstadius, S. (1957) On the regulation of bilateral symmetry in plutei with exchanged meridional halves and in giant plutei. *Journal of Embryology and Experimental Morphology*, 5, 60–73.

Hörstadius, S. (1973) *Experimental Embryology of Echinoderms*. Oxford, Clarendon Press.

Hörstadius, S. (1975) Isolation and transplant experiments. In Czihak, G. (Ed.) *The Sea Urchin Embryo: Biochemistry and Morphogenesis*. Berlin, Springer.

Horton, G. E., Letcher, B. H., Bailey, M. M. & Kinnison, M. T. (2009) Atlantic salmon (*Salmo salar*) smolt production: the relative importance of survival and body growth. *Canadian Journal of Fisheries and Aquatic Sciences*, 66, 471–483.

Horwood, L. J., Darlow, B. A. & Mogridge, N. (2001) Breast milk feeding and cognitive ability at 7-8 years. *Archive of Disease in Childhood: Fetal and Neonatal Edition*, 84, F23–F27.

Hotchkiss, F. (1995) Loven's law and adult ray homologies in echinoids, ophiuroids, edrioasteroids, and an ophiocistioid (Echinodermata: Eleutherozoa). *Proceedings of the Biological Society of Washington*, 108, 401–435.

Houle, D. (1991) Genetic covariance of fitness correlates: what genetic correlations are made of and why it matters. *Evolution*, 45, 630–648.

Houle, D. (2001) Characters as the units of evolutionary change. In Wagner, G. P. (Ed.) *The Character Concept in Evolutionary Biology*. New York, Academic Press.

Houle, D. (2010) Numbering the hairs on our heads: the shared challenge and promise of phenomics. *Proceedings of the National Academy of Sciences of the United States of America*, 107 1691–1695.

Houston, A. I. & McNamara, J. M. (1999) *Models of adaptive behaviour. An approach based on state*. Cambridge, Cambridge University Press.

Houthoofd, K. & Vanfleteren, J. R. (2006) The longevity effect of dietary restriction in *Caenorhabditis elegans*. *Experimental Gerontology*, 41, 1026–1031.

Houthoofd, K., Braeckman, B. P., Lenaerts, I., Brys, K., De Vreese, A., Van Eygen, S. & Vanfleteren, J. R. (2002) Axenic growth up-regulates mass-specific metabolic rate, stress resistance, and extends lifespan in *Caenorhabditis elegans*. *Experimental Gerontology*, 37, 1371–1378.

Howell, N. (2000) *Demography of the Dobe !Kung*, 2nd edn. New York, Aldine de Gruyter.

Howie, P. W. & McNeilly, A. S. (1982) Effect of breast-feeding patterns on human birth intervals. *Journal of Reproduction and Fertility*, 65, 545–557.

Howitz, K. T., Bitterman, K. J., Cohen, H. Y., Lamming, D. W., Lavu, S., Wood, J. G., Zipkin, R. E., Chung, P., Kisielewski, A., Zhang, L. L., Scherer, B. & Sinclair, D. A. (2003) Small molecule activators of sirtuins extend *Saccharomyces cerevisiae* lifespan. *Nature*, 425, 191–196.

Hrdy, S. B. (2005) Evolutionary context of human development: The cooperative breeding model. In Carter, C. S. & Ahnert, L. (Eds.) *Attachment and Bonding: A New Synthesis*. Cambridge, MIT Press.

Hrdy, S. B. (2009) *Mothers and others: the evolutionary origins of mutual understanding*. Cambridge, Harvard University Press.

Hsin, H. & Kenyon, C. (1999) Signals from the reproductive system regulate the lifespan of *C. elegans*. *Nature*, 399, 362–366.

Hsu, A. L., Murphy, C. T. & Kenyon, C. (2003) Regulation of aging and age-related disease by DAF-16 and heat-shock factor. *Science*, 300, 1142–1145.

Hsu, H. J., LaFever, L. & Drummond-Barbose, D. (2008) Diet controls normal and tumorous germline stem cells via insulin-dependent and -independent mechanisms in *Drosophila*. *Developmental Biology*, 313, 700–712.

Huang, H. & Brown, D. D. (2000a) Overexpression of *Xenopus laevis* growth hormone stimulates growth of tadpoles and frogs. *Proceedings of the National Academy of Sciences of the United States of America*, 97, 190–194.

Huang, H. & Brown, D. D. (2000b) Prolactin is not a juvenile hormone in *Xenopus laevis* metamorphosis. *Proceedings of the National Academy of Sciences of the United States of America*, 97, 195–199.

Huang, L. Y., Miwa, S., Bengtson, D. A. & Specker, J. L. (1998a) Effect of triiodothyronine on stomach formation and pigmentation in larval striped bass (*Morone saxatilis*). *Journal of Experimental Zoology*, 280, 231–237.

Huang, L. Y., Schreiber, A. M., Soffientino, B., Bengtson, D. A. & Specker, J. L. (1998b) Metamorphosis of summer flounder (*Paralichthys dentatus*): Thyroid status and the timing of gastric gland formation. *The Journal of Experimental Zoology*, 280, 413–420.

Huang, H., Marsh-Armstrong, N. & Brown, D. D. (1999) Metamorphosis is inhibited in transgenic *Xenopus laevis* tadpoles that overexpress type III deiodinase. *Proceedings of the National Academy of Sciences of the United States of America*, 96, 962–967.

Huang, D. A. W., Sherman, B. T., Tan, Q., Kir, J., Liu, D., Bryant, D., Guo, Y., Stephens, R., Baseler, M. W., Lane, H. C. & Lempicki, R. A. (2007) DAVID Bioinformatics

Resources: expanded annotation database and novel algorithms to better extract biology from large gene lists. *Nucleic Acids Research*, 35, W169–W175.

Huang, D. A. W., Sherman, B. T. & Lempicki, R. A. (2009) Systematic and integrative analysis of large gene lists using DAVID bioinformatics resources. *Nature Protocols*, 4, 44–57.

Huey, R. B., Gilchrist, G. W., Carlson, M. L., Berrigan, D. & Serra, L. (2000) Rapid evolution of a geographic cline in size in an introduced fly. *Science*, 287, 308–309.

Hughes, R. N. (1989) *A Functional Biology of Clonal Animals.* New York, Chapman and Hall.

Hulbert, A. J. (2000) Thyroid hormones and their effects: a new perspective. *Biological Reviews of the Cambridge Philosophical Society*, 75, 519–631.

Hulbert, A. J. (2008) The links between membrane composition, metabolic rate and lifespan. *Comparative Biochemistry and Physiology A*, 150, 196–203.

Hulbert, A. J., Faulks, S. C. & Buffenstein, R. (2006) Oxidation-resistant membrane phospholipids can explain longevity differences among the longest-living rodents and similarly-sized mice. *Journal of Gerontology A*, 61, 1009–1018.

Hultmark, D. (1996) Insect lysozymes. *EXS*, 75, 87–102.

Hunt, J. H. & Amdam, G. V. (2005) Bivoltinism as an antecedent to eusociality in the paper wasp genus *Polistes*. *Science*, 308, 264–267.

Hunt, G. J., Amdam, G. V., Schlipalius, D., Emore, C., Sardesai, N., Williams, C. E., Rueppell, O., Guzman-Novoa, E., Arechavaleta-Valasco, M., Chandra, S., Fondrk, M. K., Beye, M. & Page, R. E. (2007) Behavioral genomics of honeybee foraging and nest defense. *Naturwissenschaften*, 94, 247–267.

Hunter, W. S., Croson, W. B., Bartke, A., Gentry, M. V. & Meliska, C. J. (1999) Low body temperature in long-lived Ames dwarf mice at rest and during stress. *Physiology & Behavior*, 67, 433–437.

Hurtado, A. M., Hill, K., Kaplan, H. & Hurtado, I. (1992) Trade-offs between female food acquisition and child care among Hiwi and Ache foragers. *Human Nature*, 3, 185–216.

Hutchings, J. A. & Myers, R. A. (1988) Mating success of alternative maturation phenotypes in male atlantic salmon, *Salmo salar*. *Oecologia*, 75, 169–174.

Hwangbo, D. S., Gersham, B., Tu, M. P., Palmer, M. & Tatar, M. (2004) *Drosophila* dFOXO controls lifespan and regulates insulin signalling in brain and fat body. *Nature*, 429, 562–566.

Hylemon, P. B., Zhou, H., Pandak, W. M., Ren, S., Gil, G. & Dent, P. (2009) Bile acids as regulatory molecules. *Journal of Lipid Research*, 50, 1509–1520.

Ibarguengoytia, N. R. & Casalins, L. M. (2007) Reproductive biology of the southernmost gecko *Homonota darwini*: Convergent life history patterns among southern hemisphere reptiles living in harsh environments. *Journal of Herpetology*, 41, 72–80.

Ikeya, T., Galic, M., Belawat, P., Nairz, K. & Hafen, E. (2002) Nutrient-dependent expression of insulin-like peptides from neuroendocrine cells in the CNS contributes to growth regulation in *Drosophila*. *Current Biology*, 12, 1293–1300.

Ikeya, T., Broughton, S., Alic, N., Grandison, R. & Partridge, L. (2009) The endosymbiont *Wolbachia* increases insulin/IGF-like signalling in *Drosophila*. *Proceedings of the Royal Society of London B*, 276, 3799–3807.

Ilmonen, P., Kotrschal, A. & Penn, D. J. (2008) Telomere attrition due to infection. *PLoS One*, 3, e2143.

Imler, J. L. & Eleftherianos, I. (2009) *Drosophila* as a model for studying antiviral defences. In Rolff, J. & Reynolds, S. E. (Eds.) *Infection and Immunity: Evolution, Ecology, and Mechanisms*. Oxford, Oxford University Press.

Inagaki, T., Dutchak, P., Zhao, G., Ding, X., Gautron, L., Parameswara, V., Li, Y., Goetz, R., Mohammadi, M., Esser, V., Elmquist, J. K., Gerard, R. D., Burgess, S. C., Hammer, R. E., Mangelsdorf, D. J. & Kliewer, S. A. (2007) Endocrine regulation of the fasting response by PPARalpha-mediated induction of fibroblast growth factor 21. *Cell Metabolism*, 5, 415–425.

Inagaki, T., Lin, V. Y., Goetz, R., Mohammadi, M., Mangelsdorf, D. J. & Kliewer, S. A. (2008) Inhibition of growth hormone signaling by the fasting-induced hormone FGF21. *Cell Metabolism*, 8, 77–83.

Ingram, D. K., Weindruch, R., Spangler, E. L., Freeman, J. R. & Walford, R. L. (1987) Dietary restriction benefits learning and motor performance of aged mice. *Journal of Gerontology*, 42, 78–81.

Ingram, D. K., Anson, R. M., De Cabo, R., Mamczarz, J., Zhu, M., Mattison, J., Lane, M. A. & Roth, G. S. (2004) Development of calorie restriction mimetics as a prolongevity strategy. *Annals of the New York Academy of Sciences*, 1019, 412–423.

Innocenti, P. & Morrow, E. H. (2009) Immunogenic males: a genome-wide analysis of reproduction and the cost of mating in *Drosophila melanogaster* females. *Journal of Evolutionary Biology*, 22, 964–973.

Insel, T. R. & Young, L. J. (2001) The neurobiology of attachment. *Nature Reviews Neuroscience*, 2, 129–136.

Inui, Y. & Miwa, S. (1985) Thyroid hormone induces metamorphosis of flounder larvae. *General and Comparative Endocrinology*, 60, 450–454.

Ishizaki, H. & Suzuki, A. (1994) The brain secretory peptides that control molting and metamorphosis of the

Silkmoth, *Bombyx mori*. *International Journal of Developmental Biology*, 38, 301–310.

Ishigaki, S., Abramovitz, M. & Listowsky, I. (1989) Glutathione-S-transferases are major cytosolic thyroid hormone binding proteins. *Archives of Biochemistry and Biophysics*, 273, 265–272.

Ishizuya-Oka, A. & Shimozawa, A. (1991) Induction of metamorphosis by thyroid hormone in anuran small intestine cultured organotypically *in vitro*. *In vitro Cellular & Developmental Biology*, 27A, 853–857.

Isorna, E., Obregon, M. J., Calvo, R. M., Vazquez, R., Pendon, C., Falcon, J. & Munoz-Cueto, J. A. (2009) Iodothyronine deiodinases and thyroid hormone receptors regulation during flatfish (*Solea senegalensis*) metamorphosis. *Journal of Experimental Zoology B*, 312B, 231–246.

Izawa, T., Takahashi, Y. & Yano, M. (2003) Comparative biology comes into bloom: genomic and genetic comparison of flowering pathways in rice and *Arabidopsis*. *Current Opinion in Plant Biology*, 6, 113–120.

Ja, W. W., Carvalho, G. B., Mak, E. M., De La Rosa, N. N., Fang, A. Y., Liong, J. C., Brummel, T. & Benzer, S. (2007) Prandiology of *Drosophila* and the CAFE assay. *Proceedings of the National Academy of Sciences of the United States of America*, 104, 8253–8256.

Jackson, J. B. C. & Coates, A. G. (1986) Life cycles and evolution of clonal (modular) animals. *Philosophical Transactions of the Royal Society of London B*, 313, 7–22.

Jacobs, M. W., Degnan, S. M., Woods, R., Williams, E., Roper, K., Green, K. & Degnan, B. M. (2006) The effect of larval age on morphology and gene expression during ascidian metamorphosis. *Integrative and Comparative Biology*, 46, 760–776.

Jaeckle, W. B. (1994) Multiple modes of asexual reproduction by tropical and subtropical sea star larvae: an unusual adaptation for genet dispersal and survival. *Biological Bulletin*, 186, 62–71.

Jaenike, J. (1978) An hypothesis to account for the maintenance of sex within populations. *Evolutionary Theory*, 3, 191–194.

Jaenisch, R. & Bird, A. (2003) Epigenetic regulation of gene expression: how the genome integrates intrinsic and environmental signals. *Nature Genetics*, 33 Suppl, 245–254.

Jakob, W. & Schierwater, B. (2007) Changing hydrozoan bauplans by silencing Hox-like genes. *PLoS One*, 2, e694.

James, A. C., Azevedo, R. B. R. & Partridge, L. (1997) Genetic and environmental responses to temperature of *Drosophila melanogaster* from a latitudinal cline. *Genetics*, 146, 881–890.

Jang, Y. C. & Remmen, V. H. (2009) The mitochondrial theory of aging: insight from transgenic and knock-out mouse models. *Experimental Gerontology*, 44, 256–260.

Janke, A., Erpenbeck, D., Nilsson, M. & Arnason, U. (2001) The mitochondrial genomes of the iguana (*Iguana iguana*) and the caiman (*Caiman crocodylus*): implications for amniote phylogeny. *Proceedings of the Royal Society of London B*, 268, 623–631.

Jansen, J., Friesema, E. C., Kester, M. H., Schwartz, C. E. & Visser, T. J. (2008) Genotype-phenotype relationship in patients with mutations in thyroid hormone transporter MCT8. *Endocrinology*, 149, 2184–2190.

Janson, C. H. & van Schaik, C. P. (1993) Ecological risk aversion in juvenile primates: slow and steady wins the race. In Pereira, M. E. & Fairbanks, L. A. (Eds.) *Juvenile Primates: Life History, Development and Behavior*. Oxford, Oxford University Press.

Jasienska, G. (2001) Why energy expenditure causes reproductive suppression in women. An evolutionary and bioenergetic perspective. In Ellison, P. T. (Ed.) *Reproductive Ecology and Human Evolution*. New York, Aldine de Gruyter.

Jasienska, G. (2003) Energy metabolism and the evolution of reproductive suppression in the human female. *Acta Biotheoretica*, 51, 1–18.

Jasienska, G., Nenko, I. & Jasienska, M. (2006) Daughters increase longevity of fathers, but daughters and sons equally reduce longevity of mothers. *American Journal of Human Biology*, 18, 422–425.

Jawor, J. M., McGlothlin, J. W., Casto, J. M., Greives, T. J., Snajdr, E. A., Bentley, G. E. & Ketterson, E. D. (2006) Seasonal and individual variation in response to GnRH challenge in male dark-eyed juncos (*Junco hyemalis*). *General and Comparative Endocrinology*, 149, 182–189.

Jeffery, W. R. & Swalla, B. J. (1992) Evolution of alternate modes of development in Ascidians. *BioEssays*, 14, 219–226.

Jena, B. & Patnaik, B. (1992) Changes in catalase activity and its thermolability in liver and kidneys of aging male garden lizard. *Gerontology*, 38, 252–257.

Jena, B. S., Nayak, S. B. & Patnaik, B. K. (1998) Age-related changes in catalase activity and its inhibition by manganese (II) chloride in the brain of two species of poikilothermic vertebrates. *Archives of Gerontology and Geriatrics*, 26, 119–129.

Jenkins, T. M. J. (1969) Social structure, position choice and micro-distribution of two trout species (*Salmo trutta* and *Salmo gairdneri*) resident in mountain streams. *Animal Behavior Monographs*, 2, 56–123.

Jennings, D. H. & Hanken, J. (1998) Mechanistic basis of life history evolution in anuran amphibians: thyroid gland development in the direct-developing frog, *Eleutherodactylus coqui*. *General and Comparative Endocrinology*, 111, 225–232.

Jensen, A. R. (1998) *The g factor: the science of mental ability.* New York, Praeger.

Jensen, K. H., Little, T., Skorping, A. & Ebert, D. (2006) Empirical support for optimal virulence in a castrating parasite. *PLoS Biology*, 4, 1265–1269.

Jepson, J. H., Gardner, F. H., Gorshein, D. & Hait, W. M. (1973) Current concepts of the action of androgenic steroids on erythropoiesis. *Journal of Pediatrics*, 83, 703–708.

Jjepsen, K., Hermanson, O., Onami, T. M., Gleiberman, A. S., Lunyak, V., Jepson, J. H., Gardner, F. H., Gorshein, D. & Hait, W. M. (1973) Current concepts of the action of androgenic steroids on erythropoiesis. *Journal of Pediatrics*, 83, 703–708.

Jia, K., Chen, D. & Riddle, D. L. (2004) The TOR pathway interacts with the insulin signaling pathway to regulate *C. elegans* larval development, metabolism and life span. *Development*, 131, 3897–3906.

Jiang, J. C., Jaruga, E., Repnevskaya, M. V. & Jazwinski, S. M. (2000) An intervention resembling caloric restriction prolongs lifespan and retards aging in yeast. *FASEB Journal*, 14, 2135–2137.

Jiang, Z., Castoe, T., Austin, C., Burbrink, F., Herron, M., McGuire, J., Parkinson, C. & Pollock, D. (2007) Comparative mitochondrial genomics of snakes: extraordinary substitution rate dynamics and functionality of the duplicate control region. *BMC Evolutionary Biology* 7, 123.

Jiggins, C. D., Naisbit, R. E., Coe, R. L. & Mallet, J. (2001) Reproductive isolation caused by colour pattern mimicry. *Nature*, 411, 302–305.

Joffe, T. H. (1997) Social pressures have selected for an extended juvenile period in primates. *Journal of Human Evolution*, 32, 593–605.

Johanson, U., West, J., Lister, C., Michaels, S., Amasino, R. & Dean, C. (2000) Molecular analysis of *FRIGIDA*, a major determinant of natural variation in *Arabidopsis* flowering time. *Science*, 290, 344–347.

John-Alder, H. B. & Cox, R. M. (2007) Development of sexual size dimorphism in lizards: testosterone as a bipotential growth regulator. In Daphne, J. F., Blanckenhorn, W. U. & Szekely, T. (Eds.) *Sex, Size, and Gender Roles*. Oxford, Oxford University Press.

Johnsson, J. I. & Bjornsson, B. T. (1994) Growth hormone increases growth rate, appetite and dominance in juvenile rainbow trout, *Oncorhynchus mykiss Animal Behaviour*, 48, 177–186.

Joint, I. (2006) Bacterial conversations: talking, listening and eavesdropping. A NERC discussion meeting held at the Royal Society on 7 December 2005. *Journal of the Royal Society, Interface*, 3, 459–463.

Jolivet-Jaudet, G. & Leloup-Hatey, J. (1984) Variations in aldosterone and corticosterone plasma levels during metamorphosis in *Xenopus laevis* tadpoles. *General and Comparative Endocrinology*, 56, 59–65.

Jones, D. & Jones, G. (2007) Farnesoid secretions of dipteran ring glands: What we do know and what we can know. *Insect Biochemistry & Molecular Biology*, 37, 771–798.

Jones, E. I., Ferriere, R. & Bronstein, J. L. (2009) Eco-evolutionary dynamics of mutualists and exploiters. *The American Naturalist*, 174, 780–794.

Jonsson, K. I. (1997) Capital and income breeding as alternative tactics of resource use in reproduction. *Oikos*, 78, 57–66.

Joshi, A. & Thompson, J. N. (1995) Alternative routes to the evolution of competitive ability in 2 competing species of *Drosophila*. *Evolution*, 49, 616–625.

Just, J. J., Kraus-Just, J., Check, D. A. (1981) Survey of Chordate Metamorphosis. In Gilbert, L. I. & Frieden, E. (Eds.) *Metamorphosis: A Problem in Developmental Biology*. New York, Plenum Press.

Juul, A. (2001) The effects of oestrogens on linear bone growth. *Human Reproduction Update*, 7, 303–313.

Kaeberlein, M. & Powers III, R. W., (2007) Sir2 and calorie restriction in yeast: a skeptical perspective. *Aging Research Reviews*, 6, 128–140.

Kaeberlein, M. & Shamieh, L. S. (2010) *The Role of TOR Signaling in Aging*. New York, Springer.

Kaeberlein, M., McVey, M. & Guarente, L. (1999) The SIR2/3/4 complex and SIR2 alone promote longevity in *Saccharomyces cerevisiae* by two different mechanisms. *Genes & Development*, 13, 2570–2580.

Kaeberlein, T. L., Smith, E. D., Tsuchiya, M., Welton, K. L., Thomas, J. H., Fields, S., Kennedy, B. K. & Kaeberlein, M. (2006) Lifespan extension in *Caenorhabditis elegans* by complete removal of food. *Aging Cell*, 5, 487–494.

Kaltenbach, J. (1996) Endocrinology of Amphibian Metamorphosis. In Gilbert, L. I., Tata, J. R. & Atkinson, B. G. (Eds.) *Metamorphosis: Postembryonic Reprogramming of Gene Expression in Amphibian and Insect Cells*. San Diego, Academic Press.

Kalushkov, P., Hodkova, M., Nedved, O. & Hodek, I. (2001) Effect of thermoperiod on diapause intensity in *Pyrrhocoris apterus* (Heteroptera Pyrrhocoridae). *Journal of Insect Physiology*, 47, 55–61.

Kamleh, M. A., Hobani, Y., Dow, J. A. T. & Watson, D. G. (2007) Metabolomic profiling of *Drosophila* using liquid chromatography Fourier transform mass spectrometry. *FEBS Letters*, 582, 2916–2922.

Kamm, K., Schierwater, B., Jakob, W., Dellaporta, S. L. & Miller, D. J. (2006) Axial patterning and diversification in the *Cnidaria* predate the Hox system. *Current Biology*, 16, 920–926.

Kammenga, J. E., Doroszuk, A., Riksen, J. A., Hazendonk, E., Spiridon, L., Petrescu, A. J., Tijsterman, M., Plasterk, R. H. & Bakker, J. (2007) A *Caenorhabditis elegans* wild type defies the temperature-size rule owing to a single nucleotide polymorphism in tra-3. *PLoS Genetics*, 3, e34.

Kaneko, T. (2003) Aging and the accumulation of oxidative damage to DNA. *Journal of Clinical Biochemistry and Nutrition*, 34, 51–60.

Kapahi, P., Zid, B. M., Harper, T., Koslover, D., Sapin, V. & Benzer, S. (2004) Regulation of lifespan in *Drosophila* by modulation of genes in the TOR signaling pathway. *Current Biology*, 14, 885–890.

Kaplan, H. S. & Robson, A. J. (2002) The emergence of humans: the coevolution of intelligence and longevity with intergenerational transfers. *Proceedings of the National Academy of Sciences of the United States of America*, 99, 10221–10226.

Kaplan, H., Hill, K., Lancaster, J. & Hurtado, A. M. (2000) A theory of human life history evolution: diet, intelligence, and longevity. *Evolutionary Anthropology*, 9, 156–185.

Karaolis-Danckert, N., Buyken, A. E., Sonntag, A. & Kroke, A. (2009) Birth and early life influences on the timing of puberty onset: results from the DONALD (Dortmund Nutritional and Anthropometric Longitudinally Designed) study. *American Journal of Clinical Nutrition*, 90, 1559–1565.

Karim, F. D., Guild, G. M. and Thummel, C. S. (1993) The *Drosophila* Broad-Complex plays a role in controlling ecdysone-regulated gene expression at the onset of metamorphosis. *Development*, 118, 977–988.

Kato, H., Fukuda, T., Parkison, C., McPhie, P. & Cheng, S. Y. (1989) Cytosolic thyroid hormone-binding protein is a monomer of pyruvate kinase. *Proceedings of the National Academy of Sciences of the United States of America*, 86, 7861–7865.

Katona, P. & Katona-Apte, J. (2008) The interaction between nutrition and infection. *Clinical Infectious Diseases*, 46, 1582–1588.

Katz, S. H., Hediger, M. L., Zemel, B. S. & Parks, J. S. (1985) Adrenal androgens, body fat and advanced skeletal age in puberty: new evidence for the relations of adrenarche and gonadarche in males. *Human Biology*, 57, 401–413.

Kawakami, Y., Tanda, M., Adachi, S. & Yamauchi, K. (2003) Characterization of thyroid hormone receptor alpha and beta in the metamorphosing Japanese conger eel, *Conger myriaster*. *General and Comparative Endocrinology*, 132, 321–332.

Kawakami, Y., Yokoi, K., Kumai, H. & Ohta, H. (2008) The role of thyroid hormones during the development of eye pigmentation in the Pacific bluefin tuna (*Thunnus orientalis*). *Comparative Biochemistry and Physiology B, Biochemistry & Molecular Biology*, 150, 112–116.

Kawamura, K., Shibata, T., Saget, O., Peel, D. & Peter, J. (1999) A new family of growth factors produced by the fat body and active on *Drosophila* imaginal disc cells. *Development*, 126, 211–219.

Kawecki, T. J. & Mery, F. (2006) Genetically idiosyncratic responses of *Drosophila melanogaster* populations to selection for improved learning ability. *Journal of Evolutionary Biology*, 19, 1265–1274.

Kayukawa, T., Chen, B., Miyazaki, S., Itoyama, K., Shinoda, T. & Ishikawa, Y. (2005) Expression of mRNA for the t-complex polypeptide-1, a subunit of chaperonin CCT, is upregulated in association with increased cold hardiness in *Delia antiqua*. *Cell Stress Chaperones*, 10, 204–210.

Kealy, R. D., Lawler, D. F., Ballam, J. M., Mantz, S. L., Biery, D. N., Greeley, E. H., Lust, G., Segre, M., Smith, G. K. & Stowe, H. D. (2002) Effects of diet restriction on lifespan and age-related changes in dogs. *Journal of the American Veterinary Medical Association*, 220, 1315–1320.

Keefe, M. & Able, K. W. (1993) Patterns of metamorphosis in summer flounder, *Paralichthys dentatus Journal of Fish Biology*, 42, 713–728.

Kelic, V., Obradovic, T. & Pavkovic-Lucic, S. (2007) Growth temperature, mating latency, and duration of copulation in *Drosophila melanogaster*. *Drosophila Information Service*, 90, 111–113.

Keller, L. & Genoud, M. (1997) Extraordinary lifespans in ants: a test of evolutionary theories of aging. *Nature*, 389, 958–960.

Keller, L. & Gordon, E. (2009) *The lives of ants*. Oxford, Oxford University Press.

Keller, L. & Jemielity, S. (2006) Social insects as a model to study the molecular basis of aging. *Experimental Gerontology*, 41, 553–556.

Keller, L. & Surette, M. G. (2006) Communication in bacteria: an ecological and evolutionary perspective. *Nature Reviews Microbiology*, 4, 249.

Kempf, S. C. & Page, L. R. (2005) Anti-tubulin labeling reveals ampullary neuron ciliary bundles in opisthobranch larvae and a new putative neural structure associated with the apical ganglion. *Biological Bulletin*, 208, 169–182.

Kennedy, B. K., Austriaco, N. R., JR., Zhang, J. & Guarente, L. (1995) Mutation in the silencing gene SIR4 can delay aging in *S. cerevisiae*. *Cell*, 80, 485–496.

Kenyon, C. (2010) The genetics of ageing. *Nature*, 464, 504–512.

Kenyon, C., Chang, J., Gensch, E., Rudner, A. & Tabtiang, R. (1993) A *C. elegans* mutant that lives twice as long as wild type. *Nature*, 366, 461–464.

Kerr, B., Riley, M. A., Feldman, M. W. & Bohannan, B. J. M. (2002) Local dispersal promotes biodiversity in a real-life game of rock-paper-scissors. *Nature*, 418, 171–174.

Ketterson, E. D. & Nolan, V. (1992) Hormones and life histories: an integrative approach. *The American Naturalist*, 140, 33–62.

Ketterson, E. D. & Nolan JR., V. (1999) Adaptation, exaptation, and constraint: a hormonal perspective. *The American Naturalist*, 154 (Suppl.), S4–S25.

Ketterson, E. D., Atwell, J. W. & McGlothlin, J. W. (2009) Phenotypic integration and independence: hormones, performance, and response to environmental change. *Integrative and Comparative Biology*, 49, 365–379.

Khryanin, V. N. (2007) Evolution of the pathways of sex differentiation in plants. *Russian Journal of Plant Physiology*, 54, 845–852.

Kiguchi, K. & Riddiford, L. M. (1978) A role of juvenile hormone in pupal development of the tobacco hornworm, *Manduca sexta*. *Journal of Insect Physiology*, 24, 673–680.

Kikuyama, S., Niki, K., Mayumi, M. & Kawamura, K. (1982) Retardation of thyroxine-induced metamorphosis by Amphenone B in toad tadpoles. *Endocrinologia Japonica*, 29, 659–662.

Kikuyama, S., Kawamura, K., Tanaka, S. & Yamamoto, K. (1993) Aspects of amphibian metamorphosis: hormonal control. *International Review of Cytology*, 145, 105–148.

Killian, C. E. & Wilt, F. H. (2008) Molecular aspects of biomineralization of the Echinoderm skeleton. *Chemical Reviews*, 108, 4463–4474.

Kim, S., Benguria, A., Lai, C. Y. & Jazwinski, S. M. (1999) Modulation of life-span by histone deacetylase genes in *Saccharomyces cerevisiae*. *Molecular Biology of the Cell*, 10, 3125–3136.

Kim, T., Yoon, J., Cho, H., Lee, W. B., Kim, J., Song, Y. H., Kim, S. N., Yoon, J. H., Kim-Ha, J. & Kim, Y. J. (2005) Downregulation of lipopolysaccharide response in *Drosophila* by negative crosstalk between the AP1 and NF-kappaB signaling modules. *Nature Immunology*, 6, 211–218.

Kim, L. K., Choi, U. Y., Cho, H. S., Lee, J. S., Lee, W. B., Kim, J., Jeong, K., Shim, J., Kim-Ha, J. & Kim, Y. J. (2007) Down-regulation of NF-kappaB target genes by the AP-1 and STAT complex during the innate immune response in *Drosophila*. *PLoS Biology*, 5, e238.

Kimura, M. T. (1988) Interspecific and geographic variation of diapause intensity and seasonal adaptation in the *Drosophila auraria* species complex (Diptera: Drosophilidae). *Functional Ecology*, 2, 177–183.

Kimura, K. D., Tissenbaum, H. A., Liu, Y. & Ruvkun, G. (1997) daf-2, an insulin receptor-like gene that regulates longevity and diapause in *Caenorhabditis elegans*. *Science*, 277, 942–946.

Kingsley, R. J., Afif, E., Cox, B. C., Kothari, S., Kriechbaum, K., Kuchinsky, K., Neill, A. T., Puri, A. F. & Kish, V. M.

(2003) Expression of heat shock and cold shock proteins in the gorgonian *Leptogorgia virgulata*. *Journal of Experimental Zoology A*, 296A, 98–107.

Kingsolver, J. G. (1987) Evolution and coadaptation of thermoregulatory behavior and wing pigmentation pattern in Pierid butterflies. *Evolution*, 41, 472–490.

Kingsolver, J. G. (1995a) Fitness consequences of seasonal polyphenism in western white butterflies. *Evolution*, 49, 942–954.

Kingsolver, J. G. (1995b) Viability selection on seasonally polyphenic traits - wing melanin pattern in western white butterflies. *Evolution*, 49, 932–941.

Kingsolver, J. G. (1996) Experimental manipulation of wing pigment pattern and survival in western white butterflies. *The American Naturalist*, 147, 296–306.

Kingsolver, J. G. & Pfennig, D. W. (2004) Individual-level selection as a cause of Cope's rule of phyletic size increase. *Evolution*, 58, 1608–1612.

Kington, R., Lillard, L. & Rogowski, J. (1997) Reproductive history, socioeconomic status, and self-reported health status of women aged 50 years or older. *American Journal of Public Health*, 87, 33–37.

Kinjoh, T., Kaneko, Y., Itoyama, K., Mita, K., Hiruma, K. & Shinoda, T. (2007) Control of juvenile hormone biosynthesis in *Bombyx mori*: cloning of the enzymes in the mevalonate pathway and assessment of their developmental expression in the corpora allata. *Insect Biochemistry & Molecular Biology*, 37(8), 808–818.

Kirk, D. L. (1995) Asymmetric division, cell size and germ-soma specification in *Volvox Seminars in Developmental Biology*, 6, 369–379.

Kirk, D. L. (1998) *Volvox. Molecular genetic origins of multicellularity and cellular differentiation.* New York, Cambridge University Press.

Kirk, D. L., Baran, G. J., Harper, J. F., Huskey, R. J., Huson, K. S. & Zagris, N. (1987) Stage-specific hypermutability of the reg A locus of *Volvox*, a gene regulating the germ-soma dichotomy. *Cell*, 18, 11–24.

Kirk, M., Ransick, A., McRae, S. E. & Kirk, D. L. (1993) The relationship between cell size and cell fate in *Volvox carteri*. *Journal of Cell Biology*, 123, 191–208.

Kirk, M., Stark, K., Miller, S., Muller, W., Taillon, B., Gruber, H., Schmitt, R. & Kirk, D. L. (1999) regA, a *Volvox* gene that plays a central role in germ soma differentiation, encodes a novel regulatory protein. *Development*, 126, 639–647.

Kirkwood, T. B. (1977) Evolution of aging. *Nature*, 270, 301–304.

Kirkwood, T. B. & Holliday, R. (1979) The evolution of aging and longevity. *Proceedings of the Royal Society of London B*, 205, 531–546.

Kiss, I. (1976) Prepupal larval mosaics in *Drosophila melanogaster*. *Nature*, 262, 136–138.

Kiss, I., Beaton, A. H., Tardiff, J., Fristrom, D. & Firstrom, J. W. (1988) Interactions and developmental effects of mutations in the *Broad-Complex* of *Drosophila melanogaster*. *Genetics*, 118, 247–259.

Kitajima, C., Sato, T. & Kawanishi, M. (1967) On the effect of thyroxine to promote the metamorphosis of a conger eel. *Bulletin of the Japanese Society of Scientific Fisheries*, 33, 919–922.

Kitamoto, N., Ueno, S., Takenaka, A., Tsumura, Y., Washitani, I. & Ohsawa, R. (2006) Effect of flowering phenology on pollen flow distance and the consequences for spatial genetic structure within a population of *Primula sieboldii* (Primulaceae). *American Journal of Botany*, 93, 226–233.

Klapper, W., Heidorn, K., Kuhne, K., Parwaresch, R. & Krupp, G. (1998) Telomerase activity in 'immortal' fish. *FEBS Letters*, 434, 409–412.

Klass, M. R. (1977) Aging in the nematode *Caenorhabditis elegans*: major biological and environmental factors influencing life span. *Mechanisms of Ageing and Development*, 6, 413–429.

Klass, M. R. (1983) A method for the isolation of longevity mutants in the nematode *Caenorhabditis elegans* and initial results. *Mechanisms of Ageing and Development*, 22, 279–286.

Kleemann, G. A. & Murphy, C. T. (2009) The endocrine regulation of aging in *Caenorhabditis elegans*. *Molecular and Cellular Endocrinology*, 299, 51–57.

Klein, R. G. (1984) *The human career: human biological and cultural origins*. Chicago, Chicago University Press.

Klinzing, M. S. E. & Pechenik, J. A. (2000) Evaluating whether velar lobe size indicates food limitation among larvae of the marine gastropod *Crepidula fornicata*. *Journal of Experimental Marine Biology and Ecology*, 252, 255–279.

Klok, M. D., Jakobsdottir, S. & Drent, M. L. (2007) The role of leptin and ghrelin in the regulation of food intake and body weight in humans: a review. *Obesity Reviews*, 8, 21–34.

Knauf, F., Rogina, B., Jiang, Z., Aronson, P. S. & Helfand, S. L. (2002) Functional characterization and immunolocalization of the transporter encoded by the life-extending gene Indy. *Proceedings of the National Academy of Sciences of the United States of America*, 99, 14315–14319.

Knight, T. M., Steets, J. A., Vamosi, J. C., Mazer, S. J., Burd, M., Campbell, D. R., Dudash, M. R., Johnston, M. O., Mitchell, R. J. & Ashman, T.-L. (2005) Pollen limitation of plant reproduction: pattern and process. *Annual Review of Ecology, Evolution, and Systematics*, 36, 467–497.

Knight, C. G., Zitzmann, N., Prabhakar, S., Antrobus, R., Dwek, R., Hebestreit, H. & Rainey, P. B. (2006) Unraveling adaptive evolution: how a single point mutation affects the protein coregulation network. *Nature Genetics*, 38, 1015–1022.

Knobil, E. & Neill, J. D. (1988) *The Physiology of Reproduction*. New York, Raven.

Kobuke, L., Specker, J. L. & Bern, H. A. (1987) Thyroxine content of eggs and larvae of coho salmon, *Oncorhynchus kisutch*. *Journal of Experimental Zoology*, 242, 89–94.

Koch, P. B. (1992) Seasonal polyphenism in butterflies - a hormonally controlled phenomenon of pattern-formation. *Zoologische Jahrbücher – Abteilung fur Allgemeine Zoologie und Physiologie der Tiere* 96, 227–240.

Koch, P. B., Brakefield, P. M. & Kesbeke, F. (1996) Ecdysteroids control eyespot size and wing color pattern in the polyphenic butterfly *Bicyclus anynana* (Lepidoptera: Satyridae). *Journal of Insect Physiology*, 42, 223–230.

Koch, M., Bishop, J. & Mitchell-Olds, T. (1999) Molecular systematics and evolution of *Arabidopsis* and *Arabis*. *Plant Biology*, 1, 529–537.

Kojima, S., Takahashi, Y., Kobayashi, Y., Monna, L., Sasaki, T., Araki, T. & Yano, M. (2002) Hd3a, a rice ortholog of the *Arabidopsis FT* gene, promotes transition to flowering downstream of *Hd1* under short-day conditions. *Plant & Cell Physiology*, 43, 1096–1105.

Kokko, H. & Rankin, D. J. (2006) Lonely hearts or sex in the city? Density-dependent effects in mating systems. *Philosophical Transactions of the Royal Society of London B*, 361, 319–334.

Kole, C., Quijada, P., Michaels, S. D., Amasino, R. M. & Osborn, T. C. (2001) Evidence for homology of flowering-time genes *VFR2* from *Brassica rapa* and *FLC* from *Arabidopsis thaliana*. *Theoretical and Applied Genetics*, 102, 425–430.

Komeda, Y. (2004) Genetic regulation of time to flower in *Arabidopsis thaliana*. *Annual Review of Plant Biology*, 55, 521–535.

Kondrashov, A. S. (1993) Classification of hypotheses on the advantage of amphimixis. *Journal of Heredity*, 84, 372–387.

Konopova, B. & Jindra, M. (2008) Broad Complex acts downstream of Met in juvenile hormone signaling to coordinate primitive holometabolan metamorphosis. *Development* 135, 559–568.

Konopova, B. & Jindra, M. (2007) Juvenile hormone resistance gene *Methoprene-tolerant* controls entry into metamorphosis in the beetle *Tribolium castaneum*. *Proceedings of the National Academy of Sciences of the United States of America*, 104, 10488–10493.

Konopova, B. & Jindra, M. (2009) Juvenile hormone regulates metamorphosis of holometabolous and hemimetabolous insects through Met and Kr-h1 genes. *Mechanisms of Development*, 126, S68–S69.

Konopova, B. & Zrzavy, J. (2005) Ultrastructure, development, and homology of insect embryonic cuticles. *Journal of Morphology*, 264, 339–362.

Kooi, R. E. & Brakefield, P. M. (1999) The critical period for wing pattern induction in the polyphenic tropical butterfly *Bicyclus anynana* (Satyrinae). *Journal of Insect Physiology*, 45, 201–212.

Koornneef, M., Hanhart, C. J. & Vanderveen, J. H. (1991) A genetic and physiological analysis of late flowering mutants in *Arabidopsis thaliana*. *Molecular & General Genetics*, 229, 57–66.

Koornneef, M., Alonso-Blanco, C. & Vreugdenhil, D. (2004) Naturally occurring genetic variation in *Arabidopsis thaliana*. *Annual Review of Plant Biology*, 55, 141–172.

Korte, M. S., Koolhaas, J. M., Wingfield, J. C. & McEwen, B. S. (2005) The Darwinian concept of stress: benefits of allostasis and costs of allostatic load and the trade-offs in health and disease. *Neuroscience Biobehavior Reviews*, 29, 3–38.

Kortschak, R. D., Samuel, G., Saint, R. & Miller, D. J. (2003) EST analysis of the cnidarian *Acropora millepora* reveals extensive gene loss and rapid sequence divergence in the model invertebrates. *Current Biology*, 13, 2190–2195.

Korves, T. M., K. J. Schmid, A. L. Caicedo, C. Mays, J. R. Stinchcombe, M. D. Purugganan, and J. Schmitt (2007) Fitness effects associated with the major flowering time gene FRIGIDA in *Arabidopsis thaliana* in the field. *The American Naturalist*, 169, E141–E157.

Kostál, V. (2006) Eco-physiological phases of insect diapause. *Journal of Insect Physiology*, 52, 113–127.

Kostál, V., Tollarova, M. & Dolezel, D. (2008) Dynamism in physiology and gene diapause in a heteropteran transcription during reproductive bug, *Pyrrhocoris apterus*. *Journal of Insect Physiology*, 54, 77–88.

Koufopanou, V. (1994) The evolution of soma in the Volvocales. *The American Naturalist*, 143, 907–931.

Koufopanou, V. & Bell, G. (1993) Soma and germ - an experimental approach using Volvox. *Proceedings of the Royal Society of London B*, 254, 107–113.

Kowalski, S. P., Lan, T. H., Feldmann, K. A. & Paterson, A. H. (1994) QTL mapping of naturally-occurring variation in flowering time of *Arabidopsis thaliana*. *Molecular & General Genetics*, 245, 548–555.

Kraaijeveld, A. R. & Godfray, H. C. J. (1997) Trade-off between parasitoid resistance and larval competitive ability in *Drosophila melanogaster*. *Nature*, 389, 278–280.

Kraaijeveld, A. R., Limentani, E. C. & Godfray, H. C. J. (2001) Basis of the trade-off between parasitoid resistance and larval competitive ability in *Drosophila melanogaster*. *Proceedings of the Royal Society of London B*, 268, 259–261.

Kraaijeveld, A. R., Ferrari, J. & Godfray, H. C. J. (2002) Costs of resistance in insect-parasite and insect-parasitoid interactions. *Parasitology*, 125, S71–S82.

Krain, L. P. & Denver, R. J. (2004) Developmental expression and hormonal regulation of glucocorticoid and thyroid hormone receptors during metamorphosis in *Xenopus laevis*. *Journal of Endocrinology*, 181, 91–104.

Kralj-Fiser, S., Scheiber, I. B. R., Blejec, A., Moestl, E. & Kotrschal, K. (2007) Individualities in a flock of free-roaming greylag geese: behavioral and physiological consistency over time and across situations. *Hormones and Behavior*, 51, 239–248.

Kristensen, N. P. (1999) Phylogeny of endopterygote insects, the most successful lineage of living organisms. *European Journal of Entomology* 96, 237–253.

Krug, P. J. (2009) Not my "type": larval dispersal dimorphisms and bet-hedging in Opisthobranch life histories. *Biological Bulletin*, 216, 355–372.

Krug, E. C., Honn, K. V., Battista, J. & Nicoll, C. S. (1983) Corticosteroids in serum of *Rana catesbeiana* during development and metamorphosis. *General and Comparative Endocrinology*, 52, 232–241.

Kubli, E. (2003) Sex-peptides: seminal peptides of the *Drosophila* male. *Cellular and Molecular Life Sciences*, 60, 1689–1704.

Kucharski, R., Maleszka, J., Foret, S. & Maleszka, R. (2008) Nutritional control of reproductive status in honeybees via DNA methylation. *Science*, 319, 1827–1830.

Kuiper, G. G., Klootwijk, W., Morvan-Dubois, G., Destree, O., Darras, V. M., Van Der Geyten, S., Demeneix, B. & Visser, T. J. (2006) Characterization of recombinant *Xenopus laevis* type 1 iodothyronine deiodinase: substitution of a proline residue in the catalytic center by serine (Pro132Ser) restores sensitivity to 6-propyl-2-thiouracil. *Endocrinology*, 147, 3519–3529.

Kuittinen, H., Niittyvuopio, A., Rinne, P. & Savolainen, O. (2008) Natural variation in *Arabidopsis lyrata* vernalization requirement conferred by a *FRIGIDA* indel polymorphism. *Molecular Biology and Evolution*, 25, 319–329.

Kulkarni, S. A., Singamasetty, S., Buchholz, D. R. (2010) Corticotropin-releasing factor regulates the development in the direct developing frog, Eleutherodactylus coqui. Gen. Comp. Endocrinol. 169: 225–230.

Kuningas, M., Magi, R., Westendorp, R. G., Slagboom, P. E., Remm, M. & Van Heemst, D. (2007a) Haplotypes in the human Foxo1a and Foxo3a genes; impact on disease and mortality at old age. *European Journal of Human Genetics* 15, 294–301.

Kuningas, M., Putters, M., Westendorp, R. G., Slagboom, P. E. & Van Heemst, D. (2007b) SIRT1 gene, age-related diseases, and mortality: the Leiden 85-plus study. *Journal of Gerontology A*, 62, 960–965.

Kuningas, M., Mooijaart, S. P., Van Heemst, D., Zwaan, B. J., Slagboom, P. E. & Westendorp, R. G. (2008) Genes encoding longevity: from model organisms to humans. *Aging Cell*, 7, 270–280.

Kusserow, A., Pang, K., Sturm, C., Hrouda, M., Lentfer, J., Schmidt, H. A., Technau, U., Von Haeseler, A., Hobmayer, B., Martindale, M. Q. & Holstein, T. W. (2005) Unexpected complexity of the Wnt gene family in a sea anemone. *Nature*, 433, 156–160.

Kyriazakis, I., Tolkamp, B. J. & Hutchings, M. R. (1998) Towards a functional explanation for the occurrence of anorexia during parasitic infections. *Animal Behaviour*, 56, 265–274.

Lack, D. (1947) The significance of clutch size. *Ibis*, 89, 302–352.

Lack, D. (1966). Population Studies of Birds. Clarendon Press, Oxford, England.

LaFever, L. & Drummond-Barbosa, D. (2005) Direct control of germline stem cell division and cyst growth by neural insulin in *Drosophila. Science*, 309, 1071–1073.

Lager, C. & Ellison, P. T. (1990) Effect of moderate weight loss on ovarian function assessed by salivary progesterone measurements. *American Journal of Human Biology*, 2, 303–312.

Lagios, M. D. (1982) Latimeria and the Chondrichthyes as sister taxa: A rebuttal to recent attempts at refutation. *Copeia*, 4, 942–948.

Lagueux, M., Hetru, C., Goltzene, F., Kappler, C. & Hoffmann, J. A. (1979) Ecdysone titre and metabolism in relation to cuticulogenesis in embryos of *Locusta migratoria. Journal of Insect Physiology*, 25, 709–723.

Lai, C. Q., Parnell, L. D., Lyman, R. F., Ordovas, J. A. & Mackay, T. F. C. (2007) Candidate genes affecting *Drosophila* lifespan identified by integrating microarray gene expression analysis and QTL mapping. *Mechanisms of Ageing and Development*, 128, 237–249.

Lakkis, F. G., Dellaporta, S. L. & Buss, L. W. (2008) Allorecognition and chimerism in an invertebrate model organism. *Organogenesis*, 4, 236–240.

Lakowski, B. & Hekimi, S. (1996) Determination of lifespan in *Caenorhabditis elegans* by four clock genes. *Science*, 272, 1010–1013.

Lakowski, B. & Hekimi, S. (1998) The genetics of caloric restriction in *Caenorhabditis elegans. Proceedings of the National Academy of Sciences of the United States of America*, 95, 13091–13096.

Laland, K. N., Odling-Smee, J. & Feldman, M. W. (2000) Niche construction, biological evolution, and cultural change. *Behavioral and Brain Sciences*, 23, 131–175.

Lam, T. J. (1980) Thyroxine enhances larval development and survival in *Sarotherodon* (Tilapia) *Mossambicus ruppell. Aquaculture*, 21, 287–291.

Lambert, A., Buckingham, J. A., Boysen, H. M. & Brand, M. D. (2010) Low complex I content explains the low hydrogen peroxide production rate of heart mitochondria from the long-lived pigeon, *Columa livia. Aging Cell*, 9, 78–91.

Lambrechts, L., Fellous, S. & Koella, J. C. (2006) Coevolutionary interactions between host and parasite genotypes. *Trends in Parasitology*, 22, 12–16.

Lancaster, L. T., McAdam, A. G., Wingfield, J. C. & Sinervo, B. (2007) Adaptive social and maternal induction of antipredator dorsal patterns in a lizard with alternative social strategies. *Ecology Letters*, 10, 798–808.

Lancaster, L. T., Hazard, L. C., Clobert, J. & Sinervo, B. R. (2008) Corticosterone manipulation reveals differences in hierarchical organization of multidimensional reproductive trade-offs in r-strategist and K-strategist females. *Journal of Evolutionary Biology*, 21, 556–565.

Lancaster, L. T., McAdam, A. G. & Sinervo, B. (2010) Maternal adjustment of egg size organizes alternative escape behaviors, promoting adaptive phenotypic integration. *Evolution*, 64, 1607–1621.

Lande, R. (1982) A quantitative genetic theory of life history evolution. *Ecology*, 63, 607–615.

Landry, C. R., Wittkopp, P. J., Taubes, C. H., Ranz, J. M., Clark, A. G. & Hartl, D. L. (2005) Compensatory cis-trans evolution and the dysregulation of gene expression in interspecific hybrids of *Drosophila. Genetics*, 171, 1813–1822.

Lane, M. A. (2000) Nonhuman primate models in biogerontology. *Experimental Gerontology*, 35, 533–541.

Lane, M. A., Mattison, J., Ingram, D. K. & Roth, G. S. (2002) Caloric restriction and aging in primates: Relevance to humans and possible CR mimetics. *Microscopy Research and Technique*, 59, 335–338.

Langley, E., Pearson, M., Faretta, M., Bauer, U. M., Frye, R. A., Minucci, S., Pelicci, P. G. & Kouzarides, T. (2002) Human SIR2 deacetylates p53 and antagonizes PML/p53-induced cellular senescence. *EMBO Journal*, 21, 2383–2396.

Lanzrein, B., Gentinetta, V., Abegglen, H., Baker, F. C., Miller, C. A. & Schooley, D. A. (1985) Titers of ecdysone, 20-hydroxy-ecdysone and juvenile hormone III throughout the life cycle of a hemimetabolous insect, the ovoviviparous cockroach *Nauphoeta cinerea. Experientia* 41, 913–917.

Larsen, P. L. (2001) Asking the age-old questions. *Nature Genetics*, 28, 102–104.

Larson, A., Kirk, M. & Kirk, D. L. (1992) Molecular phylogeny of the volvocine flagellates. *Molecular Biology and Evolution*, 9, 85–105.

Larsen, D. A., Swanson, P., Dickey, J. T., Rivier, J. & Dickhoff, W. W. (1998) In vitro thyrotropin-releasing activity of corticotropin-releasing hormone-family peptides in coho salmon, *Oncorhynchus kisutch. General and Comparative Endocrinology*, 109, 276–285.

Lawniczak, M. K. & Begun, D. J. (2004) A genome-wide analysis of courting and mating responses in *Drosophila melanogaster* females. *Genome*, 47, 900–910.

Lawniczak, M. K. N., Barnes, A. I., Linklater, J. R., Boone, J. M., Wigby, S. & Chapman, T. (2007) Mating and immunity in invertebrates. *Trends in Ecology & Evolution*, 22, 48–55.

Layalle, S., Arquier, N. & Leopold, P. (2008) The TOR pathway couples nutrition and developmental timing in *Drosophila*. *Developmental Cell*, 15, 568–577.

Lazar, M. A. (1993) Thyroid hormone receptors: Multiple forms, multiple possibilities. *Endocrine Reviews*, 14, 184–193.

Lazar, M. A. (2003) Thyroid hormone action: a binding contract. *Journal of Clinical Investigation*, 112, 497–499.

Lazzaro, B. P. (2008) Natural selection on the *Drosophila* antimicrobial immune system. *Current Opinion in Microbiology*, 11, 284–289.

Lazzaro, B. P. & Little, T. J. (2009) Immunity in a variable world. *Philosophical Transactions of the Royal Society of London B*, 364, 15–26.

Lazzaro, B. P., Sackton, T. B. & Clark, A. G. (2006) Genetic variation in *Drosophila melanogaster* resistance to infection: a comparison across bacteria. *Genetics*, 174, 1539–1554.

Leaf, D. S., Anstrom, J. A., Chin, J. E., Harkey, M. A., Showman, R. M. & Raff, R. A. (1987) Antibodies to a fusion protein identify a cDNA clone encoding msp 130, a primary mesenchyme-specific cell surface protein of the sea urchin. *Developmental Biology*, 121, 29–40.

Leatherland, J. F. (1994) Reflections on the thyroidology of fishes: from molecules to humankind. *Guelph Ichthyology Reviews*, 2 1–67.

Leatherland, J. F., Hilliard, R. W., Macey, D. J. & Potter, I. C. (1990) Changes in serum thyroxine and triiodothyronine concentrations during metamorphosis of the southern hemisphere lamprey *Geotria australis*, and the effect of propylthiouracil, triiodothyronine and environmental temperature on serum thyroid hormone concentrations of ammocoetes. *Fish Physiology and Biochemistry*, 8, 167–177.

LeBlanc, S. A. (2003) *Constant Battles: The Myth of the Peaceful, Noble Savage*. New York, St. Martin's Press.

Le Corre, V., Roux, F. & Reboud, X. (2002) DNA polymorphism at the *FRIGIDA* gene in *Arabidopsis thaliana*: Extensive nonsynonymous variation is consistent with local selection for flowering time. *Molecular Biology and Evolution*, 19, 1261–1271.

Lee, R. D. (2003) Rethinking the evolutionary theory of aging: transfers, not births, shape senescence in social species. *Proceedings of the National Academy of Sciences of the United States of America*, 100, 9637–9642.

Lee, R. C. & Ambros, V. (2001) An extensive class of small RNAs in *Caenorhabditis elegans*. *Science*, 294, 862–864.

Lee, S. J. & Kenyon, C. (2009) Regulation of the longevity response to temperature by thermosensory neurons in *Caenorhabditis elegans*. *Current Biology*, 19, 715–722.

Lee, S. S., Kennedy, S., Tolonen, A. C. & Ruvkun, G. (2003) DAF-16 target genes that control *C. elegans* life-span and metabolism. *Science*, 300, 644–647.

Lee, K. S., You, K. H., Choo, J. K., Han, Y. M. & Yu, K. (2004) *Drosophila* short neuropeptide F regulates food intake and body size. *Journal of Biological Chemistry*, 279, 50781–50789.

Lee, G. D., Wilson, M. A., Zhu, M., Wolkow, C. A., De Cabo, R., Ingram, D. K. & Zou, S. (2006a) Dietary deprivation extends lifespan in *Caenorhabditis elegans*. *Aging Cell*, 5, 515–524.

Lee, P. N., Pang, K., Matus, D. Q. & Martindale, M. Q. (2006b) A WNT of things to come: Evolution of Wnt signaling and polarity in Cnidarians. *Seminars in Cell & Developmental Biology*, 17, 157–167.

Lee, E. J., Oh, B., Lee, J. Y., Kimm, K., Lee, S. H. & Baek, K. H. (2008a) A novel single nucleotide polymorphism of INSR gene for polycystic ovary syndrome. *Fertility and Sterility*, 89, 1213–1220.

Lee, K. P., Simpson, S. J., Clissold, F. J., Brooks, R., Ballard, J. W. O., Taylor, P. W., Soran, N. & Raubenheimer, D. (2008b) Lifespan and reproduction in *Drosophila*: New insights from nutritional geometry. *Proceedings of the National Academy of Sciences of the United States of America*, 105, 2498–2503.

Lee, K. S., Kwon, O. Y., Lee, J. H., Kwon, K., Min, K. J., Jung, S. A., Kim, A. K., You, K. H., Tatar, M. & Yu, K. (2008c) *Drosophila* short neuropeptide F signalling regulates growth by ERK-mediated insulin signalling. *Nature Cell Biology*, 10, 468–475.

Lee, K. A., Wikelski, M., Robinson, W. D., Robinson, T. R. & Klasing, K. C. (2008d) Constitutive immune defences correlate with life history variables in tropical birds. *Journal of Animal Ecology*, 77, 356–363.

Lee, J. M., Kaciroti, N., Appugliese, D., Corwyn, R. F., Bradley, R. H. & Lumeng, J. C. (2010) Body mass index and pubertal initiation in boys. *Archives of Pediatrics & Adolescent Medicine*, 164, 139–144.

Lefranc, A. & Bundgaard, J. (2000) The influence of male and female body size on copulation duration and fecundity in *Drosophila melanogaster*. *Hereditas*, 132, 243–247.

Legan, S. K., Rebrin, I., Mockett, R. J., Radyuk, S. N., Klichko, V. I., Sohal, R. S. & Orr, W. C. (2008) Overexpression of glucose-6-phosphate dehydrogenase extends the lifespan of *Drosophila melanogaster*. *Journal of Biological Chemistry*, 283, 32492–32499.

Le Goff, G., Boundy, S., Daborn, P. J., Yen, J. L., Sofer, L., Lind, R., Sabourault, C., Madi-Ravazzi, L. & Ffrench-Constant, R. H. (2003) Microarray analysis of cytochrome P450 mediated insecticide resistance in *Drosophila*. *Insect Biochemistry & Molecular Biology*, 33, 701–708.

Lehtinen, M. K., Yuan, Z., Boag, P. R., Yang, Y., Villen, J., Becker, E. B., Dibacco, S., De La Iglesia, N., Gygi, S., Blackwell, T. K. & Bonni, A. (2006) A conserved MST-FOXO signaling pathway mediates oxidative-stress responses and extends life span. *Cell*, 125, 987–1001.

Leidy, L. E. (1994) Biological aspects of menopause: across the lifespan. *Annual Review of Anthropology*, 23, 231–253.

Leigh, S. R. (2004) Brain growth, cognition, and life history in primate and human evolution. *American Journal of Primatology*, 62, 139–164.

Leips, J. & Mackay, T. F. C. (2000) Quantitative trait loci for lifespan in *Drosophila melanogaster*: Interactions with genetic background and larval density. *Genetics*, 155, 1773–1788.

Leloup-Hatey, J., Buscaglia, M., Jolivet-Jaudet, G. & Leloup, J. (1990) Interrenal function during the metamorphosis in anuran amphibia. *Fortschritte der Zoologie*, 38, 139–154.

Lemaitre, B. & Hoffmann, J. (2007) The host defense of *Drosophila melanogaster*. *Annual Review of Immunology*, 25, 697–743.

Lengfeld, T., Watanabe, H., Simakov, O., Lindgens, D., Gee, L., Law, L., Schmidt, H. A., Özbek, S., Bode, H. & Holstein, T. W. (2009) Multiple Wnts are involved in Hydra organizer formation and regeneration. *Developmental Biology*, 330, 186–199.

Leone, D. V., Quinlan, R. J., Hayden, R., Stewart, J. & Flinn, M. V. (2004) Long-term implications for growth of prenatal and early postnatal environment. *American Journal of Human Biology*, 16, 212–213.

Leonelli, S. (2007) *Arabidopsis*, the botannical *Drosophila*: from mouse cress to model organism. *Endeavour*, 31, 34–38.

Leroi, A. M. (2001) Molecular signals versus the Loi de Balancement. *Trends in Ecology & Evolution*, 16, 24–29.

Lescai, F., Blanche, H., Nebel, A., Beekman, M., Sahbatou, M., Flachsbart, F., Slagboom, E., Schreiber, S., Sorbi, S., Passarino, G. & Franceschi, C. (2009) Human longevity and 11p15.5: a study in 1321 centenarians. *European Journal of Human Genetics* 17, 1515–1519.

Lessells, C. M. (2008) Neuroendocrine control of life histories: what do we need to know to understand the evolution of phenotypic plasticity? *Philosophical Transactions of the Royal Society of London B*, 363, 1589–1598.

Levin, L. A. & Bridges, T. S. (1995) Pattern and Diversity in Reproduction and Development. In McEdward, L. (Ed.) *Ecology of Marine Invertebrate Larvae*. Boca Raton, CRC Press.

Levine, M. & Davidson, E. H. (2005) Gene regulatory networks for development. *Proceedings of the National Academy of Sciences of the United States of America*, 102, 4936–4942.

Levitan, D. R. (2000) Optimal egg size in marine invertebrates: theory and phylogenetic analysis of the critical relationship between egg size and development time in echinoids. *The American Naturalist*, 156, 175–192.

Levy, O., Appelbaum, L., Leggat, W., Gothlif, Y., Hayward, D. C., Miller, D. J. & Hoegh-Guldberg, O. (2007) Light-responsive cryptochromes from a simple multicellular animal, the coral *Acropora millepora*. *Science*, 318, 467–470.

Levy, Y. Y. & Dean, C. (1998) The transition to flowering. *Plant Cell*, 10, 1973–1989.

Lewontin, R. (1974) *The Genetic Basis of Evolutionary Change*. New York, Columbia University Press.

Liang, V. C., Sedgwick, T. & Shi, Y. B. (1997) Characterization of the *Xenopus* homolog of an immediate early gene associated with cell activation: sequence analysis and regulation of its expression by thyroid hormone during amphibian metamorphosis. *Cell Research*, 7, 179–193.

Liang, B., Moussaif, M., Kuan, C. J., Gargus, J. J. & Sze, J. Y. (2006) Serotonin targets the DAF-16/FOXO signaling pathway to modulate stress responses. *Cell Metabolism*, 4, 429–440.

Libert, S., Zwiener, J., Chu, X., Vanvoorhies, W., Roman, G. & Pletcher, S. D. (2007) Regulation of *Drosophila* lifespan by olfaction and food-derived odors. *Science*, 315, 1133–1137.

Libert, S., Chao, Y., Zwiener, J. & Pletcher, S. D. (2008) Realized immune response is enhanced in long-lived puc and chico mutants but is unaffected by dietary restriction. *Molecular Immunology*, 45, 810–817.

Licht, P., Papkoff, H., Farmer, S. W., Muller, C. H., Tsui, H. W. & Crews, D. (1977) Evolution of gonadotropin structure and function. *Recent Progress in Hormone Research*, 33, 169–248.

Lieps, J. & Travis, J. (1994) Metamorphic responses to changing food levels in two species of hylid frogs. *Ecology*, 75, 1345–1356.

Lim, M. M., Wang, Z., Olazabal, D. E., Ren, X., Terwilliger, E. F. & Young, L. J. (2004) Enhanced partner preference in a promiscuous species by manipulating the expression of a single gene. *Nature*, 429, 754–757.

Lin, K., Dorman, J. B., Rodan, A. & Kenyon, C. (1997) daf-16: An HNF-3/forkhead family member that can function to double the life-span of *Caenorhabditis elegans*. *Science*, 278, 1319–1322.

Lin, Y. J., Seroude, L. & Benzer, S. (1998) Extended life-span and stress resistance in the *Drosophila* mutant methuselah. *Science*, 282, 943–946.

Lin, K., Hsin, H., Libina, N. & Kenyon, C. (2001) Regulation of the *C. elegans* longevity protein DAF-16 by insulin/IGF-1 and germline signaling. *Nature Genetics*, 28, 139–145.

Linklater, J. R., Wertheim, B., Wigby, S. & Chapman, T. (2007) Ejaculate depletion patterns evolve in response to experimental manipulation of sex ratio in *Drosophila melanogaster*. *Evolution*, 61, 2027–2034.

Linnen, C., Tatar, M. & Promislow, D. E. L. (2001) Cultural artifacts: A comparison of senescence in natural, lab-adapted and artificially selected lines of *Drosophila melanogaster*. *Evolutionary Ecology Research*, 3, 877–888.

Lintlop, S. P. & Youson, J. H. (1983) Concentration of triiodothyronine in the sera of the sea lamprey, *Petromyzon marinus*, and the brook lamprey, *Lampetra lamottenii*, at various phases of the life cycle. *General and Comparative Endocrinology*, 49, 187–194.

Lipson, S. F. & Ellison, P. T. (1996) Comparison of salivary steroid profiles in naturally occurring conception and non-conception cycles. *Human Reproduction*, 11, 2090–2096.

Lirman, D. (2000) Fragmentation in the branching coral *Acropora palmata* (Lamarck): growth, survivorship, and reproduction of colonies and fragments. *Journal of Experimental Marine Biology and Ecology*, 251, 41–57.

Little, T. J., Colegrave, N., Sadd, B. M. & Schmid-Hempel, P. (2008) Studying immunity at the whole organism level. *BioEssays*, 30, 404–405.

Liu, Y. W., Lo, L. J. & Chan, W. K. (2000) Temporal expression and T3 induction of thyroid hormone receptors alpha 1 and beta 1 during early embryonic and larval development in zebrafish, *Danio rerio*. *Molecular and Cellular Endocrinology*, 159, 187–195.

Liu, X., Jiang, N., Hughes, B., Bigras, E., Shoubridge, E. & Hekimi, S. (2005) Evolutionary conservation of the *clk-1*-dependent mechanism of longevity: loss of *mclk1* increases cellular fitness and lifespan in mice. *Genes & Development*, 19, 2424–2434.

Liu, W.-C., Lo, W.-T., Purcell, J. & Chang, H.-H. (2009) Effects of temperature and light intensity on asexual reproduction of the scyphozoan, *Aurelia aurita* (L.) in Taiwan. *Hydrobiologia*, 616, 247–258.

Lively, C. M. & Dybdahl, M. F. (2000) Parasite adaptation to locally common host genotypes. *Nature*, 405, 679–681.

Lively, C. M., Dybdahl, M. F., Jokela, J., Osnas, E. E. & Delph, L. F. (2004) Host sex and local adaptation by parasites in a snail-trematode interaction. *The American Naturalist*, 164 Suppl 5, S6–S18.

Lochmiller, R. L. & Dabbert, C. B. (1993) Immunocompetence, environmental stress, and the regulation of animal populations. *Trends in Comparative Biochemistry & Physiology*, 1, 823–855.

Lochmiller, R. L. & Deerenberg, C. (2000) Trade-offs in evolutionary immunology: just what is the cost of immunity? *Oikos*, 88, 87–98.

Lopez, S. & Dominguez, C. A. (2003) Sex choice in plants: facultative adjustment of the sex ratio in the perennial herb *Begonia gracilis*. *Journal of Evolutionary Biology*, 16, 1177–1185.

Loudet, O., Chaillou, S., Krapp, A. & Daniel-Vedele, F. (2003) Quantitative trait loci analysis of water and anion contents in interaction with nitrogen availability in *Arabidopsis thaliana*. *Genetics*, 163, 711–722.

Lounibos, L. P., Van Dover, C. & O'Meara, G. F. (1982) Fecundity, autogeny and the larval environment of the pitcher-plant mosquito, *Wyeomyia smithii*. *Oecologia*, 55, 160–164.

Love, A. C., Andrews, M. E. & Raff, R. A. (2007) Gene expression patterns in novel animal appendage: the sea urchin pluteus arm. *Evolution & Development*, 9, 51–68.

Lovejoy, A. O. (1981) The origin of man. *Science*, 211, 341–350.

Lowe, C. J., Issel-Tarver, L. & Wray, G. A. (2002) Gene expression and larval evolution: changing roles of distal-less and orthodenticle in echinoderm larvae. *Evolution & Development*, 4, 111–123.

Luckinbill, L. S., Arking, R. A., Clare, M. J., Cirocco, W. C. & Buck, S. A. (1984) Selection for delayed senescence in *Drosophila melanogaster*. *Evolution*, 38, 996–1004.

Luckinbill, L. S., Riha, V., Rhine, S. & Gurdzein, T. A. (2001) The role of glucose-6-phosphate dehydrogenase in the evolution of longevity of *Drosophila melanogaster*. *Heredity*, 65, 29–38.

Lukhtanov, V. A., Kandul, N. P., Plotkin, J. B., Dantchenko, A. V., Haig, D. & Al, E. (2005) Reinforcement of pre-zygotic isolation and karyotype evolution in Agrodiaetus butterflies. *Nature*, 436, 385–389.

Lumme, J. (1981) Localization of the genetic unit controlling the photoperiodic adult diapause in *Drosophila littoralis*. *Hereditas*, 94, 241–244.

Lumme, J. & Keranen, L. (1978) Photoperiodic diapause in *Drosophila lummei* Hackman is controlled by an X-chromosomal factor. *Hereditas*, 89, 261–262.

Lumme, J. & Lakovaara, S. (1983) Seasonality and diapause in Drosophilids. In Ashburner, M., Carson, H. L. & Thompson, J. N. J. (Eds.) *Genetics and Biology of Drosophila*. London, Academic Press.

Lundholm, J. T. & Aarssen, L. W. (1994) Neighbor effects on gender variation in *Ambrosia artemisiifolia*. *Canadian Journal of Botany*, 72, 794–800.

Luo, J., Nikolaev, A. Y., Imai, S., Chen, D., Su, F., Shiloh, A., Guarente, L. & Gu, W. (2001) Negative control of p53 by Sir2alpha promotes cell survival under stress. *Cell*, 107, 137–148.

Lutz, P. L., Prentice, H. M. & Milton, S. L. (2003) Is turtle longevity linked to enhanced mechanisms for surviving brain anoxia and reoxygenation? *Experimental Gerontology*, 38, 797–800.

Lynn, S. E., Walker, B. G. & Wingfield, J. C. (2005) A phylogenetically controlled test of hypotheses for behavioral insensitivity to testosterone in birds. *Hormones and Behavior*, 47, 170–177.

Lyytinen, A., Brakefield, P. M., Lindstrom, L. & Mappes, J. (2004) Does predation maintain eyespot plasticity in *Bicyclus anynana*? *Proceedings of the Royal Society of London B*, 271, 279–283.

Mabee, P. M., Olmstead, K. L. & Cubbage, C. C. (2000) An experimental study of intraspecific variation, developmental timing, and heterochrony in fishes. *Evolution*, 54, 2091–2106.

Mabee, P. M., Crotwell, P. L., Bird, N. C. & Burke, A. C. (2002) Evolution of median fin modules in the axial skeleton of fishes. *Journal of Experimental Zoology*, 294, 77–90.

MacDonald, K. & Hershberger, S. L. (2005) Theoretical issues in the study of evolution and development. In Burgess, R. L. & MacDonald, K. (Eds.) *Evolutionary Perspectives on Human Development*. Thousand Oaks, Sage Press.

Mace, R. (2000) Evolutionary ecology of human life history. *Animal Behaviour*, 59, 1–10.

Mack, P. D., Kapelnikov, A., Heifetz, Y. & Bender, M. (2006) Mating-responsive genes in reproductive tissues of female *Drosophila melanogaster*. *Proceedings of the National Academy of Sciences of the United States of America*, 103, 10358–10363.

Mackay, T. F. C., Roshina, N. V., Leips, J. W. & Pasyukova, E. G. (2006) Complex genetic architecture of *Drosophila* longevity. In Masaro, E. J. & Austad, S. N. (Eds.) *Handbook of the Biology of Aging*. Burlington, Elsevier Press.

Mackay, T. F. C., Stone, E. A. & Ayroles, J. F. (2009) The genetics of quantitative traits: challenges and prospects. *Nature Reviews Genetics*, 10, 565–577.

MacKenzie, D. S., Jones, R. A. & Miller, T. C. (2009) Thyrotropin in teleost fish. *General and Comparative Endocrinology*, 161, 83–89.

Madhavan, M. M. & Schneiderman, H. A. (1977) Histological ananlysis of the dynamics of growth of imaginal discs and histoblast nests during larval development of *Drosophila melanogaster*. *Roux's Archives of Developmental Biology*, 183, 269–305.

Magie, C. R. & Martindale, M. Q. (2008) Cell-cell adhesion in the Cnidaria: insights into the evolution of tissue morphogenesis. *Biological Bulletin*, 214, 218–232.

Mair, W. & Dillin, A. (2008) Aging and survival: the genetics of lifespan extension by dietary restriction. *Annual Review of Biochemistry*, 77, 727–754.

Mair, W., Goymer, P., Pletcher, S. D. & Partridge, L. (2003) Demography of dietary restriction and death in *Drosophila*. *Science*, 301, 1731–1733.

Mair, W., Sgro, C. M., Johnson, A. P., Chapman, T. & Partridge, L. (2004) Lifespan extension by dietary restriction in female *Drosophila melanogaster* is not caused by a reduction in vitellogenesis or ovarian activity. *Experimental Gerontology*, 39, 1011–1019.

Mair, W., Piper, M. D. & Partridge, L. (2005) Calories do not explain extension of lifespan by dietary restriction in *Drosophila*. *PLoS Biology*, 3, e223.

Majhi, S., Jena, B. S. & Patnaik, B. K. (2000) Effect of age on lipid peroxides, lipofuscin and ascorbic acid contents of the lungs of male garden lizard. *Comparative Biochemistry and Physiology C*, 126, 293–298.

Maklakov, A. A., Simpson, S. J., Zajitschek, F., Hall, M. D., Dessmann, J., Clissold, F., Raubenheimer, D., Bonduriansky, R. & Brooks, R. C. (2008) Sex-specific fitness effects of nutrient intake on reproduction and lifespan. *Current Biology*, 18, 1062–1066.

Malamuth, N. M. (1996) Sexually explicit media, gender differences, and evolutionary theory. *Journal of Communication*, 46, 8–31.

Mangel, M. & Clark, C. (1988) *Dynamic modeling in behavioral ecology*. Princeton, Princeton University Press.

Manier, M. K., Seyler, C. M. & Arnold, S. J. (2007) Adaptive divergence within and between ecotypes of the terrestrial garter snake, *Thamnophis elegans*, assessed with F-St-Q(ST) comparisons. *Journal of Evolutionary Biology*, 20, 1705–1719.

Manzon, L. A. (2006) Cloning and developmental expression of sea lamprey (Petromyzon marinus) thyroid hormone and retinoid X receptors. Unpublished Ph.D. *Dissertation*, University of Toronto.

Manzon, R. G. & Denver, R. J. (2004) Regulation of pituitary thyrotropin gene expression during *Xenopus* metamorphosis: negative feedback is functional throughout metamorphosis. *Journal of Endocrinology*, 182, 273–285.

Manzon, R. G. & Youson, J. H. (1997) The effects of exogenous thyroxine (T4) or triiodothyronine (T3), in the presence and absence of potassium perchlorate, on the incidence of metamorphosis and on serum T4 and T3 concentrations in larval sea lamprey (*Petromyzon marinus*). *General and Comparative Endocrinology*, 106, 211–220.

Manzon, R. G., Eales, J. G. & Youson, J. H. (1998) Blocking of KClO4-induced metamorphosis in premetamorphic sea lampreys by exogenous thyroid hormones (TH): Effects of KClO4 and TH on serum TH concentrations and intestinal thyroxine outer-ring deiodination. *General and Comparative Endocrinology*, 112, 54–62.

Manzon, R. G., Holmes, J. A. & Youson, J. H. (2001) Variable effects of goitrogens in inducing precocious metamorphosis in sea lampreys (*Petromyzon marinus*). *Journal of Experimental Zoology*, 289, 290–303.

Mappes, T., Koivula, M., Koskela, E., Oksanen, T. A., Savolainen, T. & Sinervo, B. (2008) Frequency and density-dependent selection on life history strategies – A field experiment. *PLoS One*, 3, e1687.

Marden, J. H., Rogina, B., Montooth, K. L. & Helfand, S. L. (2003) Conditional trade-offs between aging and organismal performance of Indy long-lived mutant flies. *Proceedings of the National Academy of Sciences of the United States of America*, 100, 3369–3373.

Margulis, L. (1981) *Symbiosis in cell evolution*, Freeman, San Francisco.

Markow, T. A. & O'Grady, P. (2008) Reproductive ecology of *Drosophila*. *Functional Ecology*, 22, 747–759.

Markow, T. A. & O'Grady, P. M. (2005) Evolutionary genetics of reproductive behavior in *Drosophila*: Connecting the dots. *Annual Review of Genetics*, 39, 263–291.

Marlowe, F. W. (2003) A critical period for provisioning by Hadza men: Implications for pair bonding. *Evolution and Human Behavior*, 24, 217–229.

Marnett, L. J. & Plastaras, J. P. (2001) Endogenous DNA damage and mutation. *Trends in Genetics*, 17, 214–221.

Maróy, P., Kaufmann, G. & Dübendorfer, A. (1988) Embryonic ecdysteroids of *Drosophila melanogaster*. *Journal of Insect Physiology*, 34, 633–637.

Marshall, D. J. & Keough, M. J. (2003) Variation in the dispersal potential of non-feeding invertebrate larvae: the desperate larva hypothesis and larval size. *Marine Ecology Progress Series*, 255, 145–153.

Marshall, W. A. & Tanner, J. M. (1986) Puberty. In Falkner, F. & Tanner, J. M. (Eds.) *Human Growth*. New York, Plenum.

Marsh-Armstrong, N., Huang, H., Berry, D. L. & Brown, D. D. (1999) Germ-line transmission of transgenes in *Xenopus laevis*. *Proceedings of the National Academy of Sciences of the United States of America*, 96, 14389–14393.

Marsh-Armstrong, N., Cai, L. & Brown, D. D. (2004) Thyroid hormone controls the development of connections between the spinal cord and limbs during *Xenopus laevis* metamorphosis. *Proceedings of the National Academy of Sciences of the United States of America*, 101, 165–170.

Martin, I. & Grotewiel, M. S. (2006) Oxidative damage and age-related functional declines. *Mechanisms of Ageing and Development*, 127, 411–423.

Martin, L. B., Weil, Z. M. & Nelson, R. J. (2008) Seasonal changes in vertebrate immune activity: mediation by physiological trade-offs. *Philosophical Transactions of the Royal Society of London B*, 363, 321–339.

Martin-Smith, K. M., Armstrong, J. D., Johnsson, J. I. & Bjornsson, B. T. (2004) Growth hormone increases growth and dominance of wild juvenile Atlantic salmon without affecting space use. *Journal of Fish Biology*, 65, 156–172.

Masuda-Nakagawa, L. M., Gröer, H., Aerne, B. L. & Schmid, V. (2000) The Hox-like gene Cnox2-Pc is expressed at the anterior region in all life cycle stages of the jellyfish *Podocoryne carnea*. *Development, Genes and Evolution*, 210, 151–156.

Mathias, D. M., Jacky, L., Bradshaw, W. E. & Holzapfel, C. M. (2007) Quantitative trait loci associated with photoperiodic response and stage of diapause in the pitcher-plant mosquito, *Wyeomyia smithii*. *Genetics*, 176, 391–402.

Matkovic, V., Ilich, J. Z., Skugor, M., Badenhop, N. E., Goel, P., Clairmont, A., Klisovic, D., Nahhas, R. W. & Landoll, J. D. (1997) Leptin is inversely related to age at menarche in human females. *Journal of Clinical Endocrinology and Metabolism*, 82, 3239–3245.

Matsuda, H., Paul, B. D., Choi, C. Y., Hasebe, T. & Shi, Y. B. (2009) Novel functions of protein arginine methyltransferase 1 in thyroid hormone receptor-mediated transcription and in the regulation of metamorphic rate in *Xenopus laevis*. *Molecular and Cellular Biology*, 29, 745–757.

Matteo, S. & Rissman, E. F. (1984) Increased sexual activity during the midcycle portion of the human menstrual cycle. *Hormones and Behavior*, 18, 249–255.

Mattison, J. A., Lane, M. A., Roth, G. S. & Ingram, D. K. (2003) Calorie restriction in rhesus monkeys. *Experimental Gerontology*, 38, 35–46.

Matus, D. Q., Thomsen, G. H. & Martindale, M. Q. (2006) Dorso/ventral genes are asymmetrically expressed and involved in germ-layer demarcation during cnidarian gastrulation. *Current Biology*, 16, 499–505.

May, R. M. & Anderson, R. M. (1990) Parasite-host coevolution. *Parasitology*, 100 Suppl, S89–S101.

Maynard Smith, J. (1958) The effects of temperature and of egg-laying on the longevity of *Drosophila subobscura*. *Journal of Experimental Biology*, 35, 832–842.

Maynard Smith, J. (1982) *Evolution and the Theory of Games*. Cambridge, Cambridge University Press.

Maynard Smith, J., Burian, R., Kauffman, S., Alberch, P., Campbell, J., Goodwin, B., Lande, R., Raup, D. & Wolpert, L. (1985) Developmental constraints and evolution. *Quarterly Review of Biology*, 60, 265–287.

Mazur, A. (1992) Testosterone and chess competition. *Social Psychology Quarterly*, 55, 70–77.

McAlister, J. S. (2007) Egg size and the evolution of phenotypic plasticity in larvae of the echinoid genus *Strongylocentrotus*. *Journal of Experimental Marine Biology and Ecology*, 352, 306–316.

McAlister, J. S. (2008) Evolutionary responses to environmental heterogeneity in central american echinoid larvae: plasticity versus constant phenotypes. *Evolution*, 62, 1358–1372.

McBrayer, Z., Ono, H., Shimell, M., Parvy, J., Beckstead, R., Warren, J., Thummel, C., Dauphinvillemant, C., Gilbert, L. & Oconnor, M. (2007) Prothoracicotropic hormone regulates developmental timing and body size in *Drosophila*. *Developmental Cell*, 13, 857–871.

McCarroll, S. A., Murphy, C. T., Zou, S., Pletcher, S. D., Chin, C. S., Jan, Y. N., Kenyon, C., Bargmann, C. I. & Li, H. (2004) Comparing genomic expression patterns across species identifies shared transcriptional profile in aging. *Nature Genetics*, 36, 197–204.

McCart, C., Buckling, A. & Ffrench-Constant, R. H. (2005) DDT resistance in flies carries no cost. *Current Biology*, 15, R587–R589.

McCart, C. & Ffrench-Constant, R. H. (2008) Dissecting the insecticide-resistance-associated cytochrome P450 gene Cyp6g1. *Pest Management Science*, 64, 639–645.

McCauley, D. W. (1997) Serotonin plays an early role in the metamorphosis of the hydrozoan *Phialidium gregarium*. *Developmental Biology*, 190, 229–240.

McCay, C. M., Dilly, W. E. & Crowell, M. F. (1929) Growth rates of brook trout reared uopn purified rations, upon skim milk diets, and upon combinations of cereal grains. *Journal of Nutrition*, 1, 233–246.

McCay, C. M., Crowell, M. F. & Maynard, L. A. (1935) The effect of retarded growth upon the length of lifespan and upon the ultimate body size. *Nutrition*, 5, 63–79.

McClain, C. R. & Rex, M. A. (2001) The relationship between dissolved oxygen concentration and maximum size in deep-sea turrid gastropods: an application of quantile regression. *Marine Biology*, 139, 681–685.

McCormick, S. D. (1996) Effects of growth hormone and insulin-like growth factor I on salinity tolerance and gill Na+, K+-ATPase in Atlantic salmon (*Salmo salar*): Interaction with cortisol. *General and Comparative Endocrinology*, 101, 3–11.

McCormick, S. D. (2001) Endocrine control of osmoregulation in teleost fish. *American Zoologist*, 41, 781–794.

McCormick, S. D., Sakamoto, T., Hasegawa, S. & Hirano, T. (1991) Osmoregulatory actions of insulin-like growth factor-I in rainbow trout (*Oncorhynchus mykiss*). *Journal of Endocrinology*, 130, 87–92.

McCormick, S. D., Hansen, L. P., Quinn, T. P. & Saunders, R. L. (1998) Movement, migration, and smolting of Atlantic salmon (*Salmo salar*). *Canadian Journal of Fisheries and Aquatic Sciences*, 55, 77–92.

McCue, M. D. (2008) Fatty acid analyses may provide insight into the progression of starvation among squamate reptiles. *Comparative Biochemistry and Physiology A*, 151, 239–246.

McDiarmid, R. W. & Altig, R. (1999) *Tadpoles: The biology of anuran larvae*. Chicago and London, University of Chicago Press.

McDowell, J. M. & Simon, S. A. (2006) Recent insights into R gene evolution. *Molecular Plant Pathology*, 7, 437–448.

McEdward, L. R. (1996) Experimental manipulation of parental investment in Echinoid Echinoderms. *American Zoologist*, 36, 169–179.

McEdward, L. R. (1997) Reproductive strategies of marine benthic invertebrates revisited: facultative feeding by planktotrophic larvae. *The American Naturalist*, 150, 48–72.

McEdward, L. R. & Herrera, J. C. (1999) Body form and skeletal morphometrics during larval development of the sea urchin *Lytechinus variegatus* Lamarck. *Journal of Experimental Marine Biology and Ecology*, 232, 151–176.

McEdward, L. R. & Miner, B. G. (2001) Larval and life-cycle patterns in Echinoderms. *Canadian Journal of Zoology-Revue Canadienne de Zoologie*, 79, 1125–1170.

McElwee, J. J., Schuster, E., Blanc, E., Thomas, J. H. & Gems, D. (2004) Shared transcriptional signature in *Caenorhabditis elegans* dauer larvae and long-lived *daf-2* mutants implicates detoxification system in longevity assurance. *Journal of Biological Chemistry*, 279, 44533–44543.

McElwee, J. J., Schuster, E., Blanc, E., Piper, M. D., Thomas, J. H., Patel, D. S., Selman, C., Withers, D. J., Thornton, J. M., Partridge, L. & Gems, D. (2007) Evolutionary conservation of regulated longevity assurance mechanisms. *Genome Biology*, 8, R132.

McGarrigle, D. & Huang, X. Y. (2007) Methuselah antagonist extends life span. *Nature Chemical Biology*, 3, 371–372.

McGlothlin, J. W. & Ketterson, E. D. (2008) Hormone-mediated suites as adaptations and evolutionary constraints. *Philosophical Transactions of the Royal Society of London B*, 363, 1611–1620.

McGlothlin, J. W., Jawor, J. M. & Ketterson, E. D. (2007) Natural variation in a testosterone-mediated trade-off between mating effort and parental effort. *The American Naturalist*, 170, 864–875.

McGlothlin, J. W., Jawor, J. M., Greives, T. J., Casto, J. M., Phillips, J. L. & Ketterson, E. D. (2008) Hormones and honest signals: males with larger ornaments elevate testosterone more when challenged. *Journal of Evolutionary Biology*, 21, 39–48.

McGraw, L. A., Gibson, G., Clark, A. G. & Wolfner, M. F. (2004) Genes regulated by mating, sperm, or seminal proteins in mated female *Drosophila melanogaster*. *Current Biology*, 14, 1509–1514.

McGuire, M. & Gruter, M. (2003) Prostitution: An evolutionary perspective. In Somit, A. & Peterson, S. (Eds.) *Human Nature and Public Policy: An Evolutionary Approach*. New York, Palgrave McMillan.

McKay, J. K., Richards, J. H. & Mitchell-Olds, T. (2003) Genetics of drought adaptation in *Arabidopsis thaliana*: I. Pleiotropy contributes to genetic correlations among ecological traits. *Molecular Ecology*, 12, 1137–1151.

McKay, J. K., Richards, J. H., Nemali, K. S., Sen, S., Mitchell-Olds, T., Boles, S., Stahl, E. A., Wayne, T., Juenger, T. E. & Rausher, M. (2008) Genetics of drought adaptation in *Arabidopsis thaliana* II. QTL analysis of a new mapping population, Kas-1 x Tsu-1. *Evolution*, 62, 3014–3026.

McKean, K. A. & Nunney, L. (2001) Increased sexual activity reduces male immune function in *Drosophila melanogaster*. *Proceedings of the National Academy of Sciences of the United States of America*, 98, 7904–7909.

McKean, K. A. & Nunney, L. (2005) Bateman's principle and immunity: Phenotypically plastic reproductive strategies predict changes in immunological sex differences. *Evolution*, 59, 1510–1517.

McKean, K. A. & Nunney, L. (2008) Sexual selection and immune function in *Drosophila melanogaster*. *Evolution*, 62, 386–400.

McKean, K. A., Yourth, C. P., Lazzaro, B. P. & Clark, A. (2008) The evolutionary costs of immunological maintenance and deployment. *BMC Evolutionary Biology*, 8, 76.

McKenzie, J. A. (2001) Pesticide resistance. In Fox, C. W., Roff, D. A. & Daphne, J. F. (Eds.) *Evolutionary Ecology*. Oxford, Oxford University Press.

McLachlan, A. J. & Allen, D. F. (1987) Male mating success in Diptera – advantages of small size. *Oikos*, 48, 11–14.

McNabb, F. M. A. (2007) The hypothalamic-pituitary-thyroid (HPT) axis in birds and its role in bird development and reproduction. *Critical Reviews in Toxicology*, 37, 163–193.

McNeilly, A. S., Glasier, A., Jonassen, J. & Howie, P. W. (1982) Evidence for direct inhibition of ovarian function by prolactin. *Journal of Reproduction and Fertility*, 65, 559–569.

Medawar, P. B. (1952) *An Unsolved Problem in Biology*. London, Lewis.

Meddle, S. L., Maney, D. L. & Wingfield, J. C. (1999) Effects of N-methyl-D-aspartate on luteinizing hormone release and fos-like immunoreactivity in the male white-crowned sparrow (*Zonotrichia leucophrys gambelii*). *Endocrinology*, 140, 5922–5928.

Meeuwis, R., Michielsen, R., Decuypere, E. & Kuhn, E. R. (1989) Thyrotropic activity of the ovine corticotropin releasing factor in the chick embryo. *General and Comparative Endocrinology*, 76, 357–363.

Meinhardt, H. (2009) Models for the generation and interpretation of gradients. *Cold Spring Harbor Perspectives in Biology*, 1, a001362.

Meissner, M., Stark, K., Cresnar, B., Kirk, D. L. & Schmitt, R. (1999) *Volvox* germline-specific genes that are putative targets of RegA repression encode chloroplast proteins. *Current Genetics*, 36, 363–370.

Melendez, A., Talloczy, Z., Seaman, M., Eskelinen, E. L., Hall, D. H. & Levine, B. (2003) Autophagy genes are essential for dauer development and life-span extension in *C. elegans*. *Science*, 301, 1387–1391.

Melzer, D., Frayling, T. M., Murray, A., Hurst, A. J., Harries, L. W., Song, H., Khaw, K., Luben, R., Surtees, P. G., Bandinelli, S. S., Corsi, A. M., Ferrucci, L., Guralnik, J. M., Wallace, R. B., Hattersley, A. T. & Pharoah, P. D. (2007) A common variant of the p16(INK4a) genetic region is associated with physical function in older people. *Mechanisms of Ageing and Development*, 128, 370–377.

Mendel, C. M., Weisiger, R. A., Jones, A. L. & Cavalieri, R. R. (1987) Thyroid hormone binding proteins in plasma facilitate uniform distribution of thyroxine within tissues: A perfused rat liver study. *Endocrinology*, 120, 1742–1749.

Merchant, S. S., Prochnik, S. E., Vallon, O., Harris, E. H., Karpowicz, S. J., Witman, G. B., Terry, A., Salamov, A., Fritz-Laylin, L. K., Marechal-Drouard, L., Marshall, W. F., Qu, L. H., Nelson, D. R., Sanderfoot, A. A., Spalding, M. H., Kapitonov, V. V., Ren, Q. H., Ferris, P., Lindquist, E., Shapiro, H., Lucas, S. M., Grimwood, J., Schmutz, J., Cardol, P., Cerutti, H., Chanfreau, G., Chen, C. L., Cognat, V., Croft, M. T., Dent, R., Dutcher, S., Fernandez, E., Fukuzawa, H., Gonzalez-Balle, D., Gonzalez-Halphen, D., Hallmann, A., Hanikenne, M., Hippler, M., Inwood, W., Jabbari, K., Kalanon, M., Kuras, R., Lefebvre, P. A., Lemaire, S. D., Lobanov, A. V., Lohr, M., Manuell, A., Meir, I., Mets, L., Mittag, M., Mittelmeier, T., Moroney, J. V., Moseley, J., Napoli, C., Nedelcu, A. M., Niyogi, K., Novoselov, S. V., Paulsen, I. T., Pazour, G., Purton, S.,

Ral, J. P., Riano-Pachon, D. M., Riekhof, W., Rymarquis, L., Schroda, M., Stern, D., Umen, J., Willows, R., Wilson, N., Zimmer, S. L., Allmer, J., Balk, J., Bisova, K., Chen, C. J., Elias, M., Gendler, K., Hauser, C., Lamb, M. R., Ledford, H., Long, J. C., Minagawa, J., Page, M. D., Pan, J. M., Pootakham, W., Roje, S., Rose, A., Stahlberg, E., Terauchi, A. M., Yang, P. F., Ball, S., Bowler, C., Dieckmann, C. L., Gladyshev, V. N., Green, P., Jorgensen, R., Mayfield, S., Mueller-Roeber, B., Rajamani, S., Sayre, R. T., Brokstein, P., et al. (2007) The *Chlamydomonas* genome reveals the evolution of key animal and plant functions. *Science*, 318, 245–251.

Merritt, T. J. S., Sezgin, E., Zhu, C.-T. & Eanes, W. F. (2006) Triglyceride pools, flight and activity variation at the *Gpdh* locus in *Drosophila melanogaster*. *Genetics*, 172, 293–304.

Metcalfe, N. B., Huntingford, F. A., Graham, W. D. & Thorpe, J. E. (1989) Early social-status and the development of life history strategies in Atlantic Salmon. *Proceedings of the Royal Society of London B*, 236, 7–19.

Metcalfe, N. B., F. A. Huntingford, J. E. Thorpe, and C. E. Adams. 1990. The effects of social status on life-history variation in juvenile salmon. *Canadian Journal of Zoology - Revue Canadienne de la. Zoologie*, 68, 2630–2636.

Metcalfe, N. B., and J. E. Thorpe. 1990. Determinants of geographical variation in the age of seaward migrating salmon, *Salmo salar*. *Journal of Animal Ecology*, 59, 135–145.

Metcalf, C. J. E. & Pavard, S. (2007) Why evolutionary biologists should be demographers. *Trends in Ecology & Evolution*, 22, 205–212.

Metcalf, V. J., George, P. M. & Brennan, S. O. (2007) Lungfish albumin is more similar to tetrapod than to teleost albumins: Purification and characterisation of albumin from the Australian lungfish, *Neoceratodus forsteri*. *Comparative Biochemistry and Physiology B*, 147, 428–437.

Metcalf, C. J. E., Rose, K. E., Childs, D. Z., Sheppard, A. W., Grubb, P. J. & Rees, M. (2008) Evolution of flowering decisions in a stochastic, density-dependent environment. *Proceedings of the National Academy of Sciences of the United States of America*, 105, 10466–10470.

Meunier, N., Belgacem, Y. H. & Martin, J. R. (2007) Regulation of feeding behaviour and locomotor activity by takeout in *Drosophila*. *Journal of Experimental Biology*, 210, 1424–1434.

Meyer, E., Davies, S., Wang, S., Willis, B. L., Abrego, D., Juenger, T. E. & Matz, M. V. (2009) Genetic variation in responses to a settlement cue and elevated temperature in the reef-building coral *Acropora millepora*. *Marine Ecology Progress Series*, 392, 81–92.

Mezentseva, N., Kumaratilake, J. & Newman, S. (2008) The brown adipocyte differentiation pathway in birds: An evolutionary road not taken. *BMC Biology*, 6, 17.

Michaels, S. D. & Amasino, R. M. (1999) *FLOWERING LOCUS C* encodes a novel MADS domain protein that acts as a repressor of flowering. *Plant Cell*, 11, 949–956.

Michaels, S. D., He, Y. H., Scortecci, K. C. & Amasino, R. M. (2003) Attenuation of *FLOWERING LOCUS C* activity as a mechanism for the evolution of summer-annual flowering behavior in *Arabidopsis*. *Proceedings of the National Academy of Sciences of the United States of America*, 100, 10102–10107.

Michaels, S. D., Bezerra, I. C. & Amasino, R. M. (2004) *FRIGIDA*-related genes are required for the winter-annual habit in *Arabidopsis*. *Proceedings of the National Academy of Sciences of the United States of America*, 101, 3281–3285.

Michod, R. E. (2006) The group covariance effect and fitness trade-offs during evolutionary transitions in individuality. *Proceedings of the National Academy of Sciences of the United States of America*, 103, 9113–9117.

Michod, R. E. & Nedelcu, A. M. (2003) On the reorganization of fitness during evolutionary transitions in individuality. *Integrative and Comparative Biology*, 43, 64–73.

Michod, R. E., Viossat, Y., Solari, C. A., Hurand, M. & Nedelcu, A. M. (2006) Life history evolution and the origin of multicellularity. *Journal of Theoretical Biology*, 239, 257–272.

Middleton, C. A., Nongthomba, U., Parry, K., Sweeney, S. T., Sparrow, J. C. & Elliott, C. J. H. (2006) Neuromuscular organization and aminergic modulation of contractions in the *Drosophila* ovary. *BMC Biology*, 4, 1–14.

Millar, J. G., Chaney, J. D. & Mulla, M. S. (1992) Identification of oviposition attractants for *Culex quinquefasciatus* from fermented bermuda grass infusions. *Journal of the American Mosquito Control Association*, 8, 11–17.

Miller, G. E. (2000) *The mating mind: how sexual choice shaped the evolution of human nature*. New York, Doubleday.

Miller, S. E. & Hadfield, M. G. (1990) Developmental arrest during larval life and life-span extension in a marine Mollusc. *Science*, 248, 356–358.

Miller, A. E. & Heyland, A. (2010) Endocrine interactions between plants and animals: Implications of exogenous hormone sources for the evolution of hormone signaling. *General and Comparative Endocrinology*, 166, 455–461.

Miller, R. A., Buehner, G., Chang, Y., Harper, J. M., Sigler, R. & Smith-Wheelock, M. (2005) Methionine-deficient diet extends mouse lifespan, slows immune and lens aging, alters glucose, T4, IGF-I and insulin levels, and increases hepatocyte MIF levels and stress resistance. *Aging Cell*, 4, 119–125.

Mills, S. M., Hazard, L., Lancaster, L., Mappes, J., Miles, D. B., Oksanen, T. A. & Sinervo, B. (2008) Gonadotropin hormone modulation of testosterone, immune function, performance and behavioral trade-offs among male morphs of the lizard, *Uta stansburiana*. *The American Naturalist*, 171, 339–357.

Milo, R., Shen-Orr, S., Itzkovitz, S., Kashtan, N., Chklovskii, D. & Alon, U. (2002) Network motifs: simple building blocks of complex networks. *Science*, 298, 824–829.

Min, K. J. & Tatar, M. (2006) Restriction of amino acids extends lifespan in *Drosophila melanogaster*. *Mechanisms of Ageing and Development*, 127, 643–646.

Min, K. J., Hogan, M. F., Tatar, M. & O'Brien, D. M. (2006) Resource allocation to reproduction and soma in *Drosophila*: A stable isotope analysis of carbon from dietary sugar. *Journal of Insect Physiology*, 52, 763–770.

Min, K. J., Flatt, T., Kulaots, I. & Tatar, M. (2007) Counting calories in *Drosophila* diet restriction. *Experimental Gerontology*, 42, 247–251.

Min, K. J., Yamamoto, R., Buch, S., Pankratz, M. & Tatar, M. (2008) *Drosophila* lifespan control by dietary restriction independent of insulin-like signaling. *Aging Cell*, 7, 199–206.

Minakuchi, C., Zhou, X. & Riddiford, L. M. (2008) Krüppel homolog 1 (Kr-h1) mediates juvenile hormone action during metamorphosis of *Drosophila melanogaster*. *Mechanisms of Development*, 124, 91–105.

Minakuchi, C., Namiki, T. & Shinoda, T. (2009) Krüppel homolog 1, an early juvenile hormone-response gene downstream of Methoprene-tolerant, mediates its anti-metamorphic action in the red flour beetle *Tribolium castaneum*. *Developmental Biology*, 325, 341–350.

Minasian, L. L. J. & Mariscal, R. N. (1979) Characteristics and regulation of fission activity in clonal cultures of the cosmopolitan sea anemone, *Haliplanella luciae* (Verrill) *Biological Bulletin*, 157, 478–493.

Miner, B. G. (2005) Evolution of feeding structure plasticity in marine invertebrate larvae: a possible trade-off between arm length and stomach size. *Journal of Experimental Marine Biology and Ecology*, 315, 117–125.

Miner, B. (2007) Larval feeding structure plasticity during pre-feeding stages of echinoids: not all species respond to the same cues. *Journal of Experimental Marine Biology and Ecology*, 343, 158–165.

Miner, B. G. & Vonesh, J. R. (2004) Effects of fine grain environmental variability on morphological plasticity. *Ecology Letters*, 7, 794–801.

Miron, M., Lasko, P. & Sonenberg, N. (2003) Signaling from Akt to FRAP/TOR targets both 4E-BP and S6K in *Drosophila melanogaster*. *Molecular and Cellular Biology*, 23, 9117–9126.

Mirth, C. (2005) Ecdysteroid control of metamorphosis in the differentiating adult leg structures of *Drosophila melanogaster*. *Developmental Biology*, 278, 163–174.

Mirth, C. K. & Riddiford, L. M. (2007) Size assessment and growth control: how adult size is determined in insects. *BioEssays*, 29, 344–355.

Mirth, C., Truman, J. W. & Riddiford, L. M. (2005) The role of the prothoracic gland in determining critical weight for metamorphosis in *Drosophila melanogaster*. *Current Biology*, 15, 1796–1807.

Misra, R. K. & Reeve, C. R. (1964) Clines in body dimensions in populations of *Drosophilia subobscura*. *Genetical Research*, 5, 240–256.

Mitchell, B. D., Hsueh, W. C., King, T. M., Pollin, T. I., Sorkin, J., Agarwala, R., Schaffer, A. A. & Shuldiner, A. R. (2001) Heritability of lifespan in the Old Order Amish. *American Journal of Medical Genetics*, 102, 346–352.

Mitchell-Olds, T. & Schmitt, J. (2006) Genetic mechanisms and evolutionary significance of natural variation in *Arabidopsis*. *Nature*, 441, 947–952.

Mitchell-Olds, T., Willis, J. H. & Goldstein, D. B. (2007) Which evolutionary processes influence natural genetic variation for phenotypic traits? *Nature Reviews Genetics*, 8, 845–856.

Mittler, R. (2002) Oxidative stress, antioxidants and stress tolerance. *Trends in Plant Science*, 7, 405–410.

Miura, K., Oda, M., Makita, S. & Chinzei, Y. (2005) Characterization of the *Drosophila* Methoprene-tolerant gene product. *FEBS Journal*, 272, 1169–1178.

Miwa, S. & Inui, Y. (1987) Effects of various doses of thyroxine and triiodothyronine on the metamorphosis of flounder (*Paralichthys olivaceus*). *General and Comparative Endocrinology*, 67, 356–363.

Miwa, S. & Inui, Y. (1991) Thyroid hormone stimulates the shift of erythrocyte populations during metamorphosis of the flounder. *Journal of Experimental Zoology*, 259, 222–228.

Miwa, S., Tagawa, M., Inui, Y. & Hirano, T. (1988) Thyroxine surge in metamorphosing flounder larvae. *General and Comparative Endocrinology*, 70, 158–163.

Miwa, S., Yamano, K. & Inui, Y. (1992) Thyroid hormone stimulates gastric development in flounder larvae during metamorphosis. *Journal of Experimental Zoology*, 261, 424–430.

Miyata, S., Begun, J., Troemel, E. R. & Ausubel, F. M. (2008) DAF-16-dependent suppression of immunity during reproduction in *Caenorhabditis elegans*. *Genetics*, 178, 903–918.

Moberg, F. & Folke, C. (1999) Ecological goods and services of coral reef ecosystems. *Ecological Economics*, 29, 215–233.

Mockett, R. J., Cooper, T. M., Orr, W. C. & Sohal, R. S. (2006) Effects of caloric restriction are species-specific. *Biogerontology*, 7, 157–160.

Moczek, A. P. (2009) Developmental plasticity and the origins of diversity: a case study on horned beetles. In Ananthakrishnan, T. N. & Whitman, D. W. (Eds.) *Phenotypic Plasticity in Insects*. Plymouth, Science Publishers Inc. pp. 81–134.

Mohamed, S. A., Rottmann, O. & Pirchner, F. (2001) Components of heterosis for growth traits and litter size in line crosses of mice after long-term selection. *Journal of Animal Breeding and Genetics*, 118, 263–270.

Mok, F. S. Y., Thiyagarajan, V. & Qian, P. Y. (2009) Proteomic analysis during larval development and metamorphosis of the spionid polychaete *Pseudopolydora vexillosa*. *Proteome Science*, 7, 44.

Moll, J., Zahn, R., De Oliveira-Souza, R., Krueger, F. & Grafman, J. (2005) The neural basis of human moral cognition. *Nature Reviews Neuroscience*, 6, 799–809.

Moller, H. (1984) Reduction of a larval herring population by jellyfish predator. *Science*, 224, 621–622.

Momose, T., Derelle, R. & Houliston, E. (2008) A maternally localised Wnt ligand required for axial patterning in the cnidarian *Clytia hemisphaerica*. *Development*, 135, 2105–2113.

Monaghan, P. (2008) Early growth conditions, phenotypic development and environmental change. *Philosophical Transactions of the Royal Society of London B*, 363, 1635–1645.

Monaghan, P. & Haussmann, M. F. (2006) Do telomere dynamics link lifestyle and lifespan? *Trends in Ecology & Evolution*, 21, 47–53.

Monaghan, P., Metcalfe, N. B. & Torres, R. (2009) Oxidative stress as a mediator of life history trade-offs: mechanisms, measurements and interpretation. *Ecology Letters*, 12, 75–92.

Monastirioti, M. (2003) Distinct octopamine cell population residing in the CNS abdominal ganglion controls ovulation in *Drosophila melanogaster*. *Developmental Biology*, 264, 38–49.

Mongold, J. A. & Lenski, R. E. (1996) Experimental rejection of a nonadaptive explanation for increased cell size in *Escherichia coli*. *Journal of Bacteriology*, 178, 5333–5334.

Monte, E., Alonso, J. M., Ecker, J. R., Zhang, Y., Li, X., Young, J., Austin-Phillips, S. & Quail, P. H. (2003) Isolation and characterization of *PHYC* mutants in *Arabidopsis* reveals complex crosstalk between phytochrome signaling pathways. *Plant Cell*, 15, 1962–1980.

Monteiro, A. & Podlaha, O. (2009) Wings, horns, and butterfly eyespots: how do complex traits evolve? *PLoS Biology*, 7, 209–215.

Mooijaart, S. P., Kuningas, M., Westendorp, R. G., Houwing-Duistermaat, J. J., Slagboom, P. E., Rensen, P. C. & Van Heemst, D. (2007) Liver X receptor alpha associates with human life span. *Journal of Gerontology A*, 62, 343–349.

Moore, T., Beltran, L., Carbajal, S., Strom, S., Traag, J., Hursting, S. D. & Digiovanni, J. (2008) Dietary energy balance modulates signaling through the Akt/mammalian target of rapamycin pathways in multiple epithelial tissues. *Cancer Prevention Research*, 1, 65–76.

Moran, N. A. (1994) Adaptation and constraint in the complex life cycles of animals. *Annual Review of Ecology and Systematics*, 25, 573–600.

Moret, Y. & Schmid-Hempel, P. (2001) Immune defence in bumble-bee offspring. *Nature*, 414, 506.

Morgan, S. G. (1995) Life and death in the plankton: larval mortality and adaptation. In McEdward, L. (Ed.) *Ecology of Marine Invertebrate Larvae*. Boca Raton, CRC Press.

Morgan, A. D. (2008) The effect of food availability on phenotypic plasticity in larvae of the temperate sea cucumber *Australostichopus mollis*. *Journal of Experimental Marine Biology and Ecology*, 363, 89–95.

Morgan, M. B. & Snell, T. W. (2002) Characterizing stress gene expression in reef-building corals exposed to the mosquitoside dibrom. *Marine Pollution Bulletin*, 44, 1206–1218.

Morgan, N. C., Le Cren, E. D. & Lowe-McConell, R. H. (1980) Secondary production. *The Functioning of Freshwater Ecosystems*. Cambridge, Cambridge University Press.

Mori, J., Suzuki, S., Kobayashi, M., Inagaki, T., Komatsu, A., Takeda, T., Miyamoto, T., Ichikawa, K. & Hashizume, K. (2002) Nicotinamide adenine dinucleotide phosphate-dependent cytosolic T(3) binding protein as a regulator for T(3)-mediated transactivation. *Endocrinology*, 143, 1538–1544.

Mori, A., Romero-Severson, J., Black, W. C. & Severson, D. W. (2008) Quantitative trait loci determining autogeny and body size in the Asian tiger mosquito (*Aedes albopictus*). *Heredity*, 101, 75–82.

Morin, P. P., Hara, T. J. & Eales, J. G. (1997) Thyroid function and olfactory responses to L-alanine during induced smoltification in Atlantic salmon, *Salmo salar*. *Canadian Journal of Fisheries and Aquatic Sciences*, 54, 596–602.

Morvan-Dubois, G., Demeneix, B. A. & Sachs, L. M. (2008) *Xenopus laevis* as a model for studying thyroid hormone signalling: From development to metamorphosis. *Molecular and Cellular Endocrinology*, 293, 71–79.

Moseley, J. L., Chang, C. W. & Grossman, A. R. (2006) Genome-based approaches to understanding phosphorus deprivation responses and PSR1 control in *Chlamydomonas reinhardtii*. *Eukaryotic Cell*, 5, 26–44.

Moshitzky, P., Fleischmann, I., Chaimov, N., Saudan, P., Klauser, S., Kubli, E. & Applebaum, S. W. (1996) Sex-peptide activates juvenile hormone biosynthesis in the *Drosophila melanogaster* corpus allatum. *Archives of Insect Biochemistry and Physiology*, 32, 363–374.

Moss, E. G. (2007) Heterochronic genes and the nature of developmental time. *Current Biology*, 17, R425–R434.

Moss, C., Burke, R. D. & Thorndyke, M. C. (1994) Immunocytochemical localization of the neuropeptide-

S1 and serotonin in larvae of the starfish *Pisaster ochraceus* and *Asterias rubens*. *Journal of the Marine Biological Association of the United Kingdom*, 74, 61–71.

Motola, D. L., Cummins, C. L., Rottiers, V., Sharma, K., Sunino, K., Xu, E., Auchus, R., Antebi, A. & Mangelsdorf, M. (2006) Identification of DAF-12 ligands that govern dauer formation and reproduction in *C. elegans*. *Cell*, 124, 1209–1223.

Mougi, A. & Nishimura, K. (2006) Evolution of the maturation rate collapses competitive coexistence. *Journal of Theoretical Biology*, 241, 467–476.

Mousseau, T. A. (1997) Ectotherms follow converse to Bergann's Rule. *Evolution*, 51, 630–632.

Mousseau, T. A. & Fox, C. W. (1998) *Maternal Effects as Adaptations*. Oxford, Oxford University Press.

Moynihan, K. A., Grimm, A. A., Plueger, M. M., Bernal-Mizrachi, E., Ford, E., Cras-Meneur, C., Permutt, M. A. & Imai, S. (2005) Increased dosage of mammalian Sir2 in pancreatic beta cells enhances glucose-stimulated insulin secretion in mice. *Cell Metabolism*, 2, 105–117.

Muehlenbein, M. P. (2008) Adaptive variation in testosterone levels in response to immune activation: empirical and theoretical perspectives. *Social Biology*, 53, 13–23.

Muehlenbein, M. P. & Bribiescas, R. G. (2005) Testosterone-mediated immune functions and male life histories. *American Journal of Human Biology*, 17, 527–558.

Muehlenbein, M. P., Campbell, B. C., Phillippi, K. M., Murchison, M. A., Richards, R. J., Svec, F. & Myers, L. (2001) Reproductive maturation in a sample of captive male baboons. *Journal of Medical Primatology*, 30, 273–282.

Muehlenbein, M. P., Algier, J., Cogswell, F., James, M. & Krogstad, D. (2005) The reproductive endocrine response to *Plasmodium vivax* infection in Hondurans. *American Journal of Tropical Medicine and Hygiene*, 73, 178–187.

Muehlenbein, M. P., Jordan, J. L., Bonner, J. Z. & Swartz, A. M. (2010) Towards quantifying the usage costs of human immunity: altered metabolic rates and hormone levels during acute immune activation in men. American Journal of Human Biology, 22, 546–556.

Mueller, J. L., Page, J. L. & Wolfner, M. F. (2007) An ectopic expression screen reveals the protective and toxic effects of *Drosophila* seminal fluid proteins. *Genetics*, 175, 777–783.

Mugat, B., Brodu, V., Kejzlarova-Lepesant, J., Antoniewski, C., Bayer, C., Fristrom, J. & Lepesant, J. (2000) Dynamic expression of broad-complex isoforms mediates temporal control of an ecdysteroid target gene at the onset of *Drosophila* metamorphosis. *Developmental Biology*, 227, 104–117.

Muller, M., Den Tonkelaar, I., Thijssen, J. H. H., Grobbee, D. E. & Van Der Schouw, Y. T. (2003) Endogenous sex hormones in men aged 40–80 years. *European Journal of Endocrinology*, 149, 583–589.

Mullen, L. M., Lightfoot, M. E. & Goldsworthy, G. J. (2004) Induced hyperlipaemia and immune challenge in locusts. *Journal of Insect Physiology*, 50, 409–417.

Munch, D., Amdam, G. V. & Wolschin, F. (2008) Aging in a eusocial insect: molecular and physiological characteristics of lifespan plasticity in the honey bee. *Functional Ecology*, 22, 407–421.

Munroe, R. L. & Munroe, R. H. (1997) A comparative anthropological perspective. In Berry, J. W., Poortinga, Y. H. & Pandey, J. (Eds.) *Handbook of Cross Cultural Psychology*, 2nd edn. Boston, Allyn and Bacon.

Murakami, H. & Murakami, S. (2007) Serotonin receptors antagonistically modulate *Caenorhabditis elegans* longevity. *Aging Cell*, 6, 483–488.

Murdock, G. P. (1967) *Ethnographic Atlas*. Pittsburgh, University of Pittsburgh Press.

Murphy, C. T., McCarroll, S. A., Bargmann, C. I., Fraser, A., Kamath, R. S., Ahringer, J., Li, H. & Kenyon, C. (2003) Genes that act downstream of DAF-16 to influence the lifespan of *Caenorhabditis elegans*. *Nature*, 424, 277–283.

Musset, L., Le Bras, J. & Clain, J. (2007) Parallel evolution of adaptive mutations in plasmodium falciparum mitochondrial DNA during atovaquone-proguanil treatment. *Molecular Biology and Evolution*, 24, 1582–1585.

Mutti, N. S., Wolschin, F., Dolezal, A. G., Mutti, J. S., Gill, K. S. & Amdam, G. V. (submitted, under review) IIS and TOR nutrient-signaling pathways act via juvenile hormone to influence honey bee caste fate.

Myers, R. A. (1984) Demographic consequences of precocious maturation of Atlantic salmon (*Salmo salar*). *Canadian Journal of Fisheries and Aquatic Sciences*, 41, 1349–1353.

Nahmad, M., Glass, L. & Abouheif, E. (2008) The dynamics of developmental system drift in the gene network underlying wing polyphenism in ants: a mathematical model. *Evolution & Development*, 10, 360–374.

Nakajima, Y. (1986) Development of the nervous system of sea urchin embryos: formation of ciliary bands and the appearance of two types of ectoneural cells in the pluteus. *Development, Growth & Differentiation*, 28, 243–249.

Nakajima, Y., Kaneko, H., Murray, G. & Burke, R. D. (2004) Divergent patterns of neural development in larval echinoids and asteroids. *Evolution & Development*, 6, 95–104.

Nakano, H., Hibino, T., Oji, T., Hara, Y. & Amemiya, S. (2003) Larval stages of the living sea lily (stalked crinoid Echinoderm). *Nature*, 421, 158–160.

Nakano, H., Murabe, N., Amemiya, S. & Nakajima, Y. (2006) Nervous system development of the sea cucumber *Stichopus japonicus*. *Developmental Biology*, 292, 205–212.

Nanji, M., Hopper, N. A. & Gems, D. (2005) LET-60 RAS modulates effects of insulin/IGF-1 signaling on development and aging in *Caenorhabditis elegans*. *Aging Cell*, 4, 235–245.

Nappi, A. J. & Ottaviani, E. (2000) Cytotoxicity and cytotoxic molecules in invertebrates. *BioEssays*, 22, 469–480.

Napp-Zinn, K. (1985) *Arabidopsis thaliana*. In Halevy, A. (Ed.) *CRC Handbook of Flowering*. Boca Raton, FL, CRC.

Nedelcu, A. M. (2009) Environmentally induced responses co-opted for reproductive altruism. *Biology Letters*, 5, 805–808.

Nedelcu, A.M. and Michod, R.E. (2004) Evolvability, modularity, and individuality during the transition to multicellularity in volvocalean green algae. In Schlosser, G. & Wagner, G. (Eds.) *Modularity in Development and Evolution*. Chicago, University of Chicago Press.

Nedelcu, A. M. & Michod, R. E. (2006) The evolutionary origin of an altruistic gene. *Molecular Biology and Evolution*, 23, 1460–1464.

Nelson, R. J. (2005) *An Introduction to Behavioral Endocrinology*. Sunderland, Sinauer.

Nelson, J. S. (2006) *Fishes of the World*. New Jersy, John Wiley and Sons, Inc.

Nelson, E. R. & Habibi, H. R. (2009) Thyroid receptor subtypes: Structure and function in fish. *General and Comparative Endocrinology*, 161, 90–96.

Nelson, C. M., Ihle, K. E., Fondrk, M. K., Page, R. E. & Amdam, G. V. (2007) The gene vitellogenin has multiple coordinating effects on social organization. *PLoS Biology*, 5, e62.

Nemoto, S., Fergusson, M. M. & Finkel, T. (2005) SIRT1 functionally interacts with the metabolic regulator and transcriptional coactivator PGC-1α. *Journal of Biological Chemistry*, 280, 16456–16460.

Nepomnaschy, P. A., Welch, K. B., McConnel, D. S., Strassmann, B. I. & England, B. G. (2004) Stress and female reproductive function: a study of daily variations in cortisol, gonadotropins, and gonadal steroids in a rural Mayan population. *American Journal of Human Biology*, 16, 523–532.

Nepomnaschy, P. A., Welch, K. B., McConnell, D. S., Low, B. S., Strassmann, B. I. & England, B. G. (2006) Cortisol levels and very early pregnancy loss in humans. *Proceedings of the National Academy of Sciences of the United States of America*, 103, 3938–3942.

Neretti, N., Wang, P. Y., Brodsky, A. S., Nguyen, H. H., White, K. P., Rogina, B. & Helfand, S. L. (2009) Long-lived Indy induces reduced mitochondrial reactive oxygen species production and oxidative damage. *Proceedings of the National Academy of Sciences of the United States of America*, 106, 2277–2282.

Ness, R. B., Harris, T., Cobb, J., Flegal, K. M., Kelsey, J. L., Balanger, A., Stunkard, A. J. & Dagostino, R. B. (1993) Number of pregnancies and the subsequent risk of cardiovascular disease. *New England Journal of Medicine*, 328, 1528–1533.

Nesse, R. M. & Williams, G. C. (1996) *Why We Get Sick: The New Science of Darwinian Medicine*. New York, Vintage Books.

Neukirch, A. (1982) Dependence of the life-span of the Honeybee (*Apis mellifica*) upon flight performance and energy-consumption. *Journal of Comparative Physiology B*, 146, 35–40.

Niehrs, C. (2004) Synexpression groups: genetic modules and embryonic development. In Schlosser, G. & Wagner, G. P. (Eds.) *Modularity in Development and Evolution*. Chicago, University of Chicago Press.

Nielsen, C. (1998) Origin and evolution of animal life cycles. *Biological Reviews of the Cambridge Philosophical Society*, 73, 125–155.

Nielsen, R., Bustamante, C., Clark, A. G., Glanowski, S., Sackton, T. B., Hubisz, M. J., Fledel-Alon, A., Tanenbaum, D. M., Civello, D., White, T. J., J Sninsky, J., Adams, M. D. & Cargill, M. (2005) A scan for positively selected genes in the genomes of humans and chimpanzees. *PLoS Biology*, 3, e170.

Nieman, D. C. (1999) Nutrition, exercise and immune system function. *Clinics in Sports Medicine*, 18, 537–548.

Nieman, D. C. (2008) Immunonutrition support for athletes. *Nutrition Reviews*, 66, 310–320.

Niinuma, K., Yamamoto, K. & Kikuyama, S. (1991) Changes in plasma and pituitary prolactin levels in toad (*Bufo japonicus*) larvae during metamorphosis. *Zoological Science*, 8, 97–101.

Nijhout, H. F. (1979) Stretch-induced molting in *Oncopeltus fasciatus*. *Journal of Insect Physiology*, 25, 277–281.

Nijhout, H. F. (1994) *Insect Hormones*. Princeton, Princeton University Press.

Nijhout, H. F. (1999) Control mechanisms of polyphenic development in insects. *Bioscience*, 49, 181–192.

Nijhout, H. F. (2003) Development and evolution of adaptive polyphenisms. *Evolution & Development*, 5, 9–18.

Nijhout, H. F., Davidowitz, G. & Roff, D. A. (2006) A quantitative analysis of the mechanism that controls body size in *Manduca sexta*. *Journal of Biology*, 5(5) 16.

Nijhout, H. F. & Williams, C. F. (1974a) Control of moulting and metamorphosis in the tobacco hornworm, *Manduca sexta* (L.): cessation of juvenile hormone secretion as a trigger for pupation. *Journal of Experimental Biology*, 61, 493–501.

Nijhout, H. F. & Williams, D. W. (1974b) Control of moulting and metamorphosis in the tobacco hornworm, *Manduca sexta* (L.): growth of the las instar larva and the decision to pupate. *Journal of Experimental Biology*, 61, 481–491.

Nilsen, K.-A., Frederick, M., Fondrk, K. M., Smedal, B. & Hartfelder, G. V. (submitted) Dual role of fat body insulin/insulin-like growth factor signaling in honey bee behavioral physiology.

Nishiwaki, K., Kubota, Y., Chigira, Y., Roy, S. K., Suzuki, M., Schvarzstein, M., Jigami, Y., Hisamoto, N. & Matsumoto, K. (2004) An NDPase links ADAM protease glycosylation with organ morphogenesis in *C. elegans*. *Nature Cell Biology*, 6, 31–37.

Nordborg, M. & Bergelson, J. (1999) The effect of seed and rosette cold treatment on germination and flowering time in some *Arabidopsis thaliana* (Brassicaceae) ecotypes. *American Journal of Botany*, 86, 470–475.

Norris, D. O. (2007) *Vertebrate Endocrinology*. 4th edition New York, Academic Press.

Novak, V. J. A. (1966) *Insect Hormones*. London, Methuen and Co Ltd.

Nowak, M. A., Kamarova, N. L. & Niyogi, P. (2001) Evolution of universal grammar. *Science*, 291, 114–118.

Nugegoda, D., Walfor, J. & Lam, T. H. (1994) Thyroid hormones in early development of seabass (*Lates calcarifer*) larvae. *Journal of Aquaculture in the Tropics*, 9 279–290.

Nunez, J., Celi, F. S., Ng, L. & Forrest, D. (2008) Multigenic control of thyroid hormone functions in the nervous system. *Molecular and Cellular Endocrinology*, 287, 1–12.

Nunney, L. (2007) Pupal period and adult size in *Drosophila melanogaster*: a cautionary tale of contrasting correlations between two sexually dimorphic traits. *Journal of Evolutionary Biology*, 20, 141–151.

Nussey, D. H., Wilson, A. J. & Brommer, J. E. (2007) The evolutionary ecology of individual phenotypic plasticity in wild populations. *Journal of Evolutionary Biology*, 20, 831–844.

Nylin, S. (1992) Seasonal plasticity in life history traits – growth and development in *Polygonia c-album* (Lepidoptera, Nymphalidae). *Biological Journal of the Linnean Society*, 47, 301–323.

Nylin, S. & Gotthard, K. (1998) Plasticity in life history traits. *Annual Review of Entomology*, 43, 63–83.

Nylin, S. & Wahlberg, N. (2008) Does plasticity drive speciation? Host-plant shifts and diversification in nymphaline butterflies (Lepidoptera : Nymphalidae) during the tertiary. *Biological Journal of the Linnean Society*, 94, 115–130.

Nylin, S., Wickman, P. O. & Wiklund, C. (1989) Seasonal plasticity in growth and development of the speckled wood butterfly, *Pararge aegeria* (Satyrinae). *Biological Journal of the Linnean Society*, 38, 155–171.

Oberdoerffer, P., Michan, S., McVay, M., Mostoslavsky, R., Vann, J., Park, S. K., Hartlerode, A., Stegmuller, J., Hafner, A., Loerch, P., Wright, S. M., Mills, K. D., Bonni, A., Yankner, B. A., Scully, R., Prolla, T. A., Alt, F. W. & Sinclair, D. A. (2008) SIRT1 redistribution on chromatin promotes genomic stability but alters gene expression during aging. *Cell*, 135, 907–918.

O'Brien, D. M., Fogel, M. L. & Boggs, C. L. (2002) Renewable and nonrenewable resources: Amino acid turnover and allocation to reproduction in Lepidoptera. *Proceedings of the National Academy of Sciences of the United States of America*, 99, 4413–4418.

O'Brien, D. M., Min, K. J., Larsen, T. & Tatar, M. (2008) Use of stable isotopes to examine how dietary restriction extends *Drosophila* lifespan. *Current Biology*, 18, R155–R156.

O'Connell, J. F., Hawkes, K. & Blurton Jones, N. G. (1999) Grandmothering and the evolution of Homo erectus. *Journal of Human Evolution*, 36, 461–485.

Odell, W. D. & Parker, L. N. (1985) Control of adrenal androgen production. *Endocrine Research*, 10, 617–630.

Oeppen, J. & Vaupel, J. W. (2002) Broken limits to life expectancy. *Science*, 296, 1029–1031.

Oetting, A. & Yen, P. M. (2007) New insights into thyroid hormone action. *Best Practice & Research. Clinical Endocrinology & Metabolism*, 21, 193–208.

Oftedal, O. T. & Iverson, S. J. (1995) Comparative analysis of nonhuman milks: phylogenetic variation in the gross composition of milks. In Jensen, R. G. (Ed.) *Handbook of Milk Composition*. San Diego, Academic Press.

Ogawa, A., Streit, A., Antebi, A. & Sommer, R. J. (2009) A conserved endocrine mechanism controls the formation of dauer and infective larvae in nematodes. *Current Biology*, 19, 67–71.

Oh, S. W., Mukhopadhyay, A., Dixit, B. L., Raha, T., Green, M. R. & Tissenbaum, H. A. (2006) Identification of direct DAF-16 targets controlling longevity, metabolism and diapause by chromatin immunoprecipitation. *Nature Genetics*, 38, 251–257.

Ohtsu, T., Kimura, M. T. & Hori, S. H. (1992) Energy storage during reproductive diapause in the *Drosophila melanogaster* species group. *Journal of Comparative Physiology B*, 162, 203–208.

Oikarinen, A. & Lumme, J. (1979) Selection against photoperiodic reproductive diapause in *Drosophila littoralis*. *Hereditas*, 90, 119–125.

Ojeda, S. R. (2004) The anterior pituitary and hypothalamus. In Griffin, J. E. & Ojeda, S. R. (Eds.) *Textbook of Endocrine Physiology*, 5th edn. New York, Oxford University Press.

Oksanen, T. A., Koskela, E. & Mappes, T. (2002) Hormonal manipulation of offspring number: maternal effort and reproductive costs. *Evolution*, 56, 1530–1537.

Oldham, S. & Hafen, E. (2003) Insulin/IGF and target of rapamycin signaling: a TOR de force in growth control. *Trends in Cell Biology*, 13, 79–85.

Oldham, S., Bohni, R., Stocker, H., Brogiolo, W. & Hafen, E. (2000) Genetic control of size in *Drosophila*. *Philosophical Transactions of the Royal Society of London B*, 355, 945–952.

Oliver, B., Perrimon, N. & Mahowald, A. P. (1987) The ovo locus is required for sex-specific germ line maintenance in *Drosophila*. *Genes & Development*, 1, 913–923.

Oliver, K. M., Russell, J. A., Moran, N. A. & Hunter, M. S. (2003) Facultative bacterial symbionts in aphids confer resistance to parasitic wasps. *Proceedings of the National Academy of Sciences of the United States of America*, 100, 1803–1807.

Oliver, K. M., Moran, N. A. & Hunter, M. S. (2005) Variation in resistance to parasitism in aphids is due to symbionts not host genotype. *Proceedings of the National Academy of Sciences of the United States of America*, 102, 12795–12800.

Olsen, K. M., Womack, A., Garrett, A. R., Suddith, J. I. & Purugganan, M. D. (2002) Contrasting evolutionary forces in the *Arabidopsis thaliana* floral developmental pathway. *Genetics*, 160, 1641–1650.

Olsen, K. M., Halldorsdottir, S. S., Stinchcombe, J. R., Weinig, C., Schmitt, J. & Purugganan, M. D. (2004) Linkage disequilibrium mapping of *Arabidopsis CRY2* flowering time alleles. *Genetics*, 167, 1361–1369.

Olsson, M. & Shine, R. (2002) Growth to death in lizards. *Evolution*, 56, 1867–1870.

Olsson, M., Wilson, M., Isaksson, C., Uller, T. & Mott, B. (2008a) Carotenoid intake does not mediate a relationship between reactive oxygen species and bright colouration: experimental test in a lizard. *Journal of Experimental Biology*, 211, 1257–1261.

Olsson, M., Wilson, M., Uller, T., Mott, B., Isaksson, C., Healey, M. & Wanger, T. (2008b) Free radicals run in lizard families. *Biology Letters*, 4, 186–188.

Olsson, M., Wilson, M., Uller, T., Mott, B. & Isaksson, C. (2009) Variation in levels of reactive oxygen species is explained by maternal identity, sex and body-size-corrected clutch size in a lizard. *Naturwissenschaften*, 96, 25–29.

Oostra, V., de Jong, M. A., Invergo, B. M., Kesbeke, F. H. N., Wende, F., Brakefield, P. M., and Zwaan, B. J. (2010).

Translating environmental gradients into discontinuous reaction norms via hormone signaling in a polyphenic butterfly. Proceedings of the Royal Society of London B print September 8, 2010, doi: 10.1098/rspb.2010.1560.

Optim, O., Amore, G., Minokawa, T., McClay, D. R. & Davidson, E. H. (2004) SpHnf6, a transcription factor that executes multiple function in sea urchin embryogensis. *Developmental Biology*, 273, 226–243.

Orentreich, N., Matias, J. R., Defelice, A. & Zimmerman, J. A. (1993) Low methionine ingestion by rats extends life span. *Journal of Nutrition*, 123, 269–274.

Orgogozo, V., Broman, K. W. & Stern, D. L. (2006) High-resolution quantitative trait locus mapping reveals sign epistasis controlling ovariole number between two *Drosophila* species. *Genetics*, 173, 197–205.

O'Rourke, E. J., Soukas, A. A., Carr, C. E. & Ruvkun, G. (2009) *C. elegans* major fats are stored in vesicles distinct from lysosome-related organelles. *Cell Metabolism*, 10, 430–435.

Orozco, A. & Valverde, R. (2005) Thyroid hormone deiodination in fish. *Thyroid*, 15, 799–813.

Orr, H. A. & Coyne, J. A. (1992) The genetics of adaptation: a reassessment. *The American Naturalist*, 140, 725–742.

Ostrowski, E. A., Woods, R. J. & Lenski, R. E. (2008) The genetic basis of parallel and divergent phenotypic responses in evolving populations of *Escherichia coli*. *Proceedings of the Royal Society of London B*, 275, 277–284.

Ouellet, J., Li, S. & Roy, R. (2008) Notch signalling is required for both dauer maintenance and recovery in *C. elegans*. *Development*, 135, 2583–2592.

Overgaard, J., Malmendal, A., J.G., S., Bundy, J. G., Loeschcke, V., Nielsen, N. C. & Holmstrup, M. (2007) Metabolomic profiling of rapid cold hardening and cold shock in *Drosophila melanogaster*. *Journal of Insect Physiology*, 53, 1218–1232.

Ozaki, Y., Okumura, H., Kazeto, Y., Ikeuchi, T., Ijiri, S., Nagae, M., Adachi, S. & Yamauchi, K. (2000) Developmental changes in pituitary-thyroid axis, and formation of gonads in leptocephali and glass eels of *Anguilla* spp. *Fisheries Science*, 66, 1115–1122.

Ozsolak, F., Platt, A. R., Jones, D. R., Reifenberger, J. G., Sass, L. E., McInerney, P., Thompson, J. F., Bowers, J., Jarosz, M., & Milos, P. M. (2009). Direct RNA sequencing. *Nature*, 461, 814–818.

Paaby, A. B. & Schmidt, P. S. (2008) Functional significance of allelic variation at *methuselah*, an aging gene in *Drosophila*. *PLoS One*, 3, e1987.

Paaby, A., Blacket, M. J., Hoffmann, A. A. & Schmidt, P. S. (2010) Identification of a candidate adaptive polymorphism for *Drosophila* life history by parallel independent clines on two continents. *Molecular Ecology*, 19, 760–774.

Packer, C., Tatar, M. & Collin, A. (1998) Reproductive cessation in female mammals. *Nature*, 392, 807–810.

Padmanabhan, S., Mukhopadhyay, A., Narasimhan, S. D., Tesz, G., Czech, M. P. & Tissenbaum, H. A. (2009) A PP2A regulatory subunit regulates *C. elegans* insulin/IGF-1 signaling by modulating AKT-1 phosphorylation. *Cell*, 136, 939–951.

Page, L. R. (2002) Comparative structure of the larval apical sensory organ in Gastropods and hypotheses about function and developmental evolution. *Invertebrate Reproduction & Development*, 41, 193–200.

Page, R. E. J., & Amdam, G. V. (2007) The making of a social insect: developmental architectures of social design. *BioEssays*, 29, 334–343.

Page, R. E. J. & Fondrk, M. K. (1995) The effects of colony level selection o the social organization of honey-bee (*Apis mellifera*) colonies – colony level components of pollen hoarding. *Behavioral Ecology and Sociobiology*, 36, 135–144.

Page, R. E. J., Scheiner, R., Erber, J. & Amdam, G. V. (2006) The development and evolution of division of labor and foraging specialization in a social insect (*Apis mellifera* L.). *Current Topics in Developmental Biology*, 74, 253–286.

Paitz, R. T., Haussmann, M. F., Boden, R. M., Janzen, F. J. & Vleck, C. (2004) Long telomeres may minimize the effect of aging in the Painted Turtle. *Integrative and Comparative Biology*, 44, 617.

Palopoli, M. F., Rockman, M. V., Tinmaung, A., Ramsay, C., Curwen, S., Aduna, A., Laurita, J. & Kruglyak, L. (2008) Molecular basis of the copulatory plug polymorphism in *Caenorhabditis elegans*. *Nature*, 454, 1019–1022.

Pamplona, R. & Barja, G. (2007) Highly resistant macromolecular components and low rate of generation of endogenous damage: Two key traits of longevity. *Aging Research Reviews*, 6, 189–210.

Pan, K. Z., Palter, J. E., Rogers, A. N., Olsen, A., Chen, D., Lithgow, G. J. & Kapahi, P. (2007) Inhibition of mRNA translation extends lifespan in *Caenorhabditis elegans*. *Aging Cell*, 6, 111–119.

Panowski, S. H., Wolff, S., Aguilaniu, H., Durieux, J. & Dillin, A. (2007) PHA-4/Foxa mediates diet-restriction-induced longevity of *C. elegans*. *Nature*, 447, 550–555.

Panter-Brick, C., Lotstein, D. S. & Ellison, P. T. (1993) Seasonality of reproductive function and weight loss in rural Nepali women. *Human Reproduction*, 8, 684–690.

Panza, F., D'Introno, A., Capurso, C., Colacicco, A. M., Seripa, D., Pilotto, A., Santamato, A., Capurso, A. & Solfrizzi, V. (2007) Lipoproteins, vascular-related genetic factors, and human longevity. *Rejuvenation Research*, 10, 441–458.

Papaceit, M., Sanantonio, J. & Prevosti, A. (1991) Genetic-analysis of extra sex combs in the hybrids between *Drosophila subobscura* and *D madeirensis*. *Genetica*, 84, 107–114.

Papadopoulou, E. & Grumet, R. (2005) Brassinosteriod-induced femaleness in cucumber and relationship to ethylene production. *HortScience*, 40, 1763–1767.

Papadopoulou, E., Little, H. A., Hammar, S. A. & Grumet, R. (2005) Effect of modified endogenous ethylene production on sex expression, bisexual flower development and fruit production in melon (*Cucumis melo* L.). *Sexual Plant Reproduction*, 18, 131–142.

Paris, M. & Laudet, V. (2008) The history of a developmental stage: Metamorphosis in chordates. *Genesis*, 46, 657–672.

Paris, M., Escriva, H., Schubert, M., Brunet, F., Brtko, J., Ciesielski, F., Roecklin, D., Vivat-Hannah, V., Jamin, E. L., Cravedi, J. P., Scanlan, T. S., Renaud, J. P., Holland, N. D. & Laudet, V. (2008) Amphioxus postembryonic development reveals the homology of chordate metamorphosis. *Current Biology*, 18, 825–830.

Parthasarathy, R., Tan, A., Bai, H. & Palli, S. R. (2008) Transcription factor broad suppresses precocious development of adult structures during larval-pupal metamorphosis in the red flour beetle, *Tribolium castaneum*. *Mechanisms of Development*, 125, 299–313.

Partridge, L. & Fowler, K. (1992) Direct and correlated responses to selection on age at reproduction in *Drosophila melanogaster*. *Evolution*, 46, 76–91.

Partridge, L. & Gems, D. (2006) Beyond the evolutionary theory of aging, from functional genomics to evo-gero. *Trends in Ecology & Evolution*, 21, 334–340.

Partridge, L. & Harvey, P. H. (1988) The ecological context of life history evolution. *Science*, 241, 1449–1455.

Partridge, L., Barrie, B., Fowler, K. & French, V. (1994) Evolution and development of body-size and cell-size in *Drosophila melanogaster* in response to temperature. *Evolution*, 48, 1269–1276.

Partridge, L., Barrie, B., Barton, N. H., Fowler, K. & French, V. (1995) Rapid laboratory evolution of adult life history traits in *Drosophila melanogaster* in response to temperature. *Evolution*, 49, 538–544.

Partridge, L., Langelan, R., Fowler, K., Zwaan, B. & French, V. (1999) Correlated responses to selection on body size in *Drosophila melanogaster*. *Genetical Research*, 74, 43–54.

Partridge, L., Green, A., & Fowler, K. (1989) Effects of egg-production and of exposure to males on female survival in *Drosophila melanogaster*. *Journal of Insect Physiology*, 33, 745–749.

Partridge, L., Gems, D. & Withers, D. J. (2005a) Sex and death: What is the connection? *Cell*, 120, 461–472.

Partridge, L., Piper, M. D. & Mair, W. (2005b) Dietary restriction in *Drosophila*. *Mechanisms of Ageing and Development*, 126, 938–950.

Patel, A., Fondrk, M. K., Kaftanoglu, O., Emore, C., Hunt, G., Frederick, K. & Amdam, G. V. (2007) The making of a queen: TOR pathway is a key player in diphenic caste development. *PLoS One*, 2, e509.

Patel, D. S., Fang, L. L., Svy, D. K., Ruvkun, G. & Li, W. (2008) Genetic identification of HSD-1, a conserved steroidogenic enzyme that directs larval development in *Caenorhabditis elegans*. *Development*, 135, 2239–2249.

Paul, V. J. & Ritson-Williams, R. (2008) Marine chemical ecology. *Natural Product Reports*, 25, 662–695.

Paul, B. D., Fu, L., Buchholz, D. R. & Shi, Y. B. (2005a) Coactivator recruitment is essential for liganded thyroid hormone receptor to initiate amphibian metamorphosis. *Molecular and Cellular Biology*, 25, 5712–5724.

Paul, M. J., Freeman, D. A., Park, J. H. & Dark, J. (2005b) Neuropeptide Y induces torpor-like hypothermia in Siberian hamsters. *Brain Research*, 1055, 83–92.

Paul, B. D., Buchholz, D. R., Fu, L. & Shi, Y. B. (2007) Srcp300 coactivator complex is required for thyroid hormone-induced amphibian metamorphosis. *Journal of Biological Chemistry*, 282, 7472–7481.

Pearse, V. B. (2002) Prodigies of propagation: the many modes of clonal replication in boloceroidid sea anemones (Cnidaria, Anthozoa, Actiniaria). *Invertebrate Reproduction & Development*, 41, 201–213.

Pearse, J. S., Pearse, V. B. & Newberry, A. T. (1989) Telling sex from growth: dissolving Maynard Smith's paradox. *Bulletin of Marine Science*, 45, 433–446.

Pearson, K. J., Baur, J. A., Lewis, K. N., Peshkin, L., Price, N. L., Labinskyy, N., Swindell, W. R., Kamara, D., Minor, R. K., Perez, E., Jamieson, H. A., Zhang, Y., Dunn, S. R., Sharma, K., Pleshko, N., Woollett, L. A., Csiszar, A., Ikeno, Y., Le Couteur, D., Elliott, P. J., Becker, K. G., Navas, P., Ingram, D. K., Wolf, N. S., Ungvari, Z., Sinclair, D. A. & De Cabo, R. (2008) Resveratrol delays age-related deterioration and mimics transcriptional aspects of dietary restriction without extending life span. *Cell Metabolism*, 8, 157–168.

Pecasse, F., Beck, Y., Ruiz, C. & Richards, G. (2000) Kruppel-homolog, a stage-specific modulator of the prepupal ecdysone response, is essential for *Drosophila* metamorphosis. *Developmental Biology*, 221, 53–67.

Pechenik, J. A., Estrella, M. S. & Hammer, K. (1996) Food limitation stimulates metamorphosis of competent larvae and alters postmetamorphic growth rate in the marine gastropod *Crepidula fornicata*. *Marine Biology*, 127, 267–275.

Peck, L. S. & Maddrell, S. H. (2005) Limitation of size by hypoxia in the fruit fly *Drosophila melanogaster*. *Journal of Experimental Zoology A*, 303, 968–975.

Pedra, J. H., Festucci-Buselli, R. A., Sun, W. L., Muir, W. M., Scharf, M. E. & Pittendrigh, B. R. (2005) Profiling of abundant proteins associated with dichlorodiphenyl-trichloroethane resistance in *Drosophila melanogaster*. *Proteomics*, 5, 258–269.

Peel, D. J. & Milner, M. J. (1992) The response of *Drosophila* imaginal disk cell-lines to ecdysteroids. *Roux's Archives of Developmental Biology*, 202, 23–35.

Peeters, A. V., Beckers, S., Verrijken, A., Mertens, I., Roevens, P., Peeters, P. J., Van Hul, W. & Van Gaal, L. F. (2008) Association of SIRT1 gene variation with visceral obesity. *Human Genetics*, 124, 431–436.

Peled-Kramar, M., Hamilton, P. & Wilt, F. H. (2002) Spicule matrix protein LSM34 is essential for biomineralization of the sea urchin spicule. *Experimental Cell Research*, 272, 56–61.

Pelosi, L., Kuhn, L., Guetta, D., Garin, J., Geiselmann, J., Lenski, R. E. & Schneider, D. (2006) Parallel changes in global protein profiles during long-term experimental evolution in *Escherichia coli*. *Genetics*, 173, 1851–1869.

Pelz, H.-J., Rost, S., Hunerberg, M., Fregin, A., Heiberg, A.-C., Baert, K., MacNicoll, A. D., Prescott, C. V., Walker, A.-S., Oldenburg, J. & Muller, C. R. (2005) The genetic basis of resistance to anticoagulants in rodents. *Genetics*, 170, 1839–1847.

Penaz, M. (2001) A general framework of fish ontogeny: a review of the ongoing debate. *Folia Zoologica*, 50, 241–256.

Pener, M. P. (1972) The corpus allatum in adult acridids: the interrelation of its functions and possible correlations with the life cycle. In Hemming, C. F. & Taylor, T. H. C. (Eds.) *Proceedings of the International Study Conference on the Current and Future Problems of Acridology*. London: Centre for Overseas Pest Research.

Pener, M. P. (1992) Environmental cues, endocrine factors, and reproductive diapause in male insects. *Chronobiology International*, 9, 102–113.

Peng, J., Zipperlen, P. & Kubli, E. (2005) *Drosophila* sex-peptide stimulates female innate immune system after mating via the Toll and Imd pathways. *Current Biology*, 15, 1690–1694.

Perez, V. I., Bokov, A., Van Remmen, H., Mele, J., Ran, Q., Ikeno, Y. & Richardson, A. (2009) Is the oxidative stress theory of aging dead? *Biochimica et Biophysica Acta*, 1790, 1005–1014.

Perls, T. T., Wilmoth, J., Levenson, R., Drinkwater, M., Cohen, M., Bogan, H., Joyce, E., Brewster, S., Kunkel, L. & Puca, A. (2002) Life-long sustained mortality advantage of siblings of centenarians. *Proceedings of the National Academy of Sciences of the United States of America*, 99, 8442–8447.

Perron, F. E. (1986) Life history consequences of differences in developmental mode among gastropods in the genus Conus *Bulletin of Marine Science*, 39, 485–497.

Peter, R. E., Yu, K. L., Marchant, T. A. & Rosenblum, P. M. (1990) Direct neural regulation of the teleost adenohypophysis. *Journal of Experimental Zoology*, 4, 84–89.

Peterson, C. C., Walton, B. M. & Bennett, A. F. (1999) Metabolic costs of growth in free-living garter snakes and the energy budgets of ectotherms. *Functional Ecology*, 13, 500–507.

Pfannschmidt, T., Brautigam, K., Wagner, R., Dietzel, L., Schroter, Y., Steiner, S. & Nykytenko, A. (2009) Potential regulation of gene expression in photosynthetic cells by redox and energy state: approaches towards better understanding. *Annals of Botany*, 103, 599–607.

Pfennig, D. (1990) The adaptive significance of an environmentally-cued developmental switch in an anuran tadpole. *Oecologia*, 85, 101–107.

Pfennig, D. W. (1992a) Polyphenism in spadefoot toad tadpoles as a locally adjusted evolutionarily stable strategy. *Evolution*, 46, 1408–1420.

Pfennig, D. W. (1992b) Proximate and functional causes of polyphenism in an anuran tadpole. *Functional Ecology*, 6, 167–174.

Pfennig, D. W., Reeve, H. K. & Sherman, P. W. (1993) Kin recognition and cannibalism in spadefoot toad tadpoles. *Animal Behaviour*, 46, 67–84.

Philipp, I., Aufschnaiter, R., Ozbek, S., Pontasch, S., Jenewein, M., Watanabe, H., Rentzsch, F., Holstein, T. W. & Hobmayer, B. (2009) Wnt/beta-Catenin and noncanonical Wnt signaling interact in tissue evagination in the simple eumetazoan *Hydra*. *Proceedings of the National Academy of Sciences of the United States of America*, 106, 4290–4295.

Phillips, P. C. & Arnold, S. J. (1989) Visualizing multivariate selection. *Evolution*, 43, 1209–1222.

Phillips, J. P., Frye, F., Berkovitz, A., Calle, P., Millar, R., Rivier, J. & Lasley, B. E. (1987) Exogenous GnRH overrides the endogenous annual reproductive rhythm in green iguanas, *Iguana iguana*. *Journal of Experimental Zoology A*, 241, 227–236.

Phillips, J. P., Campbell, S. D., Michaud, D., Charbonneau, M. & Hilliker, A. J. (1989) Null mutation of copper/zinc superoxide dismutase in *Drosophila* confers hypersensitivity to paraquat and reduced longevity. *Proceedings of the National Academy of Sciences of the United States of America*, 86, 2761–2765.

Picard, F., Kurtev, M., Chung, N., Topark-Ngarm, A., Senawong, T., Machado De Oliveira, R., Leid, M., McBurney, M. W. & Guarente, L. (2004) Sirt1 promotes fat mobilization in white adipocytes by repressing PPAR-γ. *Nature*, 429, 771–776.

Pickering, A. D., Griffiths, R. & Pottinger, T. G. (1987) A comparison of the effects of overhead cover on the growth, survival and hematology of juvenile Atlantic Salmon, *Salmo salar* L., Brown Trout, *Salmo trutta* L., and Rainbow Trout, *Salmo Gairdneri*. *Aquaculture*, 66, 109–124.

Pierce, S. B., Costa, M., Wisotzkey, R., Devadhar, S., Homburger, S. A., Buchman, A. R., Ferguson, K. C., Heller, J., Platt, D. M., Pasquinelli, A. A., Liu, L. X., Doberstein, S. K. & Ruvkun, G. (2001) Regulation of DAF-2 receptor signaling by human insulin and ins-1, a member of the unusually large and diverse *C. elegans* insulin gene family. *Genes & Development*, 15, 672–686.

Pigliucci, M. (2001) *Phenotypic Plasticity: Beyond Nature and Nurture*. Baltimore, Johns Hopkins Press.

Pigliucci, M. (2005) Evolution of phenotypic plasticity: where are we going now? *Trends in Ecology & Evolution*, 20, 481–486.

Pigliucci, M. (2010) Genotype–phenotype mapping and the end of the 'genes as blueprint' metaphor. *Philosophical Transactions of the Royal Society B*, 365, 557–566.

Pijpe, J., Fischer, K., Brakefield, P. & Zwaan, B. (2006) Consequences of artificial selection on pre-adult development for adult lifespan under benign conditions in the butterfly *Bicyclus anynana*. *Mechanisms of Ageing and Development*, 127, 802–807.

Pijpe, J., Brakefield, P. M. & Zwaan, B. J. (2007) Phenotypic plasticity of starvation resistance in the butterfly *Bicyclus anynana*. *Evolutionary Ecology*, 21, 589–600.

Pijpe, J., Brakefield, P. M. & Zwaan, B. J. (2008) Increased lifespan in a polyphenic butterfly artificially selected for starvation resistance. *The American Naturalist*, 171, 81–90.

Pike, I. L. (1999) Age, reproductive history, seasonality, and maternal body composition during pregnancy for nomadic Turkana of Kenya. *American Journal of Human Biology*, 11, 658–672.

Pike, I. L. (2000) Pregnancy outcome for nomadic Turkana pastoralists of Kenya. *American Journal of Physical Anthropology*, 113, 31–45.

Pinker, S. (1994) *The Language Instinct*. New York, William Morrow.

Pinto, L. Z., Bitondi, M. M. & Simoes, Z. L. (2000) Inhibition of vitellogenin synthesis in *Apis mellifera* workers by a juvenile hormone analogue, pyriproxyfen. *Journal of Insect Physiology*, 46, 153–160.

Piper, M. D. W., Selman, C., McElwee, J. J. & Partridge, L. (2005) Models of insulin signalling and longevity. *Drug Discovery Today: Disease Models*, 2, 249–256.

Piper, M. D. W., Selman, C., McElwee, J. J. & Partridge, L. (2008) Separating cause from effect: how does insulin/IGF signalling control lifespan in worms, flies and mice? *Journal of Internal Medicine*, 263, 179–191.

Pirke, K. M., Schweiger, V., Lemmel, W., Krieg, J. C. & Berger, M. (1985) The influence of dieting on the menstrual cycle of healthy young women. *Journal of Clinical Endocrinology and Metabolism*, 70, 1174–1179.

Pishcedda, A. & Chippindale, A. K. (2006) Intralocus conflict diminishes the benefits of sexual selection. *PLoS Biology*, 4, e394.

Pitnick, S. (1991) Male size influences mate fecundity and remating interval in *Drosophila melanogaster*. *Animal Behaviour*, 41, 735–745.

Pitnick, S. & Garcia-Gonzalez, F. (2002) Harm to females increases with male body size in *Drosophila melanogaster*. *Proceedings of the Royal Society of London*, 269, 1821–1828.

Plas, E., Berger, P., Hermann, M. & Pflüger, H. (2000) Effects of aging on male fertility? *Experimental Gerontology*, 35, 543–551.

Pletcher, S. D., MacDonald, S. J., Marguerie, S. R., Certa, U., Stearns, S. C. & Partridge, L. (2002) Genome-wide transcript profiles in aging and calorically restricted *Drosophila melanogaster*. *Current Biology*, 12, 712–723.

Plickert, G., Jacoby, V., Frank, U., Müller, W. A. & Mokady, O. (2006) Wnt signaling in hydroid development: Formation of the primary body axis in embryogenesis and its subsequent patterning. *Developmental Biology*, 298, 368–378.

Podolsky, R. D. & McAlister, J. S. (2005) Developmental plasticity in *Macrophiothrix* brittlestars: Are morphologically convergent larvae also convergently plastic? *Biological Bulletin*, 209, 127–138.

Poe, S. (2004) A test for patterns of modularity in sequences of developmental events. *Evolution*, 58, 1852–1855.

Poodry, C. A. & Woods, D. F. (1990) Control of the developmental timer for *Drosophila* pupariation. *Roux's Archives of Developmental Biology*, 199, 219–227.

Portmann, A. (1941) Die Tragzeiten der Primaten und die Dauer der Schwangerschaft beim Menschen: Ein Problem der vergleichenden Biologie. *Revue Suisse de Zoologie* 48, 511–518.

Postlethwait, J. H. & Shirk, P. D. (1981) Genetic and endocrine regulation of vitellogenesis in *Drosophila*. *American Zoologist*, 21, 687–700.

Potthoff, M. J., Inagaki, T., Satapati, S., Ding, X., He, T., Goetz, R., Mohammadi, M., Finck, B. N., Mangelsdorf, D. J., Kliewer, S. A. & Burgess, S. C. (2009) FGF21 induces PGC-1α and regulates carbohydrate and fatty acid metabolism during the adaptive starvation response. *Proceedings of the National Academy of Sciences of the United States of America*, 106, 10853–10858.

Powanda, M. C. & Beisel, W. R. (2003) Metabolic effects of infection on protein and energy status. *Journal of Nutrition*, 133, 322–327.

Power, D. M., Llewellyn, L., Faustino, M., Nowell, M. A., Bjornsson, B. T., Einarsdottir, I. E., Canario, A. V. M. & Sweeney, G. E. (2001) Thyroid hormones in growth and development of fish. *Comparative Biochemistry and Physiology C*, 130, 447–459.

Power, D. M., Einarsdottir, I. E., Pittman, K., Sweeney, G. E., Hildahl, J., Campinho, M. A., Silva, N., Saele, O., Galay-Burgos, M., Smaradottir, H. & Bjornsson, B. T.

(2008) The molecular and endocrine basis of flatfish metamorphosis. *Reviews in Fisheries Science*, 16, 95–111.

Prasad, N. G. & Joshi, A. (2003) What have two decades of laboratory life history evolution studies on *Drosophila melanogaster* taught us? *Journal of Genetics*, 82, 45–76.

Prentice, A. M. & Prentice, A. (1988) Energy costs of lactation. *Annual Review of Nutrition*, 8, 63–79.

Prevosti, A. (1955) Geographical variability in quantitative traits in populations of *Drosophila subobscura*. *Quantitative Biology*, 20, 294–299.

Price, G. R. (1970) Selection and covariance. *Nature*, 227, 520–521.

Priest, N. K., Galloway, L. F. & Roach, D. A. (2008) Mating frequency and inclusive fitness in *Drosophila melanogaster*. *The American Naturalist*, 171, 10–21.

Prior, J. C., Vigna, Y. M. & McKay, D. W. (1992) Reproduction for the athletic woman: New understandings of physiology and management. *Sports Medicine*, 14, 190–199.

Prochnik, S. E., Umen, J., Nedelcu, A. M., Hallmann, A., Miller, S. M., Nishii, I., Ferris, P., Kuo, A., Mitros, T., Fritz-Laylin, L. K., Hellsten, U., Chapman, J., Simakov, O., Rensing, S. A., Terry, A., Pangilinan, J., Kapitonov, V., Jurka, J., Salamov, A., Shapiro, H., Schmutz, J., Grimwood, J., Lindquist, E., Lucas, S., Grigoriev, I. V., Schmitt, R. D., Kirk, D. & Rokhsar, D. S. (2010) Genomic Analysis of Organismal Complexity in the Multicellular Green Alga Volvox carteri. Science, 329, 223–226.

Purcell, J. E. (2007) Environmental effects on asexual reproduction rates of the scyphozoan, *Aurelia labiata*. *Marine Ecology Progress Series*, 348, 183–196.

Putterill, J., Laurie, R. & MacKnight, R. (2004) It's time to flower: the genetic control of flowering time. *BioEssays*, 26, 363–373.

Qazi, M. C. B., Heifetz, Y. & Wolfner, M. F. (2003) The developments between gametogenesis and fertilization: ovulation and female sperm storage in *Drosophila melanogaster*. *Developmental Biology*, 256, 195–211.

Qin, W., Chachich, M., Lane, M., Roth, G., Bryant, M., De Cabo, R., Ottinger, M. A., Mattison, J., Ingram, D., Gandy, S. & Pasinetti, G. M. (2006) Calorie restriction attenuates Alzheimer's disease type brain amyloidosis in Squirrel monkeys (*Saimiri sciureus*). *Journal of Alzheimer's Disease*, 10, 417–422.

Quinlan, R. J., Quinlan, M. B. & Flinn, M. V. (2003) Parental investment and age at weaning in a Caribbean village. *Evolution and Human Behavior*, 24, 1–17.

Qureshi, A. I., Giles, W. H., Croft, J. B. & Stern, B. J. (1997) Number of pregnancies and risk for stroke and stroke subtypes. *Archives of Neurology*, 54, 203–206.

Råberg, L., Sim, D. & Read, A. F. (2007) Disentangling genetic variation for resistance and tolerance to infectious diseases in animals. *Science*, 318, 812–814.

Råberg, L., Graham, A. L. & Read, A. F. (2009) Decomposing health: tolerance and resistance to parasites in animals. *Philosophical Transactions of the Royal Society of London B*, 364, 37–49.

Rachinsky, A., Strambi, C., Strambi, A. & Hartfelder, K. (1990) Caste and metamorphosis: hemolymph titers of juvenile hormone and ecdysteroids in last instar honeybee larvae. *General and Comparative Endocrinology*, 79, 31–38.

Radimerski, T., Montagne, J., Rintelen, F., Stocker, H., Van Der Kaay, J., Downes, C. P., Hafen, E. & Thomas, G. (2002) dS6k-regulated cell growth is dPKB/dPI(3)K-independent, but requires dPDK1. *Nature Cell Biology*, 4, 251–255.

Raff, R. A. & Byrne, M. (2006) The active evolutionary lives of echinoderm larvae. *Heredity*, 97, 244–252.

Raff, R. A. & Kaufman, T. C. (1983) *Embryos, Genes, and Evolution: Developmental-Genetic Basis of Evolutionary Change*. Bloomington, Indiana University Press,

Raine, J. C. & Leatherland, J. F. (2003) Trafficking of L-triiodothyronine between ovarian fluid and oocytes of rainbow trout (*Oncorhynchus mykiss*). *Comparative Biochemistry and Physiology B*, 136, 267–274.

Raine, J. C., Takemura, A. & Leatherland, J. F. (2001) Assessment of thyroid function in adult medaka (*Oryzias latipes*) and juvenile rainbow trout (*Oncorhynchus mykiss*) using immunostaining methods. *Journal of Experimental Zoology*, 290, 366–378.

Raine, J. C., Cameron, C., Vijayan, M. M., Lamarre, J. & Leatherland, J. F. (2004) The effect of elevated oocyte triiodothyronine content on development of rainbow trout embryos and expression of mRNA encoding for thyroid hormone receptors. *Journal of Fish Biology*, 65, 206–226.

Ramachandran, R. K., Widramanayake, A. H., Uzman, J. A., Govindarajan, V. & Tomlinson, C. R. (1997) Disruption of gastrulation and oral-aboral ectoderm differentiation in the *Lytechinus pictus* embryo by a dominant/negative PDGF receptor. *Development*, 124, 2355–2364.

Ramsey, K. M., Mills, K. F., Satoh, A. & Imai, S. (2008) Age-associated loss of Sirt1-mediated enhancement of glucose-stimulated insulin secretion in beta cell-specific Sirt1-overexpressing (BESTO) mice. *Aging Cell*, 7, 78–88.

Rassoulzadegan, F., Fenuaux, L. & Strathmann, R. R. (1984) Effect of flavor and size on selection of food by suspension-feeding plutei. *Limnology and Oceanography*, 29, 357–361.

Rathcke, B. & Lacey, E. P. (2003) Phenological patterns of terrestrial plants. *Annual Review of Ecology and Systematics*, 16, 179–214.

Ratnieks, F. L. W. (1993) Egg-laying, egg-removal, and ovary development by workers in queenright honey bee colonies. *Behavioral Ecology and Sociobiology*, 32, 191–198.

Rauschenbach, I. Y., Gruntenko, N. E., Bownes, M., Adonieva, N. V., Terashima, J., Karpova, E. K., Faddeeva, N. V. & Chentsova, N. A. (2004) The role of juvenile hormone in the control of reproductive function in *Drosophila virilis* under nutritional stress. *Journal of Insect Physiology*, 50, 323–330.

Rebeiz, M., Pool, J. E., Kassner, V. A., Aquadro, C. F. & Carroll, S. B. (2009) Stepwise modification of a modular enhancer underlies adaptation in a *Drosophila* population. *Science*, 326, 1663–1667.

Reber-Muller, S., Studer, R., Muller, P., Yanze, N. & Schmid, V. (2001) Integrin and talin in the jellyfish *Podocoryne carnea*. *Cell Biology International*, 25, 753–769.

Reboud, X. & Bell, G. (1997) Experimantal evolution in *Chamydomonas*. III. Evolution of specialist and generalist types in environments that vary in space and time. *Heredity*, 78, 507–514.

Reddy, P. K. & Lam, T. J. (1991) Effect of thyroid hormones on hatching in the tilapia, *Oreochromis mossambicus*. *General and Comparative Endocrinology*, 81, 484–491.

Reddy, P. K., Brown, C. L., Leatherland, J. F. & Lam, T. J. (1992) Role of thyroid hormones in tilapia larvae (*Oreochromis mossambicus*). 2. Changes in the hormones and 5'-monodeiodinase activity during development. *Fish Physiology and Biochemistry*, 9, 487–496.

Reed, W. L., Clark, M. E., Parker, P. G., Raouf, S. A., Arguedas, N., Monk, D. S., Snajdr, E., Nolan, V. & Ketterson, E. D. (2006) Physiological effects on demography: A long-term experimental study of testosterone's effects on fitness. *The American Naturalist*, 167, 667–683.

Reiling, J. H. & Hafen, E. (2004) The hypoxia-induced paralogs Scylla and Charybdis inhibit growth by downregulating S6K activity upstream of TSC in *Drosophila*. *Genes & Development*, 18, 2879–2892.

Reiner, D. J., Ailion, M., Thomas, J. H. & Meyer, B. J. (2008) *C. elegans* anaplastic lymphoma kinase ortholog SCD-2 controls dauer formation by modulating TGF-beta signaling. *Current Biology*, 18, 1101–1109.

Reiter, E. O. & Rosenfeld, R. G. (1998) Normal and aberrant growth. In Larsen, P. R., Kronenberg, H. M., Melmed, S. & Polonsky, K. S. (Eds.) *Williams Textbook of Endocrinology*. 9th edn. Philadelphia, Saunders.

Reitzel, A. M. & Heyland, A. (2007) Reduction in morphological plasticity in echinoid larvae: relationship of plasticity with maternal investment and food availability. *Evolutionary Ecology Research*, 9, 109–121.

Reitzel, A. M., Burton, P. M., Krone, C. & Finnerty, J. R. (2007) Comparison of developmental trajectories in the starlet sea anemone *Nematostella vectensis*: embryogenesis, regeneration, and two forms of asexual fission. *Invertebrate Biology*, 126, 99–112.

Reitzel, A. M., Sullivan, J. C., Traylor-Knowles, N. & Finnerty, J. R. (2008) Genomic survey of candidate stress-response genes in the estuarine anemone *Nematostella vectensis*. *Biological Bulletin*, 214, 233–254.

Reitzel, A. M., Daly, M., Sullivan, J. C. & Finnerty, J. R. (2009) Comparative anatomy and histology of developmental and parasitic stages in the life cycle of the lined sea anemone *Edwardsiella lineata*. *Journal of Parasitology*, 95, 100–112.

Rembold, H., Czoppelt, C. & Rao, P. J. (1974) Effect of juvenile hormone treatment on caste differentiation in the honeybee, *Apis mellifera*. *Journal of Insect Physiology*, 20, 1193–1202.

Remy, C. & Bounhiol, J. J. (1971) Normalized metamorphosis achieved by adrenocorticotropic hormone in hypophysectomized and thyroxined *Alytes* tadpoles. *Comptes Rendus Hebdomadaires des Seances de L'Academie des Sciences. Serie D: Sciences Naturelles*, 272, 455–458.

Rewitz, K. F., Yamanaka, N., Gilbert, L. I. & O'Connor, M. B. (2009) The insect neuropeptide PTTH activates receptor tyrosine kinase torso to initiate metamorphosis. *Science*, 326, 1403–1405.

Rey, B., Sibille, B., Romestaing, C., Belouze, M., Letexier, D., Servais, S., Barre, H., Duchamp, C. & Voituron, Y. (2008) Reptilian uncoupling protein: functionality and expression in sub-zero temperatures. *Journal of Experimental Biology*, 211, 1456–1462.

Reznick, D. (1985) Costs of Reproduction - an Evaluation of the Empirical-Evidence. *Oikos*, 44, 257–267.

Reznick, D., Nunney, L. & Tessier, A. (2000) Big houses, big cars, superfleas and the costs of reproduction. *Trends in Ecology & Evolution*, 15, 421–425.

Reznick, D., Bryant, M., Roff, D., Ghalambor, C. & Ghalambor, D. (2004) Effect of extrinsic mortality on the evolution of senescence in guppies. *Nature*, 431, 1095–1099.

Rice, W. R. & Chippindale, A. K. (2001) Intersexual ontogenetic conflict. *Journal of Evolutionary Biology*, 14, 685–693.

Richard, D. S., J. M. Jones, M. R. Barbarito, S. Cerula, J. P. Detweiler, S. J. Fisher, D. M. Brannigan, & D. M. Scheswohl. (2001) Vitellogenesis in diapausing and mutant *Drosophila melanogaster*: further evidence for the relative roles of ecdysteroids and juvenile hormones. *Journal of Insect Physiology*, 47, 905–913.

Richard, D. S., Rybczynski, R., Wilson, T. G., Wang, Y., Wayne, M. L., Zhou, Y., Partridge, L., & Harshman, L. G. (2005) Insulin signaling is necessary for vitellogenesis in *Drosophila melanogaster* independent of the roles of juvenile hormone and ecdysteroids: female sterility of the chico1 insulin signaling mutation is autonomous to the ovary. *Journal of Insect Physiology*, 51, 455–464.

Richardson, R. H. & Kojima, K. I. (1965) The kinds of genetic variability in relation to selection responses in *Drosophila* fecundity. *Genetics*, 52, 583–598.

Richardson, S. J., Aldred, A. R., Leng, S. L., Renfree, M. B., Hulbert, A. J. & Schreiber, G. (2002) Developmental profile of thyroid hormone distributor proteins in a marsupial, the tammar wallaby *Macropus eugenii*. *General and Comparative Endocrinology*, 125, 92–103.

Richardson, S. J., Monk, J. A., Shepherdley, C. A., Ebbesson, L. O. E., Sin, F., Power, D. M., Frappell, P. B., Kohrle, J. & Renfree, M. B. (2005) Developmentally regulated thyroid hormone distributor proteins in marsupials, a reptile, and fish. *American Journal of Physiology. Regulatory, Integrative and Comparative Physiology*, 288, R1264–R1272.

Ricklefs, R. E. (1974) Energetics of reproduction in birds. In Paynter JR., R. A. (Ed.) *Avian Energetics*. Cambridge, Nuttall Ornithological Club.

Ricklefs, R. (2006) Embryo development and aging in birds and mammals. *Proceedings of the Royal Society of London B*, 273, 2077–2082.

Ricklefs, R. E. & Wikelski, M. (2002) The physiology / life history nexus. *Trends in Ecology & Evolution*, 17, 462–468.

Riddiford, L. M. (1976) Hormonal control of insect epidermal cell commitment *in vitro*. *Nature*, 259, 115–117.

Riddiford, L. M. (1994) Cellular and molecular actions of juvenile hormone. I. General considerations and premetamorphic actions In Evans, P. D. (Ed.) *Advances in Insect Physiology*. London, Academic Press.

Riddiford, L. M. & Ashburner, M. (1991) Effects of juvenile hormone mimics on larval development and metamorphosis of *Drosophila melanogaster*. *General and Comparative Endocrinology*, 82, 172–183.

Rincon, M., Rudin, E. & Barzilai, N. (2005) The insulin/ IGF-1 signaling in mammals and its relevance to human longevity. *Experimental Gerontology*, 40, 873–877.

Rinehart, J. P., Cikra-Ireland, R. A., Flannagan, R. D. & Denlinger, D. L. (2001) Expression of ecdysone receptor is unaffected by pupal diapause in the flesh fly, *Sarcophaga crassipalpis*, while its dimerization partner, USP, is downregulated. *Journal of Insect Physiology*, 47, 915–921.

Rinehart, J. P., Li, A., Yocum, G. D., Robich, R. M., Hayward, S. A. & Denlinger, D. L. (2007) Up-regulation of heat shock proteins is essential for cold survival during insect diapause. *Proceedings of the National Academy of Sciences of the United States of America*, 104, 11130–11137.

Rintelen, F., Stocker, H., Thomas, G. & Hafen, E. (2001) PDK1 regulates growth through Akt and S6K in *Drosophila*. *Proceedings of the National Academy of Sciences of the United States of America*, 98, 15020–15025.

Ritchie, J. W., Shi, Y. B., Hayashi, Y., Baird, F. E., Muchekehu, R. W., Christie, G. R. & Taylor, P. M. (2003) A role for thyroid hormone transporters in transcriptional regulation by thyroid hormone receptors. *Molecular Endocrinology*, 17, 653–661.

R'Kha, S., Moreteau, B., Coyne, J. A. & David, J. R. (1997) Evolution of a lesser fitness trait: egg production in the specialist *Drosophila sechelia*. *Genetical Research*, 69, 17–23.

Robert, K. A. & Bronikowski, A. M. (2010) Evolution of senescence in nature: physiological evolution in populations of garter snake with divergent life histories. *The American Naturalist*, 175, 147–159.

Robert, K. A., Brunet-Rossinni, A. & Bronikowski, A. M. (2007) Testing the 'free radical theory of aging' hypothesis: physiological differences in long-lived and short-lived Colubrid snakes. *Aging Cell*, 6, 395–404.

Robert, K., Vleck, C. & Bronikowski, A. (2009) The effects of maternal corticosterone levels on offspring behavior in fast- and slow-growth garter snakes (*Thamnophis elegans*). *Hormones and Behavior*, 55, 24–32.

Roberts, S. P. & Elekonich, M. M. (2005) Muscle biochemistry and the ontogeny of flight capacity during behavioral development in the honey bee, *Apis mellifera*. *The Journal of Experimental Biology*, 208, 4193–4198.

Roberts, M. L., Buchanan, K. L. & Evans, M. R. (2004) Testing the immunocompetence handicap hypothesis: a review of the evidence. *Animal Behaviour*, 68, 227–239.

Roberts, R. D., Kawamura, T. & Handley, C. M. (2007) Factors affecting settlement of abalone (*Haliotis iris*) larvae on benthic diatom films. *Journal of Shellfish Research*, 26, 323–334.

Robertson, A. (1968) The spectrum of genetic variation. In Lewontin, R. C. (Ed.) *Population Biology and Evolution*. Syracuse, Syracuse University Press.

Robertson, K. A. & Monteiro, A. (2005) Female *Bicyclus anynana* butterflies choose males on the basis of their dorsal UV-reflective eyespot pupils. *Proceedings of the Royal Society of London B*, 272, 1541–1546.

Robich, R. M., Rinehart, J. P., Kitchen, L. J. & Denlinger, D. L. (2007) Diapause-specific gene expression in the northern house mosquito, *Culex pipiens* L., identified by suppressive subtractive hybridization. *Journal of Insect Physiology*, 53, 235–245.

Rockman, M. V. (2008) Reverse engineering the genotype-phenotype map with natural genetic variation. *Nature*, 456, 738–44.

Rodgers, J. T., Lerin, C., Haas, W., Gygi, S. P., Spiegelman, B. M. & Puigserver, P. (2005) Nutrient control of glucose homeostasis through a complex of PGC-1α and SIRT1. *Nature*, 434, 113–118.

Rodolfo-Metalpa, R., Peirano, A., Houlbrèeue, F., Abbate, M. & Ferrier-Pagès, C. (2008) Effects of temperature, light and heterotrophy on the growth rate and budding of the temperate coral *Cladocora caespitosa*. *Coral Reefs*, 27, 17–25.

Roe, R. M., Jesudason, P., Venkatesh, K., Kallapur, V. L., Anspaugh, D. D., Majumder, C., Linderman, R. J. & Graves, D. M. (1993) Developmental role of juvenile-hormone metabolism in Lepidoptera. *American Zoologist*, 33, 375–383.

Roff, D. A. (1990) Selection for changes in the incidence of wing dimorphism in *Gryllus firmus*. *Heredity*, 65, 163–168.

Roff, D. A. (1992) *The Evolution of Life Histories: Theory and Analysis*. New York, Chapman & Hall.

Roff, D. A. (1997) *Evolutionary Quantitative Genetics*. New York, Chapman and Hall.

Roff, D. (2002) *Life History Evolution*. Sunderland, Sinauer Associates, Inc.

Roff, D. A. (2007a) A centennial celebration for quantitative genetics. *Evolution*, 61, 1017–1032.

Roff, D. A. (2007b) Contributions of genomics to life history theory. *Nature Reviews Genetics*, 8, 116–125.

Roff, D. A. & Fairbairn, D. J. (2007a) Laboratory evolution of the migratory polymorphism in the sand cricket: Combining physiology with quantitative genetics. *Physiological and Biochemical Zoology*, 80, 358–369.

Roff, D. A. & Fairbairn, D. J. (2007b) The evolution of trade-offs: where are we? *Journal of Evolutionary Biology*, 20, 433–447.

Rogina, B. & Helfand, S. L. (2004) Sir2 mediates longevity in the fly through a pathway related to calorie restriction. *Proceedings of the National Academy of Sciences of the United States of America*, 101, 15998–16003.

Rogina, B., Reenan, R. A., Nilsen, S. P. & Helfand, S. L. (2000) Extended life-span conferred by cotransporter gene mutations in *Drosophila*. *Science*, 290, 2137–2140.

Rogina, B., Helfand, S. L. & Frankel, S. (2002) Longevity regulation by *Drosophila* Rpd3 deacetylase and caloric restriction. *Science*, 298, 1745.

Roitberg, B. D. & Gordon, I. (2005) Does the Anopheles blood meal - fecundity curve, curve? *Journal of Vector Ecology*, 30, 83–86.

Rolff, J. & Siva-Jothy, M. T. (2002) Copulation corrupts immunity: A mechanism for a cost of mating in insects. *Proceedings of the National Academy of Sciences of the United States of America*, 99, 9916–9918.

Rolff, J. & Siva-Jothy, M. T. (2003) Invertebrate ecological immunology. *Science*, 301, 472–475.

Roney, J. R., Mahler, S. V. & Maestripieri, D. (2003) Behavioral and hormonal responses of men to brief interactions with women. *Evolution and Human Behavior*, 24, 365–375.

Rose, M. R. (1984) Laboratory selection of postponed senescence in *Drosophila melanogaster*. *Evolution*, 38, 1003–1010.

Rose, M. R. (1991) *Evolutionary Biology of Aging*. New York, Oxford University Press.

Rose, M. R. & Bradley, T. J. (1998) Evolutionary physiology of the cost of reproduction. *Oikos*, 83, 443–451.

Rose, M. R. & Charlesworth, B. (1981) Genetics of life history in *Drosophila melanogaster*. II. Exploratory selection experiments. *Genetics*, 97, 187–196.

Rose, M. R., Drapeau, M. D., Yazdi, P. G., Sahah, K. H., Moise, D. B., Thaker, R. R., Rauser, C. L. & Mueller, L. D. (2002) Evolution of late-life mortality in *Drosophila melanogaster*. *Evolution*, 56, 1982–1991.

Rosenberg, K. (2004) Living longer: Information revolution, population expansion, and modern human origins. *Proceedings of the National Academy of Sciences of the United States of America*, 101, 10847–10848.

Rosenberg, K. & Trevathan, W. (2002) Birth, obstetrics and human evolution. *BJOG: An International Journal of Obstetrics and Gynaecology*, 109, 1199–1206.

Rosenkilde, P. & Ussing, A. P. (1996) What mechanisms control neoteny and regulate induced metamorphosis in Urodeles? *International Journal of Developmental Biology*, 40, 665–673.

Rosenkilde, P., Mogensen, E., Centervall, G. & Jorgensen, O. S. (1982) Peaks of neuronal membrane antigen and thyroxine in larval development of the Mexican axolotl. *General and Comparative Endocrinology*, 48, 504–514.

Rosenstiel, P., Philipp, E. E. R., Schreiber, S. & Bosch, T. C. G. (2009) Evolution and function of innate immune receptors - insights from marine invertebrates. *Journal of Innate Immunity*, 1, 291–300.

Roskam, J. C. & Brakefield, P. M. (1999) Seasonal polyphenism in *Bicyclus* (Lepidoptera: Satyridae) butterflies: different climates need different cues. *Biological Journal of the Linnean Society*, 66, 345–356.

Rosta, K., Molvarec, A., Enzsoly, A., Nagy, B., Ronai, Z., Fekete, A., Sasvari-Szekely, M., Rigo, J., JR. & Ver, A. (2009) Association of extracellular superoxide dismutase (SOD3) Ala40Thr gene polymorphism with pre-eclampsia complicated by severe fetal growth restriction. *European Journal of Obstetrics, Gynecology, and Reproductive Biology*, 142, 134–138.

Roth, G. S., Ingram, D. K. & Lane, M. A. (1999) Calorie restriction in primates: will it work and how will we know? *Journal of the American Geriatrics Society*, 47, 896–903.

Rottiers, V. & Antebi, A. (2006) Control of *Caenorhabditis elegans* life history by nuclear receptor signal transduction. *Experimental Gerontology*, 41, 904–909.

Röttinger, E., Saudemont, A., Duboc, V., Bensnardeau, L., McClay, D. R. & Lepage, T. (2008) FGF signals guide migration of mesenchymal cells, control skeletal morphogenesis of the skeleton and regulate gastrulation during sea urchin development. *Development*, 135, 353–365.

Roughgarden, J. (1988) The evolution of marine life cycles. In M. W. Feldman (Ed.) *Mathematical Evolutionary Theory*. Princeton, Princeton University Press.

Rountree, D. B. & Bollenbacher, W. E. (1986) The release of the prothoracicotropic hormone in the tobacco hornworm, *Manduca sexta*, is controlled intrinsically by juvenile hormone. *Journal of Experimental Biology*, 120, 41–58.

Roux, E. A., Roux, M. & Korb, J. (2009) Selection on defensive traits in a sterile caste - caste evolution: a mechanism to overcome life history trade-offs? *Evolution & Development*, 11, 80–87.

Rowe, L. & Day, T. (2006) Detecting sexual conflict and sexually antagonistic coevolution. *Philosophical Transactions of the Royal Society of London B*, 361, 277–285.

Royet, J. & Dziarski, R. (2007) Peptidoglycan recognition proteins: pleiotropic sensors and effectors of antimicrobial defences. *Nature Reviews Microbiology*, 5, 264–277.

Rudzinska, M. A. (1951) The influence of amount of food on the reproduction rate and longevity of a sectarian (*Tokophyra infusionum*). *Science*, 113, 10–11.

Ruebenbauer, A., Schlyter, F., Hansson, B. S., Lofstedt, C. & Larsson, M. C. (2008) Genetic variability and robustness of host odor preference in *Drosophila melanogaster*. *Current Biology*, 18, 1438–1443.

Rueppell, O., Christine, S., Mulcrone, C. & Groves, L. (2007) Aging without functional senescence in honey bee workers. *Current Biology*, 17, R274–R275.

Rueppell, O., Linford, R., Gardner, P., Coleman, J. & Fine, K. (2008) Aging and demographic plasticity in response to experimental age structures in honeybees (*Apis mellifera L*). *Behavioral Ecology and Sociobiology*, 62, 1621–1631.

Ruff, C. B. (2002) Variation in human body size and shape. *Annual Review of Anthropology*, 31, 211–232.

Rulifson, E. J., Kim, S. K. & Nusse, R. (2002) Ablation of insulin-producing neurons in flies: growth and diabetic phenotypes. *Science*, 296, 1118–1120.

Ruppell, O., Pankiw, T. & Page, R. E. J. (2004) Pleiotropy, epistasis and new QTL: The genetic architecture of honey bee foraging behavior. *Journal of Heredity*, 95, 481–491.

Russell, D. W. (2003) The enzymes, regulation, and genetics of bile acid synthesis. *Annual Review of Biochemistry*, 72, 137–174.

Russell, S. J. & Kahn, C. R. (2007) Endocrine regulation of aging. *Nature Reviews Molecular Cell Biology*, 8, 681–691.

Ryan, J. & Baxevanis, A. (2007) Hox, Wnt, and the evolution of the primary body axis: insights from the early-divergent phyla. *Biology Direct*, 2, 37.

Ryan, J. F., Mazza, M. E., Pang, K., Matus, D. Q., Baxevanis, A. D., Martindale, M. Q. & Finnerty, J. R. (2007) Pre-Bilaterian origins of the Hox cluster and the Hox code: evidence from the sea anemone, *Nematostella vectensis*. *PLoS One*, 2, e153.

Rybczynski, R., Bell, S. C. & Gilbert, L. I. (2001) Activation of an extracellular signal-regulated kinase (ERK) by the insect prothoracicotropic hormone. *Molecular and Cellular Endocrinology*, 184, 1–11.

Ryu, J. H., Ha, E. M., Oh, C. T., Seol, J. H., Brey, P. T., Jin, I., Lee, D. G., Kim, J., Lee, D. & Lee, W. J. (2006) An essential complementary role of NF-kappaB pathway to microbicidal oxidants in *Drosophila* gut immunity. *EMBO Journal*, 25, 3693–3701.

Ryu, J. H., Kim, S. H., Lee, H. Y., Bai, J. Y., Nam, Y. D., Bae, J. W., Lee, D. G., Shin, S. C., Ha, E. M. & Lee, W. J. (2008) Innate immune homeostasis by the homeobox gene caudal and commensal-gut mutualism in *Drosophila*. *Science*, 319, 777–782.

Saastamoinen, M., van der Sterren, D., Vastenhout, N., Zwaan, B. J., and Brakefield, P. M. (2010). Predictive adaptive responses: Condition-dependent impact of adult nutrition and flight in the tropical butterfly, *Bicyclus anynana*. *American Naturalist* 176: 686–698.

Saccone, G., Pane, A. & Polito, L. C. (2002) Sex determination in flies, fruitflies and butterflies. *Genetica*, 116, 15–23.

Sackton, T. B., Lazzaro, B. P., Schlenke, T. A., Evans, J. D., Hultmark, D. & Clark, A. G. (2007) Dynamic evolution of the innate immune system in *Drosophila*. *Nature Genetics*, 39, 1461–1468.

Sadd, B. M. & Siva-Jothy, M. T. (2006) Self-harm caused by an insect's innate immunity. *Proceedings of the Royal Society of London B*, 273, 2571–2574.

Saito, M., Seki, M., Amemiya, S., Yamasu, K., Suyemitsu, T. & Ishihara, K. (1998) Induction of metamorphosis in the sand dollar *Peronella japonica* by thyroid hormones. *Development, Growth & Differentiation*, 40, 307–312.

Sakai, K. L. (2005) Language acquisition and brain development. *Science*, 310, 815–819.

Sakamoto, T. & Hirano, T. (1993) Expression of insulin-like growth factor-I gene in osmoregulatory organs during seawater adaptation of the salmonid fish: Possible mode of osmoregulatory action of growth hormone. *Proceedings of the National Academy of Sciences of the United States of America*, 90, 1912–1916.

Sakamoto, T., McCormick, S. D. & Hirano, T. (1993) Osmoregulatory actions of growth hormone and its mode of action in salmonids: A review. *Fish Physiology and Biochemistry*, 11, 155–164.

Salih, D. A. & Brunet, A. (2008) FoxO transcription factors in the maintenance of cellular homeostasis during aging. *Current Opinion in Cell Biology*, 20, 126–136.

Salmon, A. B., Marx, D. B. & Harshman, L. G. (2001) A cost of reproduction in *Drosophila melanogaster*: Stress susceptibility. *Evolution*, 55, 1600–1608.

Samis, K. E., Heath, K. D. & Stinchcombe, J. R. (2008) Discordant longitudinal clines in flowering time and *PHYTOCHROME C* in *Arabidopsis thaliana*. *Evolution*, 62, 2971–2983.

Sanchez-Blanco, A., Fridell, Y. W. & Helfand, S. L. (2005) Involvement of *Drosophila* uncoupling protein 5 in metabolism and aging. *Genetics*, 172, 1699–1710.

Sandrelli, F., Tauber, E., Pegoraro, M., Mazzotta, G., Cisotto, P., Landskron, J., Stanewsky, R., Piccin, A., Rosato, E., Zordan, M., Costa, R. & Kyriacou, C. P. (2007) A molecular basis for natural selection at the timeless locus in *Drosophila melanogaster*. *Science*, 316, 1898–1900.

Santos, C. R. A. & Power, D. M. (1999) Identification of transthyretin in fish (*Sparus aurata*): cDNA cloning and characterisation. *Endocrinology*, 140, 2430–2433.

Sato, Y., Buchholz, D. R., Paul, B. D. & Shi, Y. B. (2007) A role of unliganded thyroid hormone receptor in postembryonic development in *Xenopus laevis*. *Mechanisms of Development*, 124, 476–488.

Satterlee, D. G. & Johnson, W. A. (1988) Selection of Japanese quail for contrasting blood corticosterone response to immobilization. *Poultry Science*, 67, 25–32.

Satuito, C. G., Natoyama, K., Yamazaki, M., Shimizu, K. & Fusetani, N. (1999) Induction of metamorphosis in the pediveliger larvae of the mussel *Mytilus galloprovincialis* by neuroactive compounds. *Fisheries Science*, 65, 384–389.

Saunders, D. S. (2002) *Insect Clocks*. Amsterdam, Elsevier.

Saunders, D. S., Henrich, V. C. & Gilbert, L. I. (1989) Induction of diapause in *Drosophila melanogaster*: photoperiodic regulation and the impact of arrhythmic clock mutations on time measurement. *Evolution*, 86, 3748–3752.

Saunders, D. S., Richard, D. S., Applebaum, S. W., Ma, M. & Gilbert, L. I. (1990) Photoperiodic diapause in *Drosophila melanogaster* involves a block to the juvenile hormone regulation of ovarian maturation. *General and Comparative Endocrinology*, 79, 174–184.

Savage-Dunn, C. (2005) TGF-beta signaling. In The *C. elegans* Research Community (Ed.) *WormBook*, doi/10.1895/wormbook.1.7.1, http://www.wormbook.org.

Sbrenna-Micciarelli, A. (1977) Effects of farnesyl methyl ether on embryos of *Schistocerca gregaria* (Orthoptera). *Acta Embryologiae et Morphologiae Experimentali*, 3, 295–303.

Scarborough, C. L., Ferrari, J. & Godfray, H. C. J. (2005) Aphid protected from pathogen by endosymbiont. *Science*, 310, 1781.

Scarcelli, N., Cheverud, J. M., Schaal, B. A. & Kover, P. X. (2007) Antagonistic pleiotropic effects reduce the potential adaptive value of the *FRIGIDA* locus. *Proceedings of the National Academy of Sciences of the United States of America*, 104, 16986–16991.

Scarpulla, R. C. (2002) Nuclear activators and coactivators in mammalian mitochondrial biogenesis. *Biochimica et Biophysica Acta (BBA) - Gene Structure and Expression*, 1576, 1–14.

Schaller, F. & Charlet, M. (1980) Neuroendocrine control and rate of ecdysone biosynthesis in larvae of a paleopteran insect: *Aeshna cyanea* Muler. In Hoffman, J. A. (Ed.) *Progress in Ecdysone Research*. Amsterdam, Elsevier.

Schilder, R. J. & Marden, J. H. (2006) Metabolic syndrome and obesity in an insect. *Proceedings of the National Academy of Sciences of the United States of America*, 103, 18805–18809.

Schippers, M. P., Dukas, R., Smith, R. W., Wang, J., Smolen, K. & McClelland, G. B. (2006) Lifetime performance in foraging honeybees: behaviour and physiology. *Journal of Experimental Biology*, 209, 3828–3836.

Schläppi, M. R. (2006) *FRIGIDA LIKE 2* is a functional allele in Landsberg erecta and compensates for a nonsense allele of *FRIGIDA LIKE 1. Plant Physiology*, 142, 1728–1738.

Schlenke, T. A. & Begun, D. J. (2003) Natural selection drives *Drosophila* immune system evolution. *Genetics*, 164, 1471–1480.

Schlichting, C. & Pigliucci, M. (1998) *Phenotypic Evolution: A Reaction Norm Perspective*. Sunderland, Sinauer.

Schlinger, B. A. (1997) The activity and expression of aromatase in songbirds. *Brain Research*, 44, 359–364.

Schlinger, B. A. & London, S. E. (2006) Neurosteroids and the songbird model system. *Journal of Experimental Zoology A*, 305, 743–748.

Schlosser, G. & Wagner, G. P. (2004) The modularity concept in developmental and evolutionary biology. In Schlosser, G. & Wagner, G. P. (Eds.) *Modularity in Development and Evolution*. Chicago, University of Chicago Press.

Schluter, D., Clifford, E. A., Nemethy, M. & McKinnon, J. S. (2004) Parallel evolution and inheritance of quantitative traits. *The American Naturalist*, 163, 809–822.

Schmid-Hempel, P. (2003) Variation in immune defence as a question of evolutionary ecology. *Proceedings of the Royal Society of London B*, 270, 357–366.

Schmid-Hempel, P. (2005) Evolutionary ecology of insect immune defenses. *Annual Review of Entomology*, 50, 529–551.

Schmidt, H. (1970) *Anthopleura stellula* (Actiniaria: Actiniidae) and its reproduction by transverse fission. *Marine Biology*, 5, 245–255.

Schmidt, P. S. & Conde, D. R. (2006) Environmental heterogeneity and the maintenance of genetic variation for reproductive diapause in *Drosophila melanogaster. Evolution*, 60, 1602–1611.

Schmidt, P. S. & Paaby, A. B. (2008) Reproductive diapause and life history clines in North American populations of *Drosophila melanogaster. Evolution*, 62, 1204–1215.

Schmidt, P. S., Duvernell, D. D. & Eanes, W. F. (2000) Adaptive evolution of a candidate gene for aging in *Drosophila. Proceedings of the National Academy of Sciences of the United States of America*, 97, 10861–10865.

Schmidt, P. S., Matzkin, L. M., Ippolito, M. & Eanes, W. F. (2005a) Geographic variation in diapause incidence, life history traits and climatic adaptation in *Drosophila melanogaster. Evolution*, 59, 1721–1732.

Schmidt, P. S., Paaby, A. B. & Heschel, M. S. (2005b) Genetic variance for diapause expression and associated life histories in *Drosophila melanogaster. Evolution*, 59, 2616–2625.

Schmidt, P. S., Zhu, C. T., Das, J., Batavia, M., Yang, L. & Eanes, W. F. (2008) An amino acid polymorphism in the *couch potato* gene forms the basis for climatic adaptation in *Drosophila melanogaster. Proceedings of the National Academy of Sciences of the United States of America*, 105, 16207–16211.

Schmidt-Nielsen, K. (1984) *Scaling: why is animal size so important?* Cambridge, Cambridge University Press.

Schmitt, J. & Wulff, R. D. (1993) Light spectral quality, phytochrome, and plant competition. *Trends in Ecology & Evolution*, 8, 47–51.

Schneider, D. S. (2007) How and why does a fly turn its immune system off? *PLoS Biology*, 5, e247.

Schneider, D. S. & Ayres, J. S. (2008) Two ways to survive infection: what resistance and tolerance can teach us about treating infectious diseases. *Nature Reviews Immunology*, 8, 889–895.

Schneider, T. & Leitz, T. (1994) Protein kinase C in Hydrozoans – Involvement in metamorphosis of Hydractinia and in pattern-formation of Hydra. *Roux's Archives of Developmental Biology*, 203, 422–428.

Schneider, D. S., Ayres, J. S., Brandt, S. M., Costa, A., Dionne, M. S., Gordon, M. D., Mabery, E. M., Moule, M. G., Pham, L. N. & Shirasu-Hiza, M. M. (2007) *Drosophila eiger* mutants are sensitive to extracellular pathogens. *PLoS Pathogens*, 3, e41.

Schoech, S. J. & Hahn, T. P. (2008) Latitude affects degree of advancement in laying by birds in response to food supplementation: a meta-analysis. *Oecologia*, 157, 369–376.

Schoenmaker, M., De Craen, A. J., De Meijer, P. H., Beekman, M., Blauw, G. J., Slagboom, P. E. & Westendorp, R. G. (2006) Evidence of genetic enrichment for exceptional survival using a family approach: the Leiden Longevity Study. *European Journal of Human Genetics*, 14, 79–84.

Schoneberg, T., Hofreiter, M., Schulz, A. & Rompler, H. (2007) Learning from the past: evolution of GPCR functions. *Trends in Pharmacological Sciences*, 28, 117–121.

Schradin, C. & Anzenberger, G. (1999) Prolactin, the hormone of paternity. *News in Psychological Sciences*, 14, 223–231.

Schram, F. R. & Koenemann, S. (2001) Developmental genetics and arthropod evolution: part 1, on legs. *Evolution & Development*, 3, 343–354.

Schranz, E. M., Windsor, A. J., Song, B. H., Lawton-Rauh, A. & Mitchell-Olds, T. (2007) Comparative genetic mapping in *Boechera stricta*, a close relative of *Arabidopsis*. *Plant Physiology*, 144, 286–298.

Schreiber, A. M. & Specker, J. L. (1998) Metamorphosis in the summer flounder (*Paralichthys dentatus*): Stage-specific developmental response to altered thyroid status. *General and Comparative Endocrinology*, 111, 156–166.

Schreiber, A. M. & Specker, J. L. (1999a) Early larval development and metamorphosis in the summer flounder: changes in per cent whole-body water content and effects of altered thyroid status. *Journal of Fish Biology*, 55, 148–157.

Schreiber, A. M. & Specker, J. L. (1999b) Metamorphosis in the summer flounder *Paralichthys dentatus*: Changes in gill mitochondria-rich cells. *Journal of Experimental Biology*, 202, 2475–2484.

Schreiber, A. M. & Specker, J. L. (1999c) Metamorphosis in the summer flounder, *Paralichthys dentatus*: Thyroidal status influences salinity tolerance. *Journal of Experimental Zoology*, 284, 414–424.

Schulz, D. J. & Robinson, G. E. (1999) Biogenic amines and division of labor in honey bee colonies: behaviorally related changes in the antennal lobes and age-related changes in the mushroom bodies. *Journal of Comparative Physiology A*, 184, 481–488.

Schulz, D. J., Sullivan, J. P. & Robinson, G. E. (2002) Juvenile hormone and octopamine in the regulation of division of labor in honey bee colonies. *Hormones and Behavior*, 42, 222–231.

Schwartz, T. S., Murray, S. & Seebacher, F. (2008) Novel reptilian uncoupling proteins: molecular evolution and gene expression during cold acclimation. *Proceedings of the Royal Society of London B*, 275, 979–985.

Schwarz, J., Brokstein, P., Voolstra, C., Terry, A., Miller, D., Szmant, A., Coffroth, M. & Medina, M. (2008) Coral life history and symbiosis: functional genomic resources for two reef building Caribbean corals, *Acropora palmata* and *Montastraea faveolata*. *BMC Genomics*, 9, 97.

Scott, N. M., Haussmann, M. F., Elsey, R. M., Trosclair, P. L. & Vleck, C. M. (2006) Telomere length shortens with body length in *Alligator mississippiensis*. *Southeastern Naturalist*, 5, 685–692.

Sear, R., Mace, R. & McGregor, I. A. (2000) Maternal grandmothers improve the nutritional status and survival of children in rural Gambia. *Proceedings of the Royal Society of London B*, 267, 1641–1647.

Sebens, K. P. (1980) The regulation of asexual reproduction and indeterminate body size in the sea anemone *Anthopleura elegantissima*. *Biological Bulletin*, 158, 152–171.

Sebens, K. P. (1982a) Asexual reproduction in *Anthopleura elegantissima* (Anthozoa: Actiniaria): seasonality and spatial extent of clones. *Ecology*, 63, 434–444.

Sebens, K. P. (1982b) The limits to indeterminate growth: an optimal size model applied to passive suspension feeders. *Ecology*, 63, 209–222.

Sebens, K. P. (2002) Energetic constraints, size gradients, and size limits in benthic marine invertebrates. *Integrative and Comparative Biology*, 42, 853–861.

Secor, S. (2009) Specific dynamic action: a review of the postprandial metabolic response. *Journal of Comparative Physiology B*, 179, 1–56.

Seebacher, F. (2005) A review of thermoregulation and physiological performance in reptiles: what is the role of phenotypic flexibility? *Journal of Comparative Physiology B*, 175, 453–461.

Seebacher, F. & Murray, S. A. (2007) Transient receptor potential ion channels control thermoregulatory behaviour in reptiles. *PLoS One*, 2, e281.

Seehuus, S. C., Krekling, T. & Amdam, G. V. (2006a) Cellular senescence in honey bee brain is largely independent of chronological age. *Experimental Gerontology*, 41, 1117–1125.

Seehuus, S. C., Norberg, K., Gimsa, U., Krekling, T. & Amdam, G. V. (2006b) Reproductive protein protects functionally sterile honey bee workers from oxidative stress. *Proceedings of the National Academy of Sciences of the United States of America*, 103, 962–967.

Seeley, T. D. (1989) The honey bee colony as a superorganism. *American Scientist*, 77, 546–553.

Segre, A. V., Murray, A. W. & Leu, J. Y. (2006) High-resolution mutation mapping reveals parallel experimental evolution in yeast. *PLoS Biology*, 4, 1372–1385.

Sehnal, F., Svacha, P. & Zrzavy, J. (1996) Evolution of insect metamorphosis. In Gilbert, L. I., Tata, J. R. & Atkinson, B. G. (Eds.) *Metamorphosis*. San Diego, Academic Press.

Selman, C., Lingard, S., Choudhury, A. I., Batterham, R. L., Claret, M., Clements, M., Ramadani, F., Okkenhaug, K., Schuster, E., Blanc, E., Piper, M. D., Al-Qassab, H., Speakman, J. R., Carmignac, D., Robinson, I. C., Thornton, J. M., Gems, D., Partridge, L. & Withers, D. J. (2008) Evidence for lifespan extension and delayed age-related biomarkers in insulin receptor substrate 1 null mice. *FASEB Journal*, 22, 807–818.

Sempere, L. F., Sokol, N. S., Dubrovsky, E. B., Berger, E. M. & Ambros, V. (2003) Temporal regulation of microRNA expression in *Drosophila melanogaster* mediated by hormonal signals and broad-Complex gene activity. *Developmental Biology*, 259, 9–18.

Serhan, C. N. & Savill, J. (2005) Resolution of inflammation: the beginning programs the end. *Nature Immunology*, 6, 1191–1197.

Sewell, M. A., Cameron, M. J. & McArdle, B. H. (2004) Developmental plasticity in larval development in the echinometrid sea urchin *Evechinus chloroticus* with varying food ration. *Journal of Experimental Marine Biology and Ecology*, 309, 219–237.

Sezgin, E., Duvernell, D. D., Matzkin, L. M., Duan, Y., Zhu, C. T., Verrelli, B. C. & Eanes, W. F. (2004) Single-locus latitudinal clines and their relationship to temperate adaptation in metabolic genes and derived alleles in *Drosophila melanogaster. Genetics*, 168, 923–931.

Sfakianaki, A. K. & Norwitz, E. R. (2006) Mechanisms of progesterone action in inhibiting prematurity. *Journal of Maternal-Fetal & Neonatal Medicine*, 19, 763–772.

Sgrò, C. M. & Partridge, L. (1999) A delayed wave of death from reproduction in *Drosophila. Science*, 286, 2521–2524.

Sgrò, C. M. & Partridge, L. (2000) Evolutionary Responses of the Life History of Wild Caught *Drosophila melanogaster* to Two Standard Methods of Laboratory Culture. *The American Naturalist*, 156, 341–353.

Shamay-Tsoory, S. G., Tomer, R. & Aharon-Peretz, J. (2005) The neuroanatomical basis of understanding sarcasm and its relationship to social cognition. *Neuropsychology*, 19, 288–300.

Shanley, D. P. & Kirkwood, T. B. L. (2000) Calorie restriction and aging: a life history analysis. Evolution 54: 740–750.

Shanley, D. P. & Kirkwood, T. B. L. (2001) Caloric restriction, life history evolution, and bioenergetics, response to Mitteldorf. *Evolution*, 55, 1906–1906.

Shanley, D. P. & Kirkwood, T. B. (2006) Caloric restriction does not enhance longevity in all species and is unlikely to do so in humans. *Biogerontology*, 7, 165–168.

Shapiro, A. M. (1976) Seasonal polyphenism. *Evolutionary Biology*, 9, 259–333.

Shaw, W. M., Luo, S., Landis, J., Ashraf, J. & Murphy, C. T. (2007) The *C. elegans* TGF-beta dauer pathway regulates longevity via insulin signaling. *Current Biology*, 17, 1635–1645.

Sheaffer, K. L., Updike, D. L. & Mango, S. E. (2008) The target of rapamycin pathway antagonizes *pha-4/FoxA* to control development and aging. *Current Biology*, 18, 1355–1364.

Sheldon, B. C. & Verhulst, S. (1996) Ecological immunology: costly parasite defenses and trade-offs in evolutionary ecology. *Trends in Ecology & Evolution*, 11, 317–321.

Sheldon, C. C., Rouse, D. T., Finnegan, E. J., Peacock, W. J. & Dennis, E. S. (2000) The molecular basis of vernalization: The central role of *FLOWERING LOCUS C (FLC). Proceedings of the National Academy of Sciences of the United States of America*, 97, 3753–3758.

Shemshedini, L. & Wilson, T. G. (1990) Resistance to juvenile hormone and an insect growth regulator in *Drosophila* is associated with an altered cytosolic juvenile hormone binding protein. *Proceedings of the National Academy of Sciences of the United States of America*, 87, 2072–2076.

Shenk, M. A., Bode, H. R. & Steele, R. E. (1993) Expression of Cnox-2, a HOM/HOX homeobox gene in hydra, is correlated with axial pattern formation. *Development*, 117, 657–667.

Sherman, C. D. H. & Ayre, D. J. (2008) Fine-scale adaptation in a clonal sea anemone. *Evolution*, 62, 1373–1380.

Shi, Y.-B. (1999) *Amphibian Metamorphosis: From morphology to molecular biology*. New York, Wiley-Liss.

Shi, Y. B. (2009) Dual functions of thyroid hormone receptors in vertebrate development: the roles of histone-modifying cofactor complexes. *Thyroid*, 19, 987–999.

Shi, Y. B., Liang, V. C., Parkison, C. & Cheng, S. Y. (1994) Tissue-dependent developmental expression of a cytosolic thyroid hormone protein gene in Xenopus: its role in the regulation of amphibian metamorphosis. *FEBS Letters*, 355, 61–64.

Shi, Y., Hon, M. & Evans, R. M. (2002) The peroxisome proliferator-activated receptor delta, an integrator of transcriptional repression and nuclear receptor signaling. *Proceedings of the National Academy of Sciences of the United States of America*, 99, 2613–2618.

Shiao, J. C. & Hwang, P. P. (2006) Thyroid hormones are necessary for the metamorphosis of tarpon *Megalops cyprinoides* leptocephali. *Journal of Experimental Marine Biology and Ecology*, 331, 121–132.

Shick, J. M. (1991) *A Functional Biology of Sea Anemones*. London, New York, Chapman & Hall.

Shick, J. M., Hoffmann, R. J. & Lamb, A. N. (1979) Asexual reproduction, population structure, and genotype-environment interactions in sea anemones. *American Zoologist*, 19, 699–713.

Shifren, J. L., Braunstein, G. D., Simon, J. A., Casson, P. R., Buster, J. E., Redmond, G. P., Burki, R. E., Ginsburg, E. S., Rosen, R. C., Leiblum, S. R., Caramelli, K. E. & Mazer, N. A. (2000) Transdermal testosterone treatment in women with impaired sexual function after oophorectomy. *New England Journal of Medicine*, 343, 682–688.

Shimizu, K. K. & Purugganan, M. D. (2005) Evolutionary and ecological genomics of Arabidopsis. *Plant Physiology*, 138, 578–84.

Shimizu, K., Hunter, E. & Fusetani, N. (2000) Localisation of biogenic amines in larvae of *Bugula neritina* (Bryozoa: Cheilostomatida) and their effects on settlement. *Marine Biology*, 136, 1–9.

Shindo, C., Aranzana, M. J., Lister, C., Baxter, C., Nicholls, C., Nordborg, M. & Dean, C. (2005) Role of *FRIGIDA* and *FLOWERING LOCUS C* in determining variation in flowering time of *Arabidopsis*. *Plant Physiology*, 138, 1163–1173.

Shine, R. (2005) Life history evolution in reptiles. *Annual Review of Ecology, Evolution, and Systematics*, 36, 23–46.

Shine, R. & Madsen, T. (1997) Prey abundance and predator reproduction: Rats and pythons on a tropical Australian floodplain. *Ecology*, 78, 1078–1086.

Shingleton, A. W. (2005) Body-size regulation: combining genetics and physiology. *Current Biology*, 15, R825–R827.

Shingleton, A. W. (2010) The regulation of organ size in *Drosophila*: Physiology, plasticity, patterning and physical force. *Organogenesis*, 6, 1–13.

Shingleton, A. W., Das, J., Vinicius, L. & Stern, D. L. (2005) The temporal requirements for insulin signaling during development in *Drosophila*. *PLoS Biology*, 3, e289.

Shingleton, A. W., Frankino, W. A., Flatt, T., Nijhout, H. F. & Emlen, D. J. (2007) Size and shape: The developmental regulation of static allometry in insects. *BioEssays*, 29, 536–548.

Shingleton, A. W., Mirth, C. K. & Bates, P. W. (2008) Developmental model of static allometry in holometabolous insects. *Proceedings of the Royal Society of London B*, 275, 1875–1885.

Shingleton, A. W., Estep, C. M., Driscoll, M. V. & Dworkin, I. (2009) Many ways to be small: different environmental regulators of size generate distinct scaling relationships in *Drosophila melanogaster*. *Proceedings of the Royal Society of London B*, 276, 2625–2633.

Shintani, N., Nohira, T., Hikosaka, A. & Kawahara, A. (2002) Tissue-specific regulation of type III iodothyronine 5-deiodinase gene expression mediates the effects of prolactin and growth hormone in *Xenopus* metamorphosis. *Development, Growth & Differentiation*, 44, 327–335.

Shiomi, K. & Yamaguchi, M. (2008) Expression patterns of three *Par*-related genes in sea urchin embryos. *Gene Expression Patterns*, 8, 323–330.

Shoguchi, E., Yharada, Y., Numakunai, T. & Satoh, N. (2000) Expression of the Otx gene in the ciliary bands during sea cucumber embryogenesis. *Genesis*, 27, 58–63.

Short, S. M. & Lazzaro, B. P. (2010) Female and male genetic contributions to female post-mating susceptibility to infection in *Drosophila melanogaster*. *Proceedings of the Royal Society, B: Biological Sciences*, 277: 3649–3657.

Shostak, S. (1993) Studies of asexual reproduction in Cnidaria. In Adiyodi, K. G. & Adiyodi, R. G. (Eds.) *Reproductive Biology of Invertebrates*. New York, John Wiley and Sons.

Sibly, R. & Calow, P. (1986) Why breeding earlier is always worthwhile. *Journal of Theoretical Biology*, 123, 311–319.

Siedlinski, M., Van Diemen, C. C., Postma, D. S., Vonk, J. M. & Boezen, H. M. (2009) Superoxide dismutases, lung function and bronchial responsiveness in a general population. *European Respiratory Journal*, 33, 986–992.

Siefker, B., Kroiher, M. & Berking, S. (2000) Induction of metamorphosis from the larval to the polyp stage is similar in Hydrozoa and a subgroup of Scyphozoa (Cnidaria, Semaeostomeae). *Helgoland Marine Research*, 54, 230–236.

Siegal, M. & Varley, R. (2002) Neural systems involved with 'Theory of Mind'. *Nature Reviews Neuroscience*, 3, 463–471.

Silbermann, R. & Tatar, M. (2000) Reproductive costs of heat shock protein in transgenic *Drosophila melanogaster*. *Evolution*, 54, 2038–2045.

Silverin, B., Baillien, M., Foidart, A. & Balthazart, J. (2000) Distribution of aromatase activity in the brain and peripheral tissue of passerine and nonpasserine avian species. *General and Comparative Endocrinology*, 117, 34–53.

Sim, C. & Denlinger, D. L. (2008) Insulin signaling and FOXO regulate the overwintering diapause of the mosquito *Culex pipiens*. *Proceedings of the National Academy of Sciences of the United States of America*, 105, 6777–6781.

Simandle, E. T., Espinoza, R. E., Nussear, K. E. & Tracy, C. R. (2001) Lizards, lipids, and dietary links to animal function. *Physiological and Biochemical Zoology*, 74, 625–640.

Simmons, D., Shaw, J., McKenzie, A., Eaton, S., Cameron, A. J. & Zimmet, P. (2006) Is grand multiparity associated with an increased risk of dysglycaemia? *Diabetologia*, 49, 1522–1527.

Simon, A. F., Shih, C., Mack, A. & Benzer, S. (2003) Steroid control of longevity in *Drosophila melanogaster*. *Science*, 299, 1407–1410.

Simpson, P., Berreur, P. & Berreurbonnenfant, J. (1980) The initiation of pupariation in *Drosophila* – dependence on growth of the imaginal disks. *Journal of Embryology and Experimental Morphology*, 57, 155–165.

Sinervo, B. (1999) Mechanistic analysis of natural selection and a refinement of Lack's and William's principles. *The American Naturalist*, 154, S26–S42.

Sinervo, B. (2000) Adaptation, natural selection, and optimal life history allocation in the face of genetically-based trade-offs. In Mousseau, T., Sinervo, B. & Endler, J. A. (Eds.) *Adaptive Genetic Variation in the Wild*. Oxford, Oxford University Press.

Sinervo, B. & Calsbeek, R. (2003) Physiological epistasis, ontogenetic conflict and natural selection on physiology and life history. *Integrative and Comparative Biology*, 43, 419–430.

Sinervo, B. & Calsbeek, R. (2006) The developmental, physiological, neural and genetical causes and consequences of frequency dependent selection in the wild. *Annual Review of Ecology, Evolution and Systematics*, 37, 581–610.

Sinervo, B. & Clobert, J. (2003) Morphs, dispersal behavior, genetic similarity, and the evolution of cooperation. *Science*, 300, 1949–1951.

Sinervo, B. & Clobert, J. (2008) Life history strategies, multi-dimentional trade-offs and behavioural syndromes. In Danchin, E., Giraldeau, L.-A. & Cézilly, F. (Eds.) *Behavioural Ecology*. Oxford, Oxford University Press.

Sinervo, B. & DeNardo, D. F. (1996) Costs of reproduction in the wild: path analysis of natural selection and experimental tests of causation. *Evolution*, 50, 1299–1313.

Sinervo, B. & Licht, P. (1991). The physiological and hormonal control of clutch size, egg size, and egg shape in *Uta stansburiana*: Constraints on the evolution of lizard life histories. J. Exp. Zool. 257: 252–264.

Sinervo, B. & Lively, C. M. (1996) The rock-paper-scissors game and the evolution of alternative male strategies. *Nature*, 380, 240–243.

Sinervo, B. & McAdam, A. G. (2008) Maturational costs of reproduction due to clutch size and ontogenetic conflict as revealed in the invisible fraction. *Proceedings of the Royal Society of London B*, 275, 629–638.

Sinervo, B. & Miles, D. B. (2010) Hormones and behavior of reptiles. In Norris, D. O. & Lopez, K. H. (Eds.) *Hormones and Reproduction of Vertebrates*. London, Academic Press.

Sinervo, B. & Svensson, E. (1998) Mechanistic and selective causes of life history trade-offs and plasticity. *Oikos*, 83, 432–442.

Sinervo, B. & Svensson, E. (2002) Correlational selection and the evolution of genomic architecture. *Heredity*, 89, 329–338.

Sinervo, B., Doughty, P., Huey, R. B. & Zamudio, K. (1992) Allometric engineering: a causal analysis of natural selection on offspring size. *Science*, 258, 1927–1930.

Sinervo, B., Svensson, E. & Comendant, T. (2000) Density cycles and an offspring quantity and quality game driven by natural selection. *Nature*, 406, 985–988.

Sinervo, B., Bleay, C. & Adamopoulou, C. (2001) Social causes of correlational selection and the resolution of a heritable throat color polymorphism in a lizard. *Evolution*, 55, 2040–2052.

Sinervo, B., Chaine, A., Clobert, J., Calsbeek, R., Hazard, L., Lancaster, L., McAdam, A. G., Alonzo, S., Corrigan, G. & Hochberg, M. E. (2006) Color morphs and genetic cycles of greenbeard mutualism and transient altruism. *Proceedings of the National Academy of Sciences of the United States of America*, 102, 7372–7377.

Sinervo, B., Huelin, B., Surget-Groba, Y., Clobert, J., Miles, D. B., Corl, A., Chaine, A. & Davis, A. (2007) Models of density-dependent genic selection and a new Rock-Paper-Scissors social system. *The American Naturalist*, 170, 663–680.

Sinervo, B., Clobert, J., Miles, D. B., McAdam, A. G. & Lancaster, L. T. (2008) The role of pleiotropy vs signaller–receiver gene epistasis in life history trade-offs: dissecting the genomic architecture of organismal design in social systems. *Heredity*, 101, 197–211.

Sisodia, S. & Singh, B. N. (2004) Size dependent sexual selection in *Drosophila ananassae*. *Genetica*, 121, 207–217.

Siva-Jothy, M. T., Moret, Y. & Rolff, J. (2005) Insect immunity: An evolutionary ecology perspective. *Advances in Insect Physiology*, 32, 1–48.

Skorupa, D. A., Dervisefendic, A., Zwiener, J. & Pletcher, S. D. (2008) Dietary composition specifies consumption, obesity, and lifespan in *Drosophila melanogaster*. *Aging Cell*, 7, 478–490.

Skulason, S. & Smith, T. B. (1995) Resource polymorphisms in vertebrates. *Trends in Ecology & Evolution*, 10, 366–370.

Slack, C., Werz, C., Wieser, D., Alic, N., Foley, A., Stocker, H., Withers, D. J., Thornton, J. M., Hafen, E. & Partridge, L. (2010) Regulation of lifespan, metabolism and stress responses by the *Drosophila* SH2B protein, Lnk. *PLoS Genetics*, 6, e1000881.

Slatkin, M. (1981) Estimating levels of gene flow in natural populations. *Annual Review of Ecology and Systematics*, 16, 393–430.

Slotte, T., Huang, H. R., Holm, K., Ceplitis, A., Onge, K. S., Chen, J., Lagercrantz, U. & Lascoux, M. (2009) Splicing variation at a *FLOWERING LOCUS C* homeolog is associated with flowering time variation in the tetraploid *Capsella bursa-pastoris*. *Genetics*, 183, 337–345.

Sly, B. J., Snoke, M. S. & Raff, R. A. (2003) Who came first – larvae or adults? Origins of bilaterian metazoan larvae. *International Journal of Developmental Biology*, 47, 623–632.

Smith, D. C. (1987) Adult recruitment in chorus frogs: Effects of size and date at metamorphosis. *Ecology*, 68, 344–350.

Smith, B. H. (1992) Life history and the evolution of human maturation. *Evolutionary Anthropology*, 1, 134–142.

Smith, C. C. & Fretwell, S. D. (1974) The optimal balance between size and number of offspring. *The American Naturalist*, 108, 499–506.

Smith, M. M. & Krupina, N. I. (2001) Conserved developmental processes constrain evolution of lungfish dentitions. *Journal of Anatomy*, 199, 161–168.

Smith, N. & Lenhoff, H. M. (1976) Regulation of pedal laceration in a sea anemone. In Mackie, G. (Ed.) *Coelenterate Ecology and Behavior*. New York, Plenum Press.

Smith, W., Priester, J. & Morais, J. (2003) PTTH-stimulated ecdysone secretion is dependent upon tyrosine phosphorylation in the prothoracic glands of *Manduca sexta*. *Insect Biochemistry & Molecular Biology*, 33, 1317–1325.

Smith, J. J., Kenney, R. D., Gagne, D. J., Frushour, B. P., Ladd, W., Galonek, H. L., Israelian, K., Song, J., Razvadauskaite, G., Lynch, A. V., Carney, D. P., Johnson, R. J., Lavu, S., Iffland, A., Elliot, P. J., Lambert, P. D., Elliston, K. O., Jirousek, M. R., Milne, J. C. & Boss, O. (2009) Small molecule activators of SIRT1 replicate signaling pathways triggered by calorie restriction *in vivo*. *BMC Systems Biology*, 3, 31.

Soars, N. A., Prowse, T. A. A. & Byrne, M. (2009) Overview of phenotypic plasticity in echinoid larvae, 'Echinopluteus transversus' type vs. typical echinoplutei. *Marine Ecology Progress Series*, 383, 113–125.

Socha, R., Sula, J., Kodrik, D. & Gelbic, I. (1991) Hormonal control of vitellogenin synthesis in *Pyrrhocoris apterus* (L.) (Heteroptera). *Journal of Insect Physiology*, 37, 805–816.

Sodergren, E., Weinstock, G. M., Davidson, E. H., Cameron, R. A., Gibbs, R. A., Angerer, R. C., Angerer, L. M., Arnone, M. I., Burgess, D. R., Burke, R. D., Coffman, J. A., Dean, M., Elphick, M. R., Ettensohn, C. A., Foltz, K. R., Hamdoun, A., Hynes, R. O., Klein, W. H., Marzluff, W., McClay, D. R., Morris, R. L., Mushegian, A., Rast, J. P., Smith, L. C., Thorndyke, M. C., Vacquier, V. D., Wessel, G. M., Wray, G., Zhang, L., Elsik, C. G., Ermolaeva, O., Hlavina, W., Hofmann, G., Kitts, P., Landrum, M. J., MacKey, A. J., Maglott, D., Panopoulou, G., Poustka, A. J., Pruitt, K., Sapojnikov, V., Song, X., Souvorov, A., Solovyev, V., Wei, Z., Whittaker, C. A., Worley, K., Durbin, K. J., Shen, Y., Fedrigo, O., Garfield, D., Haygood, R., Primus, A., Satija, R., Severson, T., Gonzalez-Garay, M. L., Jackson, A. R., Milosavljevic, A., Tong, M., Killian, C. E., Livingston, B. T., Wilt, F. H., Adams, N., Belle, R., Carbonneau, S., Cheung, R., Cormier, P., Cosson, B., Croce, J., Fernandez-Guerra, A., Geneviere, A.-M., Goel, M., Kelkar, H., Morales, J., Mulner-Lorillon, O., Robertson, A. J., Goldstone, J. V., Cole, B., Epel, D., Gold, B., Hahn, M. E., Howard-Ashby, M., Scally, M., Stegeman, J. J., Allgood, E. L., Cool, J., Judkins, K. M., McCafferty, S. S., Musante, A. M., Obar, R. A., Rawson, A. P., Rossetti, B. J., Gibbons, I. R., Hoffman, M. P., Leone, A., Istrail, S., Materna, S. C., Samanta, M. P., Stolc, V., et al. (2006) The genome of the sea urchin *Strongylocentrotus purpuratus*. *Science*, 314, 941–952.

Soetaert, K., Muthumbi, A. & Heip, C. (2002) Size and shape of ocean margin nematodes: morphological diversity and depth-related patterns. *Marine Ecology Progress Series*, 242, 179–193.

Sohal, R. S., Mockett, R. J. & Orr, W. C. (2002) Mechanisms of aging: an appraisal of the oxidative stress hypothesis. *Free Radical Biology & Medicine*, 33, 575–586.

Solari, F. & Ahringer, J. (2000) Nurd-complex genes antagonise Ras-induced vulval development in *C. elegans*. *Current Biology*, 10, 223–226.

Solari, C. A., Kessler, J. O. & Michod, R. E. (2006a) A hydrodynamics approach to the evolution of multicellularity: Flagellar motility and germ-soma differentiation in volvocalean green algae. *The American Naturalist*, 167, 537–554.

Solari, C. A., Ganguly, S., Kessler, J. O., Michod, R. E., and Goldstein, R. E. 2006b. Multicellularity and the functional interdependence of motility and molecular transport. *Proceedings of the National Academy of Sciences of the United States of America* 103: 1353–1358.

Soller, M., Bownes, M. & Kubli, E. (1999) Control of oocyte maturation in sexually mature *Drosophila* females. *Developmental Biology*, 208, 337–351.

Soma, K. K. (2006) Testosterone and aggression: berthold, birds and beyond. *Journal of Neuroendocrinology*, 18, 543–551.

Soma, K. K., Hartman, V. N., Wingfield, J. C. & Brenowitz, E. Z. (1999) Seasonal changes in androgen receptor immunoreactivity in the song nucleus HVc of a wild bird. *Journal of Comparative Neurology*, 409, 224–236.

Soma, K. K., Schlinger, B. A., Wingfield, J. C. & Saldanha, C. J. (2003) Brain aromatase, 5 alpha-reductase, and 5 beta-reductase change seasonally in wild male song sparrows: relationship to aggressive and sexual behavior. *Journal of Neurobiology*, 56, 209–221.

Sommer, U. & Gliwicz, Z. M. (1986) Long-range vertical migration of *Volvox* in tropical Lake Cahora bassa (Mozambique). *Limnology and Oceanography*, 31, 650–653.

Song, C. & Liao, S. (2000) Cholestenoic acid is a naturally occurring ligand for liver X receptor alpha. *Endocrinology*, 141, 4180–4184.

Soong, K. & Lang, J. C. (1992) Reproductive integration in reef corals. *Biological Bulletin*, 183, 418–431.

Sorensen, J. G., Nielsen, M. M. & Loeschcke, V. (2007) Gene expression profile analysis of *Drosophila melanogaster* selected for resistance to environmental stressors. *Journal of Evolutionary Biology*, 20, 1624–1636.

Soronen, P., Laiti, M., Torn, S., Harkonen, P., Patrikainen, L., Li, Y., Pulkka, A., Kurkela, R., Herrala, A., Kaija, H., Isomaa, V. & Vihko, P. (2004) Sex steroid hormone metabolism and prostate cancer. *Journal of Steroid Biochemistry & Molecular Biology* 92, 281–286.

Sowell, R. A., Hersberger, K. E., Kaufman, T. C. & Clemmer, D. E. (2007) Examining the proteome of *Drosophila* across organism lifespan. *Journal of Proteome Research*, 6, 3637–3647.

Sower, S. A. (1998) Brain and pituitary hormones of lampreys, recent findings and their evolutionary significance. *American Zoologist*, 38, 15–38.

Sower, S. A., Freamat, M. & Kavanaugh, S. I. (2009) The origins of the vertebrate hypothalamic-pituitary-gonadal (HPG) and hypothalamic-pituitary-thyroid (HPT) endocrine systems: New insights from lampreys. *General and Comparative Endocrinology*, 161, 20–29.

Spangenberg, D. B. (1965) A study of strobilation in *Aurelia aurita* under controlled conditions. *Journal of Experimental Biology*, 160, 1–10.

Spangenberg, D. B. (1967) Iodine induction of metamorphosis in *Aurelia*. *Journal of Experimental Biology*, 165, 441–449.

Spangenberg, D. B. (1972) Thyroxine induced metamorphosis in *Aurelia*. *Journal of Experimental Biology*, 178, 183–194.

Sparkman, A. & Palacios, M. G. (2009) A test of life history theories of immune defense in two ecotypes of the garter snake, *Thamnophis elegans*. *Journal of Animal Ecology*, 78, 1242–1248.

Sparkman, A. M., Arnold, S. J. & Bronikowski, A. M. (2007) An empirical test of evolutionary theories for reproductive senescence and reproductive effort in the garter snake *Thamnophis elegans*. *Proceedings of the Royal Society of London B*, 274, 943–950.

Sparkman, A. M., Vleck, C. M. & Bronikowski, A. M. (2009) Evolutionary ecology of endocrine-mediated life history variation in the garter snake *Thamnophis elegans*. *Ecology*, 90, 720–728.

Sparks, T. C., Hammock, B. D. & Riddiford, L. M. (1983) The hemolymph juvenile-hormone esterase of *Manduca sexta* (L) – inhibition and regulation. *Insect Biochemistry*, 13, 529–541.

Specker, J. L., Eales, J. G., Tagawa, M. & Tyler, W. A. (2000) Parr-smolt transformation in Atlantic salmon: Thyroid hormone deiodination in liver and brain and endocrine correlates of change in rheotactic behavior. *Canadian Journal of Zoology-Revue Canadienne de Zoologie*, 78, 696–705.

Sperry, T. S. & Thomas, P. (1999) Characterization of two nuclear androgen receptors in Atlantic croaker: comparison of their biochemical properties and binding specificities. *Endocrinology*, 140, 1602–1611.

Spindler, S. R. (2010) Caloric restriction: from soup to nuts. *Aging Research Reviews*, 9, 324–353.

Srinivasan, J., Kaplan, F., Ajredini, R., Zachariah, C., Alborn, H. T., Teal, P. E., Malik, R. U., Edison, A. S., Sternberg, P. W. & Schroeder, F. C. (2008) A blend of small molecules regulates both mating and development in *Caenorhabditis elegans*. *Nature*, 454, 1115–1118.

Stampar, S. N., Silveira, F. L. D. & Morandini, A. C. (2007) Asexual reproduction of *Nausithoe aurea* (Cnidaria, Scyphozoa, Coronatae) induced by sterile polystyrene dishes. *Brazilian Journal of Oceanography*, 55, 231–233.

Stark, K. & Schmitt, R. (2002) Genetic control of germ-soma differentiation in *Volvox carteri*. *Protist*, 153, 99–107.

Stark, K., Kirk, D. L. & Schmitt, R. (2001) Two enhancers and one silencer located in the introns of reg A control somatic cell differentiation in *Volvox reinhardtii*. *Genes & Development*, 15, 1449–1460.

Starr, R. (1970) Control of differentiation in *Volvox*. *Developmental Biology* 4, 59–100.

Stay, B., Friedel, T., Tobe, S. S. & Mundall, E. C. (1980) Feedback control of juvenile-hormone synthesis in cockroaches - possible role for ecdysterone. *Science*, 207, 898–900.

St-Cyr, J., Derome, N. & Bernatchez, L. (2008) The transcriptomics of life history trade-offs in whitefish species pairs (*Coregonus* sp.). *Molecular Ecology*, 17, 1850–1870.

Stearns, S. C. (1976) Life history tactics: a review of the ideas. *Quarterly Review of Biology*, 51, 3–47.

Stearns, S. C. (1986) Natural selection and fitness, adaptation and constraint. In Raup, D. M. & Jablonski D. (Eds.) *Patterns and Processes in the History of Life*. Berlin, Springer.

Stearns, S. C. (1989) Trade-offs in life history evolution. *Functional Ecology*, 3, 259–268.

Stearns, S. C. (1992) *The Evolution of Life Histories*. Oxford, Oxford University Press.

Stearns, S. (1994) The evolutionary links between fixed and variable traits. *Acta Paleontologica Polonica*, 38, 215–232.

Stearns, S. C. (2000) Life history evolution: successes, limitations, and prospects. *Naturwissenschaften*, 87, 476–486.

Stearns, S. C. (2005) Issues in evolutionary medicine. *American Journal of Human Biology*, 17, 131–140.

Stearns, S. & Kaiser, M. (1996) Effects on fitness components of P-element inserts in *Drosophila melanogaster*: Analysis of trade-offs. *Evolution*, 50, 795–806.

Stearns, S. C. & Koella, J. C. (1986) The evolution of phenotypic plasticity in life-history traits: predictions of reaction norms for age and size at maturity. *Evolution*, 40, 893–913.

Stearns, S. C. & Koella, J. C. (Eds.) (2008) *Evolution in Health and Disease*. Oxford, Oxford University Press.

Stearns, S. C. & Magwene, P. (2003) The naturalist in a world of genomics. *The American Naturalist*, 161, 171–180.

Stearns, S. C. & Partridge, L. (2001) The genetics of aging in *Drosophila*. In Masoro, E. & Austad, S. (Eds.) *Handbook of the Biology of Aging*, 5th edn. New York, Academic Press.

Stearns, S. C., Ackermann, M., Doebele, M. & Kaiser, M. (2000) Experimental evolution of aging, growth and reproduction in fruit flies. *Proceedings of the National Academy of Sciences of the United States of America*, 97, 3309–3313.

Stearns, S. C., Nesse, M. N. & Haig, D. (2008) Introducing evolutionary thinking for medicine. In Stearns, S. C. & Koella, J. C. (Eds.) *Evolution in Health and Disease*. Oxford, New York, Oxford University Press.

Steenstrup, J. J. S. 1845. On the Alternation of Generations. Translated from the German version of C. H. Lorenzen by G. Busk. Ray Society, London.

Stefansson, S. O., Bjornsson, B., Ebbesson, L. O. E., McCormick (2008) Smoltification. In Finn, R. N. & Kapoor, B. G. (Eds.) *Fish Larval Physiology*. New Delhi, Enfield (NH), Science Publishers, Inc. & IBH Publishing Co. Pvt. Ltd.

Stéhlík, J., Závodská, R., Shimada, K., Sauman, I. & Kostál, V. (2008) Photoperiodic induction of diapause requires regulated transcription of timeless in the larval brain of *Chymomyza costata*. *Journal of Biological Rhythms*, 23, 129–139.

Steigenga, M. J., Hoffmann, K. H. & Fischer, K. (2006) Effects of the juvenile hormone mimic pyriproxyfen on female reproduction and longevity in the butterfly *Bicyclus anynana*. *Entomological Science*, 9, 269–279.

Steinacker, J. M., Brkic, M., Simsch, C., Nething, K., Kresz, A., Prokopchuk, O. & Liu, Y. (2005) Thyroid hormones, cytokines, physical training and metabolic control. *Hormone and Metabolic Research*, 37, 538–544.

Stephenson, T. A. (1935) *The British Sea Anemones, Volume II*. London, The Ray Society.

Stern, D. L. (2000) Perspective: Evolutionary developmental biology and the problem of variation. *Evolution*, 54, 1079–1091.

Stern, D. L. (2010) *Evolution, Development, and the Predictable Genome*. Greenwood Village, Colorado, Roberts and Company Publishers.

St Germain, D. L., Schwartzman, R. A., Croteau, W., Kanamori, A., Wang, Z., Brown, D. D. & Galton, V. A. (1994) A thyroid hormone-regulated gene in *Xenopus laevis* encodes a type III iodothyronine 5-deiodinase. *Proceedings of the National Academy of Sciences of the United States of America*, 91, 7767–7771.

Stieper, B. C., Kupershtok, M., Driscoll, M. V. & Shingleton, A. W. (2008) Imaginal discs regulate developmental timing in *Drosophila melanogaster*. *Developmental Biology*, 321, 18–26.

Stinchcombe, J. R. & Hoekstra, H. E. (2008) Combining population genomics and quantitative genetics: finding the genes underlying ecologically important traits. *Heredity*, 100, 158–170.

Stinchcombe, J. R., Dorn, L. A. & Schmitt, J. (2004) Flowering time plasticity in *Arabidopsis thaliana*: a reanalysis of Westerman & Lawrence (1970). *Journal of Evolutionary Biology*, 17, 197–207.

Stoleru, D., Nawathean, P., De La Paz Fernandez, M., Menet, J. S., Fernanda Ceriani, M. & Rosbash, M. (2007) The *Drosophila* circadian network is a seasonal timer. *Cell*, 129, 207–219.

Storey, K. B. (2006) Reptile freeze tolerance: Metabolism and gene expression. *Cryobiology*, 52, 1–16.

Storey, K. B. (2007) Anoxia tolerance in turtles: Metabolic regulation and gene expression. *Comparative Biochemistry and Physiology A*, 147, 236–276.

Storey, K. B. (2010) Out cold: Biochemical regulation of mammalian hibernation – a mini-review. *Gerontology*, 56, 220–230.

Storey, A. E., Walsh, C. J., Quinton, R. L. & Wynn-Edwards, K. E. (2000) Hormonal correlates of paternal responsiveness in new and expectant fathers. *Evolution and Human Behavior*, 21, 79–95.

Strassman, B. (1996) Energy economy in the evolution of menstruation. *Evolutionary Anthropology*, 5, 157–164.

Strathmann, R. R. (1971) The feeding behavior of planktotrophic echinoderm larvae: mechanisms, regulation, and rates of suspension feeding. *Journal of Experimental Marine Biology and Ecology*, 6, 109–160.

Strathmann, R. R. (1975) Larval feeding in echinoderms. *American Zoologist*, 15, 717–730.

Strathmann, R. R. (1978) Evolution and loss of feeding larval stages of marine invertebrates. *Evolution*, 32, 894–906.

Strathmann, R. R. (1985) Feeding and nonfeeding larval development and life history evolution in marine invertebrates. *Annual Review of Ecology and Systematics*, 16, 339–361.

Strathmann, R. R. (1990) Why life histories evolve differently in the sea. *American Zoologist*, 30, 197–207.

Strathmann, R. R. (2007) Time and extent of ciliary response to particles in a non-filtering feeding mechanism. *Biological Bulletin*, 212, 93–103.

Strathmann, R. R., Fenaux, L. & Strathmann, M. F. (1992) Heterochronic developmental plasticity in larval sea-urchins and its implications for evolution of nonfeeding larvae. *Evolution*, 46, 972–986.

Strathmann, R. R., Fenaux, L., Sewell, M. A. & Strathmann, M. F. (1993) Abundance of food affects relative size of larval and postlarval structures of a molluscan veliger. *Biological Bulletin*, 185, 232–239.

Strathmann, R. R., Foley, G. P. & Hysert, A. N. (2008) Loss and gain of the juvenile rudiment and metamorphic competence during starvation and feeding of bryozoan larvae. *Evolution & Development*, 10, 731–736.

Stuart, L. M. & Ezekowitz, R. A. (2008) Phagocytosis and comparative innate immunity: learning on the fly. *Nature Reviews Immunology*, 8, 131–141.

Su, Y., Li, E., Geiss, G. K., Longabaugh, W. J. R., Kämer, A. & Davidson, E. H. (2009) A perturbation model of the gene regulatory network for oral and aboral ectoderm specficiation in the sea urchin embryo. *Developmental Biology*, 329, 410–421.

Suarez-Lopez, P., Wheatley, K., Robson, F., Onouchi, H., Valverde, F. & Coupland, G. (2001) CONSTANS mediates between the circadian clock and the control of flowering in *Arabidopsis*. *Nature*, 410, 1116–1120.

Suh, Y., Atzmon, G., Cho, M. O., Hwang, D., Liu, B., Leahy, D. J., Barzilai, N. & Cohen, P. (2008) Functionally significant insulin-like growth factor I receptor mutations in centenarians. *Proceedings of the National Academy of Sciences of the United States of America*, 105, 3438–3442.

Sula, J., Kodrik, D. & Socha, R. (1995) Hexameric haemolymph protein related to adult diapause in the red firebug, *Pyrrhocoris apterus* (L.) (Heteroptera). *Journal of Insect Physiology*, 41, 793–800.

Sultan, S. E. (2007) Development in context: the timely emergence of eco-devo. *Trends in Ecology & Evolution*, 22, 575–582.

Sutter, N. B., Bustamante, C. D., Chase, K., Gray, M. M., Zhao, K., Zhu, L., Padhukasahasram, B., Karlins, E., Davis, S., Jones, P. G., Quignon, P., Johnson, G. S., Parker, H. G., Fretwell, N., Mosher, D. S., Lawler, D. F., Satyaraj, E., Nordborg, M., Lark, K. G., Wayne, R. K. & Ostrander, E. A. (2007) A single IGF-1 allele is a major determinant of small size in dogs. *Science*, 316, 112–115.

Suzuki, Y., Truman, J. W. and Riddiford, L. M. (2008) The role of Broad in the development of *Tribolium castaneum*: implication s for the evolution of the holometabolous insect pupa. *Development*, 135, 569–577.

Svensson, E., Sinervo, B. & Comendant, T. (2001a) Condition, genotype-by-environment interaction, and correlational selection in lizard life history morphs. *Evolution*, 55, 2053–2069.

Svensson, E., Sinervo, B. & Comendant, T. (2001b) Density-dependent competition and selection on immune function in genetic lizard morphs. *Proceedings of the National Academy of Sciences of the United States of America*, 98, 12561–12565.

Svensson, E., Sinervo, B. & Comendant, T. (2002) Mechanistic and experimental analysis of condition and reproduction in a polymorphic lizard. *Journal of Evolutionary Biology*, 15, 1034–1047.

Svensson, E., McAdam, A. G. & Sinervo, B. (2009) Intralocus sexual conflict over immune defense resolves gender load and affects sex-specific signaling in a natural lizard population. *Evolution*, 63, 3124–3135.

Swanson, R. L., De Nys, R., Huggett, M. J., Green, J. K. & Steinberg, P. D. (2006) In situ quantification of a natural settlement cue and recruitment of the Australian sea urchin *Holopneustes purpurascens*. *Marine Ecology Progress Series*, 314, 1–14.

Swanson, R. L., Marshall, D. J. & Steinberg, P. D. (2007) Larval desperation and histamine: how simple responses can lead to complex changes in larval behaviour. *The Journal of Experimental Biology*, 210, 3228–3235.

Swarup, K., Alonso-Blanco, C., Lynn, J. R., Michaels, S. D., Amasino, R. M., Koornneef, M. & Millar, A. J. (1999) Natural allelic variation identifies new genes in the *Arabidopsis* circadian system. *Plant Journal*, 20, 67–77.

Swoap, S. J. & Weinshenker, D. (2008) Norepinephrine controls both torpor initiation and emergence via distinct mechanisms in the mouse. *PLoS One*, 3, e4038.

Sze, J. Y., Victor, M., Loer, C., Shi, Y. & Ruvkun, G. (2000) Food and metabolic signalling defects in a *C. elegans* serotonin-synthesis mutant. *Nature*, 403, 560–564.

Szewczyk, N. J., Udranszky, I. A., Kozak, E., Sunga, J., Kim, S. K., Jacobson, L. A. & Conley, C. A. (2006) Delayed development and lifespan extension as features of metabolic lifestyle alteration in *C. elegans* under dietary restriction. *Journal of Experimental Biology*, 209, 4129–4139.

Tagawa, M. & Hirano, T. (1987) Presence of thyroxine in eggs and changes in its content during early development of chum salmon, *Oncorhynchus keta*. *General and Comparative Endocrinology*, 68, 129–135.

Tagawa, M. & Hirano, T. (1991) Effects of thyroid hormone deficiency in eggs on early development of the medaka, *Oryzias latipes*. *Journal of Experimental Zoology*, 257, 360–366.

Tagawa, M., Miwa, S., Inui, Y., de Jesus, E. G. & Hirano, T. (1990a) Changes in thyroid hormone concentrations during early development and metamorphosis of the flounder, *Paralichthys olivaceus*. *Zoological Science*, 7, 93–96.

Tagawa, M., Tanaka, M., Matsumoto, S. & Hirano, T. (1990b) Thyroid hormones in eggs of various freshwater, marine and diadromous teleosts and their changes during egg development. *Fish Physiology and Biochemistry*, 8, 515–520.

Taguchi, A., Wartschow, L. M. & White, M. F. (2007) Brain IRS2 signaling coordinates lifespan and nutrient homeostasis. *Science*, 317, 369–372.

Takada, Y., Ye, X. & Simon, S. (2007) The integrins. *Genome Biology*, 8, 215.

Takahashi, N., Yoshihama, K., Kikuyama, S., Yamamoto, K., Wakabayashi, K. & Kato, Y. (1990) Molecular cloning and nucleotide sequence analysis of complementary DNA for bullfrog prolactin. *Journal of Molecular Endocrinology*, 5, 281–287.

Takano, T. S. (1998) Loss of notum macrochaetae as an interspecific hybrid anomaly between *Drosophila melanogaster* and *D. simulans*. *Genetics*, 149, 1435–1450.

Takken, W. & Knols, B. G. J. (1999) Odor-mediated behavior of Afrotropical malaria mosquitoes. *Annual Review of Entomology*, 44, 131–157.

Tan, F. C. & Swain, S. M. (2006) Genetics of flower initiation and development in annual and perennial plants. *Physiologia Plantarum*, 128, 8–17.

Tanaka, K. & Truman, J.W. (2007) Molecular patterning mechanism underlying metamorphosis of the thoracic leg in *Manduca sexta*. *Developmental Biology*, 305, 539–550.

Tang, X. C., Liu, X. C., Zhang, Y., Zhu, P. & Lin, H. R. (2008) Molecular cloning, tissue distribution and expression profiles of thyroid hormone receptors during embryogenesis in orange-spotted grouper (*Epinephelus coioides*). *General and Comparative Endocrinology*, 159, 117–124.

Tata, J. R. (1958) A cellular thyroxine-binding protein fraction. *Biochimica et Biophysica Acta*, 28, 91–94.

Tata, J. R. (1996) Amphibian metamorphosis: An exquisite model for hormonal regulation of postembryonic development in vertebrates. *Development, Growth & Differentiation*, 38, 223–231.

Tata, J. R., Kawahara, A. & Baker, B. S. (1991) Prolactin inhibits both thyroid hormone-induced morphogenesis and cell death in cultured amphibian larval tissues. *Developmental Biology*, 146, 72–80.

Tatar, M. (1999) Transgenes in the analysis of lifespan and fitness. *The American Naturalist*, 154, S67–S81.

Tatar, M. (2004) The neuroendocrine regulation of *Drosophila* aging. *Experimental Gerontology*, 39, 1745–1750.

Tatar, M. & Carey, J. R. (1995) Nutrition mediates reproductive trade-offs with age-specific mortality in the beetle *Callosobruchus maculatus*. *Ecology*, 76, 2066–2073.

Tatar, M. & Yin, C. M. (2001) Slow aging during insect reproductive diapause: why butterflies, grasshoppers and flies are like worms. *Experimental Gerontology*, 36, 723–738.

Tatar, M., Chien, S. A. & Priest, N. K. (2001a) Negligible senescence during reproductive dormancy in *Drosophila melanogaster*. *The American Naturalist*, 158, 248–258.

Tatar, M., Kopelman, A., Epstein, D., Tu, M. P., Yin, C. M. & Garofalo, R. S. (2001b) A mutant *Drosophila* insulin receptor homolog that extends life-span and impairs neuroendocrine function. *Science*, 292, 107–110.

Tatar, M., Bartke, A. & Antebi, A. (2003) The endocrine regulation of aging by insulin-like signals. *Science*, 299, 1346–1351.

Tauber, M. J., Tauber, C. A. & Masaki, S. (1986) *Seasonal Adaptations of Insects*. New York, Oxford University Press.

Tauber, E., Zordan, M., Sandrelli, F., Pegoraro, M., Osterwalder, N., Breda, C., Daga, A., Selmin, A., Monger, K., Benna, C., Rosato, E., Kyriacou, C. P. & Costa, R. (2007) Natural selection favors a newly derived *timeless* allele in *Drosophila melanogaster*. *Science*, 316, 1895–1898.

Temin, G., Zander, M. & Roussel, J. P. (1986) Physiochemical (Gc-MS) measurements of juvenile hormone III titres during embryogenesis of *Locusta migratoria*. *International Journal of Invertebrate Reproduction and Development* 9, 105–112.

Tennekoon, K. H. & Karunanayake, E. H. (1993) Serum FSH, LH, and testosterone concentrations in presumably fertile men: effect of age. *International Journal of Fertility*, 38, 108–112.

Teotónio, H., Chelo, I. M., Bradic, M., Rose, M. R. & Long, A. D. (2009) Experimental evolution reveals natural selection on standing genetic variation. *Nature Genetics*, 41, 251–257.

Terashima, J. & Bownes, M. (2004) Translating available food into the number of eggs laid by *Drosophila melanogaster*. *Genetics*, 167, 1711–1719.

Terashima, J., Takaki, K., Sakurai, S. & Bownes, M. (2005) Nutritional status affects 20-hydroxyecdysone concentration and progression of oogenesis in *Drosophila melanogaster*. *Journal of Endocrinology*, 187, 69–79.

Terry, D. F., Wyszynski, D. F., Nolan, V. G., Atzmon, G., Schoenhofen, E. A., Pennington, J. Y., Andersen, S. L., Wilcox, M. A., Farrer, L. A., Barzilai, N., Baldwin, C. T. & Asea, A. (2006) Serum heat shock protein 70 level as a biomarker of exceptional longevity. *Mechanisms of Ageing and Development*, 127, 862–868.

Thibaudeau, G. & Altig, R. (1999) Endotrophic anurans: Development and Evolution. In McDiarmid, R. W. (Ed.) *Tadpoles: The Biology of Anuran Larvae*. Chicago, Chicago University Press.

Thisse, C., Degrave, A., Kryukov, G. V., Gladyshev, V. N., Obrecht-Pflumio, S., Krol, A., Thisse, B. & Lescure, A. (2003) Spatial and temporal expression patterns of selenoprotein genes during embryogenesis in zebrafish. *Gene Expression Patterns*, 3, 525–532.

Thomas, T. D. (2008) The effect of in vivo and in vitro applications of ethrel and GA(3) on sex expression in bitter melon (*Momordica charantia* L.). *Euphytica*, 164, 317–323.

Thomas, J. H., Birnby, D. A. & Vowels, J. J. (1993) Evidence for parallel processing of sensory information controlling dauer formation in *C. elegans*. *Genetics*, 134, 1105–1117.

Thorpe, J. E. (1977) Bimodal distribution of length of juvenile Atlantic Salmon (Salmo salar L.) under artificial rearing conditions. *Journal of Fish Biology*, 11, 175–184.

Thorpe, J. E. (1986) Age at first maturity in Atlantic salmon, *Salmo salar*: freshwater period influences and conflicts with smolting. In Meerburg, D. J. (Ed.) *Salmonid Age at Maturity*. Ottawa, National Research Council of Canada.

Thorpe, J. E. (1987) Smolting versus residency: developmental conflict in salmonids. *American Fisheries Society Symposium*, 1, 244–252.

Thorpe, J. E. (1994a) An alternative view of smolting in salmonids. *Aquaculture*, 121, 105–113.

Thorpe, J. E. (1994b) Reproductive strategies in Atlantic salmon, *Salmo salar* L. *Aquaculture*, 25, 77–87.

Thorpe, J. E., Mangel, M., Metcalfe, N. B. & Huntingford, F. A. (1998) Modelling the proximate basis of salmonid life history variation, with application to Atlantic salmon, *Salmo salar* L. *Evolutionary Ecology*, 12, 581–599.

Tian, D., Traw, M. B., Chen, J. Q., Kreitman, M. & Bergelson, J. (2003) Fitness costs of R-gene-mediated resistance in *Arabidopsis thaliana*. *Nature*, 423, 74–77.

Tissenbaum, H. A. & Guarente, L. (2001) Increased dosage of a *sir-2* gene extends lifespan in *Caenorhabditis elegans*. *Nature*, 410, 227–230.

Tissenbaum, H. A., Hawdon, J., Perregaux, M., Hotez, P., Guarente, L. & Ruvkun, G. (2000) A common muscarinic pathway for diapause recovery in the distantly related nematode species *Caenorhabditis elegans* and *Ancylostoma caninum*. *Proceedings of the National Academy of Sciences of the United States of America*, 97, 460–465.

Todesco, M., Balasubramanian, S., Hu, T. A., Traw, M. B., Horton, M., Epple, P., Kuhns, C., Sureshkumar, S., Schwrartz, C., Lanz, C., Laitinen, R. A. E., Huang, Y., Chory, J., Lipka, V., Borevitz, J. O., Dangl, J. L., Bergelson, J., Magnus, N. & Weigel, D. (2010) Natural allelic variation underlying a major fitness trade-off in *Arabidopsis thaliana*. *Nature*, 465, 632–636.

Toivonen, J. M. & Partridge, L. (2009) Endocrine regulation of aging and reproduction in *Drosophila*. *Molecular and Cellular Endocrinology*, 299, 39–50.

Toivonen, J. M., Walker, G. A., Martinez-Diaz, P., Bjedov, I., Driege, Y., Jacobs, H. T., Gems, D. & Partridge, L. (2007) No influence of Indy on lifespan in *Drosophila* after correction for genetic and cytoplasmic background effects. *PLoS Genetics*, 3, e95.

Tomasello, M. (1999) *The Cultural Origins of Human Cognition*. Cambridge, Harvard University Press.

Tooby, J. & Cosmides, L. (1992) The psychological foundations of culture. In Barkow, J. H., Cosmides, L. & Tooby, J. (Eds.) *The Adapted Mind*. New York, Oxford University Press.

Toomajian, C., Hu, T. T., Aranzana, M. J., Lister, C., Tang, C., Zheng, H., Zhao, K., Calabrese, P., Dean, C. & Nordborg, M. (2006) A nonparametric test reveals selection for rapid flowering in the *Arabidopsis* genome. *PLoS Biology*, 4, e137.

Toonen, R. J. & Pawlik, J. R. (2001) Foundations of gregariousness: A dispersal polymorphism among the planktonic larvae of a marine invertebrate. *Evolution*, 55, 2439–2454.

Toonen, R. J. & Tyre, A. J. (2007) If larvae were smart: a simple model for optimal settlement behavior of competent larvae. *Marine Ecology Progress Series*, 349, 43–61.

Touhara, K. & Vosshall, L. B. (2009) Sensing odorants and pheromones with chemosensory receptors. *Annual Review of Physiology*, 71, 307–332.

Tower, J. (2004) There's a problem in the furnace. *Science of Aging Knowledge Environment*, 2004, pe1.

Townsend, C. R. & Calow, P. (1981) *Physiological Ecology. An Evolutionary Approach to Resource use.* Oxford, Blackwell Scientific Publications.

Travers, S. E., Smith, M. D., Bai, J., Hulbert, S. H., Leach, J. E., Schnable, P. S., Knapp, A. K., Milliken, G. A., Fay, P. A., Saleh, A. & Garrett, K. A. (2007) Ecological genomics: making the leap from model systems in the lab to native populations in the field. *Frontiers in Ecology and the Environment*, 5, 19–24.

Trembley, A. 1744. Mémoires pour server a l'histoire d'un genre de polypes de l'eau douce, à bras en forme de cornes. Verbeek, Leiden.

Trivers, R. L. & Willard, D. E. (1973) Natural selection of parental ability to vary sex-ratio of offspring. *Science*, 179, 90–92.

Troemel, E. R., Chu, S. W., Reinke, V., Lee, S. S., Ausubel, F. M. & Kim, D. H. (2006) p38 MAPK regulates expression of immune response genes and contributes to longevity in *C. elegans*. *PLoS Genetics*, 2, e183.

Troen, A. M., French, E. E., Roberts, J. F., Selhub, J., Ordovas, J. M., Parnell, L. D. & Lai, C. Q. (2007) Lifespan modification by glucose and methionine in *Drosophila melanogaster* fed a chemically defined diet. *Age*, 29, 29–39.

True, J. R. & Carroll, S. B. (2002) Gene co-option in physiological and morphological evolution. *Annual Review of Cell and Developmental Biology*, 18, 53–80.

True, J. R. & Haag, E. S. (2001) Developmental system drift and flexibility in evolutionary trajectories. *Evolution & Development*, 3, 109–119.

Truman, J. W. (1972) Physiology of insect rhythms. 1. Circadian organization of endocrine events underlying molting cycle of larval tobacco hornworms. *Journal of Experimental Biology*, 57, 805–820.

Truman, J. W. & Riddiford, L. M. (1974) Physiology of insect rhythms. 3. The temporal organization of the endocrine events underlying pupation of the tobacco hornworm. *Journal of Experimental Biology*, 60, 371–382.

Truman, J. W. & Riddiford, L. M. (1999) The origins of insect metamorphosis. *Nature*, 401, 447–452.

Truman, J. W. & Riddiford, L. M. (2002) Endocrine insights into the evolution of metamorphosis in insects. *Annual Review of Entomology*, 47, 467–500.

Truman, J. W. & Riddiford, L. M. (2007) The morphostatic actions of juvenile hormone. *Insect Biochemistry & Molecular Biology*, 37, 761–770.

Truman, J. W., Hiruma, K., Allee, J. P., MacWhinnie, S. G. B., Champlin, D. T. & Riddiford, L. M. (2006) Juvenile hormone is required to couple imaginal disc formation with nutrition in insects. *Science*, 312, 1385–1388.

Tsong, A. E., Tuch, B. B., Li, H. & Johnson, A. D. (2006) Evolution of alternative transcriptional circuits with identical logic. *Nature*, 443, 415–420.

Tsutsui, K., Saigoh, E., Yin, H., Ubuka, T., Chowdhury, V. S., Osugi, T., Uena, K. & Sharp, P. J. (2009) A new key neurohormone controlling reproduction, gonado-trophin-lnhibitory hormone in birds: discovery, progress and prospects. *Journal of Neuroendocrinology*, 21, 271–275.

Tu, M. P. & Tatar, M. (2003) Juvenile diet restriction and the aging and reproduction of adult *Drosophila melanogaster*. *Aging Cell*, 2, 327–333.

Tu, M. P., Yin, C. M. & Tatar, M. (2005) Mutations in insulin signaling pathway alter juvenile hormone synthesis in *Drosophila melanogaster*. *General and Comparative Endocrinology*, 142, 347–356.

Tu, Q., Brown, C. T., Davidson, E. H. & Oliveri, P. (2006) Sea urchin Forkhead gene family: phylogeny and embryonic expression. *Developmental Biology*, 300, 49–62.

Tullet, J. M., Hertweck, M., An, J. H., Baker, J., Hwang, J. Y., Liu, S., Oliveira, R. P., Baumeister, R. & Blackwell, T. K. (2008) Direct inhibition of the longevity-promoting factor SKN-1 by insulin-like signaling in *C. elegans*. *Cell*, 132, 1025–1038.

Tulving, E. (2002) Episodic memory: from mind to brain. *Annual Review of Psychology*, 53, 1–25.

Turner, P., Cooper, V. & Lenski, R. (1998) Tradeoff between horizontal and vertical modes of transmission in bacterial plasmids. *Evolution*, 52, 315–329.

Turner, T. L., Levine, M. T., Eckert, M. L. & Begun, D. J. (2008) Genomic analysis of adaptive differentiation in *Drosophila melanogaster*. *Genetics*, 179, 455–473.

Turri, M. G., Henderson, N. D., Defries, J. C. & Flint, J. (2001) Quantitative trait locus mapping in laboratory mice derived from a replicated selection experiment for open-field activity. *Genetics*, 158, 1217–1226.

Tzou, P., Ohresser, S., Ferrandon, D., Capovilla, M., Reichhart, J. M., Lemaitre, B., Hoffmann, J. A. & Imler, J. L. (2000) Tissue-specific inducible expression of antimicrobial peptide genes in *Drosophila* surface epithelia. *Immunity*, 13, 737–748.

Uhlirova, M., Foy, B. D., Beaty, B. J., Olson, K. E., Riddiford, L. M. & Jindra, M. (2003) Use of Sindbus virus-mediated RNA interference to demonstrate a conserved role of Broad-Complex in insect metamorphosis. *Proceedings of the National Academy of Sciences of the United States of America*, 100, 15607–15612.

Ujvari, B. & Madsen, T. (2009) Short telomeres in hatchling snakes: erythrocyte telomere dynamics and longevity in tropical pythons. *PLoS One*, 4, e7493.

Umen, J. G. & Goodenough, U. W. (2001) Control of cell division by a retinoblastoma protein homolog in *Chlamydomonas*. *Genes & Development*, 15, 1652–1661.

Ungar, P., Grine, F. E., Teaford, M. F. & El Zaatari, S. (2006) Dental microwear and diets in early African *Homo*. *Journal of Human Evolution*, 50, 78–95.

Ungerer, M. C. & Reiseberg, L. H. (2003) Genetic architecture of a selection response in *Arabidopsis thaliana*. *Evolution*, 57, 2531–2539.

Ungerer, M. C., Linder, C. R. & Reiseberg, L. H. (2003) Effects of genetic background on response to selection in experimental populations of *Arabidopsis thaliana*. *Genetics*, 163, 277–286.

Urrutia, P. M., Okamoto, K. & Fusetani, N. (2004) Acetylcholine and serotonin induce larval metamorphosis of the Japanese short-neck clam *Ruditapes philippinarum*. *Journal of Shellfish Research*, 23, 93–100.

Uvnas-Moberg, K. (1998) Antistress pattern induced by oxytocin. *News in Psychological Sciences*, 13, 22–26.

Vacher, C., Garcia-Oroz, L. & Rubinsztein, D. C. (2005) Overexpression of yeast hsp104 reduces polyglutamine aggregation and prolongs survival of a transgenic mouse model of Huntington's disease. *Human Molecular Genetics*, 14, 3425–3433.

Vagelli, A. A. (2007) New observations on the asexual reproduction of *Aurelia aurita* (Cnidaria, Scyphozoa) with comments on its life cycle and adaptive significance. *Invertebrate Zoology*, 4, 111–127.

Valdivia, P. A., Zenteno-Savin, T., Gardner, S. C. & Aguirre, A. A. (2007) Basic oxidative stress metabolites in eastern Pacific green turtles (*Chelonia mydas agassizii*). *Comparative Biochemistry and Physiology C*, 146, 111–117.

Vale, P. F. & Little, T. J. (2009) Measuring parasite fitness under genetic and thermal variation. *Heredity*, 103, 102–109.

Van Breusegem, F., Vranova, E., Dat, J. & Inze, D. (2001) The role of active oxygen species in plant signal transduction. *Plant Science*, 161, 405–414.

Vance, R. R. (1973a) More on reproductive strategies in marine benthic invertebrates. *The American Naturalist*, 107, 353–361.

Vance, R. R. (1973b) Reproductive strategies in marine benthic invertebrates. *The American Naturalist*, 107, 339–352.

Van Den Beld, A. W., De Jong, F. H., Grobbee, D. E., Pols, H. A. & Lamberts, S. W. (2000) Measures of bioavailable serum testosterone and estradiol and their relationships with muscle strength, bone density, and body composition in elderly men. *Journal of Clinical Endocrinology and Metabolism*, 85, 3276–3282.

Vandenbergh, J. G. (1967) Effect of the presence of a male on sexual maturation of female mice. *Endocrinology*, 81, 345–349.

Vandenbergh, J. G. (1973) Acceleration and inhibition of puberty in female mice by pheromones. *Journal of Reproduction and Fertility*, 19, 411–419.

Van Den Elzen, P., Garg, S., León, L., Brigl, M., Leadbetter, E. A., Gumperz, J. E., Dascher, C. C., Cheng, T. Y., Sacks, F. M., Illarionov, P. A., Besra, G. S., Kent, S. C., Moody, D. B. & Brenner, M. B. (2005) Apolipoprotein-mediated pathways of lipid antigen presentation. *Nature*, 437, 906–910.

Van Der Have, T. M. & De Jong, G. (1996) Adult size in ectotherms: Temperature effects on growth and differentiation. *Journal of Theoretical Biology*, 183, 329–340.

Van Der Meij, L., Buunk, A. P., Van De Sande, J. P. & Alicia, S. (2008) The presence of a woman increases testosterone in aggressive dominant men. *Hormones and Behavior*, 54, 640–644.

Van Dyck, H. & Wiklund, C. (2002) Seasonal butterfly design: morphological plasticity among three developmental pathways relative to sex, flight and thermoregulation. *Journal of Evolutionary Biology*, 15, 216–225.

Van Heemst, D., Beekman, M., Mooijaart, S. P., Heijmans, B. T., Brandt, B. W., Zwaan, B. J., Slagboom, P. E. & Westendorp, R. G. (2005a) Reduced insulin/IGF-1 signaling and human longevity. *Aging Cell*, 4, 79–85.

Van Heemst, D., Mooijaart, S. P., Beekman, M., Schreuder, J., De Craen, A. J., Brandt, B. W., Slagboom, P. E. & Westendorp, R. G. (2005b) Variation in the human TP53 gene affects old age survival and cancer mortality. *Experimental Gerontology*, 40, 11–15.

Van Noordwijk, A. J. & De Jong, G. (1986) Acquisition and allocation of resources - Their influence on variation in life history tactics. *The American Naturalist*, 128, 137–142.

Van Straalen, N. M. & Hoffmann, A. A. (2000) Review of evidence for physiological costs of tolerance to toxicants. In Kammenga, J. E. & Laskowski, R. (Eds.) *Demography in Ecotoxicology*. Chichester, John Wiley.

Van Straalen, N. M. & Roelofs, D. (2006) *Introduction to Ecological Genomics*. Oxford, Oxford University Press.

Van't Land, J., Van Putten, P., Villarroel, H., Kamping, A. & Van Delden, W. (1995) Latitudinal variation in wing length and allele frequencies for Adh and a-Gpdh in populations of *Drosophila melanogaster* from Ecuador and Chile. *Drosophila Information Service*, 76.

Vaziri, H., Dessain, S. K., Ng Eaton, E., Imai, S. I., Frye, R. A., Pandita, T. K., Guarente, L. & Weinberg, R. A. (2001) hSIR2(SIRT1) functions as an Nad-dependent p53 deacetylase. *Cell*, 107, 149–159.

Vellai, T., Takacs-Vellai, K., Zhang, Y., Kovacs, A. L., Orosz, L. & Muller, F. (2003) Influence of TOR kinase on lifespan in *C. elegans*. *Nature*, 426, 620.

Vernace, V. A., Arnaud, L., Schmidt-Glenewinkel, T. & Figueiredo-Pereira, M. E. (2007) Aging perturbs 26S proteasome assembly in *Drosophila melanogaster*. *FASEB Journal*, 21, 2672–2682.

Verrelli, B. C. & Eanes, W. F. (2001a) Clinal Variation for amino acid polymorphisms at the *Pgm* locus in *Drosophila melanogaster*. *Genetics*, 157, 1649–1663.

Verrelli, B. C. & Eanes, W. F. (2001b) The functional impact of *Pgm* amino acid polymorphism on glycogen content in *Drosophila melanogaster*. *Genetics*, 159, 201–210.

Via, S., Gomulkiewicz, R., de Jong, G., Scheiner, S. M., Schlichting, C. D. & van Tienderen, P. H. (1995) Adaptive Phenotypic Plasticity: Consensus and Controversy. *Trends in Ecology and Evolution*, 10, 212–216.

Videan, E. N., Fritz, J., Heward, C. B. & Murphy, J. (2006) The effects of aging on hormone and reproductive cycles in female chimpanzees *(Pan troglodytes)*. *Comparative Medicine*, 56, 291–299.

Visser, W. E., Frieserna, E. C. H., Jansen, J. & Visser, T. J. (2008) Thyroid hormone transport in and out of cells. *Trends in Endocrinology & Metabolism*, 19, 50–56.

Viswanathan, S. R., Daley, G. Q. & Gregory, R. I. (2008) Selective blockade of microRNA processing by Lin28. *Science*, 320, 97–100.

Vize, P. D. (2009) Transcriptome analysis of the circadian regulatory network in the coral *Acropora millepora*. *Biological Bulletin*, 216, 131–137.

Vlaeminck-Guillem, V., Safi, R., Guillem, P., Leteurtre, E., Duterque-Coquillaud, M. & Laudet, V. (2006) Thyroid hormone receptor expression in the obligatory paedomorphic salamander *Necturus maculosus*. *International Journal of Developmental Biology*, 50, 553–560.

von Dassow, G., Meir, E., Munro, E. M. & Odell, G. M. (2000) The segment polarity network is a robust developmental module. *Nature*, 406, 188–192.

Von Zglinicki, T. & Sitte, N. (2003) Free Radical Production and Antioxidant Defense: A Primer. In Von Zglinicki, T. (Ed.) *Aging at the Molecular Level*. Dordrecht, Kluwer.

Vourisalo, T. & Muitkainen, P. (1999) Preface. In Vourisalo, T. & Muitkainen, P. (Eds.) *Life History Evolution in Plants*. Boston, Kluwer.

Wagner, G. P., Pavlicev, M. & Cheverud, J. M. (2007) The road to modularity. *Nature Reviews Genetics*, 8, 921–931.

Wahl, M. (1985) *Metridium senile*: dispersion and small scale colonization by the combined strategy of locomotion and asexual reproduction (laceration). *Marine Ecology Progress Series*, 26, 271–277.

Wakelin, D. & Blackwell, J. (Eds.) (1988) *Genetics of Resistance to Bacterial and Parasite Infection*. London, Taylor & Francis.

Walford, R. L., Harris, S. B. & Gunion, M. W. (1992) The calorically restricted low-fat nutrient-dense diet in Biosphere 2 significantly lowers blood glucose, total leukocyte count, cholesterol, and blood pressure in humans. *Proceedings of the National Academy of Sciences of the United States of America*, 89, 11533–11537.

Walford, R. L., Mock, D., Verdery, R. & MacCallum, T. (2002) Calorie restriction in biosphere 2: alterations in physiologic, hematologic, hormonal, and biochemical parameters in humans restricted for a 2-year period. *Journal of Gerontology A*, 57, B211–B224.

Walker, R. & Hill, K. (2003) Modeling growth and senescence in physical performance among the ache of eastern Paraguay. *American Journal of Human Biology*, 15, 196–208.

Walker, A. & Leakey, R. E. (1993) *The Skull*. Berlin, Springer.

Walkiewicz, M. & Stern, M. (2009) Increased insulin/insulin growth factor signaling advances the onset of metamorphosis in *Drosophila*. *PLoS One*, 4, e5072.

Walpita, C. N., Van Der Geyten, S., Rurangwa, E. & Darras, V. M. (2007) The effect of 3,5,3'-triiodothyronine supplementation on zebrafish (*Danio rerio*) embryonic development and expression of iodothyronine deiodinases and thyroid hormone receptors. *General and Comparative Endocrinology*, 152, 206–214.

Walpita, C. N., Crawford, A. D., Janssens, E. D. R., Van Der Geyten, S. & Darras, V. M. (2009) Type 2 iodothyronine deiodinase is essential for thyroid hormone-dependent embryonic development and pigmentation in zebrafish. *Endocrinology*, 150, 530–539.

Wang, J. & Kim, S. K. (2003) Global analysis of dauer gene expression in *C. elegans*. *Development*, 130, 1621–1634.

Wang, Y. & Levy, D. E. (2006) *C. elegans* STAT cooperates with DAF-7/TGF-beta signaling to repress dauer formation. *Current Biology*, 16, 89–94.

Wang, Y. & Tissenbaum, H. A. (2006) Overlapping and distinct functions for a *Caenorhabditis elegans* SIR2 and DAF-16/FOXO. *Mechanisms of Ageing and Development*, 127, 48–56.

Wang, Y., Salmon, A. B. & Harshman, L. G. (2001) A cost of reproduction: oxidative stress susceptibility is associated with increased egg production in *Drosophila melanogaster*. *Experimental Gerontology*, 36, 1349–1359.

Wang, T., Hung, C. C. Y. & Randall, D. J. (2006a) The comparative physiology of food deprivation: from feast to famine. *Annual Review of Physiology*, 68, 223.

Wang, Y., Jorda, M., Jones, P. L., Maleszka, R., Ling, X., Robertson, H. M., Mizzen, C. A., Peinado, M. A. & Robinson, G. E. (2006b) Functional CpG methylation system in a social insect. *Science*, 314, 645–647.

Wang, M. C., O'Rourke, E. J. & Ruvkun, G. (2008) Fat metabolism links germline stem cells and longevity in *C. elegans*. *Science*, 322, 957–960.

Wang, M., Zhang, X., Zhao, H., Wang, Q. & Pan, Y. (2009a) FoxO gene family evolution in vertebrates. *BMC Evolutionary Biology*, 9, 222.

Wang, P. Y., Neretti, N., Whitaker, R., Hosier, S., Chang, C., Lu, D., Rogina, B. & Helfand, S. L. (2009b) Long-lived Indy and calorie restriction interact to extend life span.

Proceedings of the National Academy of Sciences of the United States of America, 106, 9262–9267.

Wang, Z., Zhou, X. E., Motola, D. L., Gao, X., Suino-Powell, K., Conneely, A., Ogata, C., Sharma, K. K., Auchus, R. J., Lok, J. B., Hawdon, J. M., Kliewer, S. A., Xu, H. E. & Mangelsdorf, D. J. (2009c) Identification of the nuclear receptor Daf-12 as a therapeutic target in parasitic Nematodes. *Proceedings of the National Academy of Sciences of the United States of America*, 106, 9138–9143.

Wang, Q., Sajja, U., Rosloski, S., Humphrey, T., Kim, M. C., Bomblies, K., Weigel, D., and Grbic, V. 2007. *HUA2* caused natural variation in shoot morphology of A. thaliana. Current Biology 17: 1513–1519.

Wang, Y., Mutti, N. S., Ihle, K. E., Siegel, A., Dolezal, A. G., Kaftanoglu, O. & Amdam, G. V. (2010) Down-Regulation of Honey Bee IRS Gene Biases Behavior toward Food Rich in Protein. PLoS Genetics, 6, e1000896.

Warner, R. R. (1984) Deferred reproduction as a response to sexual selection in a coral-reef fish - a test of the life historical consequences. *Evolution*, 38, 148–162.

Warner, D., Lovern, M. & Shine, R. (2007) Maternal nutrition affects reproductive output and sex allocation in a lizard with environmental sex determination. *Proceedings of the Royal Society of London B*, 274, 883–890.

Warren, J., Yerushalmi, Y., Shimell, M., O'Connor, M., Restifo, L. & Gilbert, L. (2006) Discrete pulses of molting hormone, 20-hydroxyecdysone, during late larval development of *Drosophila melanogaster*: Correlations with changes in gene activity. *Developmental Dynamics*, 235, 315–326.

Wasser, S. K. & Barash, D. P. (1983) Reproductive suppression among female mammals: implications for biomedicine and sexual selection theory. *Quarterly Review of Biology*, 58, 513–538.

Waterhouse, R. M., Kriventseva, E. V., Meister, S., Xi, Z., Alvarez, K. S., Bartholomay, L. C., Barillas-Mury, C., Bian, G., Blandin, S., Christensen, B. M., Dong, Y., Jiang, H., Kanost, M. R., Koutsos, A. C., Levashina, E. A., Li, J., Ligoxygakis, P., MacCallum, R. M., Mayhew, G. F., Mendes, A., Michel, K., Osta, M. A., Paskewitz, S., Shin, S. W., Vlachou, D., Wang, L., Wei, W., Zheng, L., Zou, Z., Severson, D. W., Raikhel, A. S., Kafatos, F. C., Dimopoulos, G., Zdobnov, E. M. & Christophides, G. K. (2007) Evolutionary dynamics of immune-related genes and pathways in disease-vector mosquitoes. *Science*, 316, 1738–1743.

Watson, G. M. & Hessinger, D. A. (1989) Cnidocyte mechanoreceptors are tuned to the movements of swimming prey by chemoreceptors. *Science*, 243, 1589–1591.

Watson, G. M. & Mire, P. (2004) Dynamic tuning of hair bundle mechanoreceptors in a sea anemone during predation. *Hydrobiologia*, 530/531, 123–128.

Watson, R. D., Agui, N., Haire, M. E. & Bollenbacer, W. E. (1987) Juvenile hormone coordinates the regulation of the hemolymph ecdysteroid titer during pupal commitment in *Manduca sexta. Journal of Experimental Zoology*, 252, 255–263.

Watt, W. B. (1968) Adaptive significance of pigment polymorphisms in Colias butterflies. I. Variation of melanin pigment in relation to thermoregulation. *Evolution*, 22, 437–458.

Weber, K. E., Greenspan, R. J., Chicoine, D. R., Fiorentino, K., Thomas, M. H. & Knight, T. L. (2008) Microarray analysis of replicate populations selected against a wing-shape correlation in *Drosophila melanogaster. Genetics*, 178, 1093–1108.

Wedekind, C. & Folstad, I. (1994) Adaptive or nonadaptive immunosuppression by sex-hormones. *The American Naturalist*, 143, 936–938.

Weigel, D. & Nordborg, M. (2005) Natural variation in *Arabidopsis*. How do we find the causal genes? *Plant Physiology*, 138, 567–8.

Weindruch, R. (1989) Dietary restriction, tumors, and aging in rodents. *Journal of Gerontology*, 44, 67–71.

Weindruch, R. & Walford, R. L. (1982) Dietary restriction in mice beginning at 1 year of age: effect on life-span and spontaneous cancer incidence. *Science*, 215, 1415–1418.

Weinig, C. & Schmitt, J. (2004) Environmental effects on the expression of quantitative trait loci and implications for phenotypic evolution. *Bioscience*, 54, 627–635.

Weinkove D, Neufeld TP, Twardzik T, Waterfield MD, Leevers SJ. 1999. Regulation of imaginal disc cell size, cell number and organ size by *Drosophila* class I(A) phosphoinositide 3-kinase and its adaptor. *Current Biology*, 9, 1019–1029.

Wenseleers, T., Ratnieks, F. L. W., De Ribeiro, M., De A Alves, D. & Imperatriz-Fonseca, V.-L. (2005) Working class royalty: bees beat the caste system. *Biology Letters*, 1, 125–128.

Werner, J. D., Borevitz, J. O., Uhlenhaut, N. H., Ecker, J. R., Chory, J. & Weigel, D. (2005) Frigida-independent variation in flowering time of natural *Arabidopsis thaliana* accessions. *Genetics*, 170, 1197–1207.

Werren, J. H. & Charnov, E. L. (1978) Facultative sex-rations and population dynamics. *Nature*, 272, 349–350.

Wessells, R. J., Fitzgerald, E., Cypser, J. R., Tatar, M. & Bodmer, R. (2004) Insulin regulation of heart function in aging fruit flies. *Nature Genetics*, 36, 1275–1281.

West, G. B., Brown, J. H. & Enquist, B. J. (1997) A general model for the origin of allometric scaling laws in biology. *Science*, 276, 122–126.

West, G. B., Brown, J. H. & Enquist, B. J. (2001) A general model for ontogenetic growth. *Nature*, 413, 628–631.

West, G. B., Savage, V. M., Gillooly, J., Enquist, B. J., Woodruff, W. H. & Brown, J. H. (2003) Physiology: Why does metabolic rate scale with body size? *Nature*, 421, 713–713.

West-Eberhard, M. J. (1983) Sexual selection, social competition, and speciation. *Quarterly Review of Biology*, 58, 155–183.

West-Eberhard, M. J. (2003) *Developmental Plasticity and Evolution*. Oxford, Oxford University Press.

Westendorp, R. G. & Kirkwood, T. B. (1998) Human longevity at the cost of reproductive success. *Nature*, 396, 743–746.

Weyrich, P., Machicao, F., Reinhardt, J., Machann, J., Schick, F., Tschritter, O., Stefan, N., Fritsche, A. & Haring, H. U. (2008) SIRT1 genetic variants associate with the metabolic response of Caucasians to a controlled lifestyle intervention--the TULIP Study. *BMC Medical Genetics*, 9, 100.

Wheeler, W. C., Whitting, M., Wheeler, Q.M. and Carpenter, J.M. (2001) The phylogeny of the extant hexapod orders. *Cladistics*, 17, 113–169.

Wheeler, D. E., Buck, N. & Evans, J. D. (2006) Expression of insulin pathway genes during the period of caste determination in the honey bee, *Apis mellifera. Insect Biochemistry & Molecular Biology*, 15, 597–602.

White B. A., Nicoll C. S. (1981). Hormonal control of amphibian metamorphosis. In Gilbert, L. I. & Frieden, E. (Eds.) *Metamorphosis: A Problem in Developmental Biology*. New York, Plenum Press.

Whitfield, C. W., Cziko, A. M. & Robinson, G. E. (2003) Gene expression profiles in the brain predict behavior in individual honey bees. *Science*, 302, 296–299.

Whitfield, C. W., Ben-Shahar, Y., Brillet, C., Leoncini, I., Crauser, D., Leconte, Y., Rodriguez-Zas, S. & Robinson, G. E. (2006) Genomic dissection of behavioral maturation in the honey bee. *Proceedings of the National Academy of Sciences of the United States of America*, 103, 16068–16075.

Whitman, D. W. & Ananthakrishnan, T. N. (2009) *Phenotypic Plasticity of Insects*. Enfield, Science Publishers.

Wichman, H. A., Badgett, M. R., Scott, L. A., Boulianne, C. M. & Bull, J. J. (1999) Different trajectories of parallel evolution during viral adaptation. *Science*, 285, 422–424.

Wiederman, M. W. (1997) Extramarital sex: prevalence and correlates in a national survey. *Journal of Sex Research*, 34, 167–174.

Wiersma, P., Munoz-Garcia, A., Walker, A. & Williams, J. B. (2007) Tropical birds have a slow pace of life. *Proceedings of the National Academy of Sciences of the United States of America*, 104, 9340–9345.

Wigby, S. & Chapman, T. (2005) Sex peptide causes mating costs in female *Drosophila melanogaster. Current Biology*, 15, 316–321.

Wigby, S., Sirot, L. K., Linklater, J. R., Buehner, N., Calboli, F. C. F., Bretman, A., Wolfner, M. F. & Chapman, T. (2009)

Seminal fluid protein allocation and male reproductive success. *Current Biology*, 19, 751–757.

Wigglesworth, V. B. (1934) The physiology of ecdysis in Rhodnius prolixus. II. Factors controlling moulting and metamorphosis. *Quarterly Journal of Microscopic Sciences*, 77, 191–222.

Wigglesworth, V. B. (1936) The function of the corpora allatum in the growth and reproduction of *Rhodnius prolixus* (Hemiptera). *Quarterly Journal of Microscopic Sciences* 79, 91–121.

Wijngaarden, P. J., Koch, P. B. & Brakefield, P. M. (2002) Artificial selection on the shape of reaction norms for eyespot size in the butterfly *Bicyclus anynana*: direct and correlated responses. *Journal of Evolutionary Biology*, 15, 290–300.

Wikelski, M., Hau, M. & Wingfield, J. C. (1999) Social instability increases testosterone year-round in a tropical bird. *Proceedings of the Royal Society of London B*, 266, 1–6.

Wikelski, M., Hau, M. & Wingfield, J. C. (2000) Seasonality of reproduction in a neotropical rain forest bird. *Ecology*, 81, 2458–2472.

Wiklund, C., Persson, A. & Wickman, P. O. (1983) Larval estivation and direct development as alternative strategies in the speckled wood butterfly, *Pararge aegeria*, in Sweden. *Ecological Entomology*, 8, 233–238.

Wilbur, H. M. (1977) Propagule size, number, and dispersion patterns in *Ambystoma* and *Asclepias*. *The American Naturalist*, 111, 43–68.

Wilbur, H. M. & Collins, J. P. (1973) Ecological aspects of amphibian metamorphosis: Nonnormal distributions of competitive ability reflect selection for facultative metamorphosis. *Science* 182, 1305–1314.

Wilcox, A. J., Baird, D. D. & Weinberg, C. J. (1999) Time of implantation of the conceptus and loss of pregnancy. *New England Journal of Medicine*, 30, 1796–1799.

Wilczek, A. M., Roe, J. L., Knapp, M. C., Cooper, M. D., Lopez-Gallego, C., Martin, L. J., Muir, C. D., Sim, S., Walker, A., Anderson, J., Egan, J. F., Moyers, B. T., Petipas, R., Giakountis, A., Charbit, E., Coupland, G., Welch, S. M. & Schmitt, J. (2009) Effects of genetic perturbation on seasonal life history plasticity. *Science*, 323, 930–934.

Wilfert, L., Gadau, J. & Schmid-Hempel, P. (2007) The genetic architecture of immune defense and reproduction in male *Bombus terrestris* bumblebees. *Evolution*, 61, 804–815.

Willcox, S., Moltschaniwskyj, N. A. & Crawford, C. (2007) Asexual reproduction in scyphistomae of *Aurelia sp.*: Effects of temperature and salinity in an experimental study. *Journal of Experimental Marine Biology and Ecology*, 353, 107–114.

Willcox, B. J., Donlon, T. A., He, Q., Chen, R., Grove, J. S., Yano, K., Masaki, K. H., Willcox, D. C., Rodriguez, B. & Curb, J. D. (2008) FOXO3A genotype is strongly associated with human longevity. *Proceedings of the National Academy of Sciences of the United States of America*, 105, 13987–13992.

Williams, G. C. (1957) Pleiotropy, natural selection, and the evolution of senescence. *Evolution*, 11, 398–411.

Williams, G.C. (1966a). Adaptation and Natural Selection: A Critique of Some Current Evolutionary Thought. Princeton University Press, Princeton, NJ.

Williams, G. C. (1966b) Natural selection, the costs of reproduction, and a refinement of Lack's Principle. *The American Naturalist*, 100, 687–690.

Williams, T. D. (2008) Individual variation in endocrine systems: moving beyond the 'tyranny of the Golden Mean'. *Philosophical Transactions of the Royal Society of London B*, 363, 1687–1698.

Williams, E. A. & Degnan, S. M. (2009) Carry-over effect of larval settlement cue on postlarval gene expression in the marine gastropod *Haliotis asinina*. *Molecular Ecology*, 18, 4434–4449.

Williams, K. D., Busto, M., Suster, M. L., So, A. K., Ben-Shahar, Y., Leevers, S. J. & Sokolowski, M. B. (2006a) Natural variation in *Drosophila melanogaster* diapause due to the insulin-regulated PI3-kinase. *Proceedings of the National Academy of Sciences of the United States of America*, 103, 15911–15915.

Williams, P. D., Day, T., Fletcher, Q. & Rowe, L. (2006b) The shaping of senescence in the wild. *Trends in Ecology & Evolution*, 21, 458–463.

Williams, J. B., Roberts, S. P. & Elekonich, M. M. (2008) Age and natural metabolically-intensive behavior affect oxidative stress and antioxidant mechanisms. *Experimental Gerontology*, 43, 538–549.

Williams, E. A., Degnan, B. M., Gunter, H., Jackson, D. J., Woodcroft, B. J. & Degnan, S. M. (2009a) Widespread transcriptional changes pre-empt the critical pelagic-benthic transition in the vetigastropod *Haliotis asinina*. *Molecular Ecology*, 18, 1006–1025.

Williams, K. D., Schmidt, P. S. & Sokolowski, M. B. (2009b) Photoperiodism in Insects: Molecular Basis and Consequences of Diapause. In Nelson, R. J., Denlinger, D. L. & Somers, R. J. (Eds.) *Photoperiodism: The Biological Calendar*. Oxford, Oxford University Press.

Wilson, T. G. & Ashok, M. (1998) Insecticide resistance resulting from an absence of target-site gene product. *Proceedings of the National Academy of Sciences of the United States of America*, 95, 14040–14044.

Wilson, M. & Daly, M. (1985) Competitiveness, risk taking, and violence: the young male syndrome. *Ethology and Sociobiology*, 6, 59–73

Wilson, T. G. & Fabian, J. (1986) A *Drosophila melanogaster* mutant resistant to a chemical analog of juvenile hormone. *Developmental Biology*, 118, 190–201.

Wilson, R. B. & Tatchell, K. (1988) SRA5 encodes the low-Km cyclic-AMP phosphodiesterase of *Saccharomyces cerevisiae*. *Molecular and Cellular Biology*, 8, 505–510.

Wilson, K., Thorndyke, M., Nilsen, F., Rogers, A. & Martinez, P. (2005) Marine systems: moving into the genomics era. *Marine Ecology*, 26, 3–16.

Windig, J. J., Brakefield, P. M., Reitsma, N. & Wilson, J. G. M. (1994) Seasonal polyphenism in the wild: survey of wing patterns in five species of *Bicyclus* butterflies in Malawi. *Ecological Entomology*, 19, 285–298.

Wingfield, J. C. (2006) Communicative behaviors, hormone-behavior interactions, and reproduction in vertebrates. In Neill, J. D. (Ed.) *Physiology of Reproduction*. New York, Academic Press.

Wingfield, J. C. (2008) Comparative endocrinology, environment and global change. *General and Comparative Endocrinology*, 157, 207–216.

Wingfield, J. C. & Farner, D. S. (1993) Endocrinology of reproduction in wild species. In Farner, D. S., King, J. R. & Parkes, K. C. (Eds.) *Avian Biology*. London, Academic Press.

Wingfield, J. & Moore, M. C. (1987) Hormonal, social and environmental factors in the reproductive biology of free-living male birds. In Crews, D. (Ed.) *Psychobiology of Reproductive Behavior: An Evolutionary Perspective*. Englewood Cliffs, NJ, Prentice Hall.

Wingfield, J. C., Hegner, R. E., Dufty, A. M. J. & Ball, G. F. (1990) The 'challenge-hypothesis': theoretical implications for patterns of testosterone secretion, mating systems, and breeding strategies. *The American Naturalist*, 136, 829–846.

Wingfield, J. C., Hahn, T. P., Levin, R. & Honey, P. (1992) Environmental predictability and control of gonadal cycles in birds. *Journal of Experimental Zoology A*, 261, 214–231.

Wingfield, J. C., Lynn, S. E. & Soma, K. K. (2001) Avoiding the 'costs' of testosterone: ecological bases of hormone-behavior interactions. *Brain, Behavior and Evolution* 57, 239–251.

Wingfield, J. C., Meddle, S. L., Moore, I., Busch, S., Wacker, D., Lynn, S., Clark, A., Vasquez, R. A. & Addis, E. (2007) Endocrine responsiveness to social challenges in northern and southern hemisphere populations of *Zonotrichia*. *Journal of Ornithology*, 148, S435–S441.

Winston, A. L. (1987) *The Biology of the Honeybee*. Cambridge, MA, Harvard University Press.

Winther, R. G. (2001) Varieties of modules: kinds, levels, origins, and behaviors. *Journal of Experimental Zoology*, 291, 116–129.

Witte, A. V., Fobker, M., Gellner, R., Knecht, S. & Floel, A. (2009) Caloric restriction improves memory in elderly humans. *Proceedings of the National Academy of Sciences of the United States of America*, 106, 1255–1260.

Wolf, J. B. & Hager, R. (2006) A maternal–offspring coadaptation theory for the evolution of genomic imprinting. *PLoS Biology*, 4: e380.

Wolf, J. B., Brodie, E. D., Cheverud, J. M., Moore, A. J. & Wade, M. J. (1998) Evolutionary consequences of indirect genetic effects. *Trends in Ecology & Evolution*, 13, 64–69.

Wolfe, K. H., Gouy, M. L., Yang, Y. W., Sharp, P. M. & Li, W. H. (1989) Date of the monocot dicot divergence estimated from chloroplast DNA sequence data. *Proceedings of the National Academy of Sciences of the United States of America*, 86, 6201–6205.

Wolff, S., Ma, H., Burch, D., Maciel, G. A., Hunter, T. & Dillin, A. (2006) SMK-1, an essential regulator of DAF-16-mediated longevity. *Cell*, 124, 1039–1053.

Wolschin, F. & Amdam, G. V. (2007a) Plasticity and robustness of protein patterns during reversible development in the honey bee (*Apis mellifera*). *Analytical and Bioanalytical Chemistry*, 389, 1095–1100.

Wolschin, F. & Amdam, G. V. (2007b) Comparative proteomics reveal characteristics of life history transitions in a social insect. *Proteome Science*, 5, 10.

Wolschin, F., Munch, D. & Amdam, G. V. (2009) Structural and proteomic analyses reveal regional brain differences during honeybee aging. *Journal of Experimental Biology*, 212, 4027–4032.

Wolschin F, Mutti NS, Amdam GV (2011) Insulin receptor substrate influences female caste development in honeybees. *Biology Letters*, 23, 112–115.

Wong, J. M. & Shi, Y. B. (1995) Coordinated regulation of and transcriptional activation by *Xenopus* thyroid hormone and retinoid-X-receptors. *Journal of Biological Chemistry*, 270, 18479–18483.

Wood, J. (1994) *Dynamics of Human Reproduction: Biology, Biometry, Demography*. Hawthorne, Aldine de Gruyter.

Wood, J. G., Rogina, B., Lavu, S., Howitz, K., Helfand, S. L., Tatar, M. & Sinclair, D. (2004) Sirtuin activators mimic caloric restriction and delay aging in Metazoans. *Nature*, 430, 686–689.

Wood, T. E., Burke, J. M. & Reiseberg, L. H. (2005) Parallel genotypic adaptation: when evolution repeats itself. *Genetica*, 123, 157–170.

Woods, R., Schneider, D., Winkworth, C. L., Riley, M. A. & Lenski, R. E. (2006) Tests of parallel molecular evolution in a long-term experiment with *Escherichia coli*. *Proceedings of the National Academy of Sciences of the United States of America*, 103, 9107–9112.

Wootton, R. J. & Kukalova-Peck, J. (2000) Flight adaptations in Palaeozoic Palaeoptera (Insecta). *Biological*

Reviews of the Cambridge Philosophical Society 75, 129–167.

Worley, A. C., Houle, D. & Barrett, S. C. H. (2003) Consequences of hierarchical allocation for the evolution of life history traits. *The American Naturalist*, 161, 153–167.

Wrangham, R. W. & Peterson, D. (1996) *Demonic males*. New York, Houghton Mifflin Company.

Wray, G. A. (1994) Developmental evolution – new paradigms and paradoxes. *Developmental Genetics*, 15, 1–6.

Wray, G. A. (1995) Evolution of larvae and developmental modes. In McEdward, L. (Ed.) *Ecology of Marine Invertebrate Larvae*. Boca Raton, CRC Press.

Wray, G. A. (1996) Parallel evolution of nonfeeding larvae in Echinoids. *Systematic Biology*, 45, 308–322.

Wright, G. M. & Youson, J. H. (1977) Serum thyroxine concentrations in larval and metamorphosing anadromous sea lamprey, *Petromyzon marinus* L. *Journal of Experimental Zoology*, 202, 27–32.

Wright, G. M. & Youson, J. H. (1980) Transformation of the endostyle of the anadromous sea lamprey, *Petromyzon marinus* L, during metamorphosis. 2. Electron-microscopy. *Journal of Morphology*, 166, 231–257.

Wu, X. H., Hopkins, P. M., Palli, S. R. & Durica, D. S. (2004) Crustacean retinoid-X receptor isoforms: distinctive DNA binding and receptor-receptor interaction with a cognate ecdysteroid receptor. *Molecular and Cellular Endocrinology*, 218, 21–38.

Wu, Q., Zhang, Y., Xu, H. & Shen, P. (2005) Regulation of hunger-driven behaviors by neural ribosomal S6 kinase in *Drosophila*. *Proceedings of the National Academy of Sciences of the United States of America*, 102, 13289–13294.

Wu, P., Shen, Q., Dong, S., Xu, Z., Tsien, J. Z. & Hu, Y. (2008) Calorie restriction ameliorates neurodegenerative phenotypes in forebrain-specific presenilin-1 and presenilin-2 double knockout mice. *Neurobiology of Aging*, 29, 1502–1511.

Wulff, J. L. (1991) Asexual fragmentation, genotype success, and population dynamics of erect branching sponges. *Journal of Experimental Marine Biology and Ecology*, 149, 227–247.

Wyatt, G. R. (1997) Juvenile hormone in insect reproduction – a paradox? *European Journal of Entomology*, 94, 323–333.

Wykoff, D. D., Davies, J. P., Melis, A. & Grossman, A. R. (1998) The regulation of photosynthetic electron transport during nutrient deprivation in *Chlamydomonas reinhardtii*. *Plant Physiology*, 117, 129–139.

Wynne-Edwards, K. E. (2003) From dwarf hamster to daddy: The intersection of ecology, evolution, and physiology that produces paternal behavior. In Slater, P. J. B., Rosenblatt, J.

S., Snowden, C. T. & Roper, T. J. (Eds.) *Advances in the Study of Behavior*. San Diego, Academic Press.

Xu, S. (2003) Theoretical basis of the Beavis effect. *Genetics*, 165, 2259–2268.

Yamada, H., Ohta, H. & Yamauchi, K. (1993) Serum thyroxine, estradiol-17-beta, and testosterone profiles during the parr-smolt transformation of Masu Salmon, *Oncorhynchus masou*. *Fish Physiology and Biochemistry*, 12, 1–9.

Yamamoto, K. & Kikuyama, S. (1982) Radioimmunoassay of prolactin in plasma of bullfrog tadpoles. *Endocrinologia Japonica*, 29, 159–167.

Yamamoto, H., Tachibana, A., Kawaii, S., Matsumura, K. & Fusetani, N. (1996) Serotonin involvement in larval settlement of the barnacle, *Balanus amphitrite*. *Journal of Experimental Zoology*, 275, 339–345.

Yamano, K. & Miwa, S. (1998) Differential gene expression of thyroid hormone receptor alpha and beta in fish development. *General and Comparative Endocrinology*, 109, 75–85.

Yamano, K., Miwa, S., Obinata, T. & Inui, Y. (1991a) Thyroid hormone regulates developmental changes in muscle during flounder metamorphosis. *General and Comparative Endocrinology*, 81, 464–472.

Yamano, K., Tagawa, M., de Jesus, E. G., Hirano, T., Miwa, S. & Inui, Y. (1991b) Changes in whole-body concentrations of thyroid hormones and cortisol in metamorphosing conger eel. *Journal of Comparative Physiology B*, 161, 371–375.

Yamano, K., Takanoohmuro, H., Obinata, T. & Inui, Y. (1994) Effect of thyroid hormone on developmental transition of myosin light chains during flounder metamorphosis. *General and Comparative Endocrinology*, 93, 321–326.

Yamasaki, S., Fujii, N. & Takahashi, H. (2005) Hormonal regulation of sex expression in plants. *Plant Hormones*. San Diego, Elsevier Academic Press Inc.

Yamauchi, K. & Tata, J. R. (1994) Purification and characterization of a cytosolic thyroid-hormone-binding protein (CTBP) in *Xenopus* liver. *European Journal of Biochemistry / FEBS*, 225, 1105–1112.

Yamauchi, K. & Tata, J. R. (1997) Tissue-dependent and developmentally regulated cytosolic thyroid-hormone-binding proteins (CTBPs) in Xenopus. *Comparative Biochemistry and Physiology C*, 118, 27–32.

Yamauchi, K., Kasahara, T., Hayashi, H. & Horiuchi, R. (1993) Purification and characterization of a 3,5,3'-L-triiodothyronine-specific binding protein from bullfrog tadpole plasma: A homolog of mammalian transthyretin. *Endocrinology*, 132, 2254–2261.

Yan, G. & Norman, S. (1995) Infection of *Tribolium* beetles with a tapeworm: variation in susceptibility within and between species and among genetic strains. *Journal of Parasitology*, 81, 37–42.

Yan, G. & Stevens, L. (1995) Selection by parasites on components of fitness in *Tribolium* beetles: the effect of intraspecific competition. *The American Naturalist*, 146, 795–813.

Yan, G. Y., Stevens, L., Goodnight, C. J. & Schall, J. J. (1998) Effects of a tapeworm parasite on the competition of *Tribolium* beetles. *Ecology*, 79, 1093–1103.

Yanai, S., Okaichi, Y. & Okaichi, H. (2004) Long-term dietary restriction causes negative effects on cognitive functions in rats. *Neurobiology of Aging*, 25, 325–332.

Yang, H., Zhao, Z. G., Qiang, W. Y., An, L. Z., Xu, S. J. & Wang, X. L. (2004) Effects of enhanced Uv-B radiation on the hormonal content of vegetative and reproductive tissues of two tomato cultivars and their relationships with reproductive characteristics. *Plant Growth Regulation*, 43, 251–258.

Yang, J., Anzo, M. & Cohen, P. (2005) Control of aging and longevity by IGF-I signaling. *Experimental Gerontology*, 40, 867–872.

Yang, C. H., Belawat, P., Hafen, E., Jan, L. Y. & Jan, Y. N. (2008) *Drosophila* egg-laying site selection as a system to study simple decision-making processes. *Science*, 319, 1679–1683.

Yano, M., Katayose, Y., Ashikari, M., Yamanouchi, U., Monna, L., Fuse, T., Baba, T., Yamamoto, K., Umehara, Y., Nagamura, Y. & Sasaki, T. (2000) *Hd1*, a major photoperiod sensitivity quantitative trait locus in rice, is closely related to the *Arabidopsis* flowering time gene *CONSTANS*. *Plant Cell*, 12, 2473–2483.

Yaoita, Y., Shi, Y. & Brown, D. (1990) *Xenopus laevis* alpha and beta thyroid hormone receptors. *Proceedings of the National Academy of Sciences of the United States of America*, 87, 7090–7094.

Ye, Y. H., Chenoweth, S. F. & McGraw, E. A. (2009) Effective but costly, evolved mechanisms of defense against a virulent opportunistic pathogen in *Drosophila melanogaster*. *PLoS Pathogens*, 5, e1000385.

Yen, P. M. (2001) Physiological and molecular basis of thyroid hormone action. *Physiological Reviews*, 81, 1097–1142.

Yerex, R. P., Young, C. W., Donker, J. D. & Marx, G. D. (1988) Effects of selection for body size on feed efficiency and size of Holsteins. *Journal of Dairy Science*, 71, 1355–1360.

Yin, T. J. & Quinn, J. A. (1995) Tests of a mechanistic model of one hormone regulating both sexes in *Cucumis sativus* (Cucurbitaceae). *American Journal of Botany*, 82, 1537–1546.

Yoshiga, T., Georgieva, T., Dunkov, B. C., Harizanova, N., Ralchev, K. & Law, J. H. (1999) *Drosophila melanogaster* transferrin. Cloning, deduced protein sequence, expression during the life cycle, gene localization and up-regulation on bacterial infection. *European Journal of Biochemistry*, 260, 414–420.

Yoshizato, K., Kistler, A. & Frieden, E. (1975) Metal ion dependence of the binding of triiodothyronine by cytosol proteins of bullfrog tadpole tissues. *Journal of Biological Chemistry*, 250, 8337–8343.

Young, L. J. & Insel, T. R. (2002) Hormones and parental behavior. In Becker, J. B., Breedlove, S. M. & McCarthy, M. M. (Eds.) *Behavioral Endocrinology*. Cambridge, MIT Press.

Youson, J. H. (1988) First Metamorphosis. In Hoar, W.S. and Randall, D. J. (Eds.) *Fish Physiology: The Physiology of Developing Fish*. Toronto, Academic Press, Inc.

Youson, J. H. (2004) The Impact of Environmental and Hormonal Cues on the Evolution of Fish Metamorphosis. In Hall, B. K., Pearson, R. D., Muller, G. B., (Eds.), *Environment, Development, and Evolution: Toward a Synthesis*. Cambridge, London, MIT Press.

Youson, J. H. & Potter, I. C. (1979) Description of the stages in the metamorphosis of the anadromous sea lamprey, *Petromyzon marinus* L. *Canadian Journal of Zoology-Revue Canadienne de Zoologie*, 57, 1808–1817.

Youson, J. H. & Sower, S. A. (2001) Theory on the evolutionary history of lamprey metamorphosis: role of reproductive and thyroid axes. *Comparative Biochemistry and Physiology B*, 129, 337–345.

Youson, J. H., Plisetskaya, E. M. & Leatherland, J. F. (1994) Concentrations of insulin and thyroid hormones in the serum of landlocked sea lampreys (*Petromyzon marinus*) of 3 larval year classes, in larvae exposed to 2 temperature regimes, and in individuals during and after metamorphosis. *General and Comparative Endocrinology*, 94, 294–304.

Youson, J. H., Holmes, J. A. & Leatherland, J. F. (1995) Serum concentrations of thyroid hormones in KClO4-treated larval sea lampreys (*Petromyzon marinus* L) *Comparative Biochemistry and Physiology C*, 111, 265–270.

Youson, J. H., Manzon, R. G., Peck, B. J. & Holmes, J. A. (1997) Effects of exogenous thyroxine (T4) and triiodothyronine (T3) on spontaneous metamorphosis and serum T4 and T3 levels in immediately premetamorphic sea lampreys, *Petromyzon marinus*. *Journal of Experimental Zoology*, 279, 145–155.

Youson, J. H. (2007) Peripheral Endocrine Glands. I. The gastroenteropancreatic endocrine system and the thyroid gland. In McKenzie D. J., Farrell A. P., and Brauner C. J. (Eds.), *Primitive Fishes*. New York, Academic Press.

Yu, J., Vodyanik, M. A., Smuga-Otto, K., Antosiewicz-Bourget, J., Frane, J. L., Tian, S., Nie, J., Jonsdottir, G. A., Ruotti, V., Stewart, R., Slukvin, I. & Thomson, J. A. (2007a) Induced pluripotent stem cell lines derived from human somatic cells. *Science*, 318, 1917–1920.

Yu, X. J., Yan, Y. & Gu, J. D. (2007b) Attachment of the bio-fouling Bryozoan *Bugula neritina* larvae affected by inorganic and organic chemical cues. *International Biodeterioration & Biodegradation*, 60, 81–89.

Zakon, H. H. (1998) The effects of steroid hormones on electrical activity of excitable cells. *Trends in Neurosciences*, 21, 202–207.

Zamudio, K. R. & Sinervo, E. (2000) Polygyny, mate-guarding, and posthumous fertilization as alternative male mating strategies. *Proceedings of the National Academy of Sciences of the United States of America*, 97, 14427–14432.

Zavala, J. A., Patankar, K., Gase, K. & Baldwin, I. T. (2004) Constitutive and inducible trypsin proteinase inhibitor production incurs large fitness costs in *Nicotiana attenuata*. *Proceedings of the National Academy of Sciences of the United States of America*, 101, 1607–1612.

Zega, G., Pennati, R., Groppelli, S., Sotgia, C. & De Bernardi, F. (2005) Dopamine and serotonin modulate the onset of metamorphosis in the Ascidian *Phallusia mammillata*. *Developmental Biology*, 282, 246–256.

Zega, G., Pennati, R., Fanzago, A. & De Bernardi, F. (2007) Serotonin involvement in the metamorphosis of the hydroid *Eudendrium racemosum*. *International Journal of Developmental Biology*, 51, 307–313.

Zera, A. J. (2005) Intermediary metabolism and life history trade-offs: Lipid metabolism in lines of the wing-polymorphic cricket, *Gryllus firmus*, selected for flight capability vs. early age reproduction. *Integrative and Comparative Biology*, 45, 511–524.

Zera, A. J. (2009) Wing polymorphism in crickets. In Whitman, D. W. & Ananthakrishnan, T. N. (Eds.) *Phenotypic Plasticity of Insects*. Enfield, Science Publishers.

Zera, A. J. (2011) Microevolution of intermediary metabolism: Evolutionary genetics meets metabolic biochemistry. Journal of Experimental Biology, 214, 179–190.

Zera, A. J. & Brink, T. (2000) Nutrient absorption and utilization by wing and flight muscle morphs of the cricket *Gryllus firmus*: implications for the trade-off between flight capability and early reproduction. *Journal of Insect Physiology*, 46, 1207–1218.

Zera, A. J. & Harshman, L. G. (2001) The physiology of life history trade-offs in animals. *Annual Review of Ecology and Systematics*, 32, 95–126.

Zera, A. J. & Harshman, L. G. (2009) Laboratory selection studies of life history physiology in insects. In Garland, T. & Rose, M. R. (Eds.) *Experimental Evolution: Methods and Applications*. Berkeley, University of California Press.

Zera, A. J. & Larsen, A. (2001) The metabolic basis of life history variation: Genetic and phenotypic differences in lipid reserves among life history morphs of the wing-polymorphic cricket, *Gryllus firmus*. *Journal of Insect Physiology*, 47, 1147–1160.

Zera, A. J. & Zhao, Z. (2003) Life history evolution and the microevolution of intermediary metabolism: activities of lipid-metabolizing enzymes in life history morphs of a wing-dimorphic cricket. *Evolution*, 57, 568–596.

Zera, A. J. & Zhao, Z. (2004) Effect of a juvenile hormone analogue on lipid metabolism in a wing-polymorphic cricket: implications for the biochemical basis of the trade-off between reproduction and dispersal. *Biochemical and Physiological Zoology*, 77, 255–266.

Zera, A. J. & Zhao, Z. (2006) Intermediary metabolism and life history trade-offs: differential metabolism of amino acids underlies the dispersal-reproduction trade-off in a wing-polymorphic cricket. *The American Naturalist*, 167, 889–900.

Zera, A. J., Koehn, R. K. & Hall, J. G. (1985) Allozymes and biochemical adaptation. In Kerkut, G. A. & Gilbert, L. I. (Eds.) *Comprehensive Insect Physiology Biochemistry and Pharmacology*. Oxford, Pergamon.

Zera, A. J., Harshman, L. G. & Williams, T. (2007) Evolutionary endocrinology: the developing synthesis between endocrinology and evolutionary genetics. *Annual Review of Ecology and Systematics*, 38, 793–817.

Zera, A. J., Berkheim, D., Newman, S., Black, C., Klug, L., and E. Crone. 2011. Purification and characterization of cytoplasmic NADP+-isocitrate dehydrogenase, and amplification of the Nadp+-Idh gene from the wing-dimorphic cricket, Gryllus firmus. Journal of Insect Science. In Press.

Zhan, M., Yamaza, H., Sun, Y., Sinclair, J., Li, H. & Zou, S. (2007) Temporal and spatial transcriptional profiles of aging in *Drosophila melanogaster*. *Genome Research*, 17, 1236–1243.

Zhang, H., Liu, J., Li, C. R., Momen, B., Kohanski, R. A. & Pick, L. (2009) Deletion of *Drosophila* insulin-like peptides causes growth defects and metabolic abnormalities. *Proceedings of the National Academy of Sciences of the United States of America*, 106, 19617–19622.

Zhao, Z. and Zera, A. J. (2001) Enzymological and radiotracer studies of lipid metabolism in the flight-capable and flightless morphs of the wing-polymorphic cricket, *Gryllus firmus*. *Journal of Insect Physiology*, 47, 1337–2347.

Zhao, Z. & Zera, A. J. (2002) Differential lipid biosynthesis underlies a trade-off between reproduction and flight capability in a wing-polymorphic cricket. *Proceedings of the National Academy of Sciences of the United States of America*, 99, 16829–16834.

Zhao, Y., Sun, H., Lu, J., Li, X., Chen, X., Tao, D., Huang, W. & Huang, B. (2005) Lifespan extension and elevated hsp gene expression in *Drosophila* caused by histone deacetylase inhibitors. *Journal of Experimental Biology*, 208, 697–705.

Zheng, J., Edelman, S. W., Tharmarajah, G., Walker, D. W., Pletcher, S. D. & Seroude, L. (2005) Differential patterns of apoptosis in response to aging in *Drosophila*. *Proceedings of the National Academy of Sciences of the United States of America*, 102, 12083–12088.

Zhong, D. B., Pai, A. & Yan, G. Y. (2005) Costly resistance to parasitism: Evidence from simultaneous quantitative rait loci mapping for resistance and fitness in *Tribolium castaneum*. *Genetics*, 169, 2127–2135.

Zhou, B., & Riddiford, L.M. (2001) Hormonal regulation and patterning of the broad-complex in the epidermis and wing discs of the tobacco hornworm, *Manduca sexta*. *Developmental Biology*, 231, 125–137.

Zhou, B., & Riddiford, L.M. (2002) Broad specifies pupal development and mediates the 'status quo' action of juvenile hormone on the pupal-adult transformation in *Drosophila* and *Manduca*. *Development*, 129, 2259–2269.

Zhou, B., Hiruma, K., Shinoda, T. and Riddiford, L. M. (1998) Juvenile hormone prevents ecdysteroid-induced expression of broad complex RNAs in the epidermis of the tobacco hornworm. *Developmental Biology*, 203, 233–244.

Zhou, N., Wilson, K. A., Andrews, M. E., Kauffman, J. S. & Raff, R. A. (2003) Evolution of Otp-independent larval skeleton patterning in the direct-developing sea urchin, *Heliocidaris erythogramma*. *Journal of Experimental Zoology B*, 300, 58–71.

Zhou, G. L., Pennington, J. E. & Wells, M. A. (2004) Utilization of pre-existing energy stores of female *Aedes aegypti* mosquitoes during the first gonotrophic cycle. *Insect Biochemistry & Molecular Biology*, 34, 919–925.

Zhou, B., Williams, D. W., Altman, J., Riddiford, L. M. & Truman, J. W. (2009) Temporal patterns of Broad isoform expression during the development of neuronal lineages in *Drosophila*. *Neural Development*, 4, 4–39.

Zid, B. M., Rogers, A. N., Katewa, S. D., Vargas, M. A., Kolipinski, M. C., Lu, T. A., Benzer, S. & Kapahi, P. (2009) 4E-BP extends lifespan upon dietary restriction by enhancing mitochondrial activity in *Drosophila*. *Cell*, 139, 149–160.

Ziegler, T. E. & Snowdon, C. T. (1997) Role of prolactin in paternal care in a monogamous New World primate, *Saguinus oedipus*. *Annals of the New York Academy of Sciences*, 807, 599–601.

Zijlstra, W. G., Steigenga, M. J., Brakefield, P. M. & Zwaan, B. J. (2003) Simultaneous selection on two fitness-related traits in the butterfly *Bicyclus anynana*. *Evolution*, 57, 1852–1862.

Zijlstra, W. G., Steigenga, M. J., Koch, P. B., Zwaan, B. J. & Brakefield, P. M. (2004) Butterfly selected lines explore the hormonal basis of interactions between life histories and morphology. *The American Naturalist*, 163, 76–87.

Zimmer, R. K. & Butman, C. A. (2000) Chemical signaling processes in the marine environment. *Biological Bulletin*, 198, 168–187.

Zirkin, B. R. & Chen, H. (2000) Regulation of Leydig cell steroidogenic function during aging. *Biology of Reproduction*, 63, 977–981.

Zuk, M. & Stoehr, A. M. (2002) Immune defense and host life history. *The American Naturalist*, 160, S9–S22.

Zwaan, B. J. (1999) The evolutionary genetics of aging and longevity. *Heredity*, 82, 589–597.

Zwaan, B., Bijlsma, R. & Hoekstra, R. F. (1995) Direct selection on life-span in *Drosophila melanogaster*. *Evolution*, 49, 649–659.

Zwaan, B. J., Azevedo, R. B., James, A. C., Van't Land, J. & Partridge, L. (2000) Cellular basis of wing size variation in *Drosophila melanogaster*: a comparison of latitudinal clines on two continents. *Heredity*, 84, 338–347.

Index